Next-Gen Plant Based Fermented Foods: Innovations for a Sustainable Future

Sunny Dhiman • Sukhminderjit Kaur
Editors

Next-Gen Plant Based Fermented Foods: Innovations for a Sustainable Future

Springer

Editors
Sunny Dhiman
Department of Biotechnology
University Institute of Biotechnology
Chandigarh University
Mohali, Punjab, India

Sukhminderjit Kaur
Department of Biotechnology
University Institute of Biotechnology
Chandigarh University
Mohali, Punjab, India

ISBN 978-3-032-29926-0 ISBN 978-3-032-29927-7 (eBook)
https://doi.org/10.1007/978-3-032-29927-7

This Springer imprint is published by the registered company Springer Nature Switzerland AG
The registered company address is: Gewerbestrasse 11, 6330 Cham, Switzerland

Preface

The global food system is undergoing an unprecedented transformation driven by the urgent need to address climate change, food insecurity, resource depletion, and growing consumer demand for healthier and more sustainable dietary choices. Among the emerging solutions, plant-based fermented foods have gained remarkable scientific, industrial, and societal attention for their ability to combine traditional food wisdom with cutting-edge biotechnological innovation. Fermentation, one of humanity's oldest food-processing techniques, is now being reimagined through advances in precision fermentation, synthetic biology, artificial intelligence, microbial engineering, and sustainable bioprocessing. These developments are paving the way for next-generation functional foods that are nutritionally superior, environmentally sustainable, and economically viable.

The book *Next-Gen Plant Based Fermented Foods: Innovations for a Sustainable Future* has been conceptualized to provide a comprehensive and interdisciplinary perspective on the rapidly evolving landscape of plant-based fermented foods. This volume brings together contributions from researchers, academicians, and industry experts across diverse disciplines to explore the scientific foundations, technological innovations, sustainability implications, market dynamics, and future prospects of fermented plant-based food systems.

The chapters in this book collectively highlight how fermentation technologies are transforming plant-based foods into highly functional, nutritionally enhanced, and consumer-acceptable products. Beginning with discussions on engineered microbes, precision fermentation, and microbiome-guided food innovation, the book traces the evolution of fermentation from traditional practices to AI-driven bioengineering approaches. The role of postbiotics, para-probiotics, and gut health-promoting metabolites is explored in depth, emphasizing their significance beyond conventional probiotics.

Several chapters focus on nutritional enhancement through fermentation, including improved protein digestibility, vitamin biosynthesis, mineral bioavailability, and reduction of anti-nutritional factors. The integration of synthetic biology and precision fermentation for developing next-generation dairy alternatives, meat substitutes, and functional beverages demonstrates the transformative potential of modern biotechnology in reshaping future food systems.

The book also examines emerging protein sources such as mycoproteins, tempeh innovations, cultivated hybrid foods, and alternative fermentation-derived proteins, reflecting the growing global interest in sustainable protein production. Artificial intelligence and automation are highlighted as key drivers for personalized nutrition, smart fermentation systems, process optimization, and scalable biomanufacturing.

Recognizing the intimate connection between fermented foods and human health, dedicated chapters explore the gut–brain axis, psychobiotics, cognitive health, and microbiome-mediated physiological responses. Consumer perception, sensory acceptance, flavor optimization, and cultural preferences are also addressed, acknowledging that technological innovation must ultimately align with consumer expectations and market realities.

Sustainability remains a central theme throughout this book. Several chapters discuss circular bioeconomy approaches, climate-resilient agriculture, zero-waste fermentation, food waste valorization, and fermentation-driven resource recovery systems. Legal, regulatory, and policy frameworks supporting sustainable fermentation technologies and their role in achieving Sustainable Development Goals (SDGs), particularly SDG 2 (Zero Hunger) and SDG 13 (Climate Action), are critically examined.

Importantly, this book not only highlights current scientific advancements but also identifies major challenges that remain unresolved, including scalability, regulatory harmonization, functional validation, standardization, and equitable technology access. By integrating traditional fermentation knowledge with modern biotechnology and sustainability science, the book aims to inspire future research, industrial innovation, and policy development in this rapidly advancing field.

We sincerely hope that this volume will serve as a valuable resource for researchers, academicians, students, food technologists, microbiologists, biotechnologists, policymakers, entrepreneurs, and professionals working in food science, fermentation technology, sustainable agriculture, and allied disciplines. It is our aspiration that this book contributes meaningfully toward the development of resilient, sustainable, and health-oriented food systems for future generations.

We express our heartfelt gratitude to all contributing authors for their efforts and valuable insights that made this book possible. We also extend our sincere appreciation to the publisher, our reviewers, and editorial team for their continuous support and cooperation throughout the development of this book.

We hope readers will find this book informative, thought-provoking, and inspiring as we collectively move toward a more sustainable and fermentation-driven future of food.

Mohali, Punjab, India Sunny Dhiman
Mohali, Punjab, India Sukhminderjit Kaur

Contents

Chapter 1
The Next-Gen Microbiome: Role of Engineered Microbes and Precision Fermentation in Transforming Plant-Based Foods

Neeraj Narwat, Shweta Jethwani, Sukhminderjit Kaur, and Sunny Dhiman

Abstract Fermented foods made with plants have gained more interest in the marketplace, as they are considered more sustainable and better for human health than the animal-derived products. Plant-fermented foods have many benefits, but in general they are nutritionally poor and have less sensory appeals and scaling capabilities than animal products. These limitations represent a significant hurdle for creating a broader sector in the marketplace. This chapter explores the potential applications of microbiome science, engineered microbes, and precision fermentation in delivering healthier and nutrient-rich plant-based food to consumers. The chapter describes the benefits arising from recent developments in the fields of microbiome and ecology development, synthetic biology, and metabolic engineering. These developments contribute to improving bioavailability, quality of protein, vitamin synthesis, functionality of metabolite production, and textural qualities of the plant-based fermented foods. Key applications of using precision fermentation to create animal-free protein, bioactive compounds, and functional ingredients portray major regulatory, commercial, and consumer acceptance hurdles. Finally, this chapter discusses future trends and applications for using AI-assisted strain development for personalised fermented food products; it is clear that the incorporation of microbiome-guided fermentation represents a significant new means to develop sustainable, nutritionally enhanced, and consumer-acceptable plant-based foods.

Keywords Metabolic engineering · Microbiome · Functional metabolites · Precision fermentation · Phytochemicals · Microbial metabolism

---

N. Narwat · S. Jethwani · S. Kaur (✉) · S. Dhiman
Department of Biotechnology, University Institute of Biotechnology, Chandigarh University, Mohali, Punjab, India
e-mail: sukhminderjit.uibt@cumail.in

S. Dhiman, S. Kaur (eds.), *Next-Gen Plant Based Fermented Foods: Innovations for a Sustainable Future*, https://doi.org/10.1007/978-3-032-29927-7_1

1.1 Introduction

Plant-based fermented foods have received a lot of attention recently as consumers want more sustainable, nutritional, and functional meals. Consumers are progressively switching to plant-based diets because of concerns such as animal cruelty, environmental sustainability, and personal health (Venter de Villiers et al., 2024). Fermented plant-based meals address these desires by providing alternatives to typical dairy and animal products (Praveen & Brogi, 2025). Plant-based diets are recognised as having fewer environmental effects than animal-based diets. Legumes, grains, and oilseeds constitute a potential source of plant protein due to their lower environmental impact compared to animal-based proteins (with respect to resource usage) (Farid et al., 2024; Clark & Tilman, 2017). The broad-spectrum beneficial compounds contained in fermented foods produced from plants include a variety of bioactive compounds, such as peptides, phenolic compounds, phytochemicals, flavonoids, and vitamin enrichment (Khayatan et al., 2024). These compounds confer several health benefits through improved digestion, immunity, and mitigation of chronic disease development associated with their consumption. As a result of demand by health-conscious, environmentally aware, and ethical consumers, global interest in the production and purchase of plant-based functional foods is increasing. Significant geographical variation exists within the market for plant-based functional foods. According to IMARC Group (2025), the Asia-Pacific region shows one of the fastest-growing plant-based food markets globally. Recent statistical data indicates that the Asia–Pacific plant-based food market was valued at approximately USD 22.55 billion in 2025 and is projected to increase nearly USD 72.57 billion by 2035, expanding at a compound annual growth rate (CAGR) of approximately 12.4%. Moreover, the Asia–Pacific plant-based functional food ingredients market is expected to increase from USD 42.24 billion in 2024 to USD 120.90 billion by 2032, reflecting a CAGR of about 18.0%. Due to increasing health awareness, urbanisation, and shifting dietary preferences in Southeast Asian countries like Indonesia, Thailand, and Vietnam, the demand for plant-based fermented foods is experiencing rapid expansion. In Indonesia specifically, the plant-based food and beverage market was valued at approximately USD 589.2 million in 2025 and is projected to reach USD 1309.0 million by 2034, with a CAGR of 9.27%. There are many Southern Asian countries and Northern Asian countries which are experiencing rapid growth within their domestic markets across the globe (including Indonesia). For example, research conducted on commercially available plant-based jerky products sold in Indonesia revealed an extremely high customer satisfaction rating (83.8–86.8%) for the product, indicating a significant opportunity for future development and growth of plant-based functional foods in that region (Veniranda & Surya, 2022).

Traditional fermented foods based on plants, e.g., tempeh, kimchi, sauerkraut, miso, etc., provide advantages, but there are also some limitations, including limited sensory appeal (taste), poor nutrition (essential nutrient deficiency), the presence of anti-nutritional factors (substances that inhibit nutrient absorption), digestion

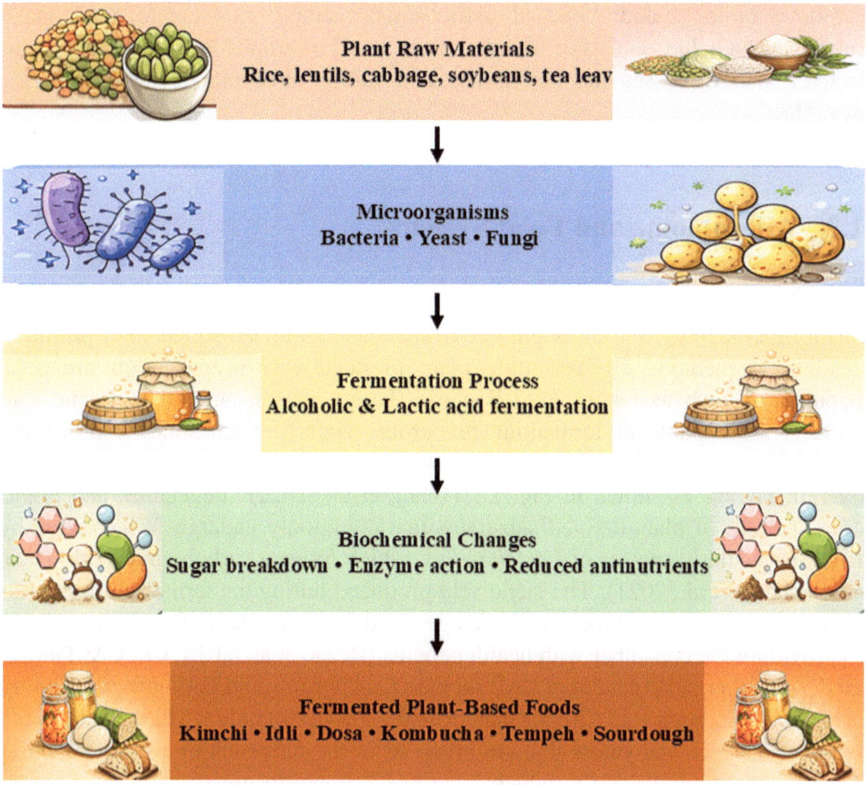

Fig. 1.1 Schematic overview of microbial fermentation in plant-based food production

problems, and lack of consumer acceptance. The majority of these products cannot replicate the flavours, textures, or functional characteristics of typical meat- or dairy-based products. However, the new advancements in the field of microbial biotechnology provide potential avenues to develop new plant-based products with improved nutritional quality and functionality. Advances in microbial ecology, synthetic biology, and precision fermentation have allowed the focused use of naturally created microorganisms to increase nutritional quality, flavour, and texture, minimise anti-nutritional chemicals, and create useful components, as shown in Fig. 1.1. Modern viewpoints on plant-based fermentation emphasise scientific knowledge of microbial interactions, advances in fermentation technology, and consumer demand for healthy and sustainable food products (Abbaspour, 2024). Fermentation-driven product development is bringing new flavours, textures, and scents to the plant-based food business, increasing customer choice and acceptability (Teng et al., 2021). This chapter basically examines the increasing importance of microbial fermentation for enhancing the nutritional value, functional properties, and sustainability of plant-based fermented products. The chapter discusses the contemporary advancement of microbial biotechnologies, including precision fermentation, synthetic biology, and AI-based optimisation methods that create nutritionally robust, sustainable, and user-friendly plant-based alternatives. Additionally, this chapter also discusses the key issues and future directions for this emerging discipline.

1.2 Microbiome and Food Systems

There is a diversity of microbes involved in plant-based fermentation that affect the quality and safety, as well as provide health benefits to fermented food products. Traditional methods of fermenting plant products rely primarily on microbial populations such as Lactic Acid Bacteria (LAB), yeasts, and fungi. Lactic acid bacteria are capable of fermenting numerous types of plant sources, producing chemotactic agents that will therefore increase the plant's nutritional value and overall flavour, as shown in Fig. 1.1 (Wang et al., 2022). Vegetables and grains are examples of plant-derived substrates that commonly undergo fermentation by LAB, particularly species of *Lactobacillus*, which ferment carbohydrates into lactic acid (Zhang et al., 2021). The lactic acid produced during the fermentation process preserves the fermented foods or beverages, and it also produces flavour in addition to providing the consumer with health benefits (Rezac et al., 2018; Clark & Tilman, 2017). Other popular genera of LAB include *Leuconostoc* and *Pediococcus* (Thierry et al., 2023; Gautam et al., 2025; Ghatani et al., 2024). Not only LAB, there are many types of bacteria and yeasts that are involved in the fermentation process, such as *Enterobacter* and *Weissella*, which have been shown to greatly contribute to the overall fermentation process (Liu et al., 2019). Yeasts play a vital role in the fermentation of plant sources due to their ability to cooperate with LAB and produce several different flavour compounds and enhance the nutritional content of the fermented product (Dimidi et al., 2019).

Filamentous fungi are also involved in the fermentation of plant-based foods such as *Aspergillus sp.* and *Rhizopus* sp. Fungi breakdown of complex carbohydrates during the fermentation and hence contribute to the final product's quality and nutritional value (Pasaribu et al., 2019). Microbial consortia consisting of LAB, yeasts and filamentous fungi metabolising the dietary fibre and other complex carbohydrates derived from plant matter may enhance the flavour or the nutritional bioavailability of fermented plants. Microbial consortia are a mixture of various microbial species that interact with one another in a synergistic nature and therefore improve fermentation process and outcome. Microbial consortia can metabolise complex plant-based compounds such as hexanal into simpler compounds, resulting in various improved flavours and textures of fermented plant products (Tang et al., 2024). The interconnections of the plant microbiome, soil microbiome, and gut microbiome show a great impact on the development of plants; improve the nutritional values; and help in recovering from various biotic and abiotic stressors which affect the quality of the raw plant material that is used for food production (Ma et al., 2025). In addition, plant-based foods and fermented food products interact with the gut microbiome of humans while being consumed, altering their variety, metabolic activity, and health-related evidence. These interactions extend from farm to gut, demonstrating the importance of microbiome management across the food value chain (van Bruggen et al., 2019). Microbial metabolism promotes important biochemical changes that improve food quality and usefulness. Microbes promote nutrient bioavailability, destroy antinutritional chemicals, synthesise vitamins and bioactive metabolites, and provide flavour, fragrance, and textural components (Iqbal et al., 2025). These functional contributions are crucial for creating nutritionally superior, sensory-optimised, and safe plant-based meals, establishing microbial metabolism as a key mechanism in microbiome-driven food innovation.

1.3 Microbiome-Driven Enhancement of Plant-Based Foods

Recently, there has been an increased interest in the fermentation of plant-based foods due to their ability to improve the nutritional value of foods from non-animal sources. The fermentation of plant-food sources can provide many advantages from a nutrition perspective, including improved nutrient availability, increased protein quality, an increased number of vitamins and minerals, and bioactive compounds being produced through a fermentative process, as shown in Fig. 1.2. There are various fermentation strategies that improve the nutritional quality of food sources. These methods improve nutrient bioavailability, increase protein level, promote synthesis of vitamins and minerals, and facilitate the production of health-promoting bioactive metabolites.

1.3.1 Bioavailability and Nutrient Release

Some components classified as antinutritional factors (ANFs), such as phytates, tannins and oxalates, can affect the absorption of minerals by the body and therefore reduce the bioavailability of nutrients from a diet that is entirely derived from plant sources. Fermentation techniques have been used to decrease or eliminate some of the ANFs from foods. The amount of nutrients released in fermented vegetables increased compared to the same vegetable that has not been fermented (Nath et al., 2022). During fermentation, microorganisms secrete various enzymes (phytases and tannases) that help to degrade phytate and tannin through enzymatic processes. This breakdown of the phytate and tannin by the enzymes creates a more soluble form of the mineral (that was previously bound to the phytate or tannin) that will now be more easily absorbed by the body after passing through the small intestine. Therefore, the nutritional benefit of a diet based only on plant sources has been significantly improved. In a study by Qi et al. (2023) the researchers investigated the

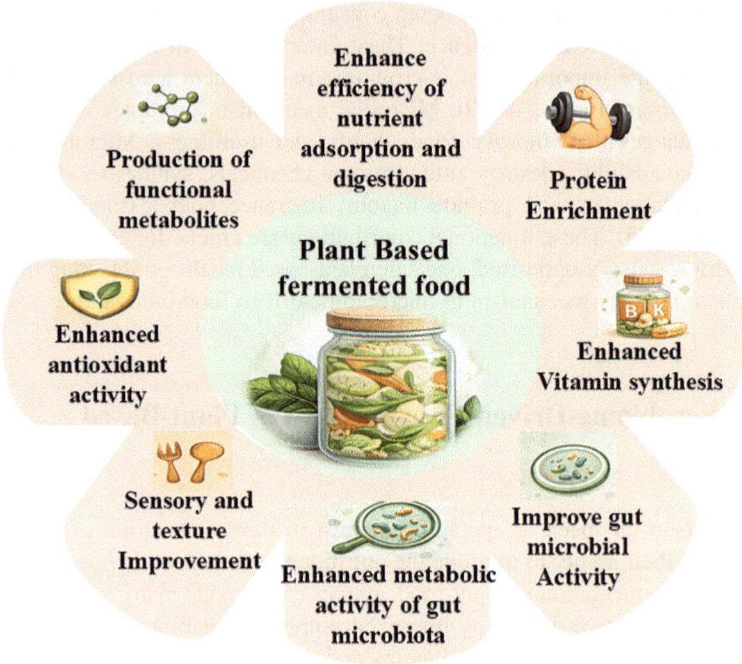

Fig. 1.2 Nutritional and functional benefits of plant-based fermented foods

fermentation of soybean meal (SBM), one of the two main sources of plant protein for both ruminants and non-ruminants, with the newly discovered probiotic strain *Bacillus licheniformis* B4. Fermentation of SBM with *B. licheniformis* B4 for 24 h resulted in a 73.3% reduction in the levels of phytic acid, along with an increase in hydrolysis from 15.9% to 25.5% and an increase in crude protein content from 44.8% to 54.3%. Fermentation of SBM also produced a reduction in the total amount of fibre in the SBM, which also contributed to an improved overall nutritional quality of the SBM.

1.3.2 Protein Enrichment and Quality Improvement

Protein quality is crucial in plant-based diets, and fermentation may dramatically increase protein quantity and digestibility. Microbial proteolysis occurs during the fermentation process, resulting in the breakdown of complex protein structures into simpler peptides and amino acids (Christensen et al., 2022). Studies have shown that microbial fermentation increases the quantities of important amino acids in plant-based diets. Filipe et al. demonstrated that solid-state fermentation of a plant feedstuff combination (25% rapeseed meal, soybean meal, rice bran, and sunflower meal) with *Aspergillus ibericus* and two strains of *Aspergillus niger* improved its nutritional profile. The protein content of all plant feedstuff rose by 5%, but the cellulose and hemicellulose levels reduced by 9–11% and 21–34%, respectively. *A. niger* CECT 2088 had the greatest enzyme activity, including cellulase (123.7 U/g), xylanase (431.8 U/g), and β-glucosidase (117.9 U/g). Fermentation, enzyme synthesis, protein increase, fibre reduction, and digestibility all showed favourable relationships according to principal component analysis (Filipe et al., 2023). Hsieh et al. (2023) found that SSF of fourth-day *Chenopodium formosanum* sprouts in a bioreactor increased antioxidant capacity, peptide production, and enzyme activity compared to standard plate fermentation. CRISPR-based metabolic engineering has achieved considerable advances in microbial production of proteinogenic amino acids and omega-3 fatty acids, providing a sustainable alternative to food inputs that were previously animal-based.

1.3.3 Enhanced Vitamin Synthesis

Fermentation is very successful in biosynthesising key vitamins, notably B vitamins, as well as increasing mineral bioavailability. Certain strains of bacteria and yeasts used in fermentation have been proven to increase the levels of B-group vitamins such as B2 (riboflavin), B6 (pyridoxine), and B12 (cobalamin), which are often lacking in plant diets (He & Howell, 2022; Lim et al., 2023). Rana et al. (2025) created vitamin B2-enriched soymilk by co-fermentation with two riboflavin-producing *Lactiplantibacillus plantarum* strains (MTCC 25432, 25,433) and *Lactobacillus acidophilus* NCIM 2902. Quan et al. (2022) investigated the effects of six LAB strains on the physicochemical, antioxidant, and sensory aspects of fermented orange juice. Sooklim et al. (2025) explored the increase of the biochemical and functional properties of rice milk by fermentation using a yeast strain, *Saccharomyces cerevisiae* RSO4, which is known for its excellent fermentative activity. *Lactobacillus* and *Propionibacterium* can be modified using genetic engineering to produce vitamin B12 during fermentation. Fermented plant-based foods, such as gallons of soy milk (dehydrated), fermented vegetables (dehydrated), and fermented nuts/seeds/legumes, also serve as appropriate non-animal food sources for the vitamin B12 (Deokar et al., 2025).

1.3.4 Production of Functional Metabolites

Fermented foods can improve not only nutritional values, but they also create new functional metabolites such as short-chain fatty acids (SCFAs), polyphenols, and bioactive peptides. Which improves the health benefits from fermented foods as well as improves fermentation functionality with plant-based products. To assist in the development of new fermented beverages, Tang et al. developed a functional beverage consisting of apple and melon; the fermentation of the two fruits showed an enhancement of antioxidant activity via total phenolic content (TPC), total flavonoid content (TFC), 1,1-diphenyl-2-picrylhydrazyl (DPPH) and 2,2′-azino-bis (3-ethylbenzthiazoline-6-sulfonic acid) (ABTS) free radical scavenging assays (Tang et al., 2023). Hsieh et al. reported the development and function of glycine-rich peptides from the fermentation of *Chenopodium formosanum* by *Rhizopus oligosporus* (Hsieh et al., 2025).

1.3.5 Microbial Exopolysaccharides and Textural Modifications

Sensorial qualities, particularly the texture and mouthfeel of plant-based fermented food items, are very crucial for customer acceptability and overall product quality. One of the most novel microbial ways to improve these sensory and functional qualities is the creation of exopolysaccharides (EPSs) by LAB and other beneficial microorganisms. *Lactobacillus delbrueckii* subsp. *bulgaricus* and *Streptococcus thermophilus*, which are used in yoghurt fermentation, create ropy and capsular EPSs that increase product viscosity and gel strength (Dan et al., 2023). Fermenting soy milk with *Weissella confusa* wild-type or sac mutant strains significantly enhanced water retention and viscosity, underscoring their potential to augment EPS production in fermented plant-based milk alternatives, thereby emulating the texture of conventional dairy products (Huang et al., 2025).

1.3.6 Gut Microbiome Interactions with Fermented Plant-Based Foods

Fermented foods are being recognised as significant regulators of gut microbiota and general gastrointestinal health. Foods rich in live microorganisms, notably LAB, have high concentrations of these beneficial bacteria during fermentation. When ingested on a regular basis, these foods may have a good impact on gut health. Continuous use of LAB-rich foods (e.g., yoghurt, kefir, kimchi, sauerkraut) has been demonstrated to improve both microbial diversity and the metabolic activity of the gut microbiota (Leeuwendaal et al., 2022). As mentioned previously, prebiotics may support a diverse microbial population in the gut and promote healthy metabolic activity and balance within the gut ecosystem and prevent or improve the symptoms associated with many gastrointestinal disorders (Fusco et al., 2023). There is also evidence that taking prebiotics has been shown to enhance the intestinal barrier function, reduce inflammation, and decrease the likelihood of developing various gastrointestinal diseases (Roy & Dhaneshwar, 2023). The aforementioned potential benefits of prebiotics demonstrate their significance in fulfilling nutritional requirements and mitigating the emergence of various gastrointestinal disorders.

Synbiotics are a mix of fermented foods and prebiotics that promote gut health by providing advantageous microbes and substrates for their development. Fermented food such as yoghurt, kefir, kimchi, and sauerkraut is rich in probiotics that improve the diversification of the gut microbiome, enhance digestion, contribute to maintaining the integrity of the gut's barrier, and improve the immune system (Dahiya & Nigam, 2022). Prebiotic substrates are non-digestible dietary fibre compounds present in foods like bananas, garlic, onions, and whole grains that promote the growth of beneficial bacteria (such as Lactobacillus and Bifidobacterium), which results in a healthy gut microbiome (Obayomi et al., 2024). Fermented foods may improve metabolic health and lower inflammation, which is a critical aspect in the management and treatment of obesity, diabetes and cardiovascular diseases. Probiotic microorganisms (e.g., *Lactobacillus* and *Bifidobacterium* species) in fermented foods influence the gut microbiota, helping to regulate metabolic activities, including insulin sensitivity, glucose homeostasis, and lipid metabolism (Nithya et al., 2023). In addition, the fermentation of prebiotic substances produces acidic waste, leading to a reduction in the number of harmful bacteria found in the colon, an increase in the number of beneficial bacteria, enhanced resilience and functional stability of the gut microbiota, and support for homeostasis in the intestines (Ford et al., 2018).

1.4 Engineered Microbes: Concepts and Tools

Genetically modified microorganisms are a crucial aspect in redoing plant-based foods using innovative fermentation techniques. Genetic engineering includes modifying microorganisms such as bacteria and fungi to improve their efficiency in producing defined food components or to improve existing production methods (Mallikarjuna & Yellamma, 2019). The most widely used recombinant DNA technology allows changes in the genetic formation of microorganisms. These modifications empower microorganisms to generate vital biomolecules, including proteins, enzymes, and other metabolites, which significantly impact the quality, safety, and nutritional content of food. A major advantage of altering the genetic makeup of microorganisms is the potential to develop entirely novel food products or enhance and modify existing ones. Synthetic biology is a multidisciplinary approach that combines molecular biology, genetics, engineering, and computer modelling to create novel systems that fulfil biological tasks (Silva & Paulillo, 2015). Several strategies have been developed throughout time to increase synthetic biology's capabilities, as shown in Fig. 1.3. Some strategies include using gene-editing technologies such as CRISPR Cas9 to change DNA to improve and simulate metabolic processes in live systems. BioBrick is another approach to synthetic biology. A BioBrick is a standard, interchangeable component used in Synbio to design biological systems and devices. The use of building blocks, which allow for the construction of new biological functionalities by combining components, has long been a topic of research since the start of the previous century with respect to the coupling of DNA in a plasmid or vector using standard assembly techniques.

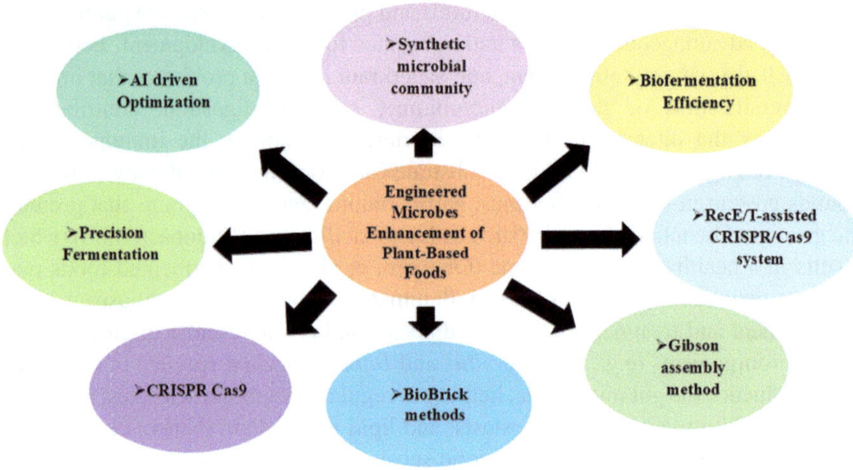

Fig. 1.3 Synthetic biology and engineered microbial approaches for enhancing plant-based foods

The BioBrick methods were used by Kildegaard et al. in (2017) to engineer the oleaginous yeast *Yarrowia lipolytica* to produce astaxanthin through de novo fermentation for nutritional applications as a food and feed additive. Similarly, Ma et al. (2016) employed the BioBrick technology in producing high levels of the proteolytic enzyme neutral protease I in both *A. oryzae* and *P. pastoris* for use in the industry. This enzyme serves multiple applications, such as textiles, food, and medicine, as it has the property of hydrolysing proteins present in soybeans and other materials, thereby creating a food product that contains flavour (Hong et al., 2004). Also, the Gibson assembly method has been successfully used for the construction or modification of food-related technological aspects. For instance, Xu et al. (2020) used the Gibson assembly method in constructing a heat-induced expression mutagenesis vector in *E. coli* for the efficient production of L-tyrosine synthase de novo. This amino acid provides nutrients and is commonly included as an ingredient in enteral and oral dietary products, as well as in nutritional food products. In addition to this, the de novo synthesis of plant-derived purine alkaloids (example: caffeine contained within coffee; theophylline, an asthma medication; theobromine obtained from cacao) was completed by McKeague et al. (2016), utilising Gibson assembly-based methodology with the eukaryote *S. cerevisiae* as a microbial expression host.

Furthermore, Costa et al. (2021) have reported an industrially applicable *S. cerevisiae* strain that utilises the CRISPR/Cas9 to generate elicit resveratrol from hydrothermally treated eucalyptus wood. More examples of CRISPR/Cas9 use in *E. coli* include the effective synthesis of anthocyanins, which are glycosylated flavonoid pigments important in the food sector and valued as natural food and beverage colourants (Cress et al., 2017). Li and colleagues knocked out the *glgc* and *gltA/ppc* genes in *Synechococcus elongatus* PCC 7942 to produce succinic acid, an essential component in the food and pharmaceutical sectors (Li et al., 2016). *S. cerevisiae* was modified to synthesise the anthocyanin pelargonidin 3-O-glucoside from glucose. Specific anthocyanin biosynthetic genes from *A. thaliana* and *Gerbera hybrida* were inserted into a *S. cerevisiae* strain that produces naringenin, a flavonoid precursor to anthocyanins (Levisson et al., 2018). Chen et al. engineered *E. coli* for improved substrate utilisation by deleting the mlc gene and adding heterologous gene constructs, resulting in L-lysine production (Chen et al., 2024). *Lactobacillus* strains have been engineered to produce bioactive peptides with antioxidant and immunomodulatory properties, enhancing the functional attributes of fermented products (Guo et al., 2023). The RecE/T-assisted CRISPR/Cas9 system achieved gene deletion efficiencies in LAB strains such as *Lactobacillus plantarum* WCFS1 and *Lactobacillus brevis* ATCC 367, which helped improve their metabolism and advance their industrial applications in the production of probiotics, bioactive compounds, and fermented food products (Xin et al., 2025).

The use of synthetic biology in the food business offers considerable potential as a solution to a variety of food challenges, as seen in Table 1.1. Currently, more than 50 enterprises operate in the food industry, providing novel synthetic goods across a broad variety of fields. From a worldwide viewpoint, these enterprises are unable to fulfil global market needs owing to their low number and concentration in the western United States. These firms' operations may be broadly classified into five categories: agricultural and crop enhancement, alternative meat production, food security solutions, tailored supplements and products, and alternative food and flavour development. The agriculture industry, which focuses on growing more robust and sustainable species as well as providing alternative foods and flavours, has the highest number of businesses. So far, fewer goods or services have been supplied in the food security sector, creating a void for future enterprises to fill (Rodrigues et al., 2023).

1.5 Precision Fermentation in Plant-Based Foods

Precision fermentation is a cutting-edge biotechnological method that uses genetically modified microbes such as yeast, fungus, and bacteria to generate particular biomolecules, including enzymes, proteins, and lipids (Boukid et al., 2023; Teng et al., 2021). Precision fermentation combines genetic engineering, synthetic biology, and various fermentation methods to allow for accurate chemical creation by microbes (Mirsalami & Mirsalami, 2025). This procedure has been extensively studied in the pharmaceutical, cosmetics, and food industries as an alternative method of producing compounds of animal or plant origin without using their natural sources (animals or plants) (Augustin et al., 2024; Knychala et al., 2024; Boukid et al., 2023). Increasingly, firms are using precision fermentation to add specific genes into microorganisms to produce large quantities of a selected protein or chemical. Following the fermentation process, the required biomolecules are extracted and known as the final product. This new method of extraction and production is much more sustainable than the old one because it can be scaled up

Table 1.1 Synthetic biology approaches and microbial hosts for nutritional enhancement in plant-based fermentation

Microbial group/ synthetic biological tool	Microorganisms	Target products	Application in plant-based foods	References
Lactic acid bacteria (LAB)	*Lactobacillus, Leuconostoc, Pediococcus*	Organic acids, EPS, vitamins	Preservation, texture improvement, probiotic effects	Rezac et al. (2018)
Yeasts	*Saccharomyces cerevisiae, Pichia pastoris*	Flavor compounds, vitamins, proteins	Flavor enhancement, functional ingredients	Dimidi et al. (2019)
Filamentous fungi	*Aspergillus, Rhizopus*	Peptides, amino acids	Improved digestibility and nutrient release	Dhiman et al. (2025)
Microbial consortia	LAB + yeasts + fungi	Bioactives, improved sensory traits	Enhanced flavor, texture, and nutrient bioavailability	Dhiman et al. (2025)
CRISPR/ Cas9 gene editing	*E. coli, S. cerevisiae*	Amino acids, pigments, polyphenols	Nutritional enrichment, natural colorants	Costa et al. (2022)
Type II-A CRISPR-Cas9	*Lacticaseibacillus rhamnosus GG*	Metabolize lactose	Milk or yogurt fermentation	Xie et al. (2025)
CRISPR-Cas9	*Saccharomyces cerevisiae*	Increase in dough volume, amino acid production	Improve the yield and taste of fermented foods	Lee et al. (2022)
CRISPR-Cas9	*Lactococcus lactis*	Folate, or vitamin B9	Increase folate biosynthesis	Fang et al. (2018)
CRISPR-Cas9	*Acetobacter pasteurianus Ab3*	Pyrroloquinoline quinone antioxidant	Optimize metabolic pathways and enhance the yields of microorganism	Han et al. (2025)
RecE/T-assisted CRISPR/ Cas9 system	*Lactobacillus plantarum WCFS1 and Lactobacillus brevis ATCC 367*	Probiotics, Bioactive compounds	Advancing in industrial applications	Xin et al. (2025)
BioBrick assembly	*Yarrowia lipolytica, Pichia pastoris*	Carotenoids, enzymes	Flavor development, functional additives	Knight (2003), Wong et al. (2017)
Gibson assembly	*E. coli, S. cerevisiae*	Amino acids, alkaloids	Nutraceuticals and fortified foods	Xu et al. (2020), McKeague et al. (2016)
Metabolic pathway engineering	Yeasts and bacteria	Vitamins, SCFAs, lipids	Precision fermentation-based foods	Nielsen et al. (2024), Peng and Wei (2025)

more easily and has less of an effect on the environment (Nielsen et al., 2024; Knychala et al., 2024; Gupta & Shukla, 2017).

Recently, precision fermentation has been changing the way we produce food by enabling the manufacture of high-quality, animal-free foods very efficiently and sustainably (Ajayeoba & Ijabadeniyi, 2025). For example, companies can now produce milk proteins without using cows to make animal-free cheese, yoghurt, and ice cream that have very similar tastes and textures to dairy products (Nielsen et al., 2024; Augustin et al., 2024; Zimberoff, 2021). Traditional fermentation processes used either wild-type or slightly modified organisms to ferment materials into food products, which indicates that it is feasible to create low-cost fermentation processes with low processing requirements as compared to other methods (Ramírez Rojas et al., 2022; Sharma et al., 2020). Precision fermentation produces very precise substances through the use of genetically modified micro-organisms (GEMs). Precision fermentation employs several synthetic biology technologies, including CRISPR-based editing, modular pathway assembly, and gene circuit design to identify and optimise biosynthetic pathways. The following processes are used to synthesise proteins, lipids, flavours (and other flavour compounds), and pigments (Peng & Wei, 2025). For instance, researchers have used CRISPR-Cas9 to edit *Lactococcus lactis* (a lactic acid bacterium used in dairy fermentation) to increase folate biosynthesis. Folate, or vitamin B9, is essential for DNA synthesis, cell division, and overall health. Additionally, a gene involved in vitamin B12 biosynthesis was introduced into *Escherichia coli* using CRISPR, enabling *E. coli* to produce vitamin B12—an important nutrient often lacking in plant-based diets (Fang et al., 2018). Lee et al. used CRISPR-Cas9 genome editing technology to improve both the quality and quantity of fermented food products using *Saccharomyces cerevisiae*. The study involved the deletion of the glucose sensor genes *RGT2, SNF3 and URE2, which help* in yeast's energy metabolism and amino acid production during rice wine production, which enhanced the overall fermentation process and increased the final product yield (Lee et al., 2022). Xie et al. aimed to enhance the commercial viability of *Lacticaseibacillus rhamnosus* GG (LGG) by using the endogenous type II-A CRISPR-Cas9 system for genome editing. Work included the construction of lactose-positive strains of LGG (MJM570), involving the deletion of a transcriptional terminator upstream of the *lacT* gene and the correction of a premature stop codon in the *lacG* gene. These genetic modifications restored lactase activity, enabling the strain to metabolise lactose and resulting in improved growth characteristics during milk or yoghurt fermentation (Xie et al., 2025).

Haryani et al. demonstrated that the CRISPR/Cas9 system can be used to reprogram the metabolic pathways of LAB strains to increase the synthesis of antimicrobial compounds. The study used a two-plasmid CRISPR/Cas9 system to knock out the lactate dehydrogenase (*LDH*) genes and shifted the carbon flux from lactic acid production to the synthesis of bacteriocin-like inhibitory substances (Haryani et al., 2024). Increasingly, researchers are emphasising the integration of multi-omics data and computer modelling, which allows the identification of metabolic bottlenecks, prediction of strain performance, and speedup of the design-build-test cycle. There are a few examples of progress in this area, including the development of gene circuits designed to create customised regulatory networks for dynamic regulation of metabolic pathways and the use of cell-free systems for rapid prototyping and testing of pathways prior to full microbial engineering. Other examples include the development of biofoundries for the rapid assembly and testing of new strains and the use of chassis organisms that are specifically designed for fermentation. Examples of chassis organisms include *Saccharomyces cerevisiae*, *Escherichia coli* (Navale et al., 2021), and *Pichia pastoris* (Gao et al., 2021).

1.6 Challenges and Limitations

1.6.1 Regulatory Challenges

Plants are fermented, and there is an increasing demand for many plant-based fermented foods due to the growing demand for sustainable, safe, healthy and innovative products by consumers (FAO, 2019). Due to differences in the way that these foods have been produced by either standard methods of fermentation or by traditional methods, regulators across the globe have faced very unique challenges related to regulating these products (FAO, 2006). When comparing these products against each other, it will be easy to see that there is an extreme variation in the regulatory framework between the various regions where emerging plant-based fermented food products are being produced. There are some regulatory variations based on the differences in legislation frameworks from region to region, the consumers' preferences in a region, and the historic approach to food safety that has been utilised by those regions (Aschemann-Witzel et al., 2021). The European Union (EU) has developed legislation to regulate innovative foods through the Innovative Food Regulation (EU) 2015/2283. This particular regulation has created a centralised approval process for innovative food products, including fermented plant-based foods (EFSA, N.D.A., 2015). Additionally, the Innovative Food Regulation (EU) 2015/2283 has created a centralised process for safety assessment of the above foods and has established requirements for the appropriate labelling of health claims and for verifying that scientific data exists to substantiate such claims. Furthermore, according to the EFSA Panel on Genetically Modified Organisms (GMO) (2011), there are several regulatory challenges that arise regarding fermented plant-based foods created through genetically modified microorganisms and their respective metabolites (Lietzow et al., 2021; Turck et al., 2024). These challenges arise from the need to ensure consumer safety while allowing for the continued development of innovative food products produced using modified microorganisms. There must be extensive safety assessments of such microorganisms used to produce food products, and the engineered microorganism must contain no potential health hazards. There have been significant additions to the list of genetic modifier microorganisms that are generally considered safe, which are now associated with opportunistic disease (Panel, 2022).

1.6.2 Scalability and Commercialization Challenges

The growing consumer interest in sustainable, healthy, and environmentally friendly food options has given rise to the growing popularity of plant-based fermented foods. However, commercialising these products and scaling them to meet consumer demand is not without challenges, including supply chain issues, technical difficulties and regulatory obstacles. Plant-based fermented products are developed from the sources of plant-based protein obtained from legumes, grains and oilseeds. However, the supply of these raw materials can be affected by factors such as seasonality, global conflict (geopolitical), and competing demands from other industries (Aswathy et al., 2024). For plant-based fermented products, fermentation is one of the most important steps, and there is a major challenge in the ability to optimise the fermentation process to obtain consistent flavour, texture, and nutritional characteristics. The rate at which lactic acid bacteria and other microorganisms metabolise the sugars present in a fermentation substrate is mainly based on the temperature, pH and composition of the substrate. Any variation in these factors can cause inconsistencies in the final product (Wang et al., 2021). The use of precision fermentation and synthetic biology in the production of new enzymes is proving effective; however, commercially scaling the production of these enzymes at an economically feasible price point continues to be a challenge (Augustin et al., 2024).

1.6.3 Consumer Perception and Market Acceptance

Recently, there has been an increased interest in the use of fermented plant-based food products for addressing consumer concerns related to health, the environment, and ethics (Aiking, 2014; Springmann et al., 2018). These fermented plant-based food products provide a viable alternative for consumers seeking sustainable, nutritious, and taste-driven food choices in lieu of animal products through the use of fermentation technology (FAO, 2019). There are some key consumer perceptions regarding plant-based fermented food products that are impacted by the sensory, nutritional, and psychological attributes of these products. Specifically, although consumers are confident that fermented plant-based food products are healthier and more sustainable, consumers still have concerns regarding the flavour, texture, and familiarity of these plant-based fermented food products versus animal-based products (Giacalone et al., 2022). The flavour, texture, and aroma/odour are all sensory factors that have a significant impact on consumer acceptance of fermented plant-based food products (Michel et al., 2021). Most fermented plant-based foods do not replicate the same sensory properties as animal-based products; for example, plant-based dairy substitutes have significantly less aroma and flavour compared to traditional dairy products (Appiani et al., 2023).

1.7 Future Perspectives

The advances in future fermentation will tend to concentrate on increasing levels of strain engineering. For example, more productive, stable and resilient next-generation strains will be developed using the advantages offered by AI and machine learning to optimise genetic changes (Yadav et al., 2024). The transition to non-food biomass feedstocks will improve sustainability and reduce costs. Innovations in bioreactor design and fermentation processes will increase scalability and energy efficiency by using renewable energy sources to reduce carbon footprints. Product diversification will go beyond food and drinks to include bioplastics, bio-based fabrics, and new flavour compounds for food applications. In addition, multifunctional microbes capable of synthesising complex substances in a single fermentation process will be created, as well as circular economy models that value industrial byproducts and waste streams as fermentation inputs (Verma et al., 2025).

Personalised fermented foods are dietary products developed using an individual's microbiome data to treat underlying physiological and health issues. The utilisation of machine learning and data-driven artificial intelligence helps scientists to create targeted bioformulations intended to correct an imbalance between the gut's microbiome and the health of the individuals, resulting in driving innovation through the development of tailored products that restore microbial balance and optimise nutrition (Patil et al., 2025). Moreover, researchers are better equipped with tools for analysing data sequencing from all studies on microbial populations via AI; thus, determining which probiotics, substrates, or bioactive components will be used within a given fermentation process (D'Urso & Broccolo, 2024). AI systems utilise data generated from multiple omics technologies (macro- and micro-omics), including metagenomic, metabolomic, and transcriptomic data, to allow for precision fermentation that meets the needs of each individual's unique microbiome profile (Walsh et al., 2023). To enhance the yield of bioactive fermentation by-products generated during the fermentative process, including short-chain fatty acids critical for maintaining gut health and decreasing inflammation (Abbaspour, 2024), AI could also be utilised to improve fermentation conditions (pH, temperature, substrate) for these products.

1.8 Conclusion

Researchers are developing new microorganisms and microbial consortia to enable a comprehensive approach to the production of innovative plant-based fermented foods. Advances in synthetic biology and precision fermentation enable the development of targeted nutrient types to provide optimal bioavailability of high-quality proteins, vitamins, and functional metabolite content while also enhancing desirable characteristics such as the flavours and textures of plant-based fermented foods. Although significant progress has been made in this area, there remain significant issues to be overcome, including regulatory issues, scalability, costs and consumers' acceptance. Advancements in microbiome-driven innovations and precision fermentation technologies within plant-based food systems have the potential to enable the development of next-generation foods that are sustainable, nutritionally enhanced, and widely accepted by consumers.

References

Abbaspour, N. (2024). Fermentation's pivotal role in shaping the future of plant-based foods: An integrative review of fermentation processes and their impact on sensory and health benefits. *Applied Food Research, 4*(2), 100468.

Aiking, H. (2014). Protein production: Planet, profit, plus people. *The American Journal of Clinical Nutrition, 100*(suppl_1), 483S–489S.

Ajayeoba, T. A., & Ijabadeniyi, O. A. (2025). Transforming food for the future: Precision fermentation as a key to sustainability, nutrition, and health.

Appiani, M., Cattaneo, C., & Laureati, M. (2023). Sensory properties and consumer acceptance of plant-based meat, dairy, fish and eggs analogs: A systematic review. *Frontiers in Sustainable Food Systems, 7*, 1268068.

Aschemann-Witzel, J., Gantriis, R. F., Fraga, P., & Perez-Cueto, F. J. (2021). Plant-based food and protein trend from a business perspective: Markets, consumers, and the challenges and opportunities in the future. *Critical Reviews in Food Science and Nutrition, 61*(18), 3119–3128.

Aswathy, M. R., Pandikkadan Sundaran, S. P. S., Sangari, T., & Samuel, S. (2024). Fermentation technology: Empowering biotechnology for a sustainable future. *Futuristic Trends in Biotechnology, 3*, 87–116.

Augustin, M. A., Hartley, C. J., Maloney, G., & Tyndall, S. (2024). Innovation in precision fermentation for food ingredients. *Critical Reviews in Food Science and Nutrition, 64*(18), 6218–6238.

Boukid, F., Ganeshan, S., Wang, Y., Tülbek, M. Ç., & Nickerson, M. T. (2023). Bioengineered enzymes and precision fermentation in the food industry. *International Journal of Molecular Sciences, 24*(12), 10156.

Chen, Y., Tan, Z., Zhang, Q., & Wang, H. (2024). Enhanced lysine production in Escherichia coli via CRISPR-mediated metabolic rewiring. *ACS Synthetic Biology, 13*, 287–298.

Christensen, L. F., García-Béjar, B., Bang-Berthelsen, C. H., & Hansen, E. B. (2022). Extracellular microbial proteases with specificity for plant proteins in food fermentation. *International Journal of Food Microbiology, 381*, 109889.

Clark, M., & Tilman, D. (2017). Comparative analysis of environmental impacts of agricultural production systems, agricultural input efficiency, and food choice. *Environmental Research Letters, 12*(6), 064016.

Costa, C. E., Møller-Hansen, I., Romani, A., Teixeira, J. A., Borodina, I., & Domingues, L. (2021). Resveratrol production from hydrothermally pretreated eucalyptus wood using recombinant industrial Saccharomyces cerevisiae strains. *ACS Synthetic Biology, 10*(8), 1895–1903.

Costa, C. E., Romaní, A., Teixeira, J. A., & Domingues, L. (2022). Resveratrol production for the valorisation of lactose-rich wastes by engineered industrial Saccharomyces cerevisiae. *Bioresource Technology, 359*, 127463.

Cress, B. F., Leitz, Q. D., Kim, D. C., Amore, T. D., Suzuki, J. Y., Linhardt, R. J., & Koffas, M. A. (2017). CRISPRi-mediated metabolic engineering of E. coli for O-methylated anthocyanin production. *Microbial Cell Factories, 16*(1), 10.

D'Urso, F., & Broccolo, F. (2024). Applications of artificial intelligence in microbiome analysis and probiotic interventions—An overview and perspective based on the current state of the art. *Applied Sciences, 14*(19), 8627.

Dahiya, D., & Nigam, P. S. (2022). Probiotics, prebiotics, synbiotics, and fermented foods as potential biotics in nutrition improving health via microbiome-gut-brain axis. *Fermentation, 8*(7), 303.

Dan, T., Hu, H., Tian, J., He, B., Tai, J., & He, Y. (2023). Influence of different ratios of lactobacillus delbrueckii subsp. bulgaricus and Streptococcus thermophilus on fermentation characteristics of yogurt. *Molecules, 28*(5), 2123.

Deokar, G. S., Pathak, V. A., Kshirsagar, S. J., Al-Asmari, F., & Nirmal, N. (2025). Enhancement of vitamin B12 in plant-based food through microbial fermentation-a sustainable food system. *Food Chemistry, 484*, 144437.

Dhiman, S., Kaur, S., Thakur, B., Singh, P., & Tripathi, M. (2025). Nutritional enhancement of plant-based fermented foods: Microbial innovations for a sustainable future. *Fermentation, 11*(6), 346.

Dimidi, E., Cox, S. R., Rossi, M., & Whelan, K. (2019). Fermented foods: Definitions and characteristics, impact on the gut microbiota and effects on gastrointestinal health and disease. *Nutrients, 11*(8), 1806.

Efsa Panel (EFSA Panel on Dietetic Products, Nutrition and Allergies), N. D. A. (2015). Guidance on the preparation and presentation of an application for authorisation of a novel food in the context of Regulation (EU) 2015/2283. *EFSA Journal, 14*, e04594.

EFSA Panel on Genetically Modified Organisms (GMO). (2011). Guidance on the risk assessment of genetically modified microorganisms and their products intended for food and feed use. *EFSA Journal, 9*(6), 2193.

Fang, H., Li, D., Kang, J., Jiang, P., Sun, J., & Zhang, D. (2018). Metabolic engineering of Escherichia coli for de novo biosynthesis of vitamin B12. *Nature Communications, 9*(1), 4917.

FAO. (2006). *Probiotics in food. Health and nutritional properties and guidelines for evaluation*. World Health Organization; Food and Agriculture Organization of the United Nations.

FAO, W. (2019). *Sustainable healthy diets–guiding principles*. Food and Agriculture Organization of the United Nations; World Health Organization; Food and Agriculture Organization of the United Nations/World Health Organization.

Farid, M. S., Anjum, R., Yang, Y., Tu, M., Zhang, T., Pan, D., et al. (2024). Recent trends in fermented plant-based analogues and products, bioactive peptides, and novel technologies-assisted fermentation. *Trends in Food Science & Technology, 149*, 104529.

Filipe, D., Vieira, L., Ferreira, M., Oliva-Teles, A., Salgado, J., Belo, I., & Peres, H. (2023). Enrichment of a plant feedstuff mixture's nutritional value through solid-state fermentation. *Animals, 13*(18), 2883.

Ford, A. C., Harris, L. A., Lacy, B. E., Quigley, E. M., & Moayyedi, P. (2018). Systematic review with meta-analysis: The efficacy of prebiotics, probiotics, synbiotics and antibiotics in irritable bowel syndrome. *Alimentary Pharmacology & Therapeutics, 48*(10), 1044–1060.

Fusco, W., Lorenzo, M. B., Cintoni, M., Porcari, S., Rinninella, E., Kaitsas, F., et al. (2023). Short-chain fatty-acid-producing bacteria: Key components of the human gut microbiota. *Nutrients, 15*(9), 2211.

Gao, J., Jiang, L., & Lian, J. (2021). Development of synthetic biology tools to engineer Pichia pastoris as a chassis for the production of natural products. *Synthetic and Systems Biotechnology, 6*(2), 110–119.

Gautam, A., Poopalarajah, R., Ahmad, A. R., Rana, B. N., Denekew, T. W., Ahn, N., et al. (2025). Ecological factors that drive microbial communities in culturally diverse fermented foods. *BMC Microbiology, 25*(1), 1–20.

Ghatani, K., Sha, S. P., Thapa, S., Sarkar, I., Sen, G., & Sen, A. (2024). Investigating bacterial diversity involved in the production of vegetable-based ethnic fermented food of North Bengal and their metabolic pathways with reverse ecology approach. *Frontiers in Sustainable Food Systems, 8*, 1322192.

Giacalone, D., Clausen, M. P., & Jaeger, S. R. (2022). Understanding barriers to consumption of plant-based foods and beverages: Insights from sensory and consumer science. *Current Opinion in Food Science, 48*, 100919.

Guo, Q., Chen, P., & Chen, X. (2023). Bioactive peptides derived from fermented foods: Preparation and biological activities. *Journal of Functional Foods, 101*, 105422.

Gupta, S. K., & Shukla, P. (2017). Sophisticated cloning, fermentation, and purification technologies for an enhanced therapeutic protein production: A review. *Frontiers in Pharmacology, 8*, 419.

Han, C., Xia, K., Ma, J., Liang, X., & Wang, W. (2025). CRISPR-Cas9 mediated construction of a food-grade pyrroloquinoline quinone high-yielding acetobacter pasteurianus Ab3 strain. *Food Bioscience, 68*, 106452.

Haryani, Y., Halid, N. A., Goh, S. G., Nor-Khaizura, M. A. R., Hatta, M. A. M., Sabri, S., et al. (2024). Efficient metabolic pathway modification in various strains of lactic acid bacteria using CRISPR/Cas9 system for elevated synthesis of antimicrobial compounds. *Journal of Biotechnology, 395*, 53–63.

He, M. T., & Howell, K. S. (2022). Vitamin-B12 enrichment in tempeh by co-culture with Propionibacterium freudenreichii during fermentation.

Hong, K. J., Lee, C. H., & Kim, S. W. (2004). Aspergillus oryzae GB-107 fermentation improves nutritional quality of food soybeans and feed soybean meals. *Journal of Medicinal Food, 7*(4), 430–435.

Hsieh, C. C., Yu, S. H., Cheng, K. W., Liou, Y. W., Hsu, C. C., Hsieh, C. W., et al. (2023). Production and analysis of metabolites from solid-state fermentation of Chenopodium formosanum (Djulis) sprouts in a bioreactor. *Food Research International, 168*, 112707.

Hsieh, C. C., Lin, H. W., Huang, H. C., Khumsupan, D., Shen, S. C., Lin, S. P., et al. (2025). Peptide from tempeh-like fermented Chenopodium formosanum counters senescence while enhancing antioxidant ability in non-replicative aging. *LWT, 222*, 117641.

Huang, W., Yang, S., Wätjen, A. P., Gumulya, Y., Fernández-Pacheco, P., Marcellin, E., et al. (2025). Isolation of an exopolysaccharide-producing Weissella confusa strain from lettuce and exploring its application as a texture modifying adjunct culture in a soy milk alternative. *International Journal of Food Microbiology, 428*, 110992.

IMARC Group. (2025). Asia-Pacific plant-based food market: Industry trends, share, size, growth, opportunity and forecast 2025–2035.

Iqbal, B., Alabbosh, K. F., Raheem, A., Ullah, I., Khan, A. R., & Memon, M. S. (2025). Microbial transformation: The role of fermentation in advancing nutritional quality and human health. *Archives of Microbiology, 207*(9), 228.

Khayatan, D., Nouri, K., Momtaz, S., Roufogalis, B. D., Alidadi, M., Jamialahmadi, T., et al. (2024). Plant-derived fermented products: An interesting concept for human health. *Current Developments in Nutrition, 8*(5), 102162.

Kildegaard, K. R., Adiego-Pérez, B., Belda, D. D., Khangura, J. K., Holkenbrink, C., & Borodina, I. (2017). Engineering of Yarrowia lipolytica for production of astaxanthin. *Synthetic and Systems Biotechnology, 2*(4), 287–294.

Knight, T. (2003). *Idempotent vector design for standard assembly of biobricks*. Defense Technical Information Center.

Knychała, M. M., Boing, L. A., Ienczak, J. L., Trichez, D., & Stambuk, B. U. (2024). Precision fermentation as an alternative to animal protein, a review. *Fermentation, 10*(6), 315.

Lee, Y. G., Kim, B. Y., Bae, J. M., Wang, Y., & Jin, Y. S. (2022). Genome-edited Saccharomyces cerevisiae strains for improving quality, safety, and flavor of fermented foods. *Food Microbiology, 104*, 103971.

Leeuwendaal, N. K., Stanton, C., O'toole, P. W., & Beresford, T. P. (2022). Fermented foods, health and the gut microbiome. *Nutrients, 14*(7), 1527.

Levisson, M., Patinios, C., Hein, S., de Groot, P. A., Daran, J. M., Hall, R. D., et al. (2018). Engineering de novo anthocyanin production in Saccharomyces cerevisiae. *Microbial Cell Factories, 17*(1), 103.

Li, M., Li, X., Zhou, Z., Wu, P., Fang, M., Pan, X., et al. (2016). Reassessment of the four yield-related genes Gn1a, DEP1, GS3, and IPA1 in rice using a CRISPR/Cas9 system. *Frontiers in Plant Science, 7*, 377.

Lietzow, J., Luckert, C., & Schäfer, B. (2021). Novel and traditional foods: Novel food regulation in the EU. In *Regulatory toxicology* (pp. 1373–1400). Springer International Publishing.

Lim, M. J., Barathikannan, K., Jeong, Y. J., Chelliah, R., Vijayalakshmi, S., Park, S. J., & Oh, D. H. (2023). Exploring the impact of fermentation on brown rice: Health benefits and value-added foods—A comprehensive meta-analysis. *Fermentation, 10*(1), 3.

Liu, Z., Peng, Z., Huang, T., Guan, Q., Li, J., Xie, M., & Xiong, T. (2019). Bacterial community dynamics and physicochemical characteristics in natural fermentation of jiang-shui, a traditional food made in Northwest China. *Journal of the Science of Food and Agriculture, 99*(7), 3391–3397.

Ma, X., Liu, Y., Li, Q., Liu, L., Yi, L., Ma, L., & Zhai, C. (2016). Expression, purification and identification of a thermolysin-like protease, neutral protease I, from Aspergillus oryzae with the Pichia pastoris expression system. *Protein Expression and Purification, 128*, 52–59.

Ma, H., Cornadó, D., & Raaijmakers, J. M. (2025). The soil-plant-human gut microbiome axis into perspective. *Nature Communications, 16*(1), 7748.

Mallikarjuna, R. S., & Yellamma, K. (2019). Genetic modification of microorganisms for enhanced production of food ingredients. *International Journal of Food Science and Technology, 54*(7), 2345–2359.

McKeague, M., Wang, Y. H., Cravens, A., Win, M. N., & Smolke, C. D. (2016). Engineering a microbial platform for de novo biosynthesis of diverse methylxanthines. *Metabolic Engineering, 38*, 191–203.

Michel, F., Hartmann, C., & Siegrist, M. (2021). Consumers' associations, perceptions and acceptance of meat and plant-based meat alternatives. *Food Quality and Preference, 87*, 104063.

Mirsalami, S. M., & Mirsalami, M. (2025). Advances in genetically engineered microorganisms: Transforming food production through precision fermentation and synthetic biology. *Future Foods, 11*, 100601.

Nath, H., Samtiya, M., & Dhewa, T. (2022). Beneficial attributes and adverse effects of major plant-based foods anti-nutrients on health: A review. *Human Nutrition & Metabolism, 28*, 200147.

Navale, G. R., Dharne, M. S., & Shinde, S. S. (2021). Metabolic engineering and synthetic biology for isoprenoid production in Escherichia coli and Saccharomyces cerevisiae. *Applied Microbiology and Biotechnology, 105*(2), 457–475.

Nielsen, M. B., Meyer, A. S., & Arnau, J. (2024). The next food revolution is here: Recombinant microbial production of milk and egg proteins by precision fermentation. *Annual Review of Food Science and Technology, 15*(2024), 173–187.

Nithya, A., Misra, S., Panigrahi, C., Dalbhagat, C. G., & Mishra, H. N. (2023). Probiotic potential of fermented foods and their role in non-communicable diseases management: An understanding through recent clinical evidences. *Food Chemistry Advances, 3*, 100381.

Obayomi, O. V., Olaniran, A. F., & Owa, S. O. (2024). Unveiling the role of functional foods with emphasis on prebiotics and probiotics in human health: A review. *Journal of Functional Foods, 119*, 106337.

Panel, E. B. (2022). Update of the list of QPS-recommended biological agents intentionally added to food or feed as notified to EFSA 15: Suitability of taxonomic units notified to EFSA until September 2021. *EFSA Journal, 20*(1), e07045.

Pasaribu, T., Laconi, E. B., & Kompiang, I. P. (2019). Evaluation of the nutrient contents of palm kernel cake fermented by microbial cocktails as a potential feedstuff for poultry. *Journal of the Indonesian Tropical Animal Agriculture, 44*(3), 295.

Patil, A., Singh, N., Patwekar, M., Patwekar, F., Patil, A., Gupta, J. K., et al. (2025). AI-driven insights into the microbiota: Figuring out the mysterious world of the gut. *Intelligent Pharmacy, 3*(1), 46–52.

Peng, B., & Wei, S. (2025). Synthetic engineering of microbes for production of terpenoid food ingredients. *Journal of Agricultural and Food Chemistry, 73*(17), 10052–10068.

Praveen, M., & Brogi, S. (2025). Microbial fermentation in food and beverage industries: Innovations, challenges, and opportunities. *Foods, 14*(1), 114.

Qi, N., Zhan, X., Milmine, J., Sahar, M., Chang, K. H., & Li, J. (2023). Isolation and characterization of a novel hydrolase-producing probiotic bacillus licheniformis and its application in the fermentation of soybean meal. *Frontiers in Nutrition, 10*, 1123422.

Quan, Q., Liu, W., Guo, J., Ye, M., & Zhang, J. (2022). Effect of six lactic acid bacteria strains on physicochemical characteristics, antioxidant activities and sensory properties of fermented orange juices. *Foods, 11*(13), 1920.

Ramírez Rojas, A. A., Swidah, R., & Schindler, D. (2022). Microbes of traditional fermentation processes as synthetic biology chassis to tackle future food challenges. *Frontiers in Bioengineering and Biotechnology, 10*, 982975.

Rana, A., Taneja, N. K., Singh, A., Dhewa, T., Kumar, V., Kumar, A., et al. (2025). Synergistic fermentation of vitamin B2 (riboflavin) bio-enriched soy milk: Optimization and techno-functional characterization of next generation functional vegan foods. *Discover Food, 5*(1), 10.

Rezac, S., Kok, C. R., Heermann, M., & Hutkins, R. (2018). Fermented foods as a dietary source of live organisms. *Frontiers in Microbiology, 9*, 1785.

Rodrigues, R. C., Pereira, H. S., Senra, R. L., Ribon, A. D. O. B., & de Oliveira Mendes, T. A. (2023). Understanding the emerging potential of synthetic biology for food science: Achievements, applications and safety considerations. *Food Chemistry Advances, 3*, 100476.

Roy, S., & Dhaneshwar, S. (2023). Role of prebiotics, probiotics, and synbiotics in management of inflammatory bowel disease: Current perspectives. *World Journal of Gastroenterology, 29*(14), 2078.

Sharma, R., Garg, P., Kumar, P., Bhatia, S. K., & Kulshrestha, S. (2020). Microbial fermentation and its role in quality improvement of fermented foods. *Fermentation, 6*(4), 106.

Silva, J. B. G., & Paulillo, L. C. M. S. (2015). Synthetic biology: Possibilities and challenges. *Revista da Biologia, 14*(1), 33–39.

Sooklim, C., Paemanee, A., Ratanakhanokchai, K., Wiwatratana, D., & Soontorngun, N. (2025). Integrated omic analysis of a new flavor yeast strain in fermented rice milk. *FEMS Yeast Research, 25*, foaf017.

Springmann, M., Wiebe, K., Mason-D'Croz, D., Sulser, T. B., Rayner, M., & Scarborough, P. (2018). Health and nutritional aspects of sustainable diet strategies and their association with environmental impacts: A global modelling analysis with country-level detail. *The Lancet Planetary Health, 2*(10), e451–e461.

Tang, Z., Wang, Q., Zhao, Z., Shen, N., Qin, Y., Lin, W., et al. (2023). Evaluation of fermentation properties, antioxidant capacity in vitro and in vivo, and metabolic profile of a fermented beverage made from apple and cantaloupe. *LWT, 179*, 114661.

Tang, N., Xing, X., Li, H., Suo, B., Wang, Y., Ai, Z., & Yang, Y. (2024). Co-culture fermentation by Saccharomycopsis fibuligera and lactic acid bacteria improves bioactivity and aroma profile of wheat bran and the bran-containing Chinese steamed bread. *Food Research International, 182*, 114179.

Teng, T. S., Chin, Y. L., Chai, K. F., & Chen, W. N. (2021). Fermentation for future food systems: Precision fermentation can complement the scope and applications of traditional fermentation. *The EMBO Reports, 22*(5), EMBR202152680.

Thierry, A., Madec, M. N., Chuat, V., Bage, A. S., Picard, O., Grondin, C., et al. (2023). Microbial communities of a variety of 75 homemade fermented vegetables. *Frontiers in Microbiology, 14*, 1323424.

Turck, D., Bohn, T., Castenmiller, J., De Henauw, S., Hirsch-Ernst, K. I., Maciuk, A., et al. (2024). Guidance on the scientific requirements for an application for authorisation of a novel food in the context of Regulation (EU) 2015/2283. *EFSA Journal, 22*(9), e8961.

van Bruggen, A. H., Goss, E. M., Havelaar, A., van Diepeningen, A. D., Finckh, M. R., & Morris, J. G., Jr. (2019). One health-cycling of diverse microbial communities as a connecting force for soil, plant, animal, human and ecosystem health. *Science of the Total Environment, 664*, 927–937.

Veniranda, V., & Surya, R. (2022). Consumer analysis of commercial plant-based jerky. In *IOP conference series: Earth and environmental science* (Vol. 998, p. 012059). IOP Publishing.

Venter de Villiers, M., Cheng, J., & Truter, L. (2024). The shift towards plant-based lifestyles: Factors driving young consumers' decisions to choose plant-based food products. *Sustainability, 16*(20), 9022.

Verma, K., Duhan, P., Pal, D., Verma, P., & Bansal, P. (2025). Precision fermentation for the next generation of food ingredients: Opportunities and challenges. *Future Foods, 12*, 100750.

Walsh, A. M., Leech, J., Huttenhower, C., Delhomme-Nguyen, H., Crispie, F., Chervaux, C., & Cotter, P. D. (2023). Integrated molecular approaches for fermented food microbiome research. *FEMS Microbiology Reviews, 47*(2), fuad001.

Wang, Y., Wu, J., Lv, M., Shao, Z., Hungwe, M., Wang, J., et al. (2021). Metabolism characteristics of lactic acid bacteria and the expanding applications in food industry. *Frontiers in Bioengineering and Biotechnology, 9*, 612285.

Wang, C., Wu, H., Liu, Z., Barrow, C., Dunshea, F., & Suleria, H. A. (2022). Bioaccessibility and movement of phenolic compounds from tomato (Solanum lycopersicum) during in vitro gastrointestinal digestion and colonic fermentation. *Food & Function, 13*(9), 4954–4966.

Wong, L., Engel, J., Jin, E., Holdridge, B., & Xu, P. (2017). YaliBricks, a versatile genetic toolkit for streamlined and rapid pathway engineering in Yarrowia lipolytica. *Metabolic Engineering Communications, 5*, 68–77.

Xie, Z., Jin, Y. S., & Miller, M. J. (2025). Exploiting endogenous Cas9-based genome editing of Lacticaseibacillus rhamnosus GG (LGG) for precision food fermentation and live biotherapeutics. *bioRxiv*. https://doi.org/10.1101/2025.01.21.634116

Xin, Y., Guo, T., & Qiao, M. (2025). Current application and future prospects of CRISPR-Cas in lactic acid bacteria: A review. *Food Research International, 209*, 116315.

Xu, S., Wang, Q., Zeng, W., Li, Y., Shi, G., & Zhou, J. (2020). Construction of a heat-inducible Escherichia coli strain for efficient de novo biosynthesis of L-tyrosine. *Process Biochemistry, 92*, 85–92.

Yadav, H., Singh, S., & Sinha, R. (2024). Fermentation technology for microbial products and their process optimization. In *Industrial microbiology and biotechnology: A new horizon of the microbial world* (pp. 35–64). Springer Nature Singapore.

Zhang, J., Song, H. S., Zhang, C., Kim, Y. B., Roh, S. W., & Liu, D. (2021). Culture-independent analysis of the bacterial community in Chinese fermented vegetables and genomic analysis of lactic acid bacteria. *Archives of Microbiology, 203*(7), 4693–4703.

Zimberoff, L. (2021). *Technically food: Inside Silicon Valley's mission to change what we eat*. Abrams.

Chapter 2
The Evolution of Fermentation: From Ancient Methods to AI-Driven Bioengineering

Kiran Kharb, Sukhminderjit Kaur, and Sunny Dhiman

Abstract Fermentation technology is thought to have been developed in the Fertile Crescent, the Middle East, and currently applied to improve safety attributes and preserve foods and their nutritional quality. Fermented food products, such as fermented dairy foods, alcoholic beverages, vegetables, fruits, and meats, have longer storage stability and low spoilage risk. With growing fermentation technology, even waste food is converted into valuable products and by-products. Fermented food will help feed the world more sustainably than traditional agricultural products. The integration of innovations in traditional, biomass, and precision fermentation technologies is driving a transformative shift in fermented food production systems. Moreover, artificial intelligence (AI) has become a revolutionary force, enabling revolution from microbial engineering to individualized fermented food products. AI decoded sensory attributes using a database and language processing, acting as a bridge between lab-scale innovation and market uptake. In this chapter, we will discuss how the evolution of fermentation happens from ancient methods to AI-driven bioengineering.

Keywords AI-driven bioengineering · Biomass fermentation · Fermented food · Precision fermentation · Traditional fermentation

2.1 Introduction

The raising concerns about the availability of sustainable and healthy natural food sources are due to the rapidly increasing global population. To save our food system and the world's natural resources, more sustainable food production methods or food processing operations must be used (Varzakas & Smaoui, 2024). Fermentation, as a powerful biotechnological tool, plays a pivotal role in developing next-generation food products by enabling the production of microbial biomass or their metabolites

K. Kharb · S. Kaur (✉) · S. Dhiman
Department of Biotechnology, University Institute of Biotechnology, Chandigarh University, Mohali, Punjab, India
e-mail: sukhminderjit.uibt@cumail.in

S. Dhiman, S. Kaur (eds.), *Next-Gen Plant Based Fermented Foods: Innovations for a Sustainable Future*, https://doi.org/10.1007/978-3-032-29927-7_2

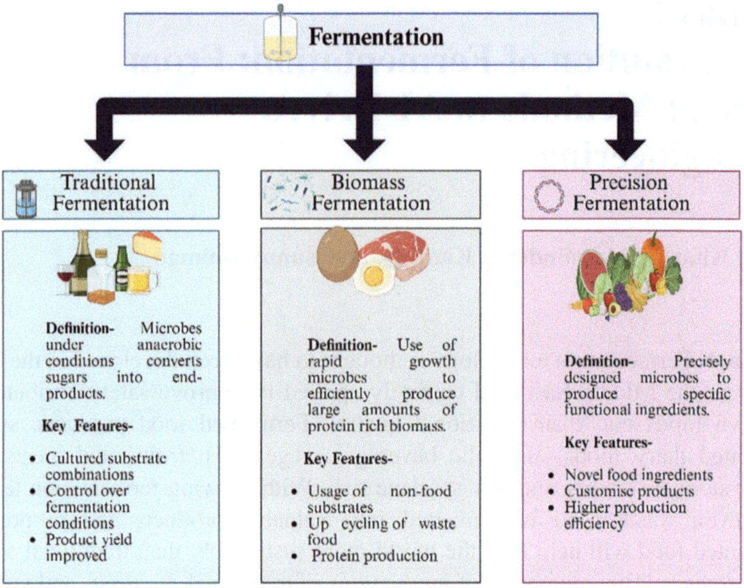

Fig. 2.1 Illustration of fermentation and its types in modern food systems

(Kamble et al., 2021). From ancient times, traditionally, plant and animal-derived food production has been on agricultural land areas. Compared to traditional farming, food produced by fermentation have less requirement of land, emit less greenhouse gas, and use less water. Fermentation, which includes traditional fermentation, biomass fermentation, and precision fermentation, has an important role in providing food to the next generation, as shown in Fig. 2.1 (Augustin et al., 2024). Additionally, fermented foods help in lowering cholesterol levels, fostering the immune system, infection protection, and reducing lactose sensitivity, among various other health benefits. Fermentation is a time-honored preservation method used for ages to prolong the shelf life of food and augment its nutritional quality (Diana Celcia et al., 2025).

From the ancient times, traditional fermentation used to preserve foods and maintain the nutritional value of foods. For the purpose of preserving food by the traditional fermentation method, substrates include fish, meat, milk, vegetables, legumes, cereals, and fruits (Mani, 2018). By this method, fermented foods are produced under controlled fermentation conditions through the action of microbial enzymes, and increase production yield. Some of the most famous fermented foods, such as yogurts, kimchi, kefir, sauerkraut, cheese, tempeh, wine, and beer are made using traditional fermentation methods (Farooqui, 2021).

Under biomass fermentation method, usage of edible microorganisms (non-food substrates) as an alternative to conventional food produced by traditional agriculture for upcycling the waste food (de Medeiros et al., 2022). Some microalgae (*Arthrospira platensis*, *Spirulina platensis*, *Chlorella vulgaris*, and *Dunaliella salina*) have been highly cultivated for protein production due to their nutritional value, metabolic versatility, and functional divergence (Linder, 2019; Vijay et al., 2025). Texturing techniques for microalgal biomass are still at an initial stage and vital for further analysis of meat analogues. For microalgae production, economic issues are the biggest challenge due to the limited scale-up and inefficient processing techniques (Prates, 2025).

Fermentation is an ancient method used for the production of various food ingredients (sweeteners, organic acids, stabilizers, colorants, thickening agents, amino acids, antioxidants, enzymes, and vitamins), provides flavour, preservation, enrichment, and textural quality to fermented food (Iram et al., 2023). Some algae (*Spirogyra sp.*, *Cosmarium sp.*, and *Cosmarium blytii*) showed their potential possibilities to be considered as novel sources of bioactive components in food, feed and biomedical applications due to their high nutritional value and antioxidant activity (Santiago-Díaz et al., 2022). Microorganisms such as fungi (*Aspergillus*, *Penicillium*, *Kluyveromyces*, and *Rhizopus*) yeast (*Saccharomyces*, *Pichia*, *Candida*, and *Yarrowia*), bacteria (*Lactobacillus*, *Bacillus*, and recombinant *Escherichia coli*) highly used in the food industry has the potential to create high-value products from industrial food waste through fermentation (Sun et al., 2022; Vivek & Venkitasamy, 2023). Fermentation can increase the extraction efficiency of compounds from the food matrix, modify the antioxidant profile, and create new bioactive compounds while requiring less water for the microorganism's growth (Marti-Quijal et al., 2021). Some examples of food waste fermentation include the production of flavonoids or carotenoids (fruit or vegetable waste), bioactive peptides (cereal processing waste), antioxidant peptides or hydrolysates (by-products of livestock and shellfish processing), glucoamylase enzymes (tea waste, rice bran, wheat bran, and corn stover), and microbial red, orange, or yellow pigments (food industry wastes like tomato waste, cottonseed meal, and fruit seeds) etc. (Ogidi & Hung, 2025; Faria et al., 2023). The bioactive compounds produced from bacterial or fungal fermentation can inhibit the growth of commonly occurring food spoilage fungi and used as natural preservatives (Deshmukh & Gaikwad, 2024).

Precision fermentation uses targeted genetic modifications in fermenting microorganisms to improve efficiency, precision, and scalability. Faster advancements in precision fermentation have enable the genetic programming of microorganisms to produce complex organic compounds (Ajayeoba & Ijabadeniyi, 2025). Using precision fermentation to produce novel food components is becoming an option for the transformation of food systems. Global needs for more sustainable food production are driving this trend. It considers the market opportunities for fermentation-derived food ingredients and issues for bringing fermentation-derived food ingredients to market (Teng et al., 2021). The technological considerations for the development and exploitation of the precision fermentation value chain including the (1) choice of the feedstock for fermentation, (2) use of synthetic biology tools to tailor microbes for efficient production in the fermentation process, (3) Downstream processing operations to recover the food component of interest, (4) scale-up of the fermentation process, and (5) characteristics of new fermentation-derived food products that are acceptable to consumers (Ledesma-Amaro & Bamezai, 2025).

Over the time, scientific research has linked the consumption of fermented foods to better health. Food compounds are broken down more easily into digestible forms during the fermentation process. It also helps to reduce the amount of toxins and pathogens in novel food products (Iqbal et al., 2025). Additionally, probiotics containing fermented food and food products are beneficial for humans in food digestion and absorb nutrients. Globally, the number of patients with noncontagious diseases (cardiovascular, type 2 diabetes, cancer, and allergies) has increased, and a diet that contains fermented foods can reduce the risk of these diseases (Ashraf et al., 2022). Globally, the food system stands at a critical point and must deliver nutritious or sustainable food and food products for rapidly growing population while the food innovation remains steady, empirical, and disintegrate. Artificial intelligence (AI) appears as a revolutionary path that links molecular composition to functional performance, chemical structure to sensory outcomes, and accelerates cross-disciplinary renovation across the entire food production channel. AI has figured out about biological complexity, streamlining procedures, and forecasting human behaviour, and has become a keystone for rebuilding the food system. Moreover, AI helps in redefining food production or consumption, offering solutions that balance planetary fermented food products without repeating the inequities of past technological revolts, from lab to fork (Lugo-Morin, 2024). This chapter elaborates the evolution of fermentation from ancient methods to AI-driven bioengineering.

2.2 Traditional Fermentation Methods from Ancient Times

Currently, all over the world, more than 5000 different fermented food products, such as cereal-based fermented foods, alcoholic/non-alcoholic beverages, fermented milk, fruits, vegetables, and meat products, have been consumed in the diet (Ray et al., 2024). Though different regions have different preparation techniques for fermented food products, because of which each fermented food product has its particular taste, texture, aroma, and other organoleptic characteristics (Sharma et al., 2020). A wide variety of fermented food products using different materials, tools, and fermentation times. Therefore, with the increasing demand of fermented food, it also raises important safety concerns, as improper fermentation techniques and inadequate hygiene can pose health risks (Singh & Kumar, 2025). Fermented commodities have been classified according to a variety of schemes. Fermented food is divided into three types: acidic, alkaline, and alcoholic fermentation, based on the main metabolic end product (Qin et al., 2022).

2.2.1 Lactic Acid Fermentation

This type of fermentation, mediated by lactic acid bacteria (LAB) and involving carbohydrate metabolism, and their final product is widely relied upon in food, cosmetics, and chemical industries. Lactic acid is a complex agent with multifunctional properties, including food preservation, maintaining pH, mosquito repellent, and skin moisturizing (Ajanaku et al., 2025). Based upon catabolic characteristics, there are two types of LAB: homofermentative and heterofermentative. Carbohydrates are converted to pyruvate during homofermentation via the Embden-Meyerhof pathway. Then, stoichiometrically, pyruvate convert into lactate. Under hetero-fermentation, through the pentose phosphor-ketolase pathway, carbohydrates are broken down into lactic acid, acetic acid, and ethanol (Rawoof et al., 2021).

Everytime microecosystem that initiates fermentation at its initial phase consists of a diversty of microorganisms. However, as fermentation proceeds, LAB species that exhibit a degree of adaptation to the respective substrate and fermentation conditions become dominant over others (Yu et al., 2020).

2.2.2 Acetic Acid Fermentation

Acetic acid fermentation is performed by acetic acid bacteria (AAB) under strictly aerobic conditions. Production of acetic acid through the oxidation of organic compounds (carbohydrates and alcohols), which is catalyzed by the dehydrogenases of the respiratory chain (Gomes et al., 2018). Acetic acid is an organic acid used a lot because of its strong preservative capability and has unique sensory effects. Acetic acid synthesis is often carried out synthetically or via microbial fermentation (*Acetobacter*, *Gluconacetobacter*, and *Gluconobacter*) (Han et al., 2024). Acetic acid is commonly utilized in the production of vinegar and pickled vegetables. The most famous products of acetic acid fermentation are kombucha and vinegar. A wide variety of raw materials have been employed in both cases, depending on their availability and local food culture (De Roos & De Vuyst, 2018). Mostly, consortial species of acetic acid fermentation are followed by alcoholic fermentation, which is caused by yeasts. The lack of studies regarding the AAB dynamics during fermentation and their physiological traits limits our knowledge and understanding of their relative fitness and the effect on the quality of the final product (Mota & Vilela, 2024).

2.2.3 Propionic Acid Fermentation

Propionic acid fermentation facilitates by propionic acid bacteria (PAB). Propionic acid production through fermentation pathways such as the acrylate, succinate, and 1,2-propanediol pathways, as well as through amino acid catabolic and biosynthetic pathways by PAB (Rachmawati & Triwibowo, 2024). In food fermentations, propionic acid production has been predominantly linked with *Propionibacterium freudenreichii* and *Pro. Acidopropionici* are produced via the Wood-Werkman cycle and the decarboxylation of succinate (Ranaei et al., 2020). Propionic acid fermentation is primarily used in the production of hard cheeses such as Emmental. Interestingly, the aforementioned species do not seem to be adapted to that specific niche, due to their inability to utilize lactose and milk proteins (Bai & Rai, 2015). Therefore, effective development of PAB is achieved through the co-existence with LAB, the metabolic activity of which provides lactate and peptides (Ray et al., 2024).

2.2.4 Alcoholic Fermentation

Alcoholic fermentation is driven by the yeast species (genus *Saccharomyces*), mainly *Saccharomyces cerevisiae*. Yeast, basically used for the production of beverages (beer, wine, and spirits) from ancient times, converts sugars to ethanol and carbon dioxide (Parapouli et al., 2020). The metabolism by yeast improves both the alcoholic flavour and content (Maicas, 2020). Sugars, mainly fructose and glucose, are catabolised by the glycolytic pathway to convert them into pyruvate, thereafter decarboxylated to acetaldehyde and reduced to ethanol (Sokra & Meta, 2025). During alcoholic fermentation, as a base, fruits and grains are used. In the case of grains, the liberation of the fermentable carbohydrates is achieved by pretreatments such as cooking, germination, or saccharification by moulds (Pawar et al., 2024).

2.2.5 Alkaline Fermentation

Alkaline fermentation is performed by species of the genus *Bacillus*, mostly *B. pumilus* and *B. cereus* (Ouoba, 2017). The raw materials employed for alkaline fermentation are seeds and grains that are rich in protein (Arbab Sakandar et al., 2023). Soybean is the most commonly employed seed; however, the use of seeds from a wide variety of domesticated and wild plants, including *Adansonia digitata L.*, *Citrullus colocynthis*, *Hibiscus sabdariffa*, and *Vigna subterranea*, has been reported (Owusu-Kwarteng et al., 2022).

2.3 Scale-Based Differentiation

Despite biochemical classification, fermentation processes can also be differentiated according to production scale, ranging from small-scale traditional methods to large-scale industrial operations. This transition involves increased fermentation process control, standardized operating conditions, and enhanced productivity to meet commercial demands.

2.3.1 Small-Scale Fermentation Process

The small-scale fermentation process is specially utilized for the production of small-scale fermented products. And the processing of the products is more or less similar to traditional processes (De Vries et al., 2018). The only difference is the requirement for small-scale fermenters, which are essential for large-scale bioreactors. They are small but highly automated and customized for the application, and maintain the pH, temperature, agitation, and aeration according to the requirements of the product (Palladino et al., 2024). There are two categories of small-scale fermenters: bench-scale and lab-scale. These fermenters, which can hold 10–100 kg of substrate, are more commonly used in labs and local fermentation units. Small-scale fermentation units lacked technologically advanced fermenters in the past, making products extremely dangerous for consumption. But, technological and scientific advancements in fermentation from traditional methods to AI-driven bioengineering, as shown in Fig. 2.2, and sophisticated versions of fermenters have made these processes more reliable and safer (Keller et al., 2024).

2.3.2 Industrial Scale Fermentation Process

The industrial fermentation process is the intentional use of fermentation technology for the large-scale production of microbial biomass or yield of primary and secondary metabolites (Gupta et al., 2022). The main feature of these industrial units is the

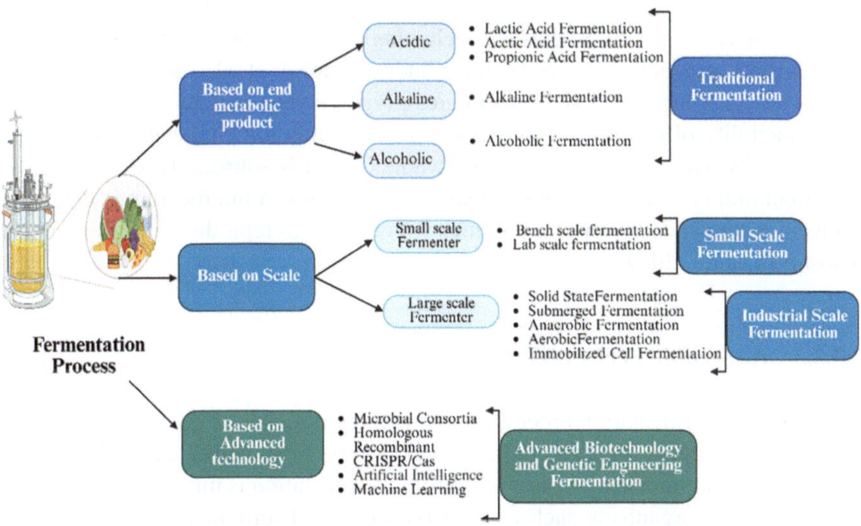

Fig. 2.2 Schematic overview of fermentation techniques, illustrating the evolution of fermentation—from ancient methods to AI-driven bioengineering

use of genetically modified microbes for the rapid production of the desired metabolites (acetic acid, citric acid, and ethanol) and industrial enzymes (lipase, invertase, and rennet) (Regassa et al., 2021). The industrial fermentation process has been categorized into four types as described below.

2.3.2.1 Solid State Fermentation

Solid-state fermentation starts with the growth of microorganisms (yeast and fungi) and product development on the surface of solid substrates, for example, mushroom cultivation and the production of mould-ripened cheeses (Sarma et al., 2024). The raw materials used in solid-state fermentation are rice or wheat bran and maize flour. These substrates may or may not be used for supplying nutrients to the microflora. Soya sauce, hamanatto, miso, tofu, and tempeh are some examples of products developed through the solid-state fermentation process (Tan et al., 2019; Šelo et al., 2021).

2.3.2.2 Submerged Fermentation

Submerged Fermentation basically uses liquid substrates (molasses and broths). During this process, substrates are utilized quite rapidly; hence need to be constantly replaced/supplemented with nutrients (Kalaiselvan et al., 2025). This fermentation technique works well for microorganisms like bacteria that need high moisture content. Additionally, this technique makes the purification of products easy. And the main purpose of submerged fermentation is to use for the extraction of secondary metabolites that must be used in liquid form. Major advantages of submerged cultivations are the well-established technological basis for scaling the processes to industrial production capacity (Hernández-Tenorio et al., 2023). Scalability causes due to controlling pH, temperature, oxygen levels, and nutrient availability. Reproducibility of submerged cultivations is frequently not challenged by medium heterogeneity, since it mostly uses dissolvable C- and N-sources. However, the use of agricultural residues and waste streams as substrates in the medium is heterogeneous in nature and may pose difficulties with respect to reproducibility (Reihani & Khosravi-Darani, 2019).

2.3.2.3 Anaerobic Fermentation

The anaerobic fermentation process is based on the principle of an oxygen-deficient fermentation process (Mishra et al., 2025). Therefore, fermenter air is changed by CO_2, N_2, H_2, or a mixture of gases. Anaerobic fermentation is important for obligate anaerobic microorganisms such as *Clostridium*, which provides them completely oxygen-deprived environment, which is toxic to them (Chauhan et al., 2025). During the anaerobic fermentation process, CO_2 and H_2 gases are created, and these gases are collected to make products like methanol and dry ice. The anaerobic conditions are essential during product recovery, depending upon the type of product. Some products require highly anaerobic conditions because of the high sensitivity of these organisms towards oxygen (Lu & Imlay, 2021).

2.3.2.4 Aerobic Fermentation

Continuous and adequate aeration is essential during the aerobic fermentation process. Some fermentation processes, like acetic acid vinegar production, may need a large amount of oxygen for developing specific aroma and pH (Mota & Vilela, 2024). Therefore, the bioreactors used for fermentation have adequate air supply. These bioreactors are generally closed or batch type, but sometimes continuous flow reactors may also be used to provide a continuous source of product generation (Kumar et al., 2025).

2.3.2.5 Immobilized Cell Fermentation

The immobilized cell fermentation process is used for the production of intracellular enzymes by immobilized cells. It is also advantageous for the synthesis of low molecular weight compounds and unstable enzymes. This type of fermentation process shows high operational activity and low operational cost (Willaert, 2018). Some examples of cell immobilization are commercial production of amino acids, e.g., *Escherichia coli* cells entrapped in polyacrylamide gel for the production of L-aspartic acid, and NADP production by *Brevibacterium* ammonia genes and yeast cells immobilized together in polyamide gel (Gombotz & Hoffman, 2019).

2.4 Advances in Biotechnology, Genetic Engineering, and Fermentation Technologies

Advancements in biotechnology and genetic engineering have significantly transformed fermentation processes. Genetic engineering techniques, like recombinant DNA technology, have enabled the design and modification of microorganisms for enhanced productivity and the production of novel compounds (Mirsalami & Mirsalami, 2025). Several other strategies are discussed below.

2.4.1 Microbial Consortia

Microbial consortia are promising, collaborative approaches to fermentation that exploit the synergy of several microbial species, complementing each other's metabolic activities (Ghosh et al., 2016). Microbial consortia distribute labour among several strains optimized for some metabolic or biosynthetic processes. This division of labour significantly increases process efficiency, reduces the metabolic burden on individual strains, and thus potentially yields more robust production capabilities (Bhatia et al., 2018).

Microbial consortia contribute significantly to the production of fermented foods, enhancing the flavour and quality of food. Umami, often referred to as a "delicious savoury taste," is part of the basic flavour profile experienced in many food items (Crosby, 2019).

However, there are still critical challenges to fully realize the power of microbial consortia on a large scale. The artificially constructed microbial communities are inherently unstable due to possible competition, mutations, and environmental variability (Santillan et al., 2025). Continued research in the optimization of such consortia for stability and productivity is a prerequisite to exploring their full potential (Bhatia et al., 2018).

2.4.2 Main Molecular Tools Used in Strain Manipulation

With the emergence of molecular tools, it became easier to engineer microbial strains to establish new metabolic pathways or improve the existing ones through genetic manipulation (Volk et al., 2022). Homologous recombination (HR) and Clustered Regularly Interspaced Short Palindromic Repeats (CRISPR/Cas) are potent techniques used for precise gene editing of microbes (Zimmermann et al., 2024). HR involves exchanging the sequences of two DNA molecules across regions of high similarity or identical sequences. This strategy takes advantage of a natural DNA repair system that is highly prevalent in bacteria and can be rationally used to introduce desired genetic material in cells, resulting in artificially modified strains (Wright et al., 2018). HR can directly manipulate bacterial genomic structures by permitting gene knockouts, substitutions, deletions, and point mutations. Additionally, genes can also receive tags, reporter devices, and nearly any possible modification. The target gene and an editing substrate, which can be either circular or linear DNA (oligonucleotides and PCR products), are typically the compounds involved in this process. Editing substrates can be introduced into the bacterial cell using methods like transformation, conjugation, or transduction (Fels et al., 2020). Compared to previous techniques, such as recombineering CRISPR/Cas has proven to be quick, marker-free, and versatile. CRISPR/Cas originated from a heritable adaptive immune system that bacteria and archaea exhibit to avoid genetic insertions, deletions, and replacements by exogenous DNA (Ghosh, 2021). This process inserts a double-strand break (DSB) at a specific target location that is highly complementary to a single-guide RNA (gRNA). Due to its benefits, CRISPR/Cas has been applied to a wide range of species, being considered a potentially universal tool for genetic manipulation (Chang, 2022). In this context, the most popular CRISPR-associated nuclease is Cas9, a single enzyme that requires only the presence of a gRNA that permits genome editing. The yeast model *S. cerevisiae* is one of the examples of microbes that underwent genetic modifications based on this system. It has been modified to produce various chemicals and amino acids and enhance tolerance to toxic compounds like ethanol (da Silva et al., 2022). Similarly, certain LAB strains can be modified to enhance food safety, starter culture vitality, and probiotic qualities. Among LAB exhibiting desirable traits for genetic modification are *L. brevis*, *L. plantarum*, *L. casei*, *L. paracasei*, and *S. thermophilus* (Moradi et al., 2020).

2.4.3 Genetic Engineering

Genetic engineering creates new microbial strains with the desired traits, such as enzyme production. Compared with plant or animal sources of enzymes, microbial enzymes have intrinsic advantages in terms of scalability, yield, and specificity (Boukid et al., 2023). Therefore, microbial sources, especially genetically modified organisms (GMOs), are a major source for producing food enzymes, since genetic modifications can be relatively easily conducted to enhance production efficiency, enzyme stability, and reduce the production of undesired by-products (Verma et al., 2019).

Gene engineering has significantly contributed to the improvement of the fermentation process by modifying microbial genomes. CRISPR-Cas9, along with other genome editing techniques, enables the precise engineering of microorganisms to utilize diverse substrates, improve stress tolerance, and enhance the production of target metabolites. These modifications not only enhance the quantity of the products but also the ability of the process to work under a wide range of fermentation conditions, thereby reducing the cost of production (Yadav et al., 2024).

The process of designing the genetically modified microorganisms (GMMs) typically starts with the selection of a host organism, as shown in Fig. 2.3. Once a suitable host has been found, scientists focus on adding specific genetic components, such as expression vectors, selection markers, and promoter sequences, all designed to ideal yield levels of the enzyme in question within the host cell. The choice of transformation method is also critical and should be tailored to ensure high integration efficiency and genetic stability through generations (Deckers et al., 2020).

Safety evaluation of the microbial strains used in food manufacturing is a legal requirement, especially when genetic modification is involved. Candidates for GMMs are required to have qualifications based on standards like the Qualified Presumption of Safety status. Further, ensure that the GMM does not produce pathogenic substances. Although rigorous, these analyses are essential to ensure

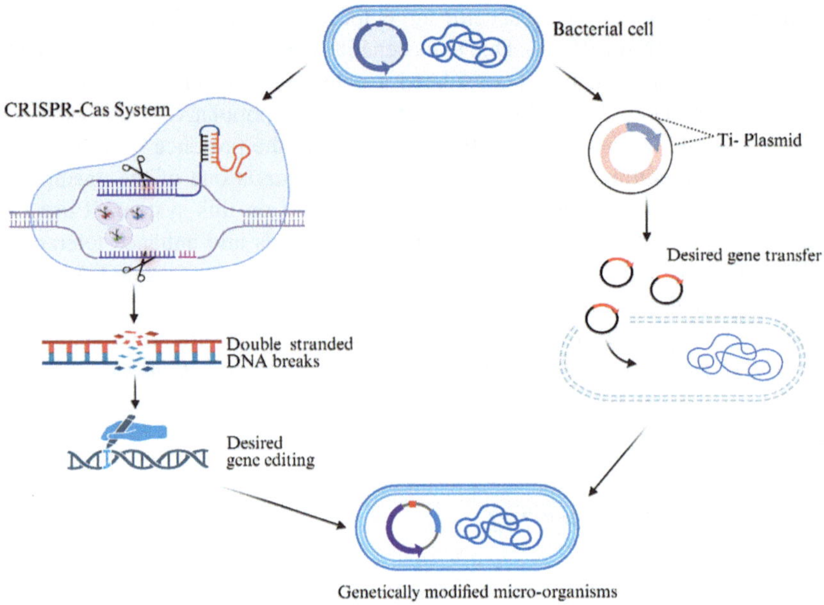

Fig. 2.3 Illustrates the genetic modification in microorganisms using the CRISPR-Cas gene editing system and Ti-plasmid-mediated gene transfer

public safety and promote consumer acceptance of enzymes produced by genetically modified microorganisms (GMMs) in large-scale food processing (Fraiture et al., 2020).

2.5 Synthetic Biology and Metabolic Engineering

Metabolic engineering and synthetic biology are two closely interrelated fields that together advance the frontiers of bioengineering to enable desired metabolite production in microorganisms, plants, and other biological systems (Yadav et al., 2025). Synthetic biology primarily involves the construction of genetic circuits, devices, and organisms through the integration of genetic parts, such as promoters, coding sequences, transcriptional factors, and binding sequences, while simultaneously bringing quantitative data together to enable predictive models of biology (Bashor & Collins, 2018). The data-driven approach becomes one of the founding stones of metabolic engineering, which is the manipulation of cellular functions in specific organisms to produce the desired products from low-cost substrates to increase yield and decrease undesirable by-products (Rymuszka & Gorczynska, 2026).

Synthetic biology promotes fermentation by developing new microbes. These engineered microbes can produce high-value compounds through the fermentation method. The control of microbial activity in fermentation by synthetic biology guarantees the reproducibility of the process and its effectiveness, even when the environmental conditions change (Saranraj et al., 2025). Such engineered microorganisms confer other beneficial characteristics to commonly known fermented foods, improving their nutritional and functional profiles (Mirsalami & Mirsalami, 2025).

Metabolic engineering is relied upon to modify the metabolic pathways of microorganisms to boost fermentation yield. Through the regulation of metabolic flux, it is feasible to boost the production of bioactive metabolites. Furthermore, this approach also reduces the formation of side products and improves the substrate conversion (Elazzazy et al., 2025). For instance, the metabolic engineering of *Lactobacillus* strains has led to the development of probiotics with enhanced potency and bioavailability, leading them to play a more significant role in the growth of the functional food market (Uhegwu & Anumudu, 2025).

However, it is often challenging to engineer the insertion of synthetic pathways into living beings. These challenges demand the thoughtful design of metabolic pathways and advanced genetic tools to ensure compatibility and minimize perturbations, which clearly indicates the further development of pathway integration strategies (Choi et al., 2019).

2.6 Artificial Intelligence and Machine Learning-Driven Fermentation Optimization

Traditional fermented foods have been examined quite successfully by using AI and ML techniques. Research on fermented foods has seen an increase in bioinformatics data, which calls for more sophisticated methods that can manage complex and high-dimensional data (Singh & Kumar, 2025). As a science that replicates and enhances human intelligence, AI is an essential part of computational technologies and optimization algorithms. AI tools are being used to monitor and optimize important production chain control points and to customize products for consumers (Okuyelu & Adaji, 2024). AI has been one of the most important tools in the optimization of fermentation processes, offering new data-driven approaches with the potential to revolutionize bioprocessing by raising conversion efficiency and reducing costs (Butean et al., 2025). Advanced algorithms, such as genetic algorithms and neural networks, can be used to develop flexible models of fermentation processes that closely mimic the actual processes (Florea et al., 2022). Consequently, these models make the online prediction of variables, such as substrate concentration and product production, possible, enabling the accurate control of fermentation conditions in order to increase both yields and operational efficiency (Tarafdar et al., 2026).

The development of functional foods is revolutionized by AI-driven fermentation optimization. It makes possible the most precise control over fermentation parameters and maximizes nutritious value in the products (Siddique et al., 2026). AI algorithms analyse large amounts of fermentation processes data, even with microbial behaviour, metabolite production, and environmental situations, to optimize processes towards desired outcomes. It ensures consistent quality, reduces nutrient wastage, and optimizes synthesis to maximize the potential for bioactive compounds (Alloun & Calvio, 2024). Similarly, AI supports the real-time monitoring and predictive modelling of adjustments in fermentation parameters for maintaining microbial balance and increasing the production of probiotics, antioxidants, and other functional components (Asar et al., 2025).

Additionally, ML models can also forecast microbial interactions and improve coculture systems that produce synergistic effects on functionality. The models also support strain selection and engineering based on the identification of microbes with specific metabolic pathways to produce targeted functional compounds (Nestor et al., 2023). The integration of AI into the fermentation process enables the producer to tailor functional food profiles. Equally, the potential to significantly reduce food waste and smooth out production flows means that AI-driven fermentation is even more in tune with nature, and is, therefore, a game-changer in modern food science (Kirwan & Zhiyong, 2025).

Using advanced feed control methodology and genetic algorithms, AI adjusts parameters by building mathematical models that optimize process regulation and predict configurations for further improvement, with reduced reliance on manual adjustments (Kanchipuram & Nadu, 2025). Furthermore, the AI-driven sensor can perform autonomous monitoring to gain direct insight into the fermentation process, even in certain scenarios where physical sensors are used. This will then allow for adaptive adjustments in these complex systems, increasing the quality of production while allowing for the data collection required for continuous optimization (Butean et al., 2025).

In the area of enzyme engineering, AI is accelerating the discovery of optimized enzymes for food processing. For example, AlphaFold, an AI-driven tool, has estimated the structures of more than 200 million proteins, ushering in the field of structural biology and enzyme design (Molla & Meseret, 2025). This potential enables the rapid and inexpensive production of enzymes that can provide innovation in food biotechnology while, in concert, tackling experimental roadblocks and accelerating production efficiencies (Mao et al., 2024).

While the promise of artificial intelligence-assisted fermentation optimization holds great potential, there are also significant barriers to the migration of data-driven frameworks from experimental settings to practical industrial applications (Bidyalakshmi et al., 2025). Laboratory investigations typically involve controlled parameters and comprehensive genetic information; however, the adaptability of these models to high-throughput processes requires further modification to adjust for natural variability among industrial substrates and conditions (Abdi et al., 2024). Although it provides practical scalability and integration of AI into commercial systems to ensure reliable industrial applications, these are under investigation (Jain, 2022).

Consequently, in the fermented food industry, it has been proven that AI is revolutionizing fermentation techniques in improving quality, scalability, and product innovation (Rawat et al., 2025). AI also plays a pivotal role in precision fermentation by enabling real-time monitoring, optimization of the process, and control over product quality (Niyigaba et al., 2025). Advanced AI aid maximises product quality, reducing waste and improving scalability. They help in the improvement of the efficiency process, consistency of guaranteed products, permitting real-time monitoring, and facilitating predictive optimization (Qudus, 2025).

2.7 Safety Issues and Risk Assessment of New Fermentation Technologies

Factors affecting the growth of pathogenic microorganisms in fermented food products include the quality of raw materials, the processing environment, and hygiene practices (Skowron et al., 2022). When good manufacturing practices are not implemented, inferior ingredients are exemplified by *Vietnamese nem chua*, a traditional raw sausage that often contains contaminants, including *Staphylococcus aureus* and *Escherichia coli* (Ngo, 2023). Food safety will also be directly impacted by the quality of water used in fermentation. Especially, in rural areas, the possibility of contamination with pathogens like *Salmonella spp.* and *E. coli* has been increased (Sehgal et al., 2024; Mkangara, 2023).

In many cases, fermented products in low-resource areas are frequently repurposed in non-sterile containers to package products, which further raises the risk of contamination. Moreover, antibiotic residues in foods from the treatment of livestock may introduce antimicrobial resistance genes to consumers and might promote multidrug-resistant bacterial strains (Jones, 2024). Under horizontal gene resistance, genes are transferred between two foodborne bacteria, demonstates the importance of the use of antibiotics and the resistant genes in both initial cultures and final products (Rossi et al., 2014).

Biogenic amines (BAs) are secondary products produced by some free amino acid-rich fermented foods. Higher levels of BAs cause signal damage and potential health threats, resulting from decarboxylation or transamination reaction (Li & Lu, 2020). Moreover, pathogen contamination in the fermented products increases health risk of exposing consumers, especially in the case of dry or semi-dry fermented sausages; therefore, more stringent safety controls and microbial surveillance throughout the production chain must be guaranteed (Osman & Sulieman, 2022). Despite being associated with various health benefits, the safety of fermented products poses a complicated problem stemming from traditional fermentation processes.

2.8 Future Directions and Opportunities in Novel Fermentation Technologies

A great knowledge of microbial genetics helps in innovation in fermentation. Basically, gene-environment interactions help to improve lactic acid production by choosing strains based on genetic markers that identify favourable metabolic traits (Li et al., 2024). The introduction of high-throughput sequencing (HTS) has allowed for an in-depth examination of the microbial communities of fermented foods, thus creating a basis for tailored strain selection and higher production efficiency (Saranyadevi et al., 2026).

Moreover, genomics has elucidated the mechanisms in *Lactobacillus* that play a role in proteolysis, which is crucial for the development of flavour and texture, by transforming amino acids into specific flavour compounds (Zang et al., 2025). Even though the use of traditional fermentation approaches to preserve food is always a risk of contamination by harmful microorganisms. This fact underlines the need for genomic research to enable the detection of such harmful microorganisms at an early stage (Kim, 2025). Synthetic biology and genomics carry huge potential in terms of engineering strains to convert non-food biomass into nutritious, safe, and consumer-friendly products through fermentation (Ajanaku et al., 2025).

As a consequence, in the future, combined scientific discoveries and fermentation knowledge will acknowledge the direction of the fermented food industry. The next wave of innovation in the sector will be driven by the use of omics technologies, precision fermentation, AI-driven fermentation control, and sustainable production methods (Paul et al., 2026). Bioprocess automation enhancement, standardized frameworks creation, and AI-based metabolic network modelling are mainly focused on improving future fermentation research. Bioprocess automation enhancement, standardized frameworks creation, and AI-based metabolic network modelling improvement are the main areas of future research in fermentation (Olodu & Enabulele, 2025). Through digital technologies, advancements in technology, and consistent regulations, the food fermentation industry can produce cost-effective, biodegradable, and healthy food products. The food fermentation industry can establish itself as a major force in environmental sustainability, nutrition improvement, and global food security by investing more in research, technology, and consumer education (Ayoub, 2026).

2.9 Conclusions

The fermentation process has been extensively used by human civilization around the world to preserve food and food products, boost the nutritional value of food, and produce unique sensory attributes. Though there are still challenges with food safety, innovative fermentation technologies have opened up new fermented food products according to consumer desires, offering various health benefits and the best nutritional profiles for food products. Besides this, the AI integrated with synthetic biology and metabolic engineering fermentation approaches opened new opportunities for improved productivity and scalability in fermented foods production. Future developments with metabolic, genomic, and synthetic biology could revolutionize fermentation processes with advanced biotechnological tools and open up new possibilities for safer, nutritionally improved products accepted by consumers globally.

References

Abdi, G., Tendulkar, R., Thatte, C., Mishra, S., Desai, B., Surve, S., Chaudhari, A., Patil, N., Jain, M., & Tarighat, M. A. (2024). Scaling up nature's chemistry: A guide to industrial production of valuable metabolites. In *Advances in metabolomics* (pp. 331–375). Springer. https://doi.org/10.1007/978-981-97-7459-3_15

Ajanaku, P. O., Olojede, A. O., Ajanaku, C. O., Egharevba, G. O., Agaja, F. O., Joseph, C. B., & Thomas, R. M. (2025). Novel fermentation techniques for improving food functionality: An overview. *Fermentation, 11*(9), 509. https://doi.org/10.3390/fermentation11090509

Ajayeoba, T. A., & Ijabadeniyi, O. A. (2025). *Transforming food for the future: Precision fermentation as a key to sustainability, nutrition, and health.* https://doi.org/10.2139/ssrn.5119686

Alloun, W., & Calvio, C. (2024). Bio-driven sustainable extraction and ai-optimized recovery of functional compounds from plant waste: A comprehensive review. *Fermentation, 10*(3), 126. https://doi.org/10.3390/fermentation10030126

Arbab Sakandar, H., Chen, Y., Peng, C., Chen, X., Imran, M., & Zhang, H. (2023). Impact of fermentation on antinutritional factors and protein degradation of legume seeds: A review. *Food Review International, 39*(3), 1227–1249. https://doi.org/10.1080/87559129.2021.1931300

Asar, R., Erenler, S., Devecioglu, D., Ispirli, H., Karbancioglu-Guler, F., Ozturk, H. I., & Dertli, E. (2025). Understanding the functionality of probiotics on the edge of artificial intelligence (ai) era. *Fermentation, 11*(5), 259. https://doi.org/10.3390/fermentation11050259

Ashraf, S. A., Elkhalifa, A. E. O., Ahmad, M. F., Patel, M., Adnan, M., & Sulieman, A. M. E. (2022). Probiotic fermented foods and health promotion. In *African fermented food products-new trends* (pp. 59–88). Springer. https://doi.org/10.1007/978-3-030-82902-5_6

Augustin, M. A., Hartley, C. J., Maloney, G., & Tyndall, S. (2024). Innovation in precision fermentation for food ingredients. *Critical Reviews in Food Science and Nutrition, 64*(18), 6218–6238. https://doi.org/10.1080/10408398.2023.2166014

Ayoub, A. (2026). Integration of artificial intelligence in food processing technologies. *Processes, 14*(3), 513. https://doi.org/10.3390/pr14030513

Bai, J. A., & Rai, V. R. (2015). Application of dairy propionic acid bacteria in fermented foods and probiotics. In *Beneficial microbes in fermentedand functional foods* (pp. 551–569). CRC Press/Taylor & Francis Group. https://doi.org/10.1201/b17912

Bashor, C. J., & Collins, J. J. (2018). Understanding biological regulation through synthetic biology. *Annual Review of Biophysics, 47*(1), 399–423. https://doi.org/10.1146/annurev-biophys-070816-033903

Bhatia, S. K., Bhatia, R. K., Choi, Y.-K., Kan, E., Kim, Y.-G., & Yang, Y.-H. (2018). Biotechnological potential of microbial consortia and future perspectives. *Critical Reviews in Biotechnology, 38*(8), 1209–1229. https://doi.org/10.1080/07388551.2018.1471445

Bidyalakshmi, T., Jyoti, B., Mansuri, S. M., Srivastava, A., Mohapatra, D., Kalnar, Y. B., Narsaiah, K., & Indore, N. (2025). Application of artificial intelligence in food processing: Current status and future prospects. *Food Engineering Reviews, 17*(1), 27–54. https://doi.org/10.1007/s12393-024-09386-2

Boukid, F., Ganeshan, S., Wang, Y., Tülbek, M. Ç., & Nickerson, M. T. (2023). Bioengineered enzymes and precision fermentation in the food industry. *International Journal of Molecular Sciences, 24*(12), 10156. https://doi.org/10.3390/ijms241210156

Butean, A., Cutean, I., Barbero, R., Enriquez, J., & Matei, A. (2025). A review of artificial intelligence applications for biorefineries and bioprocessing: From data-driven processes to optimization strategies and real-time control. *PRO, 13*(8), 2544. https://doi.org/10.3390/pr13082544

Chang, A. Y. (2022). Single-guide RNAs: Rationale and design. In *CRISPR genome surgery in stem cells and disease tissues* (pp. 47–55). Academic Press. https://doi.org/10.1016/B978-0-12-817876-8.00003-6

Chauhan, A., Arora, J., Chauhan, R., Bhardwaj, P., Ranjan, A., Basniwal, R. K., Srivastav, R., & Jindal, T. (2025). Anaerobes and methanogens microbial pioneers of oxygen-free environments. In *Anaerobes and methanogens* (pp. 1–34). CRC Press. https://doi.org/10.1201/9781003506898

Choi, K. R., Jang, W. D., Yang, D., Cho, J. S., Park, D., & Lee, S. Y. (2019). Systems metabolic engineering strategies: Integrating systems and synthetic biology with metabolic engineering. *Trends in Biotechnology, 37*(8), 817–837. https://doi.org/10.1016/j.copbio.2019.08.010

Crosby, G. (2019). *Cook, taste, learn: How the evolution of science transformed the art of cooking*. Columbia University Press.

da Silva, F. F., de Souza, É. S., Carneiro, L. M., Alves Silva, J. P., de Souza, J. V. B., & da Silva, B. J. (2022). Current ethanol production requirements for the yeast Saccharomyces cerevisiae. *International Journal of Microbiology, 2022*(1), 7878830. https://doi.org/10.1155/2022/7878830

de Medeiros, V. P., da Costa, W. K. A., da Silva, R. T., Pimentel, T. C., & Magnani, M. (2022). Microalgae as source of functional ingredients in new-generation foods: Challenges, technological effects, biological activity, and regulatory issues. *Critical Reviews in Food Science and Nutrition, 62*(18), 4929–4950. https://doi.org/10.1080/10408398.2021.1879729

De Roos, J., & De Vuyst, L. (2018). Acetic acid bacteria in fermented foods and beverages. *Current Opinion in Biotechnology, 49*, 115–119. https://doi.org/10.1016/j.copbio.2017.08.007

De Vries, H., Mikolajczak, M., Salmon, J.-M., Abecassis, J., Chaunier, L., Guessasma, S., Lourdin, D., Belhabib, S., Leroy, E., & Trystram, G. (2018). Small-scale food process engineering—Challenges and perspectives. *Innovative Food Science & Emerging Technologies, 46*, 122–130. https://doi.org/10.1016/j.ifset.2017.09.009

Deckers, M., Deforce, D., Fraiture, M.-A., & Roosens, N. H. C. (2020). Genetically modified micro-organisms for industrial food enzyme production: An overview. *Foods, 9*(3), 326. https://doi.org/10.3390/foods9030326

Deshmukh, R. K., & Gaikwad, K. K. (2024). Natural antimicrobial and antioxidant compounds for active food packaging applications. *Biomass Conversion and Biorefinery, 14*(4), 4419–4440. https://doi.org/10.1007/s13399-022-02623-w

Diana Celcia, R., Naveen, G., Gali, S. P., Meenatchi, R., & Singh, P. (2025). Prospects of ethnic fermented foods as functional foods. In *Ethnic and indigenous food technologies: Role in climate and health resilience* (pp. 159–184). Springer. https://doi.org/10.1007/978-981-96-9286-6_8

Elazzazy, A. M., Baeshen, M. N., Alasmi, K. M., Alqurashi, S. I., Desouky, S. E., & Khattab, S. M. R. (2025). Where biology meets engineering: Scaling up microbial nutraceuticals to bridge nutrition, therapeutics, and global impact. *Microorganisms, 13*(3), 566. https://doi.org/10.3390/microorganisms13030566

Faria, D. J., de Carvalho, A. P. A., & Conte-Junior, C. A. (2023). Valorization of fermented food wastes and byproducts: Bioactive and valuable compounds, bioproduct synthesis, and applications. *Fermentation, 9*(10), 920. https://doi.org/10.3390/fermentation9100920

Farooqui, A. A. (2021). Importance of fermented foods on human health. In *Gut microbiota in neurologic and visceral diseases* (pp. 69–86). Academic Press. https://doi.org/10.1016/B978-0-12-821039-0.00012-5

Fels, U., Gevaert, K., & Van Damme, P. (2020). Bacterial genetic engineering by means of recombineering for reverse genetics. *Frontiers in Microbiology, 11*, 548410. https://doi.org/10.3389/fmicb.2020.548410

Florea, A., Sipos, A., & Stoisor, M.-C. (2022). Applying AI tools for modeling, predicting and managing the white wine fermentation process. *Fermentation, 8*(4), 137. https://doi.org/10.3390/fermentation8040137

Fraiture, M.-A., Deckers, M., Papazova, N., & Roosens, N. H. C. (2020). Detection strategy targeting a chloramphenicol resistance gene from genetically modified bacteria in food and feed products. *Food Control, 108*, 106873. https://doi.org/10.1016/j.foodcont.2019.106873

Ghosh, D. (2021). *Advances in crispr/cas and related technologies*. Academic Press. https://doi.org/10.1016/bs.pmbts.2020.12.001

Ghosh, S., Chowdhury, R., & Bhattacharya, P. (2016). Mixed consortia in bioprocesses: Role of microbial interactions. *Applied Microbiology and Biotechnology, 100*(10), 4283–4295. https://doi.org/10.1007/s00253-016-7448-1

Gombotz, W. R., & Hoffman, A. S. (2019). Immobilization of biomolecules and cells on and within synthetic polymeric hydrogels. In *Hydrogels in medicine and pharmacy* (pp. 95–126). CRC Press. https://doi.org/10.1201/9780429285097

Gomes, R. J., de Fatima Borges, M., de Freitas Rosa, M., Castro-Gómez, R. J. H., & Spinosa, W. A. (2018). Acetic acid bacteria in the food industry: Systematics, characteristics and applications. *Food Technology and Biotechnology, 56*(2), 139. https://doi.org/10.17113/ftb.56.02.18.5593

Gupta, K., Bardhan, P., Saikia, D., Rather, M. A., Loying, S., Mandal, M., & Kataki, R. (2022). Microbial fermentation: Basic fundamentals and its dynamic prospect in various industrial applications. In *Industrial microbiology and biotechnology* (pp. 107–128). Springer. https://doi.org/10.1007/978-981-16-5214-1_4

Han, D., Yang, Y., Guo, Z., Dai, S., Jiang, M., Zhu, Y., Wang, Y., Yu, Z., Wang, K., Rong, C., & Yu, Y. (2024). A review on the interaction of acetic acid bacteria and microbes in food fermentation: A microbial ecology perspective. *Foods, 13*(16), 2534. https://doi.org/10.3390/foods13162534

Hernández-Tenorio, F., Sáez, A. A., Habeych-Narváez, D. I., Marín-Palacio, L. D., & Giraldo-Estrada, C. (2023). Bioingredients: An overview from biotechnology and bioeconomy. In *Selected topics in food process engineering. ICEF 2023. Food engineering series* (pp. 497–515). Springer. https://doi.org/10.1007/978-3-031-62415-5_19

Iqbal, B., Alabbosh, K. F., Raheem, A., Ullah, I., Khan, A. R., & Memon, M. S. (2025). Microbial transformation: The role of fermentation in advancing nutritional quality and human health. *Archives of Microbiology, 207*(9), 228. https://doi.org/10.1007/s00203-025-04435-8

Iram, A., Ozcan, A., Turhan, I., & Demirci, A. (2023). Production of value-added products as food ingredients via microbial fermentation. *PRO, 11*(6), 1715. https://doi.org/10.3390/pr11061715

Jain, N. S. (2022). Pioneering the future of technology: Integrating advanced cloud computing with artificial intelligence for scalable, intelligent systems. *World Journal of Advanced Research and Reviews, 13*, 847–862. https://doi.org/10.30574/wjarr.2022.13.1.0014

Jones, A. D. G. (2024). Microbial risk assessment. *Human and Ecological Risk Assessment: Theory and Practice, 2*, 779–832.

Kalaiselvan, P., Devi, N. C., Deepti, M., Devi, A. A., Akamad, K., Dheeran, P., Debbarma, S., Vadivel, D., & Rajesh, D. (2025). Solid-state fermentation—A sustainable future technology in aquafeeds? *Frontiers in Marine Science, 12*, 1669719. https://doi.org/10.3389/fmars.2025.1669719

Kamble, P., Tamhane, R., Tarnekar, I., Sushil, S., Samant, S., & Menon, S. (2021). Cellular agriculture: The way ahead for food, textile, and cosmetic industries. *International Journal of Advance Research, Ideas and Innovations in Technology, 7*(4), 1539–1546.

Kanchipuram, C., & Nadu, T. (2025). Optimization techniques and genetic algorithms in AI. *Mathematical Innovation*, (pp. 82–91).

Keller, W., Gezork, K., Nienhaus, B., & Popp, N. (2024). Modern fermentation and fermenter design. *Chemical Engineering, 131*(4), 35–39.

Kim, S.-S. (2025). Application of comparative genomics for the development of PCR primers for the detection of harmful or beneficial microorganisms in food: Mini-review. *Foods, 14*(6), 1060. https://doi.org/10.3390/foods14061060

Kirwan, C. G., & Zhiyong, F. (2025). *Smart cities and artificial intelligence: Convergent systems for planning, design, and operations*. Elsevier.

Kumar, K. V., Rishi, B., Sruthi, B., Garg, S., Gujjala, L. K. S., & Avanthi, A. (2025). Fundamentals on bioreactors, operation modes and control system mechanisms. In *Biotechnology engineering* (pp. 71–106). Elsevier. https://doi.org/10.1016/B978-0-443-31476-6.00010-5

Ledesma-Amaro, R., Bamezai, S., et al. (2025). *Sustainable use and conservation of fermentation-associated microorganisms within the agrifood system*. Food and Agriculture Organization. https://doi.org/10.4060/cd5636en

Li, B., & Lu, S. (2020). The importance of amine-degrading enzymes on the biogenic amine degradation in fermented foods: A review. *Process Biochemistry, 99*, 331–339. https://doi.org/10.1016/j.procbio.2020.09.012

Li, W., Wu, Q., Kwok, L., Zhang, H., Gan, R., & Sun, Z. (2024). Population and functional genomics of lactic acid bacteria, an important group of food microorganism: Current knowledge, challenges, and perspectives. *Food Frontiers, 5*(1), 3–23. https://doi.org/10.1002/fft2.321

Linder, T. (2019). Making the case for edible microorganisms as an integral part of a more sustainable and resilient food production system. *Food Security, 11*(2), 265–278. https://doi.org/10.1007/s12571-019-00912-3

Lu, Z., & Imlay, J. A. (2021). When anaerobes encounter oxygen: Mechanisms of oxygen toxicity, tolerance and defence. *Nature Reviews Microbiology, 19*(12), 774–785. https://doi.org/10.1038/s41579-021-00583-y

Lugo-Morin, D. R. (2024). Artificial intelligence on food vulnerability: Future implications within a framework of opportunities and challenges. *Societies, 14*(7), 106. https://doi.org/10.3390/soc14070106

Maicas, S. (2020). The role of yeasts in fermentation processes. *Microorganisms, 8*(8), 1142. https://doi.org/10.3390/microorganisms8081142

Mani, A. (2018). Food preservation by fermentation and fermented food products. *International Journal of Academic Research & Development, 1*, 51–57.

Mao, S., Jiang, J., Xiong, K., Chen, Y., Yao, Y., Liu, L., Liu, H., & Li, X. (2024). Enzyme engineering: Performance optimization, novel sources, and applications in the food industry. *Foods, 13*(23), 3846. https://doi.org/10.3390/foods13233846

Marti-Quijal, F. J., Khubber, S., Remize, F., Tomasevic, I., Roselló-Soto, E., & Barba, F. J. (2021). Obtaining antioxidants and natural preservatives from food by-products through fermentation: A review. *Fermentation, 7*(3), 106. https://doi.org/10.3390/fermentation7030106

Mirsalami, S. M., & Mirsalami, M. (2025). Advances in genetically engineered microorganisms: Transforming food production through precision fermentation and synthetic biology. *Future Foods, 11*, 100601. https://doi.org/10.1016/j.fufo.2025.100601

Mishra, M., Sharma, U., Rawat, M., Harshvardhan, Singh, S. S., & Korpole, S. (2025). Antimicrobial compounds from anaerobic microorganisms: A review of an untapped reservoir. *Applied Microbiology, 5*(3), 68. https://doi.org/10.3390/applmicrobiol5030068

Mkangara, M. (2023). Prevention and control of human salmonella enterica infections: An implication in food safety. *International Journal of Food Science, 2023*(1), 8899596. https://doi.org/10.1155/2023/8899596

Molla, A., & Meseret, G. (2025). The role of artificial intelligence in protein structural and functional prediction: Current status and future prospective. *Innovations, 6*(3), 130–138. https://doi.org/10.11648/j.innov.20250603.20

Moradi, M., Kousheh, S. A., Almasi, H., Alizadeh, A., Guimarães, J. T., Yilmaz, N., & Lotfi, A. (2020). Postbiotics produced by lactic acid bacteria: The next frontier in food safety. *Comprehensive Reviews in Food Science and Food Safety, 19*(6), 3390–3415. https://doi.org/10.1111/1541-4337.12613

Mota, J., & Vilela, A. (2024). Exploring microbial dynamics: The interaction between yeasts and acetic acid bacteria in port wine vinegar and its implications on chemical composition and sensory acceptance. *Fermentation, 10*(8), 421. https://doi.org/10.3390/fermentation10080421

Nestor, E., Toledano, G., & Friedman, J. (2023). Interactions between culturable bacteria are predicted by individual species' growth. *Msystems, 8*(2), e00836–e00822. https://doi.org/10.1128/msystems.00836-22

Ngo, H. H. T. (2023). *Evaluating implementation strategies for improved food safety in Vietnam.* Uppsala University. https://hdl.handle.net/10568/130627

Niyigaba, T., Küçükgöz, K., Kołożyn-Krajewska, D., Krolikowski, T., & Trząskowska, M. (2025). Advances in fermentation technology: A focus on health and safety. *Applied Sciences, 15*(6), 3001. https://doi.org/10.3390/app15063001

Ogidi, C. O., & Hung, Y.-T. (2025). Microbial bioprocesses of food processing wastes as a sustainable alternative source of functional bio-products for the food industries. In *Industrial water and waste treatment: Polluton control* (pp. 328–348). CRC Press. https://doi.org/10.1201/9781003628187

Okuyelu, O., & Adaji, O. (2024). AI-driven real-time quality monitoring and process optimization for enhanced manufacturing performance. *Journal of Advances in Mathematics and Computer Science, 39*(4), 81–89. https://doi.org/10.9734/JAMCS/2024/v39i41883

Olodu, B. A., & Enabulele, S. A. (2025). Fermentation technology and its industrial application. *Advances in Life Science and Biotechnology* (pp. 173–196).

Osman, O. A., & Sulieman, A. M. E. (2022). Factors influence the quality and safety of fermented sausages. In *African fermented food products: New trends* (pp. 219–233). Springer. https://doi.org/10.1007/978-3-030-82902-5_15

Ouoba, L. I. I. (2017). Traditional alkaline fermented foods: Selection of functional Bacillus starter cultures for soumbala production. In *Starter cultures in food production* (pp. 370–383). Wiley. https://doi.org/10.1002/9781118933794.ch18

Owusu-Kwarteng, J., Parkouda, C., Adewumi, G. A., Ouoba, L. I. I., & Jespersen, L. (2022). Technologically relevant Bacillus species and microbial safety of West African traditional alkaline fermented seed condiments. *Critical Reviews in Food Science and Nutrition, 62*(4), 871–888. https://doi.org/10.1080/10408398.2020.1830026

Palladino, F., Marcelino, P. R. F., Schlogl, A. E., José, Á. H. M., de Cássia Lacerda Brambilla Rodrigues, R., Fabrino, D. L., Santos, I. J. B., & Rosa, C. A. (2024). Bioreactors: Applications and innovations for a sustainable and healthy future—A critical review. *Applied Sciences, 14*(20), 9346. https://doi.org/10.3390/app14209346

Parapouli, M., Vasileiadis, A., Afendra, A.-S., & Hatziloukas, E. (2020). Saccharomyces cerevisiae and its industrial applications. *AIMS Microbiology, 6*(1), 1. https://doi.org/10.3934/microbiol.2020001

Paul, R., Krishnan, A., & Dasgupta, A. (2026). Omics, CRISPR, and artificial intelligence: The triad revolutionizing bioprocess engineering. In *Industrial microbiology and biotechnology: Trends in microbial biotechnology* (pp. 57–84). Springer. https://doi.org/10.1007/978-981-95-5378-5_3

Pawar, R., Awati, A. K., Zambare, V., Din, M. F. M., Krishnan, S., & Chelliapan, S. (2024). Saccharomyces cerevisiae: A facultative anaerobe for ethanol fermentation using organic waste. In *Anaerobes and waste conversion technologies* (pp. 249–278). Springer. https://doi.org/10.1007/978-981-97-4159-5_13

Prates, J. A. M. (2025). Unlocking the functional and nutritional potential of microalgae proteins in food systems: A narrative review. *Foods, 14*(9), 1524. https://doi.org/10.3390/foods14091524

Qin, H., Wu, H., Shen, K., Liu, Y., Li, M., Wang, H., Qiao, Z., & Mu, Z. (2022). Fermented minor grain foods: Classification, functional components, and probiotic potential. *Foods, 11*(20), 3155. https://doi.org/10.3390/foods11203155

Qudus, L. (2025). Leveraging artificial intelligence to enhance process control and improve efficiency in manufacturing industries. *International Journal of Computer Applications Technology and Research, 14*(02), 18–38. https://doi.org/10.7753/IJCATR1402.1002

Rachmawati, N., & Triwibowo, R. (2024). An overview of propionic acid as food additives in the global trades of traditional fermented products. *AIP Conference Proceedings, 2957*, 60045. https://doi.org/10.1063/5.0184008

Ranaei, V., Pilevar, Z., Khaneghah, A. M., & Hosseini, H. (2020). Propionic acid: Method of production, current state and perspectives. *Food Technology and Biotechnology, 58*(2), 115. https://doi.org/10.17113/ftb.58.02.20.6356

Rawat, S., Magar, Z. A. A., Gaikwad, P. S., Negi, T., Rai, S., & Choudhary, P. (2025). Biotechnological horizons in food processing: Future perspectives and challenges. In *Biotechnological innovations in food processing: Transforming the future of nutrition and sustainability* (pp. 661–707). Springer. https://doi.org/10.1007/978-3-031-96526-5_15

Rawoof, S. A. A., Kumar, P. S., Vo, D.-V. N., Devaraj, K., Mani, Y., Devaraj, T., & Subramanian, S. (2021). Production of optically pure lactic acid by microbial fermentation: A review. *Environmental Chemistry Letters, 19*(1), 539–556. https://doi.org/10.1007/s10311-020-01083-w

Ray, R. C., Paramithiotis, S., Thekkangil, A., Nethravathy, V., Rai, A. K., & Martin, J. G. P. (2024). Food fermentation and its relevance in the human history. In *Trending topics on fermented foods* (pp. 1–57). Springer. https://doi.org/10.1007/978-3-031-72000-0_1

Regassa, H., Bose, D., & Mukherjee, A. (2021). Review of microorganisms and their enzymatic products for industrial bioprocesses. *Industrial Biotechnology, 17*(4), 214–226. https://doi.org/10.1089/ind.2021.0002

Reihani, S. F. S., & Khosravi-Darani, K. (2019). Influencing factors on single-cell protein production by submerged fermentation: A review. *Electronic Journal of Biotechnology, 37*, 34–40. https://doi.org/10.1016/j.ejbt.2018.11.005

Rossi, F., Rizzotti, L., Felis, G. E., & Torriani, S. (2014). Horizontal gene transfer among microorganisms in food: Current knowledge and future perspectives. *Food Microbiology, 42*, 232–243. https://doi.org/10.1016/j.fm.2014.04.004

Rymuszka, A., & Gorczynska, W. (2026). Classical food fermentations as modern biotechnological platforms: Alcoholic, acetic, butyric, lactic and propionic pathways and applications. *Molecules, 31*(2), 333. https://doi.org/10.3390/molecules31020333

Santiago-Díaz, P., Rico, M., Rivero, A., & Santana-Casiano, M. (2022). Bioactive metabolites of microalgae from Canary Islands for functional food and feed uses. *Chemistry & Biodiversity, 19*(9), e202200230. https://doi.org/10.1002/cbdv.202200230

Santillan, E., Neshat, S. A., & Wuertz, S. (2025). Disturbance and stability dynamics in microbial communities for environmental biotechnology applications. *Current Opinion in Biotechnology, 93*, 103304. https://doi.org/10.1016/j.copbio.2025.103304

Saranraj, P., Ray, R. C., Kumar, N. A., Lokeshwari, B., Gayathri, K., & Sivasakthivelan, P. (2025). Microbes in action: Powering sustainable fermentation for food, pharma, and bioeconomy. *Applied Food Biotechnology, 12*(1), 1–24.e20. https://doi.org/10.22037/afb.v12i1.49836

Saranyadevi, S., Abdullahi, A., Navina, B., & others. (2026). Microbial fermentation profiling: Tools and techniques for food analysis. In *Novel food analysis* (pp. 295–320). Springer. https://doi.org/10.1007/978-1-0716-4787-5_14

Sarma, U., Chatterjee, A., & Shrestha, R. (2024). Fermented food production in the bioeconomy context. In *Trending topics on fermented foods* (pp. 59–84). Springer. https://doi.org/10.1007/978-3-031-72000-0_2

Sehgal, S., Aggarwal, S., Banik, S. P., & Kaushik, P. (2024). Impact of water contamination on food safety and related health risks. In *Microbial biotechnology in the food industry: Advances, challenges, and potential solutions* (pp. 337–363). Springer. https://doi.org/10.1007/978-3-031-51417-3_14

Šelo, G., Planinić, M., Tišma, M., Tomas, S., Koceva Komlenić, D., & Bucić-Kojić, A. (2021). A comprehensive review on valorization of agro-food industrial residues by solid-state fermentation. *Foods, 10*(5), 927. https://doi.org/10.3390/foods10050927

Sharma, R., Garg, P., Kumar, P., Bhatia, S. K., & Kulshrestha, S. (2020). Microbial fermentation and its role in quality improvement of fermented foods. *Fermentation, 6*(4), 106. https://doi.org/10.3390/fermentation6040106

Siddique, I., Farooq, M., Ullah, Q., Nabi, A., Siddique, M., Siddique, S., Naseer, M., Younis, A., et al. (2026). Future trends and direction in AI and ML for food science and bioprocess development. In *Artificial intelligence in food science: Transforming food and bioprocess development* (pp. 811–825). Academic Press/Elsevier. https://doi.org/10.1016/B978-0-443-26468-9.00080-1

Singh, A., & Kumar, S. (2025). Exploring the functionality of microbes in fermented foods: Technological advancements and future directions. *Fermentation, 11*(6), 300. https://doi.org/10.3390/fermentation11060300

Skowron, K., Budzyńska, A., Grudlewska-Buda, K., Wiktorczyk-Kapischke, N., Andrzejewska, M., Wałecka-Zacharska, E., & Gospodarek-Komkowska, E. (2022). Two faces of fermented foods—The benefits and threats of its consumption. *Frontiers in Microbiology, 13*, 845166. https://doi.org/10.3389/fmicb.2022.845166

Sokra, I., & Meta, H. (2025). Types of fermentation in food and beverage processing: Principles, microorganisms, and applications. *Journal of Agriculture and Technology, 1*(1), 114–127. https://doi.org/10.5281/zenodo.18044914

Sun, W., Shahrajabian, M. H., & Lin, M. (2022). Research progress of fermented functional foods and protein factory-microbial fermentation technology. *Fermentation, 8*(12), 688. https://doi.org/10.3390/fermentation8120688

Tan, Y. X., Mok, W. K., Lee, J., Kim, J., & Chen, W. N. (2019). Solid state fermentation of Brewers' spent grains for improved nutritional profile using Bacillus subtilis WX-17. *Fermentation, 5*(3), 52. https://doi.org/10.3390/fermentation5030052

Tarafdar, A., Das, A., Singh, B., et al. (2026). Application of artificial intelligence and machine learning in solid-state fermentation proceses. In *Current developments in solid-state fermentation* (pp. 129–144). Elsevier. https://doi.org/10.1016/B978-0-443-27624-8.00002-6

Teng, T. S., Chin, Y. L., Chai, K. F., & Chen, W. N. (2021). Fermentation for future food systems: Precision fermentation can complement the scope and applications of traditional fermentation. *EMBO Reports, 22*(5), EMBR202152680. https://doi.org/10.15252/embr.202152680

Uhegwu, C. C., & Anumudu, C. K. (2025). Probiotic potential of traditional and emerging microbial strains in functional foods: From characterization to applications and health benefits. *Microorganisms, 13*(11), 2521. https://doi.org/10.3390/microorganisms13112521

Varzakas, T., & Smaoui, S. (2024). Global food security and sustainability issues: The road to 2030 from nutrition and sustainable healthy diets to food systems change. *Foods, 13*(2), 306. https://doi.org/10.3390/foods13020306

Verma, D. K., Kimmy, G., Mahato, D. K., & Patel, A. (2019). *Genetically modified organisms produced enzymes: Multifarious applications in food manufacturing industries*. Apple Academic Press.

Vijay, A. P., Chakraborty, K., & Pai, S. K. (2025). Nutritional profiling of Chlorella salina, Dunaliella salina, and Arthrospira platensis as sustainable functional foods for human nutrition. *Journal of Food Measurement and Characterization, 19*, 4769–4782. https://doi.org/10.1007/s11694-025-03289-4

Vivek, K., & Venkitasamy, C. (2023). Role and applications of fungi in food and fermentation technology. In *Fungal resources for sustainable economy: Current status and future perspectives* (pp. 71–87). Springer. https://doi.org/10.1007/978-981-19-9103-5_3

Volk, M. J., Tran, V. G., Tan, S.-I., Mishra, S., Fatma, Z., Boob, A., Li, H., Xue, P., Martin, T. A., & Zhao, H. (2022). Metabolic engineering: Methodologies and applications. *Chemical Reviews, 123*(9), 5521–5570. https://doi.org/10.1021/acs.chemrev.2c00403

Willaert, R. G. (2018). Cell immobilization and its applications in biotechnology: Current trends and future prospects. In *Fermentation microbiology and biotechnology* (4th ed., pp. 323–358). CRC Press. https://doi.org/10.1201/9780429506987

Wright, W. D., Shah, S. S., & Heyer, W.-D. (2018). Homologous recombination and the repair of DNA double-strand breaks. *The Journal of Biological Chemistry, 293*(27), 10524–10535. https://doi.org/10.1074/jbc.TM118.000372

Yadav, H., Singh, S., & Sinha, R. (2024). Fermentation technology for microbial products and their process optimization. In *Industrial microbiology and biotechnology: A new horizon of the microbial world* (pp. 35–64). Springer. https://doi.org/10.1007/978-981-97-6270-5_2

Yadav, J., Marwah, H., & Kumar, C. (2025). Synthetic biology and metabolic engineering paving the way for sustainable next-gen biofuels: A comprehensive review. *Energy Advances, 4*(10), 1209–1228. https://doi.org/10.1039/D5YA00118H

Yu, A. O., Leveau, J. H. J., & Marco, M. L. (2020). Abundance, diversity and plant-specific adaptations of plant-associated lactic acid bacteria. *Environmental Microbiology Reports, 12*(1), 16–29. https://doi.org/10.1111/1758-2229.12794

Zang, J., Lin, T., Xu, C., Lin, Y., Ma, K., Zhang, C., Rui, X., Gan, D., & Li, W. (2025). Advances in lactic acid bacteria and their metabolites in fermented foods: Mechanisms of food quality enhancement, gut microbiota modulation, and future prospects. *Food Reviews International, 42*, 1657–1691. https://doi.org/10.1080/87559129.2025.2524405

Zimmermann, A., Prieto-Vivas, J. E., Voordeckers, K., Bi, C., & Verstrepen, K. J. (2024). Mutagenesis techniques for evolutionary engineering of microbes–exploiting CRISPR-Cas, oligonucleotides, recombinases, and polymerases. *Trends in Microbiology, 32*(9), 884–901. https://doi.org/10.1016/j.tim.2024.02.006

Chapter 3
Postbiotics and Para-Probiotics: The Future of Gut Health Beyond Probiotics

Neelam Anil Revankar and Pradeep Singh Negi

Abstract The global shift toward plant-based diets and sustainable food systems has accelerated the demand for innovative functional foods. Probiotics have long been the gold standard for gut health and have been a part of several functional foods. In an era where gut health is increasingly recognised as a cornerstone of overall well-being, probiotic application is often hindered by viability challenges, safety concerns in vulnerable populations (such as infants and the elderly), and strict formulation constraints. To address these limitations, microbial biotechnology has shifted towards stable alternatives, the postbiotics (bioactive microbial metabolites) and para-probiotics (inactivated microbial cells). This chapter provides a comprehensive analysis of postbiotics and para-probiotics detailing their production from diverse plant-based substrates, unique health-promoting mechanisms, and inherent advantages in stability and safety over live cultures. Despite their huge potential, a significant research gap exists regarding the standardized industrial-scale-up and long-term interactions within complex food matrices. This chapter aims to bridge this gap by examining current formulation strategies and future regulatory perspectives. The advancements in postbiotics, para-probiotics, and related microbial technologies are paving the way for safe, effective, and sustainable strategies.

Keywords Postbiotics · Para-probiotics · Gut health · Microbial metabolites · Immunomodulation

N. A. Revankar · P. S. Negi (✉)
Department of Fruit and Vegetable Technology, CSIR-Central Food Technological Research Institute, Mysuru, Karnataka, India

Academy of Scientific and Innovative Research (AcSIR), Ghaziabad, Uttar Pradesh, India
e-mail: psnegi.cftri@csir.res.in

S. Dhiman, S. Kaur (eds.), *Next-Gen Plant Based Fermented Foods: Innovations for a Sustainable Future*, https://doi.org/10.1007/978-3-032-29927-7_3

3.1 Introduction

Gut health science has evolved since the early twentieth century, when the focus was primarily on live Lactobacillus in sweetened drinks. Recent findings indicate that microbial by-products (postbiotics) and inactivated cells (para-probiotics) may provide superior outcomes compared to live probiotics (Vallejo-Cordoba et al., 2020). These postbiotics and para-probiotics are intrinsic to traditional fermented foods consumed across cultures for centuries. In Indian idly and dosa, lactic acid bacteria generate acetate and butyrate, the classic postbiotics, during rice-lentil co-fermentation, enhancing mineral absorption and gut barrier function (Kavitake et al., 2022). Korean kimchi harbors heat-stable para-probiotic *Lactobacillus* fragments that induce anti-inflammatory cytokines even after gastric transit. Japanese natto, African ogi, and Ethiopian injera is reported to yield short chain fatty acids (SCFAs), bacteriocins, and inactivated cell wall ligands, which act as functional by-products that demonstrate anti-biotic activity against harmful bacterial species (Tamang & Lama, 2022). This "beyond" category also includes synbiotics (prebiotic-probiotic combos for enhanced efficacy), pharmabiotics (microbe-derived drugs targeting specific conditions), and engineered live microbes as therapeutics. They have been reported to improve precision health and also work very well with plant-based systems (Wang & Selomulya, 2020).

While live probiotics are widely used, their efficacy is often limited by significant viability loss during storage, strain-specific survival variance in the gastrointestinal tract, and safety concerns in immune-compromised populations (Lee et al., 2023). Such risks underscore the requirement of exploring safer ways, like postbiotics and para-probiotics, which can deliver health benefits without the hazards associated with live microbial administration (O'Connor et al., 2021). Therefore, the current key challenges in biotics include viability loss during storage, strain-specific survival variance, and safety for immune-compromised populations. This chapter aims to highlight how "beyond probiotics" concepts propel gut health forward, focusing on plant-based innovations and precision nutrition contributing to overall health.

3.2 Post-biotics: Production, Mechanisms, and Benefits

3.2.1 Definition and Classification

The International Scientific Association for Probiotics and Prebiotics (ISAPP) defines postbiotics as "preparations of inanimate microorganisms and or their components that confer a health benefit on the host" which was then modified to the beneficial metabolic products, such as enzymes and SCFA, produced by probiotics, while paraprobiotics are the non-living, inactivated microbial cells or their components (Salminen et al., 2021). This distinguishes live probiotics from their derivatives, also now defined as postbiotics and para-probiotics. They are classified based

Table 3.1 Bioactive components from microbial fermentation and inactivated biomass

Category	Subcategory	Examples
Soluble metabolic factors: Microbial fermentation end-products released in extracellular environment	Short-chain fatty acids	Acetate, propionate, butyrate
	Secondary bile acids	Deoxycholic acid, lithocholic acid, ursodeoxycholic acid
	Organic acids	Lactic acid, acetic acid
	Bacteriocins and antimicrobial peptides	Nisin, pediocin
	Enzymes	β-glucuronidase, bile salt hydrolase (BSH)
	Vitamins	K2 (menaquinone), vit B12, vit B2, vit B6, folic acid (vit B9)
	Neuroactive metabolites	GABA, indole derivatives
Cell-associated components: Derived from non-viable microbial biomass	Cell wall fragments	Peptidoglycan, lipoteichoic acids
	Exopolysaccharides	EPS (e.g., kefiran, levan)
	Surface-layer proteins	S-layer proteins
	Pili and fimbriae	Type IV pili, SpaCBA pili
	Membrane vesicles	Outer membrane vesicles (OMVs)

Compiled from Urbani et al. (2025) and Amobonye et al. (2025)

on origin, composition, and complexity (Table 3.1). This classification is helpful for justifying regulatory claims and for guiding the development of new functional foods. For instance, secondary bile acids are increasingly recognized as postbiotics due to their microbial origin and host receptor-mediated effects (FXR, TGR5 signaling), despite historical classification under host metabolism (Urbani et al., 2025).

3.2.2 Mechanisms and Health Benefits

The transformative power of postbiotics comes from their interaction with the host. They offer benefits similar to probiotics but with greater stability and consistency. Recent clinical trials and observational studies show clear improvements in important biomarkers.

3.2.2.1 Effect of Postbiotics on Gut Health and Immune Regulation

Postbiotics act as natural organisers of the gut microbiome. They selectively stimulate the growth of beneficial bacteria and competitively excluding pathogens. One commercially known postbiotic is Probio-Eco®. It contains postbiotics from *Lacticaseibacillus paracasei* Zhang, *Lactiplantibacillus plantarum* P-8, and *Bifidobacterium animalis* subsp. lactis V9. This supplement supports gastrointestinal health, especially in managing chronic diarrhoea and associated symptoms. Succinate activates colonic peristalsis via GPR91. The succinate/3-IA/5-HTP triad modulates the microbiome-metabolite-neuroimmune axis by upregulating motility (increased 5-HT and MLNR) and barrier genes (upregulated claudin CLDN1 and mucin MUC2), and downregulating IL-6/STAT3 (Salminen et al., 2021). Similarly, study reported benefits after consuming Probio-Eco, containing 10 postbiotic metabolites, among 58 participants. Results showed upregulated IL-12 and TLR2 signaling and downregulated CRP. This enhances the gut–immune tone via a Th1 bias and microbial pattern recognition (Guo et al. 2024a, b). Almahbashi & Gunes Altuntas (2025) compared cell-free supernatant (CFS), thermal, and ultrasonic preparation methods for *Lactiplantibacillus plantarum*-derived postbiotics, finding CFS superior in metabolite diversity (increase in SCFAs and lactic acid) and bioactivity (DPPH scavenging, 45–60%; MIC, 50–100 μg/mL vs. *E. coli*, *S. aureus*; enhanced IL-10 immunomodulation). Ultrasonic treatment improved solubility by 20–30%, supporting CFS as best for functional food applications targeting gut health via organic acid and phenolic pathways (Almahbashi & Gunes Altuntas, 2025). Similarly, in another study, a reduction in metastatic castration-resistant prostate cancer mCRPC was observed among those treated with microbially produced bile acids ($n = 142$). The study group also reported an increased CD8+ T-cell infiltration, higher IFN-γ, reduced Treg cells, and supported retention of Faecalibacterium and Ruminococcus species. The treated subjects also showcased 2.3-fold increase in SCFAs in fecal samples. Fecal Microbiota Transplantation (FMT) enhanced tumour infiltration (higher IFN-γ) and reduced tumour volume by 35% (Jin et al., 2025), suggesting that microbially produced Bile acids not only managed tumour suppression but also mitigated gut dysbiosis. Thus, the methods mentioned have reported preservation of antioxidant and antimicrobial effects as well as immunological improvements.

γ-Aminobutyric acid (GABA) rich supernatant produced by *B. fragilis* showed an altered GABA-mediated response. Researchers found that brain signatures associated with depression were reduced (Strandwitz et al., 2019). Exopolysaccharides (EPS) are well-known postbiotics. EPS-producing *Streptococcus thermophilus* strains decreased IFNγ and increased IL-10 in the gastric mucosa. This prevented gastritis and strengthened the mucosal layer in an ASA-induced gastritis BALB/c mice model (Rodríguez et al., 2009). Similarly, EPS-*Bifidobacterium animalis* IPLA R1 treatment in high-fat induced *in-vivo* model significantly reduced the triglycerides accumulation in the liver by positively altering the gut dysbiosis caused due to high fat diet (Pandey et al., 2025; Salazar et al., 2019). EPS from lactic acid bacteria has shown a 6.5-fold improvement in the in-vitro studies of β-carotene abortion (Morifuji et al., 2020). Pili and fimbriae, which are structural appendages of bacteria, act as postbiotic ligands used to modulate mucosal immunity and epithelial adhesion after inactivation.

Several studies have reported gut-immune regulations, for example, type IV pili (a, b, and c), which also constitute microbial-associated molecular patterns (MAMPs) recognized by TLR2, TLR4, and are known to regulate the key immunogenic responses in hosts, such as IL-8 production (Lebeer et al., 2012), enhances mucosal immune surveillance, IgA production, and improves the epithelial integrity, thereby helps in maintaining the architecture of the gut (Ligthart et al., 2020). Peptides, such as nisin, pediocin, and plantaricin, act by disrupting bacterial membranes and selectively inhibiting pathogenic strains, thereby shaping the gut microbiota. Communication with host immune receptors, like TLR2, TLR4, and NOD2, triggers cytokine modulation, causes upregulation of IL-10 and the downregulation of TNF-α and IL-6, which results in anti-inflamatory induction in the host. Additionally, AMPs enhance barrier function by stimulating tight junction proteins such as occludin and claudin-1, and increasing mucin-2 secretion. These molecular markers validate AMP's role as non-viable, bioactive agents capable of modulating the gut–immune axis without requiring live bacterial colonization (Dinić et al., 2024; Parhi, 2024; Ra & Bang, 2024). Similarly, a study reported that Postbiotic Water Extract (PWE) of *Bacillus amyloliquefaciens* and *Lactobacillus plantarum* SN4 containing bioactive metabolites (378 ± 4.1 mg/g lactic acid and 40.8 ± 2.2 mg/g acetic acid) and microbial cell wall fragments restores gut health by upregulating alkaline phosphatase, PI uptake, degrading DNA, downregulating IPEC-J2 binding, lowering oxidative stress, and promoting epithelial integrity via mucin-2 and other tight junction protein expression (Tong et al., 2025).

3.2.2.2 Gut and Multi-organelle Axis

Effect on Gut-Neuro axis In Alzheimer's models, postbiotics reduced amyloid-beta (Aβ40, Aβ42) accumulation and tau hyperphosphorylation, while enhancing synaptic markers like PSD-95 and synaptophysin. Tight junction proteins (occludin, claudin-1) were upregulated, restoring the integrity of the blood-brain barrier. These effects were mediated through TLR2/NOD2 signaling and cytokine modulation (upregulation of IL-10, TNF-α, and IL-6), confirming their role in neuroimmune regulation. Postbiotics influence neuroprotection by reducing neuroinflammation and restoring mitochondrial and synaptic integrity. SCFAs such as acetate and butyrate, derived from microbial fermentation, cross the blood–brain barrier and regulate neurotransmitter synthesis (GABA, serotonin) (Braga et al., 2024).

Effect on Gut-Hepato axis Postbiotics have been reported to modulate lipid metabolism, reduce systemic inflammation, and improve endothelial function. SCFAs and microbial peptides influence AMP-activated protein kinase (AMPK) and peroxisome proliferator-activated receptors (PPARs), which regulate glucose and lipid homeostasis. Treatment with levansucrase of *Bacillus licheniformis* significantly improved metabolic markers in diabetic rats. Glycogen levels rose by 52%, while plasma glucose levels dropped by the same percentage (52%) (Dahech et al., 2011). EPS from *Bacillus subtilis* sp. has been reported to supress hyperglycemia, dyslipidemia serum LDL and triglycerides while increasing HDL, partly through gut-mediated modulation of bile acids (Ghoneim et al., 2016). Anti-inflammatory effects (downregulation of IL-1β and CRP) and improvement in insulin sensitivity were observed in metabolic syndrome models, suggesting postbiotics as adjuncts in cardiometabolic therapy (Gao et al., 2025). Bacteriocins such as nisin and plantaricin are reported to inhibit pathogenic growth, reducing lipopolysaccharide (LPS) translocation, thereby reducing hepatic TLR4 activation and subsequent NF-κB signalling (Pan et al., 2022). In murine models, postbiotics restored hepatic antioxidant enzymes (SOD and catalase) and reduced steatosis. Peptidoglycan fragments and SCFAs also modulated Kupffer cell activity, promoting IL-10 secretion and reducing TNF-α-mediated hepatocyte damage (Yilmaz, 2024). Thus, indicating an effective protection of hepatic health even in cases of lifestyle disorders induced by High fat diet and stress.

Effect on Gut-Skin axis Postbiotics are also reported to improve skin health via the gut–skin axis by restoring barrier integrity, reducing inflammation, and enhancing antioxidant defenses. They regenerate lost components of the skin barrier by upregulating tight junction proteins (occludin, claudin-1, ZO-1) and mucin-2, thereby replenishing structural cohesion and hydration (Chai et al., 2024). Biosurfactants and exopolysaccharides stimulate keratinocyte proliferation and collagen synthesis, repairing microtears and enhancing elasticity. Molecular targets include NF-κB, TLR2/TLR4, and Nrf2, which regulate cytokines (decrease in TNF-α, IL-6; and upregulation in IL-10) and antioxidant enzymes (increase in SOD, catalase) (Paz & Lio, 2024). Physical signs of recovery include reduced trans-epidermal water loss, diminished erythema and acne lesions, a smoother texture, and improved resilience against environmental stressors (Prajapati et al., 2025). These gut-multi-organ correlations underscore their potential in overall health management.

3.2.2.3 Production and Characterization of Postbiotics

Postbiotics from plant-based fermentation are gaining attention for their stability, safety, and bioactivity. This section outlines the production workflow and characterization strategies, emphasizing fermentation substrates, microbial strains, inactivation methods, and analytical validation.

Submerged fermentation (SmF) and solid-state fermentation (SSF) are both employed for plant-based postbiotic production, with SmF offering precise control of pH, dissolved oxygen, agitation and more straightforward downstream clarification, while SSF on moist plant matrices (bran, pomace, peels) yields higher metabolite concentration, lower water/energy use and better cost-efficiency, however, requires careful control of moisture gradients and heat removal; substrate preconditioning (milling, enzymatic saccharification, steam) and particle size standardization to improve porosity and fermentable sugar availability, and initial moisture for SSF is typically adjusted to 40–60% w/w to balance activity and oxygen diffusion (Prajapati et al., 2023).

Inoculum standardization (late-exponential cells, 5–10% inoculum), mono- vs co-culture strategies (Bacillus + Lactobacillus or Lactobacillus + Saccharomyces) enable metabolic cross-feeding to increase bacteriocin, SCFA and phenolic diversity (da Silva Vale et al., 2023; Tong et al., 2023). Key operating variables such as temperature, pH, water content (SSF), agitation/aeration (SmF), inoculum size, and time are best optimized using a design of experiments (Plackett–Burman for screening, RSM for response optimization). SSF conditions were optimized at 37 °C, 50% moisture and 8% inoculum, achieving a 3.6 fold increase in antibacterial activity (Tong et al., 2023). Stirred tanks or airlift bioreactors with controlled kLa suit SmF, whereas tray, packed-bed or rotating drum systems with forced ventilation and intermittent mixing are suitable for SSF as it avoids hotspots (Shree Kumar & Mohanasrinivasan, 2025). Process monitoring tracks substrate consumption, pH, organic acids (HPLC/GC), biomass proxies, CO_2/O_2 evolution, and specific bioactive markers (bacteriocin assays, EPS quantitation) with plug-flow sampling across SSF beds to map heterogeneity (Pimentel et al., 2023).

Harvesting strategies depend on targets. Generally, centrifugation/filtration is used to obtain cell-free extracts, which is followed by mild inactivation (pasteurisation, high-pressure, and pulsed electric fields) to preserves metabolites. The extraction/concentration is achieved via ultrafiltration, resin adsorption or solvent partitioning prior to drying (spray- or freeze-drying) with carriers or it is encapsulated for stability and gastric protection (Monteiro et al., 2023). Quality control is done to confirm absence of viable cells (plating, qPCR), quantify key actives (HPLC/GC-MS, LC-MS), as well as to assess antimicrobial (MIC/MBC), antioxidant (DPPH/ABTS) and immunomodulatory (cytokine assays, NO inhibition) activities. Additionally, their stability under varied pH and heat conditions, as well as simulated gastrointestinal conditions must be ensured to ensure functional retention for food applications. These parameters are important for scaling-up while preserving geometric and mass-transfer characteristics (Reynolds, kLa) for SmF, and bed porosity/heat profiles for SSF to maintain yields and process economics. These aspects are also directly linked to inactivation methods and extraction techniques (Table 3.2) indicating specific approach to preserve metabolites and bioactive components.

3.2.2.4 Strain Selection and Optimization for Yield

Microbial strain selection is a foundational step in designing effective postbiotic formulations, emphasizing the importance of specific microorganisms for their resilience and production of bioactive metabolites (Morán & Kilasoniya, 2024). These are selected depending on the carbon and nitrogen sources used and ability to

Table 3.2 Techniques for producing the postbiotics

Inactivation method	Mechanism	Retained/extracted postbiotics	Reference
Pasteurization	Moderate heat (60–80 °C) denatures cells while preserving heat-stable metabolites.	SCFAs (acetate, butyrate), bacteriocins (e.g., nisin), peptidoglycan fragments	Tong et al. (2023)
Sonication (ultrasound)	Cavitation disrupts cell membranes without high heat	Exopolysaccharides (EPS), bioactive peptides, intact cell wall structures	da Silva Vale et al. (2023)
High-pressure processing	Pressure (100–600 MPa) disrupts cells while preserving molecular integrity	EPS, SCFAs, lipoteichoic acids, antioxidant enzymes (SOD, catalase)	Pimentel et al. (2023)
Pulsed electric fields	Short bursts of high voltage permeabilize membranes	Intracellular peptides, SCFAs, bacteriocins, lipoteichoic acids	Racioppo et al. (2017)
Supercritical CO_2 (Sc-CO_2)	High-pressure CO_2 in its supercritical state induces cell disruption and enhances	Intracellular components, minimizing the risk of thermal degradation of valuable compounds	Thorakkattu et al. (2022)
Column purification technique	Octyl-Sepharose™ CL-4B column for isolation	Lipoteichoic acid from *L. johnsonii* La1 and L. acidophilus La10	Matsuguchi et al. (2003)
Cold plasma	Low-vacuum cold plasma enhanced polysaccharide (3.16%) synthesis	Maintains the structural and functional integrity of temperature-sensitive postbiotic compounds such as SCFAs and bacteriocins, thereby improving product quality	Guo et al. (2024a, b)

produce enhanced antibacterial and antioxidant activities (Hemalatha & Subathra Devi, 2022). Lactic acid bacteria (LAB) such as *Lactiplantibacillus plantarum*, *Pediococcus acidilactici*, and *Weissella cibaria* are frequently chosen due to their postbiotic potential. Bacillus species such as *B. amyloliquefaciens* and *B. subtilis* produce exopolysaccharides, which is also considered a desirable attribute for postbiotic potential (Yang et al., 2024). In yeasts, *Saccharomyces cerevisiae* and related species are preferred due to their fermentation capacity and adaptability (Bitew et al., 2024). Fungal strains such as *Xylaria* sp. and *Penicillium chrysogenum* are selected for their unique metabolite production. Further, using co-culturing strategies further enhance postbiotic metabolite yield. *Xylaria* sp. BCC 1067, the primary metabolite quantified was β-glucans, produced at about 115.84 mg/g of cell dry weight. The strain also showed vigorous antifungal activity against *Saccharomyces cerevisiae*, linked to the polyketide 19,20-epoxycytochalasin Q (ECQ), though its yield was not specified. Antibacterial activity against *Propionibacterium acnes* was reported with a MIC50 of ~1000 mg/L and a MBC of ~2000 mg/L. Strain-specific optimization strategies (tailoring media and environmental conditions) are critical for maximizing postbiotic yield (Jayasekara et al., 2022). These approaches (Table 3.3) are being used to improve yield and process economics.

Table 3.3 Strain specific optimization approaches for postbiotic yield

Microbe type	Strains	Reported optimization approaches	Effect on postbiotic yield	Citations
Bacteria	*L. plantarum*	Media, statistical design, genetics	Enhanced postbiotic	Garrote Achou et al. (2025)
	B. subtilis	Media, statistical design, genetics	Enhanced EPS yield	Yang et al. (2024)
Yeasts	*S. cerevisiae*	Co-culture	Higher bioethanol	Bitew et al. (2024)
	P. barkeri	Fermentation parameters	Enhanced postbiotic yield	Yang et al. (2024)
Fungi	*Xylaria* sp.	Media, metabolic engineering	Increased mycelial/ metabolite yield	Abdalla et al. (2024)
	P. chrysogenum	Media, metabolic engineering	Enhanced metabolite yield	Jayasekara et al. (2022)

Yield optimisation involves fine-tuning fermentation parameters and adjusting media composition. Approaches include single-factor experiments, response surface methodology, and advanced statistical designs to identify optimal conditions for water content, temperature, pH, inoculum size, and substrate ratios (Raita et al., 2023; Tong et al., 2023; Zhang et al., 2025). For example, increasing sodium succinate and sucrose while reducing yeast extract improved exopolysaccharide yield in *Bacillus licheniformis*. In yeasts, adjusting fermentation time, co-culturing, temperature, and pH significantly enhanced bioethanol and postbiotic yields (van Tatenhove-Pel et al., 2021). Among fungi, media replacement and light adjustments led to a four-fold increase in mycelial mass and metabolite production (Abdalla et al., 2024). Modern optimization uses metabolic engineering, systems biology, and machine learning to design high-yield strains and processes, balancing growth rate, product yield, and economic viability (Chen et al., 2025; Young et al., 2018). Downstream processing, scalability, and regulatory compliance are also essential for industrial application (Prajapati et al., 2023).

3.2.2.5 Advanced Analytical Techniques and Downstream Processing

Characterizing postbiotics requires advanced analytical platforms (Calvigioni et al., 2023; Weng et al., 2024). Metabolomics techniques such as GC-MS and liquid chromatography mass spectrometry (LC-MS) are widely used to quantify short-chain fatty acids (SCFAs), organic acids, and other volatile compounds produced during fermentation (Feng et al., 2022; Zhang et al., 2019). Proteomic profiling, including MALDI-TOF and LC-MS/MS, helps identify bioactive peptides and cell wall fragments with immunomodulatory potential (Assadpour & Jafari, 2019; Jacobsen et al., 2018). Downstream processing methods like centrifugation, ultrafiltration, and freeze-drying are critical for stabilizing postbiotic formulations while preserving bioactivity (Sharma et al., 2022; Wang & Selomulya, 2020).

3.3 Para-Probiotics: Production, Mechanisms, and Benefits

Para-probiotics, these inactivated microbial cells also known as "ghost" probiotics that retain potent health effects without the risks of live strains, making them a versatile factor in gut health innovation (Lee et al., 2023). Derived from beneficial microbes such as Lactobacilli, they preserve structural components for immune and barrier functions, emerging as stable alternatives in plant-based foods amid rising sustainability demands. Sonication, the use of ultrasound to disrupt cells, is an emerging inactivation method that enhances fragment release and bioactivity while maintaining safety.

3.3.1 Effect of Para-Probiotics on Gut-Health

Para-probiotics exert their effects through preserved surface molecules (MAMPs, PAMPs, TLRs) that engage host systems, often mirroring those of live probiotics but with enhanced stability, as evidenced by human trials showing immune, barrier, and anti-inflammatory benefits. Para-probiotic efficacy depends on inactivation conditions (Ashique et al., 2025). Different methods or the same method under varied parameters alter key microbial structures, such as peptidoglycan. To preserve gut-modulating potential, culturability tests should be paired with flow cytometry to assess membrane integrity and avoid excessive damage to functional components (Yolmeh et al., 2024). Consumption of Heat-killed *L. plantarum* L-137 (10 mg) 12 weeks among 78 human adults, showed improved mood and reduced cold incidence via immunomodulation (Hirose et al., 2013). Similarly another study indicated that *Firmicutes/Bacteroidetes* ratio in stool samples was significantly lower at 12 months than at control groups (Nakai et al., 2021).

Para-probiotics and gut—Immune response Another study conducted in 2008 confirmed that UV-inactivated *L. rhamnosus* downregulated IL-8 by 17-folds, and prevented Flagellin-induced NFkB nuclear translocation in CaCo2 model (Lopez et al., 2008). Similarly, a double-blind, placebo-controlled trial with heat-killed *B. bifidum* among 443 (placebo-treated) IBS patients reported that 30% of patients showed improvement according to the IBS-severity scoring system. This study indicated that non-viable probiotic consumption is a safe alternative to confer health benefits as compared to probiotics for patients exposed to dangers of translocation (Andresen et al., 2020). Similarly, in this randomized, placebo-controlled, double-blind trial, oral intake of *Lactobacillus pentosus* strain b240 (heat-killed) significantly increased secretory IgA (SIgA) secretion by 20% at week 4 in 80 healthy elderly individuals, suggesting enhanced mucosal immunity with no significant side effects. The authors suggest that the inactivated probiotics stimulate the gut-associated lymphoid tissue (GALT), leading to enhanced SIgA production in salivary glands via mucosal immune system cross-talk. This supports the concept of para-probiotic immunomodulation through non-viable bacterial components (Kotani et al., 2010). In two other clinical trials involving 15 healthy adults and 16 patients with irritable bowel syndrome (IBS), heat-inactivated *L. casei* 01 reduced postprandial glycemia via α-glucosidase inhibition and increased peptide release. At the same time, live *L. plantarum* 299v decreased IL-2, IL-6, IL-12, and TNF-α, improving intestinal permeability and IBS symptoms. Showcasing their efficacy in enzyme modulation, cytokine suppression, and barrier restoration as key gut-targeted mechanisms (Sugawara et al., 2016). In vitro studies with UV-inactivated *L. rhamnosus* demonstrated strong anti-inflammatory effects, reducing IL-8 by 17-fold and blocking NF-κB activation. Clinically, heat-killed *B. bifidum* improved IBS symptoms in 30% of patients, while heat-killed *L. pentosus* b240 boosted secretory IgA by 20% in elderly participants, enhancing mucosal immunity. Across these trials, para-probiotics consistently show both mechanistic and clinical benefits.

Para-probiotics and stress -sleep cycle In this randomized, double-blind, placebo-controlled study involving 69 medical students under chronic psychological stress, daily intake of heat-inactivated *Lactobacillus gasseri* (CP2305) (10^{10} cells/day) for 12 weeks significantly improved mental and physical health outcomes. Participants receiving CP2305 showed reduced anxiety (as measured by the STAI), improved general health (as assessed by the GHQ-28), better sleep quality, and normalised bowel habits. Physiologically, CP2305 intake led to lower salivary cortisol levels, enhanced parasympathetic activity (increased HF norm), and improved sleep structure (increased N3 non-REM sleep, delta power and downregulation of sleep latency). Authors also observed the suppression of stress-responsive microRNAs, including miR-144, miR-144*, and miR-16, suggesting modulation of oxidative stress and serotonin signalling. These findings support CP2305's para-probiotic role in regulating the gut–brain axis via neuroendocrine, immune, and epigenetic pathways (Nishida et al., 2017). Similarly, another 12-week trial reported with 49 male athletes, using heat-inactivated *Lactobacillus gasseri* CP2305, showed reduced fatigue, anxiety, and salivary chromogranin A; preserved mitochondrial gene expression; and stabilized gut microbiota diversity. Other effects included upregulated stress-resilient pathways, such as EIF2, mTOR, and integrin signaling, highlighting the para-probiotic modulation of the gut–brain–endocrine axis under physical stress (Sawada et al., 2017). Extending beyond gut health, heat-inactivated *L. gasseri* CP2305 improved anxiety, sleep quality, and bowel habits in stressed medical students, alongside reduced cortisol and modulation of stress-related miRNAs. Together, these findings highlight para-probiotics as safe, well-tolerated alternatives to live probiotics, with benefits spanning gut, immune, and neuroendocrine health.

Para-probiotics and Glycemic index In a randomised crossover trial involving 15 healthy adults, the consumption of a whey and grape juice drink containing *Lacticaseibacillus casei* 01 inactivated by ohmic heating (8 V/cm, 95 °C for 7 min) significantly inhibited digestive enzymes and improved the postprandial glycemic response. The paraprobiotic drink exhibited comparable inhibition of α glucosidase (51.8%) and α-amylase (44.2%) to its probiotic counterpart, and both formulations resulted in similar glycemic curves. The authors also discussed EPS-mediated enzyme binding, the release of bioactive peptides (<3 kDa) from whey proteins, and the preservation of cytoplasmic enzymes and metabolites that modulate carbohydrate digestion. These findings support the use of ohmic-heated paraprobiotics as safe and effective alternatives to live probiotics for glycemic control (Barros et al., 2021).

While live probiotics focus on colonization and competitive exclusion, para-probiotics act primarily through "biomimetic signaling." The studies by Kotani et al. (2010) and Nishida et al. (2017) suggest that heat-inactivated strains are particularly effective for long-term immunomodulation and stress management, whereas enzymatic modulation (as seen in the Barros 2021 study) is more dependent on the preservation of specific cell-wall-associated peptides and exopolysaccharides. Para-probiotics demonstrate consistent, multi-axis benefits by modulating immunity, stress, gut barrier function, and microbiota via preserved microbial structures and signalling. Diverse studies confirm their safety, mechanistic depth, and therapeutic promise across human models.

3.3.2 Definition and Characteristics: Production Methods

Para-probiotics, also known as ghost probiotics, are non-viable microbial cells or fragments that confer health benefits, distinct from live probiotics, as they eliminate viability while maintaining bioactivity through intact cell walls and metabolites. Production methods of para-probiotics are described in Table 3.4. Industries selecting strains for para-probiotic production typically follow an integrated workflow: initial in silico screening (taxonomic verification and whole-genome sequencing to exclude mobile AMR and virulence determinants), small-scale cultivation to assess biomass yield on target (preferably plant-based) media, systematic evaluation of inactivation modalities (heat, HPP, irradiation) with assays quantifying retention of defined effectors, in vitro functional testing of the inactivated material for required health target (Lee et al., 2023). Industrial production needs reproducible biomass yield and compositional consistency across lots. Evaluation of growth kinetics on scalable, food-grade media (including plant-based substrates), batch-to-batch reproducibility, and downstream processing behaviour (centrifugation, drying, encapsulation) for strain-specific targets and methods. Selecting strains this way prioritizes predictable, safe, and analytically traceable paraprobiotic products that translate from bench to market with fewer biosafety and cold-chain constraints while aligning with regulatory and consumer expectations (Siciliano et al., 2021).

3.3.3 Strain Selection and Optimization for Yield

Strain selection is pivotal for para-probiotic efficacy; as microbial species differ in their ability to retain functional properties post-inactivation. Commonly used strains

Table 3.4 Key inactivation methods in producing para-probiotics

Method	Mechanism of inactivation	Reference
Heat-killed (60–100 °C for 5–30 min)	Protein coagulation, DNA breakup, ribosome damage	Mady et al. (2025)
High pressure (20–180 MPa)	Membrane rupture, pH drop, protein denaturation	Tsevdou et al. (2020)
UV rays (short (200–280 nm), medium (280–320 nm), and long (320–400 nm)	DNA pyrimidine dimers	Lopez et al. (2008)
Sonication	Cavitation, cell wall rupture, free radicals	Manyatsi et al. (2024)
Ionizing radiation (Caesium Cs^{137} and Cobalt Co^{60})	Nucleic acid damage, radiolysis of water	Mañas and Pagán (2005)
Tyndallization	Repeated heat cycles (used in colic study)	Piqué et al. (2019)
Fragmented cells	Mechanical lysis	Salminen et al. (2021)
Supercritical CO_2 technology	pH drop, inactivates enzymes, disturbing the intracellular electrolytic balance	Yuk and Geveke (2011)
Dehydration methods (Vaccume drying/lyophilization)	Disruption of membranes, alterations in nucleic acids, ribosomes, and protein structures	Puttarat et al. (2021)
Pulse electric field range (2–87 kV/cm range)	Disruption of cell membrane through electroporation	Cavalcanti et al. (2023)

include *Lactobacillus acidophilus*, *L. rhamnosus* CNCM I-3698, *L. farciminis* CNCM I-3699, and *L. reuteri* DSM 17938. Live probiotics and heat-inactivated para-probiotics reduce inflammation, oxidative stress, and epithelial barrier loss in DSS induced-colitis via distinct and complementary routes. Live strains remodel community structure and produce SCFAs *in situ*, engaging GPCR signalling and altering microbial ecology, whereas para-probiotics deliver consistent MAMPs (peptidoglycan, lipoteichoic acids, surface proteins), preformed metabolites and EPS that activate TLR/NOD pathways. Para-probiotics induce IL-10, attenuate NF-κB, and restore tight-junction expression without viability-related risks. Paraprobiotics therefore, offer superior stability, standardization, and negligible potential for horizontal gene transfer, making them especially suitable for vulnerable populations and for incorporation into plant-based matrices (Kang et al., 2025).

L. acidophilus treated at 60 °C for 30 min retained adhesion and pathogen exclusion capabilities (Yolmeh et al., 2024), while autoclaving *L. rhamnosus* and *L. farciminis* enhanced co-aggregation and pathogen inhibition (Kawahara, 2010). However, strain-specific sensitivity to inactivation methods has effects on functional aspects of para-probiotics: *L. reuteri* DSM 17938 retained adhesion post-heat treatment but lost function after Lithium Chloride (LiCl) exposure, which acts by extracting surface proteins, significantly decreasing the adhesive capacity of *L. reuteri*, implicating surface proteins as critical mediators (Singh et al., 2017;

Jeong et al., 2019). Yeasts and acetic acid bacteria from fermented foods (e.g., nata de coco, kombucha) are emerging candidates, but their para-probiotic potential remains underexplored (de Almada et al., 2016). Thus, strain selection must be tightly coupled with process optimization to preserve para-probiotic bioactivity.

3.3.4 Analytical Techniques for Para-Probiotics (Metabolomics, Proteomics) and Downstream Processing

Quantifying non-viable para-probiotics requires advanced analytical techniques beyond traditional culturing. Flow cytometry using dual nucleic acid staining (SYTO and propidium iodide) enables rapid discrimination of live, damaged, and dead cells based on membrane integrity (Pane et al., 2018). Digital PCR targeting multiple copies of 16S rDNA offers high accuracy for quantifying heat-killed lactobacilli in food matrices, unaffected by DNA recovery variability. Apart from these techniques, omics-based characterisation for multi-strain detection is on trial, flowcytometry provides robustness of the cytofluorimetric analysis (Klein et al., 2013; Mora et al., 2019). Downstream processing methods, as mentioned in Table 3.4, must be validated for preserving functional structures and large-scale processing specific to strains. Most used methods include thermal treatment, which is widely used in the laboratory as well as at the industrial level (Barros et al., 2020). Heating induces thermal damage, retaining cell wall integrity while ensuring microbial death (Barros et al., 2021).

Ultrasound generates cavitation forces that shear membranes and denature proteins, while ionizing radiation and UV irradiation target nucleic acids, disrupting transcription and translation (Chandrapala et al., 2012). In a recent study by Almada et al. (2021), 42 male Wistar rats were fed postbiotics obtained through six different inactivation treatments of *L. acidophilus La-5* and *B. animalis BB-12*. Inactivation techniques included heat treatment, high-pressure processing (HPP), ultrasound, ionising radiation, UV irradiation, and Ohmic heating (Table 3.5). According to the outcome, Peptidoglycan and surface proteins were key mediators of immune signalling via TLR2 activation, leading to upregulation of IL-10. Cell wall integrity correlated with epithelial protection and cytokine balance, while DNA damage and membrane disruption varied by method, influencing downstream immune responses. Each method affects microbial physiology differently, making tailored analytical workflows necessary to ensure para-probiotic products retain their health-promoting features while meeting regulatory and industrial standards (Siciliano et al., 2021).

Table 3.5 Probiotic alteration methods and their effect on cytokine balance in Wistar rats

<table>
<tr><th>Treatment (method + parameters)</th><th>Significant biochemical effects (vs control)</th><th>Microbiota notes</th></tr>
<tr><td>Thermal (80 °C)</td><td>No significant effects</td><td rowspan="6">Dominant orders across groups:
Clostridiales 45.6–56%
Bacteroidales 31.9–44.2%
Lactobacillales 3.9–7.8%</td></tr>
<tr><td>High pH (pH: 12.5)</td><td>No significant effects</td></tr>
<tr><td>Low pH (pH: 1)</td><td>No significant effects</td></tr>
<tr><td>Ultrasound (power: 792 W; frequency: 20 kHz)</td><td>No significant effects</td></tr>
<tr><td>Irradiation (dose: 2.5 kGy/h) bifidobacterium lactis</td><td>Decreased total serum cholesterol ($p < 0.05$)</td></tr>
<tr><td>Supercritical CO_2 (pressure: 10 MPa; 40 °C): Bifidobacterium lactis</td><td>Decreased total serum cholesterol, decreased HDL-cholesterol increased albumin increased creatinine ($p < 0.05$)</td></tr>
</table>

Compiled from Almada et al. (2021)

3.4 Synergistic Applications and Formulations of Postbiotics and Para-Probiotics

Synergistic formulations pairing postbiotics and para-probiotics are being studied across food, feed, and therapeutic contexts to amplify antimicrobial, barrier and immunomodulatory effects while improving stability and safety.

Livestock improvement Studies using sheep treated with 60 mg/kg parabiotics showed a 72% drop in parasite burden, 90% fewer fecal eggs, and 96% fewer female parasites. Piglets fed *E. faecalis* parabiotics (12 mg/kg/day) had taller villi, while *L. plantarum* postbiotics improved weight gain, reduced diarrhea, and raised body weight by day 32. In broilers, postbiotics boosted beneficial bacteria, cut *E. coli* and Salmonella, and improved gut morphology, sometimes outperforming oxytetracycline. Horses treated with 30 mg/kg Cry5B parabiotics cleared fecal eggs, and antioxidant activity ranged from 17.8–90.6% (DPPH) to 27.1–99.5% (ABTS) (Corona et al., 2021) These quantitative outcomes highlight their potential to enhance animal health, performance, and disease resistance. The following section discusses the combined uses of postbiotics and para-probiotics, where co-formulation has been reported to produce measurable benefits. It highlights industry and experimental examples that illustrate outcomes relevant to food safety, animal production, and disease management.

Biological (anti-microbial and anti-aflatoxin) effect In dairy, an applied study combining both postbiotic extracts, para-probiotics and encapsulated *L. paracasei* KC39 this study showcased up to 60.56% aflatoxin reduction and significantly inhibiting mold growth. Given that 96% of infant formula samples exceeded aflatoxin limits, these bioactives provide a promising alternative to improve food safety (Mogahed Fahim et al., 2021). Together, they improve feed conversion ratio, body weight gain, and intestinal morphology, while postbiotic metabolites (COM456, COM345, COM246) boost egg yield and shell quality. Parabiotics induce immune protection, and the combination leads to better yolk pigmentation and improved fatty acid profiles. Immunologically, they suppress pro-inflammatory cytokines like TNF-α, elevate anti-inflammatory IL-10, increase IgA, IgG, and IgM, and upregulate β-defensin genes. Overall, their synergy delivers measurable health improvements, making it a safe, stable, and effective alternative for sustainable poultry production (Franco, 2024; Parhi, 2024; Sionek & Gantner, 2025).

Antimicrobial Resistance (AMR) Activity AMR is an important concern with respect to superbugs and hospital-acquired infections. Postbiotics and para-probiotics are emerging as potent anti-biotic replacements or adjuvants and are considered as one of the solutions to this long-anticipated epidemic (Ding et al., 2025). They have also been reported to act synergistically at low doses causing up to ~90% inhibition in-vitro, in MDR Klebsiella, *E. coli*, Pseudomonas, and Staphylococcus. The mechanism is based on neutralisation of resistance biomarker genes by postbiotic compounds such as bile acids, SCFAs, bacteriocins, and organic acids produced by popular probiotics. Postbiotics from *L. plantarum* RP225 and *L. salivarius* RP317 (25–100 mg/mL) combined with 1–4 μg/mL of amoxicillin or imipenem, resulted in complete eradication *K. pneumoniae* (Aghamohammad et al., 2025). They have also been reported to show down-regulation of key resistance genes like Extended Spectrum Beta-Lactamases (bla ermB, and blaKPC) in nosocomial and MDR pathogens after exposure to Lactobacillus-derived postbiotics (Li et al., 2025). Thus, making them one of the safe alternatives against MDR bacteria.

Collectively, these findings demonstrate that co-formulated postbiotic and probiotic interventions provide measurable benefits across animal species. They enhance growth performance and gut morphology and aid in toxin removal. At the same time, they support immune modulation and antioxidant protection. Importantly, in many cases, these outcomes equal or even surpass those achieved with conventional chemotherapeutics, underscoring their potential as safe and effective alternatives (Fig. 3.1).

3.5 Market Potential and Product Development

Postbiotics and paraprobiotics are being increasingly incorporated into functional foods, beverages, and cosmetics intended for applications in gut, skin, and metabolic health. These developments underscore their potential as bioactive ingredients in next-generation functional foods. The stability and health benefits of postbiotics and para-probiotics address some of the long-standing challenges of incorporating live probiotic cultures into diverse food products while maintaining their viability and efficacy (Morán & Kilasoniya, 2024). The global gut health supplement market is expanding significantly, with a compound annual growth rate (CAGR) of 8.4%, an estimated USD 14.4 billion in 2025 to USD 32.4 billion by 2035. The postbiotics and paraprobiotics are prepared to capture an increasing share of it due to their

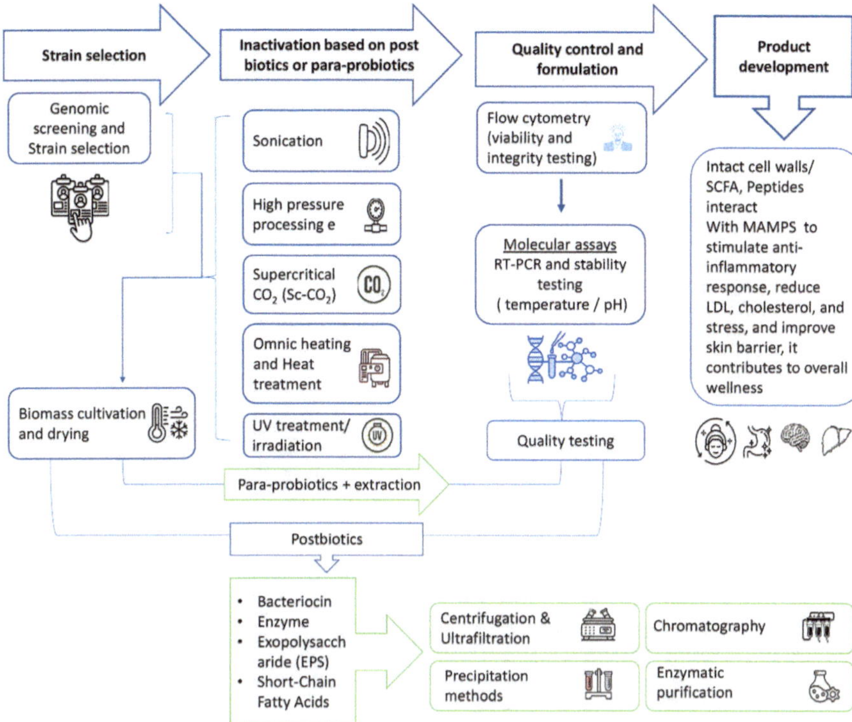

Fig. 3.1 Development of postbiotics and para-probiotics, from strain selection and microbial inactivation to quality control and product formulation, highlighting key processes and mechanisms that contribute to health benefits

unique advantages in product development (Nataraj et al., 2020; Zhang et al., 2024). This growth is fuelled by increasing consumer awareness of the link between gut health and overall well-being. Clinically, strains such as heat-treated *Lactobacillus gasseri* CP2305 have been reported to improve sleep quality and gut function in stressed individuals, while *Lactobacillus plantarum* strains have demonstrated immune and gut barrier benefits in animal and clinical studies (Gong et al., 2024). Several postbiotics and para-probiotic products (Table 3.6) have been studied in human clinical studies (Das et al., 2023; Lim et al., 2022). The growing understanding of the microbiome among consumers is increasing the demand for scientifically validated functional ingredients (Rashidinejad, 2024). The utilization of individual's specific gut microbiota may further facilitate the development of personalised nutritional products and services offering new avenues for market growth (Domenico et al., 2022).

The commercialization of postbiotics and paraprobiotics faces several barriers despite their growing market potential. A primary challenge is the lack of uniform definitions and regulatory frameworks, which creates confusion in product classification, labeling, and health claims approval, thereby slowing market entry and

Table 3.6 Commercially available postbiotics and para-probiotic products with published human/animal studies

Postbiotics and para-probiotics product name	Active postbiotic/ paraprobiotics (s)	Gut health benefit	Mechanism of action	Delivery format	Manufacturer/ distributor	Study citation (PMID)
EpiCor®	Yeast fermentate (*S. cerevisiae*)	↑ Microbial diversity, ↓ GI symptoms	↑ sIgA secretion, modulation of gut-associated lymphoid tissue (GALT)	Oral capsule	Cargill	Pinheiro et al. (2017)
Wellmune®	β-1,3/1,6-glucan (EPS)	↑ Gut barrier integrity, ↓ GI inflammation	Priming of innate immune cells, modulation of gut mucosal immunity	Oral powder	Kerry Group	Edo et al. (2025), McFarlin et al. 2024)
Mitopure®	Urolithin A (from ellagitannin metabolism)	↑ Mitochondrial health in gut epithelial cells	Induction of mitophagy via Nrf2 pathway, mitigate colitis	Oral sachet	Amazentis	Singh et al. (2019), Zhao et al. (2023)
PharmaGABA®	γ Aminobutyricacid (GABA)	↓ Stress-related gut dysbiosis	GABA-A/B receptor binding, modulation of enteric nervous system	Oral tablet	Pharma Foods International	Kanehira et al. (2011), Yoto et al. (2012)
ImmunoWall®	MOS + β-glucans + teichoic acids	↓ Pathogen colonization ↑ Gutimmune tone	Pathogen agglutination, TLR2 activation in gut epithelium in poultry	Feed additive (animal model)	ICC Brazil	Bonato et al. (2020)
Pylopass™	Non-Viable *Lactobacillus reuteri* DSMZ 17648	↓ Abundance *Helicobacter pylori*, ↑ Gut-immue tone	Low % of 13C urea breath test in humans	Solid tablets	Novozymes	Mehling and Busjahn (2013)
Lactéol® Hydro	Heat-killed *L.acidophilus* LB and zinc	↑ sIgA, TLR2 activation, ↓ Pathogen colonization	↓ Chronic diarrhea in randomised controlled human trials	Chewable tablets	Adare Pharmaceuticals SAS	Xiao et al. (2003)

consumer trust (Divsalar et al., 2026). Few of the technical difficulties include standardizing production processes with consistent bioactive compound profiles and implementation of quality control (Calvanese et al., 2025). Limited clinical evidence from robust human trials restricts the ability to substantiate health claims, affecting regulatory acceptance and consumer confidence. Additionally, consumer awareness of probiotics remains low, making effective communication and market differentiation difficult (Liang et al., 2024). Addressing these scientific, regulatory, technological, and communication challenges is essential to unlock the full commercial potential of postbiotic and paraprobiotic products in the expanding gut health market.

3.6 Regulatory Landscape and Commercialization

The regulatory landscape and commercialization pathways for postbiotics and para-probiotics exhibit significant divergence across global regions, particularly between Europe, India, and the United States, impacting their market entry and consumer availability (Morán & Kilasoniya, 2024). Their increasing appeal towards postbiotics and para-probiotics stems from potentially better safety profiles compared to live probiotics, as they avoid risks associated with viable bacteria (Homayouni-Rad et al., 2022; Nataraj et al., 2020; Wang et al., 2024). In Europe, the regulatory framework primarily categorizes many microbiome-derived preparations, including postbiotics, as novel foods under Regulation (EU) 2015 (Morán & Kilasoniya, 2024). This regulation mandates that products not widely consumed in the EU before May 1997 require pre-market authorization (Bulut et al., 2025; Lähteenmäki-Uutela et al., 2021). Consequently, inactivated cells, purified short-chain fatty acid (SCFA), and other non-conventional microbial derivatives necessitate comprehensive dossiers for authorization. This approach has led to precedent-driven reviews for substances like human-identical milk oligosaccharides and cell-culture proteins (Corona et al., 2021). As of 2024–2025, a specific regulatory category for "postbiotics"does not exist in Europe, resulting in market entry timelines that can span months to years.

The Food Safety and Standards Authority of India (FSSAI) regulates probiotics and related preparations under its consolidated food standards, specifically the Food Safety and Standards (Health Supplements, Nutraceuticals, Prebiotic and Probiotic Food) Regulations, 2022 (Valluru et al., 2025). This framework requires developers to provide compositional data, substantiated safety and efficacy claims, and human safety evidence for novel microbial ingredients. However, explicit guidance for postbiotics and para-probiotics remains interpretive rather than categorical. The developers typically submit case-specific dossiers and human safety evidence to obtain market authorization (Arora et al., 2019).

The United States offers a more flexible regulatory pathway through the Generally Recognized as Safe (GRAS) notification route and existing food additive frameworks (Oluwarinde et al., 2023). This allows inactivated probiotic ingredients and certain metabolite preparations to reach the market more rapidly, provided their safety is demonstrated "under the intended conditions of use". Since, the regulatory landscape remains under-explored, several research teams are stressing that safety assessments should not rely solely on traditional probiotic criteria (e.g. viability or colonization potential) (Roe et al., 2022), but instead focus on molecular effectors such as MAMPs, metabolites, and surface proteins that remain active after inactivation (Guglielmetti et al., 2025). Regulatory frameworks, including validated biomarkers, strain-specific genomic screening for AMR and virulence, and standardized inactivation protocols will ensure consumer acceptance and safety (Fachri et al., 2024; Guglielmetti et al., 2025; Thorakkattu et al., 2022).

Although postbiotics and para-probiotics offer promising health benefits with improved stability and safety over live probiotics, their regulatory oversight remains unorganised and underdeveloped. However, the commercial success and widespread adoption of postbiotics and para-probiotics critically depends on transparent and informed global regulatory frameworks that ensure their safety, efficacy, and appropriate labeling for consumers.

3.7 Conclusion

Postbiotics and para-probiotics offer a targeted solution beyond traditional live probiotics. In the gut, they act through complementary, well-defined pathways via soluble postbiotic metabolites such as SCFAs, peptides, EPS, and small-molecule neurotransmitter precursors, which act as signalling ligands that engage G protein-coupled receptors (GPCRs). These neurotransmitter precursors serve as crucial mediators in metabolic, immunologic, and neuro-transmissional regulation, and modulate epigenetic regulators. However, Para-probiotics, the inactivated whole cells or fragments, present conserved microbial ligands like peptidoglycan, teichoic acids, surface proteins, S-layer motifs that activate pattern-recognition receptors (TLRs, NODs). Para-probiotics act by accelerating immune responses toward regulatory cytokines (IL-10) and strengthening barrier function via tight-junction support and mucin induction. Together, they deliver bimodal signalling that reduces pro-inflammatory markers like NF-κB activation, preserves mitochondrial, and protein-synthesis pathways under stress, and immune homeostasis. Non-viable preparations such as post-biotics and para-probiotics are increasingly preferred over traditional live probiotics due to several key advantages. Infants and older adults benefit from reduced infection risk and targeted immune support without the uncertainties of live colonization, while pediatric and geriatric groups gain safe immune modulation along with enhanced barrier integrity and anti-inflammatory effects.

In plant-based matrices, these ingredients are stable, compatible with prebiotic substrates, and can be efficiently incorporated into vegan and clean-label products. These advantages make postbiotics and para-probiotics as safer, stable, and versatile alternatives for a wide range of clinical and functional food applications. The future of postbiotic and para-probiotic research lies in precision biotics, utilizing AI-driven modelling, and multi-omics to tailor microbial derivatives to specific health profiles. The research roadmap must prioritize large-scale clinical trials to validate these biotics as adjunctive therapies for metabolic and oncological care. Furthermore, their application as natural biopreservatives offers a sustainable industrial alternative to chemical additives. As global regulatory standards get harmonized, the integration of these stable microbial tools into personalized nutrition will define the next era of functional food and medicine.

References

Abdalla, S. B., Moghazy, R. M., Hamed, A. A., Abdel-Monem, M. O., El-Khateeb, M. A., & Hassan, M. G. (2024). Strain selection and adaptation of a fungal-yeast-microalgae consortium for sustainable bioethanol production and wastewater treatment from livestock wastewater. *Microbial Cell Factories, 23*(1), 288. https://doi.org/10.1186/s12934-024-02537-4

Aghamohammad, S., Sakaki, F., Haghighatshenas, Z., Khiavi, E. H. A. G., Badmasti, F., & Rohani, M. (2025). Novel antimicrobial strategy: Native postbiotics synergize with antibiotics to overcome resistance in *Klebsiella pneumoniae*. *BMC Microbiology, 25*(1), 723. https://doi.org/10.1186/s12866-025-04474-7

Almada, C. N., Almada-Érix, C. N., Roquetto, A. R., Santos-Junior, V. A., Cabral, L., Noronha, M. F., Gonçalves, A. E. S. S., dos Santos, P., dos Santos, A., Martinez, J., Lollo, P. C., Costa, W. K. A., Magnani, M., & Sant'Ana, A. S. (2021). Paraprobiotics obtained by six different inactivation processes: Impacts on the biochemical parameters and intestinal microbiota of Wistar male rats. *International Journal of Food Sciences and Nutrition, 72*(8), 1057–1070. https://doi.org/10.1080/09637486.2021.1906211

Almahbashi, A., & Gunes Altuntas, E. (2025). From preparation to bioactivity: A comparative study on preparation methods and characterization of postbiotics. *Food Science & Nutrition, 13*(5), e70294. https://doi.org/10.1002/fsn3.70294

Amobonye, A., Pillay, B., Hlope, F., Asong, S. T., & Pillai, S. (2025). Postbiotics: an insightful review of the latest category in functional biotics. *World Journal of Microbiology and Biotechnology, 41*(8), 293.

Andresen, V., Gschossmann, J., & Layer, P. (2020). Heat-inactivated *Bifidobacterium bifidum* MIMBb75 (SYN-HI-001) in the treatment of irritable bowel syndrome: A multicentre, randomised, double-blind, placebo-controlled clinical trial. *The Lancet Gastroenterology & Hepatology, 5*(7), 658–666. https://doi.org/10.1016/S2468-1253(20)30056-X

Arora, M., Kaur, N., Arora, M., & Bansal, P. (2019). Critical review on food safety standard regulations: A real scrutinizing authority or a misleading player for probiotic products. *Applied Clinical Research, Clinical Trials and Regulatory Affairs, 6*(2), 71–85. https://doi.org/10.2174/2213476X05666181011144420

Ashique, S., Debnath, B., Islam, A., Tariq, M., Taj, T., Mantry, S., Sridhar, S. B., Verma, B., Kumar, N., Adhikary, K., Saxena, D., Maqbool, M., & Hussain, M. S. (2025). Para-probiotics as novel anti-inflammatory agents: Insight into health benefits and therapeutic applications. *Current Pharmaceutical Design*. https://doi.org/10.2174/0113816128411949250909113034

Assadpour, E., & Jafari, S. M. (2019). Advances in spray-drying encapsulation of food bioactive ingredients: From microcapsules to nanocapsules. *Annual Review of Food Science and Technology, 10*, 103–131. https://doi.org/10.1146/annurev-food-032818-121641

Barros, C. P., Guimarães, J. T., Esmerino, E. A., Duarte, M. C. K., Silva, M. C., Silva, R., Ferreira, B. M., Sant'ana, A. S., Freitas, M. Q., & Cruz, A. G. (2020). Paraprobiotics and postbiotics: Concepts and potential applications in dairy products. *Current Opinion in Food Science, 32*, 1–8. https://doi.org/10.1016/j.cofs.2019.12.003

Barros, C. P., Grom, L. C., Guimarães, J. T., Balthazar, C. F., Rocha, R. S., Silva, R., Almada, C. N., Pimentel, T. C., Venâncio, E. L., Collopy Junior, I., Maciel, P. M. C., Freitas, M. Q., Esmerino, E. A., Silva, M. C., Duarte, M. C. K. H., Sant'ana, A. S., & Cruz, A. G. (2021). Paraprobiotic obtained by ohmic heating added in whey-grape juice drink is effective to control postprandial glycemia in healthy adults. *Food Research International, 140*, 109905. https://doi.org/10.1016/j.foodres.2020.109905

Bitew, D., Alemu, M., Tesfaye, A., & Andualem, B. (2024). Ethanologenic yeasts from Ethiopian fermented beverages and optimization of fermentation conditions. *Renewable and Sustainable Energy Reviews, 190*, 114090. https://doi.org/10.1016/j.rser.2023.114090

Bonato, M., Borges, L. L., Ingberman, M., Fávaro, C., Mesa, D., Caron, L. F., & Beirão, B. C. B. (2020). Effects of yeast cell wall on immunity, microbiota, and intestinal integrity of *Salmonella*-infected broilers. *Journal of Applied Poultry Research, 29*(3), 545–558. https://doi.org/10.1016/j.japr.2020.03.002

Braga, J. D., Thongngam, M., & Kumrungsee, T. (2024). Gamma-aminobutyric acid as a potential postbiotic mediator in the gut–brain axis. *npj Science of Food, 8*(1), 16. https://doi.org/10.1038/s41538-024-00253-2

Bulut, S. D., Döndaş, H. A., Celebioglu, H. U., Sansano, J. M., & Döndaş, N. Y. (2025). Recent insights about probiotics related pharmabiotics in pharmacology: Prevention and management of diseases. *Probiotics and Antimicrobial Proteins, 17*(4), 2262–2292. https://doi.org/10.1007/s12602-025-10613-3

Calvanese, C. M., Villani, F., Ercolini, D., & De Filippis, F. (2025). Postbiotics versus probiotics: Possible new allies for human health. *Food Research International, 217*, 116869. https://doi.org/10.1016/j.foodres.2025.116869

Calvigioni, M., Bertolini, A., Codini, S., Mazzantini, D., Panattoni, A., Massimino, M., Celandroni, F., Zucchi, R., Saba, A., & Ghelardi, E. (2023). HPLC-MS-MS quantification of short-chain fatty acids actively secreted by probiotic strains. *Frontiers in Microbiology, 14*, 1124144. https://doi.org/10.3389/fmicb.2023.1124144

Cavalcanti, R. N., Balthazar, C. F., Margalho, L. P., Freitas, M. Q., Sant'ana, A. S., & Cruz, A. G. (2023). Pulsed electric field-based technology for microbial inactivation in milk and dairy products. *Current Opinion in Food Science, 54*, 101087. https://doi.org/10.1016/j.cofs.2023.101087

Chai, J., Deng, F., Li, Y., Wei, X., & Zhao, J. (2024). Editorial: The gut-skin axis: Interaction of gut microbiome and skin diseases. *Frontiers in Microbiology, 15*, 1427770. https://doi.org/10.3389/fmicb.2024.1427770

Chandrapala, J., Oliver, C., Kentish, S., & Ashokkumar, M. (2012). Ultrasonics in food processing – Food quality assurance and food safety. *Trends in Food Science & Technology, 26*(2), 88–98. https://doi.org/10.1016/j.tifs.2012.01.010

Chen, C., Guo, T., Wu, D., Shu, J., Huang, N., Tian, H., Yu, H., & Ge, C. (2025). Study on *Lacticaseibacillus casei* TCS fermentation kinetic models and high-density culture strategy. *Microbiology Spectrum, 13*(7), e02590-24. https://doi.org/10.1128/spectrum.02590-24

Corona, L., Lussu, A., Bosco, A., Pintus, R., Cesare Marincola, F., Fanos, V., & Dessì, A. (2021). Human milk oligosaccharides: A comprehensive review towards metabolomics. *Children, 8*(9), 804. https://doi.org/10.3390/children8090804

da Silva Vale, A., de Melo Pereira, G. V., de Oliveira, A. C., de Carvalho Neto, D. P., Herrmann, L. W., Karp, S. G., Soccol, V. T., & Soccol, C. R. (2023). Production, formulation, and application of postbiotics in the treatment of skin conditions. *Fermentation, 9*(3), 264. https://doi.org/10.3390/fermentation9030264

Dahech, I., Belghith, K. S., Hamden, K., Feki, A., Belghith, H., & Mejdoub, H. (2011). Antidiabetic activity of Levan polysaccharide in alloxan-induced diabetic rats. *International Journal of Biological Macromolecules, 49*(4), 742–746. https://doi.org/10.1016/j.ijbiomac.2011.07.007

Das, A., Behera, R. N., Kapoor, A., & Ambatipudi, K. (2023). The potential of meta-proteomics and artificial intelligence to establish the next generation of probiotics for personalized healthcare. *Journal of Agricultural and Food Chemistry, 71*(46), 17528–17542. https://doi.org/10.1021/acs.jafc.3c03834

de Almada, C. N., Almada, C. N., Martinez, R. C. R., & Sant'Ana, A. S. (2016). Paraprobiotics: Evidences on their ability to modify biological responses, inactivation methods and perspectives on their application in foods. *Trends in Food Science & Technology, 58*, 96–114. https://doi.org/10.1016/j.tifs.2016.09.011

Ding, W., Cheng, Y., Liu, X., Zhu, Z., Wu, L., Gao, J., Lei, W., Li, Y., Zhou, X., Wu, J., Gao, Y., Ling, Z., & Jiang, R. (2025). Harnessing the human gut microbiota: An emerging frontier in combatting multidrug-resistant bacteria. *Frontiers in Immunology, 16*, 1563450. https://doi.org/10.3389/fimmu.2025.1563450

Dinić, M., Burgess, J. L., Lukić, J., Catanuto, P., Radojević, D., Marjanović, J., Verpile, R., Thaller, S. R., Gonzalez, T., Golić, N., Strahinić, I., Tomic-Canic, M., & Pastar, I. (2024). Postbiotic lactobacilli induce cutaneous antimicrobial response and restore the barrier to inhibit the intracellular invasion of Staphylococcus aureus in vitro and ex vivo. *The FASEB Journal, 38*(14), e23801. https://doi.org/10.1096/fj.202400054RR

Divsalar, E., İncili, G. K., Shi, C., Semsari, E., Hosseini, S. H., Ebrahimi Tirtashi, F., Toker, O. S., Mojgani, N., Liu, S.-Q., & Moradi, M. (2026). Challenges with the use of postbiotics/parabiotics in food industry. *Critical Reviews in Food Science and Nutrition, 66*(6), 1155–1171. https://doi.org/10.1080/10408398.2025.2541047

Domenico, M. D., Ballini, A., Boccellino, M., Scacco, S., Lovero, R., Charitos, I. A., & Santacroce, L. (2022). The-intestinal-microbiota-may-be-a-potential-theranostic-tool-for-personalized-medicine. *Journal of Personalized Medicine, 12*(4), 523. https://doi.org/10.3390/jpm12040523

Edo, G. I., Mafe, A. N., Ali, A. B. M., Akpoghelie, P. O., Yousif, E., Isoje, E. F., Igbuku, U. A., Zainulabdeen, K., Owheruo, J. O., Essaghah, A. E. A., Umar, H., Ahmed, D. S., & Alamiery, A. A. (2025). A critical review on the impacts of β-glucans on gut microbiota and human health. *The Microbe, 7*, 100394. https://doi.org/10.1016/j.microb.2025.100394

Fachri, M., Amoah, K., Huang, Y., Cai, J., Alfatat, A., Ndandala, C. B., Shija, V. M., Jin, X., Bissih, F., & Chen, H. (2024). Probiotics and paraprobiotics in aquaculture: A sustainable strategy for enhancing fish growth, health and disease prevention-a review. *Frontiers in Marine Science, 11*, 1499228. https://doi.org/10.3389/fmars.2024.1499228

Feng, X., Liu, N., Yang, Y., Feng, S., Wang, J., & Meng, Q. (2022). Isotope-coded chemical derivatization method for highly accurately and sensitively quantifying short-chain fatty acids. *Journal of Agricultural and Food Chemistry, 70*(20), 6253–6263. https://doi.org/10.1021/acs.jafc.2c01836

Franco, W. (2024). Postbiotics and parabiotics derived from bacteria and yeast: Current trends and future perspectives. *CyTA Journal of Food, 22*(1), 2425838. https://doi.org/10.1080/19476337.2024.2425838

Gao, J., Mang, Q., Sun, Y., & Xu, G. (2025). Short-chain fatty acids (SCFAs) modulate the hepatic glucose and lipid metabolism of *Coilia nasus* via the FFAR/AMPK signaling pathway in vitro. *International Journal of Molecular Sciences, 26*(8), 3654. https://doi.org/10.3390/ijms26083654

Garrote Achou, C., Cantalejo Díez, M. J., Diaz Cano, J., & Molinos Equiza, X. (2025). Evaluation of different nutritional sources in lactic acid bacteria fermentation for sustainable postbiotic production. *Foods, 14*(4), 649. https://doi.org/10.3390/foods14040649

Ghoneim, M. A. M., Hassan, A. I., Mahmoud, M. G., & Asker, M. S. (2016). Effect of polysaccharide from *Bacillus subtilis* sp. on cardiovascular diseases and atherogenic indices in diabetic rats. *BMC Complementary and Alternative Medicine, 16*(1), 112. https://doi.org/10.1186/s12906-016-1093-1

Gong, X., Zhang, J., Gan, Q., Teng, Y., Hou, J., Lyu, Y., Liu, Z., Wu, Z., Dai, R., Zou, Y., Wang, X., Zhu, D., Zhu, H., Liu, T., & Yan, Y. (2024). Advancing microbial production through artificial intelligence-aided biology. *Biotechnology Advances, 74*, 108399. https://doi.org/10.1016/j.biotechadv.2024.108399

Guglielmetti, S., Boyte, M.-E., Smith, C. L., Ouwehand, A. C., Paraskevakos, G., & Younes, J. A. (2025). Commercial and regulatory frameworks for postbiotics: An industry-oriented scientific perspective for non-viable microbial ingredients conferring beneficial physiological effects. *Trends in Food Science & Technology, 163*, 105130. https://doi.org/10.1016/j.tifs.2025.105130

Guo, S., Ma, T., Kwok, L.-Y., Quan, K., Li, B., Wang, H., Zhang, H., Menghe, B., & Chen, Y. (2024a). Effects of postbiotics on chronic diarrhea in young adults: A randomized, double-blind, placebo-controlled crossover trial assessing clinical symptoms, gut microbiota, and metabolite profiles. *Gut Microbes, 16*(1), 2395092. https://doi.org/10.1080/19490976.2024.2395092

Guo, Y., Wang, Y., Xu, X., Niu, D., Qing, Q., Wang, L., & Zhu, J. (2024b). Effects of cold plasma pretreatment on the synthesis of polysaccharide from *Pleurotus ostreatus*. *Applied Biochemistry and Biotechnology, 196*(4), 1977–1991. https://doi.org/10.1007/s12010-023-04662-z

Hemalatha, M., & Subathra Devi, C. (2022). A statistical optimization by response surface methodology for the enhanced production of riboflavin from *Lactobacillus plantarum*–HDS27: A strain isolated from bovine milk. *Frontiers in Microbiology, 13*, 982260. https://doi.org/10.3389/fmicb.2022.982260

Hirose, Y., Yamamoto, Y., Yoshikai, Y., & Murosaki, S. (2013). Oral intake of heat-killed *Lactobacillus plantarum* L-137 decreases the incidence of upper respiratory tract infection in healthy subjects with high levels of psychological stress. *Journal of Nutritional Science, 2*, e39. https://doi.org/10.1017/jns.2013.35

Homayouni-Rad, A., Pouraga, B., Abbasi, A., & Soleimani, R. A. (2022). Could-postbiotics-substitute-probiotics. *Current Nutrition & Food Science, 19*(7), 670–672. https://doi.org/10.2174/1389201023666221004125745

Jacobsen, C., García-Moreno, P. J., Mendes, A. C., Mateiu, R. V., & Chronakis, I. S. (2018). Use of electrohydrodynamic processing for encapsulation of sensitive bioactive compounds and applications in food. *Annual Review of Food Science and Technology, 9*, 525–549. https://doi.org/10.1146/annurev-food-030117-012348

Jayasekara, L. A. C. B., Poonsawad, A., Watchaputi, K., Wattanachaisaereekul, S., & Soontorngun, N. (2022). Media optimization of antimicrobial activity production and beta-glucan content of endophytic fungi *Xylaria* sp. BCC 1067. *Biotechnology Reports, 35*, e00742. https://doi.org/10.1016/j.btre.2022.e00742

Jeong, M., Kim, J. H., Yang, H., Kang, S. D., Song, S., Lee, D., Lee, J. S., Yoon Park, J. H., Byun, S., & Lee, K. W. (2019). Heat-killed *Lactobacillus plantarum* KCTC 13314BP enhances phagocytic activity and immunomodulatory effects via activation of MAPK and STAT3 pathways. *Journal of Microbiology and Biotechnology, 29*(8), 1248–1254. https://doi.org/10.4014/jmb.1905.05066

Jin, W.-B., Xiao, L., Jeong, M., Han, S.-J., Zhang, W., Yano, H., Shi, H., Arifuzzaman, M., Lyu, M., Wang, D., Tang, Y. A., Qiao, S., Yang, X., Yang, H. S., Fu, J., Sonnenberg, G. F., Collins, N., Artis, D., & Guo, C.-J. (2025). Microbiotaderived bile acids antagonize the host androgen receptor and drive anti-tumor immunity. *Cell, 188*(9), 2336–2353.e38. https://doi.org/10.1016/j.cell.2025.02.029

Kanehira, T., Nakamura, Y., Nakamura, K., Horie, K., Horie, N., Furugori, K., Sauchi, Y., & Yokogoshi, H. (2011). Relieving occupational fatigue by consumption of a beverage containing γ-amino butyric acid. *Journal of Nutritional Science and Vitaminology, 57*(1), 9–15. https://doi.org/10.3177/jnsv.57.9

Kang, Y. Y., Song, H. J., Park, S. Y., Oh, D. N., Kim, G. Y., Been, N. Y., Kim, D. Y., Lee, E. J., Nam, B.-H., & Lee, J.-M. (2025). Comparative effects of probiotics and paraprobiotics derived from *Lactiplantibacillus plantarum, Latilactobacillus sakei,* and *Limosilactobacillus reuteri* in a DSS-induced ulcerative colitis mouse model. *Journal of Microbiology and Biotechnology, 35*, e2411045. https://doi.org/10.4014/jmb.2411.11045

Kavitake, D., Suryavanshi, M. V., Kandasamy, S., Devi, P. B., Shouche, Y., & Shetty, P. H. (2022). Bacterial diversity of traditional fermented food, idli by high thorough-put sequencing. *Journal of Food Science and Technology, 59*(10), 3918–3927. https://doi.org/10.1007/s13197-022-05421-4

Kawahara, T. (2010). Inhibitory effect of heat-killed Lactobacillus strain on immunoglobulin E-mediated degranulation and late-phase immune reactions of mouse bone marrow-derived mast cells. *Animal Science Journal, 81*(6), 714–721. https://doi.org/10.1111/j.1740-0929.2010.00788.x

Klein, G., Schanstra, J. P., Hoffmann, J., Mischak, H., Siwy, J., & Zimmermann, K. (2013). Proteomics as a quality control tool of pharmaceutical probiotic bacterial lysate products. *PLoS One, 8*(6), e66682. https://doi.org/10.1371/journal.pone.0066682

Kotani, Y., Shinkai, S., Okamatsu, H., Toba, M., Ogawa, K., Yoshida, H., Fukaya, T., Fujiwara, Y., Chaves, P. H., Kakumoto, K., & Kohda, N. (2010). Oral intake of Lactobacillus pentosus strain b240 accelerates salivary immunoglobulin A secretion in the elderly: A randomized, placebo-controlled, double-blind trial. *Immunity & Ageing, 7*(1), 11. https://doi.org/10.1186/1742-4933-7-11

Lähteenmäki-Uutela, A., Rahikainen, M., Lonkila, A., & Yang, B. (2021). Alternative proteins and EU food law. *Food Control, 130*, 108336. https://doi.org/10.1016/j.foodcont.2021.108336

Lebeer, S., Claes, I., Tytgat, H. L. P., Verhoeven, T. L. A., Marien, E., von Ossowski, I., Reunanen, J., Palva, A., de Vos, W. M., De Keersmaecker, S. C. J., & Vanderleyden, J. (2012). Functional analysis of *Lactobacillus rhamnosus* GG pili in relation to adhesion and immunomodulatory interactions with intestinal epithelial cells. *Applied and Environmental Microbiology, 78*(1), 185–193. https://doi.org/10.1128/AEM.06192-11

Lee, N., Park, Y.-S., Kang, D.-K., & Paik, H.-D. (2023). Paraprobiotics: Definition, manufacturing methods, and functionality. *Food Science and Biotechnology, 32*(14), 1981–1991. https://doi.org/10.1007/s10068-023-01378-y

Li, S., Tantibhadrasapa, A., Buddhasiri, S., Boonpan, P., Sukjoi, C., Mongkolkarvin, P., Nakphaichit, M., Nitisinprasert, S., & Thiennimitr, P. (2025). Probiotic, Paraprobiotic, and postbiotic activities of *Lactiplantibacillus plantarum* KUNN19-2 against non-typhoidal salmonella serovars. *International Journal of Molecular Sciences, 26*(5), 1821. https://doi.org/10.3390/ijms26051821

Liang, D., Wu, F., Zhou, D., Tan, B., & Chen, T. (2024). Commercial probiotic products in public health: Current status and potential limitations. *Critical Reviews in Food Science and Nutrition, 64*(19), 6455–6476. https://doi.org/10.1080/10408398.2023.2169858

Ligthart, K., Belzer, C., de Vos, W. M., & Tytgat, H. L. P. (2020). Bridging bacteria and the gut: Functional aspects of type IV pili. *Trends in Microbiology, 28*(5), 340–348. https://doi.org/10.1016/j.tim.2020.02.003

Lim, H., Cankara, F., Tsai, C.-J., Keskin, O., Nussinov, R., & Gursoy, A. (2022). Artificial intelligence approaches to human-microbiome protein–protein interactions. *Current Opinion in Structural Biology, 73*, 102328. https://doi.org/10.1016/j.sbi.2022.102328

Lopez, M., Li, N., Kataria, J., Russell, M., & Neu, J. (2008). Live and ultraviolet-inactivated *Lactobacillus rhamnosus* GG decrease flagellin-induced interleukin-8 production in Caco-2 cells. *The Journal of Nutrition, 138*(11), 2264–2268. https://doi.org/10.3945/jn.108.093658

Mady, E. A., El-Husseiny, H. M., Makioka-Itaya, Y., Ishikawa, S., Inoue, R., Li, C., Yamamoto, Y., & Nagaoka, K. (2025). Therapeutic potential of heat-killed *Enterococcus faecalis* (EC-12) Paraprobiotic in reversing maternal separation-triggered social deficit in mice. *Biomedicine & Pharmacotherapy, 191*, 118513. https://doi.org/10.1016/j.biopha.2025.118513

Mañas, P., & Pagán, R. (2005). Microbial inactivation by new technologies of food preservation. *Journal of Applied Microbiology, 98*(6), 1387–1399. https://doi.org/10.1111/j.1365-2672.2005.02561.x

Manyatsi, T. S., Mousavi Khaneghah, A., & Gavahian, M. (2024). The effects of ultrasound on probiotic functionality: An updated review. *Critical Reviews in Food Science and Nutrition, 64*(31), 11643–11660. https://doi.org/10.1080/10408398.2023.2242490

Matsuguchi, T., Takagi, A., Matsuzaki, T., Nagaoka, M., Ishikawa, K., Yokokura, T., & Yoshikai, Y. (2003). Lipoteichoic acids from lactobacillus strains elicit strong tumor necrosis factor alpha-inducing activities in macrophages through toll-like receptor 2. *Clinical and Vaccine Immunology, 10*(2), 259–266. https://doi.org/10.1128/CDLI.10.2.259-266.2003

McFarlin, B. K., Bridgeman, E. A., Curtis, J. H., Vingren, J. L., & Hill, D. W. (2024). Baker's yeast beta glucan supplementation was associated with an improved innate immune mRNA expression response after exercise. *Methods, 230*, 68–79. https://doi.org/10.1016/j.ymeth.2024.07.013

Mehling, H., & Busjahn, A. (2013). Non-viable *Lactobacillus reuteri* DSMZ 17648 (Pylopass™) as a new approach to helicobacter pylori control in humans. *Nutrients, 5*(8), 3062–3073. https://doi.org/10.3390/nu5083062

Mogahed Fahim, K., Noah Badr, A., Gamal Shehata, M., Ibrahim Hassanen, E., & Ibrahim Ahmed, L. (2021). Innovative application of postbiotics, parabiotics and encapsulated *Lactobacillus plantarum* RM1 and *Lactobacillus paracasei* KC39 for detoxification of aflatoxin M1 in milk powder. *The Journal of Dairy Research, 88*(4), 429–435. https://doi.org/10.1017/S002202992100090X

Monteiro, S. S., Schnorr, C. E., & de Pasquali, M. A. B. (2023). Paraprobiotics and postbiotics—Current state of scientific research and future trends toward the development of functional foods. *Foods, 12*(12), 2394. https://doi.org/10.3390/foods12122394

Mora, D., Filardi, R., Arioli, S., Boeren, S., Aalvink, S., & de Vos, W. M. (2019). Development of omics-based protocols for the microbiological characterization of multi-strain formulations marketed as probiotics: The case of VSL#3. *Microbial Biotechnology, 12*(6), 1371–1386. https://doi.org/10.1111/1751-7915.13476

Morán, J., & Kilasoniya, A. (2024). Integration-of-postbiotics-in-food-products-through-attenuated-probiotics-a-case-study-with-lactic-acid-bacteria-in-bread. *Foods, 13*(13), 2042. https://doi.org/10.3390/foods13132042

Morifuji, M., Ichikawa, S., Kitade, M., Fukasawa, T., Asami, Y., Manabe, Y., & Sugawara, T. (2020). Exopolysaccharides from milk fermented by lactic acid bacteria enhance dietary carotenoid bioavailability in humans in a randomized crossover trial and in rats. *The American Journal of Clinical Nutrition, 111*(4), 903–914. https://doi.org/10.1093/ajcn/nqaa020

Nakai, H., Murosaki, S., Yamamoto, Y., Furutani, M., Matsuoka, R., & Hirose, Y. (2021). Safety and efficacy of using heat-killed *Lactobacillus plantarum* L-137: High-dose and long-term use effects on immune-related safety and intestinal bacterial flora. *Journal of Immunotoxicology, 18*(1), 127–135. https://doi.org/10.1080/1547691X.2021.1979698

Nataraj, B. H., Ali, S. A., Behare, P. V., & Yadav, H. (2020). Postbiotics-parabiotics: The new horizons in microbial biotherapy and functional foods. *Microbial Cell Factories, 19*(1), 168. https://doi.org/10.1186/s12934-020-01426-w

Nishida, K., Sawada, D., Kuwano, Y., Tanaka, H., Sugawara, T., Aoki, Y., Fujiwara, S., & Rokutan, K. (2017). Daily administration of paraprobiotic *Lactobacillus gasseri* CP2305 ameliorates chronic stress-associated symptoms in Japanese medical students. *Journal of Functional Foods, 36*, 112–121. https://doi.org/10.1016/j.jff.2017.06.031

O'Connor, L. E., Gahche, J. J., Herrick, K. A., Davis, C. D., Potischman, N., & Vargas, A. J. (2021). Nonfood prebiotic, probiotic, and Synbiotic use has increased in US adults and children from 1999 to 2018. *Gastroenterology, 161*(2), 476–486.e3. https://doi.org/10.1053/j.gastro.2021.04.037

Oluwarinde, B. O., Ajose, D. J., Abolarinwa, T. O., Montso, P. K., Preez, I. D., Njom, H. A., & Ateba, C. N. (2023). Safety properties of *Escherichia coli* O157:H7 specific bacteriophages: Recent advances for food safety. *Foods, 12*, 3989. https://doi.org/10.3390/foods12213989

Pan, Z., Mao, B., Zhang, Q., Tang, X., Yang, B., Zhao, J., Cui, S., & Zhang, H. (2022). Postbiotics prepared using *Lactobacillus paracasei* CCFM1224 prevent nonalcoholic fatty liver disease by modulating the gut microbiota and liver metabolism. *International Journal of Molecular Sciences, 23*(21), 13522. https://doi.org/10.3390/ijms232113522

Pandey, K., Prakash, T. V., & G., Gosai, H., Bhatia, D. and Patel, R. (2025). Microbial exopolysaccharides as postbiotics: Structure, function, and translational potential. *International Journal of Biological Macromolecules, 330*, 147956. https://doi.org/10.1016/j.ijbiomac.2025.147956

Pane, M., Allesina, S., Amoruso, A., Nicola, S., Deidda, F., & Mogna, L. (2018). Flow cytometry: Evolution of microbiological methods for probiotics enumeration. *Journal of Clinical Gastroenterology, 52*, S41. https://doi.org/10.1097/MCG.0000000000001057

Parhi, P. (2024). Relationship between antimicrobial peptides, probiotics, postbiotics, and parabiotics in the maintenance of a well-balanced gut ecosystem. In P. Baindara & S. M. Mandal (Eds.), *Evolution of antimicrobial peptides: From self-defense to therapeutic applications* (pp. 199–219). Springer Nature. https://doi.org/10.1007/978-3-031-67515-7_8

Paz, M., & Lio, P. (2024). Postbiotics and atopic dermatitis: Aiming to modulate the gut-skin axis: Postbiotics, nonviable microorganisms and their byproducts, hold potential in managing atopic dermatitis through modulation of the gut-skin axis and the neuro-immuno-cutaneous-endocrine (NICE) system. *Journal of Integrative Dermatology*. https://doi.org/10.64550/joid.9kv22073

Pimentel, T. C., Cruz, A. G., Pereira, E., Almeida da Costa, W. K., da Silva Rocha, R., de Souza Pedrosa, G. T., dos Santos Rocha, C., Alves, J. M., Alvarenga, V. O., Sant'Ana, A. S., & Magnani, M. (2023). Postbiotics: An overview of concepts, inactivation technologies, health effects, and driver trends. *Trends in Food Science & Technology, 138*, 199–214. https://doi.org/10.1016/j.tifs.2023.06.009

Pinheiro, I., Robinson, L., Verhelst, A., Marzorati, M., Winkens, B., den Abbeele, P. V., & Possemiers, S. (2017). A yeast fermentate improves gastrointestinal discomfort and constipation by modulation of the gut microbiome: Results from a randomized double-blind placebo-controlled pilot trial. *BMC Complementary and Alternative Medicine, 17*(1), 441. https://doi.org/10.1186/s12906-017-1948-0

Piqué, N., Berlanga, M., & Miñana-Galbis, D. (2019). Health benefits of heat-killed (Tyndallized) probiotics: An overview. *International Journal of Molecular Sciences, 20*(10), 2534. https://doi.org/10.3390/ijms20102534

Prajapati, N., Patel, J., Singh, S., Yadav, V. K., Joshi, C., Patani, A., Prajapati, D., Sahoo, D. K., & Patel, A. (2023). Postbiotic production: Harnessing the power of microbial metabolites for health applications. *Frontiers in Microbiology, 14*, 1306192. https://doi.org/10.3389/fmicb.2023.1306192

Prajapati, S. K., Lekkala, L., Yadav, D., Jain, S., & Yadav, H. (2025). Microbiome and postbiotics in skin health. *Biomedicine, 13*(4), 791. https://doi.org/10.3390/biomedicines13040791

Puttarat, N., Thangrongthong, S., Kasemwong, K., Kerdsup, P., & Taweechotipatr, M. (2021). Spray-drying microencapsulation using whey protein isolate and nano-crystalline starch for enhancing the survivability and stability of *Lactobacillus reuteri* TF-7. *Food Science and Biotechnology, 30*(2), 245–256. https://doi.org/10.1007/s10068-020-00870-z

Ra, Y. E., & Bang, Y.-J. (2024). Balancing act of the intestinal antimicrobial proteins on gut microbiota and health. *Journal of Microbiology, 62*(3), 167–179. https://doi.org/10.1007/s12275-024-00122-3

Racioppo, A., Corbo, M. R., Piccoli, C., Sinigaglia, M., Speranza, B., & Bevilacqua, A. (2017). Ultrasound attenuation of lactobacilli and bifidobacteria: Effect on some technological and probiotic properties. *International Journal of Food Microbiology, 243*, 78–83. https://doi.org/10.1016/j.ijfoodmicro.2016.12.011

Raita, S., Feldmane, L., Kusnere, Z., Spalvins, K., Kuzmika, I., Berzina, I., & Mika, T. (2023). Microbial carotenoids production: Strains, conditions, and yield affecting factors. *Environmental and Climate Technologies, 27*(1), 1027–1048. https://doi.org/10.2478/rtuect-2023-0075

Rashidinejad, A. (2024). The road ahead for functional foods: Promising opportunities amidst industry challenges. *Future Postharvest and Food, 1*(2), 266–273. https://doi.org/10.1002/fpf2.12022

Rodríguez, C., Medici, M., Rodríguez, A. V., Mozzi, F., & Font de Valdez, G. (2009). Prevention of chronic gastritis by fermented milks made with exopolysaccharide-producing *Streptococcus thermophilus* strains. *Journal of Dairy Science, 92*(6), 2423–2434. https://doi.org/10.3168/jds.2008-1724

Roe, A. L., Boyte, M.-E., Elkins, C. A., Goldman, V. S., Heimbach, J., Madden, E., Oketch-Rabah, H., Sanders, M. E., Sirois, J., & Smith, A. (2022). Considerations for determining safety of probiotics: A USP perspective. *Regulatory Toxicology and Pharmacology, 136*, 105266. https://doi.org/10.1016/j.yrtph.2022.105266

Salazar, N., Neyrinck, A. M., Bindels, L. B., Druart, C., Ruas-Madiedo, P., Cani, P. D., & de los Reyes-Gavilán, C. G. and Delzenne, N. M. (2019). Functional effects of EPS-producing bifidobacterium administration on energy metabolic alterations of diet-induced obese mice. *Frontiers in Microbiology, 10*, 1809. https://doi.org/10.3389/fmicb.2019.01809

Salminen, S., Collado, M. C., Endo, A., Hill, C., Lebeer, S., Quigley, E. M. M., Sanders, M. E., Shamir, R., Swann, J. R., Szajewska, H., & Vinderola, G. (2021). The International Scientific Association of Probiotics and Prebiotics (ISAPP) consensus statement on the definition and scope of postbiotics. *Nature Reviews Gastroenterology & Hepatology, 18*(9), 649–667. https://doi.org/10.1038/s41575-021-00440-6

Sawada, D., Kawai, T., Nishida, K., Kuwano, Y., Fujiwara, S., & Rokutan, K. (2017). Daily intake of Lactobacillus gasseri CP2305 improves mental, physical, and sleep quality among Japanese medical students enrolled in a cadaver dissection course. *Journal of Functional Foods, 31*, 188–197. https://doi.org/10.1016/j.jff.2017.01.042

Sharma, R., Rashidinejad, A., & Jafari, S. M. (2022). Application of spray dried encapsulated probiotics in functional food formulations. *Food and Bioprocess Technology, 15*(10), 2135–2154. https://doi.org/10.1007/s11947-022-02803-6

Shree Kumar, G. R., & Mohanasrinivasan, V. (2025). Scaling up postbiotics production: A prospective review of processes and health benefits. *Probiotics and Antimicrobial Proteins, 18*(2), 2870–2889. https://doi.org/10.1007/s12602-025-10643-x

Siciliano, R. A., Reale, A., Mazzeo, M. F., Morandi, S., Silvetti, T., & Brasca, M. (2021). Paraprobiotics: A new perspective for functional foods and nutraceuticals. *Nutrients, 13*(4), 1225. https://doi.org/10.3390/nu13041225

Singh, T. P., Kaur, G., Kapila, S., & Malik, R. K. (2017). Antagonistic activity of *Lactobacillus reuteri* strains on the adhesion characteristics of selected pathogens. *Frontiers in Microbiology, 8*, 486. https://doi.org/10.3389/fmicb.2017.00486

Singh, R., Chandrashekharappa, S., Bodduluri, S. R., Baby, B. V., Hegde, B., Kotla, N. G., Hiwale, A. A., Saiyed, T., Patel, P., Vijay-Kumar, M., Langille, M. G. I., Douglas, G. M., Cheng, X., Rouchka, E. C., Waigel, S. J., Dryden, G. W., Alatassi, H., Zhang, H.-G., Haribabu, B., et al. (2019). Enhancement of the gut barrier integrity by a microbial metabolite through the Nrf2 pathway. *Nature Communications, 10*(1), 89. https://doi.org/10.1038/s41467-018-07859-7

Sionek, B., & Gantner, M. (2025). Probiotics and Paraprobiotics—New proposal for functional food. *Applied Sciences, 15*(1), 366. https://doi.org/10.3390/app15010366

Strandwitz, P., Kim, K. H., Terekhova, D., Liu, J. K., Sharma, A., Levering, J., McDonald, D., Dietrich, D., Ramadhar, T. R., Lekbua, A., Mroue, N., Liston, C., Stewart, E. J., Dubin, M. J., Zengler, K., Knight, R., Gilbert, J. A., Clardy, J., & Lewis, K. (2019). GABA-modulating bacteria of the human gut microbiota. *Nature Microbiology, 4*(3), 396–403. https://doi.org/10.1038/s41564-018-0307-3

Sugawara, T., Sawada, D., Ishida, Y., Aihara, K., Aoki, Y., Takehara, I., Takano, K., & Fujiwara, S. (2016). Regulatory effect of paraprobiotic *Lactobacillus gasseri* CP2305 on gut environment and function. *Microbial Ecology in Health and Disease, 27*(1), 30259. https://doi.org/10.3402/mehd.v27.30259

Tamang, J. P., & Lama, S. (2022). Probiotic properties of yeasts in traditional fermented foods and beverages. *Journal of Applied Microbiology, 132*(5), 3533–3542. https://doi.org/10.1111/jam.15467

Thorakkattu, P., Khanashyam, A. C., Shah, K., Babu, K. S., Mundanat, A. S., Deliephan, A., Deokar, G. S., Santivarangkna, C., & Nirmal, N. P. (2022). Postbiotics: Current trends in food and pharmaceutical industry. *Foods, 11*(19), 3094. https://doi.org/10.3390/foods11193094

Tong, Y., Guo, H., Abbas, Z., Zhang, J., Wang, J., Cheng, Q., Peng, S., Yang, T., Bai, T., Zhou, Y., Li, J., Wei, X., Si, D., & Zhang, R. (2023). Optimizing postbiotic production through solid-state fermentation with *Bacillus amyloliquefaciens* J and *Lactiplantibacillus plantarum* SN4 enhances antibacterial, antioxidant, and anti-inflammatory activities. *Frontiers in Microbiology, 14*, 1229952. https://doi.org/10.3389/fmicb.2023.1229952

Tong, Y., Abbas, Z., Zhang, J., Wang, J., Zhou, Y., Si, D., Wei, X., & Zhang, R. (2025). Antimicrobial activity and mechanism of novel postbiotics against foodborne pathogens. *LWT, 217*, 117464. https://doi.org/10.1016/j.lwt.2025.117464

Tsevdou, M., Ouli-Rousi, M., Soukoulis, C., & Taoukis, P. (2020). Impact of high-pressure process on probiotics: Viability kinetics and evaluation of the quality characteristics of probiotic yoghurt. *Foods, 9*(3), 360. https://doi.org/10.3390/foods9030360

Urbani, G., Rondini, E., Distrutti, E., Marchianò, S., Biagioli, M., & Fiorucci, S. (2025). Phenotyping the chemical communications of the intestinal microbiota and the host: Secondary bile acids as postbiotics. *Cells, 14*(8), 595. https://doi.org/10.3390/cells14080595

Vallejo-Cordoba, B., Castro-López, C., García, H. S., González-Córdova, A. F., & Hernández-Mendoza, A. (2020). Postbiotics and paraprobiotics: A review of current evidence and emerging trends. *Advances in Food and Nutrition Research, 94*, 1–34. https://doi.org/10.1016/bs.afnr.2020.06.001

Valluru, R., Devi, K., Sanil Kumar, R., Ravichandra, V. D., Ramamoorthi, K., Pavan Kumar, K. M., Pramodini, S. R., Deepa, R., & M. (2025). Pharmacopeial specifications for dietary supplements and nutraceuticals. In *Dietary supplements and nutraceuticals* (pp. 879–902). Springer. https://doi.org/10.1007/978-981-96-8622-3_27

van Tatenhove-Pel, R. J., Zwering, E., Boreel, D. F., Falk, M., van Heerden, J. H., Kes, M. B. M. J., Kranenburg, C. I., Botman, D., Teusink, B., & Bachmann, H. (2021). Serial propagation in water-in-oil emulsions selects for *Saccharomyces cerevisiae* strains with a reduced cell size or an increased biomass yield on glucose. *Metabolic Engineering, 64*, 1–14. https://doi.org/10.1016/j.ymben.2020.12.005

Wang, Y., & Selomulya, C. (2020). Spray drying strategy for encapsulation of bioactive peptide powders for food applications. *Advanced Powder Technology, 31*(1), 409–415. https://doi.org/10.1016/j.apt.2019.10.034

Wang, S., Wang, P., Wang, D., Shen, S., Wang, S., Li, Y., & Chen, H. (2024). Postbiotics-in-inflammatory-bowel-disease-efficacy-mechanism-and-therapeutic-implications. *Journal of the Science of Food and Agriculture, 105*(2), 721–734. https://doi.org/10.1002/jsfa.13721

Weng, C.-Y. C., Suarez, C., Cheang, S. E., Couture, G., Goodson, M. L., Barboza, M., Kalanetra, K. M., Masarweh, C. F., Mills, D. A., Raybould, H. E., & Lebrilla, C. B. (2024). Quantifying gut microbial short-chain fatty acids and their isotopomers in mechanistic studies using a rapid, readily expandable LC–MS platform. *Analytical Chemistry, 96*(6), 2415–2424. https://doi.org/10.1021/acs.analchem.3c04352

Xiao, S.-D., De Zhang, Z., Lu, H., Jiang, S. H., Liu, H. Y., Wang, G. S., Xu, G. M., Zhang, Z. B., Lin, G. J., & Wang, G. L. (2003). Multicenter, randomized, controlled trial of heat-killed *Lactobacillus acidophilus* LB in patients with chronic diarrhea. *Advances in Therapy, 20*(5), 253–260. https://doi.org/10.1007/BF02849854

Yang, L., Zhu, X., Chen, Y., & Wang, J. (2024). Enhanced bacterial cellulose production in *Gluconacetobacter xylinus* by overexpression of two genes (*bscC* and *bcsD*) and a modified static culture. *International Journal of Biological Macromolecules, 260*, 129552. https://doi.org/10.1016/j.ijbiomac.2024.129552

Yilmaz, Y. (2024). Postbiotics as antiinflammatory and immune-modulating bioactive compounds in metabolic dysfunction-associated Steatotic liver disease. *Molecular Nutrition & Food Research, 68*(23), 2400754. https://doi.org/10.1002/mnfr.202400754

Yolmeh, M., Martins Alves, P. C., Dos Santos Vilela, M. M., & Sant'Ana, A. S. (2024). Impacts of ohmic-heating on the surface characteristics, membrane integrity, morphology, and chemical compounds of *Bifidobacterium animalis* paraprobiotic. *Food Bioscience, 59*, 104141. https://doi.org/10.1016/j.fbio.2024.104141

Yoto, A., Murao, S., Motoki, M., Yokoyama, Y., Horie, N., Takeshima, K., Masuda, K., Kim, M., & Yokogoshi, H. (2012). Oral intake of γ-aminobutyric acid affects mood and activities of central nervous system during stressed condition induced by mental tasks. *Amino Acids, 43*(3), 1331–1337. https://doi.org/10.1007/s00726-011-1206-6

Young, E. M., Zhao, Z., Gielesen, B. E. M., Wu, L., Benjamin Gordon, D., Roubos, J. A., & Voigt, C. A. (2018). Iterative algorithm-guided design of massive strain libraries, applied to itaconic acid production in yeast. *Metabolic Engineering, 48*, 33–43. https://doi.org/10.1016/j.ymben.2018.05.002

Yuk, H.-G., & Geveke, D. J. (2011). Nonthermal inactivation and sublethal injury of *Lactobacillus plantarum* in apple cider by a pilot plant scale continuous supercritical carbon dioxide system. *Food Microbiology, 28*(3), 377–383. https://doi.org/10.1016/j.fm.2010.09.010

Zhang, S., Wang, H., & Zhu, M.-J. (2019). A sensitive GC/MS detection method for analyzing microbial metabolites short chain fatty acids in fecal and serum samples. *Talanta, 196*, 249–254. https://doi.org/10.1016/j.talanta.2018.12.049

Zhang, L., Li, H., Song, Z., Liu, Y., & Zhang, X. (2024). Dietary-strategies-to-improve-exercise-performance-by-modulating-the-gut-microbiota. *Foods, 13*(11), 1680. https://doi.org/10.3390/foods13111680

Zhang, J., Zhang, R., Wang, J., Abbas, Z., Tong, Y., Fang, Y., Zhou, Y., Zhang, H., Li, Z., Si, D., & Wei, X. (2025). Efficient production strategy of a novel postbiotic produced by *Bacillus subtilis* and its antioxidant and anti-inflammatory effects. *Molecules, 30*(10), 2089. https://doi.org/10.3390/molecules30102089

Zhao, H., Song, G., Zhu, H., Qian, H., Pan, X., Song, X., Xie, Y., & Liu, C. (2023). Pharmacological effects of urolithin A and its role in muscle health and performance: Current knowledge and prospects. *Nutrients, 15*(20), 4441. https://doi.org/10.3390/nu15204441

Chapter 4
Enhancing Nutrient Bioavailability: Unlocking Vitamins, Minerals, and Protein Digestibility Through Fermentation

Parul Sharma, Palak Mahajan, Abhishek Kumar Singh, and Sunny Dhiman

Abstract Fermentation is a traditional bioprocessing technique that enhances nutrient bioavailability and improves the functional quality of foods. Microbial metabolism modifies food matrices, increases vitamin synthesis, reduces anti-nutritional factors, and improves mineral solubility and protein digestibility. This chapter examines how microorganisms, such as lactic acid bacteria, probiotic yeasts, and Bacillus species, enhance the nutritional profiles of cereals, legumes, vegetables, and soy-based substrates. Fermentation promotes the synthesis of vitamins, the conversion of isoflavone glycosides into bioactive aglycones, and improvements in antioxidant potential and gut health. These biochemical transformations contribute to the development of functional and nutraceutical foods. The chapter also discusses advanced fermentation technologies that enable targeted microbial selection and process optimization to maximize nutrient enrichment, safety, and sustainability. Fermentation represents a promising strategy for improving nutrient bioavailability and addressing global nutritional challenges; however, well-controlled human studies are needed to validate long-term health outcomes and optimize strain-specific applications.

Keywords Nutrient bioavailability · Food fermentation · Microbial metabolism · Protein digestibility · Functional foods · Probiotic microorganisms · Anti-nutritional factors

P. Sharma · P. Mahajan · A. K. Singh (✉)
Department of Biotechnology Engineering and Food Technology, UIE, Chandigarh University, Mohali, Punjab, India
e-mail: abhishek.uibt@cumail.in

S. Dhiman
Department of Biotechnology, University Institute of Biotechnology, Chandigarh University, Mohali, Punjab, India

S. Dhiman, S. Kaur (eds.), *Next-Gen Plant Based Fermented Foods: Innovations for a Sustainable Future*, https://doi.org/10.1007/978-3-032-29927-7_4

4.1 Introduction to Nutrient Bioavailability

Nutrients play an important role in the physiological and metabolic functions of the human body. The food or diet chart often reports the actual nutrient content of the food before consumption; however, bioavailability is a fundamentally different concept in human nutrition. The nutrient bioavailability refers to the amount of the ingested nutrients released from the food matrix, followed by its absorption through the gastrointestinal tract and then at last ultimately its availability (Dima et al., 2020). Therefore, the diet, which actually seems to be nutritionally adequate, may still cause nutritional deficiencies if its bioavailability is compromised, leading to malnutrition, hidden hunger, and even diet-related chronic diseases.

Each food product, when ingested, undergoes multiple sequential steps for nutrient utilisation, including disintegration of the food structure, digestion with enzymes and acids, intestinal absorption, transport, cellular uptake, and metabolism (Saha & Pathak, 2021). However, any lack at any of these stages can lower the bioavailability of the nutrients. For instance, in the case of vitamins, their bioavailability is influenced by solubility, stability during digestion, microbial metabolism, and interactions with other constituents, particularly dietary fats (Stacchiotti et al., 2021). Similarly, solubility, the presence of chelating agents, their chemical form, and the intestinal transport mechanism influence mineral bioavailability (Affonfere et al., 2023). Likewise, the bioavailability of the complex biopolymer proteins depends upon its digestibility and presence of the amino acids, which together are affected by the protein structures, processing conditions and presence of the antinutritional compounds (Sá et al., 2020; Rivera del Rio et al., 2022).

The structural complexity of the food matrix significantly affects nutrient bioavailability, as nutrients are either tightly bound to macromolecules or embedded within the complex matrix, thereby limiting their release during gastrointestinal digestion (Shahidi & Pan, 2022). Interactions among nutrients, for instance, competition for available sites or the formation of insoluble complexes, affect nutrient bioavailability. Moreover, modern diets have shifted towards plant-based or vegan diets, which have also intensified concerns about nutrient bioavailability. Though the plant-based foods are rich in the contents of vitamins, minerals, phytochemicals, and dietary fibre, they also contain antinutritional factors (phytates, oxalates, protease inhibitors, and lectins) which reduce the absorption of the essential minerals (zinc, magnesium, iron), leading to micronutrient deficiencies (Soni et al., 2022). Apart from all these factors, the variations among the individuals such as age, health, genetic factors, gut microbiota, secretion of enzymes impact the extent of nutrient absorption and utilization (Ling et al., 2022; Gomaa, 2020). For instance, dysbiosis of the gut microbiota affects vitamin synthesis and mineral metabolism, and reduced gastric acid secretion in elderly persons impairs the solubilization of minerals (Cristina & Lucia, 2021; Bielik & Kolisek, 2021). Thus, these challenges and complexities have led to increased concern and focus on nutrient bioavailability across the fields of food science and processing, as well as product development. In this regard, certain strategies such as soaking, germination, encapsulation, enzymatic treatment, and fermentation have been explored and reported to enhance nutrient bioavailability (Tharifkhan et al., 2021). Among these strategies, the fermentation process is an effective and sustainable approach that modifies the food matrix, reduces antinutritional factors, and even enriches food products with bioactive metabolites (Samtiya et al., 2021). The process has been used for ages across various cultures to enhance flavour and improve digestibility, thereby enabling perceived health benefits. Various studies have shown that fermentation improves the bioavailability of vitamins, minerals, and proteins through microbial and/or enzymatic processes, thereby addressing micronutrient deficiencies in a sustainable manner (Adebo et al., 2022). Also, unlike the artificial and synthetic fortification of nutrients into food products, the fermentation process leverages biological processes to naturally enhance its nutritional benefits. Therefore, this chapter will focus on the role and mechanism of fermentation in the bioavailability of vitamins, minerals, and proteins. The chapter will also highlight various microorganisms and emerging fermentation technologies employed in this arena. The fermented foods as functional and nutraceutical products, their health implication will also be discussed in detail.

4.2 Role of Fermentation in Nutrient Enhancement

Fermentation is a metabolic process that utilises microorganisms, such as lactic acid bacteria, fungi, and yeast, to break down food substrates and produce metabolites (Nagarajan et al., 2022). This process transforms the chemical composition and physical structure of the food matrix, thus affecting its biological functionality. This plays a pivotal role in improving the nutritional quality of food products by altering the food matrix, reducing antinutritional factors, biosynthesising vitamins, and producing bioactive molecules (Fig. 4.1) (Zhao et al., 2021).

The main role of the fermentation process is to modify the food matrix. Usually, nutrients are embedded or trapped within the complex matrix of macromolecules, limiting their accessibility during digestion. Fermentation supports the activity of microbial enzymes, such as amylases, pectinases, cellulases, and hemicellulases, which degrade carbohydrate polymers and other cell wall components (Siddiqui et al., 2022). This degradation thus causes structural changes that loosen the food matrix, promoting its interaction with digestive enzymes in the gastrointestinal tract and thereby facilitating the release and availability of nutrients (Sawant et al., 2025). Apart from modifying the food structure, fermentation has been reported to reduce antinutritional factors that otherwise reduce nutrient bioavailability. For example, phytates (an antinutritional factor) usually bind strongly to essential minerals, making them unavailable for absorption. However, some fermentative microorganisms produce the enzyme phytase, which hydrolyses phytate to free phosphates and releases chelated minerals, thereby increasing their bioavailability (Samtiya et al., 2021). Similarly, other antinutritional factors, such as tannins, oxalates, and protease inhibitors, are reduced by this fermentation process (Jeyakumar & Lawrence, 2022).

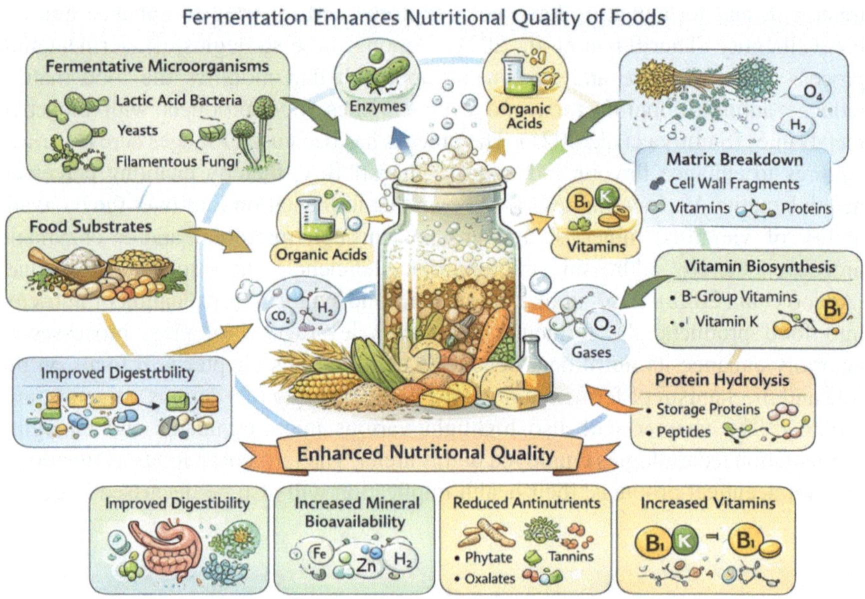

Fig. 4.1 Role of fermentation in enhancing the nutritional quality of the food

Moreover, the fermentation process has a significant role in the biosynthesis of the vitamins. For instance, upon fermentation, some yeast and lactic acid bacteria synthesizes B complex vitamins (Keyvan et al., 2025). Fermentation of soybean and dairy products has also been reported to produce Vitamin K, thereby increasing nutrient content and enhancing bioavailability, as these are in a more readily absorbable form (Kang et al., 2022). Lactic acid fermentation of food products is usually associated with the production and accumulation of the organic acid, which lowers the pH and creates conditions that improve nutrient solubility and stability, thereby supporting their absorption in the intestine (Wang et al., 2021). Apart from vitamins and minerals, which form complexes and become unavailable, proteins possess a varied mechanism. Since the proteins are biopolymers, i.e., large molecules compared to vitamins and minerals, they have their own large and complex structure, and even the binding with the protease inhibitor inhibits their digestion, causing gastric disturbances. Fermentation, in the case of protein, helps break down large molecules into peptides and free amino acids, thereby increasing their absorption and bioavailability.

Apart from these benefits, fermented foods serve as a vehicle for probiotics, modulate intestinal function by improving gut barrier integrity, regulate the secretion of digestive enzymes, and enhance nutrient absorption. Fermentation promotes the utilization of the nutrient without relying upon the chemical or synthetic fortification and this makes it an effective and attractive strategy for the development of the nutraceutical and functional foods (Sawant et al., 2025). Furthermore, fermentation also improves the sensory attributes, shelf life and safety of the food products (Shiferaw Terefe & Augustin, 2020).

4.3 Mechanisms of Bioavailability Enhancement Through Fermentation

The fermentation process involves the combination of biochemical, structural, enzymatic, and physiological mechanisms, each of which, alone or together, improves nutrient release, solubility, absorption, and metabolic utilisation, thereby increasing nutrient bioavailability (Adebo et al., 2022). In contrast to the conventional processing, fermentation is a biologically driven transformation that dismantles the nutrient utilisation barriers and generates bioactive metabolites. This makes the process a powerful and multifactorial strategy to enhance nutritional efficacy of the food products (Iqbal et al., 2025). The mechanism by which nutrient bioavailability is enhanced via fermentation is shown in Fig. 4.2.

4.3.1 Degradation of Antinutritional Factors

The key mechanism by which fermentation improves nutrient bioavailability is by degrading antinutritional compounds or factors that would otherwise form complexes, thereby altering the nutritional profile. These factors, such as tannins, phytates, oxalates, and protease inhibitors, are commonly present in plant-based diets and pose major constraints on the functionality of minerals and proteins. The fermentation process synthesizes the enzymes that degrade or metabolize these antinutritional factors and hence lower their inhibitory effects. During fermentation,

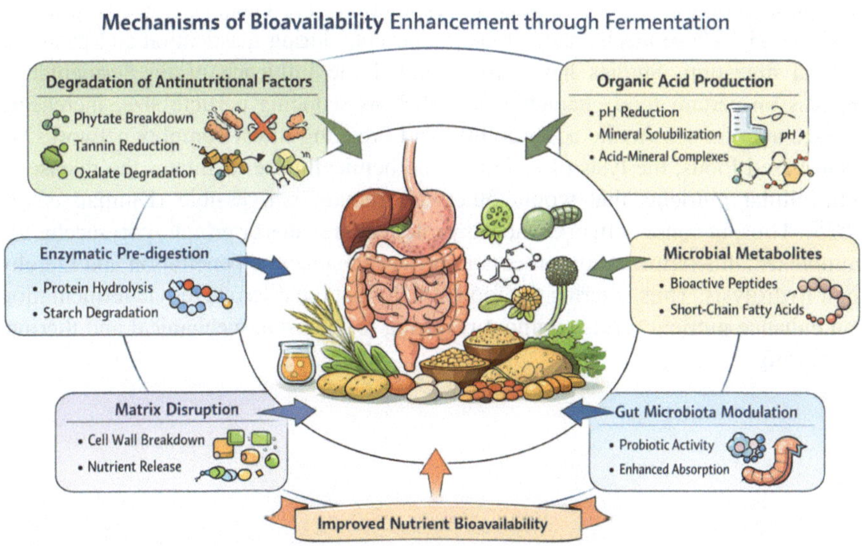

Fig. 4.2 Mechanism of nutrient enhancement through fermentation

fermentative microorganisms synthesize specific enzymes that hydrolyse or metabolize these compounds, thereby reducing their inhibitory effects. For instance, the presence of phytate in food has strong chelating properties, making minerals such as zinc, iron, magnesium, and calcium unavailable for absorption by the body (Amat et al., 2023). The amount of these inhibitors can be reduced by microbial phytase activity during fermentation, which catalyses the dephosphorylation of phytic acid, lowering its binding to minerals and thereby liberating the chelated minerals into soluble, absorbable forms (Rousta & Taherzadeh, 2025). Likewise, tannins and tannic acid form insoluble complexes with nutrients, such as proteins and minerals, and can be degraded by tannase-producing microorganisms during fermentation. This promotes the enzyme's access to dietary protein and other nutrients during digestion and absorption. Oxalate, another antinutritional compounds, which chelates with the calcium, can be degraded by the specific lactic acid bacteria, thereby improving its solubility and absorption (Arsov et al., 2024). Fermentation process reduce the activity by microbial proteolysis and forms a foundation for the improvement in the nutrient bioavailability.

4.3.2 Structural and Microstructural Modification of the Food Matrix

As discussed earlier, fermentation plays an important role in modifying the physical and structural properties of the food matrix (Zhao et al., 2021). This modification and degradation possess a crucial role in the release and accessibility of the nutrients. Upon fermentation, fermentative microbes degrade cell wall components, weaken structural integrity, disrupt nutrient encapsulation, and enhance porosity (Sawant et al., 2025). These mechanisms liberate nutrients during mastication and gastrointestinal digestion. Studies have also reported the influence of the fermentation process on certain food characteristics, such as reducing particle size, increasing surface area for enzymatic accessibility, and breaking down complex networks. In plant-based foods, the lysis of cellulose and hemicellulose facilitates the release of intracellular nutrients that would otherwise remain inaccessible (Dhiman et al., 2025). Thus, in dense networks such as protein aggregates or protein-polysaccharide complexes, this fermentation process improves the enzyme penetration and thereby their hydrolysis. Thus, altering the food structure at the micro level via fermentation can enhance nutrient release without resorting to intensive mechanical and thermal processing.

4.3.3 Enzymatic Pre-digestion of Macronutrients

The digestion of the complex food molecules is very difficult, and most of the time the product remains undigested, leading to the incomplete availability of the nutrients. To overcome this, fermentative microbes facilitate the enzymatic predigestion of macronutrients, thereby significantly enhancing their digestibility and bioavailability (Sharma et al., 2020). For instance, fermentative microorganisms include a wide array of hydrolytic enzymes, such as peptidases, proteases, amylases, hemicellulases, and cellulases, which partially break down complex biomolecules. This is highly significant for easing the digestibility of the proteins. Owing to their complex structure and higher molecular weight, protein breakdown and digestibility are major concerns for their nutritional importance. During fermentation, microbial proteolysis breaks down large, compact protein molecules into smaller peptides and even free amino acids (Kieliszek et al., 2021). This thus reduces structural complexity and exposes the peptide bonds to gastrointestinal enzymes for easier breakdown. As a result, fermented food products possess higher and more readily available protein digestibility than non-fermented ones (Shah et al., 2023). This mechanism is particularly important for plant proteins, in contrast to animal proteins, as the former are prone to anti-nutritional factors and are rigidly embedded in the cell wall. In addition to improving protein digestibility, the fermentation process also supports carbohydrate metabolism and indirectly enhances nutrient availability. Fermentation microbes, such as amylase, degrade the polysaccharide, reduce its viscosity, and facilitate its further breakdown, thereby supporting nutrient release and improved absorption (Guo et al., 2021).

4.3.4 Production of Organic Acids and Bioactive Metabolites

Mostly, the fermentation process is accompanied by the formation of certain organic acids such as citric, lactic, propionic, and acetic acids (Li et al., 2025). This thus reduces the pH of the food matrix, thereby increasing the solubility of the minerals and preventing their tendency to precipitate or form complexes. For instance, minerals such as iron and zinc are more soluble in acidic conditions, facilitating their transport across the intestinal epithelium. In addition, the development of the organic acid influences gastrointestinal physiology by stimulating gastric acid secretion and supporting protein digestion and mineral solubility (Tugnoli et al., 2020). Thus, by creating a conducive environment, the presence of organic acids serves as both a physiological driver and an enhancer of bioavailability.

Further, during fermentation, diverse bioactive metabolites are produced, which indirectly enhance nutrient bioavailability. The bioactive peptides formed during proteolytic fermentation exhibit several benefits, including antioxidant, antimicrobial, antihypertensive, and immunomodulatory functions (Chourasia et al., 2023). These peptides protect the intestinal integrity, thus supporting gut health and efficient nutrient absorption. Not only this, the formation of short-chain fatty acids (acetate, propionate, and butyrate) supports the microbial mentalism, strengthens gut health, and thereby leads to efficient nutrient absorption and utilisation (Hays et al., 2024).

4.3.5 Gut Health and Nutrient Retention

Beyond transforming the food matrix, fermentation enhances the host gut microbiota. Upon fermentation of food products, live microbes and microbial metabolites modulate the composition and activity of the gut microbiota, influencing the metabolism, synthesis, and absorption of nutrients (Vernocchi et al., 2020). Fermentation-derived substrates and metabolites stimulate the beneficial microbial population, thereby enhancing their capacity to support nutrient utilisation. Thus, this symbiotic relationship between the fermented foods and gut microbiota supports the mechanism of nutrient bioavailability (Mukherjee et al., 2024). Thus, it ensures that the nutrients remain available upon consumption.

The mechanism of fermentation-mediated nutrient bioavailability does not work solely; it involves a symbiotic relationship. Degradation of antinutritional factors, maintenance of gut health, production of organic acid and metabolites, modification of the food matrix, and pre-digestion by enzymes all work synergistically to maximize the accessibility and availability of the nutrients (Uțoiu et al., 2018).

4.4 Enhancement of Vitamin Bioavailability Through Fermentation

Fermentation is an important biotechnological method, which increases bioavailability of vitamins, vitamins B and K in particular, by using microbial synthesis and conversion processes. During fermentation, microorganisms (e.g., lactic acid bacteria (LAB), yeasts, moulds) produce vitamins B, such as folate, riboflavin, and vitamin B12, thereby enriching fermented foods that are not present in raw materials (Iqbal et al., 2025). Moreover, fermentation reduces anti-nutritional compounds, such as phytic acid, which can bind vitamins and minerals, thereby further enhancing their availability for absorption in the human gut. Preservation of sensitive vitamins during fermentation is also possible through reduced pH and the formation of anaerobic conditions, which prevent oxidation (Knez et al., 2023). To illustrate, fermented vegetables have increased antioxidant qualities and bioactive compounds that may have vitamin-like action (Tan et al., 2024). In this way, the vitamin profile and the bioavailability of vitamins in fermented products are significantly higher in contrast to the unfermented ones.

Biochemical changes are brought about by microbial fermentation during food processing. Microorganisms (including lactic acid bacteria, yeasts, and molds) help produce various vitamins (such as some vitamins of the B-complex and vitamin K) which are necessary to human health but are usually found in low concentrations or inaccessible forms in raw foods (Sawant et al., 2025). For example, fermentation promotes the activity of enzymes that decompose complex food matrices and release anti-nutritional compounds bound by phytates, thereby liberating and absorbing more micronutrients into the gastrointestinal tract (Knez et al., 2023).

Microbial processing in plant-based fermented foods, including fermented cereals, legumes, vegetables, and soymilk, can increase the bioavailability of vitamins by converting vitamin precursors into active forms and improve the overall nutritional profile. Folate, riboflavin, thiamine and cobalamin are some of the vitamins enriched in fermented food products due to biosynthetic pathways of these vitamins in lactic acid bacteria and some strains of yeast (Rekha & Vijayalakshmi, 2010; Dhiman et al., 2025). In fermentation of soy, e.g., probiotic bacteria and yeast ferment soymilk to enhance the bioavailability of isoflavone aglycones and vitamin B complex, which have a beneficial effect on the health and immune system of the intestine (Rekha & Vijayalakshmi, 2010).

Fermentation also reduces the anti-nutritional properties, such as phytates, tannins, and oxalates, which can disrupt vitamin absorption. The fermentation process results in more soluble vitamins and enhanced interaction with intestinal receptors, due to enzymatic degradation, which is more likely to promote high bioavailability (Iqbal et al., 2025). Moreover, fermentation generates bioactive peptides, short-chain fatty acids, and antioxidants that enhance gut health and systemic nutrient absorption (Iqbal et al., 2025).

Fermentation of cereal and legume products has been reported to improve the bioavailability of vitamins and minerals by modifying their nutritional composition. To give but one example, fermented cereals have elevated B vitamin levels and a higher digestible protein rate, whereas the decrease in carbohydrate and fibre content makes the nutrients more accessible. These are in part due to enzyme activities generated during fermentation, such as transamination and the production of novel substances, which enhance vitamin stability and absorption.

Fermented fruit and vegetable juices exhibit better bioavailability of vitamins and minerals because lactic acid fermentation increases the levels of phenolics and other bioactive compounds, thereby enhancing their nutritional value and functional properties (Saud et al., 2024). Vitamin metabolism is also positively affected by the presence of probiotic microorganisms introduced during fermentation, thereby enhancing microbial diversity and activity in the gut and facilitating vitamin absorption.

The new biotechnological technologies, such as precision fermentation and genome editing, have started to work on streamlining microbial strains to further maximize vitamin production and bioavailability. Such innovations offer new approaches to manufacturing nutrient-enriched fermented food products with excellent health benefits (Dhiman et al., 2025). Omics and multi-omics methods enhance existing knowledge of microbial metabolic processes, enabling precise manipulation of fermentation processes to achieve optimal vitamin bioavailability in food products.

The overall benefits of fermentation, such as the production of vitamins, the reduction of antinutritional factors, and the production of bioactive compounds, make it highly important for food security and nutrition, particularly in plant-based diets and developing economies where micronutrient deficiencies are widespread. Through microbial biotransformations, functional foods with improved vitamin bioavailability could be produced to support the health and well-being of the whole world (Sawant et al., 2025; Iqbal et al., 2025).

Bioavailability of vitamins is greatly increased by microbial fermentation in several ways by synthesizing vitamins, enzyme degradation of anti-nutrients, conversion of vitamin forms to more bioactive forms, and generation of gut health-supportive metabolites and nutrient absorbing metabolites. The combination of traditional and modern biotechnologies in fermentation processes will ensure the optimization of such processes and consequently healthy food products and the alleviation of nutritional deficiencies globally (Keyvan et al., 2025).

4.5 Mineral Bioavailability Improvement Through Fermentation

Fermentation is a traditional biotechnological process that significantly affects the nutritional value of plant-based foods by improving the bioavailability of minerals. The concept of mineral bioavailability is the ratio of minerals in food intake to those in the body, which is often reduced in plant foods due to anti-nutritional factors such as phytates. These inhibitors are degraded during fermentation, thereby enhancing mineral availability and uptake (Abbaspour, 2024).

Pseudocereals, including quinoa, canihua, and amaranth, are also good sources of minerals, including iron, zinc, and calcium, but contain phytate, a strong inhibitor of mineral absorption. It has been shown that fermentation, particularly with *Lactobacillus plantarum* 299, goes a long way toward reducing the phytate content of these grains and flours. It is shown that phytate degradation during fermentation is not solely due to the added microorganisms but also to the endogenous phytase already present in the grains. Fermentation produces lactic acid, lowering the pH and creating an environment conducive to phytate breakdown. It leads to a significant increase in mineral accessibility, with improvements of 1.7 to 4.6-fold and greater phytate-to-mineral molar ratios in fermented flours, suggesting improved estimated mineral bioavailability, especially of iron (Castro-Alba et al., 2019).

Besides, fermentation generally reduces various anti-nutritional factors, such as phytates, and enhances the proximate nutritional content of cereal-based foods. It increases the levels of crude proteins and fats and decreases the amounts of carbohydrates and crude fibre. This concomitant increase in nutritional value and the bioavailability of minerals makes fermented cereal products more nutritionally superior than their non-fermented analogues. Antioxidant activity and sensory properties are also increased by fermentation, making the foods more palatable and beneficial to health.

Microbial fermentation is not limited to nutrient enrichment but should also include altering food matrices, resulting in increased food digestibility and micronutrient accessibility. The fermentation process changes the chemical composition and cellular structure of cereals, legumes, and pseudocereals, eliminating undesirable substances such as phytate and other inhibitors. This biotransformation mediated by microorganisms causes nutrients to be more bioavailable and can also enhance food safety by decreasing the amount of pathogenic microorganisms. These advantages make fermentation an effective means to enhance plant-based protein sources and micronutrient uptake, which are essential to sustainable nutrition (Kårlund et al., 2020).

Besides its inherent phytase activity, microbial fermentation in sourdough and other lactic acid bacteria fermentations enhances mineral bioavailability by depleting phytates in wheat and other cereals. This adds to the health benefits of delivery of probiotics and a long shelf-life due to antifungal and antibacterial properties. There is also the possibility of reducing allergenic substances and enhancing overall nutrient quality by fermenting wheat and other cereal foods, thereby adding value to fermented wheat and cereal products (Sakandar et al., 2019).

The amaranth grain fermentation in particular demonstrates the dual effect of the processing techniques of popping and fermentation on the bioavailability of the minerals. The process of fermentation enhances protein, fat and ash levels and diminishes phytic acids and other mineral absorption inhibitors. It results in better levels of iron and magnesium, but when phytate levels are high at the start phytate to mineral molar ratios can be larger than recommended. Mineral bioavailability in the presence of residual inhibitors is therefore greatly increased by the fermentation process.

The products of solid-state fermentation of mixed grains using *Bacillus amyloliquefaciens* are characterized by higher essential amino acids, improved enzyme activities (amylase, protease, fibrinolysis), and improved antioxidant activity. Such biochemical and functional advantages are associated with enhanced nutrient bioavailability and the overall food quality and emphasize the use of particular strains of bacteria in the nutritional enrichment of fermentation (Heo et al., 2020). Fermented foods, particularly Lactobacillus-fermented foods, have the potential to enhance bone mineral density, increase calcium absorption, and regulate gut microbiota related to bone remodelling. These fermented diets enhance calcium absorption directly or by modulating food composition, thereby limiting diseases such as osteoporosis. The pathways associated with the gut inflammation-conditioning of lactobacilli also contribute to a broader understanding of the systemic nutritional and health benefits of fermented products.

The effects of microbial fermentation on food functional properties include the synthesis of bioactive compounds, the reduction of anti-nutritional effects, and the enhancement of sensory properties, thereby improving nutritional and functional value. The microbial fermentation is therefore important in the endeavour of addressing nutritional deficiencies and enhancing final health outcomes of the population due to enhanced uptake of mineral by consuming plant-based food (Sawant et al., 2025).

Fermentation significantly enhances mineral bioavailability in plant-based foods by decreasing phytate and other mineral absorption inhibitors, increasing nutrient composition, and altering food matrices to become more conducive to digestion and absorption. This process is mediated by various microbes, such as Lactobacillus species and Bacillus strains, which can degrade anti-nutrients and produce beneficial metabolites through enzymatic activity. The increased accessibility of minerals is then translated into health benefits, including improved access to iron and calcium and possible alleviation of bone-related diseases. Thus, fermentation is a useful, sustainable practice for improving the nutritional quality and bioavailability of minerals in cereals, pseudocereals, legumes, and mixed plant foods (Kanimozhi & Sukumar, 2025).

4.6 Enhancement of Protein Digestibility Through Fermentation

Fermentation is an age-old, more than welcome method of food processing that greatly improves the nutritional value and the digestion of proteins, especially plant-derived ones. Proteins also provide essential amino acids and bioactive compounds, but the bioavailability of plant proteins is often reduced by lower digestibility and anti-nutritional factors. Fermentation alters the structure of proteins, removes or alters these antinutrients and in other situations, produces biologically active peptides with added health value (Yousefi et al., 2025). This reaction involves the existing scientific knowledge on fermentation as a means to improve protein digestibility, its mechanisms, fermentation types, the impact of fermentation on different protein sources, and the nutritional and sustainability implications.

The effect of fermentation on protein quality is multifactorial. Biochemically, microbial activity during fermentation degrades large protein molecules into smaller peptides and free amino acids, thereby enhancing their accessibility to enzymes in the human digestive system. Proteins are cleaved by enzymes like proteases which are produced through the fermentation of microorganisms, decreasing protein size and changing the secondary and tertiary structures. This chemical modification increases solubility and exposes cleavage sites targeted by digestive enzymes, thereby enhancing digestibility (Yousefi et al., 2025). As an illustration, pea protein undergoing enzymatic hydrolysis followed by lactic acid bacterial fermentation was studied, and the molecular weight reduction by 98.79% and the decrease in the content of the second structure of sheets by 22.5% to the in vitro digestibility and 38.4% to the cellular bioavailability significantly increased relative to untreated pea protein (Kim et al., 2023).

Fermentation also helps reduce antinutritional factors (ANFs), including phytates, tannins, protease inhibitors, and complex phenolic compounds. These substances usually contain proteins or minerals, which can affect enzyme digestion and nutrient absorption. Microorganisms ferment to produce enzymes such as phytases and tannases, which break down these ANFs. An example is the lactic acid fermentation of millets that removed phytic acid by a maximum of 80.84% at the same time increasing protein digestibility and antioxidant properties (Chauhan et al., 2024). Equally, protein blends of pea and rice fermented with shiitake mycelium decreased phytate and protease inhibitors as well as enhanced digestibility and protein quality indexes like the Digestible Indispensable Amino Acid Score that classifies a protein as an excellent source of nutrients to the human diet (Clark et al., 2022). The loss of ANFs not only promotes proteolytic fermentation but also improves the bioavailability of micronutrients, which can be combined with the benefits of fermentation for nutrition (Tachie et al., 2024).

These nutritional results depend on the fermentation modality. Solid-state fermentation (SSF) has been noted to be quite effective at improving the quality of plant proteins by using the filamentous fungi and lactic acid bacteria to hydrolyze proteins and degrade large polypeptides into bioavailable peptides and get rid of ANFs. SSF enhances the amino acid composition and yields bioactive peptide profiles containing antioxidant, antihypertensive, and other health-promoting activities, potentially being even superior to submerged fermentation, because of the controlled moisture, and aeration conditions in which particular microbial activity is favoured (Feng et al., 2024). Practical dietary advantages in traditional fermentations, e.g. in African sorghum porridges, is shown by fermentation with cooking having a powerful ameliorative effect against the loss in protein digestibility caused by cooking alone by changing protein structural conformation and making them more accessible to digestive enzymes (Taylor & Taylor, 2002).

Fermentation enhances the functionality and sensory qualities of protein products beyond that of protein digestibility. Fermentation eliminates unwanted flavour and other unpleasant flavour compounds of plant protein material, making it more palatable as meat or as protein supplements. To illustrate, fermentation by shiitake mycelium positively affected the organoleptic properties of a pea and rice protein concentrate mix and nutritional value, which explains the dual role of fermentation in nutritional and sensory optimization (Clark et al., 2022). Furthermore, protein structures can also be modified through co-fermentation of whey proteins and pH shift treatment to increase the extent of solubility and digestibility as is the case with lentil protein complexes, demonstrating that fermentation can be utilized alongside other biotechnological strategies to improve protein characteristics (Alrosan et al., 2024).

Continued microbial advances are enhancing the fermentation's contribution in food sustainability and nutrition. With the development of starter culture selection, microbial strain engineering, CRISPR-based genome editing and precision fermentation, it is now possible to optimize fermentation processes to achieve the highest degree of protein digestibility, bioactive compound production, and reduction of antinutrients. These technologies can open the door to sustainability by enhancing plant-based proteins, reducing reliance on animal sources, and overcoming global nutrient shortages cost-effectively (Dhiman et al., 2025; Kårlund et al., 2020).

Fermentation enhances the digestibility of proteins by relying on enzymatic protection of protein molecules, the reduction of antinutritional factors, and alterations in protein structure, the combination of which increases nutrient bioavailability and functionality. This is especially true of plant-based proteins, which benefit from fermentation-induced reductions in the natural constraints on digestibility. The fermentation process also enhances sensory attributes, making people more willing to accept other protein sources (Çabuk et al., 2018). Moreover, the new microbial biotechnology creates opportunities for optimising fermentation for nutrition and sustainability. These combined advantages make fermentation an important tool for improving the quality of protein to meet the increasing global protein requirements in a sustainable manner.

4.7 Microorganisms Involved

Microbial fermentation is an important process of improving the bioavailability of vital nutrients like vitamins, minerals and proteins by unlocking such nutrients in complex food substrates and decreasing the antinutritional factors. This is an ancient form of biotechnology that uses a particular collection of microorganisms (mainly lactic acid bacteria (LAB), yeasts, and fungi) to alter food components, thereby making them more accessible and digestible to humans.

The first major benefit is that fermentation breaks down antinutrients like phytates and tannins, which strongly bind minerals such as iron, zinc, and calcium, thereby severely limiting their bioavailability in vegetarian foods. Fermentation by microbes breaks down these inhibitors hence enhancing the availability of minerals. For example, the reduction of tannins and phytase in fermented vegetables and legumes increases the bioavailability of zinc and iron, and the antioxidant capacity, by releasing previously complexed polyphenols (Knez et al., 2023). Equally, probiotic bacteria, including *Lactobacillus acidophilus*, *L.bulgaricus*, *L. casei*, and *L. plantarum*, along with *Saccharomyces boulardii* yeast, convert the glycoside forms of the isoflavones genistein and daidzein to aglycone forms (genistein and daidzein) in fermented soymilk, increasing the mineral bioavailability and bioaccessibility of the B vitamin complex otherwise inaccessible (Rekha & Vijayalakshmi, 2010).

Protein digestibility is also enhanced by fermentation, which partially hydrolyses proteins with microbial enzymes to release bioactive peptides with enhanced physiological functions. Fungus Pleurotus ostreatus was shown to enhance the protein content and digestibility of legumes (lentils, chickpeas and soybeans) in solid-state fermentation. Such treatment also reduces the antinutritional effects, such as phytates and tannins, and enhances antioxidant activity, thereby increasing the biological availability of proteins to the human body (Noori et al., 2025). In addition, fermented foods are produced through fermentation and contain bioactive peptides that possess health effects, such as antioxidant, antihypertensive, and immunomodulatory properties, thereby providing a functional component of nutrient bioavailability (Tachie et al., 2024).

The same advantages of fermentation are enjoyed by cereal-based fermented products, such as sourdough bread and cereal beverages. Fermentation of the lactic acid bacteria enhances the proximate composition by adding more protein and decreasing carbohydrates and crude fiber. It also reduces anti-nutritional factors, improving the absorption of minerals and antioxidant properties. Especially sourdough fermentation leads to the increase in bio-accessibility of vitamins and minerals as well as the enhancement of sensory and texture properties of bread, which demonstrates the multifaceted nature of microbial fermentation.

These advantages are primarily due to microorganisms of the lactic acid bacteria, *Saccharomyces* species, and filamentous fungi. The phytase activity allows lactic acid bacteria to enzymatically hydrolyze complex carbohydrates, and phytates and breaks down these compounds, which trap minerals and make the minerals more bioavailable. The contribution of the yeasts is the secretion of enzymes that alter the vitamin forms, enhance the production of B vitamins, and facilitate protein modification. The fermentation action of these microbial consortia improves the bioavailability of vitamins and minerals, enhances protein digestion, and improves protein functional properties (Iqbal et al., 2025).

The effects of fermentation on the bioavailability of micronutrients also include human physiological factors and microbiome factors. A healthy gut flora facilitates the absorption of vitamins and minerals, and fermented foods cannot be left out, as they provide probiotic microorganisms that help maintain a healthy microbiota, thereby indirectly promoting nutrient bioavailability. Moreover, food matrices and fibre structures that might entrap micronutrients are broken down by fermentation, releasing micronutrients into the intestinal receptors (Bielik & Kolisek, 2021).

Fermentation is a sustainable approach to plant-based protein sources, which might otherwise be poorly digested and affect mineral bioavailability due to fibre and antinutrients. Fermented legumes, cereals, and pseudocereals exhibit greater protein digestibility and mineral availability, as well as reduced levels of undesirable substances such as phytates and tannins. Such advancements positively influence nutrient absorption and enhance health by forming bioactive compounds and exerting probiotic effects (Shah et al., 2023).

More directly connected to human health, metabolites produced during fermentation processes, e.g., short-chain fatty acids, bioactive peptides, vitamins (especially B-complex), and antioxidants, have positive effects on gut health, immune system regulation, and the prevention of chronic diseases. Microbial fermentation is therefore a viable strategy to ameliorate micronutrient deficiencies worldwide and enhance overall nutritional well-being by transforming food raw materials rather than merely preserving them (Park & Mannaa, 2025).

Fermenting microorganisms orchestrate a complex series of biochemical reactions that liberate vitamins and minerals and enhance the digestibility of proteins. The anti-nutritional factors, vitamin complex bioconversion, micronutrient synthesis, and protein hydrolysis to more bioavailable and bioactive forms are broken down, bioconverted, and hydrolyzed by lactic acid bacteria, yeast and fungi respectively. It is an ancient biotechnology that improves the bioavailability of nutrients, increases gut microbiota and immune health, and promotes a sustainable food supply by rendering plant-based food and other edibles nutritious and functional to the human body (Hill et al., 2017).

4.8 Fermented Foods as Functional and Nutraceutical Products

The different health-promoting qualities of fermented foods make them valuable functional and nutraceutical products. Fermentation, an ancient food processing technique, not only enhances food sensory and shelf-life characteristics but also synthesises bioactive compounds, including peptides, probiotics, and polyphenols, with diverse health effects. They are products that exceed basic nutrition, as they help prevent diseases, modulate the immune system, support gut health, and accelerate metabolism (Dahiya & Nigam, 2022).

Bioactive peptide is a microbial fermentation product and it is short chains of amino acid with antimicrobial, antioxidant, antihypertensive and anti-inflammatory properties. Its production and action demand fermentation conditions and strains of microbes in use and further emphasizes the need in the enhancement of traditional fermentation to achieve health promoting products. Bioactive peptides fermented in food have been found to aid in reducing the risk of chronic diseases and enhance overall health and well-being, and are thus applicable in functional diets as nutraceutical agents (Fabbri et al., 2024).

Another natural source of probiotics is fermented food which consists of useful microbes that are alive and have the ability to be useful in health when taken in adequate amounts. These probiotics positively influence the gut microbiota composition and activity which improves the gut barrier integrity and controls the immune responses. They have a role in the treatment of gastrointestinal disorders such as irritable bowel syndrome, lactose intolerance, chronic diarrhea and helicobacter pylori infection. The use of probiotic fermented dairy products can be considered one of such functionalities as it provides a framework that enhances the proliferation and functionality of microbes in the gastrointestinal tract. This interplay among fermented foods, probiotics, and gut microbiota reinforces the idea that they influence the microbiome-gut-brain axis, which not only affects gut health but also neurological and psychological health through links to the vagus nerve (Kaur et al., 2022).

In addition, fermented foods deliver important bioactive compounds such as polyphenols, dietary fibre, and antioxidants. It enriches the fermented food product with these compounds when added to winemaking byproducts, such as grape pomace, which confer varied antioxidant and anti-inflammatory effects. They improve metabolic well-being, cardiovascular function, and gut health through their prebiotic action. Their use in different types of functional food, such as baked goods and dairy and plant-based yogurts, reflects the flexibility and sustainability opportunities of fermented food ingredients to enhance nutrition and prevent chronic conditions (Dahiya & Nigam, 2022).

Most functional foods contain physiologically active ingredients with non-nutritional benefits, including fermented varieties. These effects include reduced risks of cardiovascular disease, diabetes, cancer, and osteoporosis. The fermented functional foods are developed to preserve or amplify these bioactive compounds which are antioxidants, carotenoids, polyphenols, prebiotics and probiotics. Food technology and fermentation processes allow designing new products that are designed to obtain specific health benefits. These foods should help close nutritional deficits and tip the balance of disease pathology, ultimately leading to well-being and a longer life expectancy.

The scientific evidence in favour of the health merits of fermented foods and their compositional products is increasing. Antioxidant, anti-inflammatory and immune-promoting effects have been evidenced using fermented soy, nuts, berries, and cocoa, among others through bioactive compounds. The same compounds found in nutraceutical beverages can also be obtained from fermented plants and dairy products, and are effective in enhancing mental alertness and energy, and in preventing fatigue-related conditions. Nonetheless, the use and stability of these functional products and their shelf-life is key to consumer acceptance because of their sensory characteristics, including aroma and taste.

4.9 Health Implications

The health implications of plant-based fermented foods are not derived from vague wellness narratives but from specific, mechanistically defined pathways linking microbial metabolism to enhanced nutrient bioavailability and measurable physiological outcomes. Fermentation, driven by lactic acid bacteria (LAB), yeasts, and other microorganisms, fundamentally transforms the nutritional matrix of plant substrates by enzymatically degrading anti-nutritional factors, synthesising essential micronutrients, and generating bioactive peptides. This biochemical remodelling directly enhances the absorption of vitamins, minerals, and amino acids, which in turn modulates gut microbiota composition and function, leading to quantifiable systemic health benefits (Samtiya et al., 2021).

A critical mechanism underpinning these benefits is the degradation of phytates, a major anti-nutrient in cereals and legumes that chelates minerals like iron and zinc, rendering them insoluble and non-bioavailable. Microbial phytases produced during fermentation hydrolyze phytic acid, liberating bound minerals for intestinal absorption. This process has been validated in human clinical trials. A randomized cross-over study demonstrated that fractional iron absorption from fermented lentils was 2.7 times higher than from unfermented controls ($p < 0.001$), a direct consequence of phytate reduction and the concurrent production of organic acids like lactic acid, which further solubilize iron (Cakmak et al., 2025). Similarly, a stable-isotope study found that zinc bioavailability from fermented soybean (tempeh) was 42% greater than from boiled soybeans, with this improvement linked to the upregulation of the intestinal zinc transporter ZIP4. These findings establish a direct causal link between microbial activity and enhanced mineral absorption.

Fermentation also significantly improves protein quality and digestibility. The complex protein structures and protease inhibitors in raw plant materials are broken down by microbial proteases into smaller peptides and free amino acids, which are more readily digested and absorbed. In vitro pepsin-pancreatin digestion assays revealed that tempeh exhibited 92% protein hydrolysis compared to only 74% for raw soybeans. This biochemical enhancement translated into a clinically relevant outcome in a double-blind trial, in which participants consuming tempeh showed a 38% higher postprandial area under the curve (AUC) for plasma essential amino acids than those consuming unfermented soy (Shaghaghian et al., 2022). This improved protein digestibility is crucial for addressing the lower bioavailability of plant-based proteins, making fermented products a nutritionally superior alternative.

Additionally, microbial metabolism during fermentation actively synthesizes or liberates essential vitamins, particularly B-vitamins that are often scarce in plant-based diets. Strains of *Lactobacillus*, *Bifidobacterium*, and *Propionibacterium* are known to produce folate, riboflavin, and vitamin K2 (LeBlanc et al., 2013). Critically, certain fermentation processes can generate bioactive forms of vitamin B12, a nutrient otherwise absent in plant sources, thereby offering a sustainable solution to B12 deficiency in vegan and vegetarian populations (Deokar et al., 2025). This microbial biofortification directly addresses a key nutritional gap in plant-based dietary patterns.

The health consequences of these nutrient enhancements are mediated through profound effects on the gut microbiota. The consumption of plant-based fermented foods delivers live microbes and prebiotic fibers that selectively promote the growth of beneficial bacteria like *Bifidobacterium* and *Lactobacillus*. This modulation is not merely compositional but functional, as evidenced by elevated fecal concentrations of short-chain fatty acids (SCFAs). Clinical studies have shown that regular intake of fermented foods increases butyrate by 29% and acetate by 22% ($p < 0.01$) (Guse et al., 2023; Facchin et al., 2024). SCFAs serve as the primary energy source for colonic epithelial cells, enhancing gut barrier integrity and reducing systemic inflammation. This gut-centric effect translates into tangible metabolic improvements. A 12-week randomized controlled trial in prediabetic adults found that daily consumption of a fermented oat–soy beverage significantly improved insulin sensitivity, as measured by HOMA-IR (-1.4 ± 0.3 vs. -0.3 ± 0.2 in the placebo group; $p = 0.002$), and increased serum ferritin levels by 18.7 μg/L ($p = 0.008$) (Nithya et al., 2023). These outcomes were mechanistically linked to the baseline abundance of *Bifidobacterium*-associated phytase genes, demonstrating a direct pathway from microbial genetics to host nutrient status and metabolic health.

4.10 Advanced and Emerging Fermentation Technologies

A new frontier in biotechnology is high-tech, advanced fermentation technologies that enable even greater product yields, better process control, and the creation of new fermentation pathways. Food, biofuel, and pharmaceutical production have traditionally relied on fermentation, a bioconversion process that transforms substrates into useful products, which is a fundamental foundation of these industries. Recent advances introduce multidisciplinary inventions such as designs of bioreactor, real-time monitoring to the area of synthetic biology, and optimization of intelligent processes. New fermentation technologies are a broad field of innovative activities. One trend is the integration of the high-tech platforms of bioreactor which consume the real-time technology in the monitoring and control of the microbial culture with a high level of efficiency and safety. These systems use sensors, biosensors and PCR based tests to perform the rapid identification of the presence of microbial contaminants to enable the integrity of the product and safety standards. With the advent of high-throughput sequencing (HTS) technology, the accurate study of microbial communities is now feasible and, as a consequence, it is now possible to selectively manipulate fermentation consortia to add desirable product traits. In addition, predictive microbiology models have achieved high accuracy, up to 95%, in forecasting microbial dynamics to implement proactive process modifications. The parameters of fermentation are optimized with the usage of machine learning (ML) and artificial intelligence (AI) to monitor the variable parameters of the processes. These technologies may be used to predict the outcome of fermentation and detection of abnormalities and scaling. When coupled with sensor networks and blockchain technology, these innovations improve traceability, reduce spoilage, and shorten production time, thereby enhancing quality assurance in fermentation-based production chains.

Another high-tech area of development is the use of fermentation to recover sustainable energy. In particular, fermentation technologies are useful for producing hydrogen from biomass and offer a clean, renewable energy source that is essential in a low-carbon economy. Bibliometric reviews indicate that an increasing number of research groups are developing around the topic of hydrogen production via biological fermentation, driven by advances in microbiology and aligned energy policies. These articles point to the increased understanding of microbial metabolic pathways, which are the best for producing hydrogen.

Innovations in the food industry fermentation aim to enhance food safety and sensory properties. The challenges such as the development of microbial contamination and the development of biogenic amines are being minimized through advanced fermentation tanks and regulated operations. As an illustration, the benefits of coffee fermentation technologies include the methodical approach to tracking the activities of microorganisms and regulating the condition of the process in order to enrich the flavor profile, the technique is also economical and profitable. In this case, Brazil leads in the sphere of research and the innovations are aimed at making the processes more efficient, as well as less harmful to the environment. Another way industrial biotechnology is improved is through fermentation in waste-to-energy conversion pathways. Some biological methods for converting organic waste into sustainable energy sources include biogas and bioethanol produced through anaerobic digestion and fermentation. The overall performance and economic feasibility of these systems improve when combined with complementary processes, such as thermal and chemical processes. These full-cycle waste-to-energy systems align with the foundations of the circular economy and have established resource recovery as part of waste management systems.

Control of fermentation microenvironments is made easier by the development of bioreactors with real-time monitoring devices, including optical and electrochemical sensors, that reduce variability and enhance product uniformity. New designs of reactors include continuous and fed-batch, both useful in high cell density cultures and down time reduction. These technologies can be necessary to scale down fermentation processes to the production level, making them reproducible and cost-effective.

Synthetic biology and fermentation are synergistically used in emerging disciplines to engineer microbial consortia and metabolic pathways to produce novel bio-products, such as pharmaceuticals, speciality chemicals, and biofuels. Specific editing of microbial genomic and regulatory networks enables customisation of fermentation behaviour and the generation of value-added products. These metabolic engineering methods are combined with systems biology knowledge and omics technologies in order to optimize strain designs.

References

Abbaspour, N. (2024). Fermentation's pivotal role in shaping the future of plant-based foods: An integrative review of fermentation processes and their impact on sensory and health benefits. *Applied Food Research, 4*(2), 100468.

Adebo, J. A., Njobeh, P. B., Gbashi, S., Oyedeji, A. B., Ogundele, O. M., Oyeyinka, S. A., & Adebo, O. A. (2022). Fermentation of cereals and legumes: Impact on nutritional constituents and nutrient bioavailability. *Fermentation, 8*(2), 63.

Affonfere, M., Chadare, F. J., Fassinou, F. T. K., Linnemann, A. R., & Duodu, K. G. (2023). In-vitro digestibility methods and factors affecting minerals bioavailability: A review. *Food Reviews International, 39*(2), 1014–1042.

Alrosan, M., Tan, T.-C., Easa, A. M., Gammoh, S., Alu'datt, M. H., Kubow, S., Almajwal, A. M., Mahmood, A. A. R., Al-Qaisi, A., & Bawadi, H. (2024). Enhancing the quality of lentil proteins via complexation and fermentation. *Food Science and Biotechnology, 34*(1), 65–78.

Amat, T., Assifaoui, A., Schmitt, C., & Saurel, R. (2023). Importance of complex formation on nutritional properties of legume proteins. *Critical Reviews in Food Science and Nutrition, 63*(33), 12036–12058.

Arsov, A., Tsigoriyna, L., Batovska, D., Armenova, N., Mu, W., Zhang, W., & Petrova, P. (2024). Bacterial degradation of antinutrients in foods: The genomic insight. *Foods, 13*(15), 2408.

Bielik, V., & Kolisek, M. (2021). Bioaccessibility and bioavailability of minerals in relation to a healthy gut microbiome. *International Journal of Molecular Sciences, 22*(13), 6803.

Çabuk, B., Nosworthy, M. G., Stone, A. K., Korber, D. R., Tanaka, T., House, J. D., & Nickerson, M. T. (2018). Effect of fermentation on protein digestibility of pea protein concentrate. *Food Technology and Biotechnology, 56*(2), 257–264.

Cakmak, H., Düven, G., Gultekin Subasi, B., Ilyasoglu Buyukkestelli, H., Bravo-Núñez, Á., Ozyurt, V. H., Sogut, E., & Simsek, S. (2025). Plant-based fermented foods: Quality, metabolites, and health-related functions. *Food Bioscience, 69*, 106960.

Castro-Alba, V., Lazarte, C. E., Perez-Rea, D., Carlsson, N.-G., Almgren, A., Bergenståhl, B., & Granfeldt, Y. (2019). Fermentation of pseudocereals to improve mineral accessibility. *Journal of the Science of Food and Agriculture, 99*(11), 5239–5248.

Chauhan, D., Gujral, H. S., & Kaler, R. S. S. (2024). Modification of millets by lactic acid fermentation. *International Journal of Food Science and Technology, 59*, 4336–4347.

Chourasia, R., Chiring Phukon, L., Abedin, M. M., Padhi, S., Singh, S. P., & Rai, A. K. (2023). Bioactive peptides in fermented foods: A review. *Systems Microbiology and Biomanufacturing, 3*(1), 88–109.

Clark, A. J., Soni, B. K., Sharkey, B., et al. (2022). Shiitake mycelium fermentation improves plant proteins. *LWT, 156*, 113065.

Cristina, N. M., & Lucia, D. A. (2021). Nutrition and healthy aging. *Nutrients, 13*(12), 4337.

Dahiya, D., & Nigam, P. S. (2022). Probiotics, prebiotics, synbiotics and fermented foods in nutrition. *Fermentation, 8*(7), 303.

Deokar, G. S., Pathak, V. A., Kshirsagar, S. J., Al-Asmari, F., & Nirmal, N. (2025). Enhancement of vitamin B12 in plant-based food through microbial fermentation. *Food Chemistry, 484*, 144437.

Dhiman, S., Kaur, S., Thakur, B., Singh, P., & Tripathi, M. (2025). Nutritional enhancement of plant-based fermented foods. *Fermentation, 11*, 346.

Dima, C., Assadpour, E., Dima, S., & Jafari, S. M. (2020). Bioavailability of nutraceuticals. *Comprehensive Reviews in Food Science and Food Safety, 19*(3), 954–994.

Fabbri, L. P., Cavallero, A., Vidotto, F., & Gabriele, M. (2024). Bioactive peptides from fermented foods. *Foods, 13*, 3369.

Facchin, S., Bertin, L., Bonazzi, E., Lorenzon, G., De Barba, C., Barberio, B., Zingone, F., Maniero, D., Scarpa, M., Ruffolo, C., Angriman, I., & Savarino, E. V. (2024). Short-chain fatty acids and human health: From metabolic pathways to current therapeutic implications. *Life, 14*(5), 559. https://doi.org/10.3390/life14050559

Feng, X., Ng, K., Ajlouni, S., Zhang, P., & Fang, Z. (2024). Solid-state fermentation on plant proteins. *Food Reviews International, 40*(9), 2580–2617.

Gomaa, E. Z. (2020). Human gut microbiota in health and disease. *Antonie Van Leeuwenhoek, 113*(12), 2019–2040.

Guo, Y., Chen, X., Gong, P., Chen, F., Cui, D., & Wang, M. (2021). In vitro digestion and fermentation of polysaccharides. *International Journal of Food Science and Technology, 56*(10), 4970–4982.

Guse, K., Sharma, A., Weyenberg, E., Davison, S., Ma, Y., Choi, Y., Johnson, A. J., Chen, C., & Gomez, A. (2023). Regular consumption of lacto-fermented vegetables has greater effects on the gut metabolome compared with the microbiome. *Gut Microbiome, 4*, e11. https://doi.org/10.1017/gmb.2023.9

Hays, K. E., Pfaffinger, J. M., & Ryznar, R. (2024). Gut microbiota and short-chain fatty acids. *Gut Microbes, 16*(1), 2393270.

Heo, S. J., Kim, A.-J., Park, M.-J., Kang, K., & Soung, D. Y. (2020). Functional properties of fermented mixed grains. *Foods, 9*, 1693.

Hill, D., Sugrue, I., Arendt, E., Hill, C., Stanton, C., & Ross, R. P. (2017). Recent advances in microbial fermentation for dairy and health. *F1000Research, 6*, 751.

Iqbal, B., Alabbosh, K. F., Raheem, A., Ullah, I., Khan, A. R., & Memon, M. S. (2025). Microbial transformation and nutritional quality. *Archives of Microbiology, 207*(9), 228.

Jeyakumar, E., & Lawrence, R. (2022). Microbial fermentation for reduction of antinutritional factors. In *Current developments in biotechnology and bioengineering* (pp. 239–260). Elsevier.

Kang, M. J., Baek, K. R., Lee, Y. R., Kim, G. H., & Seo, S. O. (2022). Production of vitamin K by microorganisms. *Microorganisms, 10*(3), 554.

Kanimozhi, N. V., & Sukumar, M. (2025). Probiotic fermentation and phytochemical bioavailability. *Food Wellness, 1*(1), 100018.

Kårlund, A., Gómez-Gallego, C., Korhonen, J., Palo-oja, O.-M., El-Nezami, H., & Kolehmainen, M. (2020). Harnessing microbes for sustainable development. *Nutrients, 12*, 1020.

Kaur, H., Kaur, G., & Ali, S. A. (2022). Dairy-based probiotic functional foods. *Fermentation, 8*, 425.

Keyvan, E., Adesemoye, E., Champomier-Vergès, M.-C., et al. (2025). Vitamins formed by microorganisms in fermented foods. *Frontiers in Nutrition, 12*, 1653666.

Kieliszek, M., Pobiega, K., Piwowarek, K., & Kot, A. M. (2021). Proteolytic enzymes of lactic acid bacteria. *Molecules, 26*(7), 1858.

Kim, H. D., Lee, K. S., Lee, K. E., Suh, H. J., & Kim, B.-Y. (2023). Improved digestibility of pea protein via fermentation. *Food Science and Biotechnology, 33*(3), 607–615.

Knez, E., Kadac-Czapska, K., & Grembecka, M. (2023). Fermentation and nutritional quality of vegetables and legumes. *Life, 13*(3), 655.

LeBlanc, J. G., Milani, C., de Giori, G. S., Sesma, F., van Sinderen, D., & Ventura, M. (2013). Bacteria as vitamin suppliers to their host: A gut microbiota perspective. *Current Opinion in Biotechnology, 24*(2), 160–168. https://doi.org/10.1016/j.copbio.2012.08.005

Li, Y., Zhang, L., Xu, X., Lin, S., Xie, J., & Li, J. (2025). Organic acids in fermented foods. *Journal of Food Science, 90*(12), e70745.

Ling, Z., Liu, X., Cheng, Y., Yan, X., & Wu, S. (2022). Gut microbiota and aging. *Critical Reviews in Food Science and Nutrition, 62*(13), 3509–3534.

Mukherjee, A., Breselge, S., Dimidi, E., Marco, M. L., & Cotter, P. D. (2024). Fermented foods and gastrointestinal health. *Nature Reviews Gastroenterology & Hepatology, 21*(4), 248–266.

Nagarajan, M., Rajasekaran, B., & Venkatachalam, K. (2022). Microbial metabolites in fermented foods. *International Food Research Journal, 29*(3), 466–486.

Nithya, A., Misra, S., Panigrahi, C., Dalbhagat, C. G., & Mishra, H. N. (2023). Probiotic potential of fermented foods. *Food Chemistry Advances, 3*, 100381.

Noori, S. M. A., Hojjati, M., & Sorourian, R. (2025). Solid-state fermentation of legumes. *Food Science & Nutrition, 13*(11), e70783.

Park, I., & Mannaa, M. (2025). Fermented foods as functional systems. *Foods, 14*(13), 2292.

Rekha, C. R., & Vijayalakshmi, G. (2010). Bioconversion in fermented soymilk. *Journal of Applied Microbiology, 109*(4), 1198–1208.

Rivera del Rio, A., Boom, R. M., & Janssen, A. E. (2022). Processing conditions and digestibility of plant proteins. *Foods, 11*(6), 870.

Rousta, N., & Taherzadeh, M. J. (2025). Measuring phytic acid in fermented matrices. *JSFA Reports., 6*(1), 4–13.

Sá, A. G. A., Moreno, Y. M. F., & Carciofi, B. A. M. (2020). Food processing and plant protein digestibility. *Critical Reviews in Food Science and Nutrition, 60*(20), 3367–3386.

Saha, S. K., & Pathak, N. N. (2021). Digestion and metabolism of nutrients. In *Fundamentals of animal nutrition* (pp. 219–246). Springer.

Sakandar, H. A., Hussain, R., Kubow, S., Sadiq, F. A., Huang, W., & Imran, M. (2019). Sourdough bread: A cereal fermented product. *Journal of Food Processing and Preservation, 43*, e13883. https://doi.org/10.1111/jfpp.13883

Samtiya, M., Aluko, R. E., Puniya, A. K., & Dhewa, T. (2021). Enhancing micronutrients bioavailability. *Fermentation, 7*(2), 63.

Saud, S., Xiaojuan, T., & Fahad, S. (2024). Fermentation metabolism in fruit and vegetable juices. *Food Chemistry: X, 21*, 101209.

Sawant, S. S., Park, H.-Y., Sim, E.-Y., Kim, H.-S., & Choi, H.-S. (2025). Microbial fermentation in food. *Fermentation, 11*, 15.

Shaghaghian, S., McClements, D. J., Khalesi, M., Garcia-Vaquero, M., & Mirzapour-Kouhdasht, A. (2022). Digestibility of plant-based proteins in meat analogues. *Trends in Food Science & Technology, 129*, 646–656.
Shah, A. M., Tarfeen, N., Mohamed, H., & Song, Y. (2023). Fermented foods and gut microbiota. *Fermentation, 9*(2), 118.
Shahidi, F., & Pan, Y. (2022). Food matrix and phytochemical bioaccessibility. *Critical Reviews in Food Science and Nutrition, 62*(23), 6421–6445.
Sharma, R., Garg, P., Kumar, P., Bhatia, S. K., & Kulshrestha, S. (2020). Microbial fermentation and quality improvement. *Fermentation, 6*(4), 106.
Shiferaw Terefe, N., & Augustin, M. A. (2020). Fermentation and food functionality. *Critical Reviews in Food Science and Nutrition, 60*(17), 2887–2913.
Siddiqui, A. J., Singh, R., Jahan, S., et al. (2022). Enzymes in food fermentations. In *African fermented food products* (pp. 101–133). Springer.
Soni, K., Samtiya, M., Krishnan, V., & Dhewa, T. (2022). Antinutritional factors and bioavailability. In *Conceptualizing plant-based nutrition* (pp. 157–179). Springer.
Stacchiotti, V., Rezzi, S., Eggersdorfer, M., & Galli, F. (2021). Gut microbiota and fat-soluble vitamins. *Critical Reviews in Food Science and Nutrition, 61*(19), 3211–3232.
Tachie, C. Y. E., Onuh, J. O., & Aryee, A. N. A. (2024). Nutritional benefits of fermented food proteins. *Journal of the Science of Food and Agriculture, 104*(3), 1223–1233.
Tan, X., Cui, F., Wang, D., Lv, X., Li, X., & Li, J. (2024). Fermented vegetables: Health benefits and solutions. *Foods, 13*, 38.
Taylor, J., & Taylor, J. R. N. (2002). Fermentation and sorghum protein digestibility. *International Journal of Food Science and Nutrition, 37*(2), 129–137.
Tharifkhan, S. A., Perumal, A. B., Elumalai, A., Moses, J. A., & Anandharamakrishnan, C. (2021). Enzymes and nutrient bioavailability in millets. *Journal of the Science of Food and Agriculture, 101*(12), 4869–4878.
Tugnoli, B., Giovagnoni, G., Piva, A., & Grilli, E. (2020). Organic acids and intestinal health. *Animals, 10*(1), 134.
Uțoiu, E., Matei, F., Toma, A., et al. (2018). Fermented bee pollen and health benefits. *Nutrients, 10*(10), 1365.
Vernocchi, P., Del Chierico, F., & Putignani, L. (2020). Gut microbiota metabolism and food components. *International Journal of Molecular Sciences, 21*(10), 3688.
Wang, Y., Wu, J., Lv, M., et al. (2021). Metabolism characteristics of lactic acid bacteria. *Frontiers in Bioengineering and Biotechnology, 9*, 612285.
Yousefi, N., Shokrollahi Yancheshmeh, B., & Gernaey, K. V. (2025). Fermentation-based processing on protein modification. *Foods, 14*(20), 3461.
Zhao, Y. S., Eweys, A. S., Zhang, J. Y., Zhu, Y., Bai, J., Darwesh, O. M., & Xiao, X. (2021). Fermentation and antioxidant activity of plant-based foods. *Antioxidants, 10*(12), 2004.

Chapter 5
Compositional Transformation During Fermentation: Improvement of Total Antioxidant Capacity in Selected Fermented Foods

Marta Cuenca, Mussa Makran, Stefania Filice, Guadalupe Garcia-Llatas, Mónica Gandía, Antonio Cilla, and Amparo Gamero

Abstract Fermentation is an ancient method widely practiced across cultures, capable of preserving foods, enhancing their nutritional profile and generating bioactive compounds that contribute to overall health. This process recovers the bioavailability of vitamins and minerals, reduces antinutritional factors and supports gut microbiota balance and digestive function. Fermentation can increase antioxidant capacity of the fermented products compared to their raw counterparts, attributed to breaking down of compounds, generating new antioxidants, facilitating their release and reducing antinutrients. As a proof of concept, this study analyses the total antioxidant capacity (TAC) of three fermented foods—black garlic, fermented lemon and kombucha—using three widely recognized analytical methods: TEAC (Trolox Equivalent Antioxidant Capacity), ORAC (Oxygen Radical Absorbance Capacity) and total polyphenol (TP) content. Both homemade (HM) and industrially (IN) fermented samples were examined, including extractable and non-extractable

M. Cuenca · M. Makran
Nutrition and Food Science Area, Faculty of Pharmacy and Food Sciences, University of Valencia, Valencia, Spain

Food Technology Area, Faculty of Pharmacy and Food Sciences, University of Valencia, Valencia, Spain
e-mail: marta.cuenca@uv.es; mussa.makran@uv.es

S. Filice
Microtarians SIS, Société d'Impact Societal, Luxembourg, Luxembourg
e-mail: Stefania.filice@microtarians.com

G. Garcia-Llatas · A. Cilla (✉)
Nutrition and Food Science Area, Faculty of Pharmacy and Food Sciences, University of Valencia, Valencia, Spain
e-mail: guadalupe.garcia@uv.es; antonio.cilla@uv.es

M. Gandía · A. Gamero
Food Technology Area, Faculty of Pharmacy and Food Sciences, University of Valencia, Valencia, Spain
e-mail: monica.gandia@uv.es; amparo.gamero@uv.es

S. Dhiman, S. Kaur (eds.), *Next-Gen Plant Based Fermented Foods: Innovations for a Sustainable Future*, https://doi.org/10.1007/978-3-032-29927-7_5

compounds, to explore potential differences related to the type of fermentation process applied. Regarding black garlic, the HM sample demonstrated significantly higher TEAC values, whereas the IN sample performed better in ORAC measurements, with no significant differences observed in TP between both. In the case of fermented lemon, the HM sample exhibited a notably superior antioxidant capacity in ORAC and TP compared to the IN counterpart, while TEAC results were similar. Finally, the IN kombucha sample showed markedly higher values across all three methods, suggesting a more potent antioxidant profile relative to the HM version. Overall, comparison with non-fermented products used as controls from literature values suggests that fermentation (HM or IN) enhances antioxidant activity relative to non-fermented counterparts. This enhancement may be attributed to the release or transformation of antioxidant compounds during fermentation, as well as a potential reduction of antinutritional factors, as reported in previous studies. These findings contribute to the comparative characterization of fermented foods and provide a foundation for future studies to assess their bioavailability and functional effects under physiologically relevant conditions.

Keywords Antioxidant capacity · Black garlic · Fermentation · Kombucha · Lemon · ORAC · TEAC · Total polyphenols

Abbreviations

DW	Dry Weight
GAE	Gallic Acid Equivalent
HM	Homemade
IN	Industrial
ORAC	Oxygen Radical Absorbance Capacity
TE	Trolox Equivalent
TEAC	Trolox Equivalent Antioxidant Capacity
TP	Total Polyphenols

5.1 Introduction

5.1.1 Origin of Fermentation and Technological Use

The fermentation process has very ancient origins, dating back to earlier than 7000 B.C., when Koji (grains fermented with mold and/or soy) was produced in China. Over the centuries, the technique was gradually refined and improved in different parts of the world, including Georgia, Iran, Babylon, Egypt, Mexico, and Sudan, among others. In ancient Egypt, early fermentation specialists developed highly advanced methods for grain fermentation and produced an acidic beverage

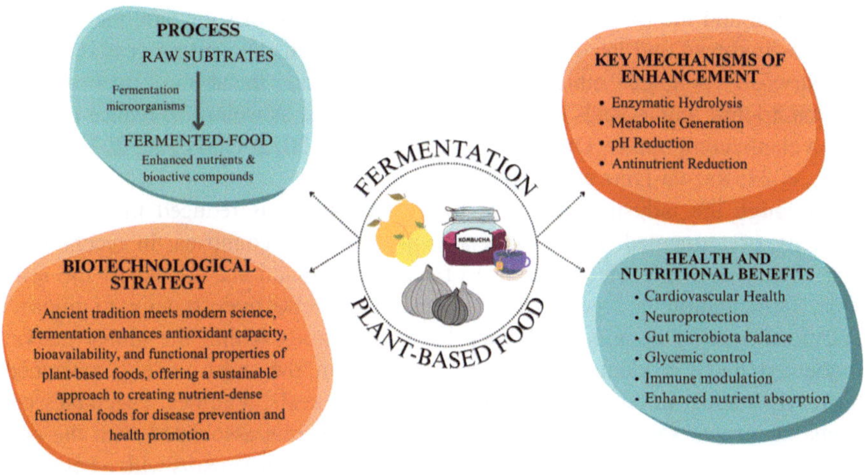

Fig. 5.1 Fermentation of plant-based foods: processes, mechanisms and health benefits

known as boza or boozah, which was obtained from lightly baked germinated grain. This fermented drink was elaborated by leaving mixtures of water and flour on warm days for extended periods, allowing natural yeasts present in flour and milk (added as flavouring) to act. Over time, fermentation has spread and became established as a technological practice on a global scale, being today an essential metabolic process carried out by microorganisms that not only preserves food but also transforms it into superior nutritional matrices with unique functional properties (Emerald et al., 2016; Mohamed, 2020).

Fermentation is a natural biological process that can significantly enhance antioxidant capacity, favourably modify the nutritional profile and generate bioactive compounds that provide tangible benefits for human health (Liu et al., 2024). Moreover, several fermented products can also protect the gastrointestinal system against pathogens, promote the absorption of bioactive molecules and minerals and support to the balance of the gut microbiota (Fig. 5.1) (Marco et al., 2017; Lemos Junior et al., 2020).

Moreover, food fermentation can be conducted either at the household level or through industrial processes and these two approaches may not always produced equivalent results. Homemade (HM) fermentation typically relies on environmental conditions, and the natural microbiota present in the raw material, which can lead to greater variability in flavour, texture, or nutritional composition. In contrast, industrial (IN) fermentation is conducted under more controlled parameters (like temperature, humidity, or the use of starter cultures) which may allow to produce more uniform and stable products. These differences in process control could influence the final concentration of bioactive compounds and, consequently, the functional properties of the fermented food (Maurice et al., 2022; Stoenescu & Stănică, 2025).

Fermentation is fundamental in the production of various foods, including bakery and pastry products, fermented beverages and milk or meat derivatives. Its application and the results obtained depend on the type of substrate, the microorganisms involved and local traditions. Additionally, this ancient process is used in industry to produce ethanol, biofuels, enzymes, metabolic intermediates, micronutrients, amino acids and aromatic compounds, as well as to reduce undesirable volatile compounds (Diez-Ozaeta & Astiazaran, 2022; Guo et al., 2023; Zheng et al., 2023; Razola-Díaz et al., 2024).

Although the industrialization of food production has reduced the traditional diversity of fermented foods, especially in developed countries, in recent years there has been a notable resurgence of fermented functional foods. At the same time, fermentation is recognised as a potentially effective strategy to improve nutritional quality in regions affected by malnutrition. Within the global market, fermented dairy products, such as yogurt, remain the most widely consumed; however, interest in new alternatives made from cereals, legumes, vegetables and fruits has grown significantly. Among the main benefits associated with the consumption of fermented foods is the introduction of microorganisms with a beneficial effect on the gut microbiota (Lang et al., 2014; Lemos Junior et al., 2020).

5.1.2 Fermentation and Its Relevance in Modern Healthy Diets

The increased consumption of ultra-processed foods high in fat, sugar and salt has been associated with the rise of non-communicable chronic diseases, generating growing interest in more balanced diets. Although many plant-based alternatives have emerged as healthier options, it is important to note that some of them also fall into the category of ultra-processed products. Beyond health considerations, the current shift toward plant-based diets is additionally driven by concerns related to environmental sustainability and animal welfare, which strongly influence consumer behaviour and food choices. In this context, plant-based foods have gained prominence as healthy alternatives, although they often present sensory and nutritional limitations that hinder consumer acceptance (Marco et al., 2017).

Fermentation can be considered a biotechnological strategy that may help overcome these limitations by providing characteristic flavour, aroma and texture, while potentially enhancing the nutritional value of plant-based foods (Reyes-Jurado et al., 2023). As a result, fermented products represent a promising avenue for the development of functional foods capable of contributing to health promotion and the prevention of diseases associated with modern dietary habits (Baruah et al., 2022).

As shown in Fig. 5.1, the fermentation process can be seen as a multi-step transformation that involves three main stages: (i) preparation of the substrate and introduction of the microbes, (ii) active fermentation with metabolic changes and (iii) production of bioactive compounds that offer health benefits. This figure depicts how fermentation connects raw plant materials to functional foods, serving as a useful approach to improve both the taste and nutritional value of alternatives.

5.1.3 Fundamental Health and Nutritional Benefits

The health and nutritional benefits of fermentation (Fig. 5.1) can be grouped into two fundamental areas: increased antioxidant capacity and improved nutritional profile. Microbial enzymes act on plant cell walls and bound compounds, releasing phenolic compounds and generating new bioactive molecules. Fermentation also reduces antinutritional factors, such as phytates and tannins, while enriching vitamins and minerals. Together, these changes contribute to better gut health, enhanced immune function, and potential disease prevention (Liu et al., 2024).

5.1.3.1 Effects on Antioxidant Capacity

Fermentation constitutes a natural process that enables an increase in antioxidant capacity through multiple biochemical pathways. Under the action of selected microorganisms (mainly lactic acid bacteria, yeasts and molds), specific enzymes such as β-glucosidases and phytases can be secreted, which may hydrolyse conjugated phenolic compounds present in plants, transforming insoluble polyphenols into soluble forms and thereby enhancing their free radical–scavenging ability (Wei et al., 2025). In addition, the metabolic activity of these microorganisms can previously activate inert metabolites or could generate new bioactive molecules, such as peptides, free phenolic acids and volatile compounds (Septembre-Malaterre et al., 2018; Leonard et al., 2021; Zhao et al., 2021; Erskine et al., 2023).

For example, fermented celery juice undergoes a significant increase in total polyphenols, flavonoids and vitamin C, along with an elevation of superoxide dismutase, a key antioxidant enzyme that helps reduce oxidative stress induced by high-fat diets (Liu et al., 2024). Similarly, the fermentation of pomegranate juice with lactic acid bacteria results in higher antioxidant capacity compared to the unfermented product (Ye et al., 2025). Consistent with these findings, recent studies have also shown that the fermentation of different plant-based beverages, including matrices such as tiger nut, carob and rice, enhances their antioxidant capacity and enhances the release and bioaccessibility of phenolic compounds during gastrointestinal digestion (Vitali et al., 2025a, b).

Overall, the regular consumption of fermented foods may help reduce the risk of cardiovascular diseases by limiting LDL lipoprotein oxidation, improve glycaemic control by decreasing oxidative stress, and might exert neuroprotective effects by reducing lipid peroxidation in the nervous system. Additionally, these foods could modulate immune responses, prevent liver damage and may support digestive and anti-inflammatory functions through the synergistic action of probiotics and prebiotics. Recent studies have shown that fermentation can enhance antiproliferative effects against colon cancer cells, though this response is matrix-dependent (Fig. 5.1) (Verni et al., 2019; Leonard et al., 2021; Zhao et al., 2021; Zheng et al., 2023; Vitali et al., 2025a, b, c).

5.1.3.2 Effects on Nutritional Profile

5.1.3.2.1 Vitamin Enrichment

Fermentation may increase the vitamin content of fruits and vegetables through two complementary mechanisms: (i) the direct synthesis of vitamins by fermenting microorganisms, especially B-complex vitamins and (ii) the stabilization of pre-existing vitamins due to the acidic and reduced environment generated during the process. Various species of the genus *Bifidobacterium* and lactic acid bacteria can produce several vitamins, significantly increasing the total micronutrient content in the food. Although vitamin C is particularly sensitive to oxidation, fermentation can contribute to its preservation by lowering the pH, limiting oxidative enzymatic activity and producing antioxidant metabolites that stabilize its structure during refrigerated storage (Diez-Ozaeta & Astiazaran, 2022; Liu et al., 2024; Lim & Abd Rahim, 2025).

5.1.3.2.2 Mineral Enrichment and Reduction of Antinutritional Factors

Fermentation significantly improves the bioavailability of essential minerals such as iron, zinc, and calcium, mainly by breaking down phytic acid, a compound that forms insoluble complexes with these micronutrients. Phytase, an enzyme produced by various lactic acid bacteria, can hydrolyse phytate and release minerals in more readily absorbable forms. At the same time, other species, such as *Bifidobacterium* spp. and certain yeasts, participate in the transformation of inorganic minerals into bioavailable organic forms. In addition to increasing mineral availability, fermentation can help to reduce other antinutritional factors (Fig. 5.1). Tannins (polyphenolic compounds that interfere with protein digestion) may be modified by microbial action, generating bioactive derivatives with antioxidant properties, such as ellagic acid. Moreover, some microorganisms could reduce pesticide residues through adsorption or metabolic degradation processes, contributing to improved food safety, although this elimination is not complete. Altogether, these transformations, mediated by enzymatic activity, pH modification and microbial metabolism, enhance both the nutritional quality and safety of fermented foods (Lazarte et al., 2015; Liu et al., 2024).

5.1.4 Selected Fermented Ingredients

Although fermentation has been used for centuries to preserve foods, scientific knowledge about certain products remains limited, particularly regarding their antioxidant capacity (Liu et al., 2024). Therefore, this chapter focused on three commonly consumed ingredients in their usual forms (garlic, lemon and tea) with the aim of analysing their fermented versions: black garlic, fermented lemon and kombucha. However, sufficient information is still lacking on the effects of fermentation and the factors that determine the quality of HM versus IN versions.

5.2 Materials and Methods

5.2.1 *Ingredients*

Three ingredients were selected for the study, with the aim of analysing and comparing different fermentation procedures (HM vs. IN) to evaluate whether artisanal variability or the consistency of industrial processing leads to greater enhancement of antioxidant content. As a reference, literature values corresponding to the unfermented versions of each ingredient were used. The HM samples were prepared by Microtarians SIS (Luxembourg) following specific fermentation protocols adapted to each food.

5.2.1.1 Black Garlic

Black garlic has gained popularity in Western markets. Several studies suggest that this vegetable exhibits higher antioxidant capacity than its fresh counterpart, due to the accumulation of polyphenols during its aging process, which generally involves subjecting whole fresh garlic bulbs to controlled heat and humidity for several weeks, promoting non-enzymatic browning and the development of its characteristic dark colour, soft texture and sweet flavour (Tahir et al., 2022).

The unfermented control for comparison purposes was based on literature data from raw fresh garlic. The HM variant was prepared using a *sous-vide* technique: whole local garlic cloves were sprinkled with Na_2CO_3, vacuum-sealed and fermented by means of their endogenous, thermotolerant microbial microbiota (Tahir et al., 2022) at 55 °C for 5 weeks. The IN sample corresponded to commercial black garlic (Azafran Bio—Schwarzer Knoblauch, Germany), produced on an industrial scale under patented conditions (Fig. 5.2).

5.2.1.2 Fermented Lemon

Traditional in various Mediterranean and North African regions, it is widely known that lactic acid fermentation transforms lemon phenolic components, but there is little information on the differences between HM and IN fermentation (Hsieh et al., 2021).

The HM sample consisted of organic lemons cut into quarters, layered with coarse non-iodized salt in airtight jars and spontaneously fermented at room temperature (22–27 °C) for 5 weeks by naturally occurring lactic acid bacteria (Fig. 5.3). The IN variant was a commercial fermented lemon product (La Natura—eingelegte Zitronen, Germany).

Fig. 5.2 Fermented black garlic samples. (**a**) Homemade; (**b**) Industrial

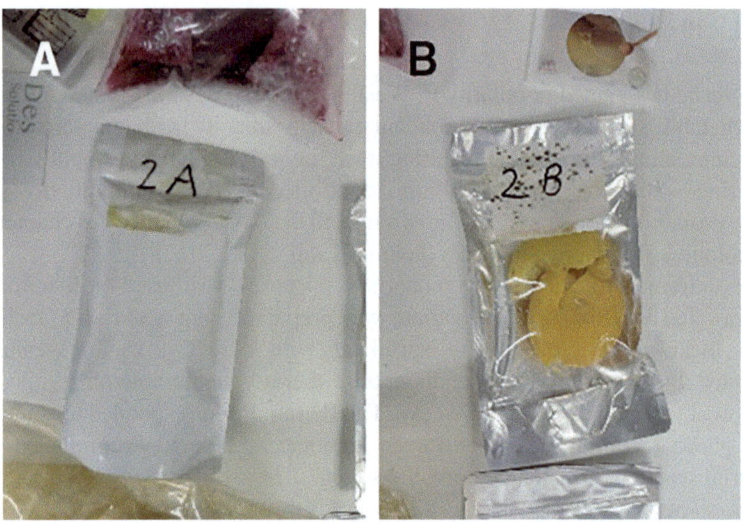

Fig. 5.3 Fermented lemon samples. (**a**) Homemade; (**b**) Industrial

5.2.1.3 Kombucha

Kombucha, in turn, has experienced significant commercial growth in recent years, although its chemical composition depends on the type of symbiotic culture of bacteria and yeasts (SCOBY) used, the tea substrate and the fermentation conditions (Antolak et al., 2021; Kluz et al., 2022).

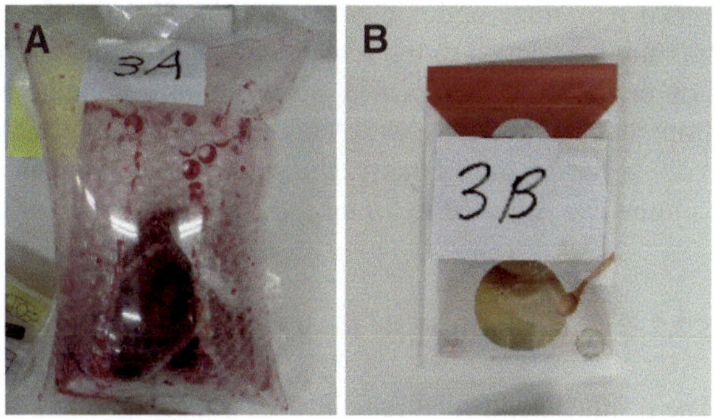

Fig. 5.4 Kombucha samples. (**a**) Homemade; (**b**) Industrial

The control consisted of the sweetened tea infusion prior to fermentation. The HM kombucha was prepared by infusing water with oak leaves and sea buckthorn berries, adding sugar and inoculating a 45-year-old of SCOBY, fermenting at 22–27 °C for 5 days until reaching a pH of 4.5 (Fig. 5.4). The IN sample corresponded to a commercial product (Natura Probio® Original Kombucha Scoby, Germany).

5.2.2 *Antioxidant Extraction Procedure*

All fermented samples were immediately frozen at—40 °C, lyophilized for 4–5 days and ground into a homogeneous powder. This standardization allowed analytical results to be expressed on a dry weight (DW) basis, eliminating variability associated with moisture content and facilitating the quantitative comparison of antioxidant capacity. Freeze-dried and ground samples were used for antioxidant extraction. Extractable (aqueous-organic extracts) and non-extractable (acid hydrolysis) antioxidants were obtained as described by Hartzfeld et al. (2002); Durazzo et al. (2017, 2019); and Cuenca-Ortolá et al. (2023).

5.2.2.1 Aqueous-Organic Extract

For the extraction of extractable antioxidants in the aqueous-organic extract, the samples first underwent a process of crushing, freeze-drying and grinding to ensure proper homogenization. Subsequently, between 3 and 5.5 g of the sample were weighed into glass tubes, to which 20 mL of a methanol:water (50:50, v/v) solution acidified with 2 N HCl was added until a pH of 2.0 was reached. The mixture was stirred at room temperature using a vortex and a universal platform shaker (IKA, KS260) for 1 h at 100 rpm, promoting the extraction of the target compounds.

After agitation, the suspension was centrifuged (Eppendorf Centrifuge, S810R. Hamburg, Germany) at 978 *g* for 10 min. The resulting supernatant was carefully recovered using a glass Pasteur pipette and transferred to a 50 mL Pyrex tube.

To maximize antioxidant extraction, 20 mL of an acetone:water (70:30, v/v) mixture was added to the remaining solid residue and the agitation and centrifugation processes were repeated under the same conditions. The supernatants obtained from both extractions were combined and subjected to a final centrifugation at 1227 *g* for 15 min. Finally, the resulting supernatant was collected, and the total volume was adjusted to 50 mL, yielding the final extract ready for analysis.

5.2.2.2 Acid Hydrolysis Residue

The residual solids that remained after the extraction of the aqueous-organic extract were dried overnight at 25 °C in an oven (Emmert, ULE 500 AO). The samples were weighed before and after drying to ensure a constant weight, with variations not exceeding 0.05 g. Once dried, 20 mg of the residue was mixed with 2 mL of methanol and 0.2 mL of concentrated sulfuric acid (18 M). The mixture was vortexed for 1 min and subsequently was heated in a thermoblock (Stuart block heater SBH200D. UK) at 85 °C for 20 h, with intermittent stirring every 1–2 h.

Following the heating process, the sample was centrifuged at 3000 *g* for 10 min. The resulting supernatant was carefully collected and transferred to a clean 10 mL glass tube. The remaining residue was then subjected to two washing steps using minimal volumes of approximately 1 mL of Milli-Q water. Each wash was followed by centrifugation at 3000 *g* for 10 min. The supernatants from both washes were combined with the initial extract and the final volume was adjusted to 5 mL with Milli-Q water.

5.2.3 Determination of Antioxidant Capacity and Total Polyphenols

5.2.3.1 TEAC

The determination of antioxidant capacity using the Trolox Equivalent Antioxidant Capacity (TEAC) method was conducted following the procedure described by Re et al. (1999). This assay evaluated the ability of the samples to scavenge free radicals present in the medium. The greenish blue $ABTS^+$ cation radical was generated through the reaction between 2,2′-azinobis (3-ethylbenzothiazoline-6-sulfonic acid) (ABTS) and potassium persulfate ($K_2S_2O_8$). This radical is a stable, water-soluble compound characterized by a distinct absorption spectrum. Upon adding the samples to the radical solution, a decrease in absorbance occurred, allowing the antioxidant capacity to be determined based on the sample's ability to reduce the radical concentration. The results were expressed relative to 6-hydroxy-2,5,7,8-tetramethylchroman-2-carboxylic acid (Trolox).

The $ABTS^+$ stock solution was prepared by dissolving 7 mM ABTS and 140 mM $K_2S_2O_8$ and allowing the mixture to react in darkness at room temperature for 12–16 h, ensuring stability for up to 2 days. Before analysis, the ABTS solution was diluted approximately 1/100 with ethanol and adjusted to an absorbance of 0.70 ± 0.02 at 734 nm at 30 °C using a thermostat-controlled spectrophotometer (PerkinElmer UV-VIS Lambda 365, PerkinElmer, Waltham, MA, USA). A 5 mM Trolox stock solution was prepared in volumetric flasks.

For the analysis, 2 mL of the ABTS radical solution was first added to the cuvette to measure its initial absorbance, followed by the addition of 100 μL of the sample or standard solution. After 3 min, the absorbance was measured again. The analysis was performed in triplicate, and the antioxidant capacity was expressed as Trolox equivalents (TE) in μM per 100 g of DW. An appropriate dilution had been previously optimized to achieve an inhibition percentage within the range of 20–80%.

5.2.3.2 ORAC

The Oxygen Radical Absorbance Capacity (ORAC) method was developed based on the ability of antioxidant compounds to scavenge oxygen radicals (Ou et al., 2001). This ability was assessed by measuring the reduction in fluorescence of fluorescein, caused by oxidative damage from free radicals generated through the thermal decomposition of 2,2-azinobis (2-amidinopropane) dihydrochloride (AAPH). The antioxidant capacity of the samples was determined by evaluating the time and percentage of protection against fluorescence loss, using Trolox as the standard antioxidant. The reaction was conducted using a Multilabel Plate Counter VICTOR^3 1420 (PerkinElmer, Waltham, MA, USA), equipped with a fluorescence filter for excitation at 485 nm and emission at 535 nm. The assays were conducted at 37 °C, with sodium fluorescein and AAPH at concentrations of 0.015 and 120 mg/mL, respectively. The samples were diluted in a 75 mM phosphate buffer, pH 7.4. In a white multi-well plate, 80 μL of fluorescein, 40 μL of AAPH and 80 μL of the diluted sample, Trolox (standard), or phosphate buffer (blank) were added. Fluorescence was measured every 5 min for 90 min until it dropped below 5% of the initial value. The results were expressed as TE μM per 100 g DW.

5.2.3.3 Total Polyphenols

To determine total polyphenols (TP), the Folin-Ciocalteu reagent was used, which was prepared by combining phosphotungstic acid and phosphomolybdic acid. In the presence of an excess of this reagent, polyphenols reduced the acids to blue tungsten and molybdenum oxides. When the samples were treated with the reagent in the presence of sodium carbonate (Na_2CO_3), the absorbance was directly proportional to the polyphenol concentration (Singleton & Rossi, 1965).

The method included centrifuging the non-extractable polyphenol samples at 3000 *g* for 10 min to remove any suspended particles formed during freezing. Then, 100 μL of the sample (whether extractable or non-extractable antioxidants) was mixed with 3 mL of a 2% (w/v) sodium carbonate solution and 100 μL of a 50% (w/v) Folin-Ciocalteu reagent. The mixture was allowed to react for 1 h at room temperature in the dark, after which the absorbance was measured at 765 nm using a spectrophotometer (PerkinElmer Lambda 365 UV-VIS). Quantification was conducted using a calibration curve based on gallic acid standards within a range of 0–300 mg/L. The results were expressed as gallic acid equivalents (GAE) per 100 g of DW to allow for comparison with existing literature.

5.2.4 Statistical Analysis

The results were expressed as mean ± standard deviation ($n = 4$). For the specific comparison between pairs of HM and IN fermentation samples within each food type (black garlic, lemon and kombucha), the Student's *t-test* for independent samples was used. All statistical tests were applied with a significance level of $p < 0.05$. The analyses were performed using GraphPad Prism version 8 Software Boston, MA, USA.

5.2.5 Study Design and Key Limitations

This study employed a single-batch design, with one HM and one IN batch per food matrix. The four replicates per sample ($n = 4$) represent analytical replicates within each batch, not biological replicates across independent fermentation events. This design limits inference about process variability, particularly for HM fermentation where batch-to-batch differences are typically more pronounced.

Non-fermented control values were obtained from literature sources, which may differ in cultivar, extraction methodology, and analytical conditions from the present study. Additionally, HM and IN fermentations differed not only in production scale but also in substrate composition (particularly evident in kombucha), microbial community structure, and processing parameters (time, temperature, starter culture use). These inherent differences, while reflecting real-world fermentation practices, complicate direct causal attribution of observed differences solely to the HM vs. IN distinction.

5.3 Results

5.3.1 Proof of Concept: Total Antioxidant Capacity in Selected Fermented Foods (Black Garlic, Fermented Lemon, Kombucha)

The total antioxidant capacity (TAC) of the selected ingredients after fermentation was evaluated using ORAC TEAC and TP assays (Fig. 5.5), showing significant differences depending on the matrix and the type of fermentation. Both extractable and non-extractable antioxidant compounds contributed to the TAC across all samples.

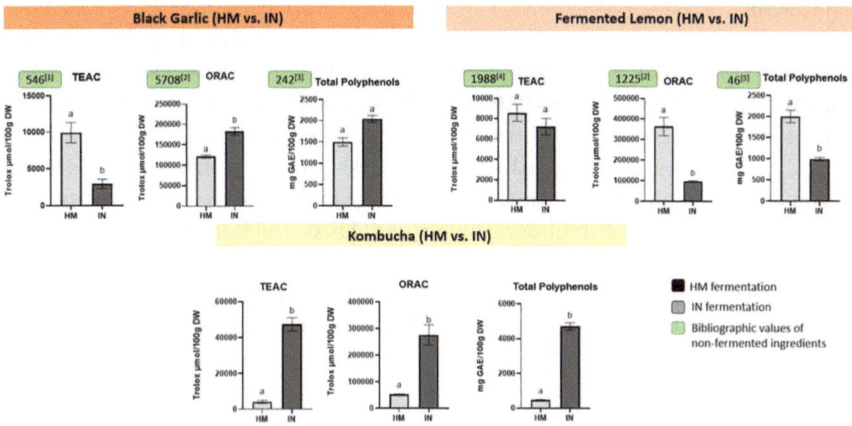

Fig. 5.5 Comparison of total antioxidant capacity (TEAC and ORAC) and total polyphenols in selected fermented foods (black garlic, fermented lemon and kombucha) obtained homemade or industrially. Different letters (a, b) in the same method indicate significant differences between the samples ($n = 4$ replicates per sample) (*t tests*) ($p < 0.05$). DW: Dry Weight; GAE: Gallic acid equivalent; ORAC: Oxygen radical absorbance Capacity; TEAC: Trolox equivalent antioxidant capacity. Bibliographic values for non-fermented ingredients correspond to the following references: [1] Najman et al. (2020); [2] USDA (2010); [3] Phenol-Explorer (2025); [4] Zhou et al. (2017);[5] Calderón Cuadrado (2020)

Black Garlic

In TEAC assay, the HM sample (10,773.3 ± 1250.2 µmol TE/100 g DW) showed a higher overall value than the IN sample (3528.4 µmol TE/100 g DW), while the raw sample presented a value of 546.3 ± 4.6 µmol TE/100 g DW (Najman et al., 2020). Conversely, the IN sample showed significantly different values, with higher measurements in ORAC (182,587.8 ± 7054.6 µmol TE/100 g DW) compared to its HM counterpart (121,659.4 ± 7434.0 µmol TE/100 g DW), both markedly exceeding the value reported for fresh garlic (5708.0 µmol TE/100 g DW) (USDA, 2010). Similarly, for TP, no significant differences were observed between the two fermentation types (1499.0 ± 79.5 and 2040.3 ± 60.2 mg GAE/100 g DW for HM and IN, respectively), although both values were substantially higher than that of fresh garlic (242.1 ± 2.4 mg GAE/100 g DW) (Phenol-explorer, 2025).

Fermented Lemon

The HM fermentation showed significantly different values compared to IN, with higher ORAC (363,294.6 ± 32,485.4 vs. 98,518.3 ± 2840.4 µmol TE/100 g DW, respectively) and TP assays (1910.9 ± 122.2 vs. 993.5 ± 30.3 mg GAE/100 g DW, respectively), though both were higher than their fresh counterparts (1225.0 µmol TE/100 g DW (USDA, 2010) and 46.8 mg GAE/100 g DW (Calderón Cuadrado, 2020)). In contrast, in the TEAC assay, no significant differences were observed between the two fermentation types (8314.0 ± 737.7 and 6734.1 ± 737.3 µmol TE/100 g DW, respectively), while fresh lemon juice reached a value of 1988 µmol TE/100 g DW (Zhou et al., 2017).

Kombucha

In the case of kombucha, the IN fermentation showed significantly different values compared to HM, with higher results across all three assays: TEAC (48,180.5 ± 2796.6 vs. 3449.6 ± 720.8 μmol TE/100 g DW), ORAC (276,547.0 ± 33,247.9 vs. 51,680.8 ± 1524.3 μmol TE/100 g DW) and TP (4717.2 ± 173.2 vs. 483.1 ± 10.1 mg GAE/100 g DW). Notably, both HM and IN kombucha samples demonstrated substantially higher antioxidant capacity compared to unfermented tea (bibliographic values: TEAC 2700.0 μmol TE/100 g DW (Zhou et al., 2022), ORAC 1280.0 μmol TE/100 g DW (Sun et al., 2015) and TP 250.5 mg GAE/100 g DW (Kim et al., 2023), confirming that fermentation significantly enhances the antioxidant profile of the tea-based beverage.

5.4 Discussion

The results of this study demonstrate that fermentation increases the TAC of the analysed ingredients, although with variations depending on the food matrix and the type of process employed. This suggests that the mechanisms responsible for the enhancement of antioxidant activity are influenced by the characteristics of the food and the specific fermentation conditions.

5.4.1 Black Garlic

In black garlic, fermentation significantly increased TAC compared to the raw version. Both HM and IN showed increases associated with the accumulation of polyphenols. The TEAC assay, which evaluates electron transfer, showed higher values in HM. The homemade sample was fermented using the sous-vide technique at 55 °C for 5 weeks, and exhibited compounds with higher reducing capacity, though the specific mechanisms underlying this difference were not directly evaluated. Conversely, the ORAC assay, which measures the capacity to neutralize oxygen radicals, showed higher values in IN, demonstrating a different analytical response pattern. These contrasting results between methods suggest that the garlic matrix responds differently to HM and IN processing conditions, resulting in distinct antioxidant capacity. Both fermentation types exceeded the antioxidant capacity of fresh garlic (Najman et al., 2020).

5.4.2 Fermented Lemon

In fermented lemon, the HM sample showed higher ORAC and TP values than the IN sample, though TEAC values were similar between both (Fig. 5.5), indicating method-dependent differences. This pattern may be associated with the specific conditions of artisanal fermentation with lactic acid bacteria at room temperature for 5 weeks, which could facilitate the degradation of cell walls and the release of phenolic compounds. Additionally, the pH reduction caused by organic acids may promote the stability and synthesis of antioxidant metabolites (Hsieh et al., 2021). The longer fermentation period of the HM sample may have promoted a greater degree of microbial transformation of lemon phenolic compounds compared with the IN process, although the specific mechanisms responsible for these differences were not directly evaluated.

5.4.3 Kombucha

In kombucha, the IN sample exhibited higher TAC values than HM across all assays. The HM sample, prepared with a 45-year-old SCOBY over 5 days, showed lower antioxidant values compared to IN, which was produced using a commercial kit with standardized SCOBY composition and controlled microbial parameters. The SCOBY plays a key role in transforming tea compounds during fermentation. Previous studies have shown that controlled fermentation conditions, including consistent temperature, fermentation time, and SCOBY composition, can influence the final antioxidant profile through microbial metabolic activity (Zhou et al., 2022), though the specific biochemical pathways were not evaluated in the present study. This suggests that process standardization may contribute to different antioxidant profiles.

5.4.4 General Patterns and Mechanisms

A common pattern among the three products is that both HM and IN increase TAC compared to their original counterparts, confirming that fermentation enriches bioactive compounds. The variability between HM and IN is associated with the food matrix, the origin of the culture and environmental conditions, indicating that the relationship between fermentation approach and antioxidant capacity is matrix-dependent rather than reflecting an inherent superiority of either method.

5.4.5 Study Limitations

This study presents several limitations that should be considered when interpreting the results. First, the comparison with literature values for the original form introduces methodological heterogeneity, since reference studies may use different ingredient varieties, cultivation conditions and/or extraction methods. Second, although four replicates per sample were performed ($n = 4$), the use of a single batch per condition (HM and IN) limits the ability to estimate within-group variability, a relevant factor in HM-type fermentations, where differences between samples tend to be more pronounced.

Furthermore, the absence of evaluations using simulated digestion or *in vivo* assays limits the interpretation of the results in terms of actual functional effects, as processes such as intestinal absorption, hepatic metabolism and tissue distribution can significantly alter bioactivity. These effects cannot be assessed with the assays employed here—TEAC and ORAC for TAC and TP.

5.5 Conclusions

Both HM and IN significantly outperformed the non-fermented versions in all three products analysed, with observed increases in extractable and non-extractable antioxidant compounds. However, the magnitude of the effect varied depending on the food: the HM method was more effective in fermented lemon, IN in kombucha and in black garlic, both produced distinct antioxidant profiles.

This variability suggests that the enrichment mechanisms depend on factors such as the initial composition of the food, the microbial community involved and process conditions (temperature, time, pH). In kombucha, the characteristics of the SCOBY culture also play a critical role.

Differences were also observed between the analytical methods used (TEAC, ORAC and TP). Some products stood out in one assay but not in others, reflecting the diversity of antioxidant mechanisms present. This underscores the importance of using complementary analytical approaches to achieve a more complete characterization of antioxidant potential in fermented foods.

These findings contribute to a better comparative understanding of fermented food products. Future research should focus on assessing bioavailability and functional effects under conditions that more closely reflect real physiological contexts, to better understand the functional significance of the observed antioxidant improvements.

Funding Information This work and project was financially supported by PRIMA 2022 Program Horizon Europe (MEDIET4ALL: A Transnational movement to support the sustainable transition towards a healthy and Eco-friendly Agri-Food System through the promotion of MEDIET and its lifestyle in modern society) (PCI2023-143402) co-financed by the State Research Agency—AEI—, the Ministry of Science and Innovation and the European Union. Marta Cuenca holds a research staff contract in the aforementioned project (CPI-23-620).

References

Antolak, H., Piechota, D., & Kucharska, A. (2021). Kombucha tea—A double power of bioactive compounds from tea and symbiotic culture of bacteria and yeasts (SCOBY). *Antioxidants, 10*(10), 1541. https://doi.org/10.3390/antiox10101541

Baruah, R., Ray, M., & Halami, P. M. (2022). Preventive and therapeutic aspects of fermented foods. *Journal of Applied Microbiology, 132*(5), 3476–3489. https://doi.org/10.1111/jam.15444

Calderón Cuadrado, S. I. (2020). *Compuestos fenólicos y capacidad antioxidante de la pasta casera de cúrcuma, jengibre, ajos, pimienta negra y jugo de limón* [Undergraduate thesis in Nutrition]. Universidad César Vallejo.

Cuenca-Ortolá, M., Alegría, A., & Cilla, A. (2023). Valencian paella: Synergistic antioxidant properties of a complete dish versus its isolated ingredients. *Biology and Life Sciences Forum, 26*(1), 55. https://doi.org/10.3390/foods2023-15057

Diez-Ozaeta, I., & Astiazaran, O. J. (2022). Fermented foods: An update on evidence-based health benefits and future perspectives. *Food Research International, 156*, 111133. https://doi.org/10.1016/j.foodres.2022.111133

Durazzo, A., Lisciani, S., Camilli, E., Gabrielli, P., Marconi, S., Gambelli, L., Aguzzi, A., Lucarini, M., Maiani, G., Casale, G., & Marletta, L. (2017). Nutritional composition and antioxidant properties of traditional Italian dishes. *Food Chemistry, 218*, 70–77. https://doi.org/10.1016/j.foodchem.2016.08.120

Durazzo, A., Lucarini, M., Santini, A., Camilli, E., Gabrielli, P., Marconi, S., Lisciani, S., Aguzzi, A., Gambelli, L., Novellino, E., & Marletta, L. (2019). Antioxidant properties of four commonly consumed popular Italian dishes. *Molecules, 24*(8), 1543. https://doi.org/10.3390/molecules24081543

Emerald, M., Rajauria, G., & Kumar, V. (2016). Novel fermented grain-based products. *Food Engineering Series, 12*, 12263–12277. https://doi.org/10.1007/978-3-319-42457-6_12

Erskine, E., Ozkan, G., Lu, B., & Capanoglu, E. (2023). Effects of fermentation process on the antioxidant capacity of fruit byproducts. *ACS Omega, 8*(5), 4543–4553. https://doi.org/10.1021/acsomega.2c07602

Guo, Q., Chen, P., & Chen, X. (2023). Bioactive peptides derived from fermented foods: Preparation and biological activities. *Journal of Functional Foods, 101*, 105422. https://doi.org/10.1016/j.jff.2023.105422

Hartzfeld, P. W., Forkner, R., Hunter, M. D., & Hagerman, A. E. (2002). Determination of hydrolyzable tannins (gallotannins and ellagitannins) after reaction with potassium iodate. *Journal of Agricultural and Food Chemistry, 50*(7), 1785–1790. https://doi.org/10.1021/jf0111155

Hsieh, C. Y., Hsieh, S. L., Ciou, J. Y., Huang, Y. W., Leang, J. Y., Chen, M. H., & Hou, C. Y. (2021). Lemon juice bioactivity in vitro increased with lactic acid fermentation. *International Journal of Food Properties, 24*(1), 28–40. https://doi.org/10.1080/10942912.2020.1861008

Kim, H., Hur, S., Lim, J., Jin, K., Yang, T., Keehm, I., Kim, S. W., Kim, T., & Kim, D. (2023). Enhancement of the phenolic compounds and antioxidant activities of Kombucha prepared using specific bacterial and yeast. *Food Bioscience, 56*, 103431. https://doi.org/10.1016/j.fbio.2023.103431

Kluz, M. I., Pietrzyk, K., Pastuszczak, M., Kacaniova, M., Kita, A., Kapusta, I., Zaguła, G., Zagrobelna, E., Struś, K., Marciniak-Lukasiak, K., Stanek-Tarkowska, J., Timar, A. V., & Puchalski, C. (2022). Microbiological and physicochemical composition of various types of homemade kombucha beverages using alternative kinds of sugars. *Foods, 11*(10), 1523. https://doi.org/10.3390/FOODS11101523

Lang, J. M., Eisen, J. A., & Zivkovic, A. M. (2014). The microbes we eat: Abundance and taxonomy of microbes consumed in a day's worth of meals for three diet types. *PeerJ, 2014*(2), e659. https://doi.org/10.7717/peerj.659

Lazarte, C. E., Vargas, M., & Granfeldt, Y. (2015). Zinc bioavailability in rats fed a plant-based diet: A study of fermentation and zinc supplementation. *Food and Nutrition Research, 59*, 27796. https://doi.org/10.3402/fnr.v59.27796

Lemos Junior, W. J. F., Guerra, A. F., Tarrah, A., da Silva Duarte, V., Giacomini, A., Luchese, R. H., & Corich, V. (2020). Safety and stability of two potentially probiotic lactobacillus strains after in vitro gastrointestinal transit. *Probiotics and Antimicrobial Proteins, 12*(2), 657–666. https://doi.org/10.1007/s12602-019-09565-2

Leonard, W., Zhang, P., Ying, D., Adhikari, B., & Fang, Z. (2021). Fermentation transforms the phenolic profiles and bioactivities of plant-based foods. *Biotechnology Advances, 49*, 107763. https://doi.org/10.1016/j.biotechadv.2021.107763

Lim, E. J., & Abd Rahim, M. H. (2025). Fermentation for enhanced biosynthesis and bioavailability of micronutrients, prebiotics, bioactive peptides and functional fatty acids in food products. *International Journal of Advanced Research in Food Science and Agriculture Technology, 4*(1), 9–27. https://doi.org/10.37934/fsat.4.1.927a

Liu, L., Li, G., Cui, L., Cai, R., Yuan, Y., Gao, Z., Yue, T., & Wang, Z. (2024). The health benefits of fermented fruits and vegetables and their underlying mechanisms. *Comprehensive Reviews in Food Science and Food Safety, 23*(6), e70072. https://doi.org/10.1111/1541-4337.70072

Marco, M. L., Heeney, D., Binda, S., Cifelli, C. J., Cotter, P. D., Foligné, B., Gänzle, M., Kort, R., Pasin, G., Pihlanto, A., Smid, E. J., & Hutkins, R. (2017). Health benefits of fermented foods: Microbiota and beyond. *Current Opinion in Biotechnology, 44*, 94–102. https://doi.org/10.1016/j.copbio.2016.11.010

Maurice, B., Saint-Eve, A., Pernin, A., Leroy, P., & Souchon, I. (2022). How different are industrial, artisanal and homemade soft breads? *Foods, 11*(10), 1484. https://doi.org/10.3390/foods11101484

Mohamed, E. (2020). The ancient Egyptian bread and fermentation. *Microbial Biosystems, 5*(1), 52–53. https://doi.org/10.21608/MB.2020.99522

Najman, K., Sadowska, A., & Hallmann, E. (2020). Influence of thermal processing on the bioactive, antioxidant and physicochemical properties of conventional and organic agriculture black garlic (*Allium sativum* l.). *Applied Sciences, 10*(23), 1–17. https://doi.org/10.3390/app10238638

Ou, B., Hampsch-Woodill, M., & Prior, R. L. (2001). Development and validation of an improved oxygen radical absorbance capacity assay using fluorescein as the fluorescent probe. *Journal of Agricultural and Food Chemistry, 49*(10), 4619–4626. https://doi.org/10.1021/jf010586o

Phenol-Explorer. (2025). *Phenol-Explorer*. Retrieved October 15, 2024, from http://phenol-explorer.eu/

Razola-Díaz, M. d. C., De Montijo-Prieto, S., Guerra-Hernández, E. J., Jiménez-Valera, M., Ruiz-Bravo, A., Gómez-Caravaca, A. M., & Verardo, V. (2024). Fermentation of orange peels by lactic acid bacteria: Impact on phenolic composition and antioxidant activity. *Foods, 13*(8), 1212. https://doi.org/10.3390/foods13081212

Re, R., Pellegrini, N., Proteggente, A., Pannala, A., Yang, M., & Rice-Evans, C. (1999). Antioxidant activity applying an improved ABTS radical cation decolorization assay. *Free Radical Biology & Medicine, 26*, 1231–1237.

Reyes-Jurado, F., Soto-Reyes, N., Dávila-Rodríguez, M., Lorenzo-Leal, A. C., Jiménez-Munguía, M. T., Mani-López, E., & López-Malo, A. (2023). Plant-based milk alternatives: Types, processes, benefits and characteristics. *Food Reviews International, 39*(4), 2320–2351. https://doi.org/10.1080/87559129.2021.1952421

Septembre-Malaterre, A., Remize, F., & Poucheret, P. (2018). Fruits and vegetables, as a source of nutritional compounds and phytochemicals: Changes in bioactive compounds during lactic fermentation. *Food Research International, 104*, 86–99. https://doi.org/10.1016/j.foodres.2017.09.031

Singleton, V. L., & Rossi, J. A. (1965). Colorymetry of total phenolics with phosphomolybdic phosphotungstic acid reagents. *American Society for Enology and Viticulture, 16*, 144–158. https://doi.org/10.5344/ajev.1965.16.3.144

Stoenescu, A.-M., & Stănică, F. (2025). Comparative characterization of phenolic profile and antioxidant activity in homemade and commercial jujube vinegar. *Beverages, 11*(3), 83. https://doi.org/10.3390/beverages11030083
Sun, T.-Y., Li, J.-S., & Chen, C. (2015). Effects of blending wheatgrass juice on enhancing phenolic compounds and antioxidant activities of traditional kombucha beverage. *Journal of Food and Drug Analysis, 23*(4), 709–718. https://doi.org/10.1016/j.jfda.2015.01.009
Tahir, Z., Saeed, F., Nosheen, F., Ahmed, A., & Anjum, F. M. (2022). Comparative study of nutritional properties and antioxidant activity of raw and fermented (black) garlic. *International Journal of Food Properties, 25*(1), 116–127. https://doi.org/10.1080/10942912.2022.2026954
USDA (U.S. Department of Agriculture), Nutrient Data Laboratory, Beltsville Human Nutrition Research Center, Agricultural Research Service. (2010). *Oxygen radical absorbance capacity (ORAC) of selected foods, release 2*. U.S. Department of Agriculture.
Verni, M., Verardo, V., & Rizzello, C. G. (2019). How fermentation affects the antioxidant properties of cereals and legumes. *Foods, 8*(9), 362. https://doi.org/10.3390/foods8090362
Vitali, M., Gandía, M., Garcia-Llatas, G., González-Sarrias, A., Vallejo, F., Cilla, A., & Gamero, A. (2025a). Modulation of antioxidant capacity, nutritional composition, probiotic viability after digestion and sensory attributes of planted-based beverages through lactic acid fermentation. *Foods, 14*, 1447. https://doi.org/10.3390/foods14091447
Vitali, M., Gandía, M., Garcia-Llatas, G., González-Sarrias, A., Vallejo, F., Cilla, A., & Gamero, A. (2025b). Impact of fermentation and *in vitro* gastrointestinal digestion on the bioaccessibility of phytochemicals and antioxidant capacity of tiger nut, carob and rive beverages. *Food Chemistry, 495*, 146562. https://doi.org/10.1016/j.foodchem.2025.146562
Vitali, M., Makran, M., Gandía, M., Cilla, A., & Gamero, A. (2025c). Differential modulation of cancer cell proliferation by fermented plant-based beverages: A comparative study of tiger nut, carob and rice beverages in colorectal adenocarcinoma cells. *Foods, 14*(17), 3072. https://doi.org/10.3390/foods14173072
Wei, L., Van Beeck, W., Hanlon, M., Dicaprio, E., & Marco, M. L. (2025). Lacto-fermented fruits and vegetables: Bioactive components and effects on human health. *Annual Review of Food Science and Technology, 16*, 289–314. https://doi.org/10.1146/annurev-food-052924
Ye, Z., Sun, S., Chen, H. Y., Ji, C., & Zhou, Z. (2025). Metabolomic and bioactivity analysis of plum pulp fermented with *Lactiplantibacillus*, *Lacticaseibacillus* and *Bifidobacterium* strains. *Food Chemistry: X, 30*, 102969. https://doi.org/10.1016/j.fochx.2025.102969
Zhao, Y. S., Eweys, A. S., Zhang, J. Y., Zhu, Y., Bai, J., Darwesh, O. M., Zhang, H. B., & Xiao, X. (2021). Fermentation affects the antioxidant activity of plant-based food material through the release and production of bioactive components. *Antioxidants, 10*(12), 2004. https://doi.org/10.3390/antiox10122004
Zheng, Z., Wei, L., Zhu, M., Qian, Z., Liu, J., Zhang, L., & Xu, Y. (2023). Effect of lactic acid bacteria co-fermentation on antioxidant activity and metabolomic profiles of a juice made from wolfberry and longan. *Food Research International, 174*, 113547. https://doi.org/10.1016/j.foodres.2023.113547
Zhou, T., Zhang, Y., Xu, D., Wang, F., Zhou, Y., Zheng, J., Li, Y., Zhang, J., & Li, H. (2017). Protective effects of lemon juice on alcohol-induced liver injury in mice. *BioMed Research International, 2017*, 7463571. https://doi.org/10.1155/2017/7463571
Zhou, D.-D., Saimaiti, A., Luo, M., Huang, S.-Y., Xiong, R.-G., Shang, A., Gan, R.-Y., & Li, H.-B. (2022). Fermentation with tea residues enhances antioxidant activities and polyphenol contents in kombucha beverages. *Antioxidants, 11*(1), 155. https://doi.org/10.3390/antiox11010155

Chapter 6
Fermented Foods and the Gut–Brain Axis: Plant-Based Fermentation, Mental Health, and Cognitive Function

Titilayo Adenike Ajayeoba, Ibukunoluwa Adenike Akintunde, Opeyemi Titilayo Lala, Eunice Damilola Wilkie, and Christianah Oluwayomi Olaoye

Abstract Plant-based fermentation represents a dynamic biological process that transforms food substrates into bioactive agents capable of modulating the gut–brain axis. This bidirectional system integrates gastrointestinal activity with the central nervous system through interconnected neural, endocrine, and immune pathways. Fermented plant foods, such as kimchi, tempeh, and indigenous varieties, function as natural synbiotics, delivering viable microorganisms and metabolites—including short-chain fatty acids (SCFAs), gamma-aminobutyric acid (GABA), and transformed polyphenols. These compounds influence neuro-inflammation, synaptic plasticity, and stress responsiveness by interacting with the vagus nerve and systemic signaling routes. While preclinical models provide robust evidence for the role of microbial metabolites in enhancing cognitive performance and attenuating anxiety-like behaviors, human evidence remains predominantly associative. This synthesis bridges traditional dietary knowledge with contemporary neuroscience, situating fermented foods within a "neuro-ecology" framework where their benefits are shaped by broader dietary patterns and lifestyle synergies. Future research must prioritize rigorous, longitudinal trials to clarify causal relationships and address the translational gap in human clinical outcomes.

Keywords Cognitive function · Gut–brain axis · Mental health · Microbial metabolites · Neuroplasticity · Plant-based fermentation

---

T. A. Ajayeoba (✉) · I. A. Akintunde · O. T. Lala
Food Science and Nutrition Program, Faculty of Science, Adeleke University, Ede, Osun, Nigeria
e-mail: ajayeoba.titilayo@adelekeuniversity.edu.ng

E. D. Wilkie
Department of Microbiology, Faculty of Science, Adeleke University, Ede, Osun, Nigeria

C. O. Olaoye
Department of Animal Nutrition, College of Animal Science and Livestock Production, Federal University of Agriculture, Abeokuta, Nigeria

S. Dhiman, S. Kaur (eds.), *Next-Gen Plant Based Fermented Foods: Innovations for a Sustainable Future*, https://doi.org/10.1007/978-3-032-29927-7_6

6.1 Introduction

Across human history, fermented plant foods have played a central role in everyday diets, long preceding scientific understanding of their biological and physiological relevance. In East Asia, kimchi developed as both a culinary tradition and a practical strategy for preserving vegetables through controlled microbial transformation. In Southeast Asia, tempeh became a protein-rich staple produced through the fermentation of soybeans. Similarly, in many agrarian societies, sourdough bread relied on wild yeasts and lactic acid bacteria to transform diverse grains into foods with enhanced digestibility and storage stability (Ashaolu & Reale, 2020; Lys, 2025). Although these practices originated from indigenous and experience-based knowledge systems, sustained empirical observation and cultural transmission reflect adaptive and functional relationships between human populations and microbial life.

In recent decades, modern science has reexamined traditionally fermented foods through the lenses of microbiology, nutrition, and neuroscience. Foods once categorized as artisanal or culturally specific are now investigated in controlled experimental settings. Their microbial communities, fermentation-derived metabolites, and physiological effects are systematically characterized. This shift from conventional practice to laboratory-based analysis represents a broader conceptual transition: fermentation is increasingly understood as a dynamic biological process with potential implications for systemic health, including brain function and mental well-being (Zhang et al., 2025).

Plant-based fermentation extends beyond the simple cultivation or ingestion of beneficial microorganisms. It is an active and dynamic process in which microbial proliferation and metabolite production transform food substrates and influence their interactions with human physiology. During fermentation, microorganisms metabolize dietary fibers, polyphenols, and other bioactive compounds that may not be readily accessible to the human digestive system. These microbial activities modify nutrient bioavailability and influence host metabolic signaling pathways, thereby reshaping the functional relationship between diet and physiology. Within this framework, fermented plant foods function not merely as passive nutritional inputs but as active components within interconnected biological systems (Zhang et al., 2025; Kasarello et al., 2023).

6.1.1 Research Gap and Objective

Despite growing interest in fermented foods and the gut–brain axis, significant gaps remain in understanding how plant-based fermentation specifically influences cognitive and mental health outcomes. Much of the current literature focuses either on isolated probiotic strains or on general dietary patterns, leaving the complex, food-matrix–dependent effects of traditionally fermented plant foods insufficiently explored. Moreover, the integration of cultural dietary knowledge with mechanistic models of microbial ecology and neurobiology remains limited.

The objective of this chapter is to bridge this gap by synthesizing traditional dietary practices with contemporary scientific evidence. It aims to establish a conceptual and empirical framework for understanding fermented plant foods as biologically active agents at the intersection of nutrition, microbial ecology, and mental health. In doing so, the chapter situates fermented plant foods within emerging models of gut–brain interaction and cognitive well-being.

6.1.2 How the Gut Talks to the Brain

Recent research on fermented foods has reinforced the significance of the gut–brain axis as a bidirectional and integrative system linking the gastrointestinal tract and the central nervous system. This axis operates through interconnected neural, endocrine, immune, and microbial pathways, through which gut activity can influence brain function, while psychological states modulate gastrointestinal physiology (Kasarello et al., 2023; Park et al., 2025). Together, these mechanisms reflect a dynamic and interdependent neurobiological network.

Neural communication is mediated primarily by the vagus nerve, which transmits sensory information from the gut to the brain and coordinates autonomic responses related to digestion, stress reactivity, and emotional regulation (Xing et al., 2026). Endocrine and immune pathways further extend this communication through gut-derived hormones, neuroactive compounds, and immune mediators that influence appetite, mood, inflammation, and cognitive processes (Kasarello et al., 2023; Pires et al., 2024). Within this integrated system, the gut microbiota serves as a central regulatory component by producing metabolites that interact with neural, endocrine, and immune signaling pathways, thereby shaping behavioral and cognitive outcomes.

Plant-based fermented foods are relevant within this framework because they provide viable microorganisms and fermentation-derived metabolites capable of modulating the composition and functional activity of the gut microbiota. Their metabolic byproducts, including short-chain fatty acids and transformed polyphenols, are implicated in pathways associated with neuro-inflammation, stress responsiveness, and synaptic plasticity (Dhiman et al., 2025). Through these mechanisms, fermented foods may influence processes linked to brain health.

Importantly, the gut–brain axis does not reduce mental health outcomes to a single nutrient, microorganism, or food source. Rather, it emphasizes the complex interplay among diet, microbial ecology, host metabolism, genetic predisposition, lifestyle factors, and environmental influences (Patil & Mehdi, 2025). Within this multifactorial model, fermented foods represent one potential dietary component that may contribute to mental well-being through interacting biological pathways.

6.2 Biological Pathways of Communication

Communication along the gut–brain axis is mediated through multiple interconnected biological pathways, each contributing distinct yet overlapping signals that influence brain function and behavior. Sensory inputs arising from gut motility, immune activation, and microbial activity are transmitted primarily via the vagus nerve to brain regions involved in emotional regulation, stress responsiveness, and cognitive processing. Alterations in gut microbiota composition can modulate the intensity and quality of these signals, underscoring the integrative role of gut physiology in shaping central nervous system activity within the gut–brain axis (Kasarello et al., 2023; Xing et al., 2026). Figure 6.1 illustrates gut–brain communication through the vagus nerve as a central neural conduit linking the gastrointestinal tract to the brain.

Table 6.1 also shows that gut–brain communication is not a simple, one-way process. Instead, it operates through a distributed, multi-layered network of overlapping metabolic, immune, endocrine, and neural pathways. Fermented plant foods are influence several of these pathways by influencing microbial communities, their metabolic byproducts, and neurochemical signaling outside the brain, thus, the evidence highlights the importance of understanding diet–brain relationships within broader biological and lifestyle contexts, rather than as direct cause-and-effect links.

6.2.1 Microbiota as a Cognitive and Emotional Modulator

Recent reports suggest that the gut microbiota plays a modulatory role in emotional regulation and cognitive performance. Greater microbial diversity has been associated with indicators of psychological resilience, including lower perceived stress, more favorable mood profiles, and improved overall functioning. Experimental studies in animal models further support this relationship, demonstrating that alterations in microbial composition can induce changes in anxiety-like behavior, stress responsivity, and cognitive performance (Homer et al., 2023).

Emerging evidence from human studies also indicates associations among diet, gut microbiome composition, and mental health outcomes. Diets rich in dietary fiber and fermented foods have been linked to improved emotional well-being and cognitive function. These associations are often accompanied by increased microbial diversity and enhanced metabolic activity within the gut, suggesting that the microbiota may mediate, at least in part, the translation of dietary inputs into psychological and cognitive effects (Patil & Mehdi, 2025; Dhiman et al., 2025; Berthoud et al., 2021).

However, the distinction between correlation and causation remains critical. Many human studies are observational and therefore cannot determine whether microbiome alterations directly contribute to improvements in mental health or merely reflect broader lifestyle patterns. Factors such as overall diet quality, physical

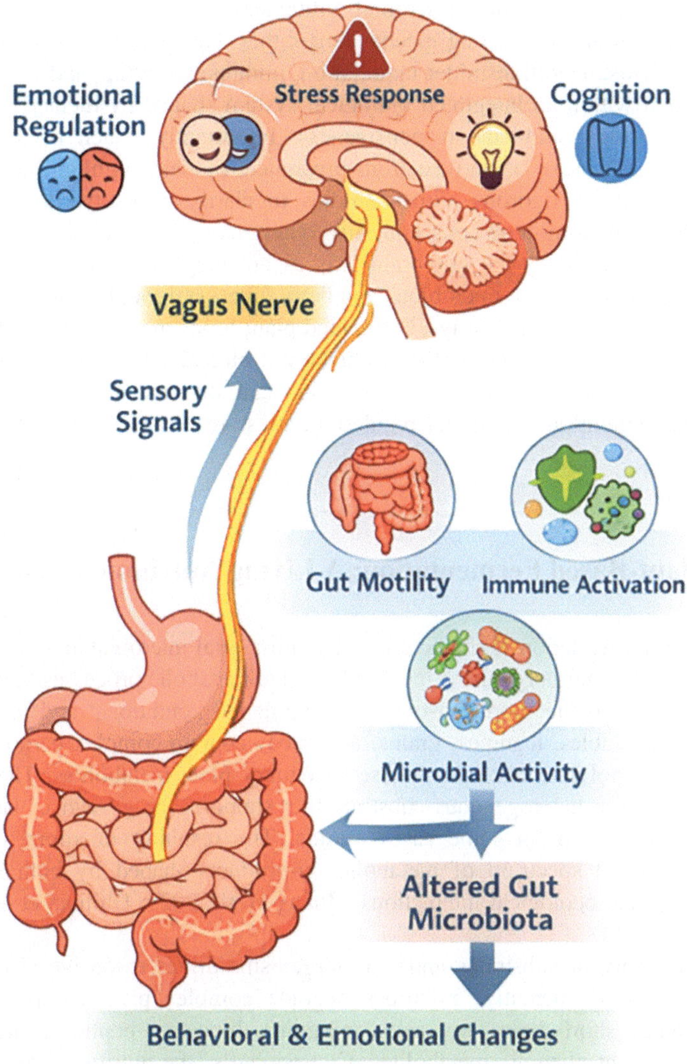

Fig. 6.1 Vagus nerve–mediated gut–brain communication: neural integration of microbial, immune, and gastrointestinal signals

activity, sleep behavior, and socioeconomic status may simultaneously influence both microbial composition and psychological outcomes. Even within intervention studies, multiple variables often change concurrently, limiting the ability to isolate the specific effects of dietary components (Sood et al., 2025). Furthermore, substantial interindividual variability in baseline microbiota composition may result in heterogeneous emotional and cognitive responses to the same dietary intervention (Magzal et al., 2023). Collectively, these considerations underscore the need for rigorously controlled, mechanistic human studies to clarify causal pathways and individual responsiveness.

Table 6.1 illustrates that microbial influences on neurophysiology operate through layered and interacting mechanisms rather than isolated pathways. Short-chain fatty acids, neurotransmitter modulation, immune signaling, and vagal activation function within an integrated biological system shaped by dietary substrates, microbial ecology, and host responsiveness. Importantly, much of the current evidence is associative or derived from preclinical models, highlighting the need for mechanistically rigorous human studies to clarify causality, dose–response relationships, and long-term neurocognitive outcome.

Figure 6.2 illustrates the gut–brain axis as an integrated network connecting fermented plant foods, gut microbiota, and brain function through neural, metabolic, immune, and endocrine pathways. Fermented plant foods modulate microbial composition and activity, leading to the production of bioactive metabolites and neuroactive compounds. These signals converge via vagal, immune, and metabolic routes to influence cognition, emotional regulation, stress responses, and overall mental health.

6.3 Plant-Based Fermentation: A Living Nutritional System

Plant-based fermentation is characterized by substantial microbial and biochemical diversity, as the microorganisms involved and the metabolic processes they initiate vary according to the food substrate, environmental conditions, and preparation methods. Vegetables, legumes, grains, and seeds contain complex carbohydrates, dietary fibers, polyphenols, and diverse secondary plant metabolites. These compounds provide a heterogeneous substrate that supports the growth of lactic acid bacteria, yeasts, and, in some cases, molds. The resulting microbial consortia generate a broad spectrum of metabolic by-products shaped by both substrate composition and ecological interactions (Zhang et al., 2025; Dhiman et al., 2025; Guo et al., 2025).

This diversity of substrates and microorganisms drives extensive biochemical transformation. Fermentative microbes degrade complex plant compounds into organic acids, short-chain fatty acids, vitamins, bioactive peptides, and lower-molecular-weight phenolic metabolites. Consequently, the nutritional and physiological value of plant-based fermented foods resides not only in their original composition but also in the metabolites produced through microbial activity (Lim & Abd Rahim, 2025). For example, fermentation—followed by colonic metabolism—can convert poorly absorbed, fiber-bound polyphenols into smaller, more bioavailable compounds through enzymatic biotransformation. These metabolites may exhibit greater systemic persistence and can modulate immune, metabolic, inflammatory, and neural signaling pathways at physiologically relevant concentrations.

Through these mechanisms, fermented plant foods generate bioactive compounds that contribute to gut barrier integrity, immune homeostasis, metabolic regulation, and neurophysiological function. Via host–microbe signaling pathways, these

Table 6.1 Microbial metabolites and neurotransmitter modulation in the gut–brain axis: sources, pathways, and neurophysiological relevance

Category	Microbial metabolite/neuroactive factor	Primary dietary precursors	Representative microbial sources	Primary biological targets	Neurophysiological and mental health relevance	Key limitations and analytical considerations	References
Microbial metabolites	Acetate (SCFA)	Dietary fibers; resistant starches	*Bifidobacterium* spp.; *Akkermansia* spp.	Intestinal epithelium; immune cells; peripheral metabolic tissues	Supports gut barrier integrity and modulates systemic inflammation implicated in mood and cognitive regulation	Effects on the brain are largely indirect and mediated through immune and metabolic pathways rather than direct central action	La Torre et al. (2021), Guo et al. (2025)
	Propionate (SCFA)	Dietary fibers	*Bacteroides* spp.; *Veillonella* spp.	Immune cells; hepatic metabolism	Regulates inflammatory tone and metabolic processes associated with cognitive energy balance	Demonstrates dose-dependent and context-specific effects; excessive levels may exert divergent outcomes	Rekha et al. (2024)
	Butyrate (SCFA)	Fermentable fibers	*Faecalibacterium prausnitzii*; *Roseburia* spp.	Colonocytes; microglia (indirectly)	Exhibits anti-inflammatory properties, epigenetic modulation (e.g., HDAC inhibition), and associations with neuroprotective pathways	Limited direct penetration of the central nervous system; mechanistic evidence in humans remains limited	De Cassia Alpino et al. (2024)
	Polyphenol-derived metabolites	Polyphenol-rich plant foods (vegetables, legumes, grains)	Diverse fermentative consortia	Antioxidant pathways; immune signaling cascades	Reduce oxidative stress and may support synaptic plasticity and neuroadaptive environments	High interindividual variability in microbial metabolism complicates reproducibility and causal attribution	Kumar Singh et al. (2019), Fiore et al. (2025)
		Dietary tryptophan			Influence serotonin biosynthesis and	Multiple competing metabolic pathways	Lin et al. (2022)

(continued)

Table 6.1 (continued)

Category	Microbial metabolite/ neuroactive factor	Primary dietary precursors	Representative microbial sources	Primary biological targets	Neurophysiological and mental health relevance	Key limitations and analytical considerations	References
	Tryptophan-derived metabolites		*Lactobacillus* spp.; *Clostridium* spp.	Enteroendocrine cells; immune cells	immune–brain communication	(e.g., kynurenine route) limit mechanistic specificity	
Neurotransmitter modulation	Serotonin (5-HT) regulation	Tryptophan	Indirect microbial modulation	Enteroendocrine cells; vagal afferents	Modulates mood, stress responsiveness, and cognitive flexibility	Peripheral serotonin does not cross the blood–brain barrier; central effects are indirect	Margoob et al. (2024)
	Gamma-aminobutyric acid (GABA)	Glutamate; fermented substrates	*Lactobacillus* spp.; *Bifidobacterium* spp.	Enteric nervous system; vagal pathways	Associated with anxiolytic signaling and stress attenuation	Evidence remains predominantly preclinical; translational validity requires further confirmation	Kurhaluk et al. (2025)
	Dopaminergic pathway modulation	Tyrosine; phenylalanine	Microbial community-level signaling effects	Mesolimbic and reward-related circuits	Potential influence on motivation and reward-related behaviors	Mechanistic pathways remain insufficiently defined and largely correlative	Hamamah et al. (2022)
Integrative signaling mechanisms	Immune-mediated signaling	Diet–microbe interactions	Community-level microbial dynamics	Cytokines; systemic inflammatory mediators	Provides a mechanistic link between gut microbial activity and affective or cognitive states	Highly context-dependent; influenced by host genetics, environment, and baseline inflammatory status	Agirman et al. (2021), Rusch et al. (2023)
	Vagal nerve activation	Microbial metabolites; gut sensory activity	Microbiota–host interface	Brainstem nuclei; limbic structures	Enables rapid gut-to-brain signaling affecting emotional and cognitive processing	Not all microbial effects are vagally mediated; relative contribution of vagal versus humoral routes remains under investigation	Berthoud et al. (2021)

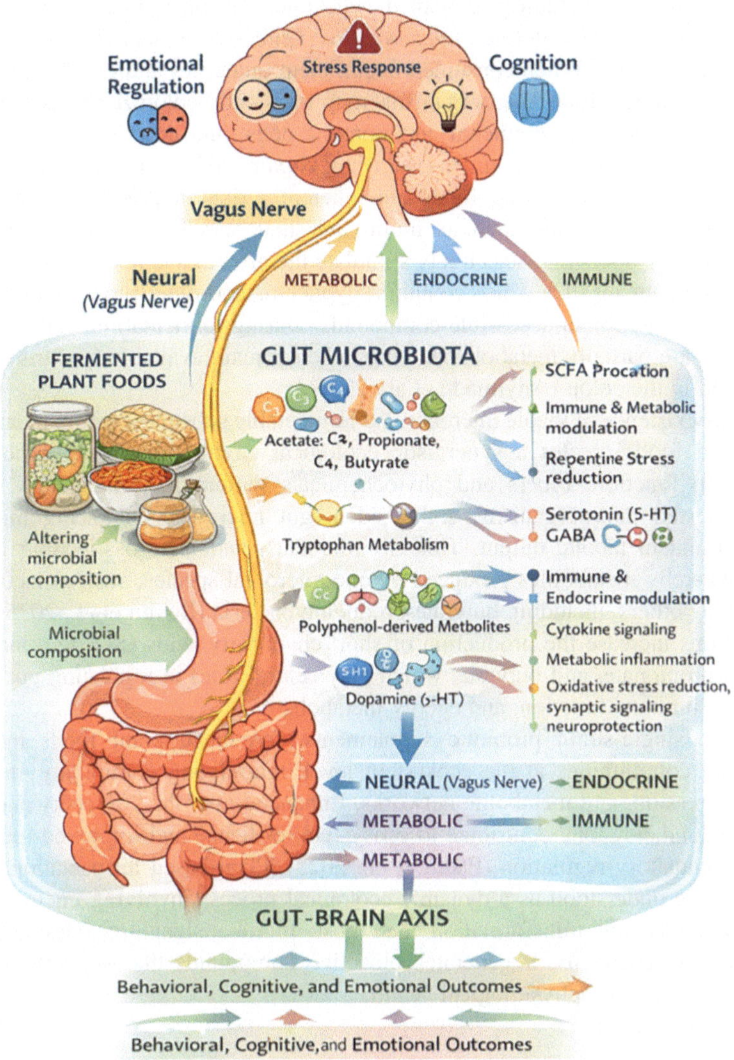

Fig. 6.2 Integrated gut–brain axis pathways linking fermented plant foods, microbial metabolism, and brain function

metabolites may influence inflammatory tone, energy metabolism, stress responsiveness, cognitive processes, and mood. Owing to their compositional characteristics, plant-based fermented foods are often described naturally as "synbiotics," as they combine elements of both probiotics and prebiotics within a single food matrix (Dahiya & Nigam, 2022).

Probiotics are defined as live microorganisms that, when administered in adequate amounts, confer a health benefit on the host. Many traditional plant-based fermented foods—including fermented vegetables, legumes, and grains—contain viable lactic acid bacteria (e.g., *Lactobacillus*, *Leuconostoc*, and *Pediococcus*) and occasionally yeasts. When consumed in sufficient quantities and capable of surviving gastric transit, these microorganisms may transiently colonize the gastrointestinal tract, interact with epithelial and immune cells, and contribute to metabolic activity within the intestinal lumen (Dahiya & Nigam, 2022; Casertano et al., 2022).

Concurrently, plant-based foods are inherently rich in prebiotics, defined as non-digestible substrates such as inulin, resistant starch, oligosaccharides, and diverse dietary fibers that selectively stimulate the growth and activity of beneficial gut microorganisms. They also contain polyphenols and other phytochemicals that function as microbiota-accessible compounds. During fermentation, some of these substrates are partially metabolized, while others remain available for further transformation in the colon (Anyimadu et al., 2026).

The coexistence of viable microbes and fermentable substrates within plant-based fermented foods creates a synergistic ecological environment that supports gut ecosystem function. Fibers and phytochemicals within the food matrix sustain both ingested microorganisms and resident gut microbes, enhancing microbial survival and metabolic output. This environment promotes cross-feeding interactions, whereby metabolites produced by one microbial species, such as lactate, are utilized by others, including butyrate-producing bacteria (Boven et al., 2025). Such interactions increase the production of short-chain fatty acids (SCFAs), including acetate, propionate, and butyrate, which play central roles in regulating gut barrier integrity, immune function, and energy metabolism.

Unlike single-strain probiotic supplements, fermented plant foods influence microbial communities at the ecosystem level, modulating community structure, gene expression, and metabolite networks. Enhanced microbial diversity and functional redundancy may contribute to ecosystem resilience and improved metabolic and inflammatory regulation (Padhi et al., 2025). Plant-based fermentation should therefore be understood as a dynamic ecological process involving microbial succession, enzymatic transformation, and host–microbe signaling. Through these integrated mechanisms, it generates bioactive compounds that support immune, metabolic, and neurophysiological health.

6.4 Microbial Metabolites, Neurochemistry, and Mental Health

The relationship between fermented plant foods and mental health is hypothesized to operate primarily through modulation of the gut microbiome rather than through direct psychoactive effects on the central nervous system. In contrast to pharmacological agents such as antidepressants or stimulants, which act directly on neural receptors or neurotransmitter systems, fermentation-derived compounds are thought to influence emotional and cognitive processes indirectly by altering immune activity, metabolic signaling, and gut–brain communication pathways. This distinction is important when interpreting mechanistic and clinical evidence, as most effects appear to be mediated through systemic and regulatory processes rather than acute neurochemical action.

6.4.1 Fermentation-Derived Compounds and Brain Function

Short-chain fatty acids (SCFAs), including acetate, propionate, and butyrate, are among the most extensively studied microbial metabolites relevant to gut–brain signaling. Produced through the fermentation of dietary fibers, SCFAs contribute to intestinal barrier integrity and the modulation of systemic inflammation—both processes implicated in neuropsychiatric vulnerability (Silva et al., 2020). Preclinical studies demonstrate that butyrate, in particular, can influence microglial activation and epigenetic pathways associated with depressive-like behaviors and cognitive decline. However, translation to human populations remains limited. Although small amounts of SCFAs may cross the blood–brain barrier, most proposed effects are indirect and mediated through immune and metabolic regulation rather than direct central activity. Clinical studies examining SCFA concentrations in relation to mood outcomes have produced heterogeneous results, and causal pathways remain incompletely defined.

Fermentation also enhances the bioavailability of polyphenols. Many plant polyphenols are poorly absorbed in their native forms; microbial metabolism converts them into smaller, bioactive metabolites with greater systemic availability. Preclinical research suggests that these compounds exert antioxidant and anti-inflammatory effects and may support synaptic signaling and neuroplasticity (Di Meo et al., 2020). In human studies, however, evidence largely derives from dietary pattern analyses rather than controlled trials isolating fermented sources of polyphenols. Consequently, it is difficult to determine whether observed psychological benefits are attributable specifically to fermentation-derived metabolites or to broader dietary quality.

Gamma-aminobutyric acid (GABA), synthesized by certain lactic acid bacteria during fermentation, has also received attention. Although microbially derived GABA does not readily cross the blood–brain barrier, animal models indicate that it may influence stress-related behaviors through enteric nervous system signaling, vagal pathways, or modulation of stress hormones (Liwinski et al., 2023). Human evidence is more limited and often based on small-scale interventions or surrogate biomarkers, underscoring the need for larger, well-controlled trials to establish clinical relevance.

Overall, preclinical models provide mechanistic plausibility for fermentation-derived metabolites in modulating neurobiological pathways. In contrast, clinical data remain associative or preliminary, highlighting a translational gap between experimental findings and demonstrated mental health outcomes in humans.

6.4.2 Depression, Anxiety, and Stress-Related Disorders

Animal studies consistently report that fermented foods or specific microbial strains can attenuate anxiety- and depressive-like behaviors, often accompanied by measurable shifts in gut microbial composition and metabolite production. These findings support the concept that microbial modulation may influence stress-related neurobiology. However, such models typically involve controlled microbial exposures and standardized environments that differ substantially from the complexity of human dietary patterns.

In human populations, research is predominantly observational. Higher intake of fermented foods has been associated with improved mood, lower perceived stress, and reduced social anxiety, particularly within plant-rich dietary patterns (Sarma et al., 2026; Shawky et al., 2025). Nevertheless, these associations are vulnerable to confounding by overall diet quality, lifestyle factors, and socioeconomic variables. Intervention studies, while growing in number, often involve small sample sizes, short durations, or combined dietary modifications, limiting the ability to attribute effects specifically to fermented plant foods.

Collectively, current evidence suggests that fermented plant foods may contribute to mental health through modulation of interconnected immune, metabolic, and neuroendocrine pathways. However, the strength of evidence differs markedly between preclinical and clinical domains. While mechanistic animal data are robust, human evidence remains suggestive rather than definitive. Rigorous, long-term randomized controlled trials incorporating microbiome, metabolomic, and neurocognitive endpoints are necessary to clarify causality and clinical significance.

6.5 Cognitive Function, Memory, and Neuroplasticity

Interest in the relationship between diet and mental health has led to increasing investigation of fermented plant foods and their potential influence on cognitive function and brain adaptability (Anih et al., 2025). These foods do not directly stimulate neurons or acutely alter brain chemistry. Rather, they are proposed to influence the gut microbiome, which in turn modulates immune signaling, stress hormone regulation, and neural communication through metabolic and vagal pathways. Through these indirect mechanisms, fermented foods may contribute to physiological conditions associated with cognitive resilience, including reduced inflammation, improved metabolic balance, and regulated host–microbe signaling.

6.5.1 Fermented Foods and Cognitive Performance

It has been reported that regular consumption of fermented plant foods may support cognitive performance indirectly by modulating inflammatory tone, metabolic regulation, and microbial diversity. Nevertheless, causal relationships in humans have not been firmly established.

Preclinical studies provide the strongest support for a link between fermented food intake and cognitive outcomes. In animal models, consumption of fermented foods or specific microbial strains has been associated with improvements in learning and memory. These changes are accompanied by shifts in gut microbial composition, reduced neuroinflammation, and increased expression of synaptic plasticity markers, including brain-derived neurotrophic factor (BDNF) (Murciano-Brea et al., 2021). Such findings support mechanistic connections among microbial metabolites, immune modulation, and neural signaling pathways.

Human evidence is more limited and less consistent. Some observational and small-scale clinical studies report associations between fermented food consumption and improved cognitive performance, particularly in older adults (Sarubbo et al., 2022). Proposed mechanisms include reductions in systemic inflammation, improvements in glucose metabolism, and increased microbial diversity—factors independently associated with cognitive health. However, most studies are correlational, and controlled intervention trials remain relatively scarce.

6.5.2 Neuroplasticity and Microbial Metabolites

It has been proposed that regular consumption of SCFA-rich fermented foods, including IFFs, may support prefrontal–hippocampal plasticity through epigenetic and anti-inflammatory mechanisms. Comparisons with Mediterranean or ketogenic dietary interventions suggest possible overlap in BDNF-related pathways, but direct evidence from controlled human studies is insufficient.

The gut–brain axis has been implicated in the modulation of neuroplasticity—the brain's capacity to reorganize and adapt—through peripheral mediators such as short-chain fatty acids (SCFAs) and other microbial metabolites (Murciano-Brea et al., 2021). SCFAs, produced during dietary fiber fermentation, interact with G-protein-coupled receptors (GPR41/43) on enteric neurons and vagal afferents. Experimental models indicate that SCFAs may also cross the blood–brain barrier in small amounts, inhibit histone deacetylases (HDACs), increase BDNF expression, and suppress NF-κB–mediated neuroinflammation. Rodent studies demonstrate SCFA-associated hippocampal neurogenesis and improved cognitive performance (Murciano-Brea et al., 2021; Al Noman et al., 2025).

In contrast, human evidence remains preliminary. Existing randomized trials, often limited to short-term high-fiber interventions (4–12 weeks), report modest increases in circulating BDNF but rely on relatively insensitive cognitive assessments, such as the Montreal Cognitive Assessment (MoCA), which may not detect subtle synaptic adaptations. Furthermore, substantial inter-individual variability—shaped by baseline microbiota diversity, APOE genotype, chronotype, colonic transit time, and mucosal dynamics—can produce marked differences in SCFA production, complicating causal interpretation.

Indigenous fermented foods (IFFs), widely consumed in African and Asian dietary patterns, represent a dietary source of SCFAs. These spontaneously fermented foods can produce measurable concentrations of SCFAs through the activity of lactic acid bacteria and yeasts (Guo et al., 2025; Rekha et al., 2024). Observational studies in Nigerian cohorts link IFF consumption to increased microbial diversity. However, direct evidence connecting IFF intake to measurable neuroplastic changes is currently lacking, largely due to the absence of neuroimaging endpoints and rigorous dietary or genetic controls.

6.5.3 Spontaneous Versus Controlled Fermentation

Spontaneous fermentation, typical of many traditional IFFs, relies on ambient microbiota and can generate variable SCFA profiles and diverse bioactive compounds. Advantages include low cost, cultural integration, and potential broad-spectrum microbial exposure. However, variability in batch composition, potential contamination, and inconsistent SCFA concentrations (5–20% variability) may limit the reproducibility of neuroplasticity-related outcomes.

Controlled fermentation using defined starter cultures (e.g., *Lactobacillus plantarum* and *Saccharomyces cerevisiae*) improves standardization of SCFA production and enhances microbiological safety (Gao et al., 2025). Human trials suggest that controlled fermentations may yield more consistent improvements in mental health outcomes, such as reductions in anxiety symptoms. However, these findings remain preliminary and do not yet establish direct neuroplastic effects.

Controlled fermentation may strengthen causal inference in future studies by producing predictable metabolite profiles suitable for mechanistic investigation of neuroplastic pathways. Nonetheless, potential trade-offs include higher production costs and reduced microbial diversity compared with traditional spontaneous fermentation.

Overall, while preclinical evidence strongly supports a role for fermentation-derived metabolites in neuroplastic processes, clinical data remain limited and heterogeneous. Robust, long-term human trials incorporating neuroimaging, metabolomics, and genetic stratification are required to clarify translational relevance. Table 6.2 systematically maps microbial mechanisms to brain health outcomes, highlighting evidence gaps and intervention pathways.

Table 6.2 Fermented foods, cognitive function, and neuroplasticity: evidence, limitations, and translational implications

Domain	Cognitive or neurobiological focus	Proposed microbial or metabolic pathways	Key evidence base	Critical limitations	Translational implications	References
Cognitive performance	Attention and overall function	Reduced systemic inflammation; SCFA-mediated metabolic regulation; improved cerebral signaling	Observational human studies; animal intervention models	Predominantly correlational data; confounding lifestyle variables	Supports inclusion of fermented plant foods within dietary patterns aimed at maintaining cognitive flexibility, rather than as isolated interventions	Chen et al. (2024), Porras-García et al. (2023)
Cognitive performance	Memory and learning	Enhanced synaptic signaling; microbial modulation of neurotrophic factors (e.g., BDNF)	Animal studies; limited human associative data	Limited human mechanistic evidence; short-term assessments	Suggests potential role in cognitive resilience strategies, particularly when combined with cognitive engagement and physical activity	Gravesteijn et al. (2022)
Aging and cognition	Age-related cognitive decline	Maintenance of microbial diversity; attenuation of neuroinflammation; polyphenol-derived neuroprotection	Animal aging models; observational studies in older adults	Lack of long-term randomized human trials	Indicates possible dietary support for healthy cognitive aging, requiring integration into broader geriatric nutrition frameworks	Nazzaro et al. (2025), Wu et al. (2021)
Neuroplasticity	Synaptic adaptability	Microbial metabolites influencing neurotrophic signaling and immune–brain interactions	Preclinical studies; emerging hypotheses	Translational uncertainty; indirect measures of plasticity	Encourages hypothesis-driven research linking diet, microbiota, and neural adaptability	Basha et al. (2025)
Neuroplasticity	Brain-derived neurotrophic regulation	SCFAs and fermented polyphenol metabolites affecting gene expression	Experimental animal data	Limited direct human evidence;	Informs future development of biomarker-guided nutritional interventions	Yuan et al. (2025)

(continued)

Table 6.2 (continued)

Domain	Cognitive or neurobiological focus	Proposed microbial or metabolic pathways	Key evidence base	Critical limitations	Translational implications	References
				biomarker variability		
Methodological considerations	Cognitive testing sensitivity	Behavioral and neuropsychological assessment tools	Cross-sectional and short-term trials	Insufficient sensitivity to subtle dietary effects	Highlights need for refined cognitive outcome measures in nutrition research	Favieri et al. (2019), Brkić et al. (2026)
Research design	Longitudinal effects	Microbiota–brain dynamics over time	Sparse cohort studies	Lack of long-term, controlled human data	Emphasizes priority for longitudinal and life-course studies to inform clinical translation	Thriene and Michels (2023)

6.6 Fermentation, Plant-Based Diets, and Lifestyle Synergy

Fermented foods are often discussed as though they act independently; however, their potential effects on mental health and cognition cannot be separated from the broader dietary and lifestyle contexts in which they are consumed. Fermented plant foods operate within interconnected biological and behavioral systems shaped by overall diet quality, physical activity, sleep patterns, and stress exposure. An integrative perspective is therefore necessary to understand their contribution within synergistic dietary patterns rather than as isolated interventions.

6.6.1 Dietary Patterns Over Isolated Foods

Health outcomes rarely depend on a single food item. Evidence from nutritional epidemiology and microbiome research consistently indicates that overall dietary patterns exert stronger and more sustained effects on microbial diversity and function. Fermented plant foods appear to be most beneficial when consumed within plant-rich, fiber-dense diets abundant in vegetables, legumes, whole grains, nuts, and seeds (Borrego-Ruiz & Borrego, 2025). Such dietary patterns provide the fermentable substrates required to sustain resilient microbial ecosystems.

Within this framework, fermented foods function as ecological enhancers rather than primary drivers of microbial change. They introduce viable microorganisms and bioactive metabolites into a gut environment already supported by fiber and polyphenols, thereby facilitating microbial integration and metabolic activity associated with immune regulation and neurochemical balance. In contrast, when consumed within low-fiber, highly processed dietary patterns, their effects may be attenuated, underscoring the importance of dietary coherence (Munteanu & Schwartz, 2024).

Lifestyle factors also influence gut–brain dynamics. Sleep disruption has been associated with reduced microbial diversity and increased systemic inflammation, whereas regular physical activity promotes microbial richness and metabolic flexibility. Chronic psychological stress may compromise gut barrier integrity and microbial stability, potentially diminishing dietary benefits. Although fermented plant foods may be incorporated into broader health-promoting routines, their effects should be considered within the context of these interacting behavioral determinants (Patil & Mehdi, 2025).

6.6.2 Personalized Nutrition and Microbial Individuality

While general dietary recommendations provide an evidence-based foundation, substantial inter-individual variability in microbiome composition and functional responsiveness is increasingly recognized. Baseline microbial diversity, genetic background, early-life exposures, and long-term dietary habits all influence individual responses to fermented foods. Consequently, identical dietary interventions may yield heterogeneous microbial and neuropsychological outcomes (Sarkar et al., 2021).

This variability has stimulated interest in precision nutrition. Advances in sequencing technologies and computational modeling enable detailed profiling of microbial communities, raising the possibility of tailoring fermented food recommendations to individual microbiome characteristics. In theory, such approaches could optimize metabolic and neurocognitive responses.

However, the translational application of precision nutrition presents scientific and ethical challenges. Current predictive models remain probabilistic, and causal pathways linking specific microbial taxa to cognitive outcomes are not fully established. Overreliance on individualized microbiome profiling may overstate scientific certainty and divert attention from well-supported population-level principles, such as dietary diversity and adequate fiber intake.

Commercial microbiome testing also raises concerns regarding cost, data privacy, and equitable access. Limited accessibility may exacerbate existing health disparities. Moreover, excessive emphasis on dietary personalization risks framing mental health as solely a matter of individual dietary choice, while overlooking structural determinants such as food availability, socioeconomic stress, and living conditions.

A balanced approach integrates individual variability with shared dietary foundations. Fermented plant foods, commonly incorporated into Nigerian dietary patterns, can be included in everyday meals without exclusivity. When combined with balanced nutrition and supportive lifestyle practices, fermentation may contribute to mental and cognitive health in a culturally relevant and practically accessible manner.

6.7 Ethical, Cultural, and Sustainability Dimensions

As scientific inquiry increasingly examines the potential role of fermented plant foods in mental health and cognitive function, these findings must be situated within broader ethical, cultural, and sustainability frameworks. The translation of laboratory-based discoveries into commercial products and public health recommendations is not value-neutral. Market forces, intellectual property regimes, cultural representation, and ecological constraints influence which foods are promoted, who benefits economically, and who retains access to traditional dietary practices. A critical perspective is therefore necessary to contextualize scientific claims within social and structural realities.

6.7.1 Fermentation in the Era of Mental Health Marketing

Growing interest in the gut–brain axis has stimulated the commercialization of products labeled as "psychobiotics" or "brain-boosting" fermented foods. These include probiotic beverages, fortified yogurts, and dietary supplements marketed as capable of reducing anxiety or enhancing cognitive performance. Although such claims often reference legitimate research on probiotics, short-chain fatty acids (SCFAs), and gamma-aminobutyric acid (GABA)–producing microbes, the evidentiary basis is frequently extrapolated beyond the scope of existing data (Casertano et al., 2022; Chowdhury, 2024).

Across cited studies, a clear discrepancy emerges between mechanistic plausibility and clinical confirmation. Preclinical models demonstrate that specific microbial strains or metabolites can influence stress reactivity and neuroinflammatory pathways. However, human trials remain limited in size, duration, and methodological rigor. Many reported effects are modest, context-dependent, and influenced by baseline microbiota composition and overall diet quality. Consequently, marketing narratives that imply direct or universal mental health benefits risk overstating causal certainty. Such simplifications may generate unrealistic consumer expectations and erode public trust when anticipated outcomes are not achieved.

Furthermore, the commercialization of fermented foods raises concerns regarding cultural appropriation and equity. Many globally marketed fermented products—including kimchi, sauerkraut, tempeh, and injera—originate from longstanding Indigenous and community-based food traditions. These foods were historically developed for preservation, nutritional adequacy, and communal resilience rather than therapeutic branding. Yet, once incorporated into global markets, they are often repackaged as premium functional products, sometimes priced beyond the reach of the communities that developed them. This shift may obscure cultural origins and detach foods from their historical and social contexts (Rani et al., 2025).

Ethically responsible translation of research into practice therefore requires transparent communication of scientific limitations, acknowledgment of cultural heritage, and avoidance of exploitative commercialization. Researchers and industry stakeholders share responsibility in ensuring that claims remain proportionate to evidence and that traditional knowledge systems are appropriately credited.

6.7.2 Fermented Plant Foods, Sustainability, and Brain Health

Beyond commercial considerations, fermentation has important implications for sustainability and food equity. Traditional plant-based fermentation methods typically require minimal external energy inputs, utilize locally available crops, and reduce food waste through preservation. In Nigeria, for example, fermented foods such as ogi, iru, ugba, and wara are produced using accessible techniques that enhance nutrient bioavailability and microbial safety. These practices contribute to food sovereignty by enabling communities to maintain control over production, preparation, and distribution systems (Woodhill et al., 2022).

From a public health perspective, reliable access to affordable and culturally familiar foods may indirectly support psychological well-being by strengthening food security and social cohesion. However, structural constraints—including economic instability, urbanization, limited infrastructure, and shifts toward ultra-processed dietary patterns—can reduce access to traditional fermented foods, potentially exacerbating nutritional and health disparities.

Modern industrial fermentation offers advantages in standardization, safety control, and scalability. Nevertheless, reliance on defined starter cultures and monocultures may reduce microbial diversity compared with spontaneous fermentation systems (Zhang et al., 2025). Given that microbial diversity is frequently associated with ecosystem resilience, this shift warrants critical evaluation. A balanced approach is required—one that enhances quality control and accessibility while preserving microbial heterogeneity and cultural heritage.

Overall, the intersection of fermentation, mental health, and sustainability extends beyond biological mechanisms. Ethical marketing practices, equitable access, preservation of traditional knowledge, and ecologically responsible production systems are integral to ensuring that potential health benefits are realized in socially and scientifically responsible ways.

6.8 Future Directions: Where Fermentation Meets Neuroscience

Advances in analytical technologies are reshaping investigation into the relationship between fermented foods and brain function. Contemporary research increasingly moves beyond examining diet, microbial composition, or neural outcomes in isolation. Instead, multi-omics approaches—integrating genomics, metabolomics, transcriptomics, and proteomics—enable a systems-level understanding of how fermentation-derived compounds may influence host physiology (Kumar Singh et al., 2019; Wen et al., 2024). These integrative frameworks enhance the capacity to trace mechanistic pathways from microbial genes present in fermented foods to the metabolites they generate, and ultimately to measurable changes in host neurobiology and behavior. For example, targeted metabolomic profiling following consumption of traditional fermented foods such as ogi or iru could identify circulating metabolites associated with stress reactivity, memory performance, or affective regulation. However, such applications remain largely exploratory and require rigorous validation.

An additional frontier involves coupling microbiome profiling with neuroimaging modalities, including functional magnetic resonance imaging (fMRI) and electroencephalography (EEG) (Bacon et al., 2024). This integrative approach allows for the examination of associations between microbial dynamics and neural activation patterns implicated in emotional regulation, attention, and executive function. Longitudinal designs that overlay microbiome trajectories onto neuroimaging data may clarify whether fermentation-derived metabolites, such as short-chain fatty acids, correspond with functional changes in prefrontal–limbic circuitry. At present, however, most available studies are cross-sectional, limiting causal inference.

Prospective cohort studies and randomized controlled trials remain essential to strengthen translational validity. Long-term investigations that concurrently assess dietary intake, microbiota composition, circulating metabolites, inflammatory markers, neuroimaging outcomes, and genetic variants would improve understanding of causal pathways linking fermented foods to cognitive resilience and stress-related psychopathology. Incorporating stratification by baseline microbiota diversity and host genetic polymorphisms (e.g., APOE status) may further clarify inter-individual variability in response.

6.8.1 Future Research Roadmap

To advance the field in a scientifically robust manner, the following priorities are proposed:

1. **Standardized Intervention Trials:**
   Conduct well-powered, long-duration randomized controlled trials comparing traditional and controlled fermentation products, with clearly defined microbial composition and metabolite profiles.
2. **Multi-Omics Integration:**
   Integrate microbiome sequencing, metabolomics, and host transcriptomics to establish mechanistic links between fermentation-derived metabolites and neurobiological pathways.
3. **Neuroimaging and Cognitive Endpoints:**
   Incorporate sensitive neuroimaging markers (e.g., hippocampal volume, functional connectivity) and domain-specific cognitive assessments rather than global screening tools alone.
4. **Longitudinal and Life-Course Designs:**
   Implement cohort studies examining cumulative exposure to fermented foods across developmental stages, aging, and stress vulnerability.
5. **Contextual and Cultural Frameworks:**
   Evaluate fermented foods within whole dietary patterns and sociocultural contexts to avoid reductionist interpretations.
6. **Equity and Accessibility Research:**
   Investigate how socioeconomic, environmental, and infrastructural factors influence access to fermented foods and potential mental health benefits.

Collectively, this emerging interdisciplinary paradigm—sometimes conceptualized as a form of "neuroecology"—positions the brain within a dynamic ecosystem shaped by diet, microbial ecology, immune regulation, cultural practices, and environmental exposures. Figure 6.3 illustrates this emerging neuro-ecological framework, integrating fermented plant foods, microbial dynamics, and brain function within a systems-level perspective. Within this systems-based framework, fermented foods may be understood not merely as nutritional components but as ecological modulators capable of influencing metabolic homeostasis and neural adaptability. Nevertheless, translating this conceptual model into clinical practice will require methodological rigor, cautious interpretation, and sustained cross-disciplinary collaboration.

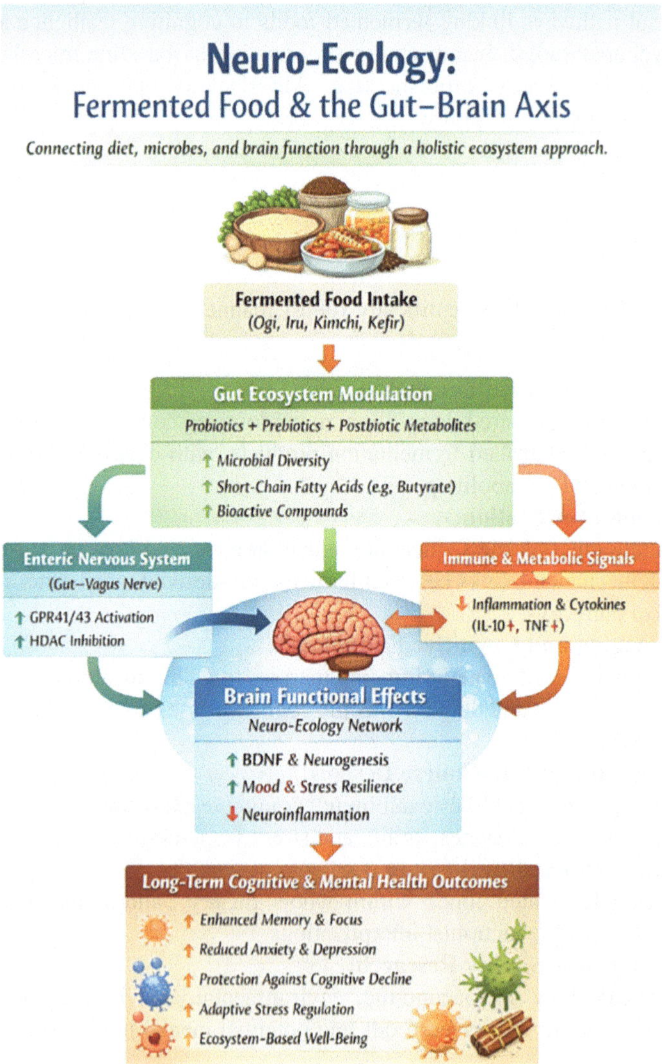

Fig. 6.3 Neuro-ecology fermented food and the gut–brain axis

References

Agirman, G., Yu, K. B., & Hsiao, E. Y. (2021). Signaling inflammation across the gut-brain axis. *Science, 374*(6571), 1087–1092.

Al Noman, A., Alhudhaibi, A. M., Afroza, M., Tonni, S. D., Shehab, H. M., Jahan Iba, N., et al. (2025). Neuroplasticity and the microbiome: How microorganisms influence brain change. *Frontiers in Microbiology, 16*, 1629349.

Anih, D. C., Tatah, V. S., Ugwuoke, K. C., & Boyi, R. H. N. (2025). Epigenetic modulation by fermented plant-based functional foods: A new frontier in neurodegenerative disease prevention. *Singapore Journal of Scientific Research, 15*(1), 36–54.

Anyimadu, C. F., Agarwal, A., Kaur, N., Tripathi, A. D., Deokar, G. S., Al-Asmari, F., et al. (2026). Effects of food components, processing technologies, additives, and processed food on gut health. *Food Reviews International, 2026*, 1–29.

Ashaolu, T. J., & Reale, A. (2020). A holistic review on Euro-Asian lactic acid bacteria fermented cereals and vegetables. *Microorganisms, 8*(8), 1176.

Bacon, E. J., He, D., D'Avilla, A. N. A., Wang, L., Li, H., Yao-Digba, P. D. Z., et al. (2024). Neuroimage analysis using artificial intelligence approaches: A systematic review. *Medical & Biological Engineering & Computing, 62*(9), 2599–2627.

Basha, S., Ks, P., Chattopadhyay, A., Ramakrishna Pai, A., & Kishore Mahato, K. (2025). Emerging insights into dairy products and Alzheimer's disease: Exploring the potential neuroprotective effects. *Critical Reviews in Food Science and Nutrition, 66*, 1–28.

Berthoud, H.-R., Albaugh, V. L., & Neuhuber, W. L. (2021). Gut-brain communication and obesity: Understanding functions of the vagus nerve. *The Journal of Clinical Investigation, 131*(10), e143770.

Borrego-Ruiz, A., & Borrego, J. J. (2025). Integrating plant-based diets into schools for a healthier and more sustainable future: A contemporary overview. *Future, 3*(4), 22.

Boven, L., Akkerman, R., & de Vos, P. (2025). Sustainable diets with plant-based proteins require considerations for prevention of proteolytic fermentation. *Critical Reviews in Food Science and Nutrition, 65*(14), 2829–2839.

Brkić, D., Concetti, C., Rémond-Derbez, N., & Hauser, J. (2026). Relationship between nutrition, brain, cognition, learning, and behavior in school age children: Systematic evidence and future opportunities. *Nutrition Reviews*, nuaf280. https://doi.org/10.1093/nutrit/nuaf280

Casertano, M., Fogliano, V., & Ercolini, D. (2022). Psychobiotics, gut microbiota and fermented foods can help preserving mental health. *Food Research International, 152*, 110892.

Chen, J.-H., Yin, X., He, H., Lu, L. W., Wang, M., Liu, B., & Cheng, K.-W. (2024). Potential neuroprotective benefits of plant-based fermented foods in Alzheimer's disease: An update on preclinical evidence. *Food & Function, 15*(8), 3920–3938.

Chowdhury, S. R. (2024). *Psychobiotics and gut–brain axis—A new paradigm for improving mental health* (pp. 146–158). CRC Press.

Dahiya, D., & Nigam, P. S. (2022). Probiotics, prebiotics, synbiotics, and fermented foods as potential biotics in nutrition improving health via microbiome-gut-brain axis. *Fermentation, 8*(7), 303.

De Cassia Alpino, G., Pereira-Sol, G. A., demourae Dias, M., Silva De Aguiar, A., & do Carmo Gouveia Peluzio, M. (2024). Beneficial effects of butyrate on brain functions: A view of epigenetic. *Critical Reviews in Food Science and Nutrition, 64*(12), 3961–3970.

Dhiman, S., Kaur, S., Thakur, B., Singh, P., & Tripathi, M. (2025). Nutritional enhancement of plant-based fermented foods: Microbial innovations for a sustainable future. *Fermentation, 11*(6), 346.

Di Meo, F., Valentino, A., Petillo, O., Peluso, G., Filosa, S., & Crispi, S. (2020). Bioactive polyphenols and neuromodulation: Molecular mechanisms in neurodegeneration. *International Journal of Molecular Sciences, 21*(7), 2564.

Favieri, F., Forte, G., & Casagrande, M. (2019). The executive functions in overweight and obesity: A systematic review of neuropsychological cross-sectional and longitudinal studies. *Frontiers in Psychology, 10*, 2126.

Fiore, M., Terracina, S., & Ferraguti, G. (2025). Brain neurotrophins and plant polyphenols: A powerful connection. *Molecules, 30*(12), 2657.

Gao, Y., Liu, Y., Ma, T., Liang, Q., Sun, J., Wu, X., et al. (2025). Fermented dairy products as precision modulators of gut microbiota and host health: Mechanistic insights, clinical evidence, and future directions. *Food, 14*(11), 1946.

Gravesteijn, E., Mensink, R. P., & Plat, J. (2022). Effects of nutritional interventions on BDNF concentrations in humans: A systematic review. *Nutritional Neuroscience, 25*(7), 1425–1436.

Guo, B., Zhang, J., Zhang, W., Chen, F., & Liu, B. (2025). Gut microbiota-derived short chain fatty acids act as mediators of the gut–brain axis targeting age-related neurodegenerative disorders: A narrative review. *Critical Reviews in Food Science and Nutrition, 65*(2), 265–286.

Hamamah, S., Aghazarian, A., Nazaryan, A., Hajnal, A., & Covasa, M. (2022). Role of microbiota-gut-brain axis in regulating dopaminergic signaling. *Biomedicine, 10*(2), 436.

Homer, B., Judd, J., Mohammadi Dehcheshmeh, M., Ebrahimie, E., & Trott, D. J. (2023). Gut microbiota and behavioural issues in production, performance, and companion animals: A systematic review. *Animals, 13*(9), 1458.

Kasarello, K., Cudnoch-Jedrzejewska, A., & Czarzasta, K. (2023). Communication of gut microbiota and brain via immune and neuroendocrine signaling. *Frontiers in Microbiology, 14*, 1118529.

Kumar Singh, A., Cabral, C., Kumar, R., Ganguly, R., Kumar Rana, H., Gupta, A., et al. (2019). Beneficial effects of dietary polyphenols on gut microbiota and strategies to improve delivery efficiency. *Nutrients, 11*(9), 2216.

Kurhaluk, N., Kołodziejska, R., Kamiński, P., & Tkaczenko, H. (2025). Integrative neuroimmune role of the parasympathetic nervous system, vagus nerve and gut microbiota in stress modulation: A narrative review. *International Journal of Molecular Sciences, 26*(23), 11706.

La Torre, D., Verbeke, K., & Dalile, B. (2021). Dietary fibre and the gut–brain axis: Microbiota-dependent and independent mechanisms of action. *Gut Microbiome, 2*, e3.

Lim, E. J., & Abd Rahim, M. H. (2025). Fermentation for enhanced biosynthesis and bioavailability of micronutrients, prebiotics, bioactive peptides and functional fatty acids in food products. *International Journal of Advanced Research in Food Science and Agriculture Technology, 4*(1), 9–27.

Lin, J., Sun-Waterhouse, D., & Cui, C. (2022). The therapeutic potential of diet on immune-related diseases: Based on the regulation on tryptophan metabolism. *Critical Reviews in Food Science and Nutrition, 62*(31), 8793–8811.

Liwinski, T., Lang, U. E., Brühl, A. B., & Schneider, E. (2023). Exploring the therapeutic potential of gamma-aminobutyric acid in stress and depressive disorders through the gut–brain axis. *Biomedicine, 11*(12), 3128.

Lys, I. M. (2025). The role of lactic fermentation in ensuring the safety and extending the shelf life of African indigenous vegetables and its economic potential. *Applied Research, 4*(1), e202400131.

Magzal, F., Turroni, S., Fabbrini, M., Barone, M., Vitman Schorr, A., Ofran, A., & Tamir, S. (2023). A personalized diet intervention improves depression symptoms and changes microbiota and metabolite profiles among community-dwelling older adults. *Frontiers in Nutrition, 10*, 1234549.

Margoob, M., Kouser, S., & Jan, N. (2024). Serotonin: The link between gut microbiome and brain. In *Serotonin-neurotransmitter and hormone of brain, bowels and blood*. IntechOpen.

Munteanu, C., & Schwartz, B. (2024). Interactions between dietary antioxidants, dietary Fiber and the gut microbiome: Their putative role in inflammation and cancer. *International Journal of Molecular Sciences, 25*(15), 8250.

Murciano-Brea, J., Garcia-Montes, M., Geuna, S., & Herrera-Rincon, C. (2021). Gut microbiota and neuroplasticity. *Cells, 10*(8), 2084.

Nazzaro, F., Coppola, F., Fratianni, F., Abdalrazeq, M., Ombra, M. N., De Giulio, B., et al. (2025). Polyphenols bioactive metabolites, and their anti-biofilm and neuroprotective potential. *Food, 14*(22), 3976.

Padhi, S., Sarkar, P., Sahoo, D., & Rai, A. K. (2025). Potential of fermented foods and their metabolites in improving gut microbiota function and lowering gastrointestinal inflammation. *Journal of the Science of Food and Agriculture, 105*(8), 4058–4069.

Park, J. C., Chang, L., Kwon, H.-K., & Im, S.-H. (2025). Beyond the gut: Decoding the gut–immune–brain axis in health and disease. *Cellular & Molecular Immunology, 22*(11), 1287–1312.

Patil, S., Mehdi, S. S., & Mehdi, S. (2025). The gut-brain axis and mental health: How diet shapes our cognitive and emotional well-being. *Cureus, 17*(7), e88420. https://doi.org/10.7759/cureus.88420

Pires, L., Gonzalez-Paramás, A. M., Heleno, S. A., & Calhelha, R. C. (2024). Gut microbiota as an endocrine organ: Unveiling its role in human physiology and health. *Applied Sciences, 14*(20), 9383.

Porras-García, E., Fernández-Espada Calderón, I., Gavala-González, J., & Fernández-García, J. C. (2023). Potential neuroprotective effects of fermented foods and beverages in old age: A systematic review. *Frontiers in Nutrition, 10*, 1170841.

Rani, J., Gulia, V., Sangwan, A., Dhull, S. S., & Mandzhieva, S. (2025). Synergies of traditional ecological knowledge in biodiversity conservation: A paradigm for sustainable food security. In *Ecologically mediated development: Promoting biodiversity conservation and food security* (pp. 27–49). Springer.

Rekha, K., Venkidasamy, B., Samynathan, R., Nagella, P., Rebezov, M., Khayrullin, M., et al. (2024). Short-chain fatty acid: An updated review on signaling, metabolism, and therapeutic effects. *Critical Reviews in Food Science and Nutrition, 64*(9), 2461–2489.

Rusch, J. A., Layden, B. T., & Dugas, L. R. (2023). Signalling cognition: The gut microbiota and hypothalamic-pituitary-adrenal axis. *Frontiers in Endocrinology, 14*, 1130689.

Sarkar, A., Yoo, J. Y., Valeria Ozorio Dutra, S., Morgan, K. H., & Groer, M. (2021). The association between early-life gut microbiota and long-term health and diseases. *Journal of Clinical Medicine, 10*(3), 459.

Sarma, U., Shrestha, R., Nethravathy, V., & Rayan, L. A. (2026). Fermented foods, microbiota, and mental health. In *Fermented foods and beverages: Functionalities and nutritional aspects* (pp. 263–275). CABI GB.

Sarubbo, F., Cavallucci, V., & Pani, G. (2022). The influence of gut microbiota on neurogenesis: Evidence and hopes. *Cells, 11*(3), 382.

Shawky, E., Surendran, S., & El-Khair, R. M. A. (2025). Fermented vegetables as a source of Psychobiotics: A review of the evidence for mental health benefits. *Probiotics and Antimicrobial Proteins, 2025*, 1–15.

Silva, Y. P., Bernardi, A., & Frozza, R. L. (2020). The role of short-chain fatty acids from gut microbiota in gut-brain communication. *Frontiers in Endocrinology, 11*, 508738.

Sood, R., Kilpatrick, L. A., Keefer, L. A., & Church, A. (2025). Biopsychosocial and environmental factors that impact brain-gut-microbiome interactions in obesity. *Clinical Gastroenterology and Hepatology, 24*(1), 10–20.

Thriene, K., & Michels, K. B. (2023). Human gut microbiota plasticity throughout the life course. *International Journal of Environmental Research and Public Health, 20*(2), 1463.

Wen, L., Yang, L., Chen, C., Li, J., Fu, J., Liu, G., et al. (2024). Applications of multi-omics techniques to unravel the fermentation process and the flavor formation mechanism in fermented foods. *Critical Reviews in Food Science and Nutrition, 64*(23), 8367–8383.

Woodhill, J., Kishore, A., Njuki, J., Jones, K., & Hasnain, S. (2022). Food systems and rural wellbeing: Challenges and opportunities. *Food Security, 14*(5), 1099.

Wu, M., Luo, Q., Nie, R., Yang, X., Tang, Z., & Chen, H. (2021). Potential implications of polyphenols on aging considering oxidative stress, inflammation, autophagy, and gut microbiota. *Critical Reviews in Food Science and Nutrition, 61*(13), 2175–2193.

Xing, T., Ozkaya, K. S., Nassrallah, Z., & Travagli, R. A. (2026). The vagus connection: Exploring the neurobiology of brain-gut communication. *Journal of Neurophysiology, 135*(1), 261–272.
Yuan, H., Zhou, J., Li, H., Kang, S., & Park, S. (2025). A systems biology approach to memory health: Integrating network pharmacology, gut microbiota, and multi-omics for health functional foods. *International Journal of Molecular Sciences, 26*(14), 6698.
Zhang, Y., Zhu, X., Wang, N., Liu, X., Wang, L., & Ning, K. (2025). Synergy of traditional practices and modern technology: Advancing the understanding and applications of microbial resources and processes in fermented foods. *Trends in Food Science & Technology, 157*, 104891.

Chapter 7
Fermentation for Personalized Nutrition—AI-Driven Approaches to Tailor Fermented Foods to Individual Needs

I. Muthuvel, D. Priyadharshini, Sathish Kumar Ponnaiah, P. S. Kavitha, and K. S. Vijay Selvaraj

Abstract Artificial Intelligence (AI) is increasingly being used to support fermentation design and control for personalised feeding. This chapter explores how machine learning (ML) methods can integrate data at individual level (genetics, gut microbiomes and metabolic markers) with data from the fermentation process to predict and optimise the composition of products. We summarise the relevant artificial intelligence approaches for (i) predicting microbial dynamics and metabolite formation, (ii) rational selection of starter cultures and (iii) monitoring and control of the bioreactor in real time. Two concept case studies illustrate how optimisation and closed loop control can adapt fermented foods to specific functional results

Keywords Artificial intelligence (AI) · Machine learning (ML) · Fermentation design · Personalized feeding · Gut microbiomes · Microbial dynamics · Metabolism · Bioreactor control · Starter cultures and closed loop control

I. Muthuvel (✉)
Department of Fruit Science, Horticultural College and Research Institute for Women, Tamil Nadu Agricultural University, Trichy, Tamil Nadu, India
e-mail: im74@tnau.ac.in

D. Priyadharshini
Department of Fruit Science, Horticultural College and Research Institute, Tamil Nadu Agricultural University, Coimbatore, Tamil Nadu, India

S. K. Ponnaiah
Plant Science, 155, Woodard Hall, Department of Agriculture, School of Agriculture, Fisheries and Human Sciences, University of Arkansas at Pine Bluff, Pine Bluff, AR, USA
e-mail: ponniahs@uapb.edu

P. S. Kavitha
Department of Planning and Monitoring, Tamil Nadu Agricultural University, Coimbatore, Tamil Nadu, India
e-mail: kavitha.ps@tnau.ac.in

K. S. Vijay Selvaraj
Vegetable Research Station, Tamil Nadu Agricultural University, Palur, Tamil Nadu, India

S. Dhiman, S. Kaur (eds.), *Next-Gen Plant Based Fermented Foods: Innovations for a Sustainable Future*, https://doi.org/10.1007/978-3-032-29927-7_7

7.1 Introduction: The New Paradigm of Personalized Fermentation

Personalised nutrition is a shift from general dietary advice to interventions that are tailored to each person's biological and physiological profile. As a biotechnology process, fermentation offers a unique platform for the production of customised food. However, the complexity of human variability requires sophisticated computational approaches. The convergence of artificial intelligence (AI) and fermentation science is increasingly being used to foster a world of personalised nutrition, moving away from the one size fits all approach that has long been the dominant approach in the food sector. This chapter explores how these areas can be combined in a synergistic way to engineer the next generation of hyperpersonalised fermented plant-based foods. AI-based methodologies are essential to address the bio-complexity of individual gut microbiomes, genetic predispositions and metabolic responses (Kumar et al., 2022). Despite the strong interest in fermented foods and precision nutrition, most commercial products and fermentation protocols are optimized for population averages rather than for individual differences in genetic make-up, microbiological function and metabolic status. This chapter summarises the biological justification for personalisation, data and modelling requirements for AI-enabled fermentation design and practical constraints (validation, safety and regulation) that determine whether personalised fermentation can be translated into trustworthy products (Singh & Kumar, 2025).

7.1.1 From Traditional Practice to Precision Food Design

Fermentation is one of the oldest and most important biotechnology processes in the world, used for millennia to preserve, enhance and enhance food. Traditional fermented foods such as kimchi, kefir and sourdough are the result of empirical, often spontaneous microbial processes that are passed down from generation to generation (Ray et al., 2024b). While these practices have sustained populations and spawned diverse food cultures, they are by their nature variable and designed for mass production, not individual convenience. The current shift is towards the concept of precision food design, a paradigm in which fermentation is no longer an artisanal craft, but a tightly controlled, data-driven engineering discipline (Xiao et al., 2025). Artificial intelligence facilitates this transformation. Using machine learning algorithms based on multi-omics, AI is transforming fermentation from a mass production process into a targeted, programmable platform for personalised food creation.

7.1.2 *The Limitations of Generic Fermented Foods in Modern Nutrition*

Despite their documented benefits, commercially available fermented foods are formulated for the average consumer, a statistical abstraction that fails to take into account the profound biodiversity of the human population. The therapeutic potential of fermented food is not homogenous; it is modulated by the individual's unique gut microbiomes, which vary significantly from person to person (Gao et al., 2025). In addition, genetic polymorphisms affect how individuals metabolise fermentation-derived bioactive compounds such as peptides and short chain fatty acids (SFLs). Standardised yoghurt or kombucha may have beneficial effects in some people, but may be less beneficial or ineffective in others depending on their specific medical conditions such as underlying inflammatory disorders, insulin sensitivity or specific nutritional deficiencies (Gaur, 2024). This general approach therefore represents a significant obstacle to harnessing fermented foods' full potential for targeted health improvements, and highlights the critical need for personalisation (Montazeri-Najafabady, 2025).

7.1.3 *Defining AI-Driven Personalization in the Fermentation Lifecycle*

Artificial intelligence-driven personalisation in fermentation is a holistic, intelligent framework that intervenes throughout the entire life cycle of a product. It is moving from static formulas to a dynamic system of continuous optimization. Artificial intelligence technologies, in particular machine learning and predictive analytics, have shown enormous potential for the analysis of large-scale, highly dimensional health data. This can be defined in three main stages: personalised strain selection and consortium design: artificial intelligence models analyze the data on the microbiome of an individual to predict which probiotic or microbial strains will colonise or modify their gut environment, leading to the design of personalised starter cultures (D'Urso & Broccolo, 2024). AI-optimised parameter control: Machine learning algorithms, in particular reinforcement learning (RL) algorithms, dynamically control fermentation parameters (e.g. temperature, pH, dissolved oxygen, substrate feeding) in real time in a smart bioreactor (Butean et al., 2025). This ensures that the process is optimised for maximum bioavailability of the desired bioactive substances (e.g. postbiotics for a specific individual) and minimisation of by-products.

Tailored Product Output The end product is not only food but also nutritional intervention. Its composition—the bioactivity, peptide, vitamin, and prebiotic profiles are carefully designed to match the genomics, metabolism, and health data of

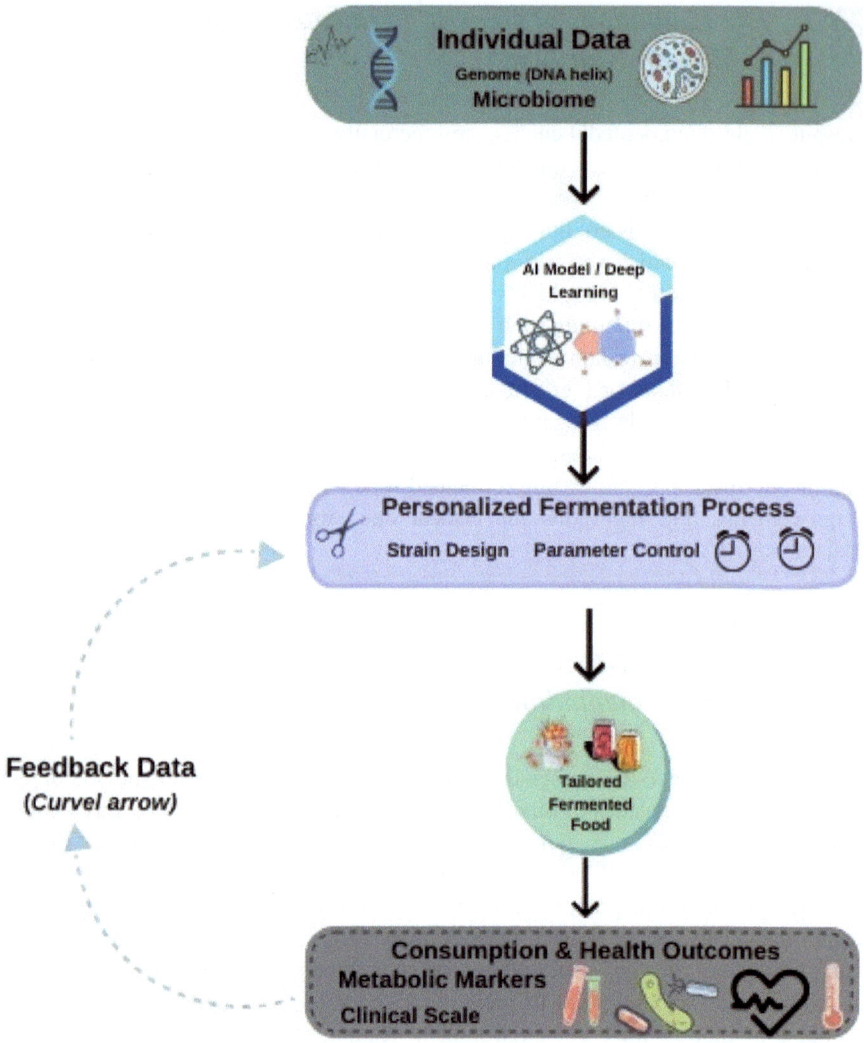

Fig. 7.1 The closed-loop framework for AI-driven personalization of fermented foods. (Source: Authors; created using Canva)

the recipient (Kussmann et al., 2023). It creates a closed-loop system where the food is as unique as the person consuming it (Fig. 7.1).

7.1.4 *Objective and Scope of the Chapter*

The main objective of this chapter is to provide a comprehensive framework for understanding and implementing the approach of artificial intelligence in food personalisation. Our aim is to synthesise in a critical way the current state of affairs in this emerging field, from its scientific basis to its technological implementation and ethical implications. The scope of this chapter includes a detailed review of the suite of tools for artificial intelligence in fermentation science, including predictive modelling and the integration of multi-omics data (Ferrocino et al., 2023). It will present concrete frameworks and case studies for adapting fermented foods, discuss the roadmap for implementation from laboratory to consumer, and address important translational challenges and ethical issues such as data protection and regulatory obstacles. By outlining the future perspectives, this chapter aims to establish artificial intelligence as not only a tool but also an essential ingredient in achieving truly personalized, efficient and sustainable nutrition through the use of fermentation technologies (Ajayeoba & Ijabadeniyi, 2025).

7.2 The Scientific Foundation for Personalization

The basic premise of personalized nutrition is based on the established understanding that individuals react differently to the same food. This inter-individual variability, caused by complex interactions of genetic, microbiological and physiological factors, limits the scope of the universal diet approach in a fundamental way (Lampe et al., 2013). A one size fits all model for the production and consumption of fermented food does not take account of this biodiversity and ultimately limits its therapeutic potential (Bell et al., 2018). This section identifies the main scientific drivers why that require a case-by-case approach and identifies fermentation bio-plasticity as a powerful tool for designing interventions that address variability at its root.

7.2.1 *Key Drivers of Individual Nutritional Needs*

7.2.1.1 Genomic and Phenotypic Variation

The genetic blueprint of an individual has a profound influence on its food needs and metabolic responses. Single nucleotide polymorphisms (SNPs) in genes coding for enzymes involved in the metabolism of food can significantly alter their activity, resulting in different phenotypic outcomes when the same food is consumed (Corella & Ordovas, 2005). For example, common variants of the *MTHFR* gene may interfere with the metabolism of folate, thereby increasing the need for dietary folate in specific bioavailable forms (Molloy, 2002). Similarly, in addition to these classic examples (Nikam et al., 2025), variations in taste receptor genes (e.g., TAS2R38) may affect sensory perception and food preference, and lipid metabolism genes (e.g., *APOE*) may modulate the response to dietary fats. This genetic layer of complexity means that a fermented food serving as a bioactive compound source for one genotype may be metabolically inert or even problematic for another, which is the proposed justification for a genotypic-based formulation (Laforest-Lapointe & Arrieta, 2017). Table 7.1 shows genetic polymorphisms and their consequences on the design of fermented foods.

7.2.1.2 The Gut Microbiome as a Dynamic Interface

The human gut microbiome, a diverse and dynamic ecosystem of trillions of microorganisms, is a critical intermediate organ between food and the physiology of the host (EFSA NDA Panel, 2016). This community encodes a large repertoire of enzymes that humans lack, allowing the biotransformation of food ingredients and fermented foods to produce unique metabolites that have a directly contribute to health and well-being (Park & Mannaa, 2025). The composition and function of this microbial community is highly individualistic, shaped by genetic predisposition, exposure at an early age, diet and environment (Deokar et al., 2025). This is clearly illustrated by the term enterotype or variation of the community status; the same fermented food, such as kimchi, which is rich in specific *Lactobacillus* strains, will react differently in a gut dominated by Bacteroides and a gut dominated by *Prevotella* strains. The resulting metabolic output the production of short-chain fatty acids, vitamins, and other bioactive substances is therefore not characteristic of the food itself, but is the result of interaction between microbial communities and the foodstuffs themselves (Elechi et al., 2023). This dynamic places the gut microbiome not only as a passenger, but as a primary and controllable target for personalised dietary interventions (EFSA NDA Panel, 2016).

7.2.1.3 Metabolic Health and Inflammatory Markers

The underlying metabolic states create a physiological environment which profoundly changes the demands and responses of the diet. These states are not static but fluid landscapes, shaped by diet, lifestyle and health choices. Key markers such as insulin sensitivity, baseline systemic inflammation (e.g., reactive C-reactive protein levels) (Margină et al., 2020). This context is defined by lipid profiles and cardiac health. Individuals with low-grade chronic inflammation and insulin resistance may benefit greatly from a fermented food prepared with high levels of anti-inflammatory peptides, such as certain bioactive peptides, or short-chain fatty acids butyrate (Kopczyńska & Kowalczyk, 2024). Conversely, individuals with specific dyslipidaemia may require a product that is optimized for the metabolism of bile salts or the assimilation of cholesterol by specific microbial strains. Standardized,

Table 7.1 Genetic polymorphisms and their implications for fermented food design

Gene	Key polymorphism	Implication for fermented food design	References
LCT (Lactase)	rs4988235	For lactose intolerant (CC genotype), design lactose-free plant-based ferments or use cultures that fully metabolize lactose.	Chin et al. (2019)
MTHFR	rs1801133 (C677T)	For impaired folate metabolism, select strains that enhance folate (B9) bioavailability in fermented cereals/plant-milks.	Rai (2017)
TAS2R38 (Bitter Taste)	rs713598	For "supertasters," optimize fermentation to reduce bitter compounds in kimchi/cruciferous ferments for improved palatability.	Saddam (2019)
HLA-DQ2/ DQ8 (celiac risk)	HLA risk haplotypes (commonly typed as DQ2/DQ8 rather than single rs)	For gluten-sensitive/at-risk consumers: design gluten-free ferments (millets, rice, sorghum), avoid barley/wheat starters, and prevent cross-contamination; if using sourdough-style processes, ensure validated gluten reduction and regulatory compliance.	Smigoc Schweiger et al. (2016)
ALDH2	rs671	For reduced acetaldehyde metabolism (common in some populations), avoid/limit ethanol-containing ferments and consider non-alcoholic ferments with controlled yeast activity.	Chang et al. (2023)
FUT2 (secretor status)	rs601338	Non-secretors can show different gut-microbiome patterns; consider ferments with documented probiotic efficacy and prebiotic substrates	Parmar et al. (2012)
APOE	ε2/ε3/ε4 (defined by rs429358/rs7412)	For cardiometabolic-focused fermented foods: emphasize low saturated fat, favor plant-based ferments, include soluble fiber and cholesterol-lowering dietary patterns;	Rasmussen et al. (2015)
CYP1A2	rs762551	Caffeine sensitivity affects acceptance of tea/coffee ferments (kombucha variants, cascara ferments). Offer low-caffeine options or clearly label caffeine; adjust formulations for stimulant-sensitive users.	Denden et al. (2016)
NAT2 (acetylation status)	NAT2 variants (multiple rs define slow/fast acetylators)	If using ferments rich in biogenic amines or specific polyphenols, consider histamine/amine management (starter selection, hygiene, time/temp) and offer low-amine versions for sensitive consumers.	Gutierrez-Virgen et al. (2023)

generic fermented foods lack the specificity to address these subtle but critical physiological differences (Singh & Kumar, 2025). The ever-changing target of metabolic health requires an equally adaptive and responsive diet that can be adapted dynamically through intelligent food design.

7.2.2 *The Plasticity of Fermentation: A Tool for Manipulating Food Matrices*

In view of the biological variables described, fermentation offers a unique, malleable biological platform for the development of food matrixes with intended functionalities. The process uses the metabolic versatility of microbes to directly and predictably change the composition, structure and bioavailability of food ingredients, providing a direct means of fighting individual differences and creating tailor-made nutritional solutions.

7.2.2.1 Bioavailability Enhancement of Nutrients

By fermentation, the bioavailability of basic micronutrients and minerals can be increased significantly, effectively overcoming some genetic or physiological absorption constraints. This is achieved by microbial biosynthesis and by enzymatic degradation of the anti-nutrients. For example, selected strains of *Lactobacillus and Bifidobacterium* have been shown to synthesise and increase the bioavailability of B vitamins such as folate and riboflavin in fermented cereals and plant milks, with direct benefit to people with genetic polymorphisms affecting the metabolism of folate (Hiraoka & Kagawa, 2017). In addition, the enzymatic reduction of phytic acid—a powerful chelator of minerals such as iron, zinc, and calcium—in cereals and beans is a quintessential example of the power of fermentation. Lactic acid bacteria and yeast express phytases that break down phytic acid, freeing the bound minerals and making them more available for absorption in the gut (Brouns, 2021). This is particularly important for individuals with higher mineral requirements or who eat a plant-based diet, as it will personalise the bioavailability of the food in terms of minerals.

7.2.2.2 Generation of Bioactive Compounds (Postbiotics, Bioactive Peptides)

This is where fermentation interacts directly and strongly with the drivers described in Sect. 7.2.1. Microbial metabolism during fermentation produces a complex bioactive compound group collectively called postbiotics, which includes short chain fatty acids (SCFA), bacteriocins, functional proteins and exopolysaccharides

(EPS). Butyrate and other SCFA produced by microbial fermentation of dietary fibres are critical for intestinal barrier integrity and immunological modulation and have systemic anti-inflammatory properties, which are directly relevant to metabolic health markers as described in Sect. 7.2.1.3 (Tian, 2021). At the same time, the proteolytic activity of the starter cultures releases encoded bioactive peptides from the parent proteins. These peptides may have a variety of target effects, such as angiotensin converting enzyme (ACE) inhibitors, antioxidant, mineral binding and opioid-like effects (Bhat et al., 2015). By selecting specific microbial strains and controlling fermentation parameters, the proteolytic profile can be directed towards peptide release supporting a specific physiological function, e.g. the production of antihypertensive peptides in pre-hypertensive individuals. It should be noted that the production of specific bioactive peptides is highly process-dependent and dependent on factors such as substrate composition, microbial activity and fermentation conditions, and therefore cannot be guaranteed without a precise control and verification.

7.2.2.3 Detoxification and Anti-nutrient Reduction

In addition to increasing nutrients, fermentation is a powerful biological detoxification process that reduces the impact of different nutrients and naturally occurring toxins that can impair food safety and nutritional integrity. This is particularly important for personalised nutrition, as individuals vary in their susceptibility to these compounds depending on their genetic background, gut microbiology and physiological status (Kolodziejczyk et al., 2019). For example, lactic acid bacteria have been shown to significantly reduce cyanogenic glycosides in cassava by enzymatic hydrolysis, thereby converting potentially toxic compounds to less toxic ones (Lei et al., 1999). Similarly, fermentation effectively degrades oligosaccharides causing flatulence in legumes such as raffinose and stachyose by microbial alpha-galactosidase activity, thereby improving gastrointestinal tolerance in patients with reduced susceptibility to these substances (Baucells et al., 2000). In addition to the anti-nutrients, some fungal starter cultures and lactic acid bacteria in contaminated grain may adsorb or biodegrade mycotoxins, thereby reducing their bioavailability and potentially causing toxicity (Dawlal, 2018). In addition, glucosinolates in cruciferous vegetables, which may affect thyroid function in susceptible individuals, are modulated by fermentation, either by degradation or by conversion to minor goitrogenic metabolites (Oliphant & Allen-Vercoe, 2019). By pre-digesting or converting these compounds, fermentation creates a safer and more palatable food matrix, and it also personalises the food by removing or reducing ingredients that would otherwise have an adverse effect on susceptible populations.

7.3 The AI Toolbox for Fermentation Science

Personalisation of fermented food based on the biological bases in Sect. 7.2 is computationally impracticable by conventional methods alone. The sheer complexity of multi-omics data, non-linear microbial dynamics and high-dimensional parameter space necessitate an in-depth algorithmic approach (Mukherjee et al., 2025). This section defines the basic machine learning techniques that make up the AI toolkit and describes the rich and diverse data ecosystems that power them and provide the infrastructure for accurate inference.

7.3.1 Overview of Relevant Machine Learning Techniques

7.3.1.1 Supervised Learning for Predictive Modelling

Supervised learning algorithms are designed to learn a mapping function from input variables to known results, which is a precondition for predicting fermentation tasks (Khaleghi et al., 2021). In this context, input variables may include the initial substrate composition, the size of the inoculum, temperature, pH and genetic information of the strain, as well as input variables such as the yield of metabolites (e.g. lactate, ethanol, specific bioactive peptides, biomass concentrations or the purity of the product) may be present (Raveschot et al., 2018). Random forest models, for example, can reliably handle a wide range of data types to predict the final gamma-aminobutyric acid (GABA) concentration in fermented soybean paste, depending on the proportion of starter cultures and the length of fermentation (Yee et al., 2024). Neural networks, including deep learning architectures, capture more complex nonlinear relationships. They can model the dynamic interaction between parameters in a time series and predict how small temperature changes affect the full metabolic activity of the microbial consortium, thus predicting the optimal time to harvest postbiotic fermentation (Ray et al., 2024a).

7.3.1.2 Unsupervised Learning for Pattern Discovery in Microbiome Data

Unlike supervised learning, unsupervised techniques reveal hidden patterns and patterns in data without any predetermined labels. This is particularly useful for the analysis of high-dimensional data obtained from gut microbiomes of consumers. Clustering algorithms such as k-measure or hierarchical clustering can process 16S rRNA or metagenomic sequencing data to identify different microbial enterotypes or populations in a cohort of individuals (Zhou et al., 2014). This allows stratification of the population into groups which can respond in a homogeneous way to a particular fermentation product. For example, analysis may show that individuals with the

Prevotella enterotype show a greater increase in the production of beneficial short-chain fatty acids when consuming a particular type of fermented fruit drink compared to individuals with the Bacteroides enterotype (Chen, 2016). Diminishing techniques such as t-SNE or UMAP can visualize these complex microbial environments and help identify new, clinically relevant sub-groups for which personalized fermentation products can be developed.

7.3.1.3 Reinforcement Learning (RL) for Dynamic Process Optimization

Fermentation is a dynamic, non-stationary process in which the state of the system (e.g. nutrient content, microbial density, concentration of metabolites) changes over time. Reinforcement learning (RL) is suitable for this environment, as RL agents can learn control behaviour by interacting with the process by sensing, controlling actions (e.g. temperature adjustment or nutrient pulsing) and a defined reward function (Priyadharshini et al., 2025; Gil et al., 2025). The agent observes the current fermentation state (via sensor data), takes action (e.g. adjusts temperature, adds nutrient pulses) and is rewarded (e.g. increases in yield of the target compound, stability). Through repeated experiments, often in a simulated environment, the agent learns the policy the strategy for what to do in each situation to maximize the cumulative reward over the fermentation cycle (Uthravalli, 2025). This allows closed loop control, which can react more flexibly to drift and unexpected microbial behaviour than fixed point-based controllers (Wang et al., 2020) and guide the process to a personalised result that can be unique for each batch, something that pre-programmed controllers cannot do. Table 7.2 Tabulated machine learning techniques for fermentation science.

7.3.2 Data Ecosystems for AI Models

The performance and predictive power of the machine learning techniques described above depend directly on the volume, diversity and accuracy of the data on which they are trained. Building a comprehensive data ecosystem is therefore a precondition for AI-based personalisation.

7.3.2.1 Multi-omics Data Integration (Genomics, Metabolomics, Proteomics)

Basic biological insights come from layered, multi-omics data. Microbial genomics (starter cultures and consumer gut microbiomes) provides a blueprint of the functional potential. Metabolomics profiles the actual small molecule fermentation outputs (in both food and consumer biofluids) and provides a functional information

Table 7.2 Machine learning (ML) techniques for fermentation science

ML technique	Primary function	Example application	Inputs → Outputs	References
Random forest	Predictive modelling	Forecasting final GABA concentration in fermented soybean paste.	Inputs: Substrate composition, time, temperature → Output: Metabolite yield, key influencing parameters.	SiYing and HuangHui (2025)
Long short-term memory (LSTM) network	Time-series forecasting	Predicting the optimal harvest window for a postbiotic-rich fermentation.	Inputs: Time-series sensor data (pH, temp) → Output: Forecasted trajectory of key bioactives.	Choi and Lee (2018)
Reinforcement learning	Dynamic process control	Real-time optimization of temperature and nutrient feed in a smart bioreactor.	Inputs: Real-time sensor state → Output: Control actions to maximize a reward (e.g., target compound yield).	Butean et al. (2025)
Multi-layer perceptron (MLP)	Supervised learning or regression	Predicting alcohol and substrate concentration in white wine fermentation.	Inputs: Temp, time, initial substrate, biomass	Khosravi and D'Aiello (2025)
Genetic algorithms (GA)	Heuristic optimization	Optimizing hyperparameters of neural networks for bioreactor control when hardware sensors are unavailable.	Inputs: Population size, mutation rate	Ismail et al. (2025)
Convolutional neural networks (CNN)	Image/pattern recognition	Real-time monitoring of microbial morphology and cell count via microscopic image analysis.	Inputs: Microscopic images →Output: Cell density and viability status.	Abeyrathna et al. (2022)
XGBoost (extreme gradient boosting)	Ensemble learning/ prediction	Predicting the flavor profile and sensory scores of fermented dairy products.	Inputs: Chemical composition, pH, starter culture type	Ling and Heldman (2025)
Transfer learning	Knowledge adaptation	Adapting a model trained on lab-scale fermentation to predict outcomes in large-scale industrial bioreactors.	Inputs: Lab-scale data features	Chen and Xia (2024)

(continued)

Table 7.2 (continued)

ML technique	Primary function	Example application	Inputs → Outputs	References
Principal component analysis (PCA)	Dimensionality reduction	Identifying key volatile markers responsible for consumer acceptability in shrimp paste.	Inputs: Multi-variate volatile profiles	Lin et al. (2025)
Gompertz-ANN hybrid	Kinetic modeling	Predicting the production of solvents (isobutanol, ethyl acetate) by *S. cerevisiae*.	Inputs: Fermentation time, temperature	Wang et al. (2025)

(Chen et al., 2017). Proteomics identifies the enzymes and bioactive peptides expressed that link genetic potential to metabolic reality. The main challenge and opportunity lies in data fusion integrating these disparate, highly dimensional data sets to create a single model. For example, a neural network may be designed to take genomic data from starter cultures and the gut microbiomes of consumers as input to predict the resulting proteomic and metabolomic profile of the fermented food and its subsequent impact on the urinary metabolome of the consumer (Gao et al., 2021).

7.3.2.2 Clinical and Real-Time Sensor Data

In order to link fermentation results to human health, artificial intelligence models need to be integrated with clinical data. This includes static data (e.g. blood lipid panel, inflammatory markers such as C-reactive protein (CRP), genotype) as well as dynamic, time-laced data (e.g. glucose monitoring, weight, blood pressure). At the same time, in the bioreactor, real-time sensor data (e.g. pH, dissolved oxygen, redox potential, and in-line spectra) provide a high-frequency stream of the state of the process. Combining this real time process data with slower, post-fermentation clinical results allows real-time agents and predictive models to learn which dynamic process models will ultimately lead to the most therapeutic product for the clinical profile (Bekatorou, 2019).

7.3.2.3 Consumer Preference and Sensory Data

To be successful, a personalized fermented food must be both nutritious and palatable. To ensure this, AI models may include consumer data. This includes hedonic (liking) scores, detailed sensory profiles (e.g. acidity, bitterness, umami), and even digital phenotyping data (e.g. images of consumption of foods). Natural language processing (NLP) can analyze open consumer feedback (Vashishtha &

Kapoor, 2023). By combining sensory data with metabolomic profiles, machine learning models can identify key compounds that drive acceptance and help to optimize fermentation parameters, achieving a balance between the intended health benefits and a sensory profile that is adapted to the preferences of the individual, ensuring long-term adherence (Ajayeoba & Ijabadeniyi, 2025; Priyadharshini et al., 2025) details artificial intelligence decodes sensory preferences using tools like convolutional neural networks (CNN) to predict flavour profiles, which directly supports your point about the integration of sensory data.

7.4 AI-Driven Frameworks for Customizing Fermented Foods

Building on the AI tools and scientific foundations set out in previous sections, this chapter explores the specific frameworks that translate computing power into personalised fermented products. This is the core of the paradigm shift as artificial intelligence (AI) moves from analytical to integrated design and control systems, enabling a new class of dynamically adjusted nutrition interventions.

7.4.1 Predictive Modelling of Microbial Dynamics and Metabolite Production

The first critical step in personalisation is the ability to predict accurately the microbial community's behaviour and metabolic output under a given set of conditions. This capability is provided by predictive modelling using supervised machine learning. Through training with historical fermentation data including initial substrate composition, environmental parameters and detailed time series measurements of microbial abundance (e.g. 16S rRNA sequencing) and concentrations of metabolites (e.g. LC-MS metabolomics) the algorithms can learn the complex non-linear relationships that underlie these biological processes.

For example, a recurrent neural network with long-term memory (LSTM) capability and an ability to process sequence data can be trained to predict the time course of a fermentation. Inputs to the model would include initial protein, carbohydrate and lipid concentrations; inoculum density of the microbial consortium; and temperature and pH time series. The output would be a forecast of the future concentration of key metabolites over time, such as specific bioactive peptides, organic acids or vitamins. This allows in-silico experiments where thousands of virtual fermentation runs can be simulated to determine the ideal starting conditions for maximising the yield of a target compound in a given set of materials, dramatically reducing the need for costly and time-consuming wet laboratory experiments (Yadav et al., 2024). In addition, these models may include kinetic parameters for metabolic flux analysis to identify enzymatic steps along the route that restrict the flow rate, which can then be used for strain engineering or optimization of the process (Priyadharshini et al., 2025). Discussion of the fusion of AI and CRISPR for the design of microbial strains (e.g. using deep learning to predict the results of gene-editing (*S. cerevisiae*) and generative models to design enzymes).

7.4.2 Design and Optimization of Personalized Starter Cultures

The traditional concept of the standard starter culture is rendered out of date in the context of individualisation. Instead, artificial intelligence allows for the rational design of custom microbial communities tailored to individual gut microbiomes and health goals. The process starts with sequencing the gut microbiomes of individuals (target environment). This metagenomic data is analysed by supervised learning (as described in Sect. 7.3.1.2) to characterise its functional capacity and to detect potential defects or dysbiosis. Subsequently, a supervised learning model, such as a Gradual Graduation Machine, is used to design the boot culture. The model is trained on a large database of microbial genomic traits, paired microbial interactions (e.g. synergies, competition, cross-feeding) and documented health outcomes. The model optimises both functional efficiency (e.g. expected *butyrogenesis*) and ecological compatibility (e.g. expected resistance to colonization and stability in native environments) (Silva Neto et al., 2023). The consortium, designed by the AI, can then be reproduced and used as a personal starter to ensure that the resulting fermented food incorporates microbes that are expected to be beneficial and perform specific functions in the body.

7.4.3 Tailoring Fermentation Parameters to Target Outcomes

Even with a customised starter culture, the fermentation process itself must be precisely tuned to ensure that the metabolism of the micro-organisms is directed towards the desired nutrient profile. Here, AI-driven optimization of process parameters becomes critical as the protocol moves from static to dynamic. The following case studies are conceptual frameworks to illustrate AI-based personalisation strategies. They are not the results of final clinical trials and no specific health outcomes are claimed without empirical verification.

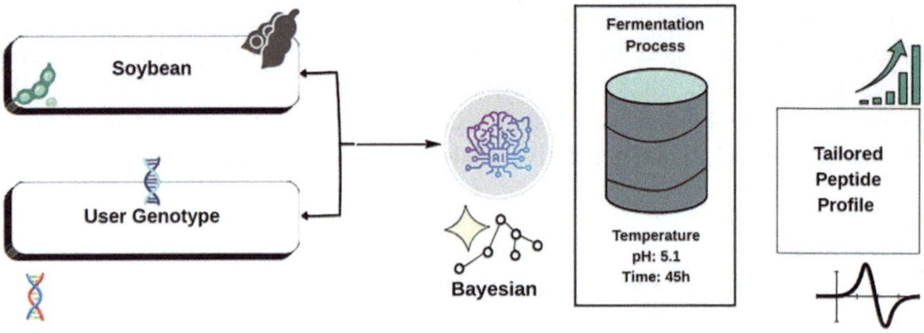

Fig. 7.2 Tempeh optimized by Tempeh.AI. (Source: Authors; created with Canva)

7.4.3.1 Case Study: AI-Optimized Tempeh for Enhanced Protein Digestibility

- **Objective**: To produce a tempeh with maximized protein digestibility and bioactive peptide profile in a patient with reduced protease activity in the pancreas, variability related to genotroph factors (as discussed in Sect. 7.2.1.1) (Fig. 7.2).
- **Data inputs**:
  - Personal data: genomic data of individuals indicating that trypsin activity has been reduced.
  - Substance data: soybean composition (protein content, lipid profile, anti-nutrients).
  - Microbiological data: genetic sequence of the strain *Rhizopus oligosporus* with particular reference to its protease enzymes (Purwandari et al., 2024)
  - Process data: historical data on temperature, humidity, fermentation time and the resulting hydrolysis of the protein (measured by degree of hydrolysis and peptide sequence by peptidomics (Sukrasno et al., 2025)).

- **AI model and workflow**: The algorithm used is the Bayesian Optimization Algorithm. The aim of the model is to find the optimal set of parameters (temperature, humidity, incubation period) to maximise the composite reward function (CPR) (Siska et al., 2026). This function balances the ultimate concentration of small, easily digestible peptides (e.g. ACE inhibitors) with the minimisation of larger, digestible peptides. The analytical unit proposes a set of parameters for the fermentation in small batches. The resulting tempeh is analysed by peptidomics and the results fed back into the Bayesian model, which updates its assumptions on parameter space and proposes a refined set of conditions for the next batch of tempehs. After several iterations, a personalized fermentation protocol is being developed for example, longer fermentation at a lower temperature, to promote the specific proteolytic activity needed for the individual (Siska et al., 2026)

- **Personalized output**: A tempeh product with a specifically adapted peptide profile that compensates for the reduced digestive capacity of the individual, ensuring optimum absorption of amino acids and delivering a specific set of antihypertensive biological active substances.

7.4.3.2 Case Study: Personalized Plant-Based Yogurt for Glycemic Control

- **Objective**: For fermentation of plant-based yogurt which produces a specific ratio of organic acids (e.g.acetate, propionate) which are known to modulate glycaemic responses and are well adapted to pre-diabetic individuals and their unique gut microbiomes.
- **Data inputs**:
  - Personal data: continuous glucose monitoring (CGM) data and gut microbial profiles.
  - Substances data: compound base of the plant (e.g. Oats, almonds, peanuts)
  - Microbial data: library of lactic acid and bifidobacterium strains with known metabolic pathways for organic acids production (Wang et al., 2021)
  - Processing data: historical fermentation data, combining strain combinations, fermentation time, temperature and final organic acid profile with postprandial glucose response in cohort studies.
- **AI model and workflow**: Multiple output random forest model is first used (Li et al., 2025) as the first choice for selecting the optimal starter culture. It takes into account individual microbiological data and the composition of the plant base and predicts which strains of *Lactobacillus* and *Bifidobacterium* will be most efficient in fermenting available sugars, thus achieving a high ratio of acetate to propionate, which is associated with a better control of the bacterial glycosaminoglycae. After selection of the culture, reinforcement learning (RL) is applied during fermentation. RL receives real-time data from in-line sensors for pH and near-infrared (NIR) to estimate the concentration of organic acids (Yang et al., 2021). Its working surface includes temperature adjustment. His reward is based on how closely his metabolic path in real time matches the required organic acid profile predicted by a random forest model.
- **Personalized output**: Plant-based yogurt with a non-standard fermentation end point determined not by time or pH alone but by the achievement of the target metabolite profile by an artificial insemination. This medicine is designed to specifically modulate the gut microbiomes of consumers and hepatic glucose production, resulting in reduced postprandial glucose spikes as confirmed by their CGM.

7.4.4 Real-Time Monitoring and Control in Smart Bioreactors

The final expression of AI-driven personalisation is the closed loop control of the fermentation process itself, which goes from batch optimization to dynamic control of the fermentation process in real time. This is done by smart bioreactors equipped with a complex set of sophisticated, often interconnected sensors. These sensors provide a continuous, multidimensional flow of data on the state of the process, including traditional variables (temperature, pH, dissolved oxygen) and advanced spectrochemical data (e.g., NIR, Raman) that can be used to determine the concentration of key metabolites and biomass.

The AI controller, which typically uses the reinforcement learning framework described in Sect. 7.3.1.3, acts as the bio-actor's 'brain'. The agent is trained, often in a high fidelity digital twin of the bioreactor, to keep the fermentation process on a path leading to the desired individualised output. For example, if the NIR (near infrared) sensor indicates a deviation from the target metabolic flux for the production of a specific postbiotic, the RL may calculate and apply a correction, such as a slight adjustment of the aeration rate or a pulsed addition of micronutrients. This creates a flexible system that can compensate for the variability of the raw material or unexpected microbial behaviour, ensuring that each small-scale fermentation consistently achieves its own unique, individual goal. This real-time AI control is the enabling technology to translate predictive models and optimized design into tangible, highly personalised fermented foods (Bertorello et al., 2024).

7.5 Implementation Roadmap: From Data to Product

The translation of the concept of AI-personalised fermentation into tangible products requires a systematic implementation path. This plan outlines the critical stages from data collection to delivery to consumers, and highlights the technological infrastructure and scaling challenges that need to be addressed.

7.5.1 A Workflow for Developing AI-Personalized Fermented Foods

The development pipeline of personalised fermented food is following an iterative, data-driven process (Fig. 7.3). It starts with data acquisition and integration, where multi-omics data (genome, metabolite, clinical biomarker, and consumer preferences) are collected from the target individual and consolidated in a secure data lake. This is followed by AI model training and validation, where machine learning models are trained on this integrated dataset as described in Sect. 7.3 to predict

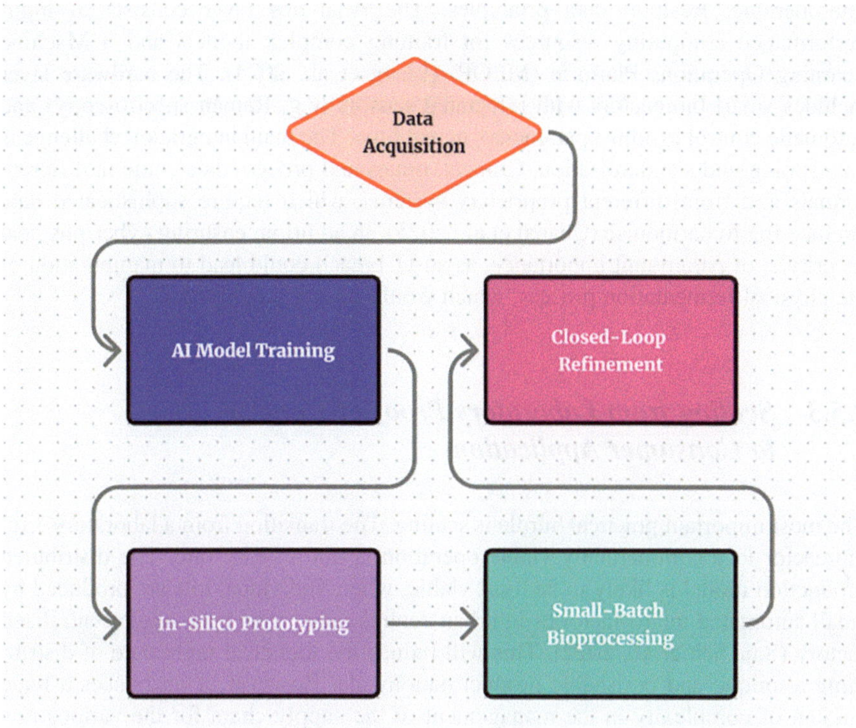

Fig. 7.3 Development workflow for AI-personalized fermented foods. (Source: Authors; created with Canva)

optimal strain combinations and fermentation parameters for desired health outcomes. The third stage is the In Silico prototype process, which simulates thousands of virtual fermentation experiments using a digital bioreactor twin to identify the most promising candidate protocols, cutting development time and costs dramatically. The fourth stage is small-scale bio-processing in smart bioreactors, where the protocols of the best candidates are physically implemented and the results are rigorously analysed. The final key step is closed loop model refinement, where the difference between predicted and observed outcomes is used to train and hone artificial intelligence models, creating a virtuous circle of increased accuracy and personalization (Marinos et al., 2024).

7.5.2 Technological Infrastructure and Integration Challenges

The backbone of this process is a solid engineering stack composed of several layers. The data layer requires secure and interoperable databases for a variety of data types, often using cloud-based platforms based on the FAIR (Findable, Accessible,

Interoperable, Resible) data principles. The Analytics layer consists of high-performance computing resources for training complex models and a Machine Learning Operations Platform (MLOP) (Liang et al., 2024). The hardware layer includes smart bioreactors with integrated sensors (e.g. Raman spectroscopy) and automatic control to adjust parameters in real time. The main integration challenge is data siloing and standardisation. Clinical, omics and process data come in different formats and from different proprietary systems, which require sophisticated data engineering to harmonise (Chahal et al., 2025). In addition, ensuring cyber-physical security is of paramount importance, as an IT breach could lead to manipulation of the physical fermentation process, which would pose a security risk.

7.5.3 Scaling from Laboratory Proof-of-Concept to Consumer Application

The most important practical hurdle is scaling. The transition from a laboratory 1 ml bioreactor to a commercially viable operation is not an easy one. The distributed production model is likely to be most viable, where individual lots are produced by small automated micro-factories in urban centres rather than by a single centralized factory (Sajadieh et al., 2022). This will reduce the logistical nightmare of distributing a unique and perishable product nationwide. However, it introduces a huge amount of complexity in the management of the supply chain for the various raw materials and culture starters and requires rigorous automated quality control at all stages. Another critical bottleneck is the economics model. The high costs of data acquisition, sequencing and small scale production risk making these products prohibitively expensive. Achieving economies of scale will depend on automation, lower cost of sensors and, where appropriate, a subscription model that reduces the cost of initial personalisation over time (Puillet et al., 2021).

7.6 Challenges and Ethical Considerations

The road to AI-personalised fermentation is full of significant non-technical obstacles which need to be tackled as rigorously as the scientific and technical challenges (Table 7.3).

7.6.1 Data Privacy, Security, and Ownership

The model is based on the most sensitive personal data imaginable: genomic sequences, intestinal microbiological profiles, and detailed medical records. This creates a high-value target for cyber attacks. Beyond safety, ownership and use

Table 7.3 Challenges, solutions, and stakeholders for AI-personalized fermentation

Challenge	Potential solution	Key stakeholders	References
Data privacy & ownership	Implement blockchain for secure, transparent consent logging; Granular user data control.	Consumers, industry, regulators, data security experts	Ameyed et al. (2021)
Regulatory hurdles	Shift from product- to process-based certification; Validate AI models and control systems.	Industry, FDA/EFSA (regulators), legal experts	Benerofe (2025)
Equity & digital divide	Develop public-private funding models; Target public health applications.	Policymakers, industry, public health agencies	Arnaout et al. (2023)
Consumer trust and transparency	Prioritize Explainable AI (XAI) for clear recommendations; Full disclosure of data use.	Industry, consumers, ethicists	Olateju et al. (2024)
Algorithmic bias	Regular auditing of AI models for demographic bias; use of diverse training datasets.	Data scientists, ethicists, minority advocacy groups	Agboola (2025)
High cost of multi-omics	Development of low-cost biosensors and surrogate biomarkers to reduce sequencing needs.	Biotech startups, research institutions, investors	Cao et al. (2025)
Interoperability of data	Establishing global standards for nutritional and genetic data exchange (e.g., Fast Healthcare Interoperability Resources (FHIR)).	Software developers, international standards orgs	Hughes and Kalra (2023)
Microbial safety risks	AI-driven real-time pathogen detection and automated "kill-switch" protocols in bioreactors.	Food safety authorities, microbiologists, engineers	Tuan et al. (2024)
Sustainability & energy	Optimization of AI model efficiency to reduce the carbon footprint of large-scale computation.	Environmental agencies, cloud providers, industry	Oyewole and Joseph (2025)
Consumer literacy	Development of intuitive AI-driven apps that simplify complex genetic/nutritional data.	Educators, UX designers, consumers	Prasad et al. (2025)
Supply chain complexity	Decentralized "Micro-factories" or 3D food printing for on-demand personalized production.	Logistics experts, manufacturers, retailers	Hachey (2018)
Long-term efficacy data	Longitudinal studies using wearable data to track the impact of personalized ferments over years.	Public health researchers, data analysts	Menassa et al. (2025)

rights are complex issues. Does an individual retain their microbiological data after sequencing? Can a company use personal data to train models for other customers? Current regulations, such as the General Data Protection Regulation (GDPR) and the Health Insurance Portability and Accountability Act (HIPAA), provide a framework but do not fully address the complexity of biomonetary data (Kumar et al., 2024). A breach could lead to genetic discrimination or individual manipulation of food. Transparent data management models, possibly based on a block chain for permanent consent, and granular user control of their data are essential to build trust and confidence (Ameyed et al., 2021).

7.6.2 Regulatory Barriers for Dynamic, Personalized Products

The existing food regulatory frameworks, such as those operated by the FDA and EFSA, are designed to deal with static, mass-produced products which cannot be reproduced from the product base (EFSA NDA Panel, 2016). How do you approve a meal that has a unique, dynamically-generated composition for each batch and each individual? The traditional notion of a standardized identity is rendered meaningless. The traditional concept of a single identity has become obsolete. Regulators may need to move from approving specific products to certifying the AI process itself, auditing algorithms, data integrity and manufacturing controls to ensure that products are safe and accurately identified (Ranjitsingh & Rao, 2025). Moreover, the justification of health claims becomes more difficult; one cannot conduct a clinical trial for each unique product variant. Regulators can accept evidence based on validated artificial intelligence predictions and surrogate endpoints, a departure from the current practice (Lesko & Atkinson Jr, 2001).

7.6.3 Ensuring Equity and Avoiding a "Digital Divide" in Nutrition

There is a strong risk that personalized AI nutrition will become a luxury only for the rich, exacerbating health inequalities. This creates a 'genetic elite' that can optimize their health, while others remain with conventional, less efficient choices. The high costs of technology, data sequencing, and custom-tailor-made production could easily make such interventions inaccessible to low-income populations (Gelband et al., 2016). Ensuring equitable access will require public sector investment, political incentives and the development of low-cost, high-impact applications that address the wide range of public health concerns rather than just cater to biohackers and affluent consumers.

7.6.4 Consumer Confidence and Transparency in AI-Driven Food Systems

Consumers are often wary of both artificial intelligence and GM microorganisms, a combination that can cause a 'yuck factor' and a high degree of resistance. The black box nature of some complex artificial intelligence models, where the reasons for decisions are unclear, is a major obstacle to trust. Companies should give priority to Explanable AI (XAI), which can provide a simple and understandable reason for recommending a particular product (e.g. a case of product recommendation). This yogurt has been formulated to support the gut microbes involved in the metabolism of glucose (Dwivedi et al., 2023). Full transparency on data usage, training algorithms and the limits of artificial intelligence recommendation is not only an ethical imperative, but also a commercial necessity for its adoption.

7.7 Future Perspectives and Conclusion

7.7.1 Emerging Trends: Integration with IoT and Home Fermentation Devices

The final expression of personalisation can be the transfer of technology from the centralised micro-factories to the household. We envisage smart, countertop fermentation devices precision fermenters that are connected to the Internet of Things (IoT). These devices could receive customised fermentation protocols via the cloud based on data from user-wearing health monitors (e.g. glucose continuous monitoring, smart rings). The device then automatically ferments and produces a weekly supply of personalised yoghurt, tempeh or kefir, tailored to the physiological state of the user in real time.

7.7.2 The Role of AI in Sustainable and Circular Food Systems

AI-assisted fermentation can be a powerful force for sustainability. It reduces waste by optimising yields and nutrient content. More fundamentally, AI can design fermentation processes that reward agricultural incremental streams and food waste in the form of personalized nutritional products. For example, artificial intelligence could design microbial consortiums and process parameters to turn spent grain or fruit pulp into a high-quality, palatable fermented drink tailored to local health needs, creating a circular-economy model of food.

7.7.3 Key Research Gaps and Avenues for Future Work

Significant research gaps remain. There are still considerable research gaps to fill. Large scale longitudinal studies are needed to confirm the health benefits of AI fermented foods as opposed to generic products. On the technological side, research into more reliable and cheaper in-line sensors is needed, as well as the development of `minimal machine learning'algorithms that can run on low-power, high-efficiency appliances for home use. Ethically, there is an urgent need for cross-disciplinary research to develop new regulatory frameworks with government agencies and to develop fair business models for personalised nutrition (Myers, 2024).

7.7.4 Concluding Remarks: The Future Is Personalized and Fermented

This chapter outlines the proposed framework: the fusion of ancient fermentation processes with cutting-edge artificial intelligence is creating a new paradigm for food. We have gone from establishing a scientific rationale based on human biological variability to an artificial intelligence toolkit that makes personalization computationally possible, to specific frameworks and ethical considerations guiding its implementation. This is not just an incremental improvement, but a transformational shift from reactive, generalised nutrition to proactive, individualised nutrition. The challenges are great, but the potential rewards a world where our food adapts dynamically to our health, is produced sustainably and is available to all are worth exploring vigorously and responsibly. The future of food is not only fermenting; it is also intelligent, adaptive, and uniquely ours.

References

Abeyrathna, D., Rauniyar, S., Sani, R. K., & Huang, P.-C. (2022). A morphological post-processing approach for overlapped segmentation of bacterial cell images. *Machine Learning and Knowledge Extraction, 4*(4), 1024–1041. https://doi.org/10.3390/make4040052

Agboola, O. K. (2025). Auditing bias in AI and machine learning-based credit algorithms: A data science perspective on fairness and ethics in FinTech. *International Journal of Technology, Management and Humanities, 11*(2), 1–11. https://doi.org/10.21590/ijtmh.11.02.10

Ajayeoba, T. A., & Ijabadeniyi, O. A. (2025). *Transforming food for the future: Precision fermentation as a key to sustainability, nutrition, and health.* https://doi.org/10.2139/ssrn.5119686

Ameyed, D., Jaafar, F., Charette-Migneault, F., & Cheriet, M. (2021). Blockchain based model for consent management and data transparency assurance. In *2021 IEEE 21st international conference on software quality, reliability and security companion (QRS-C)*. IEEE. https://doi.org/10.1109/qrs-c55045.2021.00159

Arnaout, A., Oseguera-Arasmou, M., Mishra, N., Liu, B. M., Bhattacharya, A., & Rhew, D. C. (2023). Leveraging technology in public-private partnerships: A model to address public health inequities. *Frontiers in Health Services, 3*, 1187306. https://doi.org/10.3389/frhs.2023.1187306

Baucells, F., Perez, J., Morales, J., & Gasa, J. (2000). Effect of a α-galactosidase supplementation of cereal-soya-bean-pea diets on the productive performances, digestibility and lower gut fermentation in growing and finishing pigs. *Animal Science, 71*(1), 157–164. https://doi.org/10.1017/s1357729800054989

Bekatorou, A. (2019). *Advances in vinegar production* (Vol. 1). CRC Press. https://doi.org/10.1201/9781351208475

Bell, V., Ferrão, J., Pimentel, L., Pintado, M., & Fernandes, T. (2018). One health, fermented foods, and gut microbiota. *Foods, 7*(12), 195. https://doi.org/10.3390/foods7120195

Benerofe, S. (2025). *AI governance and the verification gap: A framework for law and policy under computational intractability*. Available at SSRN 5629290. https://doi.org/10.2139/ssrn.5629290

Bertorello, S., Cei, F., Fink, D., Niccolai, E., & Amedei, A. (2024). The future exploring of gut microbiome-immunity interactions: From in vivo/vitro models to in silico innovations. *Microorganisms, 12*(9), 1828. https://doi.org/10.3390/microorganisms12091828

Bhat, Z., Kumar, S., & Bhat, H. F. (2015). Bioactive peptides of animal origin: A review. *Journal of Food Science and Technology, 52*(9), 5377–5392. https://doi.org/10.1007/s13197-015-1731-5

Brouns, F. (2021). Phytic acid and whole grains for health controversy. *Nutrients, 14*(1), 25. https://doi.org/10.3390/nu14010025

Butean, A., Cutean, I., Barbero, R., Enriquez, J., & Matei, A. (2025). A review of artificial intelligence applications for biorefineries and bioprocessing: From data-driven processes to optimization strategies and real-time control. *PRO, 13*(8), 2544. https://doi.org/10.3390/pr13082544

Cao, H., Oghenemaro, E. F., Latypova, A., Abosaoda, M. K., Zaman, G. S., & Devi, A. (2025). Advancing clinical biochemistry: Addressing gaps and driving future innovations. *Frontiers in Medicine, 12*, 1521126. https://doi.org/10.3389/fmed.2025.1521126

Chahal, C. A. A., Alahdab, F., Asatryan, B., Addison, D., Aung, N., Chung, M. K., Denaxas, S., Dunn, J., Hall, J. L., & Pamir, N. (2025). Data interoperability and harmonization in cardiovascular genomic and precision medicine. *Circulation: Genomic and Precision Medicine, 18*, e004624. https://doi.org/10.1161/circgen.124.004624

Chang, T. G., Yen, T. T., Wei, C. Y., Hsiao, T. H., & Chen, I. C. (2023). Impacts of ADH1B rs1229984 and ALDH2 rs671 polymorphisms on risks of alcohol-related disorder and cancer. *Cancer Medicine, 12*(1), 747–759. https://doi.org/10.1002/cam4.4920

Chen, T. (2016). *Modulation of gut microbiota in vitro in stools from healthy and diseased individuals using soluble plant cell wall-based dietary* [Fibers Purdue University]. https://doi.org/10.26481/dis.20191024mm

Chen, M., & Xia, J. (2024). Optimization parameters for efficient scale-up of fermentation process. In *Scale-up and chemical process for microbial production of plant-derived bioactive compounds* (pp. 29–42). Elsevier. https://doi.org/10.1016/b978-0-443-15584-0.00002-1

Chen, G., Chen, C., & Lei, Z. (2017). Meta-omics insights in the microbial community profiling and functional characterization of fermented foods. *Trends in Food Science & Technology, 65*, 23–31. https://doi.org/10.1016/j.tifs.2017.05.002

Chin, E. L., Huang, L., Bouzid, Y. Y., Kirschke, C. P., Durbin-Johnson, B., Baldiviez, L. M., Bonnel, E. L., Keim, N. L., Korf, I., & Stephensen, C. B. (2019). Association of lactase persistence genotypes (rs4988235) and ethnicity with dairy intake in a healthy US population. *Nutrients, 11*(8), 1860. https://doi.org/10.3390/nu11081860

Choi, J. Y., & Lee, B. (2018). Combining LSTM network ensemble via adaptive weighting for improved time series forecasting. *Mathematical Problems in Engineering, 2018*(1), 2470171. https://doi.org/10.1155/2018/2470171

Corella, D., & Ordovas, J. M. (2005). Single nucleotide polymorphisms that influence lipid metabolism: Interaction with dietary factors. *Annual Review of Nutrition, 25*(1), 341–390. https://doi.org/10.1146/annurev.nutr.25.050304.092656

D'Urso, F., & Broccolo, F. (2024). Applications of artificial intelligence in microbiome analysis and probiotic interventions—An overview and perspective based on the current state of the art. *Applied Sciences, 14*(19), 8627. https://doi.org/10.3390/app14198627

Dawlal, P. (2018). *Reducing fumonisin exposure in African fermented maize-based foods with lactic acid bacteria*. University of Pretoria. https://doi.org/10.3920/wmj2017.2184

Denden, S., Bouden, B., Haj Khelil, A., Ben Chibani, J., & Hamdaoui, M. (2016). Gender and ethnicity modify the association between the CYP1A2 rs762551 polymorphism and habitual coffee intake: Evidence from a meta-analysis. *Genetics and Molecular Research, 15*(2), 7487. https://doi.org/10.4238/gmr.15027487

Deokar, G. S., Pathak, V. A., Kshirsagar, S. J., Al-Asmari, F., & Nirmal, N. (2025). Enhancement of vitamin B12 in plant-based food through microbial fermentation-a sustainable food system. *Food Chemistry, 484*, 144437. https://doi.org/10.1016/j.foodchem.2025.144437

Dwivedi, R., Dave, D., Naik, H., Singhal, S., Omer, R., Patel, P., Qian, B., Wen, Z., Shah, T., & Morgan, G. (2023). Explainable AI (XAI): Core ideas, techniques, and solutions. *ACM Computing Surveys, 55*(9), 1–33. https://doi.org/10.1145/3561048

EFSA Panel on Dietetic Products, Nutrition, & Allergies. (2016). Guidance on the scientific requirements for health claims related to the immune system, the gastrointestinal tract and defence against pathogenic microorganisms. *EFSA Journal, 14*(1), 4369. https://doi.org/10.2903/j.efsa.2016.4369

Elechi, J. O. G., Sirianni, R., Conforti, F. L., Cione, E., & Pellegrino, M. (2023). Food system transformation and gut microbiota transition: Evidence on advancing obesity, cardiovascular diseases, and cancers—A narrative review. *Foods, 12*(12), 2286. https://doi.org/10.3390/foods12122286

Ferrocino, I., Rantsiou, K., McClure, R., Kostic, T., De Souza, R. S. C., Lange, L., FitzGerald, J., Kriaa, A., Cotter, P., & Maguin, E. (2023). The need for an integrated multi-OMICs approach in microbiome science in the food system. *Comprehensive Reviews in Food Science and Food Safety, 22*(2), 1082–1103. https://doi.org/10.1111/1541-4337.13103

Gao, Y., Hou, L., Gao, J., Li, D., Tian, Z., Fan, B., Wang, F., & Li, S. (2021). Metabolomics approaches for the comprehensive evaluation of fermented foods: A review. *Foods, 10*(10), 2294. https://doi.org/10.3390/foods10102294

Gao, Y., Liu, Y., Ma, T., Liang, Q., Sun, J., Wu, X., Song, Y., Nie, H., Huang, J., & Mu, G. (2025). Fermented dairy products as precision modulators of gut microbiota and host health: Mechanistic insights, clinical evidence, and future directions. *Foods, 14*(11), 1946. https://doi.org/10.3390/foods14111946

Gaur, S. S. (2024). Utilizing fermented foods in traditional therapies for addressing metabolic disorders. In *Traditional foods: The reinvented superfoods* (pp. 427–454). Springer. https://doi.org/10.1007/978-3-031-72757-3_19

Gelband, H., Sankaranarayanan, R., Gauvreau, C. L., Horton, S., Anderson, B. O., Bray, F., Cleary, J., Dare, A. J., Denny, L., & Gospodarowicz, M. K. (2016). Costs, affordability, and feasibility of an essential package of cancer control interventions in low-income and middle-income countries: Key messages from disease control priorities. *The Lancet, 387*(10033), 2133–2144. https://doi.org/10.1016/s0140-6736(15)00755-2

Gil, J. D., Chanona, E. A. D. R., Guzmán, J. L., & Berenguel, M. (2025). Reinforcement learning meets bioprocess control through behaviour cloning: Real-world deployment in an industrial photobioreactor. *arXiv preprint arXiv:2509.06853*. https://doi.org/10.1016/j.engappai.2025.113326

Gutierrez-Virgen, J. E., Pina-Pozas, M., Hernandez-Tobias, E. A., Taja-Chayeb, L., Lopez-Gonzalez, M. D. L., Meraz-Rios, M. A., & Gomez, R. (2023). NAT2 global landscape: Genetic diversity and acetylation statuses from a systematic review. *PLoS One, 18*(4), e0283726. https://doi.org/10.1371/journal.pone.0283726

Hachey, S. Q. (2018). *Prospects for sustainable micro-factory retailing in Canada: A case study of 3D printed electric vehicles*. https://doi.org/10.32014/2025.2518-170x.510

Hiraoka, M., & Kagawa, Y. (2017). Genetic polymorphisms and folate status. *Congenital Anomalies, 57*(5), 142–149. https://doi.org/10.1111/cga.12232

Hughes, N., & Kalra, D. (2023). Data standards and platform interoperability. In *Real-world evidence in medical product development* (pp. 79–107). Springer. https://doi.org/10.1007/978-3-031-26328-6_6

Ismail, U. M., Bani-Melhem, K., Khan, M. F., & Elnakar, H. (2025). Optimizing membrane bioreactor performance in wastewater treatment using machine learning and meta-heuristic techniques. *Results in Engineering, 25*, 104626. https://doi.org/10.1016/j.rineng.2025.104626

Khaleghi, M. K., Savizi, I. S. P., Lewis, N. E., & Shojaosadati, S. A. (2021). Synergisms of machine learning and constraint-based modeling of metabolism for analysis and optimization of fermentation parameters. *Biotechnology Journal, 16*(11), 2100212. https://doi.org/10.1002/biot.202100212

Khosravi, P., & D'Aiello, C. (2025). Enhancing wine fermentation: The role of AI-driven predictive modeling in flavor optimization. In *International conference on science, engineering management and information technology*. Springer. https://doi.org/10.1007/978-3-032-04225-5_5

Kolodziejczyk, A. A., Zheng, D., & Elinav, E. (2019). Diet–microbiota interactions and personalized nutrition. *Nature Reviews Microbiology, 17*(12), 742–753. https://doi.org/10.1038/s41579-019-0256-8

Kopczyńska, J., & Kowalczyk, M. (2024). The potential of short-chain fatty acid epigenetic regulation in chronic low-grade inflammation and obesity. *Frontiers in Immunology, 15*, 1380476. https://doi.org/10.3389/fimmu.2024.1380476

Kumar, P., Sinha, R., & Shukla, P. (2022). Artificial intelligence and synthetic biology approaches for human gut microbiome. *Critical Reviews in Food Science and Nutrition, 62*(8), 2103–2121. https://doi.org/10.1080/10408398.2020.1850415

Kumar, D., Gupta, S., & Parmar, P. (2024). Legal frameworks for data privacy and ownership in metagenomics. In *Genomic intelligence* (pp. 291–305). CRC Press. https://doi.org/10.1201/9781003570233

Kussmann, M., Abe Cunha, D. H., & Berciano, S. (2023). Bioactive compounds for human and planetary health. *Frontiers in Nutrition, 10*, 1193848. https://doi.org/10.3389/fnut.2023.1193848

Laforest-Lapointe, I., & Arrieta, M.-C. (2017). Patterns of early-life gut microbial colonization during human immune development: An ecological perspective. *Frontiers in Immunology, 8*, 788. https://doi.org/10.3389/fimmu.2017.00788

Lampe, J. W., Navarro, S. L., Hullar, M. A., & Shojaie, A. (2013). Inter-individual differences in response to dietary intervention: Integrating omics platforms towards personalised dietary recommendations. *Proceedings of the Nutrition Society, 72*(2), 207–218. https://doi.org/10.1017/s0029665113000025

Lei, V., Amoa-Awua, W. K. A., & Brimer, L. (1999). Degradation of cyanogenic glycosides by lactobacillus plantarum strains from spontaneous cassava fermentation and other microorganisms. *International Journal of Food Microbiology, 53*(2–3), 169–184. https://doi.org/10.1016/s0168-1605(99)00156-7

Lesko, L. J., & Atkinson, A., Jr. (2001). Use of biomarkers and surrogate endpoints in drug development and regulatory decision making: Criteria, validation, strategies. *Annual Review of Pharmacology and Toxicology, 41*(1), 347–366. https://doi.org/10.1146/annurev.pharmtox.41.1.347

Li, Y., Gu, Y., Cheng, W., Li, Z., Zhang, X., Zhao, Y., Ko, K., Liu, W., Liu, X., & Li, H. (2025). Fermentation modeling and machine learning for flavor prediction in low-sodium radish paocai with potassium chloride substitution. *npj Science of Food, 9*(1), 156. https://doi.org/10.1038/s41538-025-00528-2

Liang, P., Song, B., Zhan, X., Chen, Z., & Yuan, J. (2024). Automating the training and deployment of models in MLOps by integrating systems with machine learning. *arXiv preprint arXiv:2405.09819*. https://doi.org/10.54254/2755-2721/76/20240690

Lin, Z., Zhang, Z., Zheng, Z., Hou, R., Zhang, Y., Zheng, B., Sriboonvorakul, N., & Hu, J. (2025). Effects of hot-air drying conditions on quality attributes of meat and shell of dried shrimp. *Foods, 14*(23), 4041. https://doi.org/10.3390/foods14234041

Ling, J., & Heldman, D. R. (2025). Integration of machine learning technologies in food flavor research: Current applications, challenges, and future perspectives. *International Journal of Agricultural and Biological Engineering, 18*(3), 1–11. https://doi.org/10.25165/j.ijabe.20251803.9623

Margină, D., Ungurianu, A., Purdel, C., Tsoukalas, D., Sarandi, E., Thanasoula, M., Tekos, F., Mesnage, R., Kouretas, D., & Tsatsakis, A. (2020). Chronic inflammation in the context of everyday life: Dietary changes as mitigating factors. *International Journal of Environmental Research and Public Health, 17*(11), 4135. https://doi.org/10.3390/ijerph17114135

Marinos, G., Hamerich, I. K., Debray, R., Obeng, N., Petersen, C., Taubenheim, J., Zimmermann, J., Blackburn, D., Samuel, B. S., & Dierking, K. (2024). Metabolic model predictions enable targeted microbiome manipulation through precision prebiotics. *Microbiology Spectrum, 12*(2), e01144–e01123. https://doi.org/10.1101/2023.02.17.528811

Menassa, M., Wilmont, I., Beigrezaei, S., Knobbe, A., Arita, V. A., Bridge, L., Verschuren, W. M., Rennie, K. L., Franco, O. H., & van der Ouderaa, F. (2025). The future of healthy ageing: Wearables in public health, disease prevention and healthcare. *Maturitas, 196*, 108254. https://doi.org/10.1016/j.maturitas.2025.108254

Molloy, A. M. (2002). Folate bioavailability and health. *International Journal for Vitamin and Nutrition Research, 72*(1), 46–52. https://doi.org/10.1024/0300-9831.72.1.46

Montazeri-Najafabady, N. (2025). From one-size-fits-all to precision medicine: The promise of personalized probiotics. *Probiotics and Antimicrobial Proteins, 18*, 1–34. https://doi.org/10.1007/s12602-025-10815-9

Mukherjee, A., Abraham, S., Singh, A., Balaji, S., & Mukunthan, K. (2025). From data to cure: A comprehensive exploration of multi-omics data analysis for targeted therapies. *Molecular Biotechnology, 67*(4), 1269–1289. https://doi.org/10.1007/s12033-024-01133-6

Myers, M. C. (2024). *Data privacy laws in the United States and Germany: Implications for genomics research and personalized medicine* [University of Pittsburgh]. https://doi.org/10.1093/wentk/9780190234775.003.0020

Nikam, P., Pophaly, S., & Rizwan, M. (2025). Lactose intolerance: An overview. In *Lactose hydrolysis in dairy products* (pp. 1–27). Springer. https://doi.org/10.1007/978-3-031-78207-7_1

Olateju, O., Okon, S. U., Olaniyi, O. O., Samuel-Okon, A. D., & Asonze, C. U. (2024). *Exploring the concept of explainable AI and developing information governance standards for enhancing trust and transparency in handling customer data*. Available at SSRN. https://doi.org/10.2139/ssrn.4879025.

Oliphant, K., & Allen-Vercoe, E. (2019). Macronutrient metabolism by the human gut microbiome: Major fermentation by-products and their impact on host health. *Microbiome, 7*(1), 91. https://doi.org/10.1186/s40168-019-0704-8

Oyewole, O. O., & Joseph, J. F. (2025). Sustainable AI and green computing: Reducing the environmental impact of large-scale models with energy-efficient techniques. *International Journal of Scientific Research in Network Security and Communication, 13*(3), 19–26. https://doi.org/10.26438/ijsrnsc.v13i3.276

Park, I., & Mannaa, M. (2025). Fermented foods as functional systems: Microbial communities and metabolites influencing gut health and systemic outcomes. *Foods, 14*(13), 2292. https://doi.org/10.3390/foods14132292

Parmar, A., Alakulppi, N., Paavola-Sakki, P., Kurppa, K., Halme, L., Färkkilä, M., Turunen, U., Lappalainen, M., Kontula, K., & Kaukinen, K. (2012). Association study of FUT2 (rs601338) with celiac disease and inflammatory bowel disease in the Finnish population. *Tissue Antigens, 80*(6), 488–493. https://doi.org/10.1111/tan.12016

Prasad, G., Padhiary, M., Hoque, A., & Kumar, K. (2025). AI-driven personalized nutrition apps and platforms for enhanced diet and wellness. In *Food in the metaverse and web 3.0 era: Intersecting food, technology, and culture* (pp. 125–158). IGI Global Scientific Publishing. https://doi.org/10.4018/979-8-3693-9025-2.ch006

Priyadharshini, D., Muthuvel, I., Saraswathy, S., Kavitha, P., & Jegadeeswari, V. (2025). Precision to plate: AI-driven innovations in fermentation and hyper-personalized diets. *Frontiers in Nutrition, 12*, 1659511. https://doi.org/10.3389/fnut.2025.1659511

Puillet, L., Ducrocq, V., Friggens, N., & Amer, P. (2021). Exploring underlying drivers of genotype by environment interactions in feed efficiency traits for dairy cattle with a mechanistic model involving energy acquisition and allocation. *Journal of Dairy Science, 104*(5), 5805–5816. https://doi.org/10.3168/jds.2020-19610

Purwandari, F. A., Fogliano, V., & Capuano, E. (2024). Tempeh fermentation improves the nutritional and functional characteristics of Jack beans (Canavalia ensiformis (L.) DC). *Food & Function, 15*(7), 3680–3691. https://doi.org/10.1039/d3fo05379b

Rai, V. (2017). Association of C677T polymorphism (rs1801133) in MTHFR gene with depression. *Cellular and Molecular Biology, 63*(6), 60–67. https://doi.org/10.14715/cmb/2017.63.6.13

Ranjitsingh, L. M., & Rao, T. S. (2025). Establish legal and regulatory standards for the testing and validation of AI systems to ensure their reliability and safety in operational environments. *International Journal of System Assurance Engineering and Management, 16*(10), 3338–3353. https://doi.org/10.1007/s13198-025-02865-7

Rasmussen, K. L., Tybjærg-Hansen, A., Nordestgaard, B. G., & Frikke-Schmidt, R. (2015). Plasma levels of apolipoprotein E and risk of dementia in the general population. *Annals of Neurology, 77*(2), 301–311. https://doi.org/10.1002/ana.24326

Raveschot, C., Cudennec, B., Coutte, F., Flahaut, C., Fremont, M., Drider, D., & Dhulster, P. (2018). Production of bioactive peptides by Lactobacillus species: From gene to application. *Frontiers in Microbiology, 9*, 2354. https://doi.org/10.3389/fmicb.2018.02354

Ray, P., Faseeh, M. A., Adak, D., & Sharma, H. (2024a). Probiotics, prebiotics, and postbiotics on metabolic diseases targeting gut microbiota. In *Synbiotics in metabolic disorders* (pp. 160–172). CRC Press. https://doi.org/10.1201/9781032702438-11

Ray, R. C., Paramithiotis, S., Thekkangil, A., Nethravathy, V., Rai, A. K., & Martin, J. G. P. (2024b). Food fermentation and its relevance in the human history. In *Trending topics on fermented foods* (pp. 1–57). Springer. https://doi.org/10.1007/978-3-031-72000-0_1

Saddam, A. C. (2019). *The influence of single nucleotide polymorphisms in taste receptor gene TAS2R38 on eating behavior and body composition*. Mississippi State University. https://doi.org/10.1016/j.gene.2017.06.047

Sajadieh, S. M. M., Son, Y. H., & Noh, S. D. (2022). A conceptual definition and future directions of urban smart factory for sustainable manufacturing. *Sustainability, 14*(3), 1221. https://doi.org/10.3390/su14031221

Silva Neto, J. B., Mota, L. F., Amorim, S. T., Peripolli, E., Brito, L. F., Magnabosco, C. U., & Baldi, F. (2023). Genotype-by-environment interactions for feed efficiency traits in Nellore cattle based on bi-trait reaction norm models. *Genetics Selection Evolution, 55*(1), 93. https://doi.org/10.1186/s12711-023-00867-2

Singh, A., & Kumar, S. (2025). Exploring the functionality of microbes in fermented foods: Technological advancements and future directions. *Fermentation, 11*(6), 300. https://doi.org/10.3390/fermentation11060300

Siska, M., Pajak, E., Rosenthal, K., del Rio Chanona, A., von Lieres, E., & Helleckes, L. M. (2026). A guide to Bayesian optimization in bioprocess engineering. *Biotechnology and Bioengineering, 123*, 805–830. https://doi.org/10.1002/bit.70129

SiYing, W., & HuangHui, X. (2025). AI-driven metabolic engineering of γ-aminobutyric acid: Biosynthetic advances and industrial applications. *INNO-PRESS: Journal of Emerging Applied AI, 1*(3). https://doi.org/10.65563/jeaai.v1i3.27

Smigoc Schweiger, D., Mendez, A., Kunilo Jamnik, S., Bratanic, N., Bratina, N., Battelino, T., Brecelj, J., & Vidan-Jeras, B. (2016). High-risk genotypes HLA-DR3-DQ2/DR3-DQ2 and DR3-DQ2/DR4-DQ8 in co-occurrence of type 1 diabetes and celiac disease. *Autoimmunity, 49*(4), 240–247. https://doi.org/10.3109/08916934.2016.1164144

Sukrasno, S., Rizaldy, D., Kurniati, N. F., Raihanah, C., & Al Iman, A. (2025). Two optimized cysteine protease-aided protein hydrolysates of soybean tempeh inhibit angiotensin-converting enzyme activity in vitro. *Preventive Nutrition and Food Science, 30*(4), 379. https://doi.org/10.3746/pnf.2025.30.4.379

Tian, Q. (2021). *Regulatory effects of butyrate on vascular endothelial function* [The Chinese University of Hong Kong (Hong Kong)]. https://doi.org/10.15806/j.issn.2311-8571.2016.0025

Tuan, D. A., Uyen, P. V. N., & Masak, J. (2024). Hybrid quorum sensing and machine learning systems for adaptive synthetic biology: Toward autonomous gene regulation and precision therapies. *Preprints.org*. https://doi.org/10.20944/preprints202410.1551.v1

Uthravalli, S. H. (2025). *Dynamic, unconstrained optimization of secreted enzyme production in fed-batch fermentation using reinforcement learning* [Iowa State University]. https://doi.org/10.31274/td-20251215-238

Vashishtha, E., & Kapoor, H. (2023). Enhancing patient experience by automating and transforming free text into actionable consumer insights: A natural language processing

(NLP) approach. *International Journal of Health Sciences and Research, 13*(10), 275–288. https://doi.org/10.52403/ijhsr.20231038

Wang, B., Wang, Z., Chen, T., & Zhao, X. (2020). Development of novel bioreactor control systems based on smart sensors and actuators. *Frontiers in Bioengineering and Biotechnology, 8*, 7. https://doi.org/10.3389/fbioe.2020.00007

Wang, Y., Wu, J., Lv, M., Shao, Z., Hungwe, M., Wang, J., Bai, X., Xie, J., Wang, Y., & Geng, W. (2021). Metabolism characteristics of lactic acid bacteria and the expanding applications in food industry. *Frontiers in Bioengineering and Biotechnology, 9*, 612285. https://doi.org/10.3389/fbioe.2021.612285

Wang, Z., Wang, S., Zheng, X., Liu, W., & Shen, Z. (2025). Integrating kinetic models with physics-informed neural networks (PINNs) for predicting methane production from anaerobic co-digestion of enzyme-modified biodegradable plastics and food waste leachate. *Water, 17*(23), 3411. https://doi.org/10.3390/w17233411

Xiao, X., Fu, L., Zhang, Z., Tu, Z., Shen, N., & Fan, S. (2025). Recent advances in integrating machine learning with omics approaches in food science and nutrition research. *Journal of Agricultural and Food Chemistry, 73*, 29998–30025. https://doi.org/10.1021/acs.jafc.5c08522

Yadav, H., Singh, S., & Sinha, R. (2024). Fermentation technology for microbial products and their process optimization. In *Industrial microbiology and biotechnology: A new horizon of the microbial world* (pp. 35–64). Springer.

Yang, C., Lingli, C., Meijin, G., Xu, L., Jinsong, L., Xiaofeng, L., Zhongbing, C., Xiaojun, T., Haoyue, Z., & Xiwei, T. (2021). Application of near-infrared spectroscopy technology in the complex fermentation system to achieve high-efficiency production. *Bioresources and Bioprocessing, 8*(1), 96. https://doi.org/10.1007/978-981-97-6270-5_2

Yee, C. S., Ilham, Z., Cheng, A., Abd Rahim, M. H., Hajar-Azhari, S., Yuswan, M. H., Zaini, N. A. M., Reale, A., Di Renzo, T., & Wan, W. A. A. Q. I. (2024). Optimisation of fermentation conditions for the production of gamma-aminobutyric acid (GABA)-rich soy sauce. *Heliyon, 10*(13), e33147. https://doi.org/10.1016/j.heliyon.2024.e33147

Zhou, Y., Mihindukulasuriya, K. A., Gao, H., La Rosa, P. S., Wylie, K. M., Martin, J. C., Kota, K., Shannon, W. D., Mitreva, M., & Sodergren, E. (2014). Exploration of bacterial community classes in major human habitats. *Genome Biology, 15*(5), R66. https://doi.org/10.1186/gb-2014-15-5-r66

Chapter 8
Precision Fermentation & Synthetic Biology: Engineering Microbes to Enhance Plant-Based Dairy, Meat Substitutes, and Functional Beverages

Maxwell Wandji Nguedjo, Inelle Makamwe, Guy Roussel Nguemto Takuissu, and Hippolyte Tene Mouafo

Abstract Recent decades have been marked by rapid changes in food systems worldwide. Climate change and its detrimental effects on the environment and sustainable development have driven a shift in diets towards plant-based products. This shift reflects evolving consumer preferences, with increasing demand for products that positively impact both human health and the environment. These trends are supported by advances in microbial and plant biotechnology, along with growing ethical considerations. However, plant-based products developed as alternatives to animal-based products face several challenges, including modified sensory qualities, reduced nutritional value, cultural constraints, and the impact of processing

M. W. Nguedjo
Laboratory of Epidemiology and Nutritional Status, Centre for Food, Food Security and Nutrition Research, Institute for Medical Research and Medicinal Plant Studies, Yaoundé, Cameroon

Laboratory of Nutrition and Nutritional Biochemistry, Department of Biochemistry, University of Yaoundé I, Yaoundé, Cameroon

I. Makamwe
Laboratory of Nutrition and Nutritional Biochemistry, Department of Biochemistry, University of Yaoundé I, Yaoundé, Cameroon

Cameroon's Nutritional and Dietetic Research Center, J&A Oben Foundation, Yaoundé, Cameroon

G. R. N. Takuissu
Cameroon's Nutritional and Dietetic Research Center, J&A Oben Foundation, Yaoundé, Cameroon

Laboratory of Metabolism, Centre for Food, Food Security and Nutrition Research, Institute for Medical Research and Medicinal Plant Studies, Yaoundé, Cameroon

H. T. Mouafo (✉)
Laboratory of Development of Food Technologies, Centre for Food, Food Security and Nutrition Research, Institute for Medical Research and Medicinal Plant Studies, Yaoundé, Cameroon

S. Dhiman, S. Kaur (eds.), *Next-Gen Plant Based Fermented Foods: Innovations for a Sustainable Future*, https://doi.org/10.1007/978-3-032-29927-7_8

methods on consumer health and environmental sustainability. Synthetic biology combined with precision fermentation represents novel approaches to addressing these challenges. This chapter will examine these approaches, focusing on their applications in the production of plant-based dairy and meat substitutes, as well as in the formulation of functional beverages. The ethical and regulatory considerations related to their applicability, accessibility for low- and middle-income countries, and integration into food systems will also be discussed.

Keywords Plant-based dairy · Meat substitutes · Functional beverages · Precision fermentation · Synthetic biology · Engineered microbes · Novel foods · Regulatory framework

8.1 Introduction

A paradigm shift has occurred in the global food sector in recent decades, driven by the emergence of innovative foods experiencing rapid growth. These products include functional beverages, plant-based dairy products, and meat substitutes (Eichhorst et al., 2025). This surge in plant-based products as alternatives to animal-derived foods is due to changing consumer preferences, increasing environmental and ethical concerns, and technological advancements in the agri-food sector (Alcorta et al., 2021). These developments toward more sustainable food systems are motivated by the need to ensure food security for a growing global population through diets perceived as beneficial to both human health and the environment (Ambikapathi et al., 2022; Cheng & Truter, 2023).

Demand for plant-based beverages and other dairy alternatives, has increased markedly in recent years (Xie et al., 2023; Sharma et al., 2024). This category is now among the fastest-growing segments of the global food industry (Gajewska et al., 2024). Worldwide, plant-based dairy alternatives have entered mainstream markets, with recent projections indicating growth rate of approximately 12% and revenues exceeding USD 34.8 billion by 2026 (Fairfeild, 2022). Within this category, plant-based milk represents a major share of the market, Revenues are expected to rise from USD 14.9 billion to approximately USD 36.2 billion by 2033 (Research, 2023). The expansion plant-based beverage markets and dairy products alternatives is driven by several interconnected factors. These include rising demand for plant-based foods such as cheese substitutes, non-dairy yogurts and desserts, increased adoption of plant-based diets, greater awareness of food intolerances, and environmental concerns related to animal husbandry. Moreover, products offerings are diversified to address varied nutritional, sensory, and ethical expectations (Sharma et al., 2024; Eichhorst et al., 2025). Continued market growth is further supported by product innovation, including with enhanced formulations, functional variants, and expanded distribution across retail and food service channels (Xie et al., 2023). Furthermore, consumers increasingly view plant-based beverages, as convenient vehicles to support daily health goals, such as energy balance, immune function, digestive health, and overall well-being (Aydar et al., 2020; Kent et al., 2022).

Besides plant-based beverages, the markets for plant-based meat substitutes are also experiencing significant growth. Valued at nearly USD 1.6 billion in 2019, the global market of plant-based meat substitute is projected to reach USD 3.5 billion by 2026 (Markets and Markets, 2020). Ahmad et al. (2022) further reported that plant-based meat alternatives is expected to reach approximately USD 27.9 billion by 2025. These plant-based alternatives include textured vegetable proteins and products derived from soy, bambara groundnuts, beans, peas, and other legumes. They exhibit very high annual growth rates, exceeding those of conventional food products (The Good Food Institute, 2020). This growth is driven by positive consumer perceptions, particularly regarding health, environmental sustainability, and overall well-being (Gajewska et al., 2024). Indeed, dietary patterns in many populations are increasingly shifted towards reduced consumption of animal meat, alongside a growing demand for plant-based equivalents that provide satisfactory nutritional, health and organoleptic profiles (Sharma et al., 2024; Eichhorst et al., 2025).

Several factors including health and well-being, environmental sustainability, ethical considerations, and increasing innovation, underspin the simultaneous growth of these three categories of plant-based food products (Reyes-Jurado et al., 2021). Consumers are increasingly attentive to their dietary choices and show growing interest I how food affect their health, particularly in relation to weight management, digestive health, and the prevention of cardiometabolic diseases. Consequently, they seek alternatives perceived as healthier and more beneficial than conventional products (Xie et al., 2023). In parallel, climate and environmental concerns, especially the high carbon footprint associated with intensive animal farming, also contribute to this trend (Sharma et al., 2024; Eichhorst et al., 2025). Ethical considerations related to animal welfare further influence food choices, increasing the appeal of food products free from animal-derived ingredients. Moreover, technological advances in food formulation have enhanced product texture, taste, and functionality, while the expansion of retail channels has improved their accessibility (Irondi et al., 2025).

Despite these positive developments, currently available plant-based alternatives face significant challenges that limit their widespread adoption (Wang et al., 2022). Many products fail to effectively replicate the sensory qualities, particularly texture and taste, of their animal-based counterparts, creating a barrier for consumers accustomed to traditional diets (Perez-Cueto et al., 2022; Sharma et al., 2024). From a nutritional perspective, plant-based substitutes are often not fully equivalent to animal-derived products. Essential micronutrients such as iron, zinc, and vitamin B12 are frequently present in sufficient amounts or exhibit limited bioavailability unless extensive fortification is applied (Venter de Villiers et al., 2024). Consequently, promoting plant-based substitutes without prior appropriate dietary planning might result in adverse health. Furthermore, the use of additives and texturizers to mimic the attributes of animal-based products raises concerns regarding potential health effects (Aydar et al., 2020). To address these limitations, innovative technologies such as precision fermentation and synthetic biology, offer promising solutions (Knychala et al., 2024).

Precision fermentation is a technique that uses metabolically programmed microorganisms to produce targeted compounds, such as bioidentical animal proteins, enzymes, vitamins, lipids, minerals, and specific pigments (Boukid et al., 2023). This process relies on synthetic biology, which provides the tools needed to engineer microorganisms, and optimize metabolic pathways the industrial scale production of complex molecules (Eastham & Leman, 2024). These approaches improve production yields and efficiency while enabling the development of nutritional profiles, textures, and flavors that cannot be achieved through traditional fermentation methods (Knychala et al., 2024; Eastham & Leman, 2024). Furthermore, compounds obtained through precision fermentation are biochemically identical to those found in animal-derived products, with the added advantage of higher purity and precision, and without the need for animal farming (Eastham & Leman, 2024). Economically, the global market for precision fermentation products was valued at approximately USD 1.2 billion in 2024. This growth is driven by their increasing adoption in the meat substitute, dairy alternative, and functional food sectors.

However, despite the considerable potential of precision fermentation and synthetic biology, their application in food systems remains associated with several challenges. Key barriers to wide adoption include challenges in technological scalability such as high upfront investment in biotechnology infrastructure and the need for specialized expertise. Additional constraints arise from the ongoing complexity of microbial engineering, the need to optimize of production yields, gaps in regulatory frameworks, and persistent public concerns surrounding genetically modified organisms and products derived from advanced biotechnologies.

In this book chapter, we examine precision fermentation and synthetic biology, and their application in plant-based alternatives to animal-derived foods. We also discuss the regulatory and ethical considerations associated with these technologies, evaluate opportunities and challenges related to their large-scale implementation, particularly for developing countries, and finally, outline future prospects and research directions.

8.2 Precision Fermentation and Synthetic Biology for Food

8.2.1 Generalities on Traditional Fermentation

8.2.1.1 Definition and Classification

Fermentation can be defined as a food processing method based on the metabolic activity of microorganisms including bacteria, yeasts, and molds (Gangakhedkar et al., 2024). It is among the oldest food processing techniques, and has been used for thousands of years across diverse cultures to preserve a wide range of foods such as cereals, milk, vegetables, legumes, meat, and fish (Ajanaku et al., 2025). During fermentation, microbial enzymes transform process food components including carbohydrates, lipids, and proteins into a variety of metabolites. These include organic acids, alcohols, vitamins, flavors, biosurfactants, bacteriocins, CO_2, and other metabolites that collectively define the safety, sensory attributes, and nutritional quality of fermented foods.

Precision fermentation is characterized by several parameters, including the diversity of products generated, the variation in operating procedures, and the control of microbial activity. An operational classification helps to better understand microbial dynamics, standardize the products obtained, and identify the technological challenges associated with each approach. This classification includes:

Spontaneous fermentation Which can be defined as a process initiated by microorganisms naturally present in raw materials, equipment, or the production environment, without the deliberate addition of external inoculum (Piraine & Bochman, 2025). In this method, the development of microbial flora depends on the composition of the indigenous microflora and environmental factors such as temperature, nutrient content, and pH. These conditions favor the proliferation and dominance of the desired microorganisms, such as lactic acid bacteria or yeasts (Piraine & Bochman, 2025). Spontaneous fermentation is commonly used in the artisanal production of products such as kimchi, paocai, sauerkraut, certain fermented milks, and other traditional foods, particularly in Africa and Asia, where it is deeply rooted a practice deeply rooted in cultural and culinary traditions (Gangakhedkar et al., 2024). The process is simple, easy to implement, less expensive, and leads to products with a unique sensory profile that often reflect their local identity. However, product variability can occur between batches because microbiological dynamics are not fully controlled, and there is a risk of the growth of undesirable microorganisms.

Back-slopping It is intermediate technique between spontaneous fermentation and the use of selected starter cultures (Liao et al., 2024). In this method, a portion of a previously fermented batch is used to inoculate a fresh substrate, thereby introducing a microbial community already adapted to the raw material. Back-slopping has been traditionally applied in the production of natural leavens for baking, coffee and orange juices (Andreson et al., 2023), as well as in a range of artisanal fermented foods such as *kindirmou* and *pendidam* from the soudano-sahelian and soudano-guinean agroecological zones of northern Cameroon (Fookao et al., 2022; Mbawala et al., 2024). With repeat use, back-slopping progressively enriches the desired specific microflora and reduces the uncertainty associated with strictly spontaneous fermentation. Its advantages include shorter fermentation latency time and a greater dominance of beneficial microorganisms across successive production cycles (Liao et al., 2024). Nevertheless, microbial profile of microorganisms remains complex, which limits full product standardization.

Fermentation with starter cultures Often referred to culture-dependent fermentation, it involves the deliberate inoculation of sterilized or pretreated substrates with a known starter culture (Andreson et al., 2024; Atfaoui et al., 2025). These starters are either commercially available or derived from standardized natural leavens. This approach allows precise control over fermentation kinetics, and enables the production of foods such as yogurt, certain cheeses, kefir, *kombucha*, *natto*, dry sausages, and others, with consistent sensory and technological characteristics (Rezac et al., 2018; Kim et al., 2024). Because fermentation conditions are tightly controlled, this method supports predictable metabolic activity, optimized aroma development, and improved food safety. However, it might also lead to sensory uniformity, reduced cultural diversity in fermented foods, and high capital investments requirements.

Hybrid approaches They extend beyond traditional fermentation models by combining elements from different techniques. These strategies might involve mixed microbial consortia including yeasts, bacteria, and molds; the use of selected or genetically modified hybrid strains to achieve specific functional traits such as enhanced flavor, aroma, color, or processing efficiency, or the application of hybrid modelling tools to predict and optimize fermentation processes (Albino et al., 2024). Emerging innovations such as micro-scale ferment manufacturing, microbial encapsulation and immobilization, artificial intelligence, and omics-based microbiology, offer new opportunities to balance industrial standardization with the preservation of traditional sensory diversity (Ajanaku et al., 2025).

8.2.1.2 Advantages of Traditional Fermentation

The traditional fermentation has several advantages including an enhancement of the shelf life of food, thereby contributing to food security (Gangakhedkar et al., 2024; Ajanaku et al., 2025). Indeed, the reduction in pH due to the production of organic acids, combined with the synthesis of antimicrobial metabolites such as bacteriocins and biosurfactants, inhibits pathogenic microorganisms, extends the shelf life of food, and enhance its safety, thereby reducing food losses (Mouafo et al., 2020, 2023). In addition, fermentation enhances nutrient bioavailability including proteins and minerals, and reduces antinutritional factors such as tannins and phytates. It also enables the production of vitamins and bioactive compounds including beneficial fatty acids, peptides, and phenolic compounds, which are associated with positive effects on digestive health and the immune system. It also generates products with unique sensory profiles (flavors, textures, aroma) that are culturally valued, and strengthens the cultural identity and local food systems, thus contributing to a global gastronomic diversity (Gangakhedkar et al., 2024; Ajanaku et al., 2025).

8.2.1.3 Constraints of Traditional Fermentation

Traditional fermentation processes most often rely on complex, spontaneous, or semi-spontaneous microbial flora, which leads to significant variability among products in terms of microbial composition, aromatic profiles, and sensory quality (Galimberti et al., 2021). Indeed, traditional fermentation processes generally focused on broad transformations of raw materials such as acidification and macromolecule degradation, rather than the targeted production of specific compounds including amino acids, vitamins, bioactive peptides, particular aromas, or specific fatty acids. This generalized approach limits the development of high value-added or customized food ingredients designed to meet specific nutritional functions. Moreover, in their current form, the traditional fermentation processes are insufficient to meet the nutritional demands of growing global population or to fully support sustainable development goals related to good health and wellbeing, as well as industry, innovation, and infrastructure (Ajanaku et al., 2025). To address these limitations, the field is increasingly transitioning toward precision fermentation.

8.2.2 *Precision Fermentation*

Advances in molecular biology, next-generation sequencing, multi-omics, and bioinformatics have significantly transformed fermentation science. These developments have enabled a deeper understanding of microbial diversity, metabolic pathways, and the functional benefits of fermentation (Malomo et al., 2021; Mukherjee et al., 2022). Precision fermentation represents a key step in the evolution of modern food biotechnology, addressing major twenty-first-century challenges including population growth, environmental pressure, and the increasing demand for food sustainable, safe, and functional foods.

8.2.2.1 Definition and General Principle

Precision fermentation refers to the use of genetically modified or metabolically optimized microorganisms such as bacteria, yeasts, and molds, programmed to produce specific, and targeted food compounds in controlled environments (Gangakhedkar et al., 2024; Ajanaku et al., 2025). Unlike traditional fermentation, precision fermentation does not focus on the transformation of raw materials. Instead, it aims to produce targeted compounds (proteins, enzymes, flavors, vitamins, fatty acids) at high yields, with high purity, and precise functionality (Eastham & Leman, 2024). This technique relies on metabolic engineering, synthetic biology, and the selection of optimized strains. Its objective is to achieve precise control of metabolic pathways in order to reduce product variability and ensure stable, optimal, reproducible and efficient production of specific compounds (Ajanaku et al., 2025). Precision fermentation involves several key steps, as illustrated in Fig. 8.1. First, microbial strains are isolated and screened for their ability to produce metabolites of interest (Fig. 8.1a). Promising strains are then sequenced and genetically enhanced using CRISPR-based tools, plasmids, or gene integration to optimize growth, tolerance to environmental stresses, and target metabolite production. In the second step, these optimized strains undergo further genetic improvement by inhibiting or suppressing genes encoding for non-target compounds synthesis, and stimulating or enhancing the expression of the gene involved in the production of the desired compound. Validation assays are performed to confirm strain stability, reliability, and production efficiency (Fig. 8.1b). Following validation, the strains are cultured

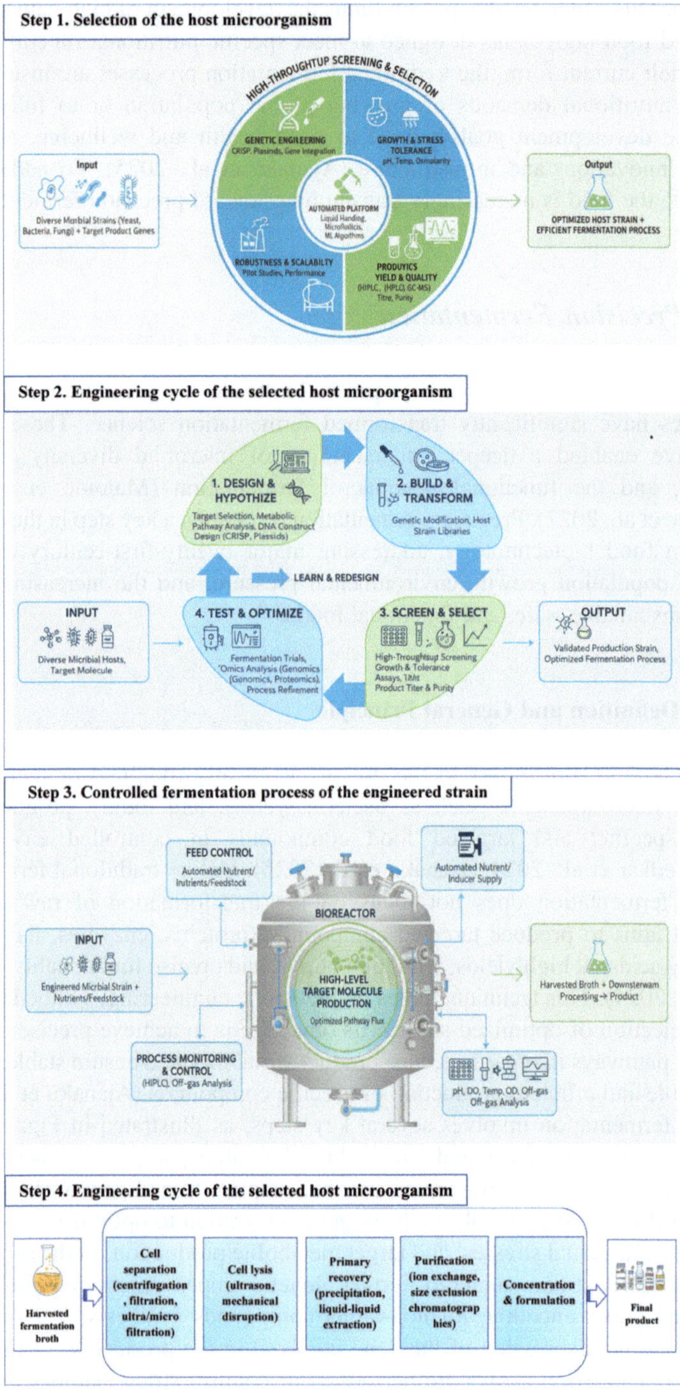

Fig. 8.1 Illustration of the different steps involved in the precision fermentation. The artificial intelligence (AI) generator (QuillBot) was partially used to generate some parts of the illustration based on the concepts

in bioreactors under controlled conditions to enable large-scale production of the compound of interest (Fig. 8.1c). The final step involved product recovery from the fermentation broth, including extraction and purification using techniques such as centrifugation, cell lysis, precipitation, and chromatography (Fig. 8.1d). Microorganisms that are most commonly used in precision fermentation processes are those generally recognized as safe (GRAS) (Gomes et al., 2018). This GRAS status is further leveraged through synthetic biology to optimize the production of target compounds and increase production yields. This approach is exemplified by molds of the genus *Trichoderma* and particularly *Trichoderma reesei*, yeasts species such as *Saccharomyces cerevisiae*, *Pichia pastoris*, *Komagaetella phaffii*, and certain bacterial species, notably *Bacillus subtilis* (Gomes et al., 2018).

8.2.2.2 Advantages of Precision Fermentation

Precision fermentation enables the high-yield and high-purity production of functional compounds with beneficial health effects, as well as vitamins, pigments, natural flavors, and other molecules that are difficult or inefficient to obtain through traditional fermentation processes (Hilgendorf et al., 2024; Silva et al., 2024a, b). As a result, it substantially expands the diversity of food products through highly controlled fermentation systems. It is noteworthy to note that ingredients derived from precision fermentation are biochemically identical to those of animal or plant origin, and exhibit a high degree of purity and uniformity, meeting food functionality requirements in term of physical, sensory, and nutritional properties (Hilgendorf et al., 2024). Furthermore, the controlled nature of the process minimizes the risk of contamination by pathogens or allergens. By enabling the cost-effective production of high-value functional compounds, precision fermentation reduces the reliance on agricultural inputs and improves the energy efficiency of bioprocesses (Pham, 2018; Silva et al., 2024a, b). Specifically, substantial reduction in the use of arable land (90%), water (96%), and energy, as well as greenhouse gas emissions (97%), have been reported. In this context, Knychala et al. (2024) highlighted reductions of 90–97% in water consumption and greenhouse gas emissions compared with conventional animal protein production systems. Furthermore, precision fermentation responds to the growing demand for ethically produced foods and helps supports the increasing adoption of vegan and vegetarian diets (Gangakhedkar et al., 2024).

8.2.3 Synthetic Biology

The traditional biological synthesis of compounds often involves processes energy-consuming processes that generate waste, and offer limited control over the chemical composition and properties of the final products (Ramachandran et al., 2025). Synthetic biology is an emerging multidisciplinary field that addresses this issue. As the science of designing living organisms, synthetic biology is experiencing rapid developments, and goes far beyond traditional descriptive approaches. It continuously integrates engineering principles to produce new or optimized biological entities for industrial, medical, and environmental applications.

8.2.3.1 Definition and Principle

Synthetic biology can be defined as the rational engineering of biology, with the aim of designing, building, and testing new or modified biological entities to perform specific targeted functions (Garner, 2021; Tebeje et al., 2021). The goal of this field is both to deepen our understanding of fundamental biological principles, and to create organisms or genetic circuits with programmed functions that address societal challenges related to energy, health, environment, and food systems (Gao et al., 2018). Industrial microbiology is the one of the primary areas applying synthetic biology, thereby driving significant advances in modern biotechnology.

Globally, synthetic biology operates through the design, modeling, computer-based simulation of biological systems, followed by *in silico* testing to program, predict, and control the behavior of genetic circuits. This approach includes the modular assembly of genes, as well as the optimization and reconstruction of metabolic pathways. The overarching objective are to improve the yield and efficiency of targeted products, compare predicted outcomes with experimental results to generate new insights and apply genome editing tools such as Clustered Regularly Interspaced Short Palindromic Repeats (CRISPR)-Cas systems, to introduce precise genetic modifications and create organisms with novel capabilities (Gilman et al., 2021; Garner, 2021; Jiang et al., 2025).

8.2.3.2 Process

Synthetic biology involves the construction of genetic circuits, devices, and microorganisms through the integration of genetic elements such as promoters, coding sequences, transcriptional factors, and binding sequences. This approach integrates quantitative data to enable predictive models of biological systems (Li et al., 2025). Synthetic biology underpins metabolic engineering, in which specific cellular functions are modified in microbial strains to enable the production of desired compounds from low-cost substrates (Janik et al., 2020; Jiang et al., 2025). Several molecular techniques are generally employed to introduce precise genetic modifications. These include gene deletion, insertion, substitution, and targeted mutagenesis of functionally relevant genes (Huang et al., 2021). Such modifications alter gene structure to modulate specific functions, thereby influencing cellular behaviors and phenotypic traits. Advances in synthetic biology has led to the development of efficient and widely applicable genome editing technologies, including zinc finger nucleases (ZFNs), transcription activator-like effector nucleases (TALENs), and clustered regularly interspaced short palindromic repeats (CRISPR) systems (Janik et al., 2020). ZFNs rely on specific recognition of DNA sequences by zinc finger proteins (ZFPs), which are coupled to the FokI endonuclease domain to form an active genome editing complex. Modifications within the FokI domain together with carboxyl and amine groups of zinc finger proteins, allow expansion target specifity (Janik et al., 2020). TALENs are generated by fusing the DNA-binding domain derived from transcription activator-like effectors (TALEs) to the catalytic domain of the FokI endonuclease (Zhang et al., 2019). CRISPRs owing to their programmable and highly precise targeting capabilities, offer a wide range of applications, including functional genomic studies, the generation of model organisms, and therapeutic gene modifications (Li et al., 2025).

Once improvements have been made to microorganisms, the resulting designs and associated data are shared within the synthetic biology community through the called Synthetic Biology Open Language (SBOL) platform. This standardized data format provides a unified method for representing biological designs *in silico*. The primary goal of SBOL is to facilitate data exchange among researchers and to establish a common standard for the visual and computational description of biological systems (Garner, 2021).

8.2.3.3 Development of Engineered Strains

The development of engineered strains involves the insertion synthetic genetic chassis into the genomes of microbial strains (Kang et al., 2019; Yu et al., 2025). An ideal chassis requires the minimum number of genes necessary to sustain essential metabolic activities. *Escherichia coli* is the most commonly used host organism for this purpose, owing to the low cost of its growth culture media, its ability to grow under easily reproducible laboratory conditions, and its short generation time of approximately 20 min (Garner, 2021). As a result, synthetic circuits can be easily introduced into the bacterial cytoplasm through plasmids-mediated transformation, after which the targeted biological effect are screened to identify and select the most efficient strains (Yang et al., 2022). In addition to *E. coli*, other well characterized and scalable microorganisms with fully sequenced genomes such as *Bacillus subtilis* and *S. cerevisiae*, are also widely used as chassis organisms (Yang et al., 2022). Among these microorganisms, *S. cerevisiae* is often considered particularly suitable for applications related to human systems because it is a eukaryote (Yang et al., 2022). However, despite extensive studies, a complete understanding of the functioning of these organisms remains limited. Consequently, some researchers have proposed the *do novo* design of chassis organisms using a bottom-up approach (Singh & Lercher, 2020). This strategy aims to enable comprehensive modeling, precise definition, and tighter control of all cellular processes within the engineered strain (Garner, 2021).

8.2.3.4 Benefits of Synthetic Biology

In food fermentation, synthetic biology enables the development of novel microbial strains with improved or newly introduced metabolic pathways (Kang et al., 2019; Yang et al., 2022). These engineered microorganisms can synthesize high value-added compounds such as vitamins, peptides, amino acids, fatty acids, and bioactive molecules, in a targeted manner and with minimal formation of by-products (Park et al., 2017). The synthetic genetic circuits introduced into these allow precise control of microbial activity during fermentation, thereby ensuring process reproducibility and efficiency, even under variable environmental conditions (Park et al., 2017; Jiang et al., 2025).

8.3 Engineering Microbes for Plant-Based Dairy Products

8.3.1 Generalities on Plant-Based Dairy Products

Plant-based dairy substitutes are water-based extracts of legumes, oil seeds, cereals or pseudo-cereals, that resemble animal milk in appearance and consistency (Plamada et al., 2023). They currently account for approximately 15% of the global dairy market, and continue to experience sustained growth (Good Food Institute, 2021). This expansion is driven by increased awareness of the environmental impacts of dairy production, ethical concerns related to livestock farming, and health factors, notably lactose intolerance, cow's milk protein allergy, and increasingly critical perceptions of the nutritional value of milk (Silva et al., 2024a, b). Historically, soy-, rice- and coconut-based beverages have been incorporated into dietary practices and culinary traditions for centuries. However, their large-scale commercial production did not emerge until the early twentieth century (Ramsing et al., 2023). More recently, the growth of veganism, and advances in fermentation technologies particularly the use lactic acid bacteria, yeasts and fungi, have enabled the fermentation of plant-based substrates. Is has led to the development of textured dairy alternatives, including, cheeses and plant-based yogurts, with sensory and nutritional properties that more closely resemble those of conventional dairy products (Nguyen, 2023). Table 8.1 summarizes examples of engineered microorganisms and their application in plant-based dairy products.

8.3.2 Challenges of Plant-Based Dairy Analogues

Milk and dairy products have traditionally occupied an important place in the human diet due to their richness in essential macronutrients and micronutrients, thereby contributing to health and overall development (Jemaa et al., 2021). In recent years, however, consumption of plant-based alternatives to milk has increased marketly (Reyes-Jurado et al., 2023). These beverages are produced from a wide range of

Table 8.1 Overview of engineered microorganisms and their application in plant-based dairy food products (Nguyen, 2023)

Name of the microorganisms	Food products
Mucor hiemalis, Neurospora crassa	Yogurt and kefir
Rhizopus stolonifer, E. coli	Processed cheese, Cheese
S. cerevisiae, Aspergillus niger	Probiotic yogurt
Saccharomyces. bayanus, S. cerevisiae	Fermented vegetables
E. faecalis, L. rhamnosus	Probiotic supplements
L. reuteri, Bifidobacterium bifidum	Functional dairy products
Aspergillus oryzae, Rhizopus oryzae	Fermented rice products (e.g., Sake)
Corynebacterium glutamicum, Brevibacterium linens	Flavor-enhanced cheeses
Bacillus subtilis, B. licheniformis	Fermented soy products (e.g., Natto)

plant sources, including legumes (soybean, pea, peanut), cereals (oats, rice), oil seeds (flax, sesame) and nuts (almond, cashew and coconut). Plant-based milk alternatives are generally characterized by a good nutritional quality (Table 8.2), including the natural absence of cholesterol, typically moderate levels of saturated fatty acids, and the presence of bioactive compounds such as polyphenols, phytosterols and antioxidants. These components are associated with a reduced risk of certain chronic diseases (Olędzki, 2025; Gerogianni et al., 2025). Furthermore, compared with conventional dairy production, plant-based beverages tend to have a low environmental footprint, requiring fewer natural resources and generating lower greenhouse gas emissions, which support their role in sustainable food systems (Poore & Nemecek, 2018).

Despite these advantages, several challenges limit their large-scale adoption. Notably, the protein content and quality of many plant-based alternatives are often lower than those of animal milk, necessitating supplementation with essential nutrients including calcium, vitamin D and vitamin B12 to ensure nutritional adequacy (Vanga & Raghavan, 2018). Sensory limitations such as grainy texture, atypical flavors and physicochemical instability leading to phase separation can also reduce consumer acceptance. As a result, technological additives such as emulsifiers, stabilizers, and taste-enhancing agents are commonly used to improve texture, stability, and overall sensory quality (Karoui & Bouaicha, 2024). In the context of growing demand, synthetic biology, metabolically engineered microorganisms and precision fermentation are increasingly viewed as essential tools for improving the nutritional composition, sensory attributes, and functional properties of plant-based milk alternatives.

8.3.3 Engineered Flavor and Aroma

Plant-based dairy products frequently exhibit undesirable flavors, often described as herbaceous, bitter or legume-like, which limit consumer acceptance. These off-flavor notes are mainly attributed to volatile compounds present in plant

Table 8.2 Average nutrients in dairy and plant-based milks (Ramsing et al., 2023)

Item, 240 mL (n, products)	Protein (g)	Calcium (mg)	Magnesium (mg)	Phosphorus (mg)	Potassium (mg)	Selenium (µg)	Zinc (mg)	Choline (mg)	Riboflavin (mg)	Vitamin D (µg)
Daily value	50	1300	420	1250	4700	55	11	550	1.3	20
Dairy milk	8.2[a]	309[b]	29	252[b]	387	4.8	1.1	44	0.33[b]	2.7[a]
Dairy milk, lactose free	8.2[a]	309[b]	29	252[b]	387	4.8	1.1	44	0.33[b]	2.7[a]
Almond milk	1.0	449[b]	15	22	163	0.2	0.1	8	0.02	2.4[a]
Cashew milk (2)	1.8	116	ND	ND	69	ND	ND	ND	ND	0.0
Coconut milk	0.5	459[b]	0	0	46	0.0	0.0	0.0	0.000	2.4[a]
Hemp milk (3)	2.3	204[a]	54[a]	199[b]	101	ND	0.6	ND	0.42[b]	2.5[a]
Oat milk (16)	2.7	248[a]	ND	170[a]	184	ND	1.0	ND	0.55[b]	2.3[a]
Pea milk (2)	7.5[a]	385[b]	0	ND	421	ND	ND	ND	ND	2.5[a]
Rice milk	0.7	288[b]	27	137[a]	66	5.4	0.3	5	0.35[b]	2.4[a]
Soy milk	6.1[a]	294[b]	33	176[a]	280	5.2	0.5	44	0.45[b]	2.7[a]

ND no data available
[a]Meets 10% daily value
[b]Meets 20% daily value

matrices, such as certain aldehydes, alcohols and ketones, notably hexanal, n-hexanol and ethyl vinyl ketone. To mitigate these sensory challenges, various pretreatment techniques have been investigated, among which fermentation has received considerable attention. Studies have shown that *Streptococcus thermophilus* NCC 1326 can enhance buttery notes in various plant-based milks, a sensory attribute particularly desirable in fermented products of plant origin. The intensification of these buttery notes contributes to sensory profiles that more closely resemble those of animal derived dairy products (Tangyu et al., 2023). Furthermore, increased intensities of cheese-like and acidic aromas have been reported in fermented pea and faba bean milks, particularly following fermentation with *Leuconostoc mesenteroides*, *Lactiplantibacillus plantarum* and *Lactobacillus johnsonii*, with reported increases of up to 83%. The formation of cheese-like aromas despite the absence of casein, the primaryc precursor of characteristic cheese aromatic compounds, highlights precision fermentation as a promising approach for the development of plant-based cheese alternatives (Tangyu et al., 2023).

8.3.4 Engineered Texture, Structure and Mouthfeel

The texture and mouthfeel of plant-based yogurts are closely related to their composition, which in turn determines the microstructure of the final product (Dhar et al., 2025; Qadir et al., 2025). Components such as proteins, hydrophilic colloids and oils, whether naturally present or added, play a key role in establishing these properties. However, their capacity to form stable gel under acidic conditions is often limited. Fermentation therefore represents an effective strategy to overcome this limitation. During fermentation, certain lactic acid bacteria synthesize exopolysaccharides (EPS) which are widely recognized for their key ability to improve the sensory and rheological properties of fermented dairy products (Korcz & Varga, 2021). The use of EPS-producing lactic acid bacteria has proven particularly effective in the manufacture of low-fat yoghurts and cheeses, because the textural and mouthfeel improvement provided by EPS can compensate for reduced lipid content (Soumya & Nampoothiri, 2021). As plant-based milks generally contain less fat than cow's milk (Silva et al., 2024a, b), incorporating EPS-producing strains is a key strategy for achieving acceptable texture in plant-based alternatives. These strains can produce EPS directly into plant matrices, resulting in increased viscosity and reduced syneresis.

8.3.5 Enhancing Nutritional Quality and Functionality

The nutritional value and organolcptic characteristics of plant based dairy beverages differ from those of products of animal origin. That is why these alternatives are often supplemented in proteins containing all essential amino acids and exhibiting good digestibility, as well as calcium, zinc, magnesium, vitamins B12, D and A, and other essential micronutrients (Ramsing et al., 2023).

Precision fermentation of soybeans using *Bacillus subtilis* MTCC5480 has been shown to significantly reduce anti-nutritional factors, such as protease inhibitors, phytic acid, urease and oxalic acids, while increasing the content and bioavailability of bioactive compounds like isoflavones and free peptides (Gong et al., 2025). It also significantly increased free amino acids levels, leading to improved protein digestibility (Gong et al., 2025). Sharma et al. (2018) noticed that fermentation of almond milk with specific lactic acid bacteria strains reduced anti-nutritional factors, improved calcium, magnesium and iron bioavailability, and contributed to improved digestion and immune-related functions.

8.4 Engineering Microbes for Plant-Based Meat Substitutes

Plant-based meat analogues have been developed to meet growing consumer expectations and contribute to a sustainable long-term protein supply. These products generally have nutritional profiles comparable to those of conventional meat, and are considered as good sources of protein (Bryngelsson et al., 2022). They are also formulated to replicate various meat products, particularly in terms of texture, color, nutritional properties and sensory characteristics (Choudhury et al., 2020).

8.4.1 Enhancing Meat-Like Fibrous Structures

Meat substitutes are mainly formulated using plant-based proteins, lipids, and binders, which are combined with coloring and flavoring compounds. Soy and peas are the most commonly used sources and are often blended with wheat flour to form fibrous structures that resemble meat. Processing techniques such as extrusion, shearing, and 3D printing are commonly applied (Sha & Xiong, 2020). However, these products are frequently described as less juicy and visually appealing than conventional meat. In addition, legumes-associated sensory attributes, including bitterness, lingering aftertaste, and astringency remain major drawbacks (Sha & Xiong, 2020; Wang et al., 2022). Several microorganisms, including *L. acidophilus*, *Lactiplantibacillus plantarum*, *Fusarium venenatum* and *Neurospora intermedia*, have demonstrated the ability to improve product texture during fermentation by forming fibrous structures comparable to those of meat. The use of *Bacillus subtilis* has also been shown to enhance the tenderness, chewiness and firmness of meat analogs (Sha & Xiong, 2020). However, a key challenge associated with these strains is identifying optimal processing conditions that prevent firmness loss and the development of undesirable viscous textures due to excessive fermentation.

8.4.2 Enhancing Sensory Attributes

Sensory attributes are a major determinants of consumer acceptance of plant-based meat analogs (Fiorentini et al., 2020). They encompass human responses to food stimuli and are primarily expressed through appearance, taste, and texture (Saintila et al., 2025). In the case of meat substitutes, these attributes largely depend on the nature of ingredients used, their formulation, and the structuring technologies implemented (Van der Weele et al., 2019).

Taste and Aroma Precision fermentation represents an effective strategy to reduce or eliminate undesirable taste and aroma associated with plant-based materials. In particular, off-flavor notes such as the characteristic beany aroma of soybean, have been significantly reduced after fermentation with *Kluyveromyces marxianus* (Hu et al., 2019), *Bacillus amyloliquefaciens* (Wang et al., 2021), *Bacillus subtilis* (Shrestha et al., 2013), and *Weissella assemble* (Tuccillo et al., 2022). These studies collectively demonstrate that fermentation inactivates lipoxygenases involved in the formation of malodorous compounds. Su et al. (2022) reported that fermentating rice bran with a mixed culture of *B. subtilis*, *S. cerevisiae*, and *Lactiplantibacillus plantarum* resulted in the denaturation of lipoxygenases and lipases, thereby limiting the formation of compounds responsible for undesirable flavors. Kale et al. (2022) proposed the addition of hydrolyzed proteins and Maillard reaction-promoting substrates to the fermentation medium to obtain plant products with meat-like aromatic profiles. Similarly, the fermentation of bean protein concentrates with *Weissella belarus*, led to the production of aromatic esters with fruity notes and a reduction in aldehydes associated with malodorous compounds (Tuccillo et al., 2022).

Texture and Mouthfeel Maung et al. (2020) demonstrated that plant-based meat analogues fermented with Bacillus subtilis exhibited increased chewability, cohesion, and firmness compared with unfermented products. Razavizadeh et al. (2021) reported that fermentation with *Lactiplantibacillus plantarum* and *L. acidophilus* improved digestibility, enhanced sensory attributes, and promoted a more pronounced fibrous structure, supporting their suitability for incorporation into meat analogues.

Flavor Specific strains of lactic acid bacteria have been used to ferment of plant-based protein ingredients for the production of meat analogues. Fermentation with *L. plantarum* P1, *L. brevis* R, *L. acidophilus* 336, and *L. acidophilus* 308, resulted in significant improvements in water- and oil-holding capacities, and a reduction in protein oxidation in soy press cake (Razavizadeh et al., 2022). Incorporation of fermented soy products at a level of 10% in meat analogue formulations further improved texture, particularly juiciness, and enhanced flavor by reducing bitterness and balancing taste (Razavizadeh et al., 2022). In another study, textured vegetable proteins produced by extrusion of soy protein, corn starch and wheat gluten, were fermented using *B. subtilis*, resulting in improvements in chewiness, hardness, integrity index, and layered structure (Maung et al., 2020). Precision fermentation has also enabled the production of targeted ingredients that enhance the sensory resemblance of plant-based meats to their animal-derived counterpart. For example, soy leghemoglobin has been produced using an engineered yeast (*Pichia pastoris*), imparting meat-like color and flavor to plant-based burgers (Ismail et al., 2020). Similarly, precision fermentation has been applied to produce fats from plant-derived substrates, with molecular structures comparable to those of animal fats (Teng et al., 2021). A precision fermentation-derived beef fat analogue produced using yeast has also been reported. Co-fermentation of wheat bran with *Propionibacterium freudenreichii* and *L. brevis* led to the production of substantial amounts of vitamin B12 (Xie et al., 2019). The authors suggested that this fermentation-derived vitamin B12 could be used to supplement plant-based meat products, as this vitamin is naturally absent in plant-based foods, and is predominantly obtained from animal-derived products.

8.5 Engineering Microbes for Functional Beverages

Functional beverages are formulated not only to quench thirst but to provide additional health benefits, such as enhanced energy, improved digestion and immune system. A functional beverage is defined as a non-alcoholic or low-alcohol drink that contains ingredients such as herbs, vitamins, minerals, amino acids or additional raw fruit or vegetables, probiotics, or other bioactive compounds (Panou & Karabagias, 2025). This category includes plant-based beverages, fortified juices, probiotic waters, adaptogenic drinks, protein beverages, and sports and hydration drinks. Figure 8.2 presents examples of functional beverages, their key functional compounds, and the associated health benefits.

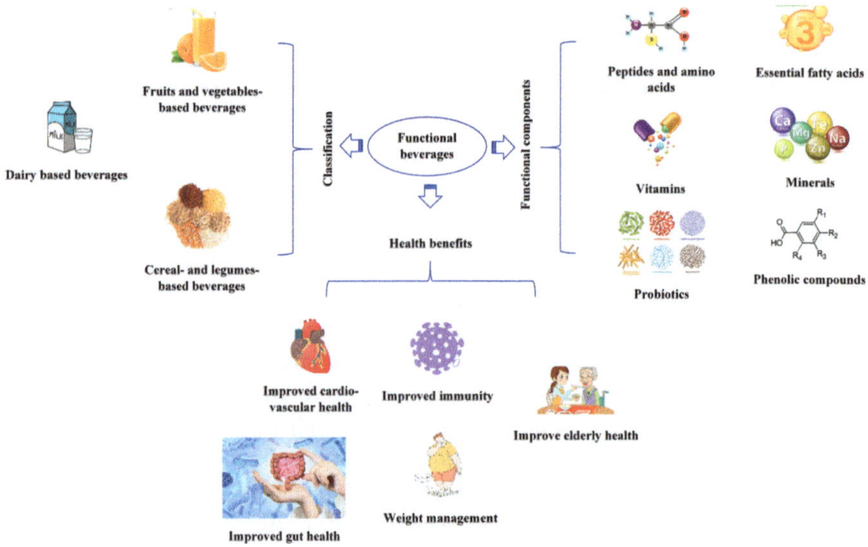

Fig. 8.2 Classification, functional components and health benefits of functional beverages

8.5.1 Gut health and Functionality as a Central Pillar of Functional Beverages

Gut health has emerged as a central driver of consumer demand and innovation in the functional beverage market, supported by growing scientific evidence linking the gut microbiota to digestive health, immune function, metabolic regulation, and the prevention of chronic diseases. Functional beverages formulated with live microorganisms, prebiotics, or postbiotics are increasingly positioned as convenient delivery systems for gut-health benefits. However, conventional strains are often unstable, exhibit variable functionality, and have health benefits that are not yet fully elucidated. To address these challenges, synthetic biology has emerged as a key technological approach (Marco et al., 2017). Its application allows enhanced functional efficacy of beverages, beyond conventional probiotic claims. It represents a promising strategy for the development of next-generation functional beverages based on scientific evidence and aligned with health and sustainable development goals.

8.5.2 Enhancing Sensory Profile and Stability of Functional Beverages

8.5.2.1 Eliminating Plant-Based Off-Flavors and Reducing Acidity

Plant-based beverages might exhibit unpleasant sensory attributes due to lipoxygenase activity. Genetic engineering of *Lactococcus lactis* and *S. cerevisiae* has enhanced the expression of aldehyde and alcohol dehydrogenases, responsible for degrading malodorous volatile aldehydes, resulting in improved sweetness and overall palatability of the beverage (Tao et al., 2022; Amadei et al., 2024). Moreover, genetically modified strains of lactic acid bacteria can synthesize fruity or buttery esters during fermentation, thereby enhancing the flavor profiles in grain- and fruit-based beverages (Li et al., 2020).

Organic acids including lactic, acetic, citric, gluconic, and malic acids, play an important role in beverage sensory attributes and preservation. However, excessive acid production, or imbalance acid ratios, can lead to harsh, sour, or vinegary flavors. Synthetic biology has enabled precise modulation of acidity in lactic acid bacteria by altering the activity of enzymes such as lactate dehydrogenase, acetolactate synthase, and malate dehydrogenase, thereby optimizing the lactic to acetic acid ratio and reducing excessive acidity (Yılmaz et al., 2024). Engineered strains of *Gluconobacter* sp. have been developed to produce gluconic acid while minimizing byproduct formation, resulting in a mild and refreshing acidity desirable in functional beverages (Goldberg & Rokem, 2019). Choi et al. (2024) engineered *Saccharomyces cerevisiae* to improve metabolic robustness and L-lactic acid production from lignocellulosic biomass. Collectively, these adjustments yield beverages with balanced acidity profiles that retain natural flavors and maintain microbial stability.

8.5.2.2 Enhancing Aroma

Aroma formation during fermentation arises from the microbial metabolism of carbohydrates, amino acids, and lipids. Advances in synthetic biology have enabled microorganisms to synthesize a wide range of aromatic compounds. Genetic modification of *Saccharomyces cerevisiae* has been shown to increase the production of volatile thiols, monoterpenes, and phenolic derivatives, thereby improving fruity, floral, or milky notes in plant-based beverages (Yuan et al., 2025). Engineered *Lactobacillus* spp. can improve the conversion of amino acids into volatile branched-chain compounds such as 3-methylbutanal and 2-methylpropanol, resulting in fermented beverages enriched in aromatic compounds and well appreciated by consumers (Lin et al., 2025).

8.5.2.3 Improving Color Stability

Color is a key quality attribute of beverages, influencing consumers' perception of freshness. Microbial engineering offers sustainable alternatives to plant-derived pigments, which are often unstable during processing and storage. Engineered *Yarrowia lipolytica*, *Escherichia coli*, and *Saccharomyces cerevisiae* have been developed to produce carotenoids (β-carotene, lycopene, astaxanthin) and anthocyanins for use as natural beverage colorants (Jing et al., 2023). In addition to improving visual appearance, these pigments enhanced the antioxidant capacity and functional properties of beverages.

8.5.2.4 Improving Mouthfeel and Texture

Plant-based beverages often exhibit an undesirable mouthfeel due to low viscosity and limited protein-water interactions. To address this limitation, microorganisms have been genetically engineered to synthesize EPS, which improve viscosity, emulsification, and stability of beverages. Genetically engineered stains of *Lactobacillus rhamnosus* and *Streptococcus thermophilus*, with altered glycosyltransferase pathways, produce EPS with tailored molecular weights, resulting in enhanced smoothness and texture in low-fat beverages (Soumya & Nampoothiri, 2021). These EPS also contribute to stabilization of plant protein and polyphenol suspensions, thereby reducing sedimentation and phase separation.

8.5.2.5 Reducing Astringency and Bitterness

Plant-based functional beverages rich in polyphenols such as cocoa, tea, hibiscus, and berries, are often characterized by bitterness and astringency resulting from tannin interactions with salivary proteins. Microbial engineering enables the production of enzymes such as tannases, glucosidases, and esterases, that convert complex tannins into less astringent and more bioavailable metabolites. Genetically modified *L. plantarum* expressing tannase has been shown to improve flavor and reduce bitterness in polyphenol-rich beverages (Lin et al., 2025). Furthermore, these microorganisms were able to release polyphenol aglycones during fermentation, thereby improving antioxidant and anti-inflammatory activities, and strengthening the functional value of the beverages.

8.5.3 Enhancing Probiotics Survival

One of the main limitations of fruit-or legume-based probiotic functional beverages is the low survival of these microorganisms during processing, storage, and gastrointestinal transit. Microbial engineering has been evidenced to improve probiotic stress tolerance through the overexpression of chaperone proteins such as *GroEL* and *DnaK*, and acid resistance systems, thereby enhancing survival in acidic beverages such as fruit juices and kombucha-type drinks (Papadimitriou et al., 2016). Genome editing using CRISPR has enabled the removal of non-essential metabolic pathways and the optimization of carbohydrate utilization, allowing probiotics to persist in low-sugar plant-based beverages (O'Connell et al., 2022). These modifications ensure the delivery of viable probiotics while preserving beverage quality (Table 8.3).

8.5.4 Enhancing Targeted Probiotic Properties

Table 8.4 summarizes the specific health-promoting functions of next-generation probiotics in functional beverages. These functional properties included production of GABA, folate, riboflavin, and vitamin B12, synthesis of ACE-inhibitory peptides, antioxidant enzymes, Indole-3-lactic acid (ILA), deconjugated bile acids, peptidoglycan fragments, and plantaricins (Table 8.4).

8.5.5 Improving Postbiotics Production for Application in Functional Beverages

Postbiotics can be defined as bioactive compounds produced by microorganisms during fermentation. These include short-chain fatty acids (acetate, propionate, and butyrate) that regulate gut barrier integrity and inflammation; bioactive peptides with antihypertensive, antioxidant, and immunomodulatory activities; EPS that improve mouthfeel, stability, and prebiotic functionality; and cell wall fragments and

Table 8.3 Engineering strategies to enhance probiotic survival in functional beverage matrices

Probiotic strain	Beverage stress factor	Engineering/adaptation strategy	Mechanism of enhanced survival	Application in functional beverages	Reference
L. plantarum	Low pH (fruit juices, kombucha)	Overexpression of acid tolerance genes (F_0F_1-ATPase, GAD system)	Maintains intracellular pH and prevents acid-induced damage	Acidic probiotic fruit beverages	Papadimitriou et al. (2016)
Bifidobacterium longum	Oxygen exposure	Upregulation of oxidative stress response genes (SOD, catalase)	Detoxifies reactive oxygen species	Non-dairy probiotic beverages	Ruiz et al. (2011)
L. rhamnosus GG	Refrigeration and storage stress	Adaptive laboratory evolution	Improved membrane integrity and cold tolerance	Chilled functional drinks	Sanders et al. (2019)
L. brevis	Low sugar availability	Engineering of carbohydrate utilization pathways	Improved metabolic flexibility	Low-calorie functional drinks	Gänzle and Follador (2012)
Bifidobacterium adolescentis	Processing heat stress	Stress protein overexpression (GroEL, DnaK)	Protection against protein denaturation	Pasteurized probiotic beverages	Papadimitriou et al. (2016)

Table 8.4 Engineering probiotic strains for targeted functional properties in functional beverages

Engineered probiotic strain	Engineering/ optimization strategy	Key bioactive produced (postbiotic/ probiotic function)	Health benefit delivered via beverage	Reference
L. brevis, *L. plantarum*	Overexpression of glutamate decarboxylase (GAD) pathway; pH and substrate optimization	GABA	Antihypertensive, anxiolytic, neuroprotective	Dhakal et al. (2012)
Lactococcus lactis	Nisin-controlled gene expression (NICE system)	Folate	Anti-inflammatory, cancer-preventive, metabolic health	LeBlanc et al. (2013)
L. plantarum	Deregulation of rib operon	Riboflavin	Energy metabolism, antioxidant defense	Capozzi et al. (2012)
Propionibacterium freudenreichii	Pathway enhancement and fermentation optimization	Cobalamin (vitamin B12)	Hematopoiesis, neurological function	Deptula et al. (2017)
L. helveticus	Enhancement of proteolytic system	ACE-inhibitory peptides	Blood pressure regulation	Martínez-Maqueda et al. (2012)
Bifidobacterium longum	Overexpression of SOD and catalase genes	Antioxidant enzymes	Reduction of oxidative stress	Plaza-Díaz et al. (2019)
L. reuteri	Optimization of tryptophan metabolism	Indole-3-lactic acid (ILA)	Gut immune modulation	Zelante et al. (2013)

enzymes that interact with host immune receptors. Table 8.5 summarizes selected examples of postbiotics produced by engineered microbial strains for functional beverages, along with their respective roles in these products.

8.5.6 Examples of Plant-Based Beverages Produced with Engineered Microbes

Table 8.6 presents some examples of plant-based beverages produced with engineered microorganisms through precision fermentation. These include plant-based kefirs, *Kombucha*-like drinks, and cereal-based fermented beverages. The different microorganisms used as ferments in the production of these plant-based beverages were specifically engineered to optimize targeted technological properties, sensory attributes, functional performance, and associated health benefits.

Table 8.5 Examples of postbiotics produced by engineered/optimized microbial strains for functional beverages

Postbiotic compound	Engineered / optimized producing strain	Engineering or optimization strategy	Role in functional beverages	Reference
Short-chain fatty acids (acetate, propionate)	*Lactobacillus plantarum*	Metabolic pathway optimization to enhance carbohydrate fermentation and SCFA output	Gut barrier reinforcement, anti-inflammatory effects, metabolic health	Liu et al. (2023)
GABA	*L. brevis*, *L. plantarum*	Overexpression of glutamate decarboxylase (GAD) system; optimized pH and glutamate supplementation	Antihypertensive, anxiolytic, neuroprotective; stress-reducing beverages	Dhakal et al. (2012)
Bioactive peptides (ACE-inhibitory peptides)	*L. helveticus*	Proteolytic system enhancement; controlled fermentation	Blood pressure regulation, cardiovascular health	Martínez-Maqueda et al. (2012)
Exopolysaccharides (EPS)	*Lactococcus lactis*, *L. rhamnosus*	EPS biosynthetic gene cluster upregulation	Improved mouthfeel, viscosity, beverage stability; prebiotic effects	Jurášková et al. (2022)
Indole-3-lactic acid (ILA)	*L. reuteri*	Tryptophan metabolism pathway enhancement	Anti-inflammatory, gut–immune modulation	Zelante et al. (2013)
Cell wall fragments (peptidoglycans, teichoic acids)	*L. casei*	Controlled heat inactivation after fermentation	Immune modulation, enhanced gut immune signaling	Żółkiewicz et al. (2020)
Organic acids (lactic, acetic acid)	*S. cerevisiae*	Redirected carbon flux toward organic acid synthesis	Flavor balance, microbial stability, controlled acidity	Walker and Stewart (2016)
Antioxidant enzymes (SOD, catalase)	*Bifidobacterium longum*	Stress-response gene overexpression	Reduction of oxidative stress, cellular protection	Plaza-Díaz et al. (2019)

8.6 Safety, Regulatory and Ethical Aspects

8.6.1 Safety of Engineered Food Microbes

Microorganisms engineered for use in plant-derived dairy substitutes, meat alternatives, and functional beverages should not pose risks beyond those considered acceptable for conventional foods. International standards including Codex Alimentarius, EFSA, and FAO/WHO, agree on a structured risk-assessment

Table 8.6 Engineered microorganisms and their technological, sensory, and functional contributions to plant-based fermented beverages

Beverage type	Engineered microorganism	Engineering/optimization strategy	Technological modification	Sensory impact	Functional/health benefit	References
Plant-based kefir	*L. plantarum*	Enhanced glycosidase and GAD pathway expression	Improved fermentation of plant sugars; stable acidification	Reduced beany flavor; smoother mouthfeel	GABA production; gut health support	Dhakal et al. (2012)
	Lc. Lactis	EPS biosynthesis gene upregulation	Increased viscosity and emulsion stability	Creamier texture; dairy-like consistency	Prebiotic and gut barrier support	Jurášková et al. (2022)
	S. cerevisiae	Modified sugar transport and ethanol pathways	Efficient fermentation with low ethanol	Mild effervescence; balanced sweetness	B-vitamin enrichment	Walker and Stewart (2016)
Kombucha-like drink	*Komagataeibacter xylinus*	Controlled cellulose and organic acid synthesis	Stable SCOBY formation; acidity control	Balanced sourness; clean finish	Antimicrobial stability	Jayabalan et al. (2014)
	S. cerevisiae	Carbon flux redirection to reduce ethanol	Controlled carbonation; low alcohol	Light carbonation; fruity notes	Reduced alcohol content	Villarreal-Soto et al. (2018)
Cereal-based beverage	*L. fermentum*	Phytase expression enhancement	Reduced phytic acid	Less astringency	Increased mineral bioavailability	Marco et al. (2017)
	L. plantarum	Stress tolerance and carbohydrate metabolism optimization	Robust fermentation of cereals	Balanced acidity	Probiotic delivery; gut health	Papadimitriou et al. (2016)
	S. cerevisiae	Amylase and maltase activity enhancement	Improved starch conversion	Mild sweetness	Improved digestibility and energy availability	Gänzle (2015)

framework for genetically modified microorganisms (GMMs) and precision fermentation products. This framework comprises hazard identification and characterization, exposure assessing, and overall risk characterization (FAO, 2024).

For engineered food microorganisms, three main product categories are relevant: (*i*) foods containing living modified cells like probiotic plant-derived yogurts or fermented drinks; (*ii*) foods containing inactivated cells or postbiotic preparations; and (*iii*) products compose solely of purified molecules produced by modified microorganisms including dairy proteins, flavor enzymes, or heme proteins, along with the producing-strain removed or inactivated. Each category presents, yet interconnected, safety considerations related to the host organism, genetic modifications, expressed products, and manufacturing process. These aspects should be assessed under realistic consumption scenarios, especially in LMICs (Jin et al., 2025).

8.6.1.1 Conceptual Framework and Product Categories

EFSA guidance on genetically modified microorganism (GMMs) categorizes products according to the presence of viable GMMs and/or recombinant DNA in the final food, as well as the extent to which novel traits are expressed (FAO, 2024). This framework can be adapted to plant-based dairy and meat alternatives.

- **Category A: Live engineered microorganisms in the final product**

  This category included engineered lactic acid bacteria or yeasts used as starter cultures or probiotics in plant-based yogurts or fermented beverages. Safety assessment should address intestinal colonization potential, metabolic activities such as D-lactate generation, and bile salt deconjugation, antimicrobial resistance (AMR), and the potential horizontal gene transfer within the gastrointestinal tract (FAO/WHO, 2024).
- **Category B: Inactive cells and postbiotic compounds**

  In these products, engineered cells are killed by heat or alternative methods, leaving only cellular fragments, metabolites, and other non-living components. Although the lack of viability reduces the risk of infection and colonization, safety evaluation remains necessary. Key considerations include the immunomodulatory effects of cell wall components, residual intracellular metabolites, and intentionally produced bioactive compounds. FAO/WHO guidelines on probiotics, along with recent consensus definitions of postbiotics, provides relevant reference for assessing these products.
- **Category C: Refined ingredients from precision fermentation**

  This category encompasses solely purified proteins or small molecules such as caseins, β-lactoglobulin, enzymes, soy leghemoglobin, and vitamins, where the producing-strain and most recombinant DNA are eliminated during downstream processing. According to FAO and various national assessments these ingredients can generally be evaluated within existing food safety frameworks, often in a manner similar to food additives or enzymes. Assessment focuses primarily on the allergenicity and toxicological properties of the novel molecules (Hilgendorf et al., 2024; Sturme et al., 2025).

8.6.1.2 Strain Selection and Genetic Characterization

A fundamental aspect of safety is selecting a secure chassis organism. The Qualified Presumption of Safety (QPS) by EFSA and the US GRAS concept formalize the extensive history of safe use associated with numerous lactic acid bacteria and baker's yeast. Isolates from QPS/GRAS species are preferred, as long as:

- Whole-genome sequencing confirms species identity and absence of virulence factors, toxins, and acquired AMR genes on mobile elements;
- The strain exhibits genetic and phenotypic stability during fermentation conditions;
- For probiotic application, strain-level characterization and comprehensive safety assessment including evidences from human studies, in accordance with FAO/WHO guidelines for probiotic evaluation (FAO/WHO, 2024).

Genetic modifications must be clearly and systematically defined in compliance with EFSA and Codex guidelines. Essential elements include:

- Full description of the inserted sequences (source, function, regulatory components);
- Evidence that no unintentional open reading frames with possible toxic or allergenic risks are formed at junctions;
- Employing marker-free end strains, avoiding antibiotic resistance markers wherever feasible;
- Proof of genetic stability across several generations and under processing conditions, supported by sequencing or reliable genotyping.

For Category C products, another consideration is whether any viable cells or recombinant DNA from the producing-strain persist in the final food. When recombinant DNA is present only at trace levels, risk assessment emphasizes the safety of the ingredient itself, whereas environmental and occupational hazards are handled through good manufacturing practices and waste management as outlined in Sect. 8.6.2.

8.6.1.3 Safety of Novel Proteins, Metabolites and Antimicrobial Resistance

The safety assessment of engineered microbes that create new proteins or significantly enhance existing ones such as milk proteins, heme proteins, enzymes, and bioactive peptides, is based on a tiered strategy (Jin et al., 2025).

(i) **Allergenicity**

- Analyzing sequences computationally using curated databases of allergens;
- Assessing digcstibility including resistance to pepsin, as a measure of possible allergenicity;

- If homology or *in vitro* data raise concerns, targeted immunological or clinical studies might be necessary.

Soy leghemoglobin produced in engineered yeast for plant-derived meat substitutes illustrates this process. Multiple toxicological tests and regulatory assessment showed that, at the suggested usage levels, the compound poses no unacceptable risk of toxicity or allergenicity (Reyes et al., 2023).

(ii) **Toxicology**

If a protein has already been evaluated and authorized as a food enzyme or additive, only limited additional testing might be required and read-across approaches can often be applied. In contrast, genuinely novel molecules such as novel peptides, and engineered vitamins, are generally subject to standard safety testing. This typically includes *in vitro* genotoxicity assays, and sub-chronic oral toxicity studies, conducted under exposure scenarios that reflect realistic consumption levels of plant-based dairy or meat alternatives (Jin et al., 2025).

(iii) **Antimicrobial resistance and horizontal gene transfer**

- Probiotic strains and production must not harbor acquired, transferable AMR genes on plasmids, transposons, or integrative elements;
- Antibiotic susceptibility profiles should be consistent with species-specific norms, and intrinsic resistance traits must be fully recorded and documented;
- For products containing live engineered microorganisms, the potential for horizontal gene transfer within the gut microbiota must be carefully considered particularly in settings with a high pre-existing burden of antimicrobial resistance even when the associated risk is considered low (ARG, 2022; WHO, 2023).

8.6.2 Regulatory Landscape

Current regulatory frameworks for novel foods, GMOs, and food additives or enzymes govern the oversight of genetically modified microorganisms and compounds obtained through precision fermentation. In relation to the safety consideration discussed in Sect. 8.6.1, three issues are of particular importance: the classification of precision fermentation products; the distinction at the system level, between GMO-producing microorganisms and non-GMO final products; and the impact of labeling, consumer perception, and trust on product adoption, particularly in low-and middle-income countries.

8.6.2.1 Novel Foods and Fundamental Regulatory Routes

A novel food is defined under European Union Regulation (EU) 2015/2283 as a food that was not consumed to a significant degree within the EU before May 15, 1997. Novel foods in the EU are subject to centralized pre-market authorization following the European Food Safety Authority (EFSA). This evaluation covers compositional analysis, toxicology, allergenicity, anticipated intake, and nutritional impacts (European Parliament and Council, 2015). Precision-fermented proteins such as animal-free dairy proteins, heme, and enzymes, are generally assessed either as novel foods or as food enzymes or additives, based on their intended technological role. Aspects of genetically modified organisms (GMOs) are regulated under Regulation (EC) No 1829/2003 on genetically modified food and feed, which applies to foods "containing, consisting of or produced from GMOs". For highly purified precision-fermented ingredients the key regulatory question is whether they are classified as "produced from" a GMO or simply "produced with" a GMO. This distinction remains under discussion and is currently addressed on a case-by-case basis by relevant authorities.

In the United States (US), most precision-fermented food ingredients are regulated under the FDA's food additive framework and GRAS notification process. Assessment focus on the safety of the final ingredient and the production method, including the production organism rather than relying on a simple GMO versus non-GMO classification. To date, several proteins and enzymes produced by microbial fermentation for plant-based or animal free applications have received no questions GRAS letters or food additives approach. Other high-income regions such as UK, Australia/New Zealand, Canada, and Singapore, typically regulate precision-fermented ingredients within existing novel food or food additives and enzyme frameworks. In some cases, these systems are supplemented by additional guidance related to cell-based food and precision fermentation, as seen in Singapore, and the UK.

In many LMICs, no specific regulatory category exists for precision fermentation. Products are therefore assessed under general food regulations, relevant biosafety or GMO legislation where available, and Codex-compliant criteria for food additives or special dietary foods. Recent FAO works on cell-based food and precision fermentation suggest that these products can generally be accommodated within existing food safety frameworks. However, it also highlights substantial gaps in risk assessment capacity, genomic characterization, and regulatory enforcement across parts of Africa, Asia, and Latin America (FAO, 2024; Sturme et al., 2025).

8.6.2.2 "GMO" Processes Versus "Non-GMO" End Products

In precision fermentation, it is essential to distinguish between two elements: (i) the production organism which is often a genetically modified microorganism (GMM), and, (ii) the final ingredient or food, which may contain no viable modified cells and little or no detectable recombinant DNA, as it is the case for purified proteins, enzymes, vitamins.

From a scientific perspective, as highlighted in Sect. 8.6.1, risk assessment focuses on the composition of the final food and the level of consumer exposure, rather than solely on whether a GMO was used during the production. Legally, process-based definitions continue to play an important role.

In the EU, foods "containing, consisting of, or produced from GMOs" are regulated under Regulation (EC) No 1829/2003, which also establishes particular labelling requirements. The regulatory treatment of highly purified precision-fermented ingredients, where recombinant DNA or genetically modified proteins are not detectable, remains under active debate and often requires product-specific interpretation.

In the US, FDA safety assessments conducted under the GRAS or Food additives pathways, primarily focus on the final product. In contrast, the USDA National Bioengineered Food Disclosure Standard (NBFDS) mandates disclosure only for food containing detectable modified genetic material. As a result, many purified precision-fermented ingredients do not require mandatory "bioengineered" labeling when modified DNA is absent, although voluntary descriptors such as "animal-free dairy protein" or "fermentation-derived" are increasingly used.

Overall, the FAO noes that no globally harmonized legal definition of precision fermentation currently exists. Regulatory approaches vary across jurisdictions, with some emphasizing the production process, particularly the use of GMO while other focus on the characteristics of the final ingredient when determining regulatory classifications (FAO, 2024; Sturme et al., 2025).

8.6.2.3 Labelling, Consumer Perception and Trust

In the European Union, genetically modified food labeling is governed by Regulation (EC) No 1829/2003, which requires foods containing or produce from genetically modified (GMOs) above defined thresholds to be clearly labeled. Regulation (EU) 2015/2283 on novel foods mandates labeling that is not misleading to consumers and that highlights any distinctive characteristics or conditions of use (European Parliament and Council, 2003, 2015).

In the US, National Bioengineered Food Disclosure Standard (NBFDS) applies only to foods containing detectable modified DNA. As a result, many precision-fermented components are not subject to mandatory "bioengineered" labeling and are instead communicated through voluntary claims, which must be truthful and not misleading under FDA and USDA guidance. Emerging private or industry-driven labels for fermentation-derived or animal-free products might further shape customer understanding although their long-term impact on trust remain uncertain (FAO, 2024).

Consumer studies indicated that many individuals are still unfamiliar with precision-fermented foods and frequently conflate them with conventional GMOs. Perceived authenticity, assessment of risks and benefits, and trust in producers and regulators authorities consistently emerge as crucial determinants of consumer acceptance. Rapid evidence reviews by the UK Food Standards Agency reported that clear explanations of the technology, visible regulatory oversight, and transparent communication of benefits such as animal welfare, environmental impact, are associated with increased willingness to try precision-fermented dairy alternatives (FSA, 2025). Similarly, cross-country studies on precision-fermented milk indicate that how information on health, environmental, and animal welfare benefits is framed can significantly influence consumers' willingness to pay. These effects are strongly moderate by levels of trust and perceived personal and societal benefits (Banovic et al., 2025).

8.6.3 *Ethical and Societal Issues*

Beyond safety and regulatory considerations, the use of engineered microorganisms in food production and components derived from precision fermentation, also raise broader ethical and societal concerns. For plant-based dairy and meat substitutes, three key aspects are particularly relevant, especially in LMICs. These include: (i) transparency and public responsibility, (ii) fairness in access and management of technology, and (iii) implications for farmers and existing food systems.

8.6.3.1 Transparency, Accountability and Public Reasoning

A substantial gap remains between the speed of technological advances and public understanding of its potential risks, benefits, and realistic timelines (Wood et al., 2023; FAO, 2024). For this reason, ethical frameworks such as Responsible Research and Innovation (RRI) advocate the development of food technologies that integrate societal values from the outset, anticipate unintended outcomes, and actively involve diverse stakeholders in decision-making processes (Gremmen, 2023). In the context of engineered microorganisms for food production and compounds derived from precision fermentation, three practical dimensions of transparency are particularly important:

- Clear and explicit terminology and labeling, avoiding obscure technical jargon and excessive promotional claims not supported by scientific evidence;
- Public access to information on environmental and nutritional impacts such as life cycle assessments and comparative nutritional profiles, especially when environmental sustainability and health benefits are key elements to product justification (Malila et al., 2024; Kossmann et al., 2025);
- Transparency regarding process characteristics, including raw materials, energy consumption, and waste management, to enable regulatory bodies, researchers, and the public to assess the faisability and viability of claimed benefits under real-world production systems.

Regarding microbial cell-based foods and precision fermentation, the FAO further emphasizes that transparency should extend beyond regulatory compliance to include structured public dialogue and socio-economic studies. Such engagement is essential to ensure that societal expectations and concerns actively influence technological development and progress, rather than merely reacting to finalized products (FAO, 2024).

8.6.3.2 Equity, Inclusivity, and Intellectual Property

Another major ethical concern associated with engineered microorganisms and precision fermentation is related to the governance and the distribution of benefits arising from these technologies. Moritz et al. (2024) observed that current patenting activity is largely concentrated in high-income countries. This reflects a strong appropriation of intellectual property rights over microbial strains, culture media, bioprocesses, and downstream applications. Intellectual property protection is essential to stimulate innovation and reduces risks for private investment. However, overly rigid regulatory frameworks can create substantial challenges. These include:

- Restrictions on technology transfer and local production capacity in low- and middle-income countries;
- Persistently high product prices, limiting access for low-income consumers and export markets;
- Limited flexibility to adapt technologies to locally available raw materials and regional contexts.

Conversely, the adoption of more inclusive technological frameworks, such as non-exclusive licenses, patent pools, open-access strain libraries, and public-private partnerships, could lower barriers to entry and support local adaptation. This approach is particularly relevant when precision fermentation delivers global public benefits such as climate change mitigation, reduced antimicrobial resistance, or improved nutritional outcomes (Gremmen, 2023; FAO, 2024; Sturme et al., 2025). Under such situations, ethical governance requires a balance between robust intellectual property protection and licensing strategies that promote equitable access and strengthen regional production capacities.

8.7 Microbial Engineering in Plant-Based Food Systems: Opportunities and Challenges

8.7.1 Relevance

Despite the well-documented benefits of plant-based diets, limitations in micronutrient bioavailability might exacerbate nutritional deficiencies when diets are poorly balanced (Ayaz et al., 2024). Microbial engineering offers considerable potential to transform agri-food systems by enabling targeted nutrient enrichment, improving food security, and supporting more sustainable production practices. It contributes to the development of more nutritionally complete plant-based food systems, aligned with global nutrition and public health goals (Steensels et al., 2019). These opportunities are particularly relevant for low- and middle-income countries, where fermentation is culturally accepted and plant-based substrates are widely, offering pathways towards affordable nutrition and strengthened food sovereignty (FAO, 2022).

8.7.2 Challenges and Adapted Models

8.7.2.1 Practical Challenges

Infrastructure and Technical Capabilities Implementing microbial engineering requires advanced infrastructures, reliable cold chain logistics, and skilled personnel. These resources are often limited in LMICs. Genetically modified microorganisms must remain functional across variable plant matrices, pH ranges, and storage conditions without generated undesirable by-products (Marco et al., 2021). Limited access to precision fermentation facilities and insufficient control of key process parameters, such as temperature, pH, and oxygen levels can further compromise product reproducibility and safety (Börner et al., 2019).

Regulatory and Biosecurity Challenges The use of genetically modified microorganisms in food is subject to strict and heterogeneous regulatory frameworks, which can restrict market access, particularly in LMICs where regulatory capacity might be limited or inconsistent (OECD, 2020). Inadequate institutional infrastructure for risk assessment, labeling, and post-market monitoring can delay adoption and commercialization (Akinbo et al., 2021), while public scepticism toward genetically modified microorganisms persists in the food sector. To strengthen consumer trust, transparency, clear labeling strategies, and the development of postbiotic approaches are increasingly recognized as essential (Siegrist & Hartmann, 2020).

Economic and Commercial Constraints The development, scale-up, and quality control of genetically modified strains involve substantial financial investment, posing significant barriers for small and medium-sized enterprises that dominate the agri-food sector in LMICs (Kumar et al., 2023). Moreover, certain products, such as functional beverages, are primarily targeted at urban or export markets, which might their affordability and accessibility for low-income populations (Galanakis, 2021).

8.7.2.2 Adapted Models for Resource-Limited Contexts

To maximize the impact of engineered microbes and precision fermentation, particularly in LMICs, innovative, context-adapted, and frugal development models are required. They include:

- The use of GRAS and food-grade microbes with minimal genetic modifications or adaptive laboratory evolution rather than complex synthetic circuits. One strategy to simplify regulatory approval and strengthen consumer confidence in the use of GRAS strains traditionally employed in food fermentation such as *L. plantarum* and *S. cerevisiae*. In this approach, genetic engineering is primarily directed towards metabolic optimization rather than the introduction of entirely new traits thereby facilitating regulatory acceptance and consumer trust (LeBlanc et al., 2013; Dagariya et al., 2025).

- Precision postbiotics: Fermentation using genetically modified strains, followed by their inactivation ensures regulatory compliance while preserving bioactive metabolites. This emerging strategy exploits engineered microorganisms to produce health-promoting metabolites during fermentation, without the presence of viable cells in the final product. This strategy improves storage stability, and simplifies regulatory approval. It is particularly suited to distribution systems in LMIC where refrigeration might be limited (Salminen et al., 2021).
- Decentralized fermentation systems: the use small-scale, modular fermentation units adapted to local substrates reduces production costs and dependence on imported inputs. When tailored to regional available raw materials such as millet, sorghum, cassava, and legumes, these systems enable decentralized production of value-added functional beverages, support rural livelihoods, and the reduction of transportation-related costs (Steensels et al., 2019).
- Open access strain platforms: Shared microbial libraries and open-access strain platforms can facilitate broader access to microbial technologies and support capacity building in resource-limited settings (FAO, 2022).

Table 8.7 summarizes key challenges related to microbial engineering in LMICs, and highlights implementation models adapted to these contexts.

8.7.3 Collaborative Models to Accelerate Adoption of Microbial Engineering

The successful implementation of microbial engineering in plant-based foods relies on coordinated collaborations among several stakeholders, such as:

- **North-South and South-South collaborations**: they are partnership between research institutions in LMICs and those in high-income countries. These collaborations facilitate technology transfer, capacity building, and strain sharing of within equitable intellectual property frameworks. In parallel, South-South collaborations are essential for exchanging solutions tailored to regional contexts should and shared challenges (FAO, 2022).
- **Public-private partnerships**: These collaborations play a crucial role in promoting pilot-scale fermentation, providing regulatory support, and developing a skilled workforce. By integrating academic innovation, industrial expertise, and public health objectives, such partnership help bridge the gap between research and commercialization. Government's involvement can further support pilot projects, subsidize infrastructure, and streamline regulatory pathways for functional foods (Galanakis, 2021).
- **Community innovation hubs**: these platforms combine traditional fermentation knowledge with innovative biotechnologies, thereby strengthening cultural acceptance and long-term sustainability (OECD, 2020). Active engagement of local producers and valorization of traditional fermentation practices facilitate

Table 8.7 Key challenges associated with microbial engineering in low- and middle-income countries and implementation models adapted to these contexts

Challenge category	Specific challenge		Adapted models for LMICs	References
Regulatory & policy	Restrictive or unclear regulations for GM microbes: delays in approval of engineered starter cultures; limits commercialization of fermented plant-based products		Use of GRAS microorganisms, precision postbiotics and ALE	OECD (2020)
Economic & cost constraints	High cost of strain development and fermentation infrastructure: Limited scalability of plant-based dairy, meat substitutes, and functional beverages		Low-cost fermentation systems Local, low-value substrates (cereals, legumes) and Small-scale modular bioreactors	Galanakis (2021), Kumar et al. (2023)
Infrastructure & capacity	Limited access to pilot plants and skilled workforce	Difficulty translating lab-scale microbial engineering to industrial production	Decentralized fermentation units University–industry shared facilities Training and technology transfer programs	Marco et al. (2021)
Consumer acceptance	Skepticism toward GM foods	Reduced market uptake of fermented plant-based products	Postbiotic-based beverages Transparency and clean-label strategies Integration with traditional fermentation	Siegrist and Hartmann (2020)
Technological robustness	Poor strain stability in plant matrices	Inconsistent fermentation performance and bioactive production	Robust native strains with minimal engineering Stress-adapted LAB and yeasts Co-culture fermentation	Börner et al. (2019)
Scalability & reproducibility	Batch-to-batch variability	Unreliable quality of functional beverages	Simplified genetic circuits Starter culture standardization Controlled fermentation protocols	Steensels et al. (2019)
Access & equity	Dependence on proprietary strains	High costs and limited local innovation	Open-source microbial platforms North–South collaborative strain development	Salminen et al. (2021), FAO (2022)

adoption, while the co-development of engineered microbial strains with small and medium-sized enterprises helps ensure that innovations remain affordable, scalable, and aligned with local market needs (Capozzi et al., 2012).

8.8 Research Gaps and Future Directions

8.8.1 Beyond Single Molecules: Multi-functional Strains and Groups

Most current applications of microbial engineering and precision fermentation focus on the production of single target molecules such as specific milk proteins, enzymes, or vitamins. A critical next step is the development of multifunctional food grade strains and microbial consortia that can collectively improve sensory attributes (taste, aroma, texture), nutritional quality (bioavailable proteins, vitamins, bioactive peptides), and product safety and shelf life through acidification, bacteriocin production, and the inhibition of spoilage and pathogen microorganisms. Achieving this goal requires the rational design of food -grade microbial chassis and consortia that comply with the safety and regulatory standards outlined in Sect. 8.6, while exploiting strain specialization to enhance robustness and resilience across various plant-based substrates.

Key remaining research gaps include:

- the lack of predictive tools to ensure the stability of microbial consortia during industrial and artisanal fermentations;
- the absence of simple, regulatory-approved guidelines for the design and validation of microbial consortia;
- limited application of these emerging approaches to indigenous fermented foods in LMICs, particularly when using local strain collections and real processing conditions instead of model systems developed primarily for high-income settings.

8.8.2 Integration with AI, Metabolic Modeling, and High-Throughput Design

Genome-scale metabolic models and machine learning are increasingly used to support strain engineering in industrial biotechnology. For food microorganisms, these tools can be adapted to address food-specific objectives. Such goals include the development of diverse aroma and flavor profiles rather than the overproduction of single target metabolite, as well as improvements in nutritional quality and digestibility, including balanced amino acid equilibrium, vitamin biosynthesis, reduction of fermentable oligosaccharides, disaccharides, monosaccharides, and polyols. In parallel, strict safety and regulatory constraints must be applied, particularly with respect to limiting biogenic amines, ethanol and unintended metabolites, and ensuring the exclusive use of food-grade genetic elements. Future research should prioritize transparent design-build-test-learn workflows for food-grade strains, with performance metrics that extend beyond productivity to include safety and regulatory compliances. Moreover, AI models should be interpretable and readily translatable into clear and understandable genetic, process modifications to support regulatory dossier preparation. Finally, the development of shared high-throughput phenotyping platforms including affordable options, and establishment of regional biofoundries would aid spread strain development, particularly in LMICs, instead of concentrating these capabilities in a limited number of high-income regions.

8.8.3 Microbiome-Aware Product Design

Section 8.5.6 highlighted the importance of plant-based diets in supporting metabolic health. Future applications of engineered microbes must explicitly account for interactions with the gut microbiome. Evidence from human and experimental studies increasingly associate microbiome composition and function to glycemic control, lipid metabolism, inflammation, and undernutrition, particularly in population exposed to high infectious and environmental burdens.

For engineered food microbes and precision-fermented components, adopting a microbiome-conscious design framework is essential. This approach involves implies formulating plant-based dairy and meat analogs along with their starter cultures, to promote beneficial function such as short-chain fatty acid production, vitamin biosynthesis, and colonization resistance. It also requires avoiding combinations of substrates and microbial metabolites that could exacerbate dysbiosis or enteric dysfunction in vulnerable populations. Finally, *in vitro* gut models and targeted human studies should be employed to characterize microbiome responses alongside conventional safety assessments prior to large-scale application.

Key priorities for future research include:

- Human intervention studies assessing the impact of foods produced using engineered microorganisms on the gut microbiota composition, metabolic activities, and relevant clinical parameters.
- Comparative studies that examine response to such foods in populations from high-income countries and LMICs, where dietary patterns, infectious disease burden, and baseline microbiota profiles differ substantially.
- Integrated approaches combining precision fermentation, gut simulation, and clinical trials to support the selection and prioritization of microbial strains and formulations for targeted nutritional programs.

Compared to current research, which largely focuses on demonstrating the capacity of individual strains to produce specific target molecules, future efforts are expected to emphasize the development of robust microbial ecosystems and food systems that perform reliability under real-world processing and consumption conditions. One effective and promising strategy to support this transition is to:

(i) **Anchor innovation within clearly defined safety and regulatory frameworks**
From the earliest stages of development, multifunctional strains and microbial consortia should be designed to comply with the safety, risk assessment, and regulatory requirements mentioned in Sect. 8.6. This approach will ensure traceability of genetic modifications, minimize the risk of antimicrobial resistance, and promote processes that are reproducible.

(ii) **Use of AI as facilitation tools rather than a hidden system**
AI and modeling should improve the design-build-test cycles while maintaining transparency to support regulatory oversight and public trust, leveraging open benchmarks and datasets whenever feasible.

(iii) **Integrate microbiome and nutritional goals**
Prior to their application in food production, the effects of engineered microorganisms on product quality, process efficiency, gut microbiota composition, metabolic markers, and nutritional health, must be carefully evaluated, particularly in populations affected by malnutrition and the resurgence of non-communicable diseases.

(iv) **Linking technological pathways to equity and transformation of food systems**
The integration of engineered microorganisms into food systems should occur within a clear and transparent transition framework aimed at: supporting smallholder farmers, strengthening local production capacities, and ensuring the availability, affordability, and cultural appropriateness of products in LMICs. Addressing these research gaps requires interdisciplinary collaborations through North-South and South-South partnerships, that integrate expertise from synthetic biology, food science, microbiology, artificial intelligence, nutrition, and the social sciences. Within this framework, engineered microorganisms and precision fermentation have the potential to evolve from specialized innovations into foundational tools for healthier, more sustainable, and more equitable food systems over the coming decade.

8.9 Conclusion

Plant-based alternatives to dairy products, meat substitutes, and functional beverages are experiencing rapidly growing demand. This trend is driven by changes in food systems and increasing concerns related to human health, food security, sustainable development, climate change, and ethical considerations. Because of their perceived benefits for human health and the environment, these products have gained consumer trust, and have contributed to a broader and more diversified food market. Despite this progress, current plant-based products present several limitations. These include shortcomings in nutritional quality, sensory attributes, cultural acceptance, and overall environmental performance. Consequently, many plant analogs, in their current forms, are not yet nutritionally adequate or functionally equivalent to foods derived from animal sources.

Precision fermentation and synthetic biology represent promising strategies to address these. These technologies enable the production of ingredients that are biologically identical to those of animal origin and exhibiting high functionalities. They offer new opportunities to improve the nutritional and sensory properties of food products while reducing reliance on complex additive systems and conventional livestock farming.

The full integration of these technologies into the design and development of plant-based foods is expected to boost the emergence of next-generation products that better meet to consumer expectations while addressing public health and sustainable development goals. Despite their substantial potential benefits for food security and nutritional resilience, the large-scale deployment of products derived from precision fermentation and synthetic biology present several challenges. These include technological, regulatory, economic, and societal barriers that must be carefully addressed. Key concerns are related to consumer trust in foods produced using engineered microorganisms, high production costs, and the lack of regulatory harmonization across regions. These challenges are particularly pronounced in LMICs, where access to advanced technological infrastructure might be limited. Addressing these disparities is essential to ensure responsible, inclusive and equitable adoption of these technologies. Future research should adopt in interdisciplinary approach that integrate food science, nutrition, biotechnology, and social and environmental sciences. Priority areas include the optimization of microbial production systems, alignment with evidence-based dietary guidelines, comprehensive assessments of health and environmental impacts, improvements in regulatory frameworks and ethical considerations, and the development of effective communication strategies to shape consumer perceptions. Through such efforts, precision fermentation and synthetic biology have the potential to play skey role in the design and development of food systems that are nutritionally adequate, sustainable, and resilient.

Funding Information This research did not receive any specific grant from funding agencies in the commercial, public, or not-for-profit sectors. The study was self-funded.

Declaration of Competing Interest The authors declare no competing interest.

References

Ahmad, M., Qureshi, S., Akbar, M. H., Siddiqui, S. A., Gani, A., Mushtaq, M., Hassan, I., & Dhull, S. B. (2022). Plantbased meat alternatives: Compositional analysis, current development and challenges. *Applied Food Research, 2*(2), 100154. https://doi.org/10.1016/j.afres.2022.100154

Ajanaku, P. O., Olojede, A. O., Ajanaku, C. O., Egharevba, G. O., Agaja, F. O., Joseph, C. B., & Thomas, R. M. (2025). Novel fermentation techniques for improving food functionality: An overview. *Fermentation, 11*(9), 509. https://doi.org/10.3390/fermentation11090509

Akinbo, O., Obukosia, S., Ouedraogo, J., Sinebo, W., Savadogo, M., Timpo, S., Mbabazi, R., Maredia, K., Makinde, D., & Ambali, A. (2021). Commercial Release of Genetically Modified

Crops in Africa: Interface Between Biosafety Regulatory Systems and Varietal Release Systems. *Frontiers in plant science, 12*, 605937. https://doi.org/10.3389/fpls.2021.605937

Albino, M., Gargalo, C. L., Albæk, M. O., Krühne, U., & Gernaey, K. V. (2024). Hybrid modeling for on-line fermentation optimization and scale-up: A review. *Processes, 12*(8), 1635. https://doi.org/10.3390/pr12081635

Alcorta, A., Porta, A., Tárrega, A., Alvarez, M. D., & Vaquero, M. P. (2021). Foods for plant-based diets: Challenges and innovations. *Foods, 10*(2), 293. https://doi.org/10.3390/foods10020293

Amadei, S., Gandolfi, I., Gottardi, D., Cevoli, C., D'Alessandro, M., Siroli, L., Lanciotti, R., & Patrignani, F. (2024). Selection of yeasts and lactic acid bacteria to improve the nutritional quality, volatile profile, and biofunctional properties of chickpea flour. *Current Research in Food Science, 11*, 101204. https://doi.org/10.1016/j.crfs.2025.101204

Ambikapathi, R., Schneider, K. R., Davis, B., Herrero, M., Winters, P., & Fanzo, J. C. (2022). Global food systems transitions have enabled affordable diets but had less favourable outcomes for nutrition, environmental health, inclusion and equity. *Nature Food, 3*, 764–779. https://doi.org/10.1038/s43016-022-00588-7

Andreson, M., Kazantseva, J., Malv, E., Kuldjärv, R., Priidik, R., & Kütt, M. (2023). Evaluation of microbial dynamics of kombucha consortia upon continuous backslopping in coffee and orange juice. *Foods, 12*, 3545. https://doi.org/10.3390/foods12193545

Andreson, M., Kazantseva, J., Kallastu, A., Jakobson, T., Sarand, I., & Kütt, M. (2024). Isolation and identification of novel non-dairy starter culture candidates from plant matrix using Backslopping propagation. *Fermentation, 10*(12), 663. https://doi.org/10.3390/fermentation10120663

ARG (Antimicrobial Resistance Collaborators). (2022). Global burden of bacterial antimicrobial resistance in 2019: A systematic analysis. *Lancet, 399*(10325), 629–655. https://doi.org/10.1016/S0140-6736(21)02724-0

Atfaoui, K., Lebrazi, S., Raffak, A., Chafai, Y., Kabous, K. E., Fadil, M., & Ouhssine, M. (2025). Impact of selected starter-based sourdough types on fermentation performance and bio-preservation of bread. *Fermentation, 11*(8), 449. https://doi.org/10.3390/fermentation11080449

Ayaz, A., Zaman, W., Radák, Z., & Gu, Y. (2024). Green strength: The role of micronutrients in plant-based diets for athletic performance enhancement. *Heliyon, 10*(12), e32803. https://doi.org/10.1016/j.heliyon.2024.e32803

Aydar, E. F., Tutuncu, S., & Ozcelik, B. (2020). Plant-based milk substitutes: Bioactive compounds, conventional and novel processes, bioavailability studies, and health effects. *Journal of Functional Foods, 70*, 103975. https://doi.org/10.1016/j.jff.2020.103975

Banovic, M., Asioli, D., & Sogari. (2025). Cross-cultural consumer valuation of precision fermentation milk: Effects of information, individual traits, and labelling preferences. *Food Quality and Preference, 131*, 105570. https://doi.org/10.1016/j.foodqual.2025.105570

Börner, R. A., Kandasamy, V., Axelsen, A. M., Nielsen, A. T., & Bosma, E. F. (2019). Genome editing of lactic acid bacteria: opportunities for food, feed, pharma and biotech. *FEMS microbiology letters, 366*(1), fny291. https://doi.org/10.1093/femsle/fny291

Boukid, F., Hassoun, A., Zouari, A., Tülbek, M. Ç., Mefleh, M., Aït-Kaddour, A., & Castellari, M. (2023). Fermentation for designing innovative plant-based meat and dairy alternatives. *Foods, 12*(5), 1005. https://doi.org/10.3390/foods12051005

Bryngelsson, S., Moshtaghian, H., Bianchi, M., & Hallström, E. (2022). Nutritional assessment of plant-based meat analogues on the Swedish market. *International Journal of Food Sciences and Nutrition, 73*(7), 889–901. https://doi.org/10.1080/09637486.2022.2078286

Capozzi, V., Russo, P., Dueñas, M. T., López, P., & Spano, G. (2012). Lactic acid bacteria producing B-group vitamins: A great potential for functional cereals products. *Applied Microbiology and Biotechnology, 96*(6), 1383–1394. https://doi.org/10.1007/s00253-012-4440-2

Cheng, J., & Truter, L. (2023). The shift towards plant-based lifestyles: Factors driving young consumers' decisions to choose plant-based food products. *Sustainability, 16*(20), 9022. https://doi.org/10.3390/su16209022

Choi, B., Tafur Rangel, A., Kerkhoven, E. J., & Nygård, Y. (2024). Engineering of *Saccharomyces cerevisiae* for enhanced metabolic robustness and L-lactic acid production from lignocellulosic biomass. *Metabolic Engineering, 84*, 23–33. https://doi.org/10.1016/j.ymben.2024.05.003

Choudhury, D., Singh, S., Seah, J. S. H., Yeo, D. C. L., & Tan, L. P. (2020). Commercialization of plant-based meat alternatives. *Trends in Plant Science, 25*(11), 1055–1058. https://doi.org/10.1016/j.tplants.2020.08.006

Dagariya, S., Bhatankar, J., Dakal, T. C., Gadi, B. R., & Giudici, P. (2025). Metabolic and evolutionary engineering of food yeasts. *Processes, 13*(6), 1852. https://doi.org/10.3390/pr13061852

Deptula, P., Chamlagain, B., Edelmann, M., Sangsuwan, P., Nyman, T. A., Savijoki, K., Piironen, V., & Varmanen, P. (2017). Food-like growth conditions support production of active vitamin B12 by Propionibacterium freudenreichii 2067 without DMBI, the lower ligand base, or cobalt supplementation. *Frontiers in Microbiology, 8*, 368. https://doi.org/10.3389/fmicb.2017.00368

Dhakal, R., Bajpai, V. K., & Baek, K. H. (2012). Production of GABA by microorganisms. *Brazilian Journal of Microbiology, 43*(4), 1230–1241. https://doi.org/10.1590/S1517-83822012000400001

Dhar, P., Sanghvi, G., Singh, R., Roopashree, R., Bhowmik, A., Sharma, M., Nayak, P. K., Nath, P. C., & Sridhar, K. (2025). Emerging trends in plant-based yogurt: Novel raw materials, bio-functionality, stability, and sustainability perspectives—A review. *Trends in Food Science & Technology, 166*, 105402. https://doi.org/10.1016/j.tifs.2025.105402

Eastham, J. L., & Leman, A. R. (2024). Precision fermentation for food proteins: Ingredient innovations, bioprocess considerations, and outlook—A mini-review. *Current Opinion in Food Science, 58*, 101194. https://doi.org/10.1016/j.cofs.2024.101194

Eichhorst, W., Shah, S., Smith, A., & Newton, P. (2025). Innovation systems for emerging food technologies: Evidence from the development of cultured proteins in Thailand. *Frontiers in Sustainable Food Systems, 9*, 1497792. https://doi.org/10.3389/fsufs.2025.1497792

European Parliament and Council. (2003). *Regulation (EC) No 1829/2003 of the European Parliament and of the Council of 22 September 2003 on genetically modified food and feed.* Official Journal of the European Union. 268, pp. 1–23.

European Parliament and Council. (2015). *Regulation (EU) 2015/2283 of 25 November 2015 on novel foods.* Official Journal of the European Union.

Fairfeild. (2022). https://www.fairfieldmarketresearch.com/report/plant-based-beverages-market/request-sample. Accessed 8 May 2024.

FAO. (2022). *Fermented foods and food systems transformation*. Food and Agriculture Organization of the United Nations.

FAO. (2024). *Cell-based food and precision fermentation: Products, safety and the future role.* Food and Agriculture Organization of the United Nations.

FAO/WHO. (2024). *Joint FAO/WHO food standards programme codex committee on nutrition and foods for special dietary uses: Discussion paper on harmonized probiotic guidelines for use in foods and food supplements.* 44th Session Dresden, Germany. Agenda Item 6.21, 5p.

Fiorentini, M., Kinchla, A. J., & Nolden, A. A. (2020). Role of sensory evaluation in consumer acceptance of plant-based meat analogs and meat extenders: A scoping review. *Foods, 9*(9), 1334. https://doi.org/10.3390/foods9091334

Fookao, A. N., Mbawala, A., Nganou, N. D., Nguimbou, R. M., & Mouafo, T. H. (2022). Improvement of the texture and dough stability of milk bread using bioemulsifiers/biosurfactants produced by lactobacilli isolated from an indigenous fermented milk (pendidam). *LWT–Food Science and Technology, 163*, 113609. https://doi.org/10.1016/j.lwt.2022.113609

FSA (Food Standards Agency). (2025). A rapid evidence review on consumer responses to precision fermentation. *FSA Research and Evidence*, 1–55. https://doi.org/10.46756/001c.136898

Gajewska, D., Szajewska, H., Włodarek, D., Weker, H., Wolnicka, K., Wiśniewska, K., Socha, P., & Hamulka, J. (2024). The role of plant-based beverages in nutrition: An expert opinion. *Nutrients, 17*(9), 1562. https://doi.org/10.3390/nu17091562

Galanakis, C. M. (2021). Functionality of Food Components and Emerging Technologies. *Foods (Basel, Switzerland), 10*(1), 128. https://doi.org/10.3390/foods10010128

Galimberti, A., Bruno, A., Agostinetto, G., Casiraghi, M., Guzzetti, L., & Labra, M. (2021). Fermented food products in the era of globalization: Tradition meets biotechnology innovations. *Current Opinion in Biotechnology, 70*, 36–41. https://doi.org/10.1016/j.copbio.2020.10.006

Gangakhedkar, P. S., Deshpande, H. W., Törős, G., Elsakhawy, T., Abdalla, N., Shaikh, A., Kovács, B., Mane, R., & Prokisch, J. (2024). Fermentation of fruits and vegetables: Bridging traditional wisdom and modern science for food preservation and nutritional value improvements. *Foods, 14*(13), 2155. https://doi.org/10.3390/foods14132155

Gänzle, M. G. (2015). Enzymatic and bacterial conversions during sourdough fermentation. *Food Microbiology, 37*, 2–10. https://doi.org/10.1016/j.fm.2013.04.007

Gänzle, M. G., & Follador, R. (2012). Metabolism of oligosaccharides and starch in lactobacilli. *Applied Microbiology and Biotechnology, 94*(6), 1369–1382. https://doi.org/10.3389/fmicb.2012.00340

Gao, X. J., Chong, L. S., Kim, M. S., & Elowitz, M. B. (2018). Programmable protein circuits in living cells. *Science, 361*, 1252–1258. https://doi.org/10.1126/science.aat5062

Garner, K. L. (2021). Principles of synthetic biology. *Essays in Biochemistry, 65*(5), 791. https://doi.org/10.1042/EBC20200059

Gerogianni, V., Mantzourani, C., Theodoropoulou, M. A., Chiou, A., & Kokotou, M. G. (2025). Phenolic compounds in plant-based milk alternatives from the greek market. *Separations, 12*(10), 282. https://doi.org/10.3390/separations12100282

Gilman, J., Walls, L., Bandiera, L., & Menolascina, F. (2021). Statistical design of experiments for synthetic biology. *ACS Synthetic Biology, 10*, 1–18. https://doi.org/10.1021/acssynbio.0c00385

Goldberg, I., & Rokem, S. J. (2019). Organic and fatty acid production, microbial. In M. T. Schmidt (Ed.), *Encyclopedia of microbiology* (4th ed., pp. 358–382). Academic Press. https://doi.org/10.1016/B978-0-12-809633-8.13083-3

Gomes, V. A. M., Souza Carmo, T., Silva Carvalho, L., Mendonça Bahia, F., & Parachin, N. S. (2018). Comparison of yeasts as hosts for recombinant protein production. *Microorganisms, 6*(2), 38. https://doi.org/10.3390/microorganisms6020038

Gong, R., Zalán, Z., Song, J., & Suo, H. (2025). Microbial fermentation of soybeans: Synergistic enhancement of bioactivity and sensory properties. *Food Chemistry: X, 30*, 102924. https://doi.org/10.1016/j.fochx.2025.102924

Good Food Institute. (2021). *U.S. retail market data for the plant-based industry* [WWW document]. https://gfi.org/marketresearch/

Gremmen, B. (2023). *Ethical issues of developing new technologies in agriculture*. Wageningen University & Research.

Hilgendorf, K., Wang, Y., Miller, M. J., & Jin, Y. S. (2024). Precision fermentation for improving the quality, flavor, safety, and sustainability of foods. *Current Opinion in Biotechnology, 86*, 103084. https://doi.org/10.1016/j.copbio.2024.103084

Hu, Y., Piao, C., Yue, C. W., Yanan, Z., Dan, W., Hansong, Y., & Xu, B. (2019). Soybean residue (okara) fermentation with the yeast *Kluyveromyces marxianus. Food Bioscience, 31*, 100439. https://doi.org/10.1016/J.FBIO.2019.100439

Huang, S., Yan, Y., Su, F., Huang, X., Xia, D., Jiang, X., Dong, Y., Lv, P., Chen, F., & Lv, Y. (2021). Research progress in gene editing technology. *Frontiers in Bioscience (Landmark Edition), 26*(10), 916–927. https://doi.org/10.52586/4997

Irondi, E. A., Aina, H. T., Imam, Y. T., Bankole, A. O., Anyiam, A. F., Elemosho, A. O., Kareem, B., & Adewumi, T. O. (2025). Plant-based milk substitutes: Sources, production, and nutritional, nutraceutical and sensory qualities. *Frontiers in Food Science & Technology, 5*, 1593870. https://doi.org/10.3389/frfst.2025.1593870

Ismail, I., Hwang, Y. H., & Joo, S. T. (2020). Meat analog as future food: A review. *Journal of Animal Science & Technology, 62*(2), 111–120. https://doi.org/10.5187/jast.2020.62.2.111

Janik, E., Niemcewicz, M., Ceremuga, M., Krzowski, L., Saluk-Bijak, J., & Bijak, M. (2020). Various aspects of a gene editing system-CRISPR-Cas9. *International Journal of Molecular Sciences, 21*(24), 9604. https://doi.org/10.3390/ijms21249604

Jayabalan, R., Malbaša, R. V., Lončar, E. S., Vitas, J. S., & Sathishkumar, M. (2014). A review on *Kombucha* tea-microbiology, composition, fermentation, beneficial effects, toxicity, and tea fungus. *Comprehensive Reviews in Food Science and Food Safety, 13*(4), 538–550. https://doi.org/10.1111/1541-4337.12073

Jemaa, M. B., Gamra, R., Falleh, H., Ksourı, R., & Bejı, R. S. (2021). Plant-based milk alternative: Nutritional profiling, physical characterization and sensorial assessment. *Current Perspectives on Medicinal & Aromatic Plants, 4*(2), 108 120. https://doi.org/10.38093/cupmap.1037118

Jiang, X., Ji, J., Yang, Q., Dou, Y., Li, Y., Yang, X., Liu, C., Dou, S., & Dong, L. (2025). Key technologies of synthetic biology in industrial microbiology. *Microorganisms, 13*(10), 2343. https://doi.org/10.3390/microorganisms13102343

Jin, J., den Besten, H. M. W., Rietjens, I. M. C. M., & Widjaja, F. (2025). Chemical and microbiological hazards arising from new plant-based foods, including precision fermentation-produced food ingredients. *Annual Review of Food Science and Technology, 16*(1), 171–194. https://doi.org/10.1146/annurev-food-111523-122059

Jing, Y., Wang, J., Gao, H., Jiang, Y., Jiang, W., Jiang, M., Xin, F., & Zhang, W. (2023). Enhanced β-carotene production in *Yarrowia lipolytica* through the metabolic and fermentation engineering. *Journal of Industrial Microbiology & Biotechnology, 50*(1), kuad009. https://doi.org/10.1093/jimb/kuad009

Jurášková, D., Ribeiro, S. C., Silva, G., & C. C. (2022). Exopolysaccharides produced by lactic acid bacteria: From biosynthesis to health-promoting properties. *Foods, 11*(2), 156. https://doi.org/10.3390/foods11020156

Kale, P., Mishra, A., & Annapure, U. S. (2022). Development of vegan meat flavour: A review on sources and techniques. *Future Foods, 5*, 100149. https://doi.org/10.1016/j.fufo.2022.100149

Kang, M., Choe, D., Kim, K., Cho, B., & Cho, S. (2019). Synthetic biology approaches in the development of engineered therapeutic microbes. *International Journal of Molecular Sciences, 21*(22), 8744. https://doi.org/10.3390/ijms21228744

Karoui, R., & Bouaicha, I. (2024). A review on nutritional quality of animal and plant-based milk alternatives: A focus on protein. *Frontiers in Nutrition, 11*, 1378556. https://doi.org/10.3389/fnut.2024.1378556

Kent, G., Kehoe, L., Flynn, A., & Walton, J. (2022). Plant-based diets: A review of the definitions and nutritional role in the adult diet. *The Proceedings of the Nutrition Society, 81*(1), 62–74. https://doi.org/10.1017/S0029665121003839

Kim, G., Baek, K., Lee, G., Lee, J., Moon, J., & Seo, S. (2024). Development of starter cultures for precision fermentation of kombucha with enriched gamma-aminobutyric acid (GABA) content. *Fermentation, 11*(1), 17. https://doi.org/10.3390/fermentation11010017

Knychala, M. M., Boing, L. A., Ienczak, J. L., Trichez, D., & Stambuk, B. U. (2024). Precision fermentation as an alternative to animal protein, a review. *Fermentation, 10*(6), 315. https://doi.org/10.3390/fermentation10060315

Korcz, E., & Varga, L. (2021). Exopolysaccharides from lactic acid bacteria: Techno-functional application in the food industry. *Trends in Food Science & Technology, 110*, 375–384. https://doi.org/10.1016/j.tifs.2021.02.014

Kossmann, H., Karslioglu, O. O., & Breunig, P. (2025). From Moo to Microbes: Pathways for precision fermentation in recombinant protein production. *Journal of Agriculture and Food Research, 22*, 102056. https://doi.org/10.1016/j.jafr.2025.102056

Kumar, R., Kumar, V., Praanjal, P., and Kumari, P.B. (2023). Genetically Modified Crops: Its Impact and Limitation in Agricultural Food Products. In Maheshwarareddy, C., Saikia, A.R., Bezboruah, M., Elumle, P., Singh, S.S. Current Trends in Agricultural Sciences & Technology. First Edition, Vital Biotech Publication, India. pp. 209–226.

LeBlanc, J. G., Milani, C., De Giori, G. S., Sesma, F., Van Sinderen, D., & Ventura, M. (2013). Bacteria as vitamin suppliers to their host: A gut microbiota perspective. *Current Opinion in Biotechnology, 24*(2), 160–168. https://doi.org/10.1016/j.copbio.2012.08.005

Li, X., Zhang, C., & Sun, L. (2020). Metabolic engineering of lactic acid bacteria for flavor enhancement in fermented foods and beverages. *Journal of Agricultural and Food Chemistry, 68*(31), 8161–8172.

Li, X., Hu, Y., & Wang, Y. (2025). CRISPR/Cas9 technology: Exploring lncRNA's own function and its role in the cancer process. *Chinese Journal of Biochemistry and Molecular Biology, 2025*(41), 364–375.

Liao, T., Li, X., Fan, L., Zhang, B., Zheng, W., Hua, J., Li, L., Mahror, N., & Cheng, L. (2024). Nature of back slopping kombucha fermentation process: Insights from the microbial succession, metabolites composition changes and their correlations. *Frontiers in Microbiology, 15*, 1433127. https://doi.org/10.3389/fmicb.2024.1433127

Lin, M., Lin, S., He, H., Yu, Y., Hu, J., & Zhou, L. (2025). *Lactiplantibacillus plantarum* in fermented beverages: Properties, mechanisms, and future prospects. *Journal of Functional Foods, 129*, 106864. https://doi.org/10.1016/j.jff.2025.106864

Liu, C., Ma, N., Feng, Y., Zhou, M., Li, H., Zhang, X., & Ma, X. (2023). From probiotics to postbiotics: Concepts and applications. *Animal Research and One Health, 1*(1), 92–114. https://doi.org/10.1002/aro2.7

Malila, Y., Owolabi, I. O., Chotanaphuti, T., Sakdibhornssup, N., Elliott, C. T., Visessanguan, W., Karoonuthaisiri, N., & Petchkongkaew, A. (2024). Current challenges of alternative proteins as future foods. *npj Science of Food, 8*, 53. https://doi.org/10.1038/s41538-024-00291-w

Malomo, A. A., Olawoye, B., Olaniran, A. F., Olaniyi, O. I., Adedayo, A., Adeniran, H. A., & Abiose, S. H. (2021). Evaluation of sugar and free amino acid during fermentation of Ogi from maize, acha and sorghum. *Food Science And Applied Biotechnology, 4*, 111–118. https://doi.org/10.30721/fsab2021.v4.i2.136

Marco, M. L., Heeney, D., Binda, S., Cifelli, C. J., Cotter, P. D., Foligné, B., Gänzle, M., Kort, R., Pasin, G., Pihlanto, A., Smid, E. J., & Hutkins, R. (2017). Health benefits of fermented foods: Microbiota and beyond. *Current Opinion in Biotechnology, 44*, 94–102. https://doi.org/10.1016/j.copbio.2016.11.010

Marco, M. L., Sanders, M. E., Gänzle, M., Arrieta, M. C., Cotter, P. D., De Vuyst, L., Hill, C., Holzapfel, W., Lebeer, S., Merenstein, D., Reid, G., Wolfe, B. E., & Hutkins, R. (2021). The International Scientific Association for Probiotics and Prebiotics (ISAPP) consensus statement on fermented foods. Nature reviews. *Gastroenterology & hepatology, 18*(3), 196–208. https://doi.org/10.1038/s41575-020-00390-5

Markets and Markets. (2020). *Meat substitutes market by source (soy protein, wheat protein, pea protein), type (concentrates, isolates, and textured), product (tofu, tempeh, seitan, and quorn), form (solid and liquid), and region—Global forecast to 2026*. Available online: https://www.marketsandmarkets.com/Market-Reports/meat-substitutes-market-979.html. Accessed 30 Nov 2020.

Martínez-Maqueda, D., Miralles, B., Recio, I., & Hernández-Ledesma, B. (2012). Antihypertensive peptides from food proteins: A review. *Food & Function, 3*(4), 350–361. https://doi.org/10.1039/C2FO10192K

Maung, T. T., Gu, B. Y., Kim, M. H., & Ryu, G. H. (2020). Fermentation of texturized vegetable proteins extruded at different moisture contents: Effect on physicochemical, structural, and microbial properties. *Food Science and Biotechnology, 29*(7), 897–907. https://doi.org/10.1007/s10068-020-00737-3

Mbawala, A., Mouafo, T. H., & Tchakouani, G. F. Y. (2024). Biovalorization of four main traditional food products and by-products produced in the Adamaoua region (Cameroon). *Journal of Nutrition Food Science and Technology, 5*(1), 1–7. https://doi.org/10.47485/2834-7854.1034

Moritz, J., Mazac, R., Ueta, M. H., Räty, N., Tuomisto, H. L., & Ryynänen, T. (2024). Prospects of justice for cellular agriculture: A just transition or reinvesting in unsustainability? *Food Ethics, 9*(2), 22. https://doi.org/10.1007/s41055-024-00156-8

Mouafo, T. H., Mbawala, A., Tanaji, K., Somashekar, D., & Ndjouenkeu, R. (2020). Improvement of the shelf life of raw ground goat meat by using biosurfactants produced by lactobacilli strains as biopreservatives. *LWT–Food Science and Technology, 133*(2020), 110071. https://doi.org/10.1016/j.lwt.2020.110071

Mouafo, T. H., Sokamte, A. T., Manet, L., Mbarga, A. J. M., Nadezdha, S., Devappa, S., & Mbawala, A. (2023). Biofilm inhibition, antibacterial and antiadhesive properties of a novel biosurfactant from *Lactobacillus paracasei* N2 against multi-antibiotics-resistant pathogens isolated from braised fish. *Fermentation, 9*, 646. https://doi.org/10.3390/fermentation9070646

Mukherjee, A., Gómez-Sala, B., O'Connor, E. M., Kenny, J. G., & Cotter, P. D. (2022). Global regulatory frameworks for fermented foods: A review. *Frontiers in Nutrition, 9*, 902642. https://doi.org/10.3389/fnut.2022.902642

Nguyen, R. (2023). Microbial diversity in Milk production and vegetables based substitutes. *Journal of Food, Microbiology, Safety & Hygiene, 8*, 223. https://doi.org/10.35248/2476-2059.23.8.223

O'Connell, A., Deo, J., Deomano, E., Wei, X., Jackson, P., Aitken, K. S., Manimekalai, R., Mohanraj, K., Hemaprabha, G., Ram, B., Viswanathan, R., & Lakshmanan, P. (2022). Combining genomic selection with genome-wide association analysis identified a large-effect QTL and improved selection for red rot resistance in sugarcane. *Frontiers in plant science, 13*, 1021182. https://doi.org/10.3389/fpls.2022.1021182

OECD. (2020). *Innovation, agricultural productivity and sustainability*. OECD Publishing.

Olędzki, R. (2025). Polyphenolic compounds in the prevention and treatment of hypertension: A review. *International Journal of Molecular Sciences, 26*(21), 10665. https://doi.org/10.3390/ijms262110665

Panou, A., & Karabagias, I. K. (2025). Composition, properties, and beneficial effects of functional beverages on human health. *Beverages, 11*, 40. https://doi.org/10.3390/beverages11020040

Papadimitriou, K., Alegría, Á., Bron, P. A., de Angelis, M., Gobbetti, M., Kleerebezem, M., Lemos, J. A., Linares, D. M., Ross, P., Stanton, C., Turroni, F., van Sinderen, D., Varmanen, P., Ventura, M., Zúñiga, M., Tsakalidou, E., & Kok, J. (2016). Stress physiology of lactic acid bacteria. *Microbiology and Molecular Biology Reviews, 80*(3), 837–890. https://doi.org/10.1128/MMBR.00076-15

Park, H. S., Jun, S. C., Han, K. H., Hong, S. B., & Yu, & J. H. (2017). Diversity, application, and synthetic biology of industrially important aspergillus fungi. In S. Sariaslani & G. M. Gadd (Eds.), *Advances in applied microbiology* (pp. 161–202). Academic Press.

Perez-Cueto, F. J., Rini, L., Faber, I., Rasmussen, M. A., Bechtold, K., Schouteten, J. J., & De Steur, H. (2022). How barriers towards plant-based food consumption differ according to dietary lifestyle: Findings from a consumer survey in 10 EU countries. *International Journal of Gastronomy and Food Science, 29*, 100587. https://doi.org/10.1016/j.ijgfs.2022.100587

Pham, P. V. (2018). Medical biotechnology: Techniques and applications. In D. Barh & V. Azevedo (Eds.), *Omics technologies and bio-engineering* (Vol. 1, pp. 449–469). Academic Press.

Piraine, R. E. A., & Bochman, M. L. (2025). Chapter 7: Spontaneous fermentation. In P. Romano, M. Ciani, & A. Capece (Eds.), *Craft Beer* (pp. 141–164). Academic Press. https://doi.org/10.1016/B978-0-443-16015-8.00007-1

Plamada, D., Teleky, B., Nemes, S. A., Mitrea, L., Szabo, K., Călinoiu, L., Pascuta, M. S., Varvara, R., Ciont, C., Martău, G. A., Simon, E., Barta, G., Dulf, F. V., Vodnar, D. C., & Nitescu, M. (2023). Plant-based dairy alternatives—A future direction to the milky way. *Foods, 12*(9), 1883. https://doi.org/10.3390/foods12091883

Plaza-Díaz, J., Ruiz-Ojeda, F. J., Gil-Campos, M., & Gil, A. (2019). Mechanisms of action of probiotics. *Advances in Nutrition, 10*(Suppl_1), S49–S66. https://doi.org/10.1093/advances/nmy063

Poore, J., & Nemecek, T. (2018). Reducing food's environmental impacts through producers and consumers. *Science (New York, N.Y.), 360*(6392), 987–992. https://doi.org/10.1126/science.aaq0216

Qadir, R., Wan Mohamad Nasir, W. N. A., Azmi, A. B., Fatima, S., Mehmood, N., & Meor Hussin, A. S. (2025). An insight into plant-based yogurts: Physicochemical, organoleptic properties and functional food aspects. *Journal of Food Composition and Analysis, 143*, 107578. https://doi.org/10.1016/j.jfca.2025.107578

Ramachandran, R., Macabenta, F., Bettencourt, G., Feng, S., Ramachandran, R., Macabenta, F., Bettencourt, G., & Feng, S. (2025). From microbes to molecules: Synthetic biology approaches for advanced materials design. *The Biochemist, 5*(2), 12. https://doi.org/10.3390/biochem5020012

Ramsing, R., Santo, R., Kim, B. F., Altema-Johnson, D., Wooden, A., Chang, K. B., Semba, R. D., & Love, D. C. (2023). Dairy and plant-based milks: Implications for nutrition and planetary health. *Current Environmental Health Reports, 10*(3), 291–302. https://doi.org/10.1007/s40572-023-00400-z

Razavizadeh, S. A., Alenčikienė, G., Šalaševičienė, A., Vaiciulyte-Funk, L., Ertbjerg, P., & Zabulionė, A. (2021). Impact of fermentation of okara on physicochemical, techno-functional, and sensory properties of meat analogues. *European Food Research & Technology, 247*, 2379–2389. https://doi.org/10.1007/s00217-021-03798-8

Razavizadeh, S., Alencikiene, G., Vaiciulyte-Funk, L., Ertbjerg, P., & Salaseviciene, A. (2022). Utilization of fermented and enzymatically hydrolyzed soy press cake as ingredient for meat analogues. *LWT Food Science & Technology, 165*, 113736. https://doi.org/10.1016/j.lwt.2022.113736

Research, P. M. (2023). *Plant based milk market*. https://www.persistencemarketresearch.com/market-research/plant-based-milk-market.asp#:~:text=The%20demand%20is%20increasing%20for,the%20end%20of%20forecast%20period. Accessed 8 May 2024.

Reyes, T. F., Agrawal, P., Chan, T., Green, R., & Matulka, R. A. (2023). The safety of soy leghemoglobin protein preparation derived from *Pichia pastoris* expressing a soy leghemoglobin gene from *Glycine max*: *In vitro* and *in vivo* studies. *Journal of Toxicology, 2023*, 7398724. https://doi.org/10.1155/2023/7398724

Reyes-Jurado, F., Soto-Reyes, N., Dávila-Rodríguez, M., Lorenzo-Leal, A., Jiménez-Munguía, M., Mani-López, E., & López-Malo, A. (2021). Plant-based milk alternatives: Types, processes, benefits, and characteristics. *Food Review International, 6*, 1–32. https://doi.org/10.1080/87559129.2021.1952421

Reyes-Jurado, F., Soto-Reyes, N., Dávila-Rodríguez, M., Lorenzo-Leal, A. C., Jiménez-Munguía, M. T., Mani-López, E., & López-Malo, A. (2023). Plant-Based Milk Alternatives: Types, Processes, Benefits, and Characteristics. *Food Reviews International, 39*(4), 2320–2351. https://doi.org/10.1080/87559129.2021.1952421

Rezac, S., Kok, C. R., Heermann, M., & Hutkins, R. (2018). Fermented foods as a dietary source of live organisms. *Frontiers in Microbiology, 9*, 1785. https://doi.org/10.3389/fmicb.2018.01785

Ruiz, L., Ruas-Madiedo, P., Gueimonde, M., de Los Reyes-Gavilán, C. G., Margolles, A., & Sánchez, B. (2011). How do bifidobacteria counteract environmental challenges? Mechanisms involved and physiological consequences. *Genes & Nutrition, 6*(3), 307–318. https://doi.org/10.1007/s12263-010-0207-5

Saintila, J., Orlando, R., Magno, E., Patricia, R., & E., Y. (2025). Health consciousness, sensory appeal, and perception of front-of-package food labels as predictors of purchase intention for unhealthy foods in Peruvian university students. *Nutrients, 17*(11), 1921. https://doi.org/10.3390/nu17111921

Salminen, S., Collado, M. C., Endo, A., Hill, C., Lebeer, S., Quigley, E. M. M., Sanders, M. E., Shamir, R., Swann, J. R., Szajewska, H., & Vinderola, G. (2021). The International Scientific Association of Probiotics and Prebiotics (ISAPP) consensus statement on the definition and scope of postbiotics. Nature reviews. *Gastroenterology & hepatology, 18*(9), 649–667. https://doi.org/10.1038/s41575-021-00440-6

Sanders, M. E., Merenstein, D. J., Reid, G., Gibson, G. R., & Rastall, R. A. (2019). Probiotics and prebiotics in intestinal health and disease: From biology to the clinic. *Nature Reviews Gastroenterology & Hepatology, 16*(10), 605–616. https://doi.org/10.1038/s41575-019-0173-3

Sha, L., & Xiong, Y. L. (2020). Plant protein-based alternatives of reconstructed meat: Science, technology, and challenges. *Trends in Food Science & Technology, 102*, 51–61. https://doi.org/10.1016/j.tifs.2020.05.022

Sharma, N. M., Vishnu, G. R., & Priya, V. (2018). Assessment of nutritional value of overnight soaked cooked rice over unsoaked cooked rice. *International Journal of Research in Pharmaceutical Sciences, 9*, 616–619. https://doi.org/10.26452/ijrps.v9i3.1230

Sharma, N., Yeasmen, N., Dubé, L., & Orsat, V. (2024). A review on current scenario and key challenges of plant-based functional beverages. *Food Bioscience, 60*, 104320. https://doi.org/10.1016/j.fbio.2024.104320

Shrestha, A. K., Dahal, N. R., & Ndungutse, V. (2013). Bacillus Fermentation of Soybean: A Review. *Journal of Food Science and Technology Nepal, 6*, 1–9. https://doi.org/10.3126/jfstn.v6i0.8252

Siegrist, M., & Hartmann, C. (2020). Consumer acceptance of novel food technologies. *Nature Food, 1*(6), 343–350. https://doi.org/10.1038/s43016-020-0094-x

Silva, A. R., Santelli, R. E., Braz, B. F., Silva, M. M., Melo, L., Lemes, A. C., & Ribeiro, B. D. (2024a). A comparative study of dairy and non-dairy milk types: Development and characterization of customized plant-based milk options. *Foods, 13*(14), 2169. https://doi.org/10.3390/foods13142169

Silva, S., Costa, E. M., & Machado, M. (2024b). Precision fermentation as a tool for sustainable cosmetic ingredient production. *Applied Sciences, 15*(17), 9246. https://doi.org/10.3390/app15179246

Singh, D., & Lercher, M. J. (2020). Network reduction methods for genome-scale metabolic models. *Cellular and Molecular Life Sciences, 77*(3), 481–488. https://doi.org/10.1007/s00018-019-03383-z

Soumya, M., & Nampoothiri, K. M. (2021). An overview of functional genomics and relevance of glycosyltransferases in exopolysaccharide production by lactic acid bacteria. *International Journal of Biological Macromolecules, 184*, 1014–1025. https://doi.org/10.1016/j.ijbiomac.2021.06.131

Steensels, J., Gallone, B., Voordeckers, K., & Verstrepen, K. J. (2019). Domestication of Industrial Microbes. *Current Biology, 29*(10), R381–R393. https://doi.org/10.1016/j.cub.2019.04.025

Sturme, M., van der Berg, J. P., & Kleter, G. (2025). Precision fermentation – With a focus on food safety. In FAO (2024) (Ed.), *Cell-based food and precision fermentation – Products, safety and the future role*. FAO. https://doi.org/10.4060/cd4448en

Su, W., Jiang, Z., Wang, C., Xu, B., Lu, Z., Wang, F., Zong, X., Jin, M., & Wang, Y. (2022). Dynamics of defatted rice bran in physicochemical characteristics, microbiota and metabolic functions during two-stage co-fermentation. *International Journal of Food Microbiology, 362*, 109489. https://doi.org/10.1016/j.ijfoodmicro.2021.109489

Tangyu, M., Fritz, M., Tan, J. P., Ye, L., Bolten, C. J., Bogicevic, B., & Wittmann, C. (2023). Flavour by design: Food-grade lactic acid bacteria improve the volatile aroma spectrum of oat milk, sunflower seed milk, pea milk, and faba milk towards improved flavour and sensory perception. *Microbial Cell Factories, 22*(1), 133. https://doi.org/10.1186/s12934-023-02147-6

Tao, A., Zhang, H., Duan, J., Xiao, Y., Liu, Y., Li, J., Huang, J., Zhong, T., & Yu, X. (2022). Mechanism and application of fermentation to remove beany flavor from plant-based meat analogs: A mini review. *Frontiers in Microbiology, 13*, 1070773. https://doi.org/10.3389/fmicb.2022.1070773

Tebeje, A., Tadesse, H., & Mengesha, Y. (2021). Synthetic bio/techno/logy and its application. *Biotechnology & Biotechnological Equipment, 35*(1), 1156–1162. https://doi.org/10.1080/13102818.2021.1960189

Teng, T. S., Chin, Y. L., Chai, K. F., & Chen, W. N. (2021). Fermentation for future food systems: Precision fermentation can complement the scope and applications of traditional fermentation. *EMBO Reports, 22*(5), e52680. https://doi.org/10.15252/embr.202152680

The Good Food Institute Plant-Based Market Overview. (2020). *U.S. retail market insights for the plant-based industry*. https://www.gfi.org/marketresearch?utm_source=linkedin&utm_medium=social&utm_campaign=SPINS2019&utm_content=staff. Accessed 30 Nov 2020.

Tuccillo, F., Wang, Y., Edelmann, M., Lampi, A. M., Coda, R., & Katina, K. (2022). Fermentation conditions affect the synthesis of volatile compounds, dextran, and organic acids by *Weissella confusa* A16 in faba bean protein concentrate. *Foods, 11*(22), 3579. https://doi.org/10.3390/foods11223579

Van der Weele, C., Feindt, P., Jan van der Goot, A., Van Mierlo, B., & Van Boekel, M. (2019). Meat alternatives: An integrative comparison. *Trends in Food Science & Technology, 88*, 505–512. https://doi.org/10.1016/j.tifs.2019.04.018

Vanga, S. K., & Raghavan, V. (2018). How well do plant based alternatives fare nutritionally compared to cow's milk? *Journal of Food Science & Technology, 55*(1), 10–20. https://doi.org/10.1007/s13197-017-2915-y

Venter de Villiers, M., Cheng, J., & Truter, L. (2024). The shift towards plant-based lifestyles: Factors driving young consumers' decisions to choose plant-based food products. *Sustainability, 16*(20), 9022. https://doi.org/10.3390/su16209022

Villarreal-Soto, S. A., Beaufort, S., Bouajila, J., Souchard, J. P., & Taillandier, P. (2018). Understanding kombucha tea fermentation: A review. *Journal of Food Science, 83*(3), 580–588. https://doi.org/10.1111/1750-3841.14068

Walker, G. M., & Stewart, G. G. (2016). *Saccharomyces cerevisiae* in fermented beverages. *Beverages, 2*(4), 30. https://doi.org/10.3390/beverages2040030

Wang, R., Dong, P., Zhu, Y., Yan, M., Liu, W., Zhao, Y., Huang, L., Zhang, D., & Guo, H. (2021). Bacterial community dynamics reveal its key bacterium, *Bacillus amyloliquefaciens* ZB, involved in soybean meal fermentation for efficient water-soluble protein production. *LWT, 135*, 110068. https://doi.org/10.1016/j.lwt.2020.110068

Wang, Y., Tuccillo, F., Lampi, A. M., Knaapila, A., Pulkkinen, M., Kariluoto, S., Coda, R., Edelmann, M., Jouppila, K., Sandell, M., Piironen, V., & Katina, K. (2022). Flavor challenges in extruded plant-based meat alternatives: A review. *Comprehensive Reviews in Food Science & Food Safety, 21*(3), 2898–2929. https://doi.org/10.1111/1541-4337.12964

WHO. (2023). *Global Antimicrobial Resistance and use surveillance system (GLASS) report: Antibiotic use data for 2022*. World Health Organization.

Wood, P., Thorrez, L., Hocquette, J. F., Troy, D., & Gagaoua, M. (2023). "Cellular agriculture": Current gaps between facts and claims regarding "cell-based meat". *Animal Frontiers: The Review Magazine of Animal Agriculture, 13*(2), 68–74. https://doi.org/10.1093/af/vfac092

Xie, C., Coda, R., Chamlagain, B., Varmanen, P., Piironen, V., & Katina, K. (2019). Co-fermentation of *Propionibacterium freudenreichii* and *lactobacillus brevis* in wheat bran for in situ production of vitamin B12. *Frontiers in Microbiology, 10*, 1541. https://doi.org/10.3389/fmicb.2019.01541

Xie, A., Dong, Y., Liu, Z., Li, Z., Shao, J., Li, M., & Yue, X. (2023). A review of plant-based drinks addressing nutrients, flavor, and processing technologies. *Foods, 12*(21), 3952. https://doi.org/10.3390/foods12213952

Yang, D., Eun, H., & Prabowo, C. P. (2022). Metabolic engineering and synthetic biology approaches for the heterologous production of aromatic polyketides. *International Journal of Molecular Sciences, 24*(10), 8923. https://doi.org/10.3390/ijms24108923

Yılmaz, B., Mortaş, H., Varlı, S. N., & Ağagündüz, D. (2024). Metabolic engineering of lactic acid bacteria and yeasts for the production of compounds with industrial applications. In E. B. Ceresino, G. Juodeikiene, S. M. Schwenninger, & J. M. Ferreira da Rocha (Eds.), *Sourdough microbiota and starter cultures for industry* (pp. 223–256). Springer. https://doi.org/10.1007/978-3-031-48604-3_8

Yu, A., Mao, J., & Xu, N. (2025). Metabolic engineering of yeasts: A key cell factory platform for advanced biomanufacturing. *Journal of Fungi, 11*(12), 863. https://doi.org/10.3390/jof11120863

Yuan, H., Li, Y., Tu, J., Wang, T., Zhu, K., Jia, H., Wu, H., & Dong, P. (2025). The impact of *S. cerevisiae* strains on the aroma profile and physicochemical properties of *Phyllanthus emblica* L. fruit wine. *Food Chemistry, X, 33*, 103388. https://doi.org/10.1016/j.fochx.2025.103388

Zelante, T., Iannitti, R. G., Cunha, C., De Luca, A., Giovannini, G., Pieraccini, G., Zecchi, R., D'Angelo, C., Massi-Benedetti, C., Fallarino, F., Carvalho, A., Puccetti, P., & Romani, L. (2013). Tryptophan catabolites from microbiota engage aryl hydrocarbon receptor and balance mucosal reactivity via interleukin-22. *Immunity, 39*(2), 372–385. https://doi.org/10.1016/j.immuni.2013.08.003

Zhang, H. X., Zhang, Y., & Yin, H. (2019). Genome editing with mRNA encoding ZFN, TALEN, and Cas9. *Molecular Therapy: The Journal of the American Society of Gene Therapy, 27*(4), 735–746. https://doi.org/10.1016/j.ymthe.2019.01.014

Żółkiewicz, J., Marzec, A., Ruszczyński, M., & Feleszko, W. (2020). Postbiotics: A step beyond pre-and probiotics. *Nutrients, 12*(8), 2189. https://doi.org/10.3390/nu12082189

Yu, C., Mao, J., & Xu, [illegible] Metabolic engineering [illegible] cell factory platform [illegible]
Yang, [illegible]

Chapter 9
Fermenting the Future of Alternative Proteins: Mycoproteins, Tempeh Innovations, Cultivated Meat and Hybrid Products

Suraj K. Panda, Olumide Joseph Olubiyo, Fernando Roberto Paz Cedeno, and Sungil Ferreira

Abstract Currently, there are significant challenges to meet the protein needs of a growing global population and this demands solutions that are nutritionally robust, environmentally responsible, and culturally acceptable. This chapter surveys fermentation-derived proteins (including mycoprotein and tempeh), cultivated meat, and animal–alternative hybrid products, framing each within production technology, nutrition and functionality, food safety, consumer acceptance, market dynamics, and regulation. It examines submerged and solid-state fermentation platforms, along with the nutritional profiles, texture attributes, and functional properties that enable diverse food applications. The chapter also assesses cultivated meat across the technological stack, cell sources, media, scaffolds, and bioreactors, together with the current global regulatory landscape and scale-up constraints. In addition, it evaluates the commercialization status of mycoproteins, the diversification of edible fungi, and the use of agro-industrial by-products as carbon and nitrogen sources for microbial fermentation. The role of animal–alternative hybrid products is analyzed as a pragmatic bridge that can pair familiar sensory qualities with improved sustainability and cost outcomes. Ultimately, this work underscores the importance of fermentation-driven proteins, hybrids, and cultivated meat in advancing towards a more sustainable protein production chain, evaluating current challenges and perspectives, as well as future opportunities.

Keywords Aerobic fermentation · Single cell protein · Submerged fermentation · Solid state fermentation · Alternative fermentation media · Biocircular economy

S. K. Panda · O. J. Olubiyo · S. Ferreira (✉)
Department of Food Science, The University of Arkansas System – Division of Agriculture (UADA), Fayetteville, AR, USA
e-mail: sunfer@uark.edu

F. R. P. Cedeno
Bioenergy Research Institute, São Paulo State University, Rio Claro, São Paulo, Brazil

S. Dhiman, S. Kaur (eds.), *Next-Gen Plant Based Fermented Foods: Innovations for a Sustainable Future*, https://doi.org/10.1007/978-3-032-29927-7_9

9.1 Introduction

The global population is projected to reach about 9.7 billion by 2050, intensifying the demand for an adequate food supply (Statista, 2025). Many underdeveloped and developing nations are still affected by nutritional disparities, contributing to poor health, food insecurity and reduced quality of life. Therefore, ensuring equitable access to safe, nutritious, and balanced foods is essential for improving the global well-being (Bayati et al., 2025; Zhang & Hu, 2025).

Protein is a vital component of the human diet. It supports muscle development and repair, improves bone health, boosts immune function, assists hormonal regulation, and strengthens cognitive performance (Aeri & Kaur, 2025; Bonjour, 2005). In the present scenario, the majority of the world's protein needs are supplied by animal-based sources. But some of these sources contribute substantially to greenhouse gas emissions and environmental degradation (Gil et al., 2024). The ratio of animal protein to alternative plant protein sources for human consumption is about 65:35 in the present market share in developed countries (Drewnowski & Hooker, 2025). If we manage to bring this ratio to 40:60 or even 50:50, humankind will be able to control the current state of environmental changes and provide protein sources for the growing population (Drewnowski & Hooker, 2025; Sampedro, 2024). This underscores the need to provide technical solutions to successfully and sustainably provide alternative proteins to balance animal and non-animal protein sources. The strategies discussed in this chapter aim to promote a diversified, multi-source protein diet that supports both human health and long-term environmental sustainability.

Fermentation derived proteins including mycoproteins, tempeh, cultured meat, and hybrid products (animal and alternative proteins combined), are promising alternatives to traditional animal-based protein source diets. These products are developed by leveraging biotechnological processes combined with advanced food processing technologies to produce high-quality, sustainable proteins (Cedeno et al., 2025; Choręziak et al., 2025). The production of high-value protein products is possible through a fermentation process under controlled environmental conditions with optimized parameters (Fletcher et al., 2024). Depending on the desired outcomes, fermentation can be applied for biomass production, functionality modification, or precision fermentation. This process results in multiple benefits, such as nutritional enhancement by elevating the protein content in the product, improving the essential amino acid profile. In plant protein, it also helps in reducing the anti-nutrient components present, causing modifications that improve digestibility and bioavailability. It is also used to develop the desired texture and flavor, improving protein functionality, as well as expanding culinary applications (Panapparambil Sooraj et al., 2025). From an environmental perspective, fermentation-based protein production has lower water and land requirements. It also emits significantly fewer greenhouse gases compared to conventional animal-based meat production (Malila et al., 2024). Moreover, it bridges traditional fermentation practices, like tempeh production, with modern technological innovations, enabling large-scale and cost-effective production. By simultaneously addressing nutritional, functional, and environmental needs, fermentation-based approaches are reshaping the future of food, redefining how proteins are produced and consumed in a more sustainable and efficient manner.

9.2 Mycoproteins: Harnessing Fungi for Sustainable Nutrition

The term "mycoprotein" is derived from Greek roots: "*myco-*" (from *mykes*, meaning fungus) and "*-protein*" (from *proteios*, meaning primary or of first importance). Mycoprotein, a fungal derived protein biomass, is considered a major innovation in sustainable food systems because of its practicality and scalability as an alternative source of protein (Wiebe, 2002). Table 9.1 presents recent studies evaluating different substrates and fungal strains evaluated for mycoprotein production. Historically, the use of fungi as a potential source of protein-rich raw material was inspired by the "Green Revolution" in the 1960s in the United Kingdom, led by Lord Rank, who sought to develop a method for producing a protein-rich food from starchy substrates. After numerous trials and the testing of over 2500 soil organisms collected from around the world, researchers at the Ranks Hovis McDougall (RHM) Research Centre identified *Fusarium venenatum* strain A3/5 (initially called

Table 9.1 Various edible fungal species employed in mycoprotein production using diverse feedstocks

Fungal strain	Fermentation type	Substrate	Highest biomass yield reported	References
Neurospora intermedia	Submerged	Soy whey and okara	14.27 g/L	Dai et al. (2025)
Pleurotus ostreatus LGAM 1123	Submerged	Glucose	26 g/L	Bakratsas et al. (2023)
Rhizopus microsporus	Submerged	Sugar cane molasses, corn steep liquor	38.34 g/L	Furlan et al. (2024)
Fusarium venenatum	Submerged	Liquefied jaggery, date extract	5.4 g/L	Prakash et al. (2014)
Rhizopus delemar	Submerged	Bread waste	22.5 g/L	Svensson et al. (2021)
Pleurotus albidus	Solid state	Brewer spent grain	22.6 g/100 g	Stoffel et al. (2019)
Lasiodiplodia theobromae	Submerged	Soybean molasses	12.44 g/L	Marchioro et al. (2025)
Agaricus bisporus	Submerged	Sugarcane extract	15 g/L	Kim et al. (2011)
Fusarium venenatum	Submerged	Glucose	9.53 g/L	Tong et al. (2024)

F. graminearum) as a viable source of mycoprotein for human consumption (Trinci, 1994; Wiebe, 2004). Over a 10-year evaluation period, RHM conducted extensive studies involving 2500 human volunteers and multiple animal trials, confirming the safety and nutritional value of the mycoprotein. In 1980, RHM received regulatory approval from the United Kingdom Ministry of Agriculture, Fisheries and Food to market mycoprotein for human consumption and its first retail product, a savory pie containing mycoprotein, was launched in January 1985 under the brand name "Quorn". This approval marked a major milestone, paving the way for the commercialization of mycoprotein as a sustainable meat substitute owing to researchers' preference for fungi over yeast or bacteria as an alternative protein source. It was primarily driven by humanity's long history of consuming fungi as food, the filamentous growth habit of fungi that provides a fibrous, meat-like texture desirable for food applications and the ease of harvesting fungal mycelia from culture broth (Trinci, 1994).

The alternative protein market, comprising of plant-based, microbial, insect-derived, and cultivated options, is experiencing robust growth as consumers increasingly seek sustainable, ethical, and health-focused substitutes for traditional animal proteins. As of 2024, the alternative protein market is valued at approximately USD 21.76 billion. This figure is primarily driven by rising awareness of environmental issues linked with livestock farming, such as high greenhouse gas emissions and resource consumption (Business Wire, 2025). This paradigm shift is particularly evident in regions like North America and Europe, where plant-based alternatives are more relevant among consumers. Conversely, Asia-Pacific is experiencing rapid adoption due to cultural preferences for diverse protein sources.

Market projections indicate strong expansion within the alternative protein sector. Estimates suggest market growth could reach up to USD 36.17 billion by 2030, reflecting a compound annual growth rate of 8.84%. This growth is largely driven by ongoing innovations in precision fermentation and cell-cultured technologies (Business Wire, 2025). Other forecasts project a compound annual growth rate of 14.1% between 2025 and 2030, with market valuations expected to exceed USD 50.22 billion by 2033. These projections highlight the influence of economic conditions and regulatory support in accelerating the adoption of sustainable foods (Grand View Research, 2025). The leaders in this rapidly expanding market include Beyond Meat, Impossible Foods, and Quorn with emerging start-ups like Meatable and BluNalu focusing on cultivated meat and seafood, respectively. These companies are increasingly investing in research and development to enhance product taste, texture, and affordability, with 2025 trends demonstrating the emergence of hybrid products that combine plant-based and fermented proteins to achieve improved sensory attributes (Yao et al., 2024). However, there are still persistent challenges, such as high production costs, limited scalability, particularly for cultivated protein technologies, and persistent consumer skepticism toward novel food innovations, that continue to pose barriers (Cedeno et al., 2025).

To this day, Marlow foods, which is currently owned by the Philippine-based company, Monde Nissin Corporation, produces mycoprotein through the continuous cultivation of *F. venenatum* under controlled conditions. After fermentation, it includes reducing biomass RNA content through a thermal process, as well as improving organoleptic properties through the incorporation of egg white and other functional ingredients as binding agents (MarketScreener, 2015; Marlow Foods, 2022). In dry weight, *F. venenatum* mycoprotein contains about 45% protein, 25% fiber with low fat and cholesterol, offering a favorable fatty acid profile and all essential amino acids, making it a nutritionally comprehensive alternative to animal-sourced proteins (Finnigan et al., 2019). However, other edible fungi like *Agaricus bisporus* (button mushroom), *Lentinula edodes* (shiitake mushroom), *Neurospora crassa* and *Pleurotus ostreatus* (oyster mushroom) are also being explored for their potential in developing mycoprotein-based meat alternatives, leveraging their distinct textural and nutritional profiles (Amara & El-Baky, 2023).

9.2.1 Production Process

Mycoprotein production from *Fusarium venenatum* has been well documented, with early foundational descriptions provided by Trinci (1994). Commercial-scale production uses submerged, continuous-flow aerobic fermentation in large, aseptic, stainless-steel bioreactors, typically pressure-cycle vessels of approximately 150,000 L capacity. In this system, *F. venenatum* is supplied with a continuous intake of glucose as the primary carbon source and ammonium as the nitrogen source, alongside vitamins and minerals to sustain rapid growth. All nutrients are maintained in excess, while process parameters, including temperature (28–30 °C), pH (~6.0), and CO_2 concentration, are tightly controlled to ensure stable biomass productivity. Under these optimized conditions, *F. venenatum* reaches a specific growth rate of 0.17–0.20 h^{-1}, yielding approximately 300–350 kg (dry weight) of fungal biomass per hour. In this continuous culture fermentation, continuous withdrawal of culture broth maintains constant fermenter volume, replacing the entire medium every 5–6 h (Trinci, 1994; Wiebe, 2002). The overall commercial mycoprotein production process using submerged fermentation is illustrated in Fig. 9.1.

Following fermentation, the harvested broth undergoes RNA reduction, a critical safety step given the naturally high RNA content (~10% dry weight) of fungal biomass. This is done to reduce purine intake and mitigate risks associated with elevated uric acid, such as gout, the biomass is subjected to controlled heat treatment that activates endogenous RNases (Denny et al., 2008). Heating is done by conventional methods. This method involves heating to 64–65 °C for 20–30 min, though more recent modifications employ rapid heating above 68 °C (optimally 72–74 °C), which improves protein retention and reduces biomass loss from ~35%–38% to ~30%–33% (Wiebe, 2002). Subsequent processing includes heating to 90 °C, centrifugation to remove released nucleotides, and concentration of the mycelia with about 20% solids. The concentrated mycelia are then blended with a binding agent such as egg albumen and mechanically aligned to form a fibrous structure that

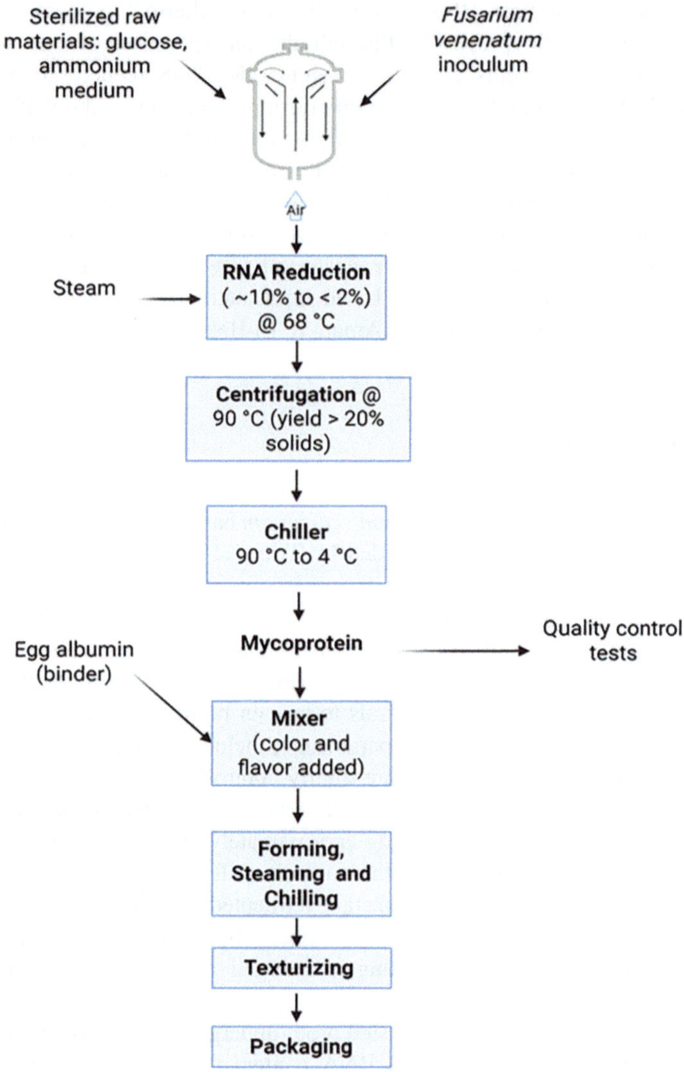

Fig. 9.1 Simplified mycoprotein production process diagram using a stirred tank bioreactor

mimics meat texture. This step enables the production of diverse food products including chunks, mince, fillets, burgers, sausages, and steaks-like products (Denny et al., 2008; Wiebe, 2002).

Recent advances in submerged fermentation (SmF) have expanded the range of substrates and organisms used for mycoprotein production beyond *F. venenatum*. SmF systems are preferred for their controllability and ease of monitoring. There are still challenges to be solved such as the high overall price of production driven by the fermentation medium, as well as reduced efficiency. Studies have demonstrated effective biomass generation using various low-cost feedstocks, including potato waste, pulse husk, date waste, fruit residues, and seaweed by-products (Khan et al., 2023). For instance, *Paradendryphiella salina* cultivated on cornmeal seawater agar has been used to convert *Macrocystis pyrifera* and *Ulva* spp. into high-yield mycoprotein biomass, achieving substantial increases in protein, essential amino acids, and ash content compared to unfermented substrates (Landeta-Salgado et al., 2021). Additional research on fungi such as *Pleurotus sapidus*, *Wolfiporia cocos*, *Lentinula edodes*, and yeasts like *Saccharomyces cerevisiae* has shown that substrate type strongly influences biomass productivity and nutritional composition. Agricultural waste, including apple pomace, beetroot molasses, spinach leaves, cucumber peel, orange peel, and banana waste, have yielded variable but often significant biomass and protein outputs, underscoring the importance of substrate composition in optimizing mycoprotein production efficiency and sustainability (Khan et al., 2023).

9.2.2 Benefits and Application of Mycoprotein

Mycoprotein derived from *Fusarium venenatum* is nutritionally robust, containing approximately 44% protein on a dry-weight basis, with a PDCAAS (Protein Digestibility-Corrected Amino Acid Score) comparable to that of milk casein (gold standard) (Finnigan et al., 2019). It provides all essential amino acids required for human nutrition. Although the concentrations of sulfur-containing amino acids are relatively modest, the overall essential amino acid profile is comparable to that of eggs and favorable to human muscle tissue. Mycoprotein also contains a substantial proportion of dietary fiber, primarily chitin and β-glucans from the fungal cell wall, while being naturally low in saturated fat, sodium, and cholesterol. This composition results in an overall nutrient profile superior to many conventional vegetarian and meat-based protein sources, supporting its suitability as an alternative to meat (Denny et al., 2008; Wiebe, 2002).

Numerous human studies have demonstrated physiological benefits associated with regular mycoprotein consumption. Daily intake over a six-week period was associated with reduced total cholesterol levels, an effect suggested to be attributable to mycoprotein's high fibre content. However, further studies are needed to substantiate these claims (Ruxton & McMillan, 2010). Mycoprotein has also been associated with enhanced satiety and reduced energy intake relative to meat-based meals, though findings vary across studies. In earlier studies, Burley et al. (1993) examined satiety responses to mycoprotein versus chicken in 18 lean, healthy male and female participants and reported that mycoprotein exhibited a greater satiating effect than other high-fiber foods evaluated in their prior research. The authors suggested that the unique fiber composition of mycoprotein may play a particularly influential role in appetite regulation. In contrast, Williamson et al. (2006) found no significant differences in appetite sensations or subsequent energy intake when comparing meals containing mycoprotein, chicken, or tofu, indicating that satiety responses may not differ substantially among these protein sources.

In recent studies, Whelehan et al. (2024) evaluated the effects of mycoprotein-enriched milk drinks in 24 randomized adults with varying BMI over a 6-h period. The findings indicated that consumption of the mycoprotein-infused beverage altered postprandial glucose kinetics. However, no significant changes were observed in overall blood glucose or insulin concentrations. These results suggest that mycoprotein may have potential to support glycaemic regulation in humans, while also emphasizing the need for further investigation into its metabolic and health benefits.

Collectively, mycoprotein's nutritional and functional attributes indicate that mycoprotein represents a high-quality, sustainable protein source with potential to support healthy dietary patterns, particularly as demand grows for nutrient-dense, low-fat, fiber-rich alternatives to traditional animal proteins. The diverse applications of mycoprotein are evident in its integration into a wide array of food products, ranging from frozen to refrigerated options, including Quorn mince, nuggets, and fillets, which benefit from its textural resemblance to meat (Zeng et al., 2023). Furthermore, its consumption provides notable environmental advantages over traditional meat production, including a 14% reduction in greenhouse gas emissions when mycoprotein is produced from lignocellulosic substrates, as well as substantially lower land and water footprints (Upcraft et al., 2021). Additional substrates investigated include soy whey, which achieved a 59.7% reduction in chemical oxygen demand (Pratama et al., 2025), as well as pea processing by-products (Souza Filho et al., 2018) and oat flour (Rousta et al., 2021). However, further studies are still required to quantify the environmental impact associated with the use of filamentous fungi cultivated on similar matrices.

In summary, the benefits of mycoprotein span through the entire food chain, from production through consumption. From a production standpoint, mycoprotein can be generated with substantially lower nutrient and energy inputs than conventional animal proteins. Also, due to higher specific growth rates, it has significantly shorter production cycles than both animal and many plant proteins. Given the anticipated increases in global demand for food and feed over the coming decades, these attributes are critical for delivering protein with reduced or minimized environmental footprints. Nutritionally, mycoprotein provides a complete essential amino acid profile and is rich in dietary fiber, so mycoprotein-containing products can supply a balanced combination of energy and macronutrients. Moreover, because mycoprotein products are not typically associated with off-flavors common to some alternative proteins (e.g., beany notes), they tend to exhibit strong consumer acceptance.

9.3 Tempeh, Enhanced Nutrition Through Solid State Fermentation

Tempeh is one of the oldest known fermented foods, where the main initial purpose is to preserve soybeans for a longer period of time. It originated in Indonesia and has been used as a staple source of protein across many South Asian nations for centuries

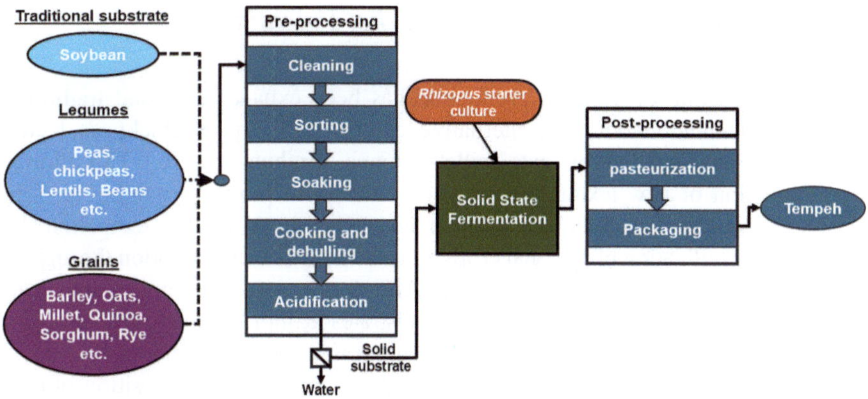

Fig. 9.2 Flow diagram depicting the tempeh production pathway, including substrate inoculation, fungal fermentation, and final product formation

(Romulo & Surya, 2021). Tempeh is produced through solid-state fermentation, where cooked and dehulled soybeans are used as a substrate and different *Rhizopus* molds, e.g., *R. oligosporus*, and *R. oryzae* are used as a starter culture (Elhalis et al., 2024). The fermentation process causes the soybeans to bind together to form a firm cake like structure. Due to the fermentation process it higher digestibility and greater bioavailability than the non-fermented grains. Overall, fermentation improves the nutritional and structural quality, as well as contributes to a unique nutty flavor (Górska et al., 2025).

The fermentation process (Fig. 9.2) by *Rhizopus* breaks down complex soy protein into more bioavailable shorter peptides and free amino acids, making Tempeh a suitable protein source with higher *in vitro* protein digestibility. The fermentation process also leads to a decrease in the anti-nutrient factors such as phytic acid, saponin, trypsin inhibitors, and tannins. Soybean has high concentration of phytates, which can bind iron, zinc, calcium, and magnesium, causing indigestion. *Rhizopus* fermentation produces phytase, an enzyme that facilitates the breakdown of phytates (an antinutritional factor), thus increasing the mineral availability for absorption during digestion (Sudarmadji & Markakis, 1977). In this sense, the overall fermentation and proteolytic process neutralises trypsin inhibitors through the production of proteases (Wang et al., 1972). The *Rhizopus* metabolism enhances bioactive compounds profiles by transforming the tannins into phenolic acids and minerals. The bitterness in soybeans is often characterised by the presence of saponins, which are oleanane-type triterpenoid glycosides. The *Rhizopus* fermentation on soybean substrates produces beta-glucosidases which break the glycosidic linkages between the sugar chains and aglycone core, converting complex glycosidic saponins into simpler less polar derivatives. These derivatives have higher antioxidant activity and increased health benefits and contributes to better nutritional quality and flavour of Tempeh (Reddy & Pierson, 1994).

9.3.1 Modern Innovations and Market Potential

The diversification of Tempeh production has been expanded beyond traditional soybean fermentation to include alternative legumes such as peas, chickpeas, lentils, beans and grains (e.g., barley, oats, millets, quinoa, sorghum, rye etc.). This allowed the production of new textures, flavors and nutritional profiles, enabling also the access to a broader consumer base through the availability of soy-free tempeh. It also promotes the utilization of regional crops and contributes to crop rotation throughout the year to produce a sustainable tempeh. After fermentation, the mycelium-bound tempeh cakes produced from different substrates exhibit distinct structural and sensory properties. These differences in graininess, cohesion, firmness, and flavor arise from variations in the inherent starch, fiber, and protein composition of each substrate. The tempeh made from chickpeas have a mild, nutty flavor and a soft texture and it is regarded as closest to soybean tempeh in overall acceptability (Abu-Salem & Abou-Arab, 2011). Erkan et al. (2020) evaluated tempeh production on various substrates including chickpea, green and red lentils, black, white and broad beans, as well as soybean. Their sensory analysis indicated that chickpea tempeh achieved higher consumer acceptance and exhibited superior protein content than other legumes such as soybean, lentils (red and green), and beans (white, black, and broad), with up to 23.5% protein (Erkan et al., 2020). Different cereal substrates such as sorghum, wheat, oats, rye, corn and barley can also be used for the production of tempeh. A study conducted on various grains found in the Kansas region, USA, for Tempeh production indicated that fermentation with *R. oligosporus* NRRL 2549 on red sorghum developed a good product quality with excellent patty structure, desirable appearance and aroma (Hachmeister & Fung, 1993). The use of locally available grains for tempeh production can create a new avenue for value-added industrial development while extending its application to a broader population.

The modified tempeh by co-fermentation and blended substrates represents innovation that is aimed at improving flavor and texture to enhance sensory appeal and acceptance among a wider consumer. A study conducted on modified tempeh, where addition of 2% w/v *S. cerevisiae* during fermentation helped in the reduction of beany flavor and contributed to an improved yeast aroma (Kustyawati, 2017). Another study investigating tempeh production using various legume combinations reported that a mixture of soybean and red kidney bean has demonstrated a greater sensory property, characterized by a harder and more compact structure. It also reflects on the production of hybrid tempeh to improve the texture preferences and product quality (Tan et al., 2024).

Tempeto represents another product innovation brought into tempeh, in which *Rhizopus* fermentation was followed by a post-fermentation sous-vide cooking, resulting in the development of malty, umami, sweet, and miso flavor profiles. Further, this approach enhanced the textures into soft, fluffy and grainy, broadening its potential to the different culinary applications (Guixer et al., 2017).

The innovation brought into functional Tempeh has further enriched the nutraceutical properties and improved health benefits. The supplementation of *Moringa oleifera* leaf powder with soybean flour during fermentation has enriched the tempeh-based biscuits with bioactive compounds, higher fiber content, multiple vitamins, minerals and higher antioxidant (Lestari et al., 2024). In another study, the optimal fortification of Ferrous fumarate into tempeh made from black-eyed peas (200 mg/100 g) and kidney beans (100 mg/100 g) has elevated the sensory impact along with reference iron content. This functional tempeh is beneficial for consumption for patients with anemia (Vulgaris et al., 2023). The development in food processing to alter the metabolites by microbial intervention has also brought innovation in the functional tempeh. The introduction of *Lactiplantibacillus plantarum* and *Pichia burtonii in* tempeh fermentation led to more than two-fold increment in the concentration of bioactive metabolites like amino acids and tyramine (Riyanto et al., 2024).

There are several food safety and quality concerns associated with the treatment, handling, fermentation, and storage of tempeh. Spoilage and food poisoning in soybean-based tempeh can result from the growth of *Bacillus cereus* (Nout et al., 1987). This risk can be minimized by acidification treatment of soybeans during the soaking process prior to the fermentation (Ashenafi & Busse, 1991). Freshly fermented tempeh is also susceptible to microbial contamination, which can introduce foodborne pathogens (Kavya et al., 2025). Major hazards include botulinum toxins produced by *Clostridium sps.*, and enterotoxins produced by *Staphylococcus sps.*, both of which may occur during improper storage conditions. These risks can be minimized by maintaining refrigeration temperature below 5 °C (Tanaka et al., 1985). Cross-contamination of the inoculum and associated mycotoxin risk can be prevented by avoiding wild unknown stater cultures and using a defined established starter culture containing only non-mycotoxin producing strains (Chin et al., 2024b). Since, tempeh is a ready-to-eat food, the implementation of microbiological safety standards, along with the safe handling practices, post processing steps such as pasteurization, appropriate packaging and refrigerated storage are crucial for the consumers safety (Ahnan-Winarno et al., 2021; Griese et al., 2013).

Tempeh has evolved beyond its traditional Indonesian roots and is now incorporated into various western cuisines, often featured in modern plant based dishes like tempeh burger, sandwiches and pizza (Purwadaria et al., 2016). It has transitioned from its original banana leaf wrapping to become a part of modern manufacturing and packaging formats including cans, plastic wraps and bottles, enabling its distribution to global markets. The health benefits of traditional tempeh are well established. Moreover, the development of modern plant-based diets like multi-source tempeh and functional tempeh further enhances its nutraceutical importance (Górska et al., 2025; Tan et al., 2024). However, it is important to follow proper product development and production to ensure that modern innovations can accommodate both traditional authenticity, safe products and long-term sustainability.

9.4 Cultivated Meat Production

Cultivated meat (CM), also known as *in-vitro* meat or lab-grown meat, refers to animal flesh produced by culturing animal cells directly, rather than by raising and slaughtering livestock. This innovative approach addresses significant environmental, ethical, and public health concerns associated with conventional animal agriculture by bypassing the need for whole animals (Godfray et al., 2018). The process typically involves isolating muscle stem cells or other progenitor cells from a living animal via biopsy, which are then proliferated in bioreactors and differentiated into muscle and fat tissues (Post, 2012).

Several studies trace the conceptual origins of cultivated meat to speculative proposals from the late nineteenth and early twentieth centuries. These early ideas were later reinforced by mid-twentieth-century advances in tissue-culture methodologies, which demonstrated the long-term maintenance of animal cells outside the living organism. Momentum increased in the 1960s, also during the "Green Revolution", when Dutch innovator Willem van Eelen, motivated by concerns related to post-war food scarcity and animal welfare, articulated the possibility of producing meat directly from cultured animal cells (Food Dive, 2023). Although he secured the first patent for cell-based meat in 1997, historical analyses note that the underlying concept had been articulated decades earlier. More recently, empirical advancements in cultivated meat research accelerated rapidly in the early 2000s. One of the earliest significant milestones occurred in 2002, when NASA successfully demonstrated the production of edible cultured fish tissue derived from goldfish cells. This research was done in an effort to address nutritional sustainability during long-duration space missions. The resulting prototype provided critical evidence for the technical feasibility of producing structured muscle tissue by differentiating bovine satellite cells within bioreactors using fetal bovine serum and scaffold-based support systems (Benjaminson et al., 2002). By 2005, academic engagement had intensified, particularly through research at Maastricht University in the Netherlands, where investigators began systematic studies aimed at optimizing the production of cultured meat. Together, these efforts paved the way for the landmark 2013 introduction of the first cultured beef burger by Mark Post, an achievement made possible through financial support from Google co-founder Sergey Brin (Food Dive, 2023).

In recent years, cultivated meat has progressed from early conceptual prototypes to limited commercial availability. Although, its broader adoption remains constrained by regulatory, economic, and scalability challenges (Zandonadi et al., 2025). Singapore continues to lead global commercialization efforts, having approved Eat Just's cultivated chicken in 2020, with products now available in both restaurants and retail outlets (Good Food Institute & Huling, 2025). Similarly, regulatory advances have also emerged in the United States, where the FDA and USDA approved cultivated chicken from Upside Foods and Good Meat in 2023. This approval enabled limited distribution in select restaurants, including Bar Crenn in San Francisco and China Chilcano in Washington, D.C., until the partnership concluded in 2024. It also supported the first U.S. retail trial of cultivated pork meatballs, conducted at Berkeley Bowl in Berkeley, California, in October 2025 (Vegconomist, 2025). Beyond these developments, Israel granted regulatory clearance for Aleph Farms' cultivated beef in 2024, while Australia and New Zealand issued approvals with strict labeling conditions (Good Food Institute & Huling, 2025). Within Europe, the sector awaits EFSA Novel Food evaluations expected by 2026. However, regulatory barriers, including Italy's ban on cultivated meat production, continue to slow broader advancement (EFSA, 2023).

Despite advancements, regulatory resistance has intensified in several regions of the United States. States including Florida, Alabama, and Texas enacted prohibitions on the sale and production of cultivated meat between 2024 and 2025, citing concerns related to food safety and the protection of conventional agricultural sectors. These legislative actions have prompted legal challenges from companies, including Upside Foods and Wildtype, on constitutional grounds (National Agricultural Law Center, 2025). Concurrently, commercial progress remains constrained by high production costs, which are estimated at $63/kg. Also limited industrial scalability, particularly due to insufficient bioreactor capacity and the overall absence of a large-scale, food-grade bioprocessing infrastructure. While investments such as ORF Genetics' €5 million Icelandic facility aim to expand bioreactor capacities and reduce these bottlenecks, current limitations continue to hinder widespread commercialization (Green Queen & Mridul, 2025). Technological innovation continues to focus on developing stable, scalable cell lines, replacing costly media components with plant-based and recombinant alternatives while improving bioreactor systems to enable efficient, large-scale cultured cell production (Gu et al., 2025).

Market projections indicate sustained interest and research and development, although consumer acceptance, labeling challenges, and regulatory delays continue to pose significant barriers (MarketsandMarkets, 2025). Nonetheless, the sector has experienced notable developments, including Eat Just Inc.'s partnership with the Halal Product Development Company to advance certification pathways and advisory support for a future halal-compliant cultivated meat. Additionally, BlueNalu has collaborated with Nutreco, a leading animal protein company, to develop and scale-up a food-grade supply chain for cell-cultured seafood production (MarketsandMarkets, 2025).

9.4.1 Technological Innovation for Culturing Meat

As previously discussed, cultivated meat relies on the production of animal muscle tissue through cellular agriculture, using stem cell technologies to generate edible products without traditional livestock farming. However, despite notable advancements, several technological hurdles persist in achieving widespread, cost-effective industrial-scale production. The overall process typically progresses from laboratory methods to large-scale bioprocessing and can be grouped into four foundational technological domains: cell sources, culture media, scaffolds, and bioreactors (Post, 2014). These areas encompass the sequential process steps of cell acquisition and banking, proliferation, differentiation and maturation using scaffolds, and ultimately harvesting. Each stage presents specific technical, economic, and regulatory obstacles; nonetheless, recent advances in biotechnology, materials science, and process engineering have improved both efficiency and commercial viability (Gu et al., 2025).

In detail, the process begins with cell acquisition and selection, in which starter cells are obtained from a specific animal source. This is typically done via biopsy of muscle, adipose, or connective tissue from species such as cattle, poultry, or fish. These cells, often satellite stem cells or induced pluripotent stem cells (iPSCs), are isolated and screened for proliferation potential, genetic stability, and the expression of relevant markers (e.g., lysine-specific demethylase 4A, estrogen-related receptor β, and stage-specific embryonic antigen-4) (Jara et al., 2023). This step is critical for establishing a consistent cell line being species specific (Xie et al., 2025). For example, bovine embryonic stem cells derived from trophoblastic bovine tissues can maintain pluripotency and sustain proliferative capacity when cultured in modified TeSR1 medium. However, goat embryonic stem cells generated using analogous methods and cultured in 2i + LiF (dual inhibition plus leukemia inhibitor factor) medium had limited proliferation ability over time. This difference in proliferation ability can be attributed to individual species' genetic markers. Therefore, it is crucial to choose a more efficient way to obtain stem cells, a storage method that can maintain their survival status alongside functional characteristics for a long time (Shi et al., 2025; Wei et al., 2017). Recent advances, most notably Clustered Regularly Interspaced Short Palindromic Repeats– CRISPR-associated protein 9 (CRISPR–Cas9) genome editing, have enabled the development of immortalized lines with enhanced proliferation and reduced senescence, thereby decreasing the frequency of required biopsies (de Bardet et al., 2023). Additionally, non-invasive sources such as umbilical cord blood or poultry feathers are being explored to reduce ethical concerns and improve scalability (Reiss et al., 2021).

After acquisition, cells are expanded into small-scale cultures and cryopreserved to establish a master cell bank. This process ensures a stable supply for subsequent production runs. During the proliferation stage, thawed cell banks are scaled up in nutrient-rich media within controlled bioreactors to achieve high cell densities; this step is resource- and media-intensive and constitutes a major cost driver in cultivated meat production (Hubalek et al., 2022).

Proliferation and differentiation involve culturing cells in nutrient media supplemented with growth factors to maintain proliferation or pluripotency, depending on the starter cell type, and to support directed differentiation toward lineages relevant for muscle or adipose tissue when required. For satellite cells and other adult stem cells (e.g., MSCs/FAPs), differentiation media commonly employ a basal formulation such as DMEM/F12 supplemented with L-glutamine, non-essential amino acids, and low concentrations of FGF-2 (Chen et al., 2011). Pluripotent stem cell media typically use a similar basal formulation but also include additional factors, such as FGF-2, EGF, TGF-β, heparin, serum or serum replacements, and extracellular matrix components, to support self-renewal and regulate cell behavior; these supplements may likewise enhance adult stem cell culture performance

(Gu et al., 2025). Fetal bovine serum (FBS) is a widely used basal growth supplement derived from the blood of bovine fetuses. It is obtained through blood coagulation followed by centrifugation, which removes fibrinogen and cellular components, yielding a serum rich in growth factors and nutrients (Tannières et al., 2024). Historically, FBS has been widely used because it supplies diverse growth factors, nutrients, and adhesion proteins. FBS can modulate processes such as the terminal differentiation of muscle progenitors. However, the animal-derived nature of FBS conflicts with the ethical and sustainability goals of cultivated meat while introducing problems like batch-to-batch variability and potential contamination by viruses or prions (Pilgrim et al., 2022). Consequently, there is an intentional shift toward serum-free and chemically defined media, driven by concerns about safety, consistency, cost, and environmental impact. In addition, the development of xeno-free formulations (e.g., Essential 8™, TeSR™, FBM™) has advanced this transition, although further optimization is required to match the performance of serum-based media for livestock cell lines (Badenes et al., 2016). In parallel, recombinant growth factors are increasingly preferred over animal-derived alternatives to improve safety, reproducibility, and alignment with industry sustainability objectives.

Maturation and scaffolding are central to producing cultivated meat with acceptable structure and sensory quality. Here, differentiated cells are seeded onto edible or removable scaffolds (e.g., collagen-, alginate-, or chitosan-based matrices) and matured in bioreactors that supply nutrients, oxygen, remove waste, and apply mechanical cues to improve texture and flavor (Woelken et al., 2024). Scaffolds (Fig. 9.3) act as extracellular-matrix analogues that provide three-dimensional support, biochemical signals, and appropriate porosity for nutrient diffusion and tissue ingrowth, and they must simultaneously satisfy practical design criteria, cell adhesion, suitable mechanical stiffness, controlled porosity, biocompatibility, and edibility or degradability (Bodiou et al., 2020; Bock et al., 2009; Derakhti et al., 2019; Jeong et al., 2022). Common scaffold systems include hydrogels, porous scaffolds (PSs), and microcarriers (Ben-Arye & Levenberg, 2019). Hydrogels emulate the hydration and mechanical properties of the extracellular matrix and can be readily molded or 3D-printed to create structured cultivated meat products. Likewise, porous scaffolds offer high porosity that facilitates efficient mass transfer, but they frequently require supplementary reinforcement to attain sufficient mechanical strength (Gu et al., 2025). Microcarriers provide high surface-area-to-volume ratios in suspension culture, enabling scalable cell expansion. Although they typically produce cellular aggregates that are more appropriate for processed or unstructured products, not whole muscle structures. Their structural properties also offer opportunities to enhance the nutritional profile of the final product (Tsai & Pacak, 2021; Yen et al., 2023). Despite these advantages, major challenges persist across scaffold formats, including ensuring biocompatibility, achieving uniform nutrient and oxygen delivery to prevent core necrosis, maintaining reproducible pore architectures, and reducing production costs, particularly for methods that remain limited to small-scale, high-cost fabrication (Gu et al., 2025; Jeong et al., 2022).

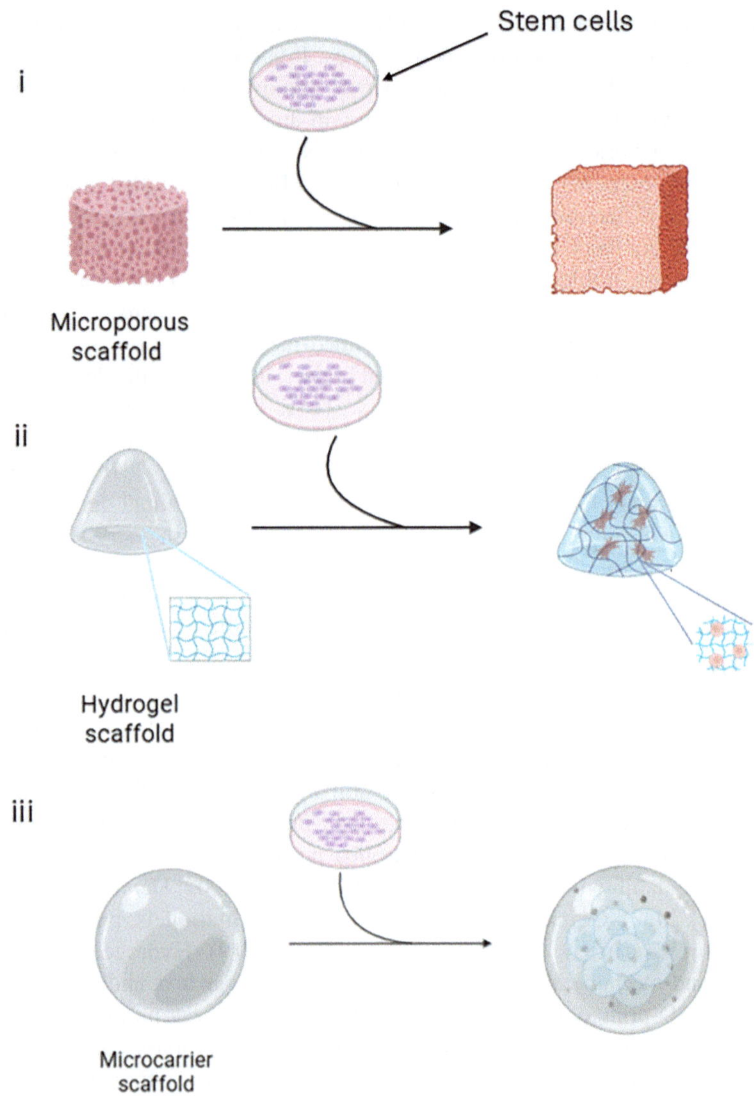

Fig. 9.3 Major types of scaffolds for stem cells. (i) Microporous (ii) Hydrogel (iii) Microcarrier scaffolds

Material selection and fabrication strategies play a critical role in determining scalability and cost-effectiveness. Animal-derived polymers such as collagen, gelatin, and silk offer native cell-adhesion motifs, generally supporting favorable cellular responses. However, their application presents challenges, including batch variability, ethical concerns, and insufficient mechanical strength (Stout et al., 2022; Zehorai et al., 2023). Several studies on scaffolding materials conclude that recombinant production methods and more sustainable extraction techniques can mitigate many of the current limitations. Nevertheless, fungal and plant-derived polysaccharides (e.g., chitosan, alginate, cellulose derivatives, beta glucomannans) and plant proteins (such as wheat gluten, pea, and soy proteins) can provide low-cost, edible scaffold alternatives, but they frequently lack native cell-adhesion motifs, as well as sufficient mechanical strength (Gu et al., 2025). These limitations can be addressed through chemical functionalization, incorporation of RGD amino acid sequences (arginine-glycine-aspartic acid motif that promotes cell adhesion), hybridization with complementary biomaterials, or the application of advanced fabrication techniques (including electrospinning, freeze-drying, and 3D printing) to modulate pore architecture and structural properties (Andersen et al., 2015; Genevro et al., 2019; Lee et al., 2024). Microcarrier geometry and pore size critically affect attachment and proliferation (optimal pores typically range from tens to a few hundred micrometers), while micropatterning and topographical cues enhance cellular alignment for fibrous muscle structure (Gu et al., 2025). Finally, electrical stimulation and conductive scaffold components can accelerate maturation and reduce reliance on costly growth factors. Integrative strategies that combine these biophysical cues with scalable, low-cost materials may enhance production efficiency. Additionally, the use of automated removal systems or biodegradable scaffold designs represents a promising approach to improve product quality while lowering the overall cost of commercial cultivated meat production (Stoppel et al., 2016).

9.4.2 Bioreactor Designs and Advancements

Bioreactors are essential technologies for scaling up cell expansion in cultivated meat production, providing controlled environments that support cell growth, delivering nutrients, exchanging gas, and removing waste. These systems integrate principles from cell biology and mechanical engineering to maintain optimal pH, dissolved oxygen, and carbon dioxide levels, all critical for sustaining high-density cultures. Oxygen delivery is particularly challenging in large-scale bioreactors due to its low solubility at processing temperatures, and therefore agitation is often employed to improve gas transfer. However, excessive agitation may introduce shear stress, which can damage cells. For this reason, bioreactors require careful optimization of mixing rates and intermittent stirring strategies to balance mass transfer and cell viability (Zhang et al., 2015; Marco et al., 2015).

In general, bioreactors for submerged processes can be classified by both medium handling and mixing methods. Batch, fed-batch, and continuous bioreactors differ in how fresh medium is supplied and conditioned medium removed, with fed-batch and continuous systems generally preferred for large-scale cultivated meat production due to improved nutrient management and automation potential (Spier et al., 2011). Different types of bioreactors offer distinct advantages (Fig. 9.4). Perfusion bioreactors, a subtypc of continuous systems, maintain cells while continuously removing waste, making them suitable for structured tissue growth. Mechanically driven bioreactors, including stirred-tank, rotating-wall, and mechanically active systems,

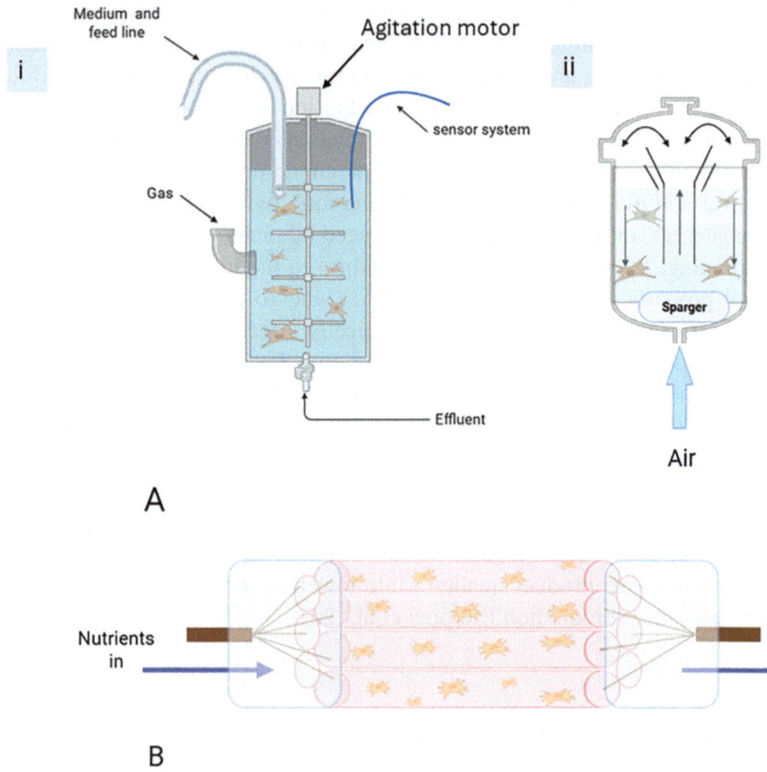

Fig. 9.4 Types of bioreactor designs. (**a**) Agitation based (i) Stirred tank bioreactor (ii) airlift bioreactor. (**b**) Hollow fiber bioreactor

use agitation or applied forces to promote nutrient circulation and tissue alignment, thereby supporting proper cell proliferation, differentiation, and maturation. Additionally, hollow-fiber and pneumatic bioreactors provide alternatives that reduce shear stress or employ gas purging, though they are less established in large-scale applications (Radtke & Herbst-Kralovetz, 2012; Werner et al., 2010).

Agitation-based (AB) and perfusion-based (PB) bioreactors differ in performance and suitability for cultivated meat products. ABs, such as stirred and wave bioreactors, generally achieve higher yields for unstructured meat but may compromise tissue integrity due to shear stress. PBs, including hollow-fiber, packed-bed, and fixed-bed systems, enable nutrient gradients, reduce shear, and support 3D tissue architecture, making them critical for structured cultivated meat, though they face challenges in scalability and cost (Gu et al., 2025; Reiss et al., 2021). Innovations such as dynamic fiber scaffolds, modular perfusion units, and hybrid systems that combine AB throughput with PB specificity aim to address these limitations, while AI-driven fluid dynamics may further optimize nutrient delivery and shear management (Li et al., 2020; Ngwa et al., 2025; Pan et al., 2023).

The choice of bioreactor also depends on the cell type and intended product. Anchorage-dependent cells, such as myocytes, require surfaces or scaffolds for proliferation and differentiation, whereas free-floating pluripotent stem cell aggregates may be cultured in suspension (Woelken et al., 2024). Consequently, co-culture approaches for multiple cell types, such as muscle and fat, are feasible but require optimization for livestock-derived cells. Perfusion rates, scaffold sizes, and shear stress must be carefully managed to maintain tissue viability. During manufacturing, bioreactor designs may need to be tailored to the specific size, shape, and structural requirements of different cultivated meat products. Understanding and controlling microcarrier movement and distribution within large-scale, high-density bioreactors is essential for optimizing culture homogeneity, mass transfer, and cell yield. Consequently, continued development and refinement of bioreactor technologies will be critical to meet industrial-scale production demands and support a diverse portfolio of cultivated meat products.

9.5 Hybrid Products

The term hybrid products, also called dual-protein or combi-foods, refers to a combination of ingredients derived from animals (e.g. meat, eggs, dairy, poultry, or seafood) with alternative non-animal proteins (Zhang et al., 2025). These alternative proteins consist of plant-based options, mycoprotein, and, in some cases, cultivated meat components (Kaplan & McClements, 2025). Combining alternative proteins with animal ingredients leverages the advantages of both ingredient categories. Hybrid products can satisfy consumer preferences for taste and texture, while meeting their nutritional requirements and reducing environmental impact and manufacturing costs. This approach allows, for example, to retain some of the sensory characteristics mostly associated with animal products, such as juiciness and chewiness, while enabling manufacturers to create products that are more affordable and sustainable, as well as having superior texture and nutrition compared to purely plant-based options (Grasso, 2024).

Hybrid products provide familiar flavors that consumers are familiar and enjoy, while simultaneously decreasing carbon emissions, conserving land and water resources. It also helps in stabilizing livestock prices that are often volatile in animal farming. When compared to exclusively plant-based foods, hybrid products present a better mouthfeel and a more complete amino acid profile due to the synergy between animal and non-animal proteins (Tkaczewska et al., 2025).

There are some hybrid products that are already gaining attention, including burgers based on beef blended with soy protein (Chin et al., 2024a), pork sausages that include pea, rice, or soy protein (Broucke et al., 2022; dos Santos et al., 2023), and some seafood items incorporating macroalgae, taking advantage of the fact that some algal species (like *Rhodomonas*, *Tetraselmis*, and *Phaeodactylum*) offer strong seafood odors and flavors, along with high amino acid profiles (Coleman et al., 2022; Jareonsin et al., 2024). Companies in Israel, Spain, Singapore, the United States,

Brazil, and Indonesia are experimenting with the so-called hybrid cultivated meat (HCM), which is a hybridization of animal cells and plant-, bacterial-, or algae-derived ingredients, including various types of binders to mimic the taste and texture of real meat (Alam et al., 2024). HCM are more practical than fully cultured meat, which still faces some cost, acceptance and scalability issues (Kaplan & McClements, 2025).

It is important to highlight that there is a trade-off between the nutritional properties and sensory attributes when designing hybrid products. Increasing the content of alternative proteins in the final product usually improves its nutritional profile by reducing saturated fat or increasing dietary fiber. On the other hand, this comes at the expense of the sensory qualities perceived by consumers, reducing attributes such as juiciness, texture, and flavor. Unfortunately, this remains a major challenge for researchers and manufacturers, and there is no "one size fits all" solution.

From a nutritional perspective, adding alternative protein to hybrid products alters their macronutrient and micronutrient profiles. For example, texturized vegetable protein (TVP) and single-cell protein (SCP) contain about 48% and 52% protein, respectively (Juhrich et al., 2025), but typically provide much more carbohydrate and dietary fiber than traditional meat. Despite this being nutritionally advantageous (due to an increase in fiber intake), they differ in digestibility compared with animal muscle proteins (Razzaq et al., 2020; Zhou et al., 2021). Lipid profiles are also affected in hybrid products. Beef and pork cuts contain a high proportion (33%–45% of total fat) of saturated fatty acids (Williams, 2007), whereas plant-based ingredients are richer in polyunsaturated fatty acids, considered beneficial for cardiovascular health (Juhrich et al., 2025). Despite this, the inclusion of alternative protein dilutes nutrients associated with meat, such as bioavailable iron, zinc, and vitamin B12.

It is known that the sensory attributes of hybrid products change according to the level of substitution of animal ingredients. For example, partially replacing animal fat with plant lipids can compromise texture; however, the use of structured oleogels has demonstrated hardness and gumminess reduction without compromising chewiness (Hernández-Nolasco et al., 2026). The content of myoglobin present in meat is responsible for its red color. Therefore, hybrid products tend to present a paler or more yellowish tone (Fraeye et al., 2020). Finally, some plant proteins may introduce off-flavors (such as beany notes), while algae-derived ingredients may present a fishy and seafood-like aroma (Coleman et al., 2022).

9.5.1 Ingredient Combinations

Different strategies have been studied to obtain hybrid products combining conventional meat, dairy, or eggs with plant proteins, mycoprotein, or fermented products. One of the most adopted strategies is to mix animal meat (beef, pork, poultry) with plant proteins such as soy, pea, lentil, or chickpea protein. Substituting a portion of meat with plant ingredients allows cost and environmental footprint reduction while improving the nutritional profile of the hybrid product. This strategy is most commonly applied to products like burgers, nuggets, and sausages, where the plant matrix enhances water retention and binding while keeping most of the original flavor and juiciness (Broucke et al., 2022; Chin et al., 2024b; dos Santos et al., 2023).

Mycoprotein provides a naturally fibrous structure, and when blended with animal meat, it can reduce costs and environmental impact. Due to its neutral taste, mycoprotein allows the natural flavors of the animal component to remain dominant in the final hybrid product (Akinsemolu & Onyeaka, 2025). Mycoprotein is suitable for formulating hybrid products such as steaks, sausages, mince, chicken, nuggets, strips, and fishless fingers, which require enhanced firmness and/or fibrous structure (Khan et al., 2023).

Among the dairy-based hybrid products, there are combinations of milk with oat, pea, or almond, which enhance nutritional value and reduce environmental footprint while maintaining creaminess and familiarity (de Brito et al., 2025b). These blends help improve amino acid content, lower saturated fat, and appeal to flexitarian consumers reluctant to adopt fully plant-based beverages. Another example is the hybrid yogurts, which blend conventional milk with coconut, soy, or pea bases (de Brito et al., 2025a). The dairy component supplies texture, protein quality, and characteristic fermented flavor, while plant components reduce cost, increase fiber content, or improve lipid profile. These hybrids have gained popularity due to improved spoonability and lower syneresis (i.e. serum separation) compared to fully plant-based yogurts (Méndez-Galarraga et al., 2025). Some processed cheeses include plant proteins (i.e., soy protein isolate, pea protein, lupin protein, oat protein) or starches (hydroxypropylated barley starches) to reduce dairy content while maintaining meltability and stretch (Genet et al., 2023; Zhang et al., 2024). Hybrid cheese can improve sustainability and decrease costs, while avoiding the sensory limitations of fully plant-based cheese alternatives.

9.5.2 Technological and Processing Challenges

One of the primary technical challenges associated with hybrid meat products is discoloration, characterized by undesirable and perceptible shifts in color attributes that diverge from the visual standards of conventional meat products. Discoloration frequently manifests through a significant decline in redness or an increase in yellowness, which can diminish consumer acceptability considerably. These colorimetric alterations are influenced not only by the native pigments present in plant proteins but also by their interactions with myoglobin within the meat matrix, as well as processing parameters, levels of protein substitution, and the pH environment of the hybrid formulations. For instance, Grasso et al. (2019) documented reductions in redness and elevated yellowness in meatballs with 15%–30% inclusion of texturized soy protein. Similar effects were observed in chicken hybrid gels containing 50% plant proteins, including soy, pea, sunflower, fava bean, and rice.

Increased yellowness indices were similarly reported in sausages with 50% rice protein (dos Santos et al., 2023) or 20% pea protein isolate or texturized pea protein (Broucke et al., 2022). The predominant factors contributing to these color shifts include the dilution of myoglobin pigment within the protein matrix and the intrinsic coloration of plant proteins. However, the extent of discoloration depends on the level of substitution. For example, hybrid pork sausages containing 0.5%–1.5% faba bean protein isolate retained color attributes similar to those of traditional formulations (Kim & Chin, 2024). Additionally, pH fluctuations in hybrid meat matrices can influence pigmentation by modulating metmyoglobin formation, thereby inducing browning and diminishing redness (Ramanathan et al., 2020).

Another key challenge in the production of hybrid products is replicating the characteristic texture of meat, which is crucial for consumer acceptance. The fibrous, chewy structure of meat is difficult to replicate with plant proteins, especially at high levels of substitution, due to inherent structural differences between animal myofibrillar proteins and plant globular proteins. Changes in texture depend on the type of plant protein and eventual protein modification, as well as the substitution percentage, with higher levels often weakening the protein network, thereby affecting gelation and water holding capacity.

At low substitution levels (<5%–10%), plant proteins can improve gel microstructure and promote cohesive gels, as seen with rice bran in chicken patties, which increased hardness (Aviles et al., 2023). Similarly, 0.5%–1.5% faba bean protein improved gel strength and surface hydrophobicity while reducing sulfhydryl groups (Kim & Chin, 2024). Different plant sources influence functionality and texture distinctly. For example, replacing 20% or 50% of chicken mince with pea protein isolate in 3D–printed chicken analogues resulted in softer textures, likely due to hydrogen bonding and fewer disulfide bonds (Wang et al., 2023). Lentil–based hybrids at just 25% substitution showed compromised texture, probably due to limited gelation capacity (Chandler & McSweeney, 2022). Texturized pea proteins also pose challenges; mixing pork with 20% of textured pea protein reduced sausage texture because the pea proteins formed large, less soluble aggregates, weakening the network and forming cavities and jelly-like exudates (Broucke et al., 2022). In pork frankfurters with 25%–75% textured pea proteins, hardness, cohesiveness, and chewiness decreased. Conversely, adding oat flour (3.5%–6.5%) to pea-meat hybrids helped counteract softening by increasing hardness, thanks to soluble fibers such as beta–glucan, which binds water and stabilizes the structure, resulting in firmer textures (Flores et al., 2023).

9.6 Consumer Perspectives and Acceptance

Understanding the perception that consumers have of hybrid food products, their attitudes, and willingness to pay is crucial for the success of these products. A complex set of motivations and concerns influences these perceptions. Consumers judge hybrid meat across several dimensions, often including perceived health advantages, sustainability, and curiosity, while common barriers relate to taste concerns, unfamiliar ingredients, processing perceptions, and pricing (Miao et al., 2023; Profeta et al., 2021; Ryder et al., 2023; Salgaonkar & Nolden, 2024; van Dijk et al., 2023). Among all the factors shaping acceptance, sensory attributes remain the strongest driver. Consumers consistently prioritize taste, texture, and appearance over more abstract benefits such as environmental impact or long-term health outcomes (Chin et al., 2024b; Realini et al., 2022, 2023). This emphasis on immediate sensory gratification underscores why creating products that mimic conventional meat is essential.

A growing number of studies have evaluated the sensory qualities of hybrid foods (Chin et al., 2024b; Melios & Grasso, 2024; Petrat-Melin & Dam, 2023). The results of a conjoint analysis of 454 U.S. consumers are consistent in showing that hybrids that most closely resemble traditional meat perform better in consumer choice scenarios. Protein source was the most influential attribute in product selection (35% relative importance), followed by price (24.5%), fat content (20.5%), and package claim (20%). Notably, hybrid hotdogs (50% beef + 50% plant protein) exhibited a negative part-worth utility (−15.9), compared with 100% plant-based (+18.6) and 100% beef (−2.6), indicating lower baseline preference. Moreover, regression analysis showed that protein ($\beta = -1.2$) and fat content ($\beta = -0.8$) were significant drivers of hybrid selection, which suggests that deviations from familiar meat-like profiles reduce consumer acceptance (Salgaonkar & Nolden, 2024).

Burgers containing 50% meat maintained desirable taste and texture, with only minor elasticity differences (Chin et al., 2024b). In contrast, when plant-based components exceed 50%, acceptance tends to decline due to off-flavors, less appealing textures, or color inconsistencies (Melios & Grasso, 2024). Although sensory appeal dominates consumer decision-making, other elements still matter. Ethical considerations, environmental benefits, and perceptions of healthiness also contribute to overall acceptance (Ryder et al., 2023). Another recurring barrier is the perception that hybrid products are "overly processed," which raises doubts about their healthfulness (Salgaonkar & Nolden, 2024). These challenges highlight opportunities for product developers to improve familiarity through clear ingredient communication, simpler formulations, and cleaner-label strategies.

In a survey of 261 Uruguayan consumers, hierarchical cluster analysis identified four distinct segments with significantly different purchase intentions ($P < 0.05$). Notably, the cluster with the highest proportion of young consumers (38% aged 17–29 years; CL4) showed a clear preference for 100% dairy products over hybrid alternatives, whereas clusters with a higher proportion of middle-aged and highly educated individuals (CL1 and CL2) demonstrated greater openness to milk–plant blends. Education level differed significantly across clusters ($P = 0.038$), and daily dairy consumption frequency was also unevenly distributed ($P = 0.005$), indicating that demographic characteristics were associated with variation in hybrid acceptance (Curutchet et al., 2024).

A qualitative study involving 28 meat-eating consumers (16 Asian and 12 non-Asian) revealed statistically significant cross-cultural differences

($P < 0.05$) in how blended products were evaluated. For example, blended beef burgers labeled "microalgae" received significantly lower desirability scores among Asian consumers (4.7/10) compared to their healthiness rating (7.1/10), indicating taste concerns despite recognized nutritional value. Similarly, non-Asian consumers assigned significantly lower healthiness scores to blended burgers with "No Monosodium Glutamate" (4.8/10 vs. 6.3/10 for Asians) and "Sustainable plant protein alternatives" (4.6/10 vs. 6.7/10), reflecting greater skepticism toward processing and ingredient claims. For conventional chicken nuggets, Asians rated desirability significantly higher than healthiness (8.2 vs. 4.5), highlighting the strong role of expected sensory pleasure in product evaluation (Miao et al., 2023). While both groups appreciated the potential health and sustainability benefits, Asian consumers prioritized flavor more heavily, whereas non-Asian consumers were more skeptical about processing and raised concerns about the naturalness of certain ingredients like microalgae.

A combination of sensory satisfaction, perceived benefits, familiarity, and price shapes the willingness to pay (WTP) and the likelihood of repeat purchases. A cross-country study involving 2477 participants in the United Kingdom, Spain, and Denmark reported that consumers are unwilling to pay a premium for hybrid burgers. However, when information about the benefits of hybrid products, the WTP rose by £10.94/kg (health information), £6.32/kg (sensory information), and £7.56/kg (convenience information) in participants from the UK. Similarly, health information increased the WTP of participants from Spain by €7.33/kg, and sensory information raised the WTP of participants from Denmark by 113.05 DKK/kg (Asioli et al., 2023). Still, price remains a critical barrier for some groups in adopting these products.

A study conducted in the UK with 802 participants identified two distinct segments: one group called "fans" (76% of the sample; $n = 610$) who were reluctant to reduce meat intake or pay extra for hybrid products, and the other group called "reducers" (24%; $n = 192$) who were more receptive to hybrids, especially when pricing aligns with the perceived added value. These differences were reflected in attitudes toward hybrid products. Reducers expressed substantially greater willingness to try (5.6 vs. 4.1) and buy (5.5 vs. 3.9) hybrid meat products on a 7-point scale. Moreover, fans preferred hybrids with a higher meat content (75% meat/25% plant), whereas reducers favored formulations with 75% plant-based ingredients (Melios & Grasso, 2024).

9.7 Market Outlook and Regulatory Considerations

The increasing interest in vegetarian and vegan diets, the diversification of protein sources, growing awareness towards climate change, resource use and carbon footprint of conventional livestock productions are the key factors supporting the rise in alternative protein diets (Davis et al., 2025; Eshel, 2020; Musicus et al., 2022;

Tilman & Clark, 2014). With the projected market growth of alternative protein, the technological advancements in fermentation processes, cultivated meat production and microbial protein technologies, can significantly improve efficiency, scalability and cost optimization of these protein products (Molfetta et al., 2022). These factors can further accelerate the mainstream adoption of alternative meat-based diets and drive a global transition in protein consumption patterns.

The rapid growth of the alternative meat sector also requires strong regulatory and ethical compliance to ensure product quality and consumer's safety. Regulatory assessment of fungus-based alternative proteins such as mycoprotein and tempeh, includes controlling reducing Ribonucleic acid (RNA) to approved levels, evaluating allergenicity, monitoring foodborne pathogens, mycotoxin risks, and ensuring the safety specifications of using various substrates or waste materials in fermentation (Finnigan et al., 2024; Hashempour-Baltork et al., 2020). Additionally, it is also crucial to conduct safety research to identify the potential adverse reactions, including the identification of potential gastrointestinal or allergic symptoms that may pose life-threatening risks among consumers (Jacobson & DePorter, 2018). A key ethical consideration for fungus-based alternative proteins is ensuring transparency to consumers regarding the use of genetically modified organisms (GMOs) in their production, as well as labelling.

The regulatory landscape for cultured meat is still emerging, evolving and adapting, and the framework differs across regions. Various components used in cultured meat, such as ingredients in growth media, cell lines, and scaffold materials are novel to the food sector. Therefore, emphasis on robust food safety guidelines is important to address issues related to the genetic stability of cell lines, microbiological risk factors and contamination with chemical residues (Bhat et al., 2019). Regulatory procedures need to be more robust for these novel foods to ensure safer products, labelling accuracy and ethical production (Zandonadi et al., 2025). Ethical consideration also arises regarding the use of embryonic or pluripotent stem cells in cultured meat production and the need to transition toward serum-free media can align with both public perception and regulatory expectations (Behera & Adhikary, 2023).

9.8 Conclusions

Fermentation systems to produce microbial proteins and cultivated meat have the potential to advance sustainable protein production, addressing both demand and nutritional needs, as well as environmental concerns. Fermented microbial proteins, particularly mycoproteins in their current commercial form, along with tempeh, hybrids, and cultivated meat, have the potential to spearhead the sustainable protein innovation for meat-like products. The opportunity to utilize diverse substrates, including agri-food by-products, positions the fermentation technology as a key component in the transition toward a more sustainable and resilient food system.

Finally, animal–alternative hybrid products provide near-term traction by delivering familiar taste, color, and texture while moderating environmental impacts and ingredient costs. However, careful formulation is needed to mitigate color shifts, off-flavor and undesirable texture, thereby aligning with and meeting consumer preferences and facilitating mainstream adoption.

By integrating fermentation technological optimization, food safety, understanding of current and future markets, as well as consumers' acceptance, the food industry will be able to ultimately provide proteins that are nutritious, affordable, and climate-friendly. The regulatory landscape worldwide will also play a pivotal role in how easy or not public policies will be implemented that facilitate or not the transition to a food system that is less reliant on animal protein, compared with the current state. Continued research, including fermentation optimization (bioreactors scale-up; high-yield, mutation-resistant, toxin-free strains), circular feedstock integration (agri-food by-products alternative media), sensory-first product design, and cost-down strategies for cellular agriculture, will be crucial to overcoming current challenges and unlocking the full potential of fermented alternative proteins in addressing global food security and environmental sustainability.

References

Abu-Salem, F. M., & Abou-Arab, E. A. (2011). Physico-chemical properties of tempeh produced from chickpea seeds. *Journal of American Science, 7*(7), 1545–1003. http://www.americanscience.org. http://www.americanscience.orgeditor@americanscience.org

Aeri, B. T., & Kaur, H. (2025). Protein quantity and quality and links with mental health in adolescents. In *Protein intake in health and disease* (pp. 449–462). https://doi.org/10.1201/B23328-33

Ahnan-Winarno, A. D., Cordeiro, L., Winarno, F. G., Gibbons, J., & Xiao, H. (2021). Tempeh: A semicentennial review on its health benefits, fermentation, safety, processing, sustainability, and affordability. *Comprehensive Reviews in Food Science and Food Safety, 20*(2), 1717–1767. https://doi.org/10.1111/1541-4337.12710

Akinsemolu, A. A., & Onyeaka, H. N. (2025). Mycoproteins as sustainable food sources: Current applications and future prospects. *Discover Applied Sciences, 7*(3). https://doi.org/10.1007/s42452-025-06614-0. Springer Nature.

Alam, A. M. M. N., Kim, C. J., Kim, S. H., Kumari, S., Lee, S. Y., Hwang, Y. H., & Joo, S. T. (2024). Trends in hybrid cultured meat manufacturing technology to improve sensory characteristics. *Food Science of Animal Resources, 44*(1), 39–50. https://doi.org/10.5851/kosfa.2023.e76. Korean Society for Food Science of Animal Resources.

Amara, A. A., & El-Baky, N. A. (2023). Fungi as a source of edible proteins and animal feed. *Journal of Fungi, 9*(1), 73. https://doi.org/10.3390/JOF9010073

Andersen, T., Auk-Emblem, P., & Dornish, M. (2015). 3D cell culture in alginate hydrogels. *Microarrays, 4*(2), 133–161. https://doi.org/10.3390/MICROARRAYS4020133

Ashenafi, M., & Busse, M. (1991). Growth of Bacillus cereus in fermenting tempeh made from various beans and its inhibition by Lactobacillus plantarum. *Journal of Applied Bacteriology, 70*(4), 329–333. https://doi.org/10.1111/J.1365-2672.1991.TB02944

Asioli, D., Banovic, M., Barone, A. M., Grasso, S., & Nayga, R. M., Jr. (2023). European consumers' valuation for hybrid meat: Does information matter? *Applied Economic Perspectives and Policy, 45*(1), 44–62. https://doi.org/10.1002/aepp.13283

Aviles, M. V., Naef, E. F., Abalos, R. A., Lound, L. H., Gómez, M. B., & Olivera, D. F. (2023). Use of a rice industry by-product as a meat replacer in a hybrid chicken patty: Technological and sensory impact. *International Journal of Gastronomy and Food Science, 31*, 100674. https://doi.org/10.1016/j.ijgfs.2023.100674

Badenes, S. M., Fernandes, T. G., Cordeiro, C. S. M., Boucher, S., Kuninger, D., Vemuri, M. C., Diogo, M. M., & Cabral, J. M. S. (2016). Defined essential 8™ medium and vitronectin efficiently support scalable xeno-free expansion of human induced pluripotent stem cells in stirred microcarrier culture systems. *PLoS One, 11*(3), e0151264. https://doi.org/10.1371/JOURNAL.PONE.0151264

Bakratsas, G., Polydera, A., Nilson, O., Kossatz, L., Xiros, C., Katapodis, P., & Stamatis, H. (2023). Single-cell protein production by Pleurotus ostreatus in submerged fermentation †. *Sustainable Food Technology*. https://doi.org/10.1039/d2fb00058j

Bayati, M., Arkia, E., & Emadi, M. (2025). Socio-economic inequality in the nutritional deficiencies among the world countries: Evidence from global burden of disease study 2019. *Journal of Health, Population, and Nutrition, 44*(1), 8. https://doi.org/10.1186/S41043-025-00739-Z

Behera, R., & Adhikary, L. (2023). Review on cultured meat: Ethical alternative to animal industrial farming. *Journal Homepage, 7*(2), 42–51. https://doi.org/10.26656/fr.2017.7(2).772

Ben-Arye, T., & Levenberg, S. (2019). Tissue engineering for clean meat production. *Frontiers in Sustainable Food Systems, 3*, 435343. https://doi.org/10.3389/FSUFS.2019.00046/FULL

Benjaminson, M. A., Gilchriest, J. A., & Lorenz, M. J. A. A. (2002). In vitro edible muscle protein production system (MPPS): Stage 1, fish. *Acta astronautica, 51*(12), 879–889. https://doi.org/10.1016/S0094-5765(02)00033-4

Bhat, Z. F., Morton, J. D., Mason, S. L., Bekhit, A. E. D. A., & Bhat, H. F. (2019). Technological, regulatory, and ethical aspects of in vitro meat: A future slaughter-free harvest. *Comprehensive Reviews in Food Science and Food Safety, 18*(4), 1192–1208. https://doi.org/10.1111/1541-4337.12473

Bock, A., Sann, H., Schulze-Horsel, J., Genzel, Y., Reichl, U., & Möhler, L. (2009). Growth behavior of number distributed adherent MDCK cells for optimization in microcarrier cultures. *Biotechnology Progress, 25*(6), 1717–1731. https://doi.org/10.1002/BTPR.262

Bodiou, V., Moutsatsou, P., & Post, M. J. (2020). Microcarriers for upscaling cultured meat production. *Frontiers in Nutrition, 7*(10), 506214. https://doi.org/10.3389/FNUT.2020.00010/FULL

Bonjour, J. P. (2005). Dietary protein: An essential nutrient for bone health. *Journal of the American College of Nutrition, 24*, 526S–536S. https://doi.org/10.1080/07315724.2005.10719501

Broucke, K., Van Poucke, C., Duquenne, B., De Witte, B., Baune, M.-C., Lammers, V., Terjung, N., Ebert, S., Gibis, M., Weiss, J., & Van Royen, G. (2022). Ability of (extruded) pea protein products to partially replace pork meat in emulsified cooked sausages. *Innovative Food Science & Emerging Technologies, 78*, 102992. https://doi.org/10.1016/j.ifset.2022.102992

Burley, V. J., Paul, A. W., & Blundell, J. E. (1993). Influence of a high-fibre food (myco-protein) on appetite: Effects on satiation (within meals) and satiety (following meals). *European Journal of Clinical Nutrition., 47*(6), 409–418. https://europepmc.org/article/med/8396025

Business Wire. (2025, May 15). *Protein alternative market focused insights 2025–2030 | Alternative proteins poised for growth as global market to hit USD 36.17 billion by 2030.* ResearchAndMarkets.com. https://www.businesswire.com/news/home/20250516308930/en/Protein-Alternative-Market-Focused-Insights-2025-2030-Alternative-Proteins-Poised-for-Growth-as-Global-Market-to-Hit-USD-36.17-Billion-by-2030%2D%2D-ResearchAndMarkets.com

Cedeno, F. R. P., Olubiyo, O. J., & Ferreira, S. (2025). From microbial proteins to cultivated meat for alternative meat-like products: A review on sustainable fermentation approaches. *Journal of Biological Engineering, 19*(1), 44. https://doi.org/10.1186/S13036-025-00509-9

Chandler, S. L., & McSweeney, M. B. (2022). Characterizing the properties of hybrid meat burgers made with pulses and chicken. *International Journal of Gastronomy and Food Science, 27*, 100492. https://doi.org/10.1016/j.ijgfs.2022.100492

Chen, G., Gulbranson, D. R., Hou, Z., Bolin, J. M., Ruotti, V., Probasco, M. D., Smuga-Otto, K., Howden, S. E., Diol, N. R., Propson, N. E., Wagner, R., Lee, G. O., Antosiewicz-Bourget, J., Teng, J. M. C., & Thomson, J. A. (2011). Chemically defined conditions for human iPSC derivation and culture. *Nature Methods, 8*(5), 424–429. https://doi.org/10.1038/nmeth.1593

Chin, S. W., Baier, S. K., Stokes, J. R., & Smyth, H. E. (2024a). Evaluating the sensory properties of hybrid (meat and plant-based) burger patties. *Journal of Texture Studies, 55*(1), e12819. https://doi.org/10.1111/jtxs.12819

Chin, X. H., Elhalis, H., Chow, Y., & Liu, S. Q. (2024b). Enhancing food safety in soybean fermentation through strategic implementation of starter cultures. *Heliyon, 10*(2), e25007. https://doi.org/10.1016/J.HELIYON.2024.E25007

Choręziak, A., Rosiejka, D., Michałowska, J., & Bogdański, P. (2025). Nutritional quality, safety and environmental benefits of alternative protein sources—An overview. *Nutrients, 17*(7), 1148. https://doi.org/10.3390/NU17071148

Coleman, B., Van Poucke, C., Dewitte, B., Ruttens, A., Moerdijk-Poortvliet, T., Latsos, C., De Reu, K., Blommaert, L., Duquenne, B., Timmermans, K., van Houcke, J., Muylaert, K., & Robbens, J. (2022). Potential of microalgae as flavoring agents for plant-based seafood alternatives. *Future Foods, 5*, 100139. https://doi.org/10.1016/j.fufo.2022.100139

Curutchet, A., Arcia, P., & Rodríguez Arzuaga, M. (2024). The future of hybrid foods: Will consumers embrace dairy-plant blends? *International Journal of Food Science and Technology, 59*(11), 8777–8785. https://doi.org/10.1111/ijfs.17473

Dai, Y., Liu, Y., Wang, Z., Han, S., Tao, Y., Wang, Z., Dong, M., Wang, D., & Xia, X. (2025). Enhanced mycoprotein production of Neurospora intermedia from soy processing byproducts: Integration of ultrasonic stimulation with repeated-batch simultaneous saccharification and culture. *Bioresource Technology, 437*, 133124. https://doi.org/10.1016/J.BIORTECH.2025.133124

Davis, U., Bobo, J., Wilson, P., Noy, P., Mansilla, R., Long, G., Welham, S., Harvey, J., Lukinova, E., Nica-Avram, G., Smith, G., Salt, D., Smith, A., & Goulding, J. (2025). Nutritional implications of alternative proteins: A commentary. *Public Health Nutrition, 28*(1), e69. https://doi.org/10.1017/S1368980025000242

de Bardet, J. C., Cardentey, C. R., González, B. L., Patrone, D., Mulet, I. L., Siniscalco, D., & Robinson-Agramonte, M. d. l. A. (2023). Cell immortalization: In vivo molecular bases and in vitro techniques for obtention. *Biotech, 12*(1), 14. https://doi.org/10.3390/BIOTECH12010014

de Brito, I. S., Tavares Filho, E. R., de Mattos, R. B. E., Pagani, M. M., Pimentel, T. C., da Cruz, A. G., Mársico, E. T., Martins, C. P. d. C., & Esmerino, E. A. (2025a). Exploring preferences and emotional responses of a dairy-based hybrid product: Insights from consumer-centric development using different health-oriented clusters. *Food Quality and Preference, 129*. https://doi.org/10.1016/j.foodqual.2025.105537

de Brito, I. S., Durço, B. B., Tavares Filho, E. R., de Mattos, R. B. E., Pagani, M. M., Mársico, E. T., da Cruz, A. G., de Carvalho Martins, C. P., & Esmerino, E. A. (2025b). How consumers make sense of hybrid dairy-based foods: A multistakeholder qualitative study. *Journal of Sensory Studies, 40*(6). https://doi.org/10.1111/joss.70083

Denny, A., Aisbitt, B., & Lunn, J. (2008). Mycoprotein and health. *Nutrition Bulletin, 33*(4), 298–310. https://doi.org/10.1111/j.1467-3010.2008.00730.x

Derakhti, S., Safiabadi-Tali, S. H., Amoabediny, G., & Sheikhpour, M. (2019). Attachment and detachment strategies in microcarrier-based cell culture technology: A comprehensive review. *Materials Science and Engineering: C, 103*, 109782. https://doi.org/10.1016/J.MSEC.2019.109782

dos Santos, M., Ribeiro, W. O., de Sousa Monteiro, J., dos Santos, B. A., Campagnol, P. C. B., & Pollonio, M. A. R. (2023). Effect of transglutaminase treatment on the structure and sensory properties of rice- or soy-based hybrid sausages. *Foods, 12*(23). https://doi.org/10.3390/foods12234226

Drewnowski, A., & Hooker, K. (2025). The protein transition: What determines the animal-to-plant (A:P) protein ratios in global diets. *Frontiers in Nutrition, 12*. https://doi.org/10.3389/fnut.2025.1518793

EFSA. (2023, May 10). *The safety of cell culture-derived food – Ready for scientific evaluation*. https://www.efsa.europa.eu/en/news/safety-cell-culture-derived-food-ready-scientific-evaluation

Elhalis, H., Chin, X. H., & Chow, Y. (2024). Soybean fermentation: Microbial ecology and starter culture technology. *Critical Reviews in Food Science and Nutrition, 64*(21), 7648–7670. https://doi.org/10.1080/10408398.2023.2188951

Erkan, S. B., Gürler, H. N., Bilgin, D. G., Germec, M., & Turhan, I. (2020). Production and characterization of tempehs from different sources of legume by Rhizopus oligosporus. *LWT, 119*, 108880. https://doi.org/10.1016/J.LWT.2019.108880

Eshel, G. (2020). How to prioritize voluntary dietary modification. *Advances in Environmental and Engineering Research*, (4), 1. https://doi.org/10.21926/AEER.2004005

Finnigan, T. J. A., Wall, B. T., Wilde, P. J., Stephens, F. B., Taylor, S. L., & Freedman, M. R. (2019). Mycoprotein: The future of nutritious nonmeat protein, a symposium review. *Current Developments in Nutrition, 3*(6). https://doi.org/10.1093/cdn/nzz021

Finnigan, T., Mach, K., & Edlin, A. (2024). Mycoprotein: A healthy new protein with a low environmental impact. In *Sustainable protein sources: Advances for a healthier tomorrow* (2nd ed., pp. 539–566). https://doi.org/10.1016/B978-0-323-91652-3.00011-3

Fletcher, A. J., Smith, N. W., Hill, J. P., & McNabb, W. C. (2024). Modeling the feasibility of fermentation-produced protein at a globally relevant scale. *Frontiers in Sustainable Food Systems, 8*, 1419259. https://doi.org/10.3389/FSUFS.2024.1419259/BIBTEX

Flores, M., Hernán, A., Salvador, A., & Belloch, C. (2023). Influence of soaking and solvent extraction for deodorization of texturized pea protein isolate on the formulation and properties of hybrid meat patties. *Journal of the Science of Food and Agriculture, 103*(6), 2806–2814. https://doi.org/10.1002/jsfa.12453

Food Dive. (2023, July 31). *From prediction to production: A timeline of cultivated meat*. https://www.fooddive.com/news/from-prediction-production-timeline-cultivated-meat-industry/689441/

Fraeye, I., Kratka, M., Vandenburgh, H., & Thorrez, L. (2020). Sensorial and nutritional aspects of cultured meat in comparison to traditional meat: Much to be inferred. *Frontiers in Nutrition, 7*. https://doi.org/10.3389/fnut.2020.00035

Furlan, O., de Oliveira, N. S., de Paula, R. C., Rosa, R. T., Michelotto, P. V., Weber, S. H., Bianchini, L. F., & Rosa, E. A. R. (2024). Pilot scale production of high-content mycoprotein using Rhizopus microsporus var. oligosporus by submerged fermentation and agro-industrial by-products. *Bioresource Technology, 413*, 131515. https://doi.org/10.1016/J.BIORTECH.2024.131515

Genet, B. M. L., Sedó Molina, G. E., Wätjen, A. P., Barone, G., Albersten, K., Ahrné, L. M., Hansen, E. B., & Bang-Berthelsen, C. H. (2023). Hybrid cheeses—Supplementation of cheese with plant-based ingredients for a tasty, nutritious and sustainable food transition. *Fermentation, 9*(7). https://doi.org/10.3390/fermentation9070667. Multidisciplinary Digital Publishing Institute (MDPI).

Genevro, G. M., de Moraes, M. A., & Beppu, M. M. (2019). Freezing influence on physical properties of glucomannan hydrogels. *International Journal of Biological Macromolecules, 128*, 401–405. https://doi.org/10.1016/J.IJBIOMAC.2019.01.112

Gil, M., Rudy, M., Duma-Kocan, P., Stanisławczyk, R., Krajewska, A., Dziki, D., & Hassoon, W. H. (2024). Sustainability of alternatives to animal protein sources, a comprehensive review. *Sustainability, 16*(17), 7701. https://doi.org/10.3390/SU16177701

Godfray, H. C. J., Aveyard, P., Garnett, T., Hall, J. W., Key, T. J., Lorimer, J., Pierrehumbert, R. T., Scarborough, P., Springmann, M., & Jebb, S. A. (2018). Meat consumption, health, and the environment. *Science (New York, N.Y.), 361*(6399). https://doi.org/10.1126/science.aam5324

Good Food Institute, & Huling, R. (2025, June 18). *Regulators approve cultivated meat sales in Australia*. https://gfi-apac.org/regulators-approve-cultivated-meat-sales-in-australia/

Górska, K., Pejcz, E., & Harasym, J. (2025). Tempeh and fermentation—Innovative substrates in a classical microbial process. *Applied Sciences (Switzerland), 15*(16). https://doi.org/10.3390/app15168888. Multidisciplinary Digital Publishing Institute (MDPI).

Grand View Research. (2025). *Alternative protein ingredients market size, share & trends report alternative protein ingredients market (2025–2030)*. https://www.grandviewresearch.com/industry-analysis/alternative-protein-market-report

Grasso, S. (2024). Opportunities and challenges of hybrid meat products: A viewpoint article. *International Journal of Food Science and Technology, 59*(11), 8693–8696. https://doi.org/10.1111/ijfs.17421

Grasso, S., Smith, G., Bowers, S., Ajayi, O. M., & Swainson, M. (2019). Effect of texturised soy protein and yeast on the instrumental and sensory quality of hybrid beef meatballs. *Journal of Food Science and Technology, 56*(6), 3126–3135. https://doi.org/10.1007/s13197-018-3552-9

Green Queen, & Mridul, A. (2025, September 10). *Iceland's ORF genetics gets $5.8M to advance cultivated meat via molecular farming*. https://www.greenqueen.com.hk/orf-genetics-funding-cultivated-lab-grown-meat-molecular-farming/

Griese, S. E., Fleischauer, A. T., MacFarquhar, J. K., Moore, Z., Harrelson, C., Valiani, A., Morrison, S. E., Sweat, D., Maillard, J. M., Griffin, D., Springer, D., Mikoleit, M., Newton, A. E., Jackson, B., Nguyen, T. A., Bosch, S., & Davies, M. (2013). Gastroenteritis outbreak associated with unpasteurized tempeh, North Carolina, USA. *Emerging Infectious Diseases, 19*(9), 1514. https://doi.org/10.3201/EID1909.130334

Gu, H., Kong, Y., Huang, D., Wang, Y., Raghavan, V., & Wang, J. (2025). *Comprehensive reviews in food science and food safety comprehensive review scaling cultured meat: Challenges and solutions for affordable mass production*. https://doi.org/10.1111/1541-4337.70221

Guixer, B., Frøst, M. B., & Flore, R. (2017). Tempeto – Expanding the scope and culinary applications of tempe with post-fermentation sousvide cooking. *International Journal of Gastronomy and Food Science, 9*, 1–9. https://doi.org/10.1016/J.IJGFS.2017.03.002

Hachmeister, K. A., & Fung, D. Y. C. (1993). Tempeh: A mold-modified indigenous fermented food made from soybeans and/or cereal grains. *Critical Reviews in Microbiology, 19*(3), 137–188. https://doi.org/10.3109/10408419309113527

Hashempour-Baltork, F., Khosravi-Darani, K., Hosseini, H., Farshi, P., & Reihani, S. F. S. (2020). Mycoproteins as safe meat substitutes. *Journal of Cleaner Production, 253*, 119958. https://doi.org/10.1016/J.JCLEPRO.2020.119958

Hernández-Nolasco, Z., Inés Acateca-Hernández, M., Real-Luna, N., & Selene Hernández-Cázares, A. (2026). *Beeswax-structured oleogels and bigels in functional meat products: Nutritional, technological, and sensory implications*. https://doi.org/10.26599/FSAP

Hubalek, S., Post, M. J., & Moutsatsou, P. (2022). Towards resource-efficient and cost-efficient cultured meat. *Current Opinion in Food Science, 47*, 100885. https://doi.org/10.1016/J.COFS.2022.100885

Jacobson, M. F., & DePorter, J. (2018). Self-reported adverse reactions associated with mycoprotein (Quorn-brand) containing foods. *Annals of Allergy, Asthma & Immunology, 120*(6), 626–630. https://doi.org/10.1016/J.ANAI.2018.03.020

Jara, T. C., Park, K., Vahmani, P., Van Eenennaam, A. L., Smith, L. R., & Denicol, A. C. (2023). Stem cell-based strategies and challenges for production of cultivated meat. *Nature Food, 4*(10), 841–853. https://doi.org/10.1038/s43016-023-00857-z

Jareonsin, S., Pumas, C., Jaitiang, D., & Uttarotai, T. (2024). Green fusion proteins: An approach to sustainable nutrition blending plant and algae-based proteins for a circular food system. *Future Foods, 10*, 100415. https://doi.org/10.1016/j.fufo.2024.100415

Jeong, D., Seo, J. W., Lee, H. G., Jung, W. K., Park, Y. H., & Bae, H. (2022). Efficient myogenic/adipogenic transdifferentiation of bovine fibroblasts in a 3D bioprinting system for steak-type cultured meat production. *Advanced Science, 9*(31), 2202877. https://doi.org/10.1002/ADVS.202202877

Juhrich, L. C., Grosse, M., Morlein, J., Bergmann, P., Zorn, H., & Gand, M. (2025). Nutritional and sensory properties of meat analogues: A current overview and future considerations. *Journal of Agricultural and Food Chemistry, 73*(4), 2236–2248. https://doi.org/10.1021/acs.jafc.4c09414. American Chemical Society.

Kaplan, D. L., & McClements, D. J. (2025). Hybrid alternative protein-based foods: Designing a healthier and more sustainable food supply. *Frontiers in Science, 3*. https://doi.org/10.3389/fsci.2025.1599300

Kavya, P., Sobhanan, A., Shinde, D., Meena, R., & Koley, T. K. (2025). Soybean-based ethnic fermented food products: Preparation methods, nutritional quality, and their health benefits. In *Ethnic and indigenous food technologies* (pp. 333–355). https://doi.org/10.1007/978-981-96-9286-6_16

Khan, R., Brishti, F. H., Arulrajah, B., Goh, Y. M., Abd Rahim, M. H., Karim, R., Hajar-Azhari, S., Kin Kit, S., Anwar, F., & Saari, N. (2023). Mycoprotein as a meat substitute: Production, functional properties, and current challenges-a review. *International Journal of Food Science and Technology, 59*(1), 522–544. https://doi.org/10.1111/IJFS.16791

Kim, G. H., & Chin, K. B. (2024). Effects of faba bean protein isolate on rheological properties of pork myofibrillar protein gels and quality characteristics of pork low-fat model sausages. *Journal of the Science of Food and Agriculture, 104*(10), 6322–6329. https://doi.org/10.1002/jsfa.13484

Kim, K., Choi, B., Lee, I., Lee, H., Kwon, S., Oh, K., & Kim, A. Y. (2011). Bioproduction of mushroom mycelium of Agaricus bisporus by commercial submerged fermentation for the production of meat analogue. *Journal of the Science of Food and Agriculture, 91*(9), 1561–1568. https://doi.org/10.1002/JSFA.4348

Kolkmann, A. M., Post, M. J., Rutjens, M. A. M., van Essen, A. L. M., & Moutsatsou, P. (2019). Serum-free media for the growth of primary bovine myoblasts. *Cytotechnology, 72*(1), 111–120. https://doi.org/10.1007/S10616-019-00361-Y

Kustyawati, M. E. (2017). *Profile of aroma compounds and acceptability of modified tempeh*. http://www.ifrj.upm.edu.my/24%20(02)%202017/(38).pdf

Landeta-Salgado, C., Cicatiello, P., & Lienqueo, M. E. (2021). Mycoprotein and hydrophobin like protein produced from marine fungi Paradendryphiella salina in submerged fermentation with green seaweed Ulva spp. *Algal Research, 56*, 102314. https://doi.org/10.1016/J.ALGAL.2021.102314

Lee, M., Park, S., Choi, B., Choi, W., Lee, H., Lee, J. M., Lee, S. T., Yoo, K. H., Han, D., Bang, G., Hwang, H., Koh, W. G., Lee, S., & Hong, J. (2024). Cultured meat with enriched organoleptic properties by regulating cell differentiation. *Nature Communications, 15*(1), 77. https://doi.org/10.1038/s41467-023-44359-9

Lestari, L. A., Rachmawati, A. I., Saskia, A., Rachmadanty, S., Azuan, Z., Daud, M., & Sholikhati, A. S. (2024). Leveraging local ingredients for better nutritional status: The role of soybean, tempeh, and moringa oleifera leaf. *Penelitian Gizi Dan Makanan (The Journal of Nutrition and Food Research), 47*(1), 33–42. https://doi.org/10.36457/PGM.V47I1.776

Li, X., Zhang, G., Zhao, X., Zhou, J., Du, G., & Chen, J. (2020). A conceptual air-lift reactor design for large scale animal cell cultivation in the context of in vitro meat production. *Chemical Engineering Science, 211*, 115269. https://doi.org/10.1016/J.CES.2019.115269

Malila, Y., Owolabi, I. O., Chotanaphuti, T., Sakdibhornssup, N., Elliott, C. T., Visessanguan, W., Karoonuthaisiri, N., & Petchkongkaew, A. (2024). Current challenges of alternative proteins as future foods. *npj Science of Food, 8*(1), 53. https://doi.org/10.1038/S41538-024-00291-W

Marchioro, M. L. K., Candeia, G. A. P. B., Bertoleti, L. M., Barbosa-Dekker, A. M., Dekker, R. F. H., & da Cunha, M. A. A. (2025). Bioprospecting the endophytic fungus, Lasiodiplodia theobromae MMPI, for the integrated production of mycoprotein and exocellular

(1→6)-β-glucan. *Fermentation, 11*(4), 166. https://doi.org/10.3390/FERMENTATION11040166/S1

Marco, C. S., Pohlscheidt, J. M., Sieblist, C., Jenzsch, Á. M., & Pohlscheidt, Á. M. (2015). Equipment characterization to mitigate risks during transfers of cell culture manufacturing processes. *Cytotechnology, 68*(4), 1381–1401. https://doi.org/10.1007/S10616-015-9899-0

MarketsandMarkets, R. (2025). *Cultured meat market size, share, trends and reports*. https://www.marketsandmarkets.com/Market-Reports/cultured-meat-market-204524444.html

MarketScreener. (2015). *Monde Nissin Corporation completed the acquisition of Marlow Foods Limited from Exponent Private Equity LLP and ICG European Fund 2006, L.P. of Intermediate Capital Group plc*. https://www.marketscreener.com/quote/stock/MONDE-NISSIN-CORPORATION-124598889/news/Monde-Nissin-Corporation-completed-the-acquisition-of-Marlow-Foods-Limited-from-Exponent-Private-Equ-38205581/

Marlow Foods. (2022, April 20). *How is quorn made? Introduction to quorn & quorn mycoprotein*. https://www.quorn.us/news/how-is-quorn-made-introduction-to-quorn-and-mycoprotein

Melios, S., & Grasso, S. (2024). Meat fans' and meat reducers' attitudes towards meat consumption and hybrid meat products in the UK: A cluster analysis. *International Journal of Food Science and Technology, 59*(12), 9394–9401. https://doi.org/10.1111/ijfs.17586

Méndez-Galarraga, M. P., Curutchet, A., & Arzuaga, M. R. (2025). Hybrid yogurts: Exploring the functional, sensory and nutritional potential of dairy-plant protein combinations. *Food Reviews International, 41*(7), 2150–2166. https://doi.org/10.1080/87559129.2025.2489033

Miao, X., Hastie, M., Ha, M., & Warner, R. (2023). Consumer response to blended beef burgers and chicken nuggets is influenced by ingredient and nutrition claims – Qualitative assessment. *Future Foods, 8*, 100247. https://doi.org/10.1016/j.fufo.2023.100247

Molfetta, M., Morais, E. G., Barreira, L., Bruno, G. L., Porcelli, F., Dugat-Bony, E., Bonnarme, P., & Minervini, F. (2022). Protein sources alternative to meat: State of the art and involvement of fermentation. *Food, 11*(14), 2065. https://doi.org/10.3390/FOODS11142065/S1

Musicus, A. A., Wang, D. D., Janiszewski, M., Eshel, G., Blondin, S. A., Willett, W., & Stampfer, M. J. (2022). Health and environmental impacts of plant-rich dietary patterns: A US prospective cohort study. *The Lancet Planetary Health, 6*(11), e892–e900. https://doi.org/10.1016/S2542-5196(22)00243-1

National Agricultural Law Center. (2025). *Alternative proteins: 2025 legislative update*. https://nationalaglawcenter.org/alternative-proteins-2025-legislative-update/

Ngwa, C. J., Chiu, K. H., Brenner, K. J., Rodrigues, C., Peršin, Z., Vajda, J., Vihar, B., Morales-Dalmau, J., Ferreira, F. C., Henkel, M., & Schillberg, S. (2025). Bioreactor parameters and systems for cultured meat production. *Future Foods, 12*, 100796. https://doi.org/10.1016/J.FUFO.2025.100796

Nout, M. J. R., Beernink, G., & Bonants-van Laarhoven, T. M. G. (1987). Growth of Bacillus cereus in soyabean tempeh. *International Journal of Food Microbiology, 4*(4), 293–301. https://doi.org/10.1016/0168-1605(87)90004-3

Pan, W., Chen, H., Wang, A., Wang, F., & Zhang, X. (2023). Challenges and strategies: Scalable and efficient production of mesenchymal stem cells-derived exosomes for cell-free therapy. *Life Sciences, 319*, 121524. https://doi.org/10.1016/J.LFS.2023.121524

Panapparambil Sooraj, I., Leech, J., O'Callaghan, T. F., & McAuliffe, O. (2025). Application and utilization of fermentation as a processing tool to mitigate protein putrefaction in plant-based diets. *Frontiers in Microbiology, 16*, 1638378. https://doi.org/10.3389/FMICB.2025.1638378/FULL

Petrat-Melin, B., & Dam, S. (2023). Textural and consumer-aided characterisation and acceptability of a hybrid meat and plant-based burger patty. *Foods, 12*(11). https://doi.org/10.3390/foods12112246

Pilgrim, C. R., McCahill, K. A., Rops, J. G., Dufour, J. M., Russell, K. A., & Koch, T. G. (2022). A review of fetal bovine serum in the culture of mesenchymal stromal cells and potential alternatives for veterinary medicine. *Frontiers in Veterinary Science, 9*, 859025. https://doi.org/10.3389/FVETS.2022.859025

Post, M. J. (2012). Cultured meat from stem cells: Challenges and prospects. *Meat Science, 92*(3), 297–301. https://doi.org/10.1016/j.meatsci.2012.04.008

Post, M. J. (2014). Cultured beef: Medical technology to produce food. *Journal of the Science of Food and Agriculture, 94*(6), 1039–1041. https://doi.org/10.1002/jsfa.6474

Prakash, P., Karthik, S., Namashivyam, R., & Narendrakumar, G. (2014). Optimization of growth parameters for elevated production of mycoprotein-Fusarium venenatum using response surface methodology. *Journal of Pure and Applied Microbiology, 8*(6), 4843–4849.

Pratama, F., Rahardja, R. T., Rachmadi, A. R., Salam, W. Q., Kho, K., Adelie, A., & Devanthi, P. V. P. (2025). Optimizing mycoprotein production by Aspergillus oryzae using soy whey as a substrate. *Journal of Fungi, 11*(5), 349. https://doi.org/10.3390/JOF11050349

Profeta, A., Baune, M.-C., Smetana, S., Broucke, K., Van Royen, G., Weiss, J., Hieke, S., Heinz, V., & Terjung, N. (2021). Consumer preferences for meat blended with plant proteins – Empirical findings from Belgium. *Future Foods, 4*, 100088. https://doi.org/10.1016/j.fufo.2021.100088

Purwadaria, H. K., Fardiaz, D., Kardono, L. B. S., & McElhatton, A. (2016). Tempe from traditional to modern practices. In *Modernization of traditional food processes and products* (pp. 145–160). https://doi.org/10.1007/978-1-4899-7671-0_13

Radtke, A. L., & Herbst-Kralovetz, M. M. (2012). Culturing and applications of rotating wall vessel bioreactor derived 3D epithelial cell models. *Journal of Visualized Experiments*, (62). https://doi.org/10.3791/3868

Ramanathan, R., Hunt, M. C., Mancini, R. A., Nair, M. N., Denzer, M. L., Suman, S. P., & Mafi, G. G. (2020). Recent updates in meat color research: Integrating traditional and high-throughput approaches. *Meat and Muscle Biology, 4*(2). https://doi.org/10.22175/mmb.9598

Razzaq, Z. U., Khan, M. K. I., Maan, A. A., & Rahman, S. u. (2020). Characterization of single cell protein from Saccharomyces cerevisiae for nutritional, functional and antioxidant properties. *Journal of Food Measurement and Characterization, 14*(5), 2520–2528. https://doi.org/10.1007/s11694-020-00498-x

Realini, C. E., Ares, G., Antúnez, L., Brito, G., Luzardo, S., del Campo, M., Saunders, C., Farouk, M. M., & Montossi, F. M. (2022). Meat insights: Uruguayan consumers' mental associations and motives underlying consumption changes. *Meat Science, 192*, 108901. https://doi.org/10.1016/j.meatsci.2022.108901

Realini, C. E., Driver, T., Zhang, R., Guenther, M., Duff, S., Craigie, C. R., Saunders, C., & Farouk, M. M. (2023). Survey of New Zealand consumer attitudes to consumption of meat and meat alternatives. *Meat Science, 203*, 109232. https://doi.org/10.1016/j.meatsci.2023.109232

Reddy, N. R., & Pierson, M. D. (1994). Reduction in antinutritional and toxic components in plant foods by fermentation. *Food Research International, 27*(3), 281–290. https://doi.org/10.1016/0963-9969(94)90096-5

Reiss, J., Robertson, S., & Suzuki, M. (2021). Cell sources for cultivated meat: Applications and considerations throughout the production workflow. *International Journal of Molecular Sciences, 22*(14), 7513. https://doi.org/10.3390/IJMS22147513

Riyanto, R. A., Fukusaki, E., & Putri, S. P. (2024). Metabolite profiling highlights the effect of microbial intervention in the soaking step of tempe. *International Journal of Food Science and Technology, 59*(10), 7414–7425. https://doi.org/10.1111/IJFS.17481

Romulo, A., & Surya, R. (2021). Tempe: A traditional fermented food of Indonesia and its health benefits. *International Journal of Gastronomy and Food Science, 26*, 100413. https://doi.org/10.1016/J.IJGFS.2021.100413

Rousta, N., Hellwig, C., Wainaina, S., Lukitawesa, L., Agnihotri, S., Rousta, K., & Taherzadeh, M. J. (2021). Filamentous fungus Aspergillus oryzae for food: From submerged cultivation to fungal burgers and their sensory evaluation—A pilot study. *Foods, 10*(11), 2774. https://doi.org/10.3390/FOODS10112774

Ruxton, C. H. S., & McMillan, B. (2010). The impact of mycoprotein on blood cholesterol levels: A pilot study. *British Food Journal, 112*(10), 1092–1101. https://doi.org/10.1108/00070701011080221

Ryder, C., Jaworska, S., & Grasso, S. (2023). Hybrid meat products and co-creation: What do consumers say, feel and think? *Frontiers in Nutrition, 10*. https://doi.org/10.3389/fnut.2023.1106079

Salgaonkar, K., & Nolden, A. A. (2024). Exploring consumer preferences and challenges in hybrid meat products: A conjoint analysis of hotdogs. *Foods, 13*(10). https://doi.org/10.3390/foods13101460

Sampedro, J. (2024). A sustainable protein ratio. *Nature Food, 5*(5), 357–358. https://doi.org/10.1038/s43016-024-00984-1

Shi, L., Khan, M. Z., Ullah, A., Liang, H., Geng, M., Akhtar, M. F., Na, J., Han, Y., & Wang, C. (2025). Advancements in stem cell applications for livestock research: A review. *Veterinary Sciences, 12*(5), 397. https://doi.org/10.3390/VETSCI12050397

Souza Filho, P. F., Nair, R. B., Andersson, D., Lennartsson, P. R., & Taherzadeh, M. J. (2018). Vegan-mycoprotein concentrate from pea-processing industry byproduct using edible filamentous fungi. *Fungal Biology and Biotechnology, 5*(1), 1–10. https://doi.org/10.1186/s40694-018-0050-9

Spier, M. R., Vandenberghe, L. P. D. S., Medeiros, A. B. P., & Soccol, C. R. (2011). Application of different types of bioreactors in bioprocesses. In *Bioreactors: Design, properties and applications*. Nova Science Publishers, Inc..

Statista. (2025). *World: Total population 1950–2100 | Statista*. https://www.statista.com/statistics/805044/total-population-worldwide/

Stoffel, F., de Oliveira Santana, W., Gregolon, J. G. N., Kist, T. B. L., Fontana, R. C., & Camassola, M. (2019). Production of edible mycoprotein using agroindustrial wastes: Influence on nutritional, chemical and biological properties. *Innovative Food Science & Emerging Technologies, 58*, 102227. https://doi.org/10.1016/J.IFSET.2019.102227

Stoppel, W. L., Kaplan, D. L., & Black, L. D. (2016). Electrical and mechanical stimulation of cardiac cells and tissue constructs. *Advanced Drug Delivery Reviews, 96*, 135–155. https://doi.org/10.1016/J.ADDR.2015.07.009

Stout, A. J., Mirliani, A. B., Rittenberg, M. L., Shub, M., White, E. C., Yuen, J. S. K., & Kaplan, D. L. (2022). Simple and effective serum-free medium for sustained expansion of bovine satellite cells for cell cultured meat. *Communications Biology, 5*(1), 466. https://doi.org/10.1038/s42003-022-03423-8

Sudarmadji, S., & Markakis, P. (1977). The phytate and phytase of soybean tempeh. *Journal of the Science of Food and Agriculture, 28*(4), 381–383. https://doi.org/10.1002/JSFA.2740280410; WGROUP:STRING:PUBLICATION

Svensson, S. E., Bucuricova, L., Ferreira, J. A., Filho, P. F. S., Taherzadeh, M. J., & Zamani, A. (2021). Valorization of bread waste to a fiber- and protein-rich fungal biomass. *Fermentation, 7*(2), 91. https://doi.org/10.3390/FERMENTATION7020091

Tan, Z. J., Bakar, A., Lim, M. F., & Sutimin, S. Y. (2024). Nutritional composition and sensory evaluation of tempeh from different combinations of beans. *Journal Homepage, 8*(2), 138–146. https://doi.org/10.26656/fr.2017.8(2).088

Tanaka, N., Kovats, S. K., & Guggisberg, J. A. (1985). Evaluation of the microbiological safety of tempeh made from unacidified soybeans. *Journal of Food Protection, 48*(5), 438–441. https://doi.org/10.4315/0362-028X-48.5.438

Tannières, M., Breugnot, D., Bon, M. C., & Grodowitz, M. J. (2024). Cultivation of monoxenous trypanosomatids: A minireview. *Journal of Invertebrate Pathology, 203*, 108047. https://doi.org/10.1016/J.JIP.2023.108047

Tilman, D., & Clark, M. (2014). Global diets link environmental sustainability and human health. *Nature, 515*(7528), 518–522. https://doi.org/10.1038/NATURE13959

Tkaczewska, J., Mungure, T., & Warner, R. (2025). Is it still meat? The effects of replacing meat with alternative ingredients on the nutritional and functional properties of hybrid products: A review. *Critical Reviews in Food Science and Nutrition, 65*(28), 5752–5771. https://doi.org/10.1080/10408398.2024.2430750

Tong, S., Hong, R., Chen, W., Chai, M., Zhang, Y., Sun, Y., Wang, Q., & Li, D. (2024). Synchronous bioproduction of betanin and mycoprotein in the engineered edible fungus Fusarium venenatum. *Journal of Agricultural and Food Chemistry, 72*(35), 19462–19469. https://doi.org/10.1021/ACS.JAFC.4C06071

Trinci, A. P. J. (1994). Evolution of the Quorn® myco-protein fungus, Fusarium graminearum A3/5. *Microbiology, 140*(9), 2181–2188. https://doi.org/10.1099/13500872-140-9-2181

Tsai, A. C., & Pacak, C. A. (2021). Bioprocessing of human mesenchymal stem cells: From planar culture to microcarrier-based bioreactors. *Bioengineering, 8*(7), 96. https://doi.org/10.3390/bioengineering8070096

Upcraft, T., Tu, W. C., Johnson, R., Finnigan, T., Van Hung, N., Hallett, J., & Guo, M. (2021). Protein from renewable resources: Mycoprotein production from agricultural residues. *Green Chemistry, 23*(14), 5150–5165. https://doi.org/10.1039/d1gc01021b

van Dijk, B., Jouppila, K., Sandell, M., & Knaapila, A. (2023). No meat, lab meat, or half meat? Dutch and Finnish consumers' attitudes toward meat substitutes, cultured meat, and hybrid meat products. *Food Quality and Preference, 108*, 104886. https://doi.org/10.1016/j.foodqual.2023.104886

Vegconomist. (2025). *Cultivated meat*. https://vegconomist.com/category/cultivated-cell-cultured-biotechnology/cultivated-meat/#:~:text=In%20August%20of%20last%20year,Products%20Inspection%20Act%20(PPIA)

Vulgaris, P., Fadly, D., Angkasa, D., & Friawan, A. (2023). Fortification of ferrous fumarate on tempeh and drinks of Vigna unguiculata. *Journal Homepage, 7*(1), 46–51. https://doi.org/10.26656/fr.2017.7(1).375

Wang, H. L., Vespa, J. B., & Hesseltine, C. W. (1972). Release of bound trypsin inhibitors in soybeans by Rhizopus oligosporus. *The Journal of Nutrition, 102*(11), 1495–1499. https://doi.org/10.1093/JN/102.11.1495

Wang, T., Kaur, L., Beniwal, A. S., Furuhata, Y., Aoyama, H., & Singh, J. (2023). Physicochemical and textural properties of 3D printed plant-based and hybrid soft meat analogs. *Plant Foods for Human Nutrition, 78*(2), 375–382. https://doi.org/10.1007/s11130-023-01068-4

Wei, Q., Xi, Q., Liu, X., Meng, K., Zhao, X., & Ma, B. (2017). Characterization of goat inner cell mass derived cells in double kinase inhibition condition. *Biochemical and Biophysical Research Communications, 483*(1), 325–331. https://doi.org/10.1016/J.BBRC.2016.12.144

Werner, S., Eibl, R., Lettenbauer, C., Röll, M., Eibl, D., De Jesus, M., Zhang, X., Stettler, M., Tissot, S., Bürki, C., Broccard, G., Kühner, M., Tanner, R., Baldi, L., Hacker, D., & Wurm, F. M. (2010). Innovative, non-stirred bioreactors in scales from milliliters up to 1000 liters for suspension cultures of cells using disposable bags and containers – A swiss contribution. *Chimia (Aarau), 64*(11), 819. https://doi.org/10.2533/chimia.2010.819

Whelehan, G., West, S., Abdelrahman, D. R., Murton, A. J., Finnigan, T. J. A., Wall, B. T., & Stephens, F. B. (2024). The addition of mycoprotein to a mixed-meal impacts postprandial glucose kinetics without altering blood glucose concentrations: A randomised controlled trial. *European Journal of Clinical Nutrition, 78*(9), 757–764. https://doi.org/10.1038/s41430-024-01470-4

Wiebe, M. (2002). Myco-protein from fusarium venenatum: A well-established product for human consumption. *Applied Microbiology and Biotechnology, 58*(4), 421–427. https://doi.org/10.1007/s00253-002-0931-x

Wiebe, M. G. (2004). Quorn™ myco-protein – Overview of a successful fungal product. *Mycologist, 18*(1), 17–20. https://doi.org/10.1017/S0269915X04001089. Cambridge University Press.

Williams, P. (2007). Nutritional composition of red meat. *Nutrition and Dietetics, 64*(Suppl. 4). https://doi.org/10.1111/j.1747-0080.2007.00197.x

Williamson, D., Geiselman, P., & Lovejoy, J. (2006). Effects of consuming mycoprotein, tofu or chicken upon subsequent eating behaviour, hunger and safety. *Appetite, 46*(1), 41–48. https://doi.org/10.1016/J.APPET.2005.10.007

Woelken, L., Weckowska, D. M., Dreher, C., & Rauh, C. (2024). Toward an innovation radar for cultivated meat: Exploring process technologies for cultivated meat and claims about their social impacts. *Frontiers in Sustainable Food Systems, 8*, 1390720. https://doi.org/10.3389/FSUFS.2024.1390720/FULL

Xie, Y., Cai, L., Ding, S., Wang, C., Wang, J., Ibeogu, I. H., Li, C., & Zhou, G. (2025). An overview of recent progress in cultured meat: Focusing on technology, quality properties, safety, industrialization, and public acceptance. *The Journal of Nutrition, 155*(3), 745–755. https://doi.org/10.1016/J.TJNUT.2025.01.010

Yao, Y., Li, C., Yuen, J. S. K., Stout, A. J., & Kaplan, D. L. (2024). Manufacture of hybrid alternative protein food products using a combination of plant-based ingredients, fermentation-derived ingredients, and animal cells. *Cellular Agriculture, 109*(12), 251–266. https://doi.org/10.1016/B978-0-443-18767-4.00032-9. Elsevier.

Yen, F. C., Glusac, J., Levi, S., Zernov, A., Baruch, L., Davidovich-Pinhas, M., Fishman, A., & Machluf, M. (2023). Cultured meat platform developed through the structuring of edible microcarrier-derived microtissues with oleogel-based fat substitute. *Nature Communications, 14*(1), 2942. https://doi.org/10.1038/s41467-023-38593-4

Zandonadi, R. P., Ramos, M. C., Elias, F. T. S., & Guimarães, N. S. (2025). Global insights into cultured meat: Uncovering production processes, potential hazards, regulatory frameworks, and key challenges—A scoping review. *Food, 14*(1). https://doi.org/10.3390/FOODS14010129/S1

Zehorai, E., Maor-Shoshani, A., Molotski, N., Dorojkin, A., Marelly, N., Dvash, T., & Lavon, N. (2023). From fertilised oocyte to cultivated meat – Harnessing bovine embryonic stem cells in the cultivated meat industry. *Reproduction Fertility and Development, 36*(2), 124–132. https://doi.org/10.1071/RD23169

Zeng, B., Nilsson, K., Teixeira, P. G., & Bergenståhl, B. (2023). Study of mycoprotein extraction methods and its functional properties. *Colloids and Surfaces A: Physicochemical and Engineering Aspects, 659*. https://doi.org/10.1016/j.colsurfa.2022.130800

Zhang, Y., & Hu, J. (2025). Nutrition challenges in less developed regions: Innovative solutions. In *Healthy food for children* (pp. 189–223). https://doi.org/10.1007/978-981-96-8512-7_7

Zhang, Y., Stobbe, P., Silvander, C. O., & Chotteau, V. (2015). Very high cell density perfusion of CHO cells anchored in a non-woven matrix-based bioreactor. *Journal of Biotechnology, 213*, 28–41. https://doi.org/10.1016/J.JBIOTEC.2015.07.006

Zhang, D., Jiang, K., Luo, H., Zhao, X., Yu, P., & Gan, Y. (2024). Replacing animal proteins with plant proteins: Is this a way to improve quality and functional properties of hybrid cheeses and cheese analogs? *Comprehensive Reviews in Food Science and Food Safety, 23*(1), 1–36. https://doi.org/10.1111/1541-4337.13262. John Wiley and Sons Inc.

Zhang, R., Farouk, M. M., Realini, C. E., & Thum, C. (2025). Hybridization in meat-based dual protein foods: Mechanisms, challenges, and consumer insights. *Comprehensive Reviews in Food Science and Food Safety, 24*(4). https://doi.org/10.1111/1541-4337.70216. John Wiley and Sons Inc.

Zhou, H., Hu, Y., Tan, Y., Zhang, Z., & McClements, D. J. (2021). Digestibility and gastrointestinal fate of meat versus plant-based meat analogs: An in vitro comparison. *Food Chemistry, 364*. https://doi.org/10.1016/j.foodchem.2021.130439

Chapter 10
Smarter Fermenters, Better Foods: Automation and AI in the Plant-Based Revolution

Victor Hugo Buttrós, Disney Ribeiro Dias, and Rosane Freitas Schwan

Abstract Artificial intelligence (AI) and automation are revolutionizing fermentation into a highly precise platform for developing plant-based foods. By incorporating smart bioreactors, real-time sensors, and machine learning models, fermentation systems can now adaptively control crucial variables such as pH, temperature, aeration, oxygen, and substrate feed. This data-driven approach enhances consistency, yield, and scalability, effectively addressing the fundamental limitations of traditional fermentation. AI models trained on historical or high-throughput data enable outcome prediction, process optimization, and the utilization of digital twins to simulate and scale up fermentation with remarkable accuracy. These tools support both biomass fermentation, where microbial cells serve as a food source, and precision fermentation, in which microorganisms are engineered or directed to produce animal-free proteins, fats, enzymes, or flavors. These innovations have paved the way for the commercial production of recombinant dairy proteins, heme analogues, essential nutrients (such as vitamin B_{12}, omega fatty acids, and amino acids), and texture-modifying enzymes that enhance the sensory qualities of plant-based foods. Biosensors and analytical tools, including electronic noses, tongues, and spectroscopic probes, facilitate real-time quality monitoring, ensuring the reproducibility of aroma, texture, and nutritional value. Automated systems enhance high-throughput screening and minimize human error, enabling continuous operation at industrial scales. Despite the rapid progress made, challenges persist, including data scarcity, a lack of interoperability and standardization, substantial initial costs, and regulatory demands for safety and traceability. Overcoming these obstacles necessitates interdisciplinary collaboration, transparent AI, and integration with synthetic

V. H. Buttrós
Department of Biology, Federal University of Lavras, Lavras, Brazil

Soil Department, Federal University of Rio Grande do Sul, Porto Alegre, Brazil

D. R. Dias
Department of Food Science, Federal University of Lavras, Lavras, Brazil

R. F. Schwan (✉)
Department of Biology, Federal University of Lavras, Lavras, Brazil
e-mail: rschwan@ufla.br

S. Dhiman, S. Kaur (eds.), *Next-Gen Plant Based Fermented Foods: Innovations for a Sustainable Future*, https://doi.org/10.1007/978-3-032-29927-7_10

biology and bioinformatics. AI-powered fermentation is empowering decentralized, modular, and adaptive food production, paving the way for a scalable and sustainable future for plant-based innovation.

Keywords Plant-based foods · Fermentation · Automation · Process analytical technology · Machine learning · Digital twins · Precision fermentation · Biomass fermentation

Abbreviations

AI	Artificial Intelligence; Computational methods that facilitate pattern recognition, prediction, and decision-making to enhance process optimization, including techniques like machine learning, deep learning, Bayesian optimization, and explainable models employed in contemporary bioprocess control and analytics.
ALCOA/ALCOA+	Attributable, Legible, Contemporaneous, Original, Accurate (+ Complete, Consistent, Enduring, Available); data-integrity principles.
BO	Bayesian Optimization; sequential ML-guided experimental optimization.
CCD	Central Composite Design; response-surface experimental design.
CFD	Computational Fluid Dynamics; simulation of mixing, flow, and mass transfer.
CIP	Cleaning in Place; automated internal cleaning of tanks and piping.
CPR	Carbon Production Rate; CO_2 evolution rate from off-gas analysis.
DCS	Distributed Control System; plant-wide automation and supervisory control.
DoE	Design of Experiments; structured experimental design framework.
DO	Dissolved Oxygen; oxygen concentration in fermentation broth.
EBR	Electronic Batch Record; digital batch documentation and genealogy.
ERP	Enterprise Resource Planning; business-level planning/scheduling system.
FAT	Factory Acceptance Test; vendor-site testing before equipment delivery.
FTIR	Fourier Transform Infrared Spectroscopy; mid-infrared PAT tool.

GC-MS	Gas Chromatography–Mass Spectrometry; analytical chemistry technique.
GMP	Good Manufacturing Practices; regulatory production standards.
HACCP	Hazard Analysis and Critical Control Points; food-safety framework.
HMI	Human–Machine Interface; operator interface for SCADA/PLC.
HPLC	High-Performance Liquid Chromatography; analytical measurement method.
HTST	High-Temperature Short-Time; continuous thermal sterilization method.
IQ	Installation Qualification; verification of correct system installation.
ISA-88	Batch Control Standard; modular recipe and phase standard.
ISA-95	Enterprise–Control Integration Standard; structure linking ERP–MES–SCADA.
kLa	Volumetric Oxygen Transfer Coefficient; metric for oxygen transfer capability.
MES	Manufacturing Execution System; batch coordination and electronic records.
MLOps	Machine Learning Operations; lifecycle management for ML models.
ML	Machine Learning; data-driven modeling approaches.
MOO	Multi-Objective Optimization; simultaneous optimization of competing goals.
MPC	Model Predictive Control; model-based constrained optimization control.
MQTT	Message Queuing Telemetry Transport; lightweight publish–subscribe protocol.
NIR	Near-Infrared Spectroscopy; an inline multivariate PAT sensor.
NSGA-II	Non-Dominated Sorting Genetic Algorithm II; evolutionary multi-objective optimizer.
OD	Optical Density; turbidity/biomass proxy.
OEM	Original Equipment Manufacturer; equipment supplier (appears indirectly).
OPC UA	Open Platform Communications—Unified Architecture; interoperability protocol.
OQ	Operational Qualification; testing that equipment functions as intended.
ORP	Oxidation–Reduction Potential (implicitly referenced in PAT contexts).
OTR	Oxygen Transfer Rate; actual oxygen delivered to cells per time.

OUR	Oxygen Uptake Rate; oxygen consumption rate measured via off-gas.
PAT	Process Analytical Technology; real-time measurement and control framework.
PID	Proportional–Integral–Derivative Control; classic feedback control loop.
PLC	Programmable Logic Controller; real-time automation controller.
PQ	Performance Qualification; proof that the system works under real process conditions.
qPCR	Quantitative Polymerase Chain Reaction; DNA-quantification method.
R&D	Research and Development; development efforts (implicit throughout).
RQ	Respiratory Quotient; CO_2 produced/O_2 consumed.
RPM	Revolutions Per Minute; impeller agitation speed.
RTD	Resistance Temperature Detector; temperature sensor (used implicitly).
SAT	Site Acceptance Test; onsite testing after installation.
SCADA	Supervisory Control and Data Acquisition; real-time supervisory automation.
SIP	Sterilization in Place; automated steam-based sterilization.
TIC/TOC	Total Inorganic/Organic Carbon (appears indirectly in PAT context).
vvm	Volumes of Gas per Volume of Liquid per Minute; aeration rate.

10.1 Introduction

Fermentation is central to plant-based foods because microorganisms can turn variable plant substrates into products that are consistent in flavor, texture, nutrition, and cost (Dias & Schwan, 2024; Samtiya et al., 2021). This chapter serves as a practical guide for making fermentation processes predictable and scalable through automation, Process Analytical Technology, a robust data foundation, and practical machine learning. The objective is to assist teams in achieving repeatability, expediting development, and ensuring consistent performance across different vessels and locations.

Key terms used throughout are introduced briefly here. Biomass fermentation treats cells as the product. Precision fermentation uses cells to produce defined molecules that enhance plant matrices, such as proteins, enzymes, lipids, vitamins, and aroma compounds (Ahirwar et al., 2023; Augustin et al., 2023). Operations in both modes depend on controlling pH, dissolved oxygen, temperature, feeding, and foam, with scale limits set by oxygen transfer, mixing, and heat removal (Montague et al., 1989; Vandenberghe et al., 2024).

Volumetric oxygen transfer coefficients (kLa) are the working metric for oxygen transfer (Badino et al., 2001). Off-gas analytics provide oxygen uptake and carbon evolution rates, summarized as the respiratory quotient, and also the profile of volatile compounds during fermentation (Mengers et al., 2022; Schmidberger et al., 2014). A soft sensor is a model that estimates an unmeasured variable from measured signals (Li et al., 2023). A digital twin is a live, data-linked model used for *in silico* scenario testing and scale-up guidance (Zhao et al., 2025). These ideas are detailed in later sections.

The chapter shows how sensing, control, data, and analytics fit together. Sensing covers pH, dissolved oxygen, temperature, off-gas, and at least one spectroscopic or capacitance probe. Control covers stable proportional–integral–derivative (PID) loops and recipe state machines with tested interlocks. Data covers a historian with clean tags, recipe, and equipment context aligned with ISA-88, and secure interfaces that support electronic batch records and audit trails consistent with ALCOA Plus.

Analytics covers a minimal set of supervised models, a small feature store, drift checks, and a calibrated twin for safe setpoint testing. Readers can skim the first pass for the architecture, then return to the sections on Process Analytical Technology (PAT), data infrastructure, machine learning, optimization and control, and digital twins to apply the methods to a specific project, whether a recombinant protein in yeast or a biomass product from filamentous fungi.

10.2 Fermentation Fundamentals for Plant-Based Applications

Biomass vs. Precision Fermentation: In alternative proteins, *biomass fermentation* grows microorganisms for use as the product itself (e.g., Quorn's mycoprotein from fungi), while *precision fermentation* uses microorganisms as "cell factories" to produce a specific ingredient (e.g., Perfect Day's dairy whey protein). Biomass methods excel for simplicity and high-volume protein output, often using natural (non-GMO) strains. In contrast, precision methods are ideal for high-value functional ingredients that require purity and consistency (such as vitamins, enzymes, and tailored fats). The two approaches are complementary, not competing: biomass fermentation efficiently yields bulk protein and nutrients, while precision fermentation provides targeted components (like heme or flavor enzymes) that nature alone cannot supply at scale (Ahirwar et al., 2023; Augustin et al., 2023; Geistlinger et al., 2024; Whittaker et al., 2020). In practice, many plant-based food formulations use both, e.g., a fungus-based protein bulk plus a precision-fermented enzyme for flavor. To clarify the key distinctions between biomass and precision fermentation, Table 10.1 summarizes their main differences side by side.

Table 10.1 Comparison between biomass fermentation and precision fermentation for food and ingredient production

Aspect	Biomass fermentation	Precision fermentation
Product of interest	Whole microbial biomass (cells) used as food ingredient or protein source	A specific target molecule produced by the microbe (purified ingredient)
Typical micro-organism traits	Often wild-type or non-modified, food-grade strains with a history of safe use	Often engineered or highly selected production strains optimized for a single product
Scale and volume tendency	Typically large-scale, high-throughput production for bulk protein or bulk biomass	Often smaller to medium scale for high-value products, can scale up when demand and economics justify
Examples	Quorn mycoprotein (filamentous fungus biomass), baker's yeast biomass, single-cell protein from yeasts/algae	Animal-free whey proteins (e.g., β-lactoglobulin), heme analogs, enzymes, vitamins, specialty fats
Key goal	Maximize cell yield and protein content, while maintaining acceptable sensory properties	Maximize target titer, productivity, and yield; ensure product quality and consistency
Downstream processing focus	Minimal to moderate (harvest, washing, texturization, flavor masking), since the biomass is the product	Substantial (clarification, concentration, purification, polishing) because the molecule is the product
Main challenges	Palatability and texture, off-flavors, nucleic acid reduction (where relevant), achieving consistent sensory quality	Purification costs, product stability and authenticity (functionality), regulatory oversight for engineered strains, process robustness and contamination control

Table Source: From Authors

Objectives and Constraints Biomass fermentation aims to maximize cell mass (since the cells *are* the food product), so high growth rate and protein content are key. Constraints include maintaining the palatability and texture of the biomass and preventing by-products that affect taste. In precision fermentation, the goal is to maximize the titer and yield of the target molecule (e.g., a protein or pigment) (Augustin et al., 2023; Bonturi, 2025). This often involves genetic engineering of the microorganism and optimizing media to channel most resources into the product (Mao et al., 2024).

Constraints here include feedstock costs, downstream purification requirements, and regulatory limits (especially when using GMOs or new additives) (Mamma, 2020; Mukherjee et al., 2022; Wilken & Nikolov, 2012). Both types of processes aim for high productivity, but precision processes may tolerate lower cell growth if the product per cell is high. On the other hand, biomass processes prioritize biomass accumulation. For instance, in precision fermentation for an enzyme, growth might be intentionally slowed down during the stationary phase to induce product formation. In contrast, in biomass fermentation (e.g., yeast for nutritional yeast), rapid exponential growth is desired (Robinson et al., 2001).

Core Unit Operations and Variables Despite their diverse end goals, the fundamental fermentation operations share commonalities. Key control variables, such as temperature, pH, and dissolved oxygen (DO), must be maintained at optimal setpoints for the specific microorganism being cultivated. Agitation and aeration are crucial operations. Mixing ensures that all cells receive nutrients and oxygen, while gas sparging provides oxygen for aerobic cultures. In large fermenters, spargers and impellers are employed to maximize the oxygen transfer rate (OTR), which is characterized by the volumetric mass-transfer coefficient kLa (Badino et al., 2001; Chai et al., 2022a; Specht, 2025).

Feeding strategy is another operation: simple *batch* fermentation provides all nutrients initially, whereas *fed-batch* adds feeds over time to extend growth, and *continuous* mode steadily feeds and harvests (Yang & Sha, 2019). In any mode, the nutrient feeding rate and timing are critical variables that affect productivity. Foam control is also a practical necessity, as it can arise from CO_2 and surfactants released during cell lysis, and is managed with mechanical foam breakers or antifoam. At scale, foam can cause contamination or sensor malfunctions, so antifoam addition and headspace sensors are used to keep it in check. If uncontrolled, foam may trigger safety valves or foul filters, it is considered a "critical process parameter" in fermentation operations (Tiso et al., 2024).

Chassis and Substrates (Including Side-Streams) The term *chassis* refers to the microorganism species used. Common chassis for plant-based fermentations include yeasts (e.g., *Saccharomyces cerevisiae* for ethanol or *Pichia pastoris/Komagataella phaffii* for recombinant proteins), filamentous fungi (e.g., *Fusarium venenatum* for Quorn's mycoprotein, or *Aspergillus* for citric acid or enzyme production), bacteria (e.g., *Bacillus* or *E. coli* for certain proteins and vitamins), and even microalgae. Each has preferred substrates and strengths: for instance, yeasts ferment simple sugars well, certain fungi can utilize complex polysaccharides, and some bacteria (like methanotrophs or H_2-oxidizers) can feed on gas substrates (Calero & Nikel, 2018; Carneiro et al., 2022; Fan et al., 2025; Geinitz et al., 2020; Liu et al., 2024; Ramírez Rojas et al., 2022).

Sourcing sustainable, low-cost *substrates* is crucial for economic viability. Here, agro-industrial side-streams are increasingly attractive. These are waste or by-products from agriculture and food processing (e.g., molasses from sugar refining, whey from cheese production, fruit pomace, lignocellulosic biomass hydrolysates, crude glycerol from biodiesel, etc.). Using such side-streams as carbon or nitrogen sources can dramatically cut costs and improve the circularity of the process (Ahlborn et al., 2019; Sharma et al., 2022; Sun et al., 2022; Tsafrakidou et al., 2024). For example, fungal mycoprotein production can use starchy waste (Majumder et al., 2024; Tsafrakidou et al., 2024) or sulfitated spent liquor as feed (He & Chen, 2020); several startups are feeding microorganisms with potato peels (Liang et al., 2014) or brewers' spent grain (Lock et al., 2023). Studies indicate that side-stream substrates reduce both cost and environmental footprint, making microbial protein more sustainable (Sharma et al., 2022). However, they introduce variability (seasonal changes in composition and impurities) that must be managed. Researchers emphasize the importance of assessing safety, such as identifying toxins in waste, and performing preprocessing steps, like enzymatic hydrolysis of fibrous waste, to render these streams usable (Raak et al., 2022). Despite the challenges, there are successful cases. For instance, Quorn's mycoprotein utilizes glucose derived from maize starch, which is an agro-refining stream. Additionally, other companies have employed whey permeate to cultivate yeast for protein production (Whittaker et al., 2020). In essence, the chassis and substrate are selected together. An ideal pairing would be a resilient microorganism capable of consuming an inexpensive feedstock (even waste or byproducts) and efficiently producing the desired protein or biomass.

Selecting the right combination of chassis and substrate isn't just about matching metabolic capabilities to affordable substrates. Some challenges need to be addressed, as already pointed. At first, by-products have variable compositions, carbon-to-nitrogen ratios, and concentrations of inhibitory compounds such as phenolics, metals, and other chemicals. This makes process control difficult and may require expensive pretreatment steps that negate cost savings. Secondly, some of the most desirable organisms for metabolic capabilities, such as certain fungi for degrading complex materials or certain bacteria for utilizing gases, often lack advanced genetic tools or safety clearances for food-grade applications.

Thirdly, the assumption that microorganisms can easily adapt to mixed waste materials isn't always accurate. Stress from salt, acid, and nutritional limitations can impair product yield or alter product quality, necessitating genetic approaches or laboratory evolution of improved stability. Lastly, claims of sustainability must be backed by a comprehensive life cycle assessment of the process. While waste utilization appears environmentally friendly, some processes could consume significant energy for cleaning, material degradation, or air addition, potentially negating some of the environmental benefits of using waste materials. Therefore, while there are opportunities to use robust organisms with affordable by-products, success hinges on overcoming several challenges that are often not fully discussed.

Scale-Up Considerations (O_2 Transfer, Mixing, Heat Removal) Scaling fermentation from the lab (shake flasks or 1–10 L bioreactors) to pilot and industrial scales (1000–100,000 L) is notoriously complex. Physical parameters do not scale linearly (Chaudhry, 2024). The biggest issue is often oxygen transfer. Large tanks have a lower surface-to-volume ratio and limited power input per unit volume, making it difficult to supply enough oxygen to a dense culture. Oxygen transfer rate depends on gas flow, agitation speed, sparger design, and medium properties. Even with vigorous stirring and aeration, big fermenters develop *gradients*: zones of lower dissolved O_2 or differing pH.

In fact, oxygen is so critical that insufficient O_2 at scale can turn an aerobic process partly anaerobic, reducing biomass yields or causing the formation of unwanted metabolites (Chen & Liu, 2020; Yang, 2014; Zabalza et al., 2009). For example, a fed-batch yeast process that was optimal in a 5 L vessel might see a 5–10% drop in biomass yield at commercial scale simply due to oxygen transfer and mixing limits. To mitigate this, researchers and technicians increase pressure (to raise O_2 solubility), use pure-oxygen sparging, or improve impeller design; however, there are practical limits, as too much power can cause shear stress or excessive heat (Enfors et al., 2001; Liu et al., 2016).

Mixing and heat removal also become bottlenecks. Fermentation is exothermic (microbial metabolism generates heat), so industrial tanks need cooling systems (jackets or internal coils) to maintain a constant temperature. The cooling duty scales with volume, but the heat transfer area grows more slowly since large tanks risk overheating if not designed with sufficient cooling capacity (Garcia-Ochoa & Gomez, 2005).

Similarly, perfect mixing is impossible on a huge scale; large reactors often have zones with nutrient or pH differences. Impellers of different types (marine impellers, Rushton turbines, etc.) are used in combination to improve circulation. Still, some mixing time of minutes is common in multi-thousand-liter tanks, versus seconds in lab-scale tanks (Amanullah et al., 2001; Larsson & Törnkvist, 1996). This can result in cells cycling between feast and famine conditions or between varying pH environments. Consequently, scale-dependent effects like these make it challenging to maintain identical conditions as in laboratory settings (Larsson & Törnkvist, 1996; Xu et al., 1999).

Engineers employ dimensionless parameters such as power-to-volume, tip speed, Reynolds number, and kLa to achieve scale similarity. However, compromises are inevitable. For instance, to maintain a constant kLa, one might increase airflow and impeller RPM at scale. However, this could result in more foam or increased shear. Consequently, scale-up is often an iterative process, with some loss of yield or productivity accepted. Additionally, strains may need to be adapted to large-scale conditions. (Badino et al., 2001; Wehrs et al., 2019).

In essence, modern plant-based fermentations leverage a rich history of fermentation science, spanning from brewing to biotechnology, but apply it in novel ways. They utilize diverse feedstocks, strive for both food functionality and efficiency, and strike a balance between biology and engineering constraints. By comprehending the fundamentals, such as pH control, dissolved oxygen (DO), feed composition, foam management, and the trade-offs between biomass production and product yield, producers can design robust fermentation processes that align with both nutritional objectives and cost targets. Continuous innovation, such as employing waste feedstocks or optimizing bioreactor designs, is driving these processes towards greater sustainability and scalability, thereby contributing to the alternative protein revolution.

10.3 Automation Architecture and Control Layers

Hardware Stack in Food-Grade Fermentation Industrial fermentation systems for food operate with a hierarchy of hardware devices to monitor and control the process. At the field level, a network of sensors and actuators is installed on the bioreactor and associated equipment. As shown in Fig. 10.1, typical sensors include: temperature probes, pH electrodes, dissolved oxygen (DO) probes, foam and level sensors, pressure transmitters, and possibly turbidity or biomass sensors (e.g., capacitance probes). These provide real-time data on the state of the culture.

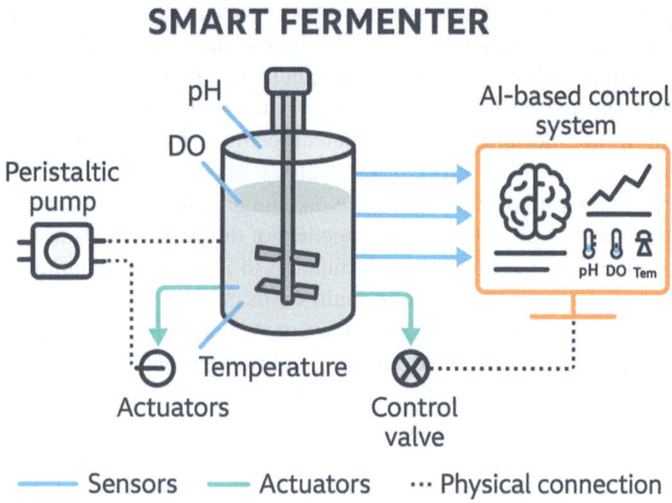

Fig. 10.1 Schematic representation of a smart fermenter, featuring a fermentation tank equipped with sensors (light blue) and actuators (green), connected to an AI-powered control system (orange). Blue arrows depict the data flow from sensors to the AI system, while green arrows represent control commands from the AI to actuators. Dashed lines illustrate the physical connections between the tank and its components, highlighting their integration within the bioreactor. (Source: From Authors)

Actuators include: pumps (for adding feed, acid/base, antifoam, etc.); motor drives for the agitator; valves for controlling gas flow or liquid addition, and heating/cooling elements (like steam valves or coolant pumps for the jacket) (Wang et al., 2020; Yee et al., 2025).

The entire vessel often sits in a skid with integrated valves and piping enabling media addition, harvest, and CIP/SIP (Cleaning/Sterilization in Place). CIP systems, for example, consist of dedicated pumps, heaters, and spray balls that automatically flush the tank with cleaning solutions and rinse it. After cleaning, SIP (Steam-In-Place) sterilizes the vessel by circulating pressurized steam, often through the jacket or via a steam inlet, to achieve aseptic conditions. All these hardware components must be food-grade (stainless steel, sanitary design) and often are automated to minimize human contact with the sterile process (Gopalakrishnan et al., 2024).

The hardware stack typically is organized around one or more PLCs (Programmable Logic Controllers) or a DCS (Distributed Control System). The PLC reads all sensor signals (usually 4–20 mA analog signals or digital inputs) and drives the actuators. It executes low-level control loops, such as proportional-integral-derivative (PID) loops, that keep temperature, pH, or DO at setpoint by manipulating actuators (e.g., adjusting heating/cooling flow, acid/base pumping, or aeration rate and stirrer speed). These are often called field control loops—they happen continuously and autonomously for tight regulation (Wang et al., 2020; Yee et al., 2025).

Above the PLC is the SCADA (Supervisory Control and Data Acquisition) layer, which, in modern systems, is often PC-based HMI (Human-Machine Interface) software. For example, a brewery might use a SCADA system in which operators can view the fermenter status graphically and change setpoints or start/stop pumps. SCADA oversees multiple PLCs and provides coordination and data logging. It is responsible for running *recipes* in batch processes: a recipe defines the sequence of steps for a fermentation (e.g., sterilize → inoculate → ramp temperature → hold → feed start → etc.) (Kakes, 2008; Pasic et al., 2019).

Using the ISA-88 batch control standard, a recipe can be programmed as phases and operations that the control system executes in sequence (De Minicis et al., 2014; Slaugenhaupt, 2025). The SCADA triggers the appropriate PLC functions at each step (for instance, "Phase 1: aerate at 1 vvm and stir 200 RPM until DO >80%, then proceed"). In effect, SCADA is the coordination layer that ensures all field devices act in concert with the process schedule.

Above SCADA, many food biotech facilities now employ a MES (Manufacturing Execution System) or similar production management software (especially if operating under GMP or large scale). The MES handles higher-level scheduling, electronic batch records, raw material inventory, and performance tracking. For fermentation, MES might store the master recipe and issue SCADA commands to initiate a batch. It also receives data (or, more often, queries a data historian) to verify that the batch followed the recipe and that critical parameters stayed in spec (Chen et al., 2020; Chen & Voigt, 2020). The MES and SCADA interface follows the ISA-95 guidelines, which provide a common vocabulary and data model for integrating enterprise systems with control systems.

In ISA-95 terms, Level 2 is SCADA/HMI control, Level 3 is MES, and Level 4 is ERP (enterprise resource planning). The MES ensures that production orders from ERP (like "make X tons of product") are executed on the plant floor and that all data (inputs, process conditions, outputs) are recorded and associated with the batch (Almada-Lobo, 2015; Chen, 2005; Hood, 2015) (Almada-Lobo, 2015; Chen, 2005; Hood, 2015). Automation and control architecture can be visualized on Fig. 10.2.

Historian and Data Links A crucial component of modern fermentation automation is the data historian, a specialized time-series database that continuously records process data from the PLC/SCADA layer. Systems like OSIsoft PI or AVEVA Historian are commonly used; they can store hundreds of tags (such as temperatures, pH, and motor speeds) at high frequency over weeks or even years. This allows for easy retrieval of trends, generation of batch reports, and provision of data for troubleshooting or optimization. The historian often serves as a bridge between SCADA and MES systems. For instance, the MES might query the historian to populate an electronic batch record with the actual temperature profile or DO readings for a batch (Vlassides et al., 2001; Yee et al., 2025).

High data integrity is crucial here, especially for regulated products. Auditors expect that process data is secure, time-stamped, and not alterable, in accordance with the ALCOA+ principles (Attributable, Legible, Contemporaneous, Original,

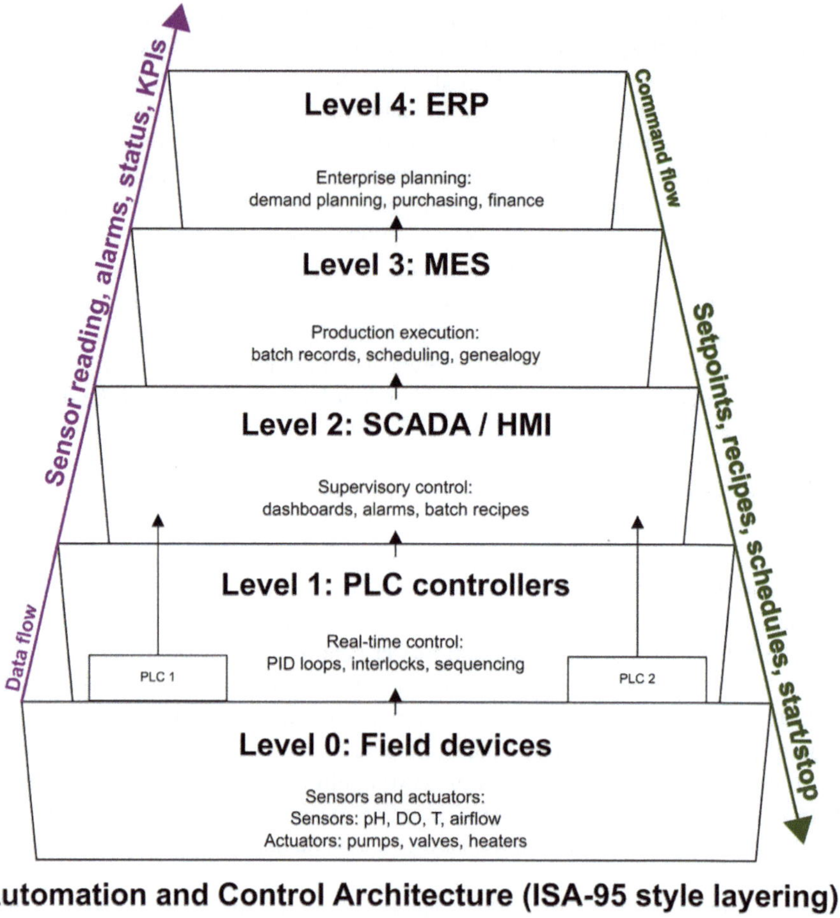

Fig. 10.2 Automation and control architecture organized according to ISA-95 style layering. The pyramid summarizes responsibilities from the physical process interface to enterprise planning: Level 0 (field devices, including sensors such as pH, dissolved oxygen, temperature, and airflow, and actuators such as pumps, valves, and heaters), Level 1 (PLC controllers executing real-time control, PID loops, interlocks, and sequencing), Level 2 (SCADA/HMI providing supervisory control via dashboards, alarms, and batch recipes), Level 3 (MES coordinating production execution through batch records, scheduling, and genealogy), and Level 4 (ERP supporting enterprise planning such as demand planning, purchasing, and finance). Upward arrows signify data flow from the process to higher layers, such as sensor readings, alarms, status, and KPIs. Conversely, downward arrows indicate command flow from higher layers to the process, including setpoints, recipes, schedules, and start/stop commands. The presence of PLC 1 and PLC 2 elements suggests that a SCADA/HMI layer can oversee multiple PLC units within the same production environment. (Source: From Authors)

Accurate, Complete, Consistent, Enduring, Available) for data integrity. That is why historians have security, audit trails, and are often validated systems. They ensure any value (like pH = 6.5 at 10:00:00) is stored with its timestamp and cannot be tampered with, which supports compliance with 21 CFR Part 11 in food/pharma settings (Alosert et al., 2022; Gokulakrishnan & Venkataraman, 2025; Leal et al., 2021; Raja et al., 2024; Rattan, 2018).

Edge vs. Cloud in Fermentation Control With the rise of Industrial IoT, some tasks in fermentation automation have moved to the cloud, but critical control remains at the *edge* (Adeleke et al., 2023). The edge refers to on-premises controllers and computers (PLCs, SCADA PCs, on-site servers) that have a direct, low-latency connection to the equipment. They handle real-time control decisions; for example, a PID loop that opens an air valve runs on a PLC and cannot depend on internet connectivity. Edge computing excels in low latency and data privacy by processing data locally (Chen et al., 2018; Wu et al., 2025). On the other hand, the cloud is used for higher-level data analysis, machine learning model training, long-term data storage, and multi-site integration. For instance, large fermentation companies might aggregate data from plants around the world to a cloud database to compare performance.

Cloud platforms can run intensive analytics (e.g., predictive modeling or feed strategy ML models) and send recommendations back to the site (Lee et al., 2025). However, the cloud is not trusted for anything that requires a millisecond response or could impact safety if delayed. In practice, a hybrid approach is common: real-time control and safety interlocks at the edge, and analytics and big data processing in the cloud (Gasset et al., 2024; Ma et al., 2019). For example, sensor readings might be streamed via protocols like MQTT to a cloud dashboard for engineers, but the actual control logic that adjusts a feed pump runs locally on the PLC. This ensures reliability. If the network fails, the fermenter remains operational. As some studies suggest, edge computing reduces latency and enhances privacy, while cloud computing provides scalability and substantial computing power (Borgosz & Dikicioglu, 2024; Jain & Yadav, 2025; Yee et al., 2025).

Control Hierarchy and Batch Operation The control hierarchy can be summarized in layers: Field devices (Level 0/1) → PLC control (Level 1/2) → SCADA supervision (Level 2) → MES coordination (Level 3) → ERP/business (Level 4). In the context of running a fermentation batch, an operator (or MES) initiates a recipe run in SCADA. SCADA then coordinates the sequential steps of the process. For each step, SCADA may invoke predefined control modules within the PLC (Chen et al., 2020; Cinar et al., 2003; Jain & Yadav, 2025). For instance, a "sterilization phase" might instruct the PLC to gradually increase the jacket steam valve to 121 °C and maintain that temperature for 30 min. The PLC executes this process using its PID loops and timers, and then reports the status back.

Once sterilization is complete, SCADA transitions to the "inoculation phase." In this phase, an operator may be prompted (via HMI) to connect the inoculum and open an air valve. Alternatively, in fully automated plants, a valve opens automatically from a seed tank. Throughout the process, the batch context is maintained. The historian tags all data with the batch ID, and any deviations (such as alarms indicating "pH out of range") are logged in batch records. This hierarchical structure ensures responsiveness at the low level (PLCs reacting within seconds) and traceability at the high level (MES tracking the batch from start to finish). Additionally, this setup enables scalability. A single SCADA/MES system can supervise multiple fermenters (each equipped with its own PLC), allowing for the orchestration of parallel or staggered batches to optimize utilization (Koulouris et al., 2021; Udugama et al., 2023a).

Finally, modern fermentation suites often incorporate OPC UA (Open Platform Communications—Unified Architecture) and/or MQTT for connectivity between layers (Masykur et al., 2020). OPC UA is an interoperability standard that allows PLC/SCADA data to be exposed uniformly to higher-level software—it has used to connect different vendors' equipment and to feed data to MES or historians in real time. MQTT is a lightweight publish/subscribe protocol popular in IoT; some plants use it to stream sensor data to cloud services or remote monitoring dashboards. Both aim to decouple systems and simplify integration. Instead of hard-coding a custom link between a fermenter and a database, a generic OPC UA server can serve all tag values, which any client (MES, historian, etc.) can read with standard calls (Adeleke et al., 2023; Braun, 2023; Pandey et al., 2025). This adherence to standards like ISA-95 and OPC UA ensures that as the plant introduces new equipment or software, they communicate effectively, simplifying and enhancing integration.

In essence, the automation architecture in contemporary fermentation facilities is layered. It comprises robust local control systems for safety and quality, overseen by supervisory systems that ensure recipes are executed as intended. Simultaneously, these systems collect extensive data. This architecture not only maintains process control (temperature, pH, etc.) but also establishes the groundwork for advanced capabilities such as electronic batch records, remote monitoring, and future plant-wide optimization.

10.4 Process Analytical Technology and Real-Time Sensing

Real-Time Sensing Overview Fermentation, a living process, demands continuous monitoring to ensure optimal performance and quality. Traditional fermentations relied on offline samples, such as daily HPLC analyses for nutrient or product levels. However, modern smart fermentation employs a comprehensive suite of in-line and on-line sensors that measure critical parameters in real time. In-line sensors reside directly within the process stream, like probes in fermenters, while on-line sensors typically involve automatic samplers that take a slipstream of broth to an analyzer. These sensors transmit data to control systems, enabling immediate adjustments (feedback control) or at least issuing alerts if deviations occur (Gerzon et al., 2022; Gomes et al., 2014).

Classic Inline Sensors—pH, DO, Temperature Every bioreactor is equipped with pH and dissolved oxygen (DO) probes inserted through ports. pH is crucial for enzyme function and nutrient uptake, and even slight deviations can stress cells or alter product properties. Modern pH sensors are usually gel-filled or solid-state electrodes that can withstand sterilization. However, they require regular calibration with standard buffers because they can drift due to fouling (such as protein coatings) and aging (Reyes et al., 2022). Do probes have historically been polarographic (Clark-type) sensors with an oxygen-permeable membrane. Newer optical DO sensors (luminescent) are also commonly used. Dissolved oxygen (DO) indicates the amount of oxygen dissolved in the broth. A drop in DO may signal that the oxygen uptake by microorganisms exceeds the available supply (Urriza-Arsuaga et al., 2019). Controllers often cascade DO control to adjust airflow or stir speed when the dissolved oxygen (DO) falls below the setpoint. For instance, to maintain DO at 30% saturation, controllers can increase airflow or add pure oxygen. These fundamental sensors, such as pH, dissolved oxygen (DO), and temperature, are robust workhorses that are widely used and typically very reliable when properly maintained. They are usually connected to transmitter units that perform signal conditioning and transmit readings to the control system (Zhou et al., 2022).

Off-Gas Analysis (CO_2/O_2) In addition to measuring inside the broth, analyzing the *exhaust gas* from the fermenter is a powerful PAT tool. Off-gas analyzers measure % O_2 and %CO_2 leaving the reactor (Hüttel & Müller, 2012). From these measurements, we can calculate the OUR (oxygen uptake rate) and CPR (carbon dioxide production rate) of the culture. Essentially, if we know the amount of O_2 we provide and the amount that is not consumed in the exhaust, the difference represents the amount of oxygen used by the cells. Similarly, the CO_2 output indicates the metabolic rate. These metrics are directly related to cell growth and substrate consumption (Klein et al., 2012).

For instance, a surge in CO_2 evolution often signifies the peak of exponential growth, while a sudden decline may indicate carbon starvation or the onset of the stationary phase. Consequently, off-gas analysis serves as a reliable proxy for biomass and growth rate. Moreover, it proves highly instrumental in process control. Numerous systems employ respiratory quotient (RQ) monitoring, defined as CO_2 produced divided by O_2 consumed. Any deviation from the expected RQ value, particularly when it exceeds 1 or falls below 1, suggests a change in metabolism, potentially indicating overflow metabolism or oxygen limitation.

In practical terms, off-gas analysis is continuous and noninvasive. Techniques employed include infrared sensors for CO_2, which detect its absorption bands, and paramagnetic or electrochemical cells for O_2. Off-gas data is often analyzed in conjunction with other parameters, making it particularly prevalent in the pharmaceutical industry and high-value fermentations (Goh et al., 2019; Schulte et al., 2025). It enables real-time yield estimation, which measures the efficiency of substrate conversion into biomass or products. Additionally, it can detect issues like a failing DO probe. For instance, if the DO probe indicates oxygen levels are normal, but off-gas analysis reveals no oxygen consumption, it suggests an underlying problem. Overall, off-gas analysis has become a crucial PAT tool for achieving precise control and comprehending fermentation kinetics.

Biomass Monitoring—Capacitance and Optical Sensors Cell density, or biomass concentration, is a crucial variable in any fermentation process. Traditionally, it has been measured offline using optical density (OD) or cell dry weight (Kiviharju et al., 2008). However, advancements in technology have enabled the measurement of cell density inline using dielectric spectroscopy sensors, commonly referred to as capacitance probes (Yardley et al., 2000). These sensors measure the permittivity of the broth at radio frequencies. Live cells, owing to their intact membranes, function as tiny capacitors, allowing for accurate cell density measurement. The measured capacitance correlates with viable cell volume. Capacitance biomass sensors (e.g., the ABER probe or Hamilton Incyte) are inserted into the vessel and provide a real-time trace of live-cell concentration (Alfano et al., 2020; Domján et al., 2022). They are handy for processes where biomass is the product (e.g., yeast propagation) or for timing induction/feeding based on growth phase.

One advantage is that, unlike optical turbidity, capacitance sees only live (polarized) cells, not debris or dead cells. It is also insensitive to gas bubbles and medium color. The readings do need calibration to cell dry weight or OD units for absolute values, but once calibrated, they offer a safe, robust way to track growth continuously (Metze et al., 2019; Morris et al., 2021). Some fermenters also use optical methods: simple near-infrared absorption or scattering sensors can track turbidity (optical density) online, though bubbles and suspensions can interfere with measurements. For example, an optical OD probe might shine NIR light and measure scatter to estimate cell density. These can be simpler but often less precise than capacitance for dense cultures (Ortega-Platas et al., 2021).

Spectroscopic Probes (NIR, Raman, FTIR) One significant advance in PAT is the use of spectroscopy to monitor broth chemistry in real time. *Near-Infrared (NIR) spectroscopy* and *Mid-Infrared (IR, including FTIR)* probes can be inserted or connected via flow cells. They measure specific wavelength absorbances that correlate to the concentrations of compounds such as sugars, organic acids, alcohol, and so on (Sales et al., 2015). Raman spectroscopy employs a laser to detect inelastically scattered light, enabling the identification of molecular "fingerprints" in various broth components. The advantage lies in the ability of a single probe to simultaneously monitor multiple nutrients and products without the need for sample withdrawal.

For instance, Raman or Near-Infrared (NIR) spectroscopy can be calibrated using chemometric models to quantify glucose, acetate, ethanol, and biomass based on their unique spectral signatures (Li et al., 2018; Lourenço et al., 2012; Ortega-Platas et al., 2021). Infrared (IR) spectroscopy is suitable for analyzing a wide range of metabolite groups, while Raman spectroscopy provides greater specificity for particular compounds. For instance, companies have employed FTIR spectroscopy to regulate glucose feeding in yeast fermentation processes, and Raman spectroscopy to monitor product titer in biotech applications. However, there are challenges to consider. The opacity of the broth can interfere with signals, as Raman's laser can be scattered by cells or particles. Additionally, the probes used in Raman spectroscopy are expensive and require robust calibration models.

Despite these challenges, advancements in these spectroscopic PAT tools are noteworthy. For instance, Raman spectroscopy has even been employed to detect the metabolic state of an individual cell in certain research studies (Classen et al., 2022; Karnachoriti et al., 2025). Many systems employ fiber optic probes. The spectrometer can be placed outside, while light is delivered through fiber optics, which offers enhanced sterility and facilitates easier integration (Metze et al., 2019). Spectroscopic PAT offers a real-time glimpse into the broth's chemical composition, a feat previously limited to lab assays. These tools facilitate closed-loop control, such as adjusting base addition when IR detects lactic acid accumulation or activating a feed pump when Raman indicates substrate depletion (Joanofarc et al., 2019; Rómoli et al., 2017).

Electronic Noses and Tongues (Flavor Monitoring) Fermentation processes in food often generate complex flavor and aroma compounds. Traditionally, assessing flavor (e.g., in a plant-based cheese ferment or a kimchi ferment) required human sensory panels or GC-MS analysis post-processing (Huang et al., 2015; Lee et al., 2024; Short et al., 2021). Now, electronic noses (E-noses) and electronic tongues (E-tongues) are emerging as PAT devices for quantifying flavor and taste volatiles in real time. An *E-nose* is essentially an array of gas sensors (conductive polymer sensors, MOSFET sensors, etc.) that respond to volatile compounds. In fermentation, an e-nose can be hooked to the fermenter off-gas line or headspace (Gliszczyńska-Świgło & Chmielewski, 2016). For example, some miso and natto producers pipe exhaust gas through an E-nose to track key aroma "milestones" and determine optimal fermentation endpoints (Rudnitskaya & Legin, 2008; Seesaard & Wongchoosuk, 2022). It is like having a consistent digital "sniffer" instead of relying solely on human smell. Similarly, an *E-tongue* has an array of chemical sensors (for pH, salt, bitter, sweet, and umami compounds) to mimic taste (Baldwin et al., 2011; Preedy, 2016; Tibaduiza et al., 2024). Some cases reported that an E-tongue successfully tracked sourness, sweetness, and umami development during kimchi fermentation, with readings matching human panel scores for ripeness. This allowed the producer to signal optimal ripeness without manual tasting (Lee et al., 2024; Moon et al., 2025). These biosensors often use AI pattern recognition of sensor array outputs to correlate with flavor profiles or quality metrics (Ciosek & Wróblewski, 2006; Singh et al., 2022). For instance, a trained model might associate a pattern of volatile signals with a "fruity aroma" score or with the concentration of diacetyl (a buttery flavor compound) (Tian et al., 2020). While still an evolving tech, E-noses and E-tongues in fermentation offer a way to monitor sensory quality *in-line*, which is extremely valuable for food applications (where taste is king). As illustrated in Fig. 10.3, they effectively digitize the senses, enabling automated decisions, such as ending fermentation when a desired flavor profile is achieved.

Microfluidics and Automated Sampling Some measurements still require actual wet-chemistry or biological assays that a probe cannot perform. For these, online analyzers with automated sampling are used. A common example is a bio monitor

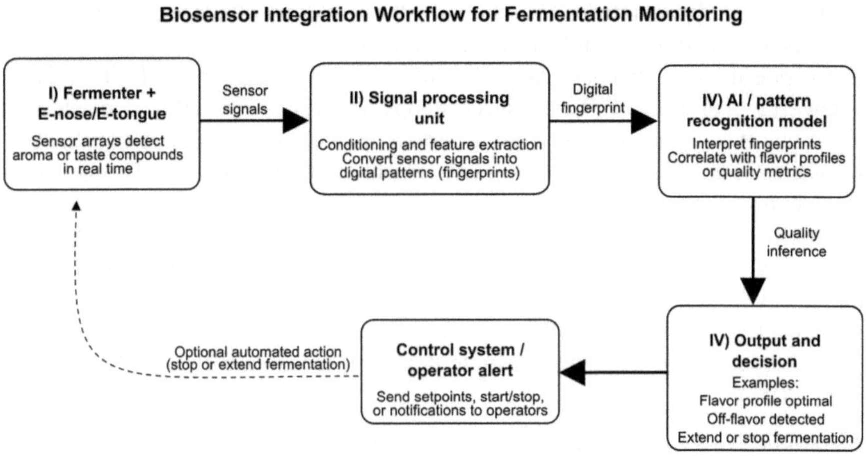

Fig. 10.3 Flowchart of biosensor integration in fermenter monitoring using electronic nose and electronic tongue sensor arrays. Volatile and taste-related compounds are detected in real time in the fermenter and converted into sensor signals, which are conditioned and transformed by a signal processing unit into digital patterns ("fingerprints"). These fingerprints are interpreted by an AI or pattern-recognition model that correlates the sensor signature with flavor profiles or quality metrics. The resulting quality inference drives an output decision (for example, "flavor profile optimal" or "off-flavor detected"), which can trigger an operator alert or feed into the control system to adjust setpoints or initiate start/stop actions. The dashed feedback loop indicates an optional automated intervention, such as stopping or extending the fermentation when a target aroma or taste profile is reached. (Source: From Authors)

for glucose or other nutrients: a microfluidic device draws a small stream of broth through a sterile filter or loop, then mixes it with reagents to measure (e.g., an enzymatic glucose assay calorimetrically or via an electrochemical biosensor) (Buchenauer et al., 2009; Heins et al., 2021; Rodrigues et al., 2024; Talaei et al., 2010). These units (e.g., YSI or flow injection analyzers) can be connected to fermenters to provide periodic readings of glucose, lactate, ammonia, and other compounds. (Qiu et al., 2014; Trojanowicz & Kołacińska, 2016).

Newer developments use *microfluidic chips* and MEMS sensors to miniaturize this. For instance, microfluidic impedance cytometry chips can count cells by flowing them through tiny channels with electrodes, counting thousands of cells per second and distinguishing by size, etc., all in a chip a few cm in size. Such a chip could be part of a loop that continuously samples the culture and reports cell count or even viability (if combined with a stain) (Ho et al., 2019; Toledo et al., 2018). Another example: microfluidic devices for *online PCR or biosensing*. There are now portable PCR machines that can take a sample of the ferment and run a quick qPCR to quantify a particular microorganism or gene (used in traditional food fermentations to monitor contamination or specific bacterial growth) (Lee, 2018). Though not continuous in the same way as a pH probe, these online analysis loops (triggered, say, every few hours) greatly enhance insight. They often feed their results to the historian or control system, just like an inline probe would.

Implementing automated sampling requires careful design to maintain sterility (steam-sterilizable sampling valves, sterile filters, etc.) and to avoid drawing too large a volume (so as not to disturb the culture significantly) (Rodrigues et al., 2024). Many high-end bioreactors now come with auto sampling systems that can periodically withdraw small samples and send them to various analyzers or even to a fraction collector for later lab analysis. This reduces the need for manual sampling and provides data with minimal delay (Hofer et al., 2020).

Calibration and Maintenance All these sensors are only helpful if they are accurate and reliable throughout a fermentation campaign, so calibration and maintenance protocols are a central part of PAT (Arroyo, 2014; Biechele et al., 2015). Drift is a common issue: for example, DO probes can drift due to membrane fouling or electrolyte depletion, and pH probes often drift due to glass and reference junction problems, especially after repeated sterilization cycles (Arroyo, 2014; Demuth et al., 2016). Spectroscopic baselines can also shift over time if window coatings, bubbles, or biofilms form at the optical interface (Mazarevica et al., 2004; Schalk et al., 2019). A strong calibration program, therefore, involves regular pre- and post-batch calibration checks.

For pH, a two-point calibration with standard buffers (typically pH 4 and 7) before each run, or at least at regular intervals, is recommended, with offline checks to quantify drift (Demuth et al., 2016; Schneider et al., 2019). DO probes are usually calibrated at 0% and 100% saturation (e.g., sodium sulfite and air-saturated water) before inoculation, and off-gas analyzers require zero and span calibration with reference gases (Arroyo, 2014). For spectroscopic sensors, robust chemometric calibration models (PLS/MLR) are built using lab reference data, and these models need periodic validation or updating when the broth matrix changes significantly (Mazarevica et al., 2004; Pollard et al., 2001; Schalk et al., 2019).

To manage fouling, many processes include cleaning cycles for sensors. Some inline sensors (e.g., turbidity probes) might have a built-in ultrasonic or mechanical wiper cleaning device. Others rely on CIP cleaning of the whole vessel. In long-running fermentations, sensor drift and biofouling can be significant. A PAT system might incorporate redundant sensors or "soft sensors" (models) to detect when a sensor is likely drifting by cross-checking (e.g., if off-gas CO_2 suggests high activity but the pH sensor shows no change when it should, maybe the pH sensor is fouled) (Brunner et al., 2019, 2021; Luttmann et al., 2012; Pérez et al., 2020). In regulated environments, traceability of sensor calibration is mandated, and every probe must be tagged with its calibration records, and any adjustments must be documented (in accordance with ALCOA data principles). Automated calibration stations exist (for example, some systems automatically pull the pH probe into a calibration chamber with buffer during a cleaning cycle). However, in food fermentation, simpler approaches such as manual calibration are common across batches (Uchil et al., 2025).

In summary, the modern PAT toolbox for fermentation is comprehensive. We have our eyes (spectroscopes), nose (e-nose), tongue (e-tongue), ears (possibly vibration or foam sensors), and hands (autosamplers) constantly monitoring the process. This shift from empirical, once-a-day monitoring to adaptive, minute-by-minute control allows us to make real-time adjustments. For instance, we can tune the feed rate on the fly if a sensor indicates nutrient depletion, or we can stop a batch early if an undesired metabolite rises. Implementing PAT requires investment and expertise (chemometric models and sensor maintenance programs), but it pays off by enhancing consistency, yield, and quality. In the food industry, this translates into reliable product flavor and safety in every batch.

10.5 Data Infrastructure for Smart Fermentation

Data Models and Time-Series Pipelines In a smart fermentation plant, data is as crucial an output as the product itself. Every sensor reading, pump status, or alarm contributes to a time-series dataset that can be analyzed and optimized (Alosert et al., 2022; Vlassides et al., 2001). A data historian, such as OSIsoft PI, Wonderware Historian, or similar software, is usually the central component of the data infrastructure. These historians collect time-stamped values from the control system at high frequency. They are specifically designed to handle time-series data, efficiently storing millions of points. Additionally, they provide query tools that allow users to retrieve trends based on batches or equipment (Yang & Ierapetritou, 2024). For practical analysis, the data must be well-modeled: using consistent tag naming conventions and contextualization. This is where standards like ISA-88 and ISA-95 help. ISA-88 provides a batch context model—e.g., it defines that we should tag data with Batch ID, step, phase, etc., which allows slicing data by batch (De Minicis et al., 2014; Slaugenhaupt, 2025). ISA-95 provides the hierarchical model (Enterprise → Site → Area → Process Cell → Unit → etc.) so each tag can be categorized (e.g. Unit = Fermentor-7, Parameter = pH) (Alosert et al., 2022). By aligning the historian data structure with ISA-95/88, integration with MES and enterprise systems becomes plug-and-play—they share a common vocabulary for equipment and process stages (Mitra & Murthy, 2021; Slaugenhaupt, 2025). For instance, an MES recipe might ask the historian: “Give me all Temperature data for Batch X during the fermentation phase,” and if the data model is standardized, that query is straightforward (Hawkins et al., 2011).

The Data Pipeline Usually Goes PLC/SCADA generates data → OPC UA or MQTT brokers transfer it → Historian stores it. Modern historians often support OPC UA (Open Platform Communications Unified Architecture) natively to collect data from many PLCs in a vendor-agnostic way (Nicolae & Korodi, 2018). In some architectures, MQTT is used: e.g., each sensor’s data can be published to an MQTT broker, which a cloud service subscribes to for streaming analytics (Masykur et al., 2020). However, for critical process control, plants keep at least one on-premises historian or buffer so that data is not lost if the cloud connection fails.

An important aspect is data enrichment with metadata. Smart data infrastructure goes beyond logging a simple "value = 6.8". It also captures contextual information such as the batch, product, calibration status, and so on. The objective is to create reusable machine-learning datasets that are rich in metadata (Mione et al., 2025; Moreddu, 2025). If we later want to train a model to predict yield, we would need not only the time series of process variables but also metadata such as the strain used, the inoculum age, the tank ID (since differences in tank geometry might be relevant), and the lot numbers of raw materials (since some lots might be subpar) (Fisher et al., 2022). Systematically capturing this metadata, often in the MES or batch records, and linking it to the sensor data is crucial. One approach is to utilize ISA-95's equipment hierarchy and ISA-88's batch records, ensuring that each data point can be traced to a specific context, such as Fermenter A, Batch #123, Product = ProteinX, Strain = ABC, and so on (Hawkins et al., 2011). This enables analysis, allowing for the comparison of batches and the extraction of features per batch (e.g., maximum OUR, time to glucose depletion, etc., for each batch).

Interoperability and Simplicity Interoperability refers to the ability of different systems, developed by various vendors or designed for distinct purposes, to share and comprehend data (Hill, 2020; Isoko et al., 2024). Using standard protocols is crucial. As mentioned, OPC UA is widely used for linking control systems to historians and higher software because it offers a universal data model with security (Jarva, 2024). In newer IoT setups, MQTT with JSON payloads might be used to send data to the cloud or to edge aggregators; it is lightweight and works well over unreliable networks (Hazra et al., 2021). The overarching idea is to avoid custom, siloed integrations, as those are brittle and hard to maintain. Instead, a fermentation plant might have all controllers expose data via OPC UA, the MES subscribes to needed tags via OPC, and a cloud analytics platform might tap into the historian's REST API or MQTT feed. This decoupling means it is possible to add a new analyzer or a digital twin software and just point it to the same standard interfaces (Kinnunen, 2024; Ladegourdie & Kua, 2022; Li et al., 2021; Llamuca et al., 2020; Sai Guptha et al., 2025).

Standards like ISA-95 not only guide structure but also define some common terminology (equipment IDs, material definitions, etc.). That helps when integrating with ERP (Enterprise Resource Planning—for example, an ERP might send a work order to MES saying "produce 1000L of Product-X using Recipe-Y"—if everything follows ISA-95, the MES knows which process cell to run Recipe-Y on, etc.) (Li et al., 2021; Llamuca et al., 2020). Simplicity also comes from using modern middleware—for instance, OPC UA can eliminate much of the custom driver coding that older OPC DA (Open Platform Communications Data Access) or proprietary protocols required. It is designed with built-in security (encryption, authentication) and a unified address space, which simplifies the maintenance of robust integrations (Schwarz & Börcsök, 2013). Similarly, MQTT is simple in concept (pub-sub), meaning adding a new data consumer does not strain the PLC (the broker handles it). Many IIoT architectures in fermentation use MQTT to push data to the cloud while the critical local loop remains unaffected (Masykur et al., 2020; Sai Guptha et al., 2025).

Governance and Data Integrity (ALCOA+) In regulated or food-safety-critical fermentation, data governance is indispensable (Gokulakrishnan & Venkataraman, 2025; Rattan, 2018). ALCOA+ principles (Attributable, Legible, Contemporaneous, Original, Accurate + Complete, Consistent, Enduring, Available) provide a checklist for data management (Alosert et al., 2022; Gokulakrishnan & Venkataraman, 2025; Leal et al., 2021; Rattan, 2018). In practice, this means: every data point is Attributable (we know the source sensor and who/what system recorded it), Contemporaneous (timestamped at the time of generation, not entered later in bulk), Original (we store raw values, or if transformed, we keep raw too), and Accurate (calibrations applied, checked against standards) (Raja et al., 2024; Uchil et al., 2025). The "Plus" elements include Complete (no cherry-picking of data; e.g., even if a sensor fails, that event is logged, not just deleting the insufficient data), Consistent (using the same units, formats, and not having contradictory duplicates), Enduring (data is kept through its retention period in a durable form), and Available (readily accessible for review or audit) (Alosert et al., 2022; Isoko et al., 2024; Leal et al., 2021).

For a sophisticated fermentation setup, the historian/MES system must have audit trails that record who made changes to a setpoint and when. It should also have secure timestamps, often synchronized with an atomic clock or NTP for accuracy. Additionally, user controls should be implemented to prevent unauthorized data modifications. Furthermore, backup strategies should be in place, such as real-time replication of historian data to an offsite server, ensuring data durability and availability even in the event of hardware failures (Aparnna et al., 2025; Short et al., 2024). In GMP manufacturing, such as enzyme production for food or supplements, electronic records must meet the trustworthiness standards set by FDA/EMA inspectors. Consequently, a well-governed data infrastructure should incorporate features like electronic signatures for critical actions (as per 21 CFR Part 11), role-based access control, and validation of spreadsheets or algorithms used for release decisions (Sahana, 2025; Short et al., 2024; Uchil et al., 2025).

For instance, if a machine learning model predicts titer, and that prediction could influence batch release, the data fed into the model and its output should be managed through a controlled process with an audit trail. While food fermentation for non-pharmaceutical purposes may not be as stringent, companies still prioritize data integrity, especially when supporting claims like an alternative protein with specific nutritional content. This is to prevent any quality issues that could lead to recalls.

Building Reusable Datasets for ML (with Batch-Aware Splits) With well-equipped instrumentation and historians, plants accumulate vast amounts of data, potentially spanning terabytes of fermentation data over several years. The subsequent step involves transforming this data into actionable intelligence, often through the use of machine learning (ML) or advanced analytics (Fisher et al., 2022; Gangadharan et al., 2021; Nguyen, 2025). To do this effectively, one must curate datasets that include not just sensor readings but all relevant context (metadata, as mentioned). For instance, a dataset for yield prediction might combine time-series (e.g., DO trend, feeding profile) with static information such as strain type, inoculum size, and batch duration (Li et al., 2023; Vlassides et al., 2001). Data engineers typically build ETL (extract, transform, load) pipelines that move data from historians or MES systems into analytics environments or data lakes, where batch records, time-series, and outcomes can be joined in a consistent schema (Dessaigne et al., 2024; Moreddu, 2025). Because each fermentation batch is an independent run, it is a natural unit for analysis; a batch-aware split means that ML models are trained and tested on different sets of whole batches (for example, 80% of past batches for training and 20% for testing) rather than on random time points, which helps prevent temporal information leakage and overly optimistic performance estimates (Gangadharan et al., 2021; Li et al., 2023). This is crucial because time points within a batch are strongly correlated, and without batch-aware splits, a model may implicitly memorize individual trajectories instead of learning generalizable patterns.

With rich metadata, one can also perform stratified splits, for instance, ensuring that the test set includes batches from different seasons, fermentors, or production lines, or that it contains examples from each strain when the goal is to test robustness across product variants (Mione et al., 2025; Moreddu, 2025). Well-designed data infrastructures make it possible to pose queries such as "return all batches of product X in the last two years with their sensor traces, quality attributes and yields" and to materialize these results as analysis-ready tables, sometimes as a wide table per batch with summary features or as a time-series database joined by batch identifier (Borgosz & Dikicioglu, 2024; Dessaigne et al., 2024). Metadata such as recipe name, raw material lots, operator identifiers, and recorded fermentation anomalies can be attached to each batch record and later reused for feature engineering, for example, to create indicators of raw material changes or flags for process deviations (Adeleke et al., 2023; Mione et al., 2025).

A concrete example is using ML to predict the optimal termination point of a food fermentation based on sensory outcomes. Here, the inputs might be time series of volatile sensor signals, pH, and sugar consumption, while the target is a sensory score generated by a trained panel or an electronic nose or tongue (Baldwin et al., 2011; Rudnitskaya & Legin, 2008; Zhao et al., 2025). If sensory scores are captured in MES and linked to batch identifiers, they can be automatically integrated into the training dataset. Alternatively, they can be imported from external sources and mapped to batches. Once a dataset and model are established, they can be reused and updated as new batches are generated. Ideally, this process should be part of an MLOps workflow that routinely retrains and monitors model performance (Gasset et al., 2024; Moreddu, 2025). The same metadata that facilitates traceability also expedites feature engineering. For instance, it enables the automatic computation of batch-level descriptors like "time to dissolved oxygen drop after feeding," "maximum respiration rate," or "duration within a target temperature window." These computations can be performed at scale when data are archived and queryable (Fisher et al., 2022; Li et al., 2023).

In essence, sophisticated fermentation data infrastructures transcend the mere storage of raw sensor values. They encapsulate these values within an interoperable context, ensuring data integrity and governance. Additionally, they meticulously organize batch histories, making them directly accessible for PAT analytics, machine learning models, and digital twins (Alosert et al., 2022; Isoko et al., 2024; Koulouris et al., 2021). By investing in such infrastructures, fermentation companies convert historical data into a reusable asset for continuous improvement, discovering better operating conditions, detecting subtle shifts such as changes in raw materials or equipment behaviour, and accelerating innovation through simulation and learning from past runs (Chai et al., 2022a; Gopalakrishnan et al., 2024; Zhao et al., 2025).

10.6 Machine Learning and Soft Sensing in Practice

Current Value-Add ML Use Cases In today's fermentation industry, machine learning (ML) and data-driven modeling are being applied to complement traditional process engineering. Some supervised learning tasks that add value include predictive modeling (e.g., forecasting final yield or product quality based on early-run data), soft sensors (estimating hard-to-measure variables in real time from easy measurements), and fault detection/classification (identifying when a fermentation is going off-track or diagnosing sensor faults) (Khaleghi et al., 2021; Wang et al., 2025). For example, an ML model might be trained to predict the remaining fermentation time or the final protein content using patterns in the first 24 h of data. This helps production planning and optimizes harvest timing (Roell et al., 2022). Another supervised case: image analysis via ML for morphology—some fermentations (e.g., fungal or algal) use vision systems to monitor morphology; ML can classify images to determine whether cells are healthy or contamination has occurred (Jeong & Kim, 2025).

Unsupervised learning is also used for tasks such as anomaly detection and multivariate analysis. Principal component analysis (PCA) or autoencoders can reduce the dimensionality of dozens of sensor signals to reveal underlying factors or detect when a batch's trajectory deviates from the normal trajectory cluster (Berahmand et al., 2024; Kwon et al., 2024; Usmani et al., 2022). Clustering algorithms can categorize batches into "high yield" and "low yield" clusters, prompting an investigation into the factors that differentiate them. Unsupervised machine learning is particularly valuable in fermentation because the processes are intricate and often non-linear. These methods can uncover subtle relationships or groupings without the need for an explicit labeled outcome for every data point (Gomersall et al., 1997; Lee et al., 2023).

Soft Sensors (Virtual Sensors) Soft sensors are a big practical win for ML in fermentation. A *soft sensor* is essentially a model (e.g., an ML regression or a simpler empirical model) that infers a difficult or impossible measurement from available data (Hua et al., 2023; Tokuyama et al., 2021). Common targets for soft sensors include product concentration, substrate concentration (e.g., glucose), biomass (if an inline probe is not available), or specific metabolite levels (e.g., acetate or methanol in cultures) (Tang et al., 2022; Wang et al., 2023). They can also estimate quality attributes (like viscosity or flavor profile) in real time. For example, measuring glucose continuously might require an expensive analyzer, but a soft sensor can use other signals (dissolved O_2 trend, feed rate, CO_2 evolution) to predict glucose concentration. Similarly, in precision fermentation making an enzyme, you might not have an on-line measurement of enzyme titer; a soft sensor model could predict titer from fermentation time, DO, pH control additions, and biomass readings, based on historical correlations (Li et al., 2023; Li et al., 2024; Pérez et al., 2020; Sun et al., 2019b; Tang et al., 2022; Wang et al., 2024).

These soft sensors are typically supervised ML models trained on historical batches, where lab measurements of the target were available occasionally (or at least at the end of each batch). They can range from simple linear models to complex neural networks (Mowbray et al., 2022a; Yuan et al., 2019). Modern trends favor the use of deep learning regression models or ensemble methods for improved accuracy in soft sensors. For instance, a random forest might take input features like area under the CO_2 curve, oxygen uptake rate, base addition rate (proxy for acid production) etc., and output an estimate of, say, lactic acid concentration (Rivera et al., 2024).

Soft sensors provide real-time estimations that can be used for control and monitoring. They essentially turn batch data and PAT signals into a new virtual signal that we would otherwise have to wait hours or days to receive from a lab. One caution: data quality and quantity for training soft sensors can be limited—fermentation datasets are often small (tens or hundreds of batches) and can have outliers or missing points (Brunner et al., 2021; Kiviharju et al., 2008; Luttmann et al., 2012; Pérez et al., 2020; Quast et al., 2013). As one paper noted, experimental datasets are sparse and noisy, which can lead to poor generalization if one is not careful. Techniques such as data augmentation (e.g., simulating variations) or hybrid models (combining first-principles with ML) can help (Yao et al., 2025). Indeed, there is research on generating synthetic time series to augment training for soft sensors (Kwon et al., 2024).

Validation Is Critical The model's predictions should be tested on historical data it has not seen (ensuring batch-level separation to avoid leakage) and, ideally, on a few new batches in shadow mode (i.e., not used for control, just for monitoring) to build trust. Once it is proven, it can be integrated into the control strategy (e.g., by feeding a soft sensor value to a PID controller or triggering an alarm) (Brunner et al., 2019; Medl et al., 2023; Pomerantsev & Rodionova, 2021; Sun et al., 2019a).

Drift monitoring is a practical necessity. A model that worked well can degrade if the process changes—e.g., a new strain or recipe might shift the relationships that the model learned. Alternatively, sensor recalibrations might introduce bias (Chandrasekaran et al., 2025). Teams set up drift checks to ensure the accuracy of their models. These checks can be as simple as comparing the soft sensor output to actual lab measurements when they become available. If the error exceeds a threshold, the model is retrained. More sophisticated checks involve monitoring the input feature distributions. If these distributions move outside the range the model was trained on, an out-of-range detection flag is raised, indicating that model predictions may be unreliable. This could happen if, for example, a much higher cell density is achieved than has been seen before (Hema et al., 2022; Kadlec et al., 2009; Kaneko & Funatsu, 2011; Kordon et al., 2003; Lu & Chiang, 2018).

Edge vs cloud deployment decisions revolve around latency and reliability. A soft sensor used for rapid control (e.g., real-time residual substrate control) should reside on the edge, near the PLC, to avoid network latency or outages affecting it (Chen et al., 2018; Wu et al., 2025). Conversely, a predictive model that just gives an end-of-batch estimate can run in the cloud asynchronously (e.g., after getting the first 12 h of data, the cloud service returns a prediction) (Kinnunen, 2024; Lee et al., 2025). Many plants use a hybrid: e.g., data is buffered locally and streamed to the cloud; a cloud ML service computes something and sends the results back down to an edge device, which interfaces with control. However, for high-frequency applications, it is usually implemented on an on-site computer (Sai Guptha et al., 2025; Vlassides et al., 2001).

Explainability for Food Quality Assurance (QA) When using ML in a food production context, *explainability* is important to satisfy QA/QC (Quality Control) teams and, potentially, regulators or auditors. Food companies are more comfortable if they understand how a model is making decisions. So, there is a preference for more interpretable models or for providing explanation tools (like feature importance or SHAP values) along with predictions (Alsaeedi et al., 2024; Hua et al., 2023; Rohart et al., 2017; Theng & Bhoyar, 2024; Zhao et al., 2023b). For example, if a soft sensor predicts "fermentation is 90% complete", QA might ask, "How do we know?" It helps if we can say, for instance, "the model heavily weights the CO_2 evolution rate, and the slowing base addition: both indicate approaching completion." Some models used are inherently interpretable (e.g., PLS—Partial Least Squares—which is popular in chemometrics and provides insight into which wavelengths or variables correlate with the quality attribute). Others, like neural nets, require post-hoc explainability tools (Andersen & Runger, 2011; Tang et al., 2022; Wang et al., 2023; Wei et al., 2022).

In highly regulated scenarios (such as producing a food ingredient under FSSC22000 or a similar standard), you might need to validate that the ML system does not adversely affect product quality decisions. Often ML is used in advisory mode, providing recommendations that humans or simple algorithms then act on. Nevertheless, as confidence grows, we see more closed-loop ML-driven control (Ng & Tan, 2025; Simethy, 2024). In such cases, simpler models (even if slightly less accurate) might be chosen because QA can validate them. For instance, a linear regression soft sensor with five inputs is easy to validate and troubleshoot ("sensor X failed, so output is off"), whereas a deep learning black box might be harder to validate. There is ongoing work on explainable AI in food to build that trust. Regulatory bodies have also guided that any AI that impacts product quality should be qualified/validated similarly to equipment, demonstrating it consistently performs as intended and has fail-safes (Chhetri, 2024; Pan, 2023; Rivera et al., 2024).

One practical approach is to use hybrid models (combining first principles and ML) that are more interpretable. Like using a known bioprocess model form (e.g., logistic growth or Monod kinetics) (Hill, 1983), but have ML adjust the parameters. This way, the structure is familiar, and one can explain predictions with terms like "the growth rate parameter has increased, indicating higher biomass yield, which is why predicted titer is higher." Another approach is to restrict the machine learning model's domain to known safe ranges and have conventional controls take over if it exceeds those limits. This approach provides quality assurance with the assurance that there are safeguards in place (Herrera-Ruiz et al., 2025; Mu'azzam et al., 2024; Oyetunde et al., 2018).

In summary, ML and soft sensing in fermentation are already delivering practical benefits: better process observation (soft sensors giving us eyes where we were blind) and early warnings/predictions (so we can intervene or optimize). The keys to success are good data (from PAT sensors; see Sect. 10.4), careful model development (to prevent overfitting given limited data), and an operational lifecycle to keep models updated and trusted (Metcalfe et al., 2025). When done well, ML becomes like an "assistant brewer" or "digital winemaker" that continuously watches over and fine-tunes the process, ensuring each batch hits its targets with minimal guesswork.

10.7 Optimization and Control with AI Tooling

Pragmatic Optimization—Blending DoE and Bayesian Methods Optimizing a fermentation process, whether it involves media composition, feeding strategy, or operating conditions, has traditionally been achieved through Design of Experiments (DoE) and expert intuition (Singh et al., 2017). DoE (e.g., factorial designs or response surface methodology) systematically tests combinations of factors to build a regression model of output (e.g., yield) as a function of the factors. This is still invaluable to map the broad design space. However, fully exploring many factors using DoE can be expensive (the number of experiments skyrockets) (Khuri & Mukhopadhyay, 2010; Reji & Kumar, 2022). Modern approaches increasingly use Bayesian optimization (BO) to accelerate this process. Bayesian optimization is essentially an AI-guided strategy for selecting the next best experiment to run, based on all data gathered so far, by balancing exploration and exploitation. It often employs a Gaussian process or other surrogate model to predict outcomes and uncertainties (Gisperg et al., 2025).

In Practice, a Pragmatic Approach Is to Blend DoE and BO: Start with a coarse DoE to identify a general region or significant factors, then use BO for fine-tuning. For example, one might perform a fractional factorial DoE on, say, five media components (glucose, nitrogen, trace elements, etc.) to see which have the most significant effects on cell growth (Rezende et al., 2018). Once key nutrients are identified, Bayesian optimization can be used to find the optimal concentrations of those nutrients to maximize growth rate or yield (Fu et al., 2012). Studies in bioprocess engineering have shown that BO can find optimal media recipes or feeding profiles with far fewer experiments than grid search, often reducing experimental load by >60% while reaching comparable or better optima (Lapierre et al., 2025). In some cases, BO was used to optimize growth media and achieved higher titers with up to 30% fewer experiments, whereas classical methods might have required many more (Cosenza et al., 2022; Narayanan et al., 2025). The strength of BO is in sequential learning: it can incorporate any nonlinear or interacting effects captured by its surrogate model, and it directs experiments to promising regions (e.g., "try a slightly higher iron concentration next, since the model predicts there might be a sweet spot") (Cosenza et al., 2022; Gisperg et al., 2025).

For fermentation processes, multi-day experiments are costly, so this efficiency is gold. Many companies use Bayesian optimization for media development and strain optimization now (Hancock, 2015). However, BO is not magic: it requires defining an objective function (which could be multi-faceted) and constraints (Gisperg et al., 2025). In a pragmatic approach, we must incorporate practical constraints into the optimization, such as "glucose cannot exceed X g/L, or it causes osmotic stress" and "DO must remain >20% or else it is oxygen-limited, invalidating results". So, each experiment recommended by an AI tool is vetted for safety and feasibility by scientists (this is the guardrail concept, more on that below) (Liu et al., 2025).

Another AI tool for optimization is evolutionary algorithms (NSGA-II) or reinforcement learning for dynamic optimizations (like time-varying feeding profiles). For example, one might use a genetic algorithm to evolve feeding schedules that maximize productivity, or reinforcement learning, where the controller "learns" an optimal feeding policy through trial-and-error simulation. These are less commonly applied than BO in fermentation at present, but they hold promise, especially when a digital twin or simulator is available to test many strategies quickly (Ahmad et al., 2024; Murugan & Subbaian, 2022; Zhao et al., 2025).

Model Predictive Control (MPC) in Bioprocessing: MPC is an advanced control technique that uses a process model to predict future outcomes and optimize control moves over a horizon (Rashedi et al., 2022). In fermentation, MPC has been used in scenarios where simple PID control is insufficient due to strong interactions or constraints, for instance, controlling the feeding rate to maximize product formation while avoiding overflow metabolism (Bolmanis et al., 2023; Mohd & Aziz, 2016). Designing an MPC for fermentation requires a dynamic model (e.g., a mechanistic model of growth/product formation or a data-driven state-space model learned from data) because fermentations are nonlinear; often, a nonlinear MPC or an adaptive MPC is needed (Chai et al., 2022b).

Is MPC Justified? It depends on the process complexity and economic payoff. In many food fermentations (such as brewing beer or making yogurt), traditional control is sufficient, and MPC might be overkill. However, in some precision fermentation (enzymes or high-valued metabolites) or high-volume processes with tight specifications, MPC can improve performance (Bolmanis et al., 2023; Chai et al., 2022b; Foss et al., 1995). Some examples might include: (1) An MPC controlling dissolved oxygen by manipulating both stirrer speed and oxygen enrichment can handle the constraint of not exceeding a certain agitator power and find an optimal combination to keep DO at setpoint more smoothly than separate PID loops fighting each other (Raja et al., 2007); (2) Fed-batch fermentations for maximizing biomass—an MPC can decide feed rate at each step to keep substrate just below inhibition level and maximize growth (Bolmanis et al., 2023). Another study has demonstrated MPC for optimizing *E. coli* feed profiles to maximize recombinant protein production, achieving better results than constant or manual profiles (Dubencovs et al., 2023).

The barrier is that developing and maintaining the model is non-trivial—cells can behave unpredictably if conditions drift from those on which the model was built, and one must update the model when strains or media change. Therefore, MPC is usually applied only when the value is high (e.g., a pharma product, or when slight yield gains are worth much money) or when a good model is readily available (e.g., in some yeast fermentations, the kinetics are well known) (Forbes et al., 2015).

Multi-Objective Trade-Offs: Fermentation optimization rarely has a single goal. We often face multiple objectives: maximize yield (g product per g substrate), maximize productivity (g/L/h), maintain good sensory qualities, *and* minimize cost. These objectives conflict—e.g., pushing for the fastest productivity might require high feed rates that lower yield or cause off-flavors; maximizing yield might mean a very slow process, which is economically suboptimal (Mauri et al., 2020; Navid et al., 2019). Thus, modern optimization uses multi-objective optimization (MOO) approaches to evaluate trade-offs and identify Pareto-optimal solutions (Petchrompo et al., 2022). For instance, one objective could be to maximize titer (product concentration), while another could be to minimize batch time. The result of MOO might be a set of solutions from which a decision-maker can choose (maybe a compromise solution that achieves 90% of the maximum titer in 70% of the minimum time) (Hernández Rodríguez et al., 2022).

In fermentation, some common trade-offs are:

- *Yield vs Rate*: A very common one. Often, if you feed the substrate slowly, cells convert it more efficiently (higher yield), but it takes longer (lower volumetric productivity). If you feed fast, you get a high production rate but waste more substrate or produce overflow metabolites (lower yield) (Hjersted & Henson, 2008).
- *Yield vs Flavor*: In some traditional ferments (beer, kombucha, etc.), maximizing yield of ethanol or acid can produce too much of certain byproducts that affect taste. So a trade-off between alcohol percentage and taste profile might exist (Rodman & Gerogiorgis, 2016).
- *Cost vs Performance*: Using richer media, which involves more expensive ingredients like yeast extract and amino acids, can enhance yield or rate, but at a higher cost. Optimization should consider the cost per unit output, not just the

biophysical yields. A multi-objective approach can include goals such as maximizing product titer and minimizing medium cost (Konzock & Nielsen, 2024).
- *Environmental objectives*: For sustainable processes, one might include minimizing waste and energy usage as objectives, in addition to the desired output (Mori et al., 2019).

AI optimization can incorporate these via weighted sums or Pareto front finding algorithms. For example, a Bayesian multi-objective optimization can handle two objectives and find a set of solutions that offer different trade-offs. Alternatively, one can use Genetic algorithms with multi-objective criteria (the NSGA-II algorithm is popular) to evolve a population of solutions that optimize multiple goals (Gisperg et al., 2025; Kumar & Yadav, 2019; Patel & Gujarathi, 2025; Zhang et al., 2025).

The result typically needs human judgment: AI will suggest "Solution A: yield 50%, rate 2 g/Lh; Solution B: yield 48%, rate 2.5 g/Lh; Solution C: yield 55%, rate 1.5 g/Lh" etc. The team then picks based on business priorities. If production volume is more important, they might pick the higher-rate option; if the substrate is expensive, they might pick the higher-yield option. In food, there are often sensory or quality targets that must be met as hard constraints (e.g., no off-flavor beyond a threshold), so optimization must be constrained by those quality metrics (which are hopefully measured or predicted via PAT/soft sensors).

Guardrails and Fail-Safes When using AI-driven optimization or control, it is crucial to embed guardrails, essentially safety and quality constraints that the AI must not violate. For instance, if an AI optimization suggests raising the temperature to increase production rate, a guardrail might enforce "do not exceed 35 °C or proteins will denature/flavor will be ruined." (Ranjan & Kravaris, 2024). These can be implemented as explicit constraints in optimization algorithms or as logic checks around an AI controller. Fail-safe mechanisms ensure that if the AI logic does something unexpected, the process does not go out of control (Herrera-Ruiz et al., 2025; Mu'azzam et al., 2024; Oyetunde et al., 2018). For example, an MPC or AI controller might have a watchdog timer, and if it outputs an actuator move that is out of normal bounds or if communication is lost, the system reverts to a known safe mode (like PID control or a steady feed rate) (Hernandez et al., 2012; Mowbray et al., 2022b). In fermentation, typical fail-safes include: pH and DO loops that will always override if those variables stray too far; emergency substrate feed shutoff if a certain metabolite accumulates (to avoid a runaway foam or byproduct scenario); or simply alarm and pause feed if predictions seem off (Mitra & Murthy, 2021; Mowbray et al., 2022b).

Especially in food manufacturing, consistency and safety trump theoretical optimality. So any AI-driven setpoints or optimizations are often introduced gradually and under observation. A conservative approach might be to have the AI recommend setpoint changes to an operator who then approves them (Metcalfe et al., 2025). Over time, as confidence builds, more automation can be allowed. But even then, companies often maintain a manual or simpler control fallback (Gasset et al., 2024). For example, if a soft sensor fails or flags high uncertainty (drift), the control system might automatically switch back to a fixed feeding profile, preventing an unseen model error from destroying a batch (Gnoth et al., 2011; Ranjan & Kravaris, 2024).

Moreover, regulatory and quality systems might require that critical control points (CCPs in HACCP terms) not be controlled solely by an "opaque" AI without human review. To comply, one might treat the AI suggestions as advisory and still keep a human in the loop for those CCP decisions, or fully validate the AI system as an approved piece of control equipment (Chhetri, 2024).

In essence, AI tools are accelerating and enhancing fermentation optimization by exploring intricate factor spaces and balancing multiple objectives that would be challenging for manual methods. The most effective approach is often a hybrid strategy: leverage AI to analyze combinations, but rely on human and domain knowledge to establish reasonable boundaries, interpret results, and implement changes safely. When executed correctly, AI-driven optimization can lead to significant yield improvements or cost reductions (e.g., reports of 10–20% yield increases with advanced media optimization, which is substantial in a commodity context). Additionally, AI-assisted control (e.g., MPC or RL-based) can push processes closer to their inherent limits (e.g., minimizing batch time without exceeding metabolite levels) while adhering to constraints. The result is a more efficient and consistent process, but always with a layer of trust but verify, ensuring that algorithms serve as tools for fermentation experts, not unchecked black boxes.

10.8 Digital Twins for Development and Scale-Up

What Is a Fermentation Digital Twin? A digital twin, in this context, is a virtual replica of the fermentation process (and potentially the equipment) that runs concurrently with the physical process. It transcends the boundaries of a mere simulation model, as it continuously receives real-time data from the plant and dynamically updates its state to accurately reflect the actual conditions of the process (Yang & Ierapetritou, 2024). Think of it as a living model that can both predict future behavior and test scenarios without risking the real system (Koulouris et al., 2021). In fermentation, one can have digital twins at different levels: an asset-level twin (e.g., a model of a single bioreactor, capturing its dynamics, mass transfer, and kinetics) and a process-line or plant-level twin (integrating upstream/downstream, utilities, and multiple tanks). Asset-level twins often include detailed equipment characteristics like vessel geometry, agitation, aeration systems, and combine that with kinetic models of the biology (Zhao et al., 2025). Line-level twins might incorporate scheduling, resource constraints, or interactions (like how upstream feed variability affects downstream purification) (Yao et al., 2023). This interaction between real-time sensing and model-based optimization is illustrated in Fig. 10.4.

Building and Calibrating the Twin To create a digital twin, you typically start with a model of the fermentation process. This can be mechanistic, e.g., differential equations for cell growth, substrate uptake, and product formation (the classic

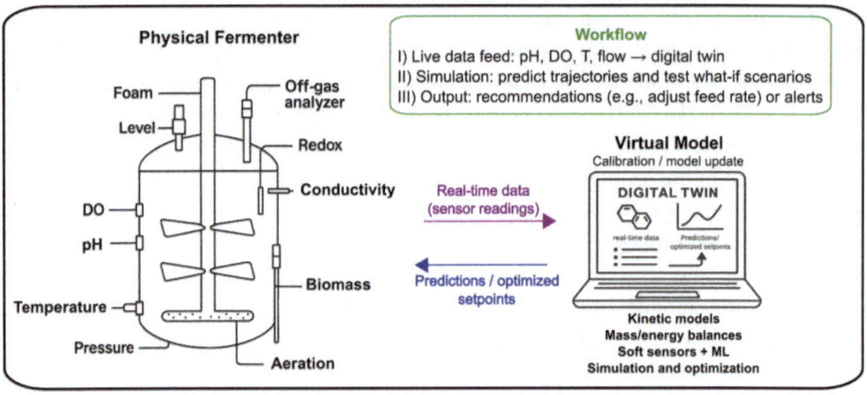

Fig. 10.4 Schematic of a fermentation digital twin linking a physical fermenter to a virtual model for development and scale-up. The physical system is monitored by in situ and off-gas measurements (for example, pH, dissolved oxygen, temperature, pressure, level, foam, redox, conductivity, biomass, aeration rate, and off-gas analysis). These measurements are streamed in real time to the digital twin, where kinetic models, mass and energy balances, and data-driven components (soft sensors and machine-learning models) are calibrated and used to simulate process behavior and test what-if scenarios. The virtual model returns predictions and optimized setpoints or recommendations (for example, feed or aeration adjustments) to the process and/or operators, forming a closed decision loop that supports real-time monitoring, optimization, and robust scale-up. (Source: From Authors)

Monod or Logistic growth models, perhaps with mass transfer equations for oxygen and heat balance equations for temperature (Luttmann et al., 2012; Lyubenova et al., 2024)). It can also involve CFD (Computational Fluid Dynamics) for a detailed understanding of mixing and oxygen distribution, especially to capture scale-dependent effects (Li, 2024). Increasingly, hybrid models that combine mechanistic and data-driven approaches are being used. Mechanistic models provide a framework and baseline, while data-driven components adjust parameters or fill gaps based on plant data (Riezzo et al., 2025).

Calibration means tuning the model parameters so that the twin behaves like the real system (Ward et al., 2021). Historically, this might involve running small-scale experiments to estimate kinetic parameters (μ_{max}, K_s, yield coefficients, etc.). Nevertheless, with the wealth of historical data, one can calibrate directly using plant data—essentially an optimization that adjusts model parameters until the model's output matches past batch profiles within an acceptable error. A digital twin might also incorporate real-time data reconciliation: as new data comes in (e.g., a current biomass reading or CO_2 evolution), the twin model state is corrected (e.g., via a Kalman filter or Bayesian update) to ensure it stays aligned. This way, even if the initial parameters are off, the twin "learns" and tracks the actual run (Dazzarola et al., 2026; Lopez et al., 2020).

It is important to calibrate under various conditions so the twin is robust. For example, calibrate it on a slow-growth batch and a fast-growth batch, or on both summer and winter data if seasonal effects are present. Often, different modes or recipes may require separate calibration, or an adaptable model that takes recipe parameters as inputs (Albino et al., 2024; Lopez et al., 2020). When moving from lab to pilot to plant scale, calibration at each scale is needed: equipment differences (mixing, k_{La}, heat removal) can be captured by adjusting parameters such as the oxygen transfer coefficient or by adding terms for gradient effects. A well-calibrated twin at a small scale can be gradually modified using known scale-up correlations (e.g., using CFD to adjust k_{La} for a large tank and calibrating it with a few pilot runs) (Moser et al., 2024; Riezzo et al., 2025).

Scenario Testing and Optimization Once you have a digital twin, one of its significant benefits is the ability to perform what-if analysis. Engineers can test different setpoint profiles or disturbances in silico without jeopardizing the product. For instance, one can simulate: "What if we increased the feed rate by 20% after hour 10—how would that affect final titer or byproduct formation?" or "If the cooling fails for 30 minutes (temperature rises), how will that impact the batch?" The twin can answer these by simulation. This helps determine optimal setpoints, and you can run many virtual experiments overnight with the twin, varying aeration, feed, pH, etc., to identify the best strategy, then implement it in the real process. This is much cheaper and faster than trying multiple approaches on the actual production line (Isoko et al., 2024; Shen et al., 2024).

Moreover, digital twins are invaluable for training and troubleshooting. Operators can be trained on the twin: e.g., practicing what to do if DO starts dropping, the twin can simulate the trajectory and let them make control moves. For troubleshooting, if an abnormal batch occurs, one can retrofit the scenario in the twin to understand why, maybe by adjusting an assumed parameter (like contamination introduced a competing reaction, or a sensor was reading wrong), you can make the twin mimic the bad batch and thus deduce the cause (Hassan et al., 2024).

Another emerging use is real-time optimization: the twin can run in parallel to the process and continuously optimize under current conditions (Goodwin et al., 2022; Viola & Chen, 2021; Zhang et al., 2021). For example, as conditions change (e.g., substrate purity), the twin could adjust and then re-optimize the feed schedule or aeration strategy to still hit targets (Bayer et al., 2021; Zhao et al., 2023a). This concept ties into Model Predictive Control, as the twin is essentially the internal model used by MPC to predict and optimize moves. Some advanced digital twin setups will actually send new setpoints to the control system in real-time (closed-loop twin) (Stevnsborg et al., 2023). For instance, the twin predicts that, given the current trajectory, the final glucose concentration will be 5 g/L (wasted). Consequently, it suggests slightly tapering the feed to ensure that more glucose is consumed by the end of the batch.

Maintaining Fidelity Over Changes One challenge is that fermentation systems are dynamic. If you alter the recipe (e.g., changing the medium composition or introducing a new feeding protocol) or switch to a different strain, the twin's model may become inaccurate. Good digital twin practice is to have a model that can be recalibrated or adapted with minimal effort. This might involve designing small experiments to update kinetic parameters for a new strain (e.g., measure its growth rate on certain substrates (Sinner et al., 2020)). Alternatively, it could leverage online learning: the twin could use the first batch of a new strain to auto-tune some parameters by comparing predicted vs observed data. Some twins incorporate machine learning components specifically for adaptability, e.g., an ML model for specific growth rate that is retrained whenever a new type of data is seen, keeping the fundamental mass-balance structure same but the "rate law" learned (Goodwin et al., 2022; Koulouris et al., 2021; Li, 2024; Riezzo et al., 2025).

We also often maintain multiple scales of twins: a high-fidelity twin at a small scale (with more sensors and more invasive measurements) and a scaled-up twin for production. The scale-up twin might simplify some things but also incorporate scale-specific factors, such as mixing time. To keep fidelity, engineers calibrate the scale-up twin using both the small-scale model and some pilot data (Udugama et al., 2023b). For example, a lab twin might assume perfect mixing; the large-scale twin might incorporate a known gradient or dispersion model to simulate imperfect mixing, and that term is tuned using data from pilot runs (where probes at different points or known scale effects are measured).

Applications in Scale-Up: When moving from development to commercial scale, the twin can simulate scale-up scenarios—e.g., *if we go from a 10 L bioreactor to 10,000 L, will oxygen limitation occur?* The twin (with CFD-derived k_{La}) can demonstrate that the dissolved oxygen (DO) concentration might decrease unless we increase the pressure or agitation. It can virtually test various mitigation strategies, such as increasing the sparger oxygen fraction, using multiple feed points, and so on (Li, 2024). This approach helps prevent costly failures in which a product performs well in a pilot phase but fails when scaled up. Companies have used digital twins to reduce the number of iterative scale-up batches. Instead of conducting five trial runs to adjust conditions, they simulate and run only one or two confirmation runs (Neubauer et al., 2020). Twins also facilitate technology transfer to new facilities by capturing all the process knowledge in the twin and then utilizing it to configure equipment at a different site (taking into account the vessel geometry and other relevant factors) or even train an inexperienced operator (Hassan et al., 2024). It essentially encapsulates the process design, making it easier to repeat elsewhere.

In essence, a digital twin is a digital counterpart to your fermentation process, allowing you to interact with it. By seamlessly integrating real plant data and employing rigorous models, it emerges as a potent tool for forecasting, optimizing, and mitigating risks associated with fermentation processes. During development, it accelerates innovation by enabling rapid testing of various ideas. In the context of scale-up, it provides valuable insights into the necessary adjustments required when transitioning to a larger scale. Moreover, in operational settings, it can even function concurrently as a "virtual advisor," predicting the batch's trajectory and enabling timely interventions if necessary. As advancements in computing and integration technologies continue, these digital twins are becoming increasingly accurate and user-friendly, propelling the industry closer to the ideal of "first-time-right" scale-ups and self-optimizing fermentation plants.

10.9 Implementation, Scale-Up, and Quality Essentials

Readiness Assessment (People, Process, Tech) Successfully implementing smart fermentation and advanced control necessitates the preparation of three crucial pillars: people, process, and technology (Elragal & Elgendy, 2024; Poch et al., 2023). People, it's crucial to evaluate the skills and training of the team, including operators, engineers, and QA personnel, to ensure they're prepared for new systems (Poch et al., 2023). For example, introducing PAT sensors and digital tools means operators need to know how to interpret new screens and alarms, and engineers might need training in data analytics or in sensor maintenance. Often, a *gap analysis* is done to identify where skill development is needed. Cultivating a culture open to data-driven decision-making is part of readiness (Elragal & Elgendy, 2024; Poch et al., 2023; Sinner et al., 2020). If the plant staff trust the new soft sensor or AI suggestion, they will use it; otherwise, they might ignore it. Thus, involving them early, such as having operators participate in the design of HMIs or in the acceptance testing of new sensors, builds ownership and understanding.

Process readiness means the fermentation workflow is well characterized and sufficiently stable to benefit from advanced control. If the baseline process is highly variable or not yet fundamentally optimized, it might be better to get that right first before layering on complex automation. Also, process documentation (SOPs, batch recipes) should be up-to-date, as the new system will formalize those, e.g., coding a recipe in MES requires clarity on steps (Gerzon et al., 2022; Simon et al., 2015).

Technology readiness involves ensuring that the infrastructure (installed sensors, automation hardware, networks) is in place and reliable. For instance, if we plan to implement an electronic batch record, we need all critical data points to be captured by the system (no more manual readings that only went on paper). Alternatively, if using cloud analytics, ensure the internet connectivity and cybersecurity measures are robust (Jaskó et al., 2020; Ji & Xu, 2025; Pietrasik et al., 2024; Thames & Schaefer, 2016).

Can companies perform a digital maturity assessment or a readiness audit similar to a checklist: Are CIP/SIP already automated or still manual? If manual, maybe automate that first as a quick win (since CIP/SIP repeat frequently and benefit from consistency). Are sensors calibrated and on a schedule? Are there data standards in place? On people: are there data scientists or engineers who can support the new ML models, or do we need to partner or hire? By evaluating these, management can plan a phased approach (Elragal & Elgendy, 2024; Simethy, 2024; Simon et al., 2015).

Modular Deployment Plan (Sensors and Controls) A smart strategy is to roll out new sensors and automation in modular increments rather than a big bang. For example, one might first install inline PAT sensors, such as a new Raman probe, on a single fermentor as a pilot project, rather than equipping all fermentors at once. This way, one vessel becomes the "model vessel" for developing methods and working out kinks (Baradez et al., 2018). Similarly, one could implement the advanced control (say, an optimized feeding algorithm or MPC) on a single line, while others run the old way, and compare results (Maruthamuthu et al., 2020). Modular deployment also refers to the architecture: using modular automation code (ISA-88 encourages modular recipe phases, equipment modules, etc.) so that pieces can be added or modified without rewriting everything (De Minicis et al., 2014; Kim et al., 2021; Slaugenhaupt, 2025). For example, adding a new pH control loop should just slot in, not break all logic.

Many facilities opt for a parallel run period: run the new system in observation mode while the old system continues to officially control, to ensure everything works as expected. Then switch gradually. During deployment, it is common to use vendor skids or packages for key tasks. For example, a modular skid for CIP/SIP might be added, containing the cleaning automation and interfacing with the central control system via I/O and recipe steps. By having CIP automated separately, it does not interfere with fermentation control logic and can even be tested independently (e.g., run CIP cycles with water to validate it) (Kiran Deokate et al., 2025).

A phased plan might look like:

Phase 1: Install historian and networking, and begin data collection (no control changes yet).

Phase 2: Add PAT sensors (pH and DO were already present; now add an off-gas analyzer, possibly a biomass capacitance probe). Use the data for monitoring, but maintain control over existing loops.

Phase 3: Introduce advanced control in one area, such as automated feed control using soft sensor readings.

Phase 4: Integrate MES and electronic batch records once the lower-level automation is stable.

Phase 5: Implement predictive models and optimize schedules, among other things.

Each module adds capability and can be validated independently.

Validation Path (FAT, SAT, IQ/OQ/PQ) In food biotechnology, especially when equipment or systems are under regulatory oversight or used for internal quality assurance, it's crucial to ensure that any new equipment or system meets the necessary qualifications (Mukherjee et al., 2022). The steps are typically:

FAT (Factory Acceptance Test): Testing equipment or systems at the vendor's site before delivery is crucial. For instance, during a fermentation system upgrade, FAT might involve simulating signals to the PLC and testing the control sequences using water or dummy loads at the vendor's facility. Similarly, when purchasing a new bioreactor or analyzer, FAT ensures that it meets the specified requirements in a controlled environment. FAT plays a vital role in identifying issues early, such as software bugs or hardware faults, and ensures that the system is built according to the agreed specifications (Gewald et al., 2015).

SAT (Site Acceptance Test): After installation at the plant, SAT tests verify that the system functions in the real environment and interfaces properly with other systems. For example, after installing the new SCADA, one would test communication with existing PLCs, test that all sensors read correctly on the new screens, and that emergency stops work, etc. Often, FAT and SAT are documented with test protocols. FAT/SAT also often cover utility connections, safety interlocks, etc., ensuring no damage in transit and proper installation (Felo et al., 2013; Shukla et al., 2023). After the mechanical and electrical components are completed, the formal IQ, OQ, and PQ stages commence (if following GMP-like validation).

IQ (Installation Qualification): Verifying that the equipment or system is installed correctly according to design and manufacturer recommendations involves checking that all components are in place, cables are labeled, software versions are recorded, instruments are calibrated, and manuals are available (Dwivedi, 2005). It essentially involves ensuring the foundation is solid. In the context of a new fermentation control system, IQ would encompass tasks such as listing all instruments, confirming their installation in the designated ports, verifying calibration certificates, identifying network nodes, and so on.

OQ (Operational Qualification): Testing that the system operates as intended across its range, typically by challenging each function and alarm in a controlled manner (Dwivedi, 2005; Jornitz, 2019). For a fermentor, OQ tests could include verifying the temperature control loop's ability to maintain the setpoint at various values, ensuring the pH control adds acid or base within specified tolerances, and confirming that alarms trigger and cause the appropriate failsafe (such as an agitator failure alarm shutting off the feed pump). These tests often involve running dummy batches with water or media and simulating different scenarios. OQ demonstrates that the automation and equipment perform as designed according to the specifications. For PAT instruments, OQ might involve running known-reference samples (e.g., for an off-gas analyzer, running a gas with a known CO_2 percentage and verifying the reading's accuracy). If using ML or soft sensors, OQ might test those models on known validation datasets to ensure they meet accuracy criteria. (This is a relatively new area in validation guidelines, but the general principle remains the same) (Brunner et al., 2019; Medl et al., 2023; Raja et al., 2024).

PQ (Performance Qualification): Confirming the system's performance under real-world conditions to produce an acceptable product is essentially the conduct of trial batches under normal operations. These batches demonstrate that the new system consistently meets the quality specifications. For food fermentation, PQ might involve three consecutive batches using the new automated process. This demonstrates that the yield, purity, and sensory attributes meet the targets. During PQ, everything is in production mode, and data from these runs is collected to show stability and control. It's like a final exam of the process and its controls. Once PQ is completed, it often blends into routine production (Dwivedi, 2005; Ghosh et al., 2022).

These steps ensure that the system is thoroughly tested and documented before going live, which is essential for obtaining quality certifications and building confidence. For instance, if an electronic batch record system is introduced, IQ would verify that all necessary data points are captured, OQ would test that records are accurately generated (e.g., by introducing a deviation and confirming its recording), and PQ would demonstrate that actual batches can be released using e-records without any issues.

Automation of Cleaning and Sterilization Ensuring robust CIP (Cleaning-In-Place) and SIP (Sanitization-In-Place) is crucial during scale-up and implementation, as contamination remains a persistent risk. Modern fermenters automate CIP/SIP sequences to guarantee reproducibility. For instance, a typical CIP cycle might involve rinsing, an alkaline wash at 80 °C, rinsing, an acid wash, rinsing, and a final rinse, all of which are precisely timed and directed by a program (Kiran Deokate et al., 2025). Automating this process eliminates operator variation, such as forgetting a step or varying contact time. During OQ/PQ, these cleaning cycles are validated, for instance, through swab tests to ensure no residual contamination. Similarly, sterilization-in-Place (SIP) automation is crucial. The system must maintain the required temperature for the specified duration throughout the vessel. Automation can manage this process, such as heating with steam, holding at 121 °C for 30 min, and then switching to cooling. Additionally, automation can control valves to perform simultaneous filter sterilization of the inlet and outlet lines, among other tasks (Miley et al., 2015; Seiberling & Ratz, 2018). For larger systems, continuous HTST media sterilization may be part of the process and is automated and integrated. The advantage is a consistent turnaround time for fermentors and easier validation (since you have electronic records of each cycle) (Agrawal, 2024).

Electronic Batch Records (EBR) and Genealogy: Implementing smart fermentation often involves transitioning away from paper records. An Electronic Batch Record (EBR) system captures all process data, operator actions, material lots, and other relevant information for each batch in an electronic document. This system typically integrates with a Manufacturing Execution System (MES). The EBR includes details such as the executed recipe steps, timestamps, any deviations or alarms, and sign-offs. A significant advantage of using an EBR is the reduction of human error, as it eliminates the risk of missed entries or illegible writing. Additionally, it enables efficient genealogy and traceability of the production process (Sousa, 2025; Yadav, 2025). Genealogy in manufacturing involves tracking the relationships between batches of input materials and the final product batches. Essentially, it determines which raw material lots were used in which fermentation processes, and which fermentation batches contributed to which final blended batch or packaged lot. With an integrated data system, you can quickly query this information if needed. For instance, if a specific lot of pea protein (used as a feed ingredient) was later found to be contaminated, you can easily identify all the fermentations that utilized it and trace the path of its product (Ren et al., 2025; Thomas et al., 2022).

For food safety and regulatory compliance, including FDA's FSMA and ISO 22000 best practices, traceability is of utmost importance. Many alternative protein companies must be prepared to conduct recalls or trace investigations. If all data is digital, it enables both forward and backward tracing. Forward tracing involves tracing from a raw lot to all products, while backward tracing involves tracing from a finished product back to all inputs (Yu et al., 2022).

EBR systems often enforce process controls. For instance, if a step requires QA approval before proceeding, such as a pH adjustment outside the normal range, the MES/EBR can mandate an electronic signature before allowing the batch to continue. This ensures that quality checks are not bypassed. Additionally, these systems time-stamp everything, which aids in maintaining data integrity, as discussed in Sect. 10.5.

Food Safety and Traceability Meets PAT/Data Food fermentations that produce consumer products need to integrate their smart tech with food safety plans (Jha & Singh, 2025; Mattarozzi et al., 2023; Yu et al., 2022). For instance, a HACCP plan (Hazard Analysis and Critical Control Points) will specify certain critical limits, such as a minimum sterilization temperature or a maximum toxin level (Rodrigues et al., 2024). PAT sensors and automation can monitor these critical control points in real-time and even log proof of compliance. For instance, the system can log that the fermenter SIP was at a minimum of 121 °C for at least 15 minutes, providing proof of compliance. If a critical limit is violated, such as the temperature not reaching the target, the system can immediately flag it for corrective action, potentially saving a batch from being released unsafely. Additionally, environmental monitoring, such as HEPA filter integrity or sterile air pressure, can also be logged (Sandle, 2024).

Traceability involves being able to answer questions about any jar of product, such as which fermentation process, cleaning cycle, sensor readings, and raw materials are associated with it. Smart fermentation setups electronically tag all materials. For instance, when adding a nutrient, an operator can scan the lot barcode, and the MES system records this information in the batch record. Later, if a contaminant is discovered in a product, the digital system quickly surfaces the commonality (maybe all problematic batches used nutrient lot XYZ, indicating that it is the source).

Regulatory bodies are increasingly expecting digital records. For instance, when it comes to novel foods or additives, regulators may audit production records to ensure consistency. Electronic records expedite audits because you can retrieve any data point or trend on demand, eliminating the need to flip through binders. However, this also means that cybersecurity and data backup are essential components of quality assurance. The data must be secure from tampering (to be trusted as accurate) and backed up (to prevent data loss in the event of a server crash and to ensure traceability). Many plants maintain a local historian and replicate their data to an off-site server for disaster recovery (Monaco, 2025; Precup et al., 2024).

In conclusion, the implementation and scale-up of smart fermentation technology are not merely technical endeavors but also require meticulous change management and compliance efforts. By carefully phasing the deployment, rigorously testing (FAT/SAT, IQ/OQ/PQ), and aligning with quality systems, the plant can modernize without compromising control. Once established, the combination of automated records, robust cleaning/sterility, and integrated data enables confident scaling—producing more product in more locations while maintaining the same high quality and full traceability. The groundwork laid in this chapter ensures that when high-tech fermentation meets the real world, it does so in a controlled, safe, and QA-approved manner, setting the stage for the facility to reap the benefits of all the advanced tools discussed earlier.

10.10 Applications and Short Case Snapshots

Precision fermentation holds great promise for producing recombinant dairy proteins such as whey and casein. Companies such as Perfect Day have engineered microorganisms, such as Trichoderma reesei, to secrete β-lactoglobulin, the primary whey protein, with remarkable accuracy. This advancement enables the production of dairy products without using animals (Geistlinger et al., 2024). Secretion is crucial because releasing the protein into the broth makes downstream purification much simpler and yields a food-grade product of the desired quality. However, a significant challenge in this process has been achieving sufficient yields and ensuring that the protein's stability matches that of the original dairy product (Cairns et al., 2021; Eastham & Leman, 2024). Early attempts in the 1990s yielded only a few milligrams per liter. However, through strain engineering and bioprocess optimization, titers have improved approximately 1000-fold, reaching around 1 g per liter in recent studies (Batt et al., 1990; Keppler et al., 2021). This progress demonstrates how bottlenecks such as low expression levels can be overcome by selecting optimal host strains (e.g., fungi with robust secretion systems) and by iterative genetic enhancements. The resulting proteins are essentially identical in nutritional and functional properties to their animal-derived counterparts. For instance, fermentation-derived whey can be used in ice cream and other products just as traditional whey is (Chai et al., 2022; Molfetta et al., 2022). This example links tools (synthetic biology and fermentation know-how) to outcomes (real dairy analogues) and shows that, although yield and quality were initial hurdles, they were overcome through improved hosts and process development. The remaining challenges involve scaling up economically, but ongoing strain improvements and large-scale fermentation facilities are addressing them, bringing costs down over time (Chai et al., 2022; Molfetta et al., 2022).

Another example is enzyme tailoring for food texture and flavor, utilizing real-time sensor feedback. Enzymes play a crucial role in food processing, influencing texture and flavor. For instance, proteases and cross-linking enzymes impact texture, while specific enzymes can generate flavor compounds. Through protein engineering, fermentation is employed to produce enzymes optimized for these tasks. For instance, an enzyme can be modified to function at a specific pH or temperature, enhancing the consistency of plant-based cheese. Fermentation allows enzymes to be adapted or engineered to exhibit increased activity or stability under specific conditions, significantly improving performance and reducing costs (Li et al., 2024; Pang et al., 2021). Developing such enzymes is hindered by a practical bottleneck:

achieving the desired sensory outcomes, such as flavor and aroma, using traditional analytical methods like GC-MS or human taste panels.

These methods are slow and expensive (Chen et al., 2025; Khalid et al., 2025). To address this issue, food technology teams are employing electronic noses and spectroscopic sensors as feedback tools throughout the development and production phases. E-nose technology utilizes an array of gas sensors to swiftly detect volatile aroma profiles. It serves as a nondestructive and rapid indicator of the production of flavor compounds during fermentation or enzymatic reactions (Preedy, 2016; Rodrigues et al., 2024; Rudnitskaya & Legin, 2008; Seesaard & Wongchoosuk, 2022). Likewise, in-line spectroscopic probes (such as FTIR or NIR) can monitor ingredients, such as sugars, acids, and other metabolites, in real time. These tools serve as eyes and ears in the bioprocess. For instance, an electronic nose can detect the headspace of a fermenting product and quantify the "cheese-like" aroma molecules being produced. This information enables engineers to make real-time adjustments to enzyme dosages or timing. Similarly, spectroscopy-based process analytics can monitor texture-related parameters, such as sugar depletion and pH changes. These analytics can be integrated into closed-loop control systems to optimize the bioprocess (Pollard et al., 2001; Sales et al., 2015). Feedback from these sensors helps steer the process toward the target flavor/texture—if the aroma profile deviates, the system can be tuned immediately. A case in point is the brewing and fermentation industry, where in-line gas sensors and IR spectrometers are already used to catch problems early and prevent off-flavors (Klein et al., 2012; Mengers et al., 2022; Schmidberger et al., 2014).

The primary challenge lies in correlating sensor readings with human perception, as an electronic nose cannot fully replicate the capabilities of a human nose. To address this, the E-nose output is calibrated against chemical analyses and sensory panels. Additionally, machine learning models are employed to interpret the sensor data, enhancing the accuracy and reliability of the E-nose's performance (Gliszczyńska-Świgło & Chmielewski, 2016; Tibaduiza et al., 2024). Companies can reliably use these high-throughput feedback tools to accelerate enzyme optimization and maintain consistent flavor. In essence, this example demonstrates the connection between advanced tools (tailored enzymes and PAT sensors) and desired outcomes (precisely tuned textures and flavors). It highlights that while measuring sensory outcomes used to be slow, real-time sensors now enable efficient process guidance.

A third snapshot highlights the production of structured fats and vitamins through fermentation, demonstrating how precision fermentation can provide both essential nutrients and novel ingredients. Many crucial vitamins are already produced on an industrial scale using microbial fermentation. For instance, most vitamins found in supplements and fortified foods, such as vitamin B_2 (riboflavin) and B_{12} (cobalamin), are produced through fermentation. This has long offered a dependable, animal-free source of nutrients. Historically, the challenge was optimizing microbial strains to produce these compounds at high levels, a problem largely addressed through decades of metabolic engineering and selection (Skoneczny & Cioch-Skoneczny, 2024; Wang et al., 2021).

Today, fermentation is not only used to produce vitamins but is also being used to develop designer fats for food applications. For instance, startup Nourish Ingredients has created an animal-free fat ingredient through precision fermentation. This ingredient enhances the flavor and cooking performance of plant-based meats. Their process involves using an oleaginous fungus (*Mortierella alpina*) as a microbial factory to produce specific lipids that replicate the taste and aroma of meat fat. By inserting the appropriate metabolic pathways (DNA instructions) into the microorganism and fermenting it, they produce an oil-based ingredient called "Tastilux." This ingredient provides a rich, meaty flavor even at a low inclusion rate of approximately 1% in the final food product. The result is a "designer fat" that mimics the sizzle, aroma, and mouthfeel of animal tallow, but is produced from sugars via fermentation rather than slaughter. (Petrie et al., 2025). A typical bottleneck for these novel fats is obtaining regulatory approval and scaling up, since they are new to the food supply.

In the case of Nourish, the company had to compile extensive safety data and navigate the GRAS (Generally Recognized as Safe) review process for flavor ingredients, a process that took several months and a large team effort to complete (Jha & Singh, 2025; Monaco, 2025; Mukherjee et al., 2022). This example links cutting-edge tools (engineered lipid pathways in microorganisms and precision fermentation) to tangible outcomes (sustainable fat ingredients with better taste). It shows that while regulatory and scale challenges exist, they can be mitigated through early safety testing and by leveraging existing large-fermentor infrastructure. Notably, the know-how from decades of microbial oil production in the chemical industry can be repurposed, as analysts suggest that insights from industrial lipid fermentations will accelerate food-grade fat fermentation. In sum, fermentation enables the consistent, sustainable production of both routine nutrients (such as vitamins) and next-generation ingredients (such as structured fats).

Finally, consider microbial biomass as a food source, specifically tailored to achieve desired textures. In biomass fermentation, the microorganism itself serves as the product. The harvested cells (or biomass) are directly used as a protein-rich ingredient. A classic example is Quorn's mycoprotein, which has been produced in the UK since the 1980s using the filamentous fungus Fusarium venenatum. Quorn's fungus is cultivated in massive 150,000-liter fermenters through a continuous process that yields approximately 300 kg of protein-rich biomass per hour (Whittaker et al., 2020). The fibrous fungal cells are then processed into meat-like food products. This was a groundbreaking achievement in its time, linking the tools of large-scale fermentation to the outcome of a new food source. However, achieving a meat-like texture from microbial biomass posed unique challenges.

One issue was morphology control. During continuous cultivation, the filamentous fungus can undergo genetic or physiological changes, resulting in the production of mutants that grow excessively "branched." Highly branched mycelium alters the texture and rheology of the biomass, making it less fibrous and doughier, and can clog the fermenter (Majumder et al., 2024; Whittaker et al., 2020; You et al., 2024). In Quorn's process, after approximately 1000 h of continuous operation, such mutants began to emerge and gradually slowed down production. This limitation restricted the duration of a single fermentation run before necessitating a restart. To address this bottleneck, producers explored various strategies to delay the appearance of these morphological mutants. For instance, they meticulously controlled growth rate, nutrient flow, and other culture conditions to favor the desired filamentous form while suppressing faster-growing variants.

By managing factors such as dilution rate and oxygen levels, they were able to lengthen production runs and maintain a more uniform, thread-like mycelial texture for longer. Another technical challenge was post-harvest processing. The fungal biomass initially contained high RNA levels, a common characteristic of single-cell protein, which required reduction for human consumption. The solution involved a heat treatment step to break down excess RNA, followed by mixing the biomass with a binding agent (originally egg albumen, now plant-based binders) to form solid, meat-like pieces (Lonchamp et al., 2022; Whittaker et al., 2020). Despite these challenges, Quorn successfully created a product with a fibrous, meat-like texture and demonstrated that it could be done at an industrial scale. This paved the way for today's generation of biomass fermentation startups.

New companies are now using mycelium (fungal biomass) to create whole-cut meat alternatives with impressive textural fidelity. For example, Meati Foods/USA (Huggins & Whiteley, 2025) and Adamo Foods/UK (Adamo, 2025) grow filamentous fungi through liquid-state fermentation, then form the mycelial network into slabs that can be sliced into steak-like cuts. The mycelium's inherent microscopic structure serves as a scaffold that remarkably resembles muscle fibers. In Adamo's case, they "brew" the mycelium in large fermentation tanks, similar to brewing beer. After a few days, the biomass is harvested as an intact mat with a naturally fibrous structure. This approach differs from traditional plant-based meats that rely on extrusion to create texture; instead, the fungus self-assembles a meat-like texture as it grows (Duque-Estrada et al., 2023). The result is a delicate yet fibrous product that can replicate, for instance, a chicken breast or a steak without the need for texturizing additives. The primary challenge in this process lies in scaling up while maintaining consistent texture. Liquid fermentation of filamentous fungi is difficult to scale because the broth becomes viscous, and oxygen transfer becomes challenging. Additionally, ensuring that every batch has the same tenderness or bite requires strict control of growth conditions (You et al., 2024). Companies mitigate these issues by employing specialized bioreactor designs, such as air-lift fermenters or other low-shear systems that facilitate the formation of filament networks by the fungus. Additionally, they closely monitor parameters like pH, dissolved oxygen, and nutrient feed rates (Zhu et al., 2025).

Additionally, extensive strain selection is performed to identify fungal strains that naturally yield the desired texture and are stable (Cervera-Gascó et al., 2023; Torres-Tiji et al., 2020). Biomass fermentation has evolved to the point where it's feasible to target specific textures. From Quorn's earlier mince-like products to the ambitious goal of whole cuts championed by today's startups, each example demonstrates the use of microbial cultivation tools to achieve a clear outcome: a novel food form. These examples highlight bottlenecks, such as genetic stability, safety, or rheological control, and propose solutions such as process controls and strain improvements.

These innovations suggest that microbial biomass can be a versatile food source, capable of tuning texture through biological and process-engineering approaches rather than solely relying on downstream processing.

10.11 Conclusions and Practical Takeaways

For teams venturing into fermentation-driven food innovation, several practical steps emerge from these examples. Begin with a small-scale approach, setting clear objectives and employing robust data collection methods. The initial step involves identifying a high-impact target molecule or functionality, such as a protein, enzyme, or flavor compound, and developing a proof of concept at a laboratory scale. Early bench experiments, conducted in shake flasks or small bioreactors, should not only validate the engineered microorganism's ability to produce the target product but also begin to map critical process parameters, including temperature, pH, and feed strategy. Investing in data infrastructure from the outset is crucial, even at a small scale. Implement sensors and maintain detailed records of each experiment. This foundational step establishes the groundwork for Quality-by-Design.

Modern fermentation startups often adopt a phased approach to scale up their operations. After validating their process in the lab, they move to benchtop fermentors (ranging from 1 to 10 liters) to optimize the process. Subsequently, they proceed to a pilot scale, typically on the order of a few hundred liters, before attempting full-scale production at thousands of liters. Each phase serves as an opportunity to refine the strain and process conditions, and each larger-scale test acts as an investment checkpoint. By phasing investments in this manner, startups can effectively mitigate risks associated with their projects. For instance, they would not construct a 50,000-liter plant until the 500-liter pilot consistently produces the desired output at expected yields. Additionally, teams can prioritize spending on elements that enhance flexibility. Instead of immediately investing in a dedicated factory, they can consider using contract fermentation facilities or shared pilot plants for intermediate-scale-up steps. This approach conserves capital and provides valuable performance data before making significant scale commitments.

In essence, prioritize learning by characterizing your microorganism and product on a limited budget. Iterate swiftly and scale up your infrastructure only when necessary. Many startups also find partnerships valuable, such as collaborating with established fermentation companies or food processors, to gain access to scale-up expertise and equipment. As demonstrated by the Nourish Ingredients example, partnering for the initial industrial manufacturing run expedited their journey to market. Therefore, a phased, data-driven approach to scale-up, coupled with strategic collaborations, is recommended. This ensures that you address problems at a smaller scale, where costs are lower, and builds confidence among investors and regulators as you progress.

Another crucial takeaway is the importance of Process Analytical Technology (PAT) and data in ensuring performance portability across various vessels and sites.

In bioprocessing, it's widely recognized that what functions effectively in a 5-liter lab reactor may behave differently in a 5000-liter tank. Factors such as mixing, oxygen transfer, and even cell physiology can undergo changes with scale. To address this challenge, teams should harness PAT tools, including sensors for pH, dissolved oxygen, biomass, metabolites, and more, in conjunction with advanced control software. This approach enables them to maintain tight process control and collect comprehensive datasets. By continuously measuring key variables and keeping them within optimal ranges, consistent product quality can be achieved even as the process scales up or is transferred to a new facility. In practice, this involves installing in-line monitors for critical parameters and, if necessary, employing soft sensors or observer models to infer difficult-to-measure attributes in real time. With robust data spanning small- and pilot-scale, models or digital twins can be developed to predict how to adjust conditions in larger tanks. For instance, if oxygen uptake rate is a critical factor, data analysis may reveal the need to increase agitation and aeration in a larger vessel to replicate the conditions of the smaller scale. PAT sensors can then confirm whether the larger vessel indeed achieves the same dissolved oxygen profile.

The ultimate objective is to make the process as "platform agnostic" as possible. A well-defined process with critical parameters can be replicated across various pieces of equipment by matching those parameter profiles. Consistent data practices are also crucial for multi-site operations. Researchers have highlighted a significant challenge: the lack of standardization in fermentation data collection. This makes it difficult to compare runs or combine datasets from different sources directly. To address this, teams should establish standard data formats and protocols (for instance, consistently record sensor readings at uniform time intervals, calibrate probes consistently, etc.). This ensures that learnings are transferable. Embracing digital tools, such as electronic lab notebooks, cloud databases, and analysis software, will facilitate easier knowledge sharing between R&D and production, or between different geographic sites. In essence, data and automation make scaling predictable. When every batch is monitored with the same rigor, any deviation can be detected and corrected, and the process can be confidently replicated elsewhere.

Furthermore, PAT goes beyond merely using sensors; it involves actively utilizing the data. This is achieved by implementing control loops that adjust feed or environmental settings in real time to keep the fermentation process on track. For instance, if an increase in exhaust CO_2 indicates a growth surge, a feedback system can automatically modify the glucose feeding rate to prevent overflow metabolism. These intelligent, data-informed controls ensure consistent performance (yield, purity, flavor profile, etc.) across vessels and even between factories, despite physical variations. The key for teams is to treat data as a valuable asset. Investing early in analytical tools and digital infrastructure is crucial. This not only enhances the reliability of daily processes but also facilitates smoother tech transfer and scale-up. By doing so, a company can strategically phase its growth, knowing when to scale up and making wise investments in proven steps. This approach enables the company to achieve reproducible results across different scales. The convergence of biology with digital technology (sensors, modeling, AI) is transforming fermentation into a more predictable and controllable engineering discipline. Embracing these approaches helps new ventures avoid costly surprises, accelerate market entry, and confidently expand production across multiple vessels and locations. Each step, from initial laboratory experiments to full-scale PAT deployment, builds on the principle that knowledge and control are the driving forces behind success in modern fermentation-based innovation.

References

Adamo. (2025). *Adamo foods*. Retrieved November 15 from https://www.adamofoods.com

Adeleke, I., Nwulu, N., & Adebo, O. A. (2023). Internet of Things (IoT) in the food fermentation process: A bibliometric review. *Journal of Food Process Engineering, 46*(5). https://doi.org/10.1111/jfpe.14321

Agrawal, R. (2024). Fermentation media and industrial sterilization. In *Textbook of industrial microbiology* (pp. 87–96). Springer.

Ahirwar, A., Rai, A., Sirotiya, V., Khandelwal, P., Singh, G., Jhadav, D., Harish, Marchand, J., Schoefs, B., Varjani, S., & Vinayak, V. (2023). Fermentation of algal biomass for its nutritional value: Perspectives and revolutions in the food industry. *Environmental Technology Reviews, 13*(1), 1–28. https://doi.org/10.1080/21622515.2023.2283097

Ahlborn, J., Stephan, A., Meckel, T., Maheshwari, G., Rühl, M., & Zorn, H. (2019). Upcycling of food industry side streams by basidiomycetes for production of a vegan protein source. *International Journal of Recycling of Organic Waste in Agriculture, 8*(S1), 447–455. https://doi.org/10.1007/s40093-019-00317-4

Ahmad, A., Yadav, A. K., Singh, A., & Singh, D. K. (2024). A comprehensive machine learning-coupled response surface methodology approach for predictive modeling and optimization of biogas potential in anaerobic co-digestion of organic waste. *Biomass and Bioenergy, 180*, 106995. https://doi.org/10.1016/j.biombioe.2023.106995

Albino, M., Gargalo, C. L., Nadal-Rey, G., Albæk, M. O., Krühne, U., & Gernaey, K. V. (2024). Hybrid modeling for on-line fermentation optimization and scale-up: a review. *Processes, 12*(8). https://doi.org/10.3390/pr12081635

Alfano, R., Pennybaker, A., Laskowski, A., & Lundeen, T. (2020). Implementation of the aber biomass probe in Pall's iCELLis nano bioreactor provides a robust and reproducible method to assess cell density. *Cytotherapy, 22*(5). https://doi.org/10.1016/j.jcyt.2020.03.309

Almada-Lobo, F. (2015). Define, integrate, implement MES with controls, ERP: Modern manufacturing execution systems (MES) can deliver the flexibility to enable manufacturers to define the solutions that meet their strategic needs. MES is what you make it: Define the MES functions needed, use standards, consider modularity, and integrate controls with MES and MES with enterprise resource planning (ERP), as needed [Article]. *Control Engineering, 62*, 38+. https://link.gale.com/apps/doc/A616448670/AONE?u=anon~fef7b5f0&sid=googleScholar&xid=b7a3aec4

Alosert, H., Savery, J., Rheaume, J., Cheeks, M., Turner, R., Spencer, C., Farid, S., & S., & Goldrick, S. (2022). Data integrity within the biopharmaceutical sector in the era of industry 4.0. *Biotechnology Journal, 17*(6). https://doi.org/10.1002/biot.202100609

Alsaeedi, A. H., Al-Mahmood, H. H. R., Alnaseri, Z. F., Aziz, M. R., Al-Shammary, D., Ibaida, A., & Ahmed, K. (2024). Fractal feature selection model for enhancing high-dimensional biological problems. *BMC Bioinformatics, 25*(1), 12. https://doi.org/10.1186/s12859-023-05619-z

Amanullah, A., McFarlane, C. M., Emery, A. N., & Nienow, A. W. (2001). Scale-down model to simulate spatial pH variations in large-scale bioreactors. *Biotechnology and Bioengineering, 73*(5), 390–399. https://doi.org/10.1002/bit.1072

Andersen, S. W., & Runger, G. C. (2011). Partitioned partial least squares regression with application to a batch fermentation process. *Journal of Chemometrics, 25*(4), 159–168.

Aparnna, V., Roy, K., Singh, S., Rose, H., Khamrui, K., & Gautam, A. K. (2025). Recent advancements in the application of blockchain, IoT, and fuzzy systems in food science and bioprocess development. In *Artificial intelligence in food science* (pp. 281–302). Elsevier.

Arroyo, A. Y. (2014). Advances in sensor and sampling technologies in fermentation and mammalian cell culture. In *Manual of industrial microbiology and biotechnology* (pp. 700–718). https://doi.org/10.1128/9781555816827.ch50

Augustin, M. A., Hartley, C. J., Maloney, G., & Tyndall, S. (2023). Innovation in precision fermentation for food ingredients. *Critical Reviews in Food Science and Nutrition, 64*(18), 6218–6238. https://doi.org/10.1080/10408398.2023.2166014

Badino, A. C., Facciotti, M. C. R., & Schmidell, W. (2001). Volumetric oxygen transfer coefficients (kLa) in batch cultivations involving non-Newtonian broths. *Biochemical Engineering Journal, 8*(2), 111–119. https://doi.org/10.1016/S1369-703X(01)00092-4

Baldwin, E. A., Bai, J., Plotto, A., & Dea, S. (2011). Electronic noses and tongues: Applications for the food and pharmaceutical industries. *Sensors, 11*(5), 4744–4766. https://doi.org/10.3390/s110504744

Baradez, M.-O., Biziato, D., Hassan, E., & Marshall, D. (2018). Application of Raman spectroscopy and univariate modelling as a process analytical technology for cell therapy bioprocessing. *Frontiers in Medicine, 5*, 47.

Batt, C. A., Rabson, L. D., Wong, D. W., & Kinsella, J. E. (1990). Expression of recombinant bovine beta-lactoglobulin in *Escherichia coli*. *Agricultural and Biological Chemistry, 54*(4), 949–955.

Bayer, B., Dalmau Diaz, R., Melcher, M., Striedner, G., & Duerkop, M. (2021). Digital twin application for model-based DoE to rapidly identify ideal process conditions for space-time yield optimization. *Processes, 9*(7). https://doi.org/10.3390/pr9071109

Berahmand, K., Daneshfar, F., Salehi, E. S., Li, Y., & Xu, Y. (2024). Autoencoders and their applications in machine learning: A survey. *Artificial Intelligence Review, 57*(2), 28.

Biechele, P., Busse, C., Solle, D., Scheper, T., & Reardon, K. (2015). Sensor systems for bioprocess monitoring. *Engineering in Life Sciences, 15*(5), 469–488. https://doi.org/10.1002/elsc.201500014

Bolmanis, E., Dubencovs, K., Suleiko, A., & Vanags, J. (2023). Model predictive control – A stand out among competitors for fed-batch fermentation improvement. *Fermentation, 9*(3), 206.

Bonturi, N. (2025). Biomass vs. precision fermentation – Understanding the technologies reshaping food innovation. *Vegconomist.*. Retrieved 04 November from https://cultivated-x.com/guest-post/guest-post-biomass-vs-precision-fermentation/

Borgosz, L., & Dikicioglu, D. (2024). Industrial internet of things: What does it mean for the bioprocess industries? *Biochemical Engineering Journal, 201*. https://doi.org/10.1016/j.bej.2023.109122

Braun, S. (2023). *Multi-agent systems to control, simulate, and optimize resilient smart factories RMIT University*. Doctorate by research, RMIT University.

Brunner, V., Klöckner, L., Kerpes, R., Geier, D. U., & Becker, T. (2019). Online sensor validation in sensor networks for bioprocess monitoring using swarm intelligence. *Analytical and Bioanalytical Chemistry, 412*(9), 2165–2175. https://doi.org/10.1007/s00216-019-01927-7

Brunner, V., Siegl, M., Geier, D., & Becker, T. (2021). Challenges in the development of soft sensors for bioprocesses: A critical review. *Frontiers in Bioengineering and Biotechnology, 9*. https://doi.org/10.3389/fbioe.2021.722202

Buchenauer, A., Hofmann, M. C., Funke, M., Büchs, J., Mokwa, W., & Schnakenberg, U. (2009). Micro-bioreactors for fed-batch fermentations with integrated online monitoring and microfluidic devices. *Biosensors and Bioelectronics, 24*(5), 1411–1416. https://doi.org/10.1016/j.bios.2008.08.043

Cairns, T. C., Zheng, X., Zheng, P., Sun, J., & Meyer, V. (2021). Turning inside out: Filamentous fungal secretion and its applications in biotechnology, agriculture, and the clinic. *Journal of Fungi, 7*(7), 535.

Calero, P., & Nikel, P. I. (2018). Chasing bacterial chassis for metabolic engineering: A perspective review from classical to non-traditional microorganisms. *Microbial Biotechnology, 12*(1), 98–124. https://doi.org/10.1111/1751-7915.13292

Carneiro, C. V. G. C., Serra, L. A., Pacheco, T. F., Ferreira, L. M. M., Brandão, L. T. D., Freitas, M. N., Trichez, D., & Almeida, J. R. (2022). Advances in Komagataella phaffii engineering for the production of renewable chemicals and proteins. *Fermentation, 8*(11). https://doi.org/10.3390/fermentation8110575

Cervera-Gascó, J., Pardo, J. E., Álvarez-Ortí, M., López-Mata, E., Zied, D. C., & Pardo-Giménez, A. (2023). An intelligent mushroom strain selection model based on their quality characteristics. *Food Bioscience, 56*, 103232.

Chai, K. F., Ng, K. R., Samarasiri, M., & Chen, W. N. (2022). Precision fermentation to advance fungal food fermentations. *Current Opinion in Food Science, 47*, 100881.

Chai, W. Y., Teo, K. T. K., Tan, M. K., & Tham, H. J. (2022a). Fermentation process control and optimization. *Chemical Engineering & Technology, 45*(10), 1731–1747. https://doi.org/10.1002/ceat.202200029

Chai, W. Y., Teo, K. T. K., Tan, M. K., & Tham, H. J. (2022b). *Model predictive control in fermentation process – A review*. AIP Conference Proceedings.

Chandrasekaran, K., Chandrasekaran, K., Jayaraman, G., & Bhatt, N. (2025). Model maintenance, monitoring, and control framework using in-situ NIR spectroscopy and offline process data for Lactococcus lactis fermentation. *Chemical Engineering Journal, 165116*.

Chaudhry, M. A. (2024). *Lessons in bioreactor scale-up: Part 1 – Exploring introductory principles*. Retrieved 05 November from https://www.bioprocessintl.com/bioreactors/lessons-in-bioreactor-scale-up-part-1-mdash-exploring-introductory-principles

Chen, D. (2005). Enterprise-control system integration – An international standard. *International Journal of Production Research, 43*(20), 4335–4357. https://doi.org/10.1080/00207540500142399

Chen, G. Q., & Liu, X. (2020). On the future fermentation. *Microbial Biotechnology, 14*(1), 18–21. https://doi.org/10.1111/1751-7915.13674

Chen, X., & Voigt, T. (2020). Implementation of the manufacturing execution system in the food and beverage industry. *Journal of Food Engineering, 278*. https://doi.org/10.1016/j.jfoodeng.2020.109932

Chen, B., Wan, J., Celesti, A., Li, D., Abbas, H., & Zhang, Q. (2018). Edge computing in IoT-based manufacturing. *IEEE Communications Magazine, 56*(9), 103–109. https://doi.org/10.1109/mcom.2018.1701231

Chen, X., Nophut, C., & Voigt, T. (2020). Manufacturing execution Systems for the Food and Beverage Industry: A model-driven approach. *Electronics, 9*(12). https://doi.org/10.3390/electronics9122040

Chen, S., Jia, Y., Duan, B., Liu, T.-L., Wang, Q., Xiao, X., Nithianandam, P., Tian, X., Yang, C., & Wu, C. (2025). A sensor-actuator–coupled gustatory interface chemically connecting virtual and real environments for remote tasting. *Science. Advances, 11*(9), eadr4797.

Chhetri, K. B. (2024). Applications of artificial intelligence and machine learning in food quality control and safety assessment. *Food Engineering Reviews, 16*(1), 1–21.

Cinar, A., Parulekar, S. J., Undey, C., & Birol, G. (2003). *Batch Fermentation*. https://doi.org/10.1201/9780203911358

Ciosek, P., & Wróblewski, W. (2006). The analysis of sensor array data with various pattern recognition techniques. *Sensors and Actuators B: Chemical, 114*(1), 85–93. https://doi.org/10.1016/j.snb.2005.04.008

Classen, J., Langer, M., Jockwer, A., & Traenkle, J. (2022). Raman spectrometric PAT models: Larger-scale, stirred-tank bioreactorslarger-scale, stirred-tank bioreactors. *BioProcess International, 20*(9), 34–38.

Cosenza, Z., Astudillo, R., Frazier, P. I., Baar, K., & Block, D. E. (2022). Multi-information source Bayesian optimization of culture media for cellular agriculture. *Biotechnology and Bioengineering, 119*(9), 2447–2458. https://doi.org/10.1002/bit.28132

Dazzarola, C., Tighe, R., Pérez-Correa, J. R., & Saa, P. A. (2026). Toward a digital twin for beer quality control: Development of a digital model integrating industrial process data and model-based fermentation descriptors. *Journal of Food Engineering, 403*, 112726.

De Minicis, M., Giordano, F., Poli, F., & Schiraldi, M. M. (2014). Recipe development process re-design with ANSI/ISA-88 batch control standard in the pharmaceutical industry. *International Journal of Engineering Business Management, 6*. https://doi.org/10.5772/59025

Demuth, C., Varonier, J., Jossen, V., Eibl, R., & Eibl, D. (2016). Novel probes for pH and dissolved oxygen measurements in cultivations from millilitre to benchtop scale. *Applied Microbiology and Biotechnology, 100*(9), 3853–3863. https://doi.org/10.1007/s00253-016-7412-0

Dessaigne, A., Briane, A., Daboussi, F., Cescut, J., & Corrales, D. C. (2024). From legacy systems to data pipelines modernization in fermentation process. In *Computer aided chemical engineering* (Vol. 53, pp. 3205–3210). Elsevier.

Dias, D. R., & Schwan, R. F. (2024). History and perspectives. In R. F. Schwan, V. K. Joshi, & D. R. Dias (Eds.), *Bioreactor technology in food processing* (pp. 1–17). CRC Press.

Domján, J., Pantea, E., Gyürkés, M., Madarász, L., Kozák, D., Farkas, A., Horváth, B., Benkő, Z., Nagy, Z. K., Marosi, G., & Hirsch, E. (2022). Real-time amino acid and glucose monitoring system for the automatic control of nutrient feeding in CHO cell culture using Raman spectroscopy. *Biotechnology Journal, 17*(5). https://doi.org/10.1002/biot.202100395

Dubencovs, K., Suleiko, A., Sile, E., Petrovskis, I., Akopjana, I., Suleiko, A., Galvanauskas, V., Tars, K., & Vanags, J. (2023). The application of adaptive model predictive control for fed-batch Escherichia coli BL21 (DE3) cultivation and biosynthesis of recombinant proteins. *Fermentation, 9*(12), 1015.

Duque-Estrada, P., Hardiman, K., Dam, A. B., Dodge, N., Aaslyng, M. D., & Petersen, I. L. (2023). Protein blends and extrusion processing to improve the nutritional quality of plant proteins. *Food & Function, 14*(16), 7361–7374.

Dwivedi, C. M. (2005). Validation of cell culture-based processes and qualification of associated equipment and facility. In *Cell culture technology for pharmaceutical and cell-based therapies* (pp. 541–570). CRC Press.

Eastham, J. L., & Leman, A. R. (2024). Precision fermentation for food proteins: Ingredient innovations, bioprocess considerations, and outlook – A mini-review. *Current Opinion in Food Science, 58*, 101194.

Elragal, A., & Elgendy, N. (2024). A data-driven decision-making readiness assessment model: The case of a Swedish food manufacturer. *Decision Analytics Journal, 10*. https://doi.org/10.1016/j.dajour.2024.100405

Enfors, S. O., Jahic, M., Rozkov, A., Xu, B., Hecker, M., Jürgen, B., Krüger, E., Schweder, T., Hamer, G., O'Beirne, D., Noisommit-Rizzi, N., Reuss, M., Boone, L., Hewitt, C., McFarlane, C., Nienow, A., Kovacs, T., Trägårdh, C., Fuchs, L., & Manelius, Å. (2001). Physiological responses to mixing in large scale bioreactors. *Journal of Biotechnology, 85*(2), 175–185. https://doi.org/10.1016/s0168-1656(00)00365-5

Fan, D., Yuan, Z., Tang, H., Ren, P., & Han, S. (2025). Metabolic engineering of Pichia pastoris as an industrial chassis enables biosynthesis of dopamine from methanol. *Bioresource Technology, 436*. https://doi.org/10.1016/j.biortech.2025.132915

Felo, M., Christensen, B., & Higgins, J. (2013). Process cost and facility considerations in the selection of primary cell culture clarification technology. *Biotechnology Progress, 29*(5), 1239–1245.

Fisher, O. J., Watson, N. J., Porcu, L., Bacon, D., Rigley, M., & Gomes, R. L. (2022). Data-driven modelling for resource recovery: Data volume, variability, and visualisation for an industrial bioprocess. *Biochemical Engineering Journal, 185*. https://doi.org/10.1016/j.bej.2022.108499

Forbes, M. G., Patwardhan, R. S., Hamadah, H., & Gopaluni, R. B. (2015). Model predictive control in industry: Challenges and opportunities. *IFAC-PapersOnLine, 48*(8), 531–538.

Foss, B. A., Johansen, T. A., & Sørensen, A. V. (1995). Nonlinear predictive control using local models – Applied to a batch fermentation process. *Control Engineering Practice, 3*(3), 389–396.

Fu, Z., Leighton, J., Cheng, A., Appelbaum, E., & Aon, J. C. (2012). Optimization of a Saccharomyces cerevisiae fermentation process for production of a therapeutic recombinant protein using a multivariate Bayesian approach. *Biotechnology Progress, 28*(4), 1095–1105.

Gangadharan, N., Sewell, D., Turner, R., Field, R., Cheeks, M., Oliver, S. G., Slater, N. K., & Dikicioglu, D. (2021). Data intelligence for process performance prediction in biologics manufacturing. *Computers & Chemical Engineering, 146*, 107226.

Garcia-Ochoa, F., & Gomez, E. (2005). Prediction of gas-liquid mass transfer coefficient in sparged stirred tank bioreactors. *Biotechnology and Bioengineering, 92*(6), 761–772. https://doi.org/10.1002/bit.20638

Gasset, A., Van Wijngaarden, J., Mirabent, F., Sales-Vallverdú, A., Garcia-Ortega, X., Montesinos-Seguí, J. L., Manzano, T., & Valero, F. (2024). Continuous process verification 4.0 application in upstream: Adaptiveness implementation managed by AI in the hypoxic bioprocess of the Pichia pastoris cell factory. *Frontiers in Bioengineering and Biotechnology, 12*. https://doi.org/10.3389/fbioe.2024.1439638

Geinitz, B., Hüser, A., Mann, M., & Büchs, J. (2020). Gas fermentation expands the scope of a process network for material conversion. *Chemie Ingenieur Technik, 92*(11), 1665–1679. https://doi.org/10.1002/cite.202000086

Geistlinger, T., Briggs, N., & Nay, K. (2024). Case study on whey protein from fermentation. In *Cellular agriculture* (pp. 323–342). https://doi.org/10.1016/b978-0-443-18767-4.00021-4

Gerzon, G., Sheng, Y., & Kirkitadze, M. (2022). Process analytical technologies – Advances in bioprocess integration and future perspectives. *Journal of Pharmaceutical and Biomedical Analysis, 207*. https://doi.org/10.1016/j.jpba.2021.114379

Gewald, D., Rostek, K., Schuster, A., & Aumann, R. (2015). From technology development to (pre-series) product–the epack hybrid. In *ASME-ORC2015: Proceedings of the 3rd international seminar on ORC power systems*.

Ghosh, G., Bodroth, R. P., & Dutta, B. (2022). Impact of process validation and equipment qualification in production of bio-therapeutics. *Journal of Advanced Biotechnology and Experimental Therapeutics, 5*(1), 163.

Gisperg, F., Klausser, R., Elshazly, M., Kopp, J., Brichtová, E. P., & Spadiut, O. (2025). Bayesian optimization in bioprocess engineering – Where do we stand today? *Biotechnology and Bioengineering, 122*(6), 1313–1325.

Gliszczyńska-Świgło, A., & Chmielewski, J. (2016). Electronic nose as a tool for monitoring the authenticity of food. A review. *Food Analytical Methods, 10*(6), 1800–1816. https://doi.org/10.1007/s12161-016-0739-4

Gnoth, S., Simutis, R., & Lübbert, A. (2011). Fermentation process supervision and strategies for fail-safe operation: A practical approach. *Engineering in Life Sciences, 11*(1), 94–106.

Goh, H.-Y., Sulu, M., Alosert, H., Lewis, G. L., Josland, G. D., & Merriman, D. E. (2019). Applications of off-gas mass spectrometry in fed-batch mammalian cell culture. *Bioprocess and Biosystems Engineering, 43*(3), 483–493. https://doi.org/10.1007/s00449-019-02242-2

Gokulakrishnan, D., & Venkataraman, S. (2025). Ensuring data integrity: Best practices and strategies in pharmaceutical industry. *Intelligent Pharmacy, 3*(4), 296–303. https://doi.org/10.1016/j.ipha.2024.09.010

Gomersall, R., Hitzmann, B., & Guthke, R. (1997). Fermentation fault management and on-line state estimation using methods of cluster analysis. *Bioprocess Engineering, 17*(2), 69–73.

Gomes, J., Chopda, V. R., & Rathore, A. S. (2014). Integrating systems analysis and control for implementing process analytical technology in bioprocess development. *Journal of Chemical Technology & Biotechnology, 90*(4), 583–589. https://doi.org/10.1002/jctb.4591

Goodwin, T., Xu, J., Celik, N., & Chen, C.-H. (2022). Real-time digital twin-based optimization with predictive simulation learning. *Journal of Simulation, 18*(1), 47–64. https://doi.org/10.1080/17477778.2022.2046520

Gopalakrishnan, A. V., Suresh, H., Mariappan, G., Kanagaraja, A., & Raman, P. (2024). Industrial microbial bioprocess development: A comprehensive overview. In *Industrial microbiology and biotechnology* (pp. 89–154). https://doi.org/10.1007/978-981-97-6270-5_4

Hancock, C. (2015). Role of microbial fermentation on degradation of folic acid in micronutrient fortified corn masa prior to baking. *Journal of Undergraduate Research, 2015*(1), 247.

Hassan, M., Montague, G., Iqbal, M. Z., & Fahey, J. (2024). Virtual reality-based bioreactor digital twin for operator training. *Digital Chemical Engineering, 11*. https://doi.org/10.1016/j.dche.2024.100147

Hawkins, W., Brandl, D., & Boyes, W. (2011). *ISA 88 and ISA 95 in the life science industries* (Vol. 4). Momentum Press.

Hazra, A., Adhikari, M., Amgoth, T., & Srirama, S. N. (2021). A comprehensive survey on interoperability for IIoT: Taxonomy, standards, and future directions. *ACM Computing Surveys, 55*(1), 1–35. https://doi.org/10.1145/3485130

He, Q., & Chen, H. (2020). Increased efficiency of butanol production from spent sulfite liquor by removal of fermentation inhibitors. *Journal of Cleaner Production, 263*. https://doi.org/10.1016/j.jclepro.2020.121356

Heins, A. L., Hoang, M. D., & Weuster-Botz, D. (2021). Advances in automated real-time flow cytometry for monitoring of bioreactor processes. *Engineering in Life Sciences, 22*(3–4), 260–278. https://doi.org/10.1002/elsc.202100082

Hema, P., Sathish, E., Maheswari, M., Khosla, A., Pant, B., & Ambethkar, M. R. (2022). Robust soft sensor systems for industry: Evaluated through real-time case study. *Measurement: Sensors, 24*, 100542.

Hernández Rodríguez, T., Sekulic, A., Lange-Hegermann, M., & Frahm, B. (2022). Designing robust biotechnological processes regarding variabilities using multi-objective optimization applied to a biopharmaceutical seed train design. *Processes, 10*(5), 883.

Hernandez, J., Medina, R., & Hernandez, M. (2012). *Instrumentation and design of a supervisory system for an anaerobic biodigestor*. In *2012 IEEE international symposium on alternative energies and energy quality (SIFAE)*.

Herrera-Ruiz, J. F., Fontalvo, J., & Prado-Rubio, O. A. (2025). Hybrid modeling for bioprocesses: Architectures, applications, and perspectives. *Engineering Reports, 7*(12), e70502.

Hill, D. (1983). Simplified monod kinetics of methane fermentation of animal wastes. *Agricultural Wastes, 5*(1), 1–16.

Hill, C. (2020). *Using lab informatics and instrumentation to optimize upstream bioprocessing*.

Hjersted, J. L., & Henson, M. A. (2008). Optimization of fed-batch Saccharomyces cerevisiae fermentation using dynamic flux balance models. *Biotechnology Progress, 22*(5), 1239–1248. https://doi.org/10.1021/bp060059v

Ho, P., Westerwalbesloh, C., Kaganovitch, E., Grünberger, A., Neubauer, P., Kohlheyer, D., & Lieres, E. V. (2019). Reproduction of large-scale bioreactor conditions on microfluidic chips. *Microorganisms, 7*(4). https://doi.org/10.3390/microorganisms7040105

Hofer, A., Kroll, P., Barmettler, M., & Herwig, C. (2020). A reliable automated sampling system for on-line and real-time monitoring of CHO cultures. *Processes, 8*(6). https://doi.org/10.3390/pr8060637

Hood, G. W. (2015). Industry standard models for design/discussion of manufacturing systems. In *2015 5th Australian control conference (AUCC)*.

Hua, L., Zhang, C., Sun, W., Li, Y., Xiong, J., & Nazir, M. S. (2023). An evolutionary deep learning soft sensor model based on random forest feature selection technique for penicillin fermentation process. *ISA Transactions, 136*, 139–151.

Huang, X.-C., Guo, C.-F., Yuan, Y.-H., Luo, X.-X., & Yue, T.-L. (2015). Detection of medicinal off-flavor in apple juice with artificial sensing system and comparison with test panel evaluation and GC–MS. *Food Control, 51*, 270–277. https://doi.org/10.1016/j.foodcont.2014.11.037

Huggins, T., & Whiteley, J. (2025). *Meati foods*. Retrieved November 15 from https://www.meati.com

Hüttel, S., & Müller, R. (2012). Methods to optimize myxobacterial fermentations using off-gas analysis. *Microbial Cell Factories, 11*(1). https://doi.org/10.1186/1475-2859-11-59

Isoko, K., Cordiner, J. L., Kis, Z., & Moghadam, P. Z. (2024). Bioprocessing 4.0: A pragmatic review and future perspectives. *Digital Discovery, 3*(9), 1662–1681.

Jain, V., & Yadav, A. (2025). Automation and advanced control systems in bio-manufacturing. In *Industrial applications for bioprocessing and biomanufacturing* (pp. 143–172). https://doi.org/10.4018/979-8-3373-2873-7.ch006

Jarva, A. (2024). *OPC UA PubSub in industrial systems integration*. Tampere University.

Jaskó, S., Skrop, A., Holczinger, T., Chován, T., & Abonyi, J. (2020). Development of manufacturing execution systems in accordance with industry 4.0 requirements: A review of standard- and ontology-based methodologies and tools. *Computers in Industry, 123*. https://doi.org/10.1016/j.compind.2020.103300

Jeong, J., & Kim, S. (2025). Application of machine learning for quantitative analysis of industrial fermentation using image processing. *Food Science and Biotechnology, 34*(2), 373–381.

Jha, P., & Singh, A. K. (2025). Regulatory compliance and food safety standards. In *Engineering solutions for sustainable food and dairy production: Innovations and techniques in food processing and dairy engineering* (pp. 463–487). Springer.

Ji, T., & Xu, X. (2025). Exploring the integration of cloud manufacturing and cyber-physical systems in the era of industry 4.0 – An OPC UA approach. *Robotics and Computer-Integrated Manufacturing, 93*. https://doi.org/10.1016/j.rcim.2024.102927

Joanofarc, X., Nivedhika, D., Patnaik, S. K., & Panda, R. C. (2019). Closed-loop performance and analysis of a real time non-linear bioreactor process. In *2019 2nd international conference on power and embedded drive control (ICPEDC)*.

Jornitz, M. W. (2019). Filter manufacturer's quality assurance and qualifications. In *Filtration and purification in the biopharmaceutical industry* (3rd ed., pp. 277–293). CRC Press.

Kadlec, P., Gabrys, B., & Strandt, S. (2009). Data-driven soft sensors in the process industry. *Computers & Chemical Engineering, 33*(4), 795–814.

Kakes, E. (2008). In B. McNeil & L. M. Harvey (Eds.), *Practical fermentation technology* (pp. 289–322). https://doi.org/10.1002/9780470725306

Kaneko, H., & Funatsu, K. (2011). A soft sensor method based on values predicted from multiple intervals of time difference for improvement and estimation of prediction accuracy. *Chemometrics and Intelligent Laboratory Systems, 109*(2), 197–206.

Karnachoriti, M., Chatzipetrou, M., Touloupakis, E., Kontos, A. G., & Zergioti, I. (2025). Raman spectroscopy as a tool for real-time nutrient monitoring in bioreactor cultivation of microalgae. *Journal of Raman Spectroscopy, 56*(9), 817–826. https://doi.org/10.1002/jrs.6841

Keppler, J. K., Heyse, A., Scheidler, E., Uttinger, M. J., Fitzner, L., Jandt, U., Heyn, T. R., Lautenbach, V., Loch, J. I., Lohr, J., Kieserling, H., Günther, G., Kempf, E., Grosch, J.-H., Lewiński, K., Jahn, D., Lübbert, C., Peukert, W., Kulozik, U., & Biedendieck, R. (2021). Towards recombinantly produced milk proteins: Physicochemical and emulsifying properties of engineered whey protein beta-lactoglobulin variants. *Food Hydrocolloids, 110*. https://doi.org/10.1016/j.foodhyd.2020.106132

Khaleghi, M. K., Savizi, I. S. P., Lewis, N. E., & Shojaosadati, S. A. (2021). Synergisms of machine learning and constraint-based modeling of metabolism for analysis and optimization of fermentation parameters. *Biotechnology Journal, 16*(11), 2100212.

Khalid, Z., Chen, Y., Liu, X., Noureen, B., Chen, Y., Wang, M., Ma, Y., Du, L., & Wu, C. (2025). Recent advances and unaddressed challenges in biomimetic olfactory-and taste-based biosensors: Moving towards integrated, AI-powered, and market-ready sensing systems. *Sensors, 25*(22), 7000.

Khuri, A. I., & Mukhopadhyay, S. (2010). Response surface methodology. *Wiley Interdisciplinary Reviews: Computational Statistics, 2*(2), 128–149.

Kim, E. J., Kim, J. H., Kim, M.-S., Jeong, S. H., & Choi, D. H. (2021). Process analytical technology tools for monitoring pharmaceutical unit operations: A control strategy for continuous process verification. *Pharmaceutics, 13*(6). https://doi.org/10.3390/pharmaceutics13060919

Kinnunen, H. (2024). *Cloud communication of factory automation system with opc ua.* Aalto University.

Kiran Deokate, K. D., Shrikrishna Baokar, S. B., & Rajendra Patil, R. P. (2025). Comprehensive overview of cleaning-in-place (CIP) systems and their validation in pharmaceutical manufacturing. *International Journal of Pharmaceutical Research and Applications, 10*(4), 440–447. https://doi.org/10.35629/4494-1004440447

Kiviharju, K., Salonen, K., Moilanen, U., & Eerikäinen, T. (2008). Biomass measurement online: The performance of in situ measurements and software sensors. *Journal of Industrial Microbiology & Biotechnology, 35*(7), 657–665. https://doi.org/10.1007/s10295-008-0346-5

Klein, T., Schneider, K., & Heinzle, E. (2012). A system of miniaturized stirred bioreactors for parallel continuous cultivation of yeast with online measurement of dissolved oxygen and off-gas. *Biotechnology and Bioengineering, 110*(2), 535–542. https://doi.org/10.1002/bit.24633

Konzock, O., & Nielsen, J. (2024). TRYing to evaluate production costs in microbial biotechnology. *Trends in Biotechnology, 42*(11), 1339–1347. https://doi.org/10.1016/j.tibtech.2024.04.007

Kordon, A. K., Smits, G. F., Kalos, A. N., & Jordaan, E. M. (2003). Robust soft sensor development using genetic programming. *Data Handling in Science and Technology, 23*, 69–108.

Koulouris, A., Misailidis, N., & Petrides, D. (2021). Applications of process and digital twin models for production simulation and scheduling in the manufacturing of food ingredients and products. *Food and Bioproducts Processing, 126*, 317–333. https://doi.org/10.1016/j.fbp.2021.01.016

Kumar, H., & Yadav, S. P. (2019). Hybrid NSGA-II based decision-making in fuzzy multi-objective reliability optimization problem. *SN Applied Sciences, 1*(11). https://doi.org/10.1007/s42452-019-1512-2

Kwon, H. J., Shiu, J. H., Yamakawa, C. K., & Rivera, E. C. (2024). Enhancing fermentation process monitoring through data-driven modeling and synthetic time series generation. *Bioengineering, 11*(8), 803.

Ladegourdie, M., & Kua, J. (2022). Performance analysis of OPC UA for industrial interoperability towards industry 4.0. *IoT, 3*(4), 507–525. https://doi.org/10.3390/iot3040027

Lapierre, F. M., Mattaliano, P., Raith, D., Castillo-Cota, M., Bermeitinger, J., & Huber, R. (2025). Multi-cycle high-throughput growth media optimization using batch Bayesian optimization. *Journal of Chemical Technology & Biotechnology, 100*(8), 1571–1583.

Larsson, G., & Törnkvist, M. (1996). Rapid sampling, cell inactivation and evaluation of low extracellular glucose concentrations during fed-batch cultivation. *Journal of Biotechnology, 49*(1–3), 69–82. https://doi.org/10.1016/0168-1656(96)01534-9

Leal, F., Chis, A. E., Caton, S., González-Vélez, H., García-Gómez, J. M., Durá, M., Sánchez-García, A., Sáez, C., Karageorgos, A., Gerogiannis, V. C., Xenakis, A., Lallas, E., Ntounas, T., Vasileiou, E., Mountzouris, G., Otti, B., Pucci, P., Papini, R., Cerrai, D., & Mier, M. (2021). Smart pharmaceutical manufacturing: Ensuring end-to-end traceability and data integrity in medicine production. *Big Data Research, 24*. https://doi.org/10.1016/j.bdr.2020.100172

Lee, N. Y. (2018). A review on microscale polymerase chain reaction based methods in molecular diagnosis, and future prospects for the fabrication of fully integrated portable biomedical devices. *Microchimica Acta, 185*(6). https://doi.org/10.1007/s00604-018-2791-9

Lee, J. A., Kim, H. U., Na, J.-G., Ko, Y.-S., Cho, J. S., & Lee, S. Y. (2023). Factors affecting the competitiveness of bacterial fermentation. *Trends in Biotechnology, 41*(6), 798–816.

Lee, M., Kim, D., Hwang, I. M., & Chang, J. Y. (2024). Exploring flavor perception through metabolite profiling and sensory approaches during starter kimchi fermentation. *Food Bioscience, 61*. https://doi.org/10.1016/j.fbio.2024.104477

Lee, J., Byun, J., & Kim, S. (2025). Scalable cloud-based relay architecture for blockchain logging from distributed IoT sensors in fermentation monitoring. *Food Science and Biotechnology*. https://doi.org/10.1007/s10068-025-02024-5

Li, J. (2024). *Digital twin based on computational fluid dynamics and cell kinetic model of a large-scale bioreactor for vanillin production optimization*. National University of Singapore.

Li, M., Ebel, B., Chauchard, F., Guédon, E., & Marc, A. (2018). Parallel comparison of in situ Raman and NIR spectroscopies to simultaneously measure multiple variables toward real-time monitoring of CHO cell bioreactor cultures. *Biochemical Engineering Journal, 137*, 205–213. https://doi.org/10.1016/j.bej.2018.06.005

Li, C., Mantravadi, S., Schou, C., Nielsen, H., Madsen, O., & Møller, C. (2021). An ISA-95 based middle data layer for data standardization – Enhancing systems interoperability for factory automation. In *Advances in automotive production technology – Theory and application* (pp. 187–194). https://doi.org/10.1007/978-3-662-62962-8_22

Li, L., Li, N., Wang, X., Zhao, J., Zhang, H., & Jiao, T. (2023). Multi-output soft sensor modeling approach for penicillin fermentation process based on features of big data. *Expert Systems with Applications, 213*. https://doi.org/10.1016/j.eswa.2022.119208

Li, S., Du, D., Wang, J., & Wei, Z. (2024). Application progress of intelligent flavor sensing system in the production process of fermented foods based on the flavor properties. *Critical Reviews in Food Science and Nutrition, 64*(12), 3764–3793.

Liang, S., McDonald, A. G., & Coats, E. R. (2014). Lactic acid production with undefined mixed culture fermentation of potato peel waste. *Waste Management, 34*(11), 2022–2027. https://doi.org/10.1016/j.wasman.2014.07.009

Liu, W.-C., Gong, T., Wang, Q.-H., Liang, X., Chen, J.-J., & Zhu, P. (2016). Scaling-up fermentation of Pichia pastoris to demonstration-scale using new methanol-feeding strategy and increased air pressure instead of pure oxygen supplement. *Scientific Reports, 6*(1). https://doi.org/10.1038/srep18439

Liu, X., Tang, K., & Hu, J. (2024). Application of cyanobacteria as chassis cells in synthetic biology. *Microorganisms, 12*(7). https://doi.org/10.3390/microorganisms12071375

Liu, C., Li, F., Zhang, P., & Balasubramanian, P. (2025). Augmented machine learning with limited data for hydrogen yield prediction in wastewater dark fermentation. *npj Clean Water, 8*(1), 101.

Llamuca, J. D., Garcia, C. A., Naranjo, J. E., Rosero, C., Alvarez-M, E., & Garcia, M. V. (2020). Integrating ISA-95 and IEC-61499 for distributed control system monitoring. In *Information and communication technologies of Ecuador (TIC.EC)* (pp. 66–80). https://doi.org/10.1007/978-3-030-35740-5_5

Lock, T. J., Mah, S. H., & Lai, Z. W. (2023). Versatile applications of brewer's spent grain: Solid-state fermentation and nutritional added value. *Applied Biochemistry and Biotechnology, 196*(8), 5508–5532. https://doi.org/10.1007/s12010-023-04769-3

Lonchamp, J., Stewart, K., Munialo, C. D., Evans, L., Akintoye, M., Gordon, S., Clegg, P. S., Willoughby, N., & Euston, S. R. (2022). Mycoprotein as novel functional ingredient: Mapping of functionality, composition and structure throughout the Quorn fermentation process. *Food Chemistry, 396*, 133736.

Lopez, P. C., Udugama, I. A., Thomsen, S. T., Roslander, C., Junicke, H., Mauricio-Iglesias, M., & Gernaey, K. V. (2020). Towards a digital twin: A hybrid data-driven and mechanistic digital shadow to forecast the evolution of lignocellulosic fermentation. *Biofuels, Bioproducts and Biorefining, 14*(5), 1046–1060.

Lourenço, N. D., Lopes, J. A., Almeida, C. F., Sarraguça, M. C., & Pinheiro, H. M. (2012). Bioreactor monitoring with spectroscopy and chemometrics: A review. *Analytical and Bioanalytical Chemistry, 404*(4), 1211–1237. https://doi.org/10.1007/s00216-012-6073-9

Lu, B., & Chiang, L. (2018). Semi-supervised online soft sensor maintenance experiences in the chemical industry. *Journal of Process Control, 67*, 23–34.

Luttmann, R., Bracewell, D. G., Cornelissen, G., Gernaey, K. V., Glassey, J., Hass, V. C., Kaiser, C., Preusse, C., Striedner, G., & Mandenius, C. F. (2012). Soft sensors in bioprocessing: A status report and recommendations. *Biotechnology Journal, 7*(8), 1040–1048. https://doi.org/10.1002/biot.201100506

Lyubenova, V., Ignatova, M., Zoteva, D., & Roeva, O. (2024). Model-based adaptive control of bioreactors – A brief review. *Mathematics, 12*(14). https://doi.org/10.3390/math12142205

Ma, Z., Xiao, M., Xiao, Y., Pang, Z., Poor, H. V., & Vucetic, B. (2019). High-reliability and low-latency wireless communication for Internet of Things: Challenges, fundamentals, and enabling technologies. *IEEE Internet of Things Journal, 6*(5), 7946–7970. https://doi.org/10.1109/jiot.2019.2907245

Majumder, R., Miatur, S., Saha, A., & Hossain, S. (2024). Mycoprotein: Production and nutritional aspects: A review. *Sustainable Food Technology, 2*(1), 81–91. https://doi.org/10.1039/d3fb00169e

Mamma, D. (2020). Food wastes: Feedstock for value-added products. *Fermentation, 6*(2). https://doi.org/10.3390/fermentation6020047

Mao, S., Jiang, J., Xiong, K., Chen, Y., Yao, Y., Liu, L., Liu, H., & Li, X. (2024). Enzyme engineering: Performance optimization, novel sources, and applications in the food industry. *Food, 13*(23). https://doi.org/10.3390/foods13233846

Maruthamuthu, M. K., Rudge, S. R., Ardekani, A. M., Ladisch, M. R., & Verma, M. S. (2020). Process analytical technologies and data analytics for the manufacture of monoclonal antibodies. *Trends in Biotechnology, 38*(10), 1169–1186. https://doi.org/10.1016/j.tibtech.2020.07.004

Masykur, F., Prasetyo, A., Widaningrum, I., Cobantoro, A. F., & Setyawan, M. B. (2020). Application of message queuing telemetry transport (MQTT) protocol in the internet of things to monitor mushroom cultivation. In *2020 7th international conference on information technology, computer, and electrical engineering (ICITACEE).*

Mattarozzi, M., Laski, E., Bertucci, A., Giannetto, M., Bianchi, F., Zoani, C., & Careri, M. (2023). Metrological traceability in process analytical technologies and point-of-need technologies for food safety and quality control: Not a straightforward issue. *Analytical and Bioanalytical Chemistry, 415*(1), 119–135.

Mauri, M., Gouzé, J.-L., De Jong, H., & Cinquemani, E. (2020). Enhanced production of heterologous proteins by a synthetic microbial community: Conditions and trade-offs. *PLoS Computational Biology, 16*(4), e1007795.

Mazarevica, G., Diewok, J., Baena, J. R., Rosenberg, E., & Lendl, B. (2004). On-line fermentation monitoring by mid-infrared spectroscopy. *Applied Spectroscopy, 58*(7), 804–810. https://doi.org/10.1366/0003702041389229

Medl, M., Rajamanickam, V., Striedner, G., & Newton, J. (2023). Development and validation of an artificial neural-network-based optical density soft sensor for a high-throughput fermentation system. *Processes, 11*(1), 297.

Mengers, H. G., Zimmermann, M., & Blank, L. M. (2022). Using off-gas for insights through online monitoring of ethanol and baker's yeast volatilome using SESI-Orbitrap MS. *Scientific Reports, 12*(1). https://doi.org/10.1038/s41598-022-16554-z

Metcalfe, B., Acosta-Pavas, J. C., Robles-Rodriguez, C. E., Georgakilas, G. K., Dalamagas, T., Aceves-Lara, C. A., Daboussi, F., Koehorst, J. J., & Corrales, D. C. (2025). Towards a machine learning operations (MLOps) soft sensor for real-time predictions in industrial-scale fed-batch fermentation. *Computers & Chemical Engineering, 194*. https://doi.org/10.1016/j.compchemeng.2024.108991

Metze, S., Ruhl, S., Greller, G., Grimm, C., & Scholz, J. (2019). Monitoring online biomass with a capacitance sensor during scale-up of industrially relevant CHO cell culture fed-batch processes in single-use bioreactors. *Bioprocess and Biosystems Engineering, 43*(2), 193–205. https://doi.org/10.1007/s00449-019-02216-4

Miley, B., Riley, J., & Zelmanovich, Y. (2015). Large-scale fermentation systems: Hygienic design principles. *Chemical Engineering, 122*(11), 59.

Mione, F. M., Luna, M. F., Kaspersetz, L., Neubauer, P., Martinez, E. C., & Cruz Bournazou, M. N. (2025). A property graph schema for automated metadata capture, reproducibility and knowledge discovery in high-throughput bioprocess development. *Digital Discovery, 4*(9), 2401–2422. https://doi.org/10.1039/d5dd00070j

Mitra, S., & Murthy, G. S. (2021). Bioreactor control systems in the biopharmaceutical industry: A critical perspective. *Systems Microbiology and Biomanufacturing, 2*(1), 91–112. https://doi.org/10.1007/s43393-021-00048-6

Mohd, N., & Aziz, N. (2016). Performance and robustness evaluation of nonlinear autoregressive with exogenous input model predictive control in controlling industrial fermentation process. *Journal of Cleaner Production, 136*, 42–50.

Molfetta, M., Morais, E. G., Barreira, L., Bruno, G. L., Porcelli, F., Dugat-Bony, E., Bonnarme, P., & Minervini, F. (2022). Protein sources alternative to meat: State of the art and involvement of fermentation. *Food, 11*(14), 2065.

Monaco, A. (2025). *Regulating food innovations in the EU: An analysis of the novel food regulation.*

Montague, G. A., Morris, A. J., & Ward, A. C. (1989). Fermentation monitoring and control: A perspective. *Biotechnology and Genetic Engineering Reviews, 7*(1), 147–188. https://doi.org/10.1080/02648725.1989.10647858

Moon, S. Y., Jeong, C. H., Hwang, I. M., Kim, D., Lee, M. E., Jeong, S.-G., & Kim, H. M. (2025). Modulating kimchi taste and fermentation with Latilactobacillus sakei, Leuconostoc mesenteroides, and Weissella koreensis starters. *Food Research International, 220*. https://doi.org/10.1016/j.foodres.2025.117055

Moreddu, R. (2025). Integrated data-driven biotechnology research environments. *Database, 2025*. https://doi.org/10.1093/database/baaf064

Mori, M., Marinari, E., & De Martino, A. (2019). A yield-cost tradeoff governs Escherichia coli's decision between fermentation and respiration in carbon-limited growth. *npj Systems Biology and Applications, 5*(1). https://doi.org/10.1038/s41540-019-0093-4

Morris, C., Madhavarao, C. N., Yoon, S., & Ashraf, M. (2021). Single in-line biomass probe detects CHO cell growth by capacitance and bacterial contamination by conductivity in bioreactor. *Biotechnology Journal, 16*(12). https://doi.org/10.1002/biot.202100126

Moser, A., Appl, C., Pörtner, R., Baganz, F., & Hass, V. C. (2024). A new concept for the rapid development of digital twin Core models for bioprocesses in various reactor designs. *Fermentation, 10*(9). https://doi.org/10.3390/fermentation10090463

Mowbray, M., Kay, H., Kay, S., Caetano, P. C., Hicks, A., Mendoza, C., Lane, A., Martin, P., & Zhang, D. (2022a). Probabilistic machine learning based soft-sensors for product quality prediction in batch processes. *Chemometrics and Intelligent Laboratory Systems, 228*, 104616.

Mowbray, M., Petsagkourakis, P., del Rio-Chanona, E. A., & Zhang, D. (2022b). Safe chance constrained reinforcement learning for batch process control. *Computers & Chemical Engineering, 157*. https://doi.org/10.1016/j.compchemeng.2021.107630

Mu'azzam, K., da Silva, F. V. S., Murtagh, J., & Gallagher, M. J. S. (2024). A roadmap for model-based bioprocess development. *Biotechnology Advances, 73*, 108378.

Mukherjee, A., Gómez-Sala, B., O'Connor, E. M., Kenny, J. G., & Cotter, P. D. (2022). Global regulatory frameworks for fermented foods: A review. *Frontiers in Nutrition, 9*. https://doi.org/10.3389/fnut.2022.902642

Murugan, C., & Subbaian, S. (2022). Multi-objective optimization for enhanced ethanol production during whey fermentation. In *2022 international conference on power, energy, control and transmission systems (ICPECTS).*

Narayanan, H., Hinckley, J. A., Barry, R., Dang, B., Wolffe, L. A., Atari, A., Tseng, Y.-Y., & Love, J. C. (2025). Accelerating cell culture media development using Bayesian optimization-based iterative experimental design. *Nature Communications, 16*(1). https://doi.org/10.1038/s41467-025-61113-5

Navid, A., Jiao, Y., Wong, S. E., & Pett-Ridge, J. (2019). System-level analysis of metabolic trade-offs during anaerobic photoheterotrophic growth in Rhodopseudomonas palustris. *BMC Bioinformatics, 20*(1), 233.

Neubauer, P., Anane, E., Junne, S., & Cruz Bournazou, M. N. (2020). Potential of integrating model-based design of experiments approaches and process analytical technologies for bioprocess scale-down. In *Digital twins* (pp. 1–28). https://doi.org/10.1007/10_2020_154

Ng, W. L., & Tan, J. S. (2025). Machine learning in cultivated meat: Enhancing sustainability, efficiency, quality, and scalability across the production pipeline. *Food and Bioprocess Technology*, 1–22.

Nguyen, X. D. (2025). *Data analytics and machine learning applications in fermentation processes and molecular property prediction.*

Nicolae, A., & Korodi, A. (2018). Node-red and OPC UA based lightweight and low-cost historian with application in the water industry. In *2018 IEEE 16th international conference on industrial informatics (INDIN).*

Ortega-Platas, A., Corro-Herrera, V., Aguilar-Uscanga, M. G., & Gómez-Rodríguez, J. (2021). Real-time monitoring of xylitol production in a bioreactor by Candida tropicalis IEC5-ITV using near-infrared spectroscopy (NIRS). *Revista Mexicana de Ingeniería Química, 20*(3), 1–10. https://doi.org/10.24275/rmiq/Bio2420

Oyetunde, T., Bao, F. S., Chen, J.-W., Martin, H. G., & Tang, Y. J. (2018). Leveraging knowledge engineering and machine learning for microbial bio-manufacturing. *Biotechnology Advances, 36*(4), 1308–1315.

Pan, Z. (2023). Machine learning for real-time optimization of bioprocessing parameters: Applications and improvements. *Artificial Intelligence and Machine Learning Review, 4*(3), 30–42.

Pandey, S., Chaudhary, M., & Tóth, Z. (2025). An investigation on real-time insights: Enhancing process control with IoT-enabled sensor networks. *Discover Internet of Things, 5*(1). https://doi.org/10.1007/s43926-025-00124-6

Pang, C., Yin, X., Zhang, G., Liu, S., Zhou, J., Li, J., & Du, G. (2021). Current progress and prospects of enzyme technologies in future foods. *Systems Microbiology and Biomanufacturing, 1*(1), 24–32.

Pasic, F., Wohlers, B., Dziwok, S., Becker, M., & Heinrich, M. (2019). *A KPI-based condition monitoring system for the beer brewing process.* In *2019 24th IEEE international conference on emerging technologies and factory automation (ETFA).*

Patel, S. P., & Gujarathi, A. M. (2025). Multi-objective optimization of industrial batch Balhimycin antibiotic and fed-batch lysine biochemical processes. *Process Biochemistry, 155*, 93–116. https://doi.org/10.1016/j.procbio.2025.05.003

Pérez, P. A. L., Lopez, R. A., & Femat, R. (2020). *Control in bioprocessing: Modeling, estimation and the use of soft sensors*. Wiley.

Petchrompo, S., Coit, D. W., Brintrup, A., Wannakrairot, A., & Parlikad, A. K. (2022). A review of Pareto pruning methods for multi-objective optimization. *Computers & Industrial Engineering, 167*, 108022.

Petrie, J., Tahchy, A. E., Singh, S., Vecilla, E., Adler, L., & Leupin, K. (2025). *Nourish ingredients.* Retrieved November 21 from https://www.nourishingredients.com

Pietrasik, M., Wilbik, A., & Grefen, P. (2024). The enabling technologies for digitalization in the chemical process industry. *Digital Chemical Engineering, 12.* https://doi.org/10.1016/j.dche.2024.100161

Poch, A., Rodriguez, L., Nettleton, D., & Mc Nerney, O. (2023). *Roadmap for the adoption of PAT and digital twin technologies in the European Pharmaceutical Industry: Barriers and opportunities.*

Pollard, D., Buccino, R., Connors, N., Kirschner, T., Olewinski, R., Saini, K., & Salmon, P. (2001). Real-time analyte monitoring of a fungal fermentation, at pilot scale, using in situ mid-infrared spectroscopy. *Bioprocess and Biosystems Engineering, 24*(1), 13–24. https://doi.org/10.1007/s004490100226

Pomerantsev, A. L., & Rodionova, O. Y. (2021). New trends in qualitative analysis: Performance, optimization, and validation of multi-class and soft models. *TrAC Trends in Analytical Chemistry, 143*, 116372.

Precup, G., Marini, E., Zakidou, P., Beneventi, E., Consuelo, C., Fernández-Fraguas, C., Garcia Ruiz, E., Laganaro, M., Magani, M., & Mech, A. (2024). Novel foods, food enzymes, and food additives derived from food by-products of plant or animal origin: Principles and overview of the EFSA safety assessment. *Frontiers in Nutrition, 11*, 1390734.

Preedy, V. R. (2016). *Electronic noses and tongues in food science*. Academic.

Qiu, J., Arnold, M. A., & Murhammer, D. W. (2014). On-line near infrared bioreactor monitoring of cell density and concentrations of glucose and lactate during insect cell cultivation. *Journal of Biotechnology, 173*, 106–111. https://doi.org/10.1016/j.jbiotec.2014.01.009

Quast, C., Pruesse, E., Yilmaz, P., Gerken, J., Schweer, T., Yarza, P., Peplies, J., & Glöckner, F. O. (2013). The SILVA ribosomal RNA gene database project: Improved data processing and web-based tools. *Nucleic Acids Research, 41*(Database issue), D590–596. https://doi.org/10.1093/nar/gks1219

Raak, N., Struck, S., Jaros, D., Hernando, I., Gülseren, İ., Michalska-Ciechanowska, A., Foschino, R., Corredig, M., & Rohm, H. (2022). Blending side streams. A potential solution to reach a resource efficient, circular, zero-waste food system. *Future Foods, 6*. https://doi.org/10.1016/j.fufo.2022.100207

Raja, R. P., Badwe, A. S., Patwardhan, S. C., & Noronha, S. B. (2007). Adaptive model predictive control of dissolved oxygen in a bioreactor. *IFAC Proceedings Volumes, 40*(5), 53–58.

Raja, J. R., Kella, A., & Narayanasamy, D. (2024). The essential guide to computer system validation in the pharmaceutical industry. *Cureus*. https://doi.org/10.7759/cureus.67555

Ramírez Rojas, A. A., Swidah, R., & Schindler, D. (2022). Microbes of traditional fermentation processes as synthetic biology chassis to tackle future food challenges. *Frontiers in Bioengineering and Biotechnology, 10*. https://doi.org/10.3389/fbioe.2022.982975

Ranjan, R., & Kravaris, C. (2024). Model-predictive fault-tolerant control of safety-critical processes based on dynamic safe set. *Journal of Process Control, 144*. https://doi.org/10.1016/j.jprocont.2024.103329

Rashedi, M., Khodabandehlou, H., Demers, M., Wang, T., & Garvin, C. (2022). Model predictive controller design for bioprocesses based on machine learning algorithms. *IFAC-PapersOnLine, 55*(7), 45–50.

Rattan, A. K. (2018). Data integrity: History, issues, and remediation of issues. *PDA Journal of Pharmaceutical Science and Technology, 72*(2), 105–116. https://doi.org/10.5731/pdajpst.2017.007765

Reji, M., & Kumar, R. (2022). Response surface methodology (RSM): An overview to analyze multivariate data. *Indian Journal of Microbiology Research, 9*, 241–248.

Ren, L., Dong, J., Zeng, X., Yang, L., & Wang, Y. (2025). Digital genealogy: AIGC-driven evolution of digital twin for future smart manufacturing. *IEEE Transactions on Automation Science and Engineering*.

Reyes, S. J., Durocher, Y., Pham, P. L., & Henry, O. (2022). Modern sensor tools and techniques for monitoring, controlling, and improving cell culture processes. *Processes, 10*(2). https://doi.org/10.3390/pr10020189

Rezende, C. A., Atta, B. W., Breitkreitz, M. C., Simister, R., Gomez, L. D., & McQueen-Mason, S. J. (2018). Optimization of biomass pretreatments using fractional factorial experimental design. *Biotechnology for Biofuels, 11*(1), 206.

Riezzo, L., Kay, H., Feng, Y., Jing, K., & Zhang, D. (2025). Accelerating bioprocess digital twin development by integrating hybrid modelling with transfer learning. *Chemical Engineering Journal, 511*, 162018.

Rivera, E. C., Yamakawa, C. K., Rossell, C. E., Nolasco, J., Jr., & Kwon, H. J. (2024). Prediction of intensified ethanol fermentation of sugarcane using a deep learning soft sensor and process analytical technology. *Journal of Chemical Technology & Biotechnology, 99*(1), 207–216.

Robinson, T., Singh, D., & Nigam, P. (2001). Solid-state fermentation: A promising microbial technology for secondary metabolite production. *Applied Microbiology and Biotechnology, 55*(3), 284–289. https://doi.org/10.1007/s002530000565

Rodman, A. D., & Gerogiorgis, D. I. (2016). Multi-objective optimisation of flavour and processing time in beer fermentation via dynamic simulation. In *26th European symposium on computer aided process engineering* (pp. 1033–1038). https://doi.org/10.1016/b978-0-444-63428-3.50177-6

Rodrigues, R. C., Canettieri, E. V., Prata, A. M. R., Martinez, E. A., José, Á. H. M., Palladino, F., da Silva, S. P., & Pessoa, A. (2024). Analytical tools and operational requirements for bioreactors. In R. F. Schwan, V. K. Joshi, & D. R. Dias (Eds.), *Bioreactor technology in food processing* (pp. 59–83). CRC Press.

Roell, G. W., Sathish, A., Wan, N., Cheng, Q., Wen, Z., Tang, Y. J., & Bao, F. S. (2022). A comparative evaluation of machine learning algorithms for predicting syngas fermentation outcomes. *Biochemical Engineering Journal, 186*, 108578.

Rohart, F., Gautier, B., Singh, A., & Lê Cao, K.-A. (2017). mixOmics: An R package for 'omics feature selection and multiple data integration. *PLoS Computational Biology, 13*(11), e1005752. https://doi.org/10.1371/journal.pcbi.1005752

Rómoli, S., Serrano, M., Rossomando, F., Vega, J., Ortiz, O., & Scaglia, G. (2017). Neural network-based state estimation for a closed-loop control strategy applied to a fed-batch bioreactor. *Complexity, 2017*, 1–16. https://doi.org/10.1155/2017/9391879

Rudnitskaya, A., & Legin, A. (2008). Sensor systems, electronic tongues and electronic noses, for the monitoring of biotechnological processes. *Journal of Industrial Microbiology & Biotechnology, 35*(5), 443–451. https://doi.org/10.1007/s10295-007-0298-1

Sahana, V. (2025). Enhancing data integrity and regulatory compliance in biomanufacturing using pas X manufacturing execution system. *International Journal of Applied Mathematics, 38*(7s), 134–145. https://doi.org/10.12732/ijam.v38i7s.471

Sai Guptha, V. D. P., Anamalamudi, S., Madupuri, R. P., & Sivarajan, S. (2025). Interoperability between OPC-UA and MQTT in hybrid industrial IoT(IIoT) environments. In *2025 17th international conference on COMmunication systems and NETworks (COMSNETS)*.

Sales, K. C., Rosa, F., Sampaio, P. N., Fonseca, L. P., Lopes, M. B., & Calado, C. R. C. (2015). In situ near-infrared (NIR) versus high-throughput mid-infrared (MIR) spectroscopy to monitor biopharmaceutical production. *Applied Spectroscopy, 69*(6), 760–772. https://doi.org/10.1366/14-07588

Samtiya, M., Aluko, R. E., Puniya, A. K., & Dhewa, T. (2021). Enhancing micronutrients bioavailability through fermentation of plant-based foods: A concise review. *Fermentation, 7*(2). https://doi.org/10.3390/fermentation7020063

Sandle, T. (2024). *Biocontamination control for pharmaceuticals and healthcare*. Elsevier.

Schalk, R., Heintz, A., Braun, F., Iacono, G., Rädle, M., Gretz, N., Methner, F.-J., & Beuermann, T. (2019). Comparison of Raman and mid-infrared spectroscopy for real-time monitoring of yeast fermentations: A proof-of-concept for multi-channel photometric sensors. *Applied Sciences, 9*(12). https://doi.org/10.3390/app9122472

Schmidberger, T., Gutmann, R., Bayer, K., Kronthaler, J., & Huber, R. (2014). Advanced online monitoring of cell culture off-gas using proton transfer reaction mass spectrometry. *Biotechnology Progress, 30*(2), 496–504. https://doi.org/10.1002/btpr.1853

Schneider, M. Y., Carbajal, J. P., Furrer, V., Sterkele, B., Maurer, M., & Villez, K. (2019). Beyond signal quality: The value of unmaintained pH, dissolved oxygen, and oxidation-reduction potential sensors for remote performance monitoring of on-site sequencing batch reactors. *Water Research, 161*, 639–651. https://doi.org/10.1016/j.watres.2019.06.007

Schulte, A., Brockmann, J., Müller, N., Anderlei, T., & Büchs, J. (2025). A new approach to off-gas analysis for shaken bioreactors showing high CTR and RQ accuracy. *Journal of Biological Engineering, 19*(1). https://doi.org/10.1186/s13036-025-00480-5

Schwarz, M. H., & Börcsök, J. (2013). A survey on OPC and OPC-UA: About the standard, developments and investigations. In *2013 XXIV international conference on information, communication and automation technologies (ICAT)*.

Seesaard, T., & Wongchoosuk, C. (2022). Recent progress in electronic noses for fermented foods and beverages applications. *Fermentation, 8*(7). https://doi.org/10.3390/fermentation8070302

Seiberling, D. A., & Ratz, A. J. (2018). Engineering considerations for CIP/SIP systems. In *Sterile pharmaceutical products* (pp. 135–219). Routledge.

Sharma, V., Tsai, M.-L., Nargotra, P., Chen, C.-W., Kuo, C.-H., Sun, P.-P., & Dong, C.-D. (2022). Agro-industrial food waste as a low-cost substrate for sustainable production of industrial enzymes: A critical review. *Catalysts, 12*(11). https://doi.org/10.3390/catal12111373

Shen, Z., Arrano-Vargas, F., & Konstantinou, G. (2024). What-if scenario testing through power system digital twins. In *2024 IEEE 22nd international conference on industrial informatics (INDIN)*.

Short, E. C., Kinchla, A. J., & Nolden, A. A. (2021). Plant-based cheeses: A systematic review of sensory evaluation studies and strategies to increase consumer acceptance. *Foods, 10*(4). https://doi.org/10.3390/foods10040725

Short, A. R., Orfanoudakis, T. G., & Leligou, H. C. (2024). *PLCBlox: Using blockchain-based audit trails to generate secure PLC commands*.

Shukla, S. S., Pandey, R. K., Gidwani, B., & Kalyani, G. (2023). Objectives of qualification. In *Pharmaceutical calibration, validation and qualification: A comprehensive approach* (pp. 49–61). Springer.

Simethy, G. (2024). *Implementation and assessment of advanced reinforcement learning methods for control of chemical processes*. Hochschule für Angewandte Wissenschaften Hamburg.

Simon, L. L., Pataki, H., Marosi, G., Meemken, F., Hungerbühler, K., Baiker, A., Tummala, S., Glennon, B., Kuentz, M., Steele, G., Kramer, H. J. M., Rydzak, J. W., Chen, Z., Morris, J., Kjell, F., Singh, R., Gani, R., Gernaey, K. V., Louhi-Kultanen, M., & Chiu, M.-S. (2015). Assessment of recent process analytical technology (PAT) trends: A multiauthor review. *Organic Process Research & Development, 19*(1), 3–62. https://doi.org/10.1021/op500261y

Singh, V., Haque, S., Niwas, R., Srivastava, A., Pasupuleti, M., & Tripathi, C. (2017). Strategies for fermentation medium optimization: An in-depth review. *Frontiers in Microbiology, 7*, 2087.

Singh, H., Singh, G., Kaur, N., & Singh, N. (2022). Pattern-based colorimetric sensor array to monitor food spoilage using automated high-throughput analysis. *Biosensors and Bioelectronics, 196*. https://doi.org/10.1016/j.bios.2021.113687

Sinner, P., Daume, S., Herwig, C., & Kager, J. (2020). Usage of digital twins along a typical process development cycle. In *Digital twins* (pp. 71–96). https://doi.org/10.1007/10_2020_149

Skoneczny, S., & Cioch-Skoneczny, M. (2024). Production of vitamins in a bioreactor. *Bioreactor Technology in Food Processing, 424*.

Slaugenhaupt, R. (2025). *Using ISA-88 to enhance MES implementation*. ISA interchange. Retrieved 4 November from https://blog.isa.org/using-isa-88-to-enhance-mes-implementation

Sousa, M. R. S. (2025). *Industry 4.0 maturity model and DES-based validation for digital*.

Specht, L. (2025). *The science of fermentation*. Good Food Institute. Retrieved 5 November from https://gfi.org/science/the-science-of-fermentation/

Stevnsborg, M., Selle, K., Barton, R., Prado-Rubio, O. A., Gargalo, C., Gernaey, K. V., Gilleskie, G., & Huusom, J. K. (2023). Digital twins in pilot scale fermentation: Non-linear state estimation for improving induction timing. *Computer Aided Chemical Engineering, 52*, 2637–2642. https://doi.org/10.1016/b978-0-443-15274-0.50419-4

Sun, Y., Du, N., Sun, Q.-Y., Chen, X.-G., & Yang, J.-W. (2019a). Research and application of biological potency soft sensor modeling method in the industrial fed-batch chlortetracycline fermentation process. *Cluster Computing, 22*(Suppl 3), 6019–6030.

Sun, Y., Tang, L., Sun, Q., Wang, M., Han, X., & Chen, X. (2019b). Study on the prediction of the contamination symptoms in the fermentation process of chlortetracycline based on soft sensor modeling method. *Technology and Health Care, 27*(1_suppl), 205–215.

Sun, W., Shahrajabian, M. H., & Lin, M. (2022). Research progress of fermented functional foods and protein factory-microbial fermentation technology. *Fermentation, 8*(12). https://doi.org/10.3390/fermentation8120688

Talaei, S., Frey, O., Psoma, S., van der Wal, P. D., & de Rooij, N. F. (2010). Smart SU-8 pillars implemented in a microfluidic bioreactor for continuous measurement of glucose. *Procedia Engineering, 5*, 448–451. https://doi.org/10.1016/j.proeng.2010.09.143

Tang, H., Yang, Z., Xu, F., Wang, Q., & Wang, B. (2022). Soft sensor modeling method based on improved KH-RBF neural network bacteria concentration in marine alkaline protease fermentation process. *Applied Biochemistry and Biotechnology, 194*(10), 4530–4545.

Thames, L., & Schaefer, D. (2016). Software-defined cloud manufacturing for industry 4.0. *Procedia CIRP, 52*, 12–17. https://doi.org/10.1016/j.procir.2016.07.041

Theng, D., & Bhoyar, K. K. (2024). Feature selection techniques for machine learning: A survey of more than two decades of research. *Knowledge and Information Systems, 66*(3), 1575–1637.

Thomas, A., Arvind, A., & Geetha, A. (2022). A study of genealogy and tracking systems for automotive manufacturing industries. In *2022 international conference on automation, computing and renewable systems (ICACRS)*.

Tian, H., Yu, B., Yu, H., & Chen, C. (2020). Evaluation of the synergistic olfactory effects of diacetyl, acetaldehyde, and acetoin in a yogurt matrix using odor threshold, aroma intensity, and electronic nose analyses. *Journal of Dairy Science, 103*(9), 7957–7967. https://doi.org/10.3168/jds.2019-17495

Tibaduiza, D., Anaya, M., Gómez, J., Sarmiento, J., Perez, M., Lara, C., Ruiz, J., Osorio, N., Rodriguez, K., Hernandez, I., & Sanchez, C. (2024). Electronic tongues and noses: A general overview. *Biosensors, 14*(4). https://doi.org/10.3390/bios14040190

Tiso, T., Demling, P., Karmainski, T., Oraby, A., Eiken, J., Liu, L., Bongartz, P., Wessling, M., Desmond, P., Schmitz, S., Weiser, S., Emde, F., Czech, H., Merz, J., Zibek, S., Blank, L. M., & Regestein, L. (2024). Foam control in biotechnological processes – Challenges and opportunities. *Discover Chemical Engineering, 4*(1). https://doi.org/10.1007/s43938-023-00039-0

Tokuyama, K., Shimodaira, Y., Kodama, Y., Matsui, R., Kusunose, Y., Fukushima, S., Nakai, H., Tsuji, Y., Toya, Y., & Matsuda, F. (2021). Soft-sensor development for monitoring the lysine fermentation process. *Journal of Bioscience and Bioengineering, 132*(2), 183–189.

Toledo, J., Ruiz-Díez, V., Pfusterschmied, G., Schmid, U., & Sánchez-Rojas, J. L. (2018). Flow-through sensor based on piezoelectric MEMS resonator for the in-line monitoring of wine fermentation. *Sensors and Actuators B: Chemical, 254*, 291–298. https://doi.org/10.1016/j.snb.2017.07.096

Torres-Tiji, Y., Fields, F. J., & Mayfield, S. P. (2020). Microalgae as a future food source. *Biotechnology Advances, 41*, 107536.

Trojanowicz, M., & Kołacińska, K. (2016). Recent advances in flow injection analysis. *The Analyst, 141*(7), 2085–2139. https://doi.org/10.1039/c5an02522b

Tsafrakidou, P., Zagklis, D., & Bekatorou, A. (2024). Bioreactor technology in waste utilization and management. In R. F. Schwan, V. K. Joshi, & D. R. Dias (Eds.), *Bioreactor technology in food processing* (pp. 597–622).

Uchil, D. D., Jacob, A., & Kumar, B. S. H. (2025). Standard operating procedures (SOPs) and auditing in bioprocess industries. In *Biotechnology engineering* (pp. 181–204). https://doi.org/10.1016/b978-0-443-31476-6.00004-x

Udugama, I. A., Hall, K. K., Bayer, C., Young, B. R., & Walmsley, T. G. (2023a). Real-time control and scheduling decision making in fermentation-based food process operations. *Computer Aided Chemical Engineering, 52*, 2381–2386. https://doi.org/10.1016/b978-0-443-15274-0.50378-4

Udugama, I. A., Kelton, W., & Bayer, C. (2023b). Digital twins in food processing: A conceptual approach to developing multi-layer digital models. *Digital Chemical Engineering, 7*. https://doi.org/10.1016/j.dche.2023.100087

Urriza-Arsuaga, I., Ielasi, G., Bedoya, M., & Orellana, G. (2019). Luminescence-based sensors for bioprocess applications. In *Fluorescence in industry* (pp. 1–38). https://doi.org/10.1007/4243_2019_10

Usmani, U. A., Happonen, A., & Watada, J. (2022). A review of unsupervised machine learning frameworks for anomaly detection in industrial applications. In *Science and information conference*.

Vandenberghe, L. P. D. S., Martínez-Burgos, W. J., Herrmann, L. W., Mores, S., de Mello, A. F. M., de Oliveira Penha, R., de Souza Kirnev, P. C., Magalhães, A. I., & Soccol, C. R.

(2024). Scale-up technology of bioreactors. In R. F. Schwan, V. K. Joshi, & D. R. Dias (Eds.), *Bioreactor technology in food processing* (pp. 123–150). CRC Press.

Viola, J., & Chen, Y. (2021). Parallel self optimizing control framework for digital twin enabled smart control engineering. In *2021 IEEE 1st international conference on digital twins and parallel intelligence (DTPI)*.

Vlassides, S., Ferrier, J. G., & Block, D. E. (2001). Using historical data for bioprocess optimization: Modeling wine characteristics using artificial neural networks and archived process information. *Biotechnology and Bioengineering, 73*(1), 55–68. https://doi.org/10.1002/1097-0290(20010405)73:1<55::aid-bit1036>3.0.co;2-5

Wang, B., Wang, Z., Chen, T., & Zhao, X. (2020). Development of novel bioreactor control systems based on smart sensors and actuators. *Frontiers in Bioengineering and Biotechnology, 8*, 7. https://doi.org/10.3389/fbioe.2020.00007

Wang, Y., Liu, L., Jin, Z., & Zhang, D. (2021). Microbial cell factories for green production of vitamins. *Frontiers in Bioengineering and Biotechnology, 9*, 661562.

Wang, K., Zhao, W., Lin, L., Wang, T., Wei, P., Ledesma-Amaro, R., Zhang, A. H., & Ji, X. J. (2023). A robust soft sensor based on artificial neural network for monitoring microbial lipid fermentation processes using Yarrowia lipolytica. *Biotechnology and Bioengineering, 120*(4), 1015–1025.

Wang, B., Wei, J., Zhang, L., Jiang, H., Jin, C., & Huang, S. (2024). Soft sensor modeling method for Pichia pastoris fermentation process based on substructure domain transfer learning. *BMC Biotechnology, 24*(1), 104.

Wang, Z.-Z., Zeng, D.-W., Zhu, Y.-F., Zhou, M.-H., Kondo, A., Hasunuma, T., & Zhao, X.-Q. (2025). Fermentation design and process optimization strategy based on machine learning. *Biodesign Research, 7*(1), 100002.

Ward, R., Choudhary, R., Gregory, A., Jans-Singh, M., & Girolami, M. (2021). Continuous calibration of a digital twin: Comparison of particle filter and Bayesian calibration approaches. *Data-Centric Engineering, 2*, e15.

Wehrs, M., Tanjore, D., Eng, T., Lievense, J., Pray, T. R., & Mukhopadhyay, A. (2019). Engineering robust production microbes for large-scale cultivation. *Trends in Microbiology, 27*(6), 524–537. https://doi.org/10.1016/j.tim.2019.01.006

Wei, Q., Liu, G., Zhang, C., Sun, J., & Zhang, Y. (2022). Identification of characteristic volatile compounds and prediction of fermentation degree of pomelo wine using partial least squares regression. *LWT, 154*, 112830.

Whittaker, J. A., Johnson, R. I., Finnigan, T. J. A., Avery, S. V., & Dyer, P. S. (2020). The biotechnology of Quorn mycoprotein: Past, present and future challenges. In *Grand challenges in fungal biotechnology* (pp. 59–79). https://doi.org/10.1007/978-3-030-29541-7_3

Wilken, L. R., & Nikolov, Z. L. (2012). Downstream processing of transgenic plant systems: Protein recovery and purification strategies. In *Molecular farming in plants: Recent advances and future prospects* (pp. 217–257). https://doi.org/10.1007/978-94-007-2217-0_11

Wu, L., Xiao, G., Huang, D., Zhang, X., Ye, D., & Weng, H. (2025). Edge computing-based machine vision for non-invasive and rapid soft sensing of mushroom liquid strain biomass. *Agronomy, 15*(1). https://doi.org/10.3390/agronomy15010242

Xu, B., Jahic, M., & Enfors, S. O. (1999). Modeling of overflow metabolism in batch and fed-batch cultures of Escherichia coli. *Biotechnology Progress, 15*(1), 81–90. https://doi.org/10.1021/bp9801087

Yadav, A. (2025). Bio-manufacturing. *Industrial Applications for Bioprocessing and Biomanufacturing, 143*.

Yang, X. (2014). Scale-up of microbial fermentation process. In *Manual of industrial microbiology and biotechnology* (pp. 669–675). https://doi.org/10.1128/9781555816827.ch47

Yang, C. D. D. O., & Ierapetritou, M. (2024). Towards digital twin for biopharmaceutical processes: Concept. *Biopharmaceutical Manufacturing: Progress, Trends and Challenges, 11*, 179.

Yang, Y., & Sha, M. (2019). *A beginner's guide to bioprocess modes–batch, fed-batch, and continuous fermentation* (Vol. 408, pp. 1–16). Eppendorf Inc.

Yao, B., Xu, W., Shen, T., Ye, X., & Tian, S. (2023). Digital twin-based multi-level task rescheduling for robotic assembly line. *Scientific Reports, 13*(1), 1769.

Yao, Z., Jiang, Q., & Gu, X. (2025). Regularized decoupling transformer-based variable selection and soft sensor modeling for fermentation processes. *IEEE Transactions on Instrumentation and Measurement.*

Yardley, J. E., Kell, D. B., Barrett, J., & Davey, C. L. (2000). On-line, real-time measurements of cellular biomass using dielectric spectroscopy. *Biotechnology and Genetic Engineering Reviews, 17*(1), 3–36. https://doi.org/10.1080/02648725.2000.10647986

Yee, C. S., Zahia-Azizan, N. A., Abd Rahim, M. H., Mohd Zaini, N. A., Raja-Razali, R. B., Ushidee-Radzi, M. A., Ilham, Z., & Wan-Mohtar, W. A. A. Q. I. (2025). Smart fermentation technologies: Microbial process control in traditional fermented foods. *Fermentation, 11*(6), 323.

You, X.-Y., Ding, Y., Bu, Q.-Y., Wang, Q.-H., & Zhao, G.-P. (2024). Nutritional, textural, and sensory attributes of protein bars formulated with mycoproteins. *Food, 13*(5), 671.

Yu, Z., Jung, D., Park, S., Hu, Y., Huang, K., Rasco, B. A., Wang, S., Ronholm, J., Lu, X., & Chen, J. (2022). Smart traceability for food safety. *Critical Reviews in Food Science and Nutrition, 62*(4), 905–916.

Yuan, X., Gu, Y., Wang, Y., Yang, C., & Gui, W. (2019). A deep supervised learning framework for data-driven soft sensor modeling of industrial processes. *IEEE Transactions on Neural Networks and Learning Systems, 31*(11), 4737–4746.

Zabalza, A., van Dongen, J. T., Froehlich, A., Oliver, S. N., Faix, B., Gupta, K. J., Schmälzlin, E., Igal, M., Orcaray, L., Royuela, M., & Geigenberger, P. (2009). Regulation of respiration and fermentation to control the plant internal oxygen concentration. *Plant Physiology, 149*(2), 1087–1098. https://doi.org/10.1104/pp.108.129288

Zhang, Y., Huang, J., Liu, X., & Ni, Z. (2021). Digital twin-based process optimization system research for micro-assembly products. In *2021 international conference on computer, control and robotics (ICCCR).*

Zhang, Y., Yu, H., Hou, Y., & Zhu, N. (2025). Multi-objective optimization research based on NSGA-II and experimental study of triplex-tube phase change thermal energy storage system. *Energies, 18*(8). https://doi.org/10.3390/en18082129

Zhao, B., Li, X., Sun, W., Qian, J., Liu, J., Gao, M., Guan, X., Ma, Z., & Li, J. (2023a). BioDT: An integrated digital-twin-based framework for intelligent biomanufacturing. *Processes, 11*(4). https://doi.org/10.3390/pr11041213

Zhao, L., Tan, K., Wang, X., Ding, J., Liu, Z., Ma, H., & Han, B. (2023b). Hyperspectral feature selection for SOM prediction using deep reinforcement learning and multiple subset evaluation strategies. *Remote Sensing, 15*(1), 127. https://www.mdpi.com/2072-4292/15/1/127

Zhao, S., Jiao, T., Adade, S. Y.-S. S., Wang, Z., Ouyang, Q., & Chen, Q. (2025). Digital twin for predicting and controlling food fermentation: A case study of kombucha fermentation. *Journal of Food Engineering, 393.* https://doi.org/10.1016/j.jfoodeng.2025.112467

Zhou, X., Wang, J., Huang, L., Li, D., & Duan, Q. (2022). Modelling and controlling dissolved oxygen in recirculating aquaculture systems based on mechanism analysis and an adaptive PID controller. *Computers and Electronics in Agriculture, 192.* https://doi.org/10.1016/j.compag.2021.106583

Zhu, J., Chen, W., Chen, Y., Amin, F. R., Li, Y., Lu, M., & Li, D. (2025). CFD optimization of an air lift fermenter for *Fusarium venenatum* fermentation. *Bioresource Technology Reports, 29,* 102024.

Chapter 11
Novel Starter Cultures and Probiotics for Functional Plant-Based Fermentations

Nádia Nara Batista, Ana Paula Pereira Bressani, Cintia Lacerda Ramos, Disney Ribeiro Dias, and Rosane Freitas Schwan

Abstract The demand for plant-based foods is growing, and in this context, non-dairy fermented foods have emerged as promising functional and sustainable alternatives, offering improved sensory properties, enhanced nutritional value, and a lower environmental impact. However, these products may present limitations, including physical instability, off-flavors, antinutrient content, nutritional imbalances, standardization issues, and challenges to consumer acceptance. Focusing primarily on spontaneous fermentations, starter culture selection, and potential probiotic microorganisms, this chapter aims to provide a comprehensive and critical overview of the fermentation of non-dairy foods and emerging technologies that support the development of safer, more stable, and sustainable functional products.

Keywords Plant-based fermentation · Starter cultures · Probiotics · Lactic acid bacteria · Functional foods · Spontaneous fermentation · Microbial ecology

11.1 Introduction

The global plant-based industry is projected to expand from $14 billion (€12.3 billion) to nearly $44 billion (€38.6 billion) by 2035, reflecting a 12% Compound Annual Growth Rate over the period (Bambridge-Sutton, 2025). Although this rapid expansion demonstrates strong consumer demand and commercial potential, it also exposes important technological, safety, and regulatory challenges. These limitations include issues related to physical stability, off-flavors, antinutrient content, and

N. N. Batista · R. F. Schwan (✉)
Federal University of Lavras, Department of Biology, Lavras, MG, Brazil
e-mail: rschwan@ufla.br

A. P. P. Bressani · D. R. Dias
Federal University of Lavras, Department of Food Science, Lavras, MG, Brazil

C. L. Ramos
Federal University of Jequitinhonha and Mucuri Valleys, Department of Basic Science, Diamantina, MG, Brazil

S. Dhiman, S. Kaur (eds.), *Next-Gen Plant Based Fermented Foods: Innovations for a Sustainable Future*, https://doi.org/10.1007/978-3-032-29927-7_11

nutritional imbalance, which compromise product quality, standardization, and consumer acceptance. These challenges highlight the need for an integrated approach connecting food technology, sensory science, nutrition, and dietetics.

In this context, non-dairy fermented foods have emerged as promising functional and sustainable alternatives, resulting in improved sensory properties, enhanced nutritional value, and lower environmental impact (Anjos et al., 2024; Tachie et al., 2023). Spontaneous fermentation processes, conducted by native microbial communities (primarily lactic acid bacteria and yeasts), result in products with different chemical and sensory characteristics. The characterization of microbial communities provides the scientific evidence necessary to select starter cultures and probiotics, thereby ensuring the standardization, safety, and predictability required for industrial-scale fermentation (Gänzle et al., 2024).

However, the selection of starter culture for the development of fermented plant-based products requires careful attention. The strain must adapt to plant matrices and maintain viability throughout processing and storage. Furthermore, aspects such as the generation of bioactive compounds, the degradation of anti-nutritional factors, and the improvement of nutrient bioavailability are considered fundamental to ensuring functional benefits.

To reduce these challenges, emerging non-thermal technologies, such as ultrasound, high-pressure processing (HPP), and pulsed electric fields (PEF), have been investigated as alternatives to conventional heat treatments. These technologies preserve the nutritional and sensory quality of the products. Consequently, they enhance microbiological and chemical stability and extend shelf life (Galván-D'Alessandro & Carciochi, 2018; Pizarro-Oteíza et al., 2025).

This chapter provides an integrated and critical overview of the fermentation of non-dairy foods, encompassing spontaneous and controlled processes, selection of starter and probiotic cultures, and emerging technologies that support the development of safer, more stable, and sustainable functional products.

11.2 Starter Culture and Probiotics: Sensory and Functional Impact of Fermentation

Spontaneously fermented foods are an important source of microorganisms that can act in technological, nutritional, and sensorial aspects. These microbial communities enhance shelf life, safety, flavor complexity, and, at times, the nutritional and functional value of foods (Gänzle et al., 2024). Several studies have investigated the biotechnological properties of microorganisms to select suitable starter or probiotic cultures for the development of fermented food, including plant-based fermented foods (Agarbati et al., 2024; Buatong et al., 2022; Contreras et al., 2025; Fonseca et al., 2021a; Gungor et al., 2024; Liu et al., 2023; Ogura et al., 2025; Resende et al., 2025; Wang et al., 2020).

Starter cultures are microbial preparations containing high cell densities from one or more selected strains, introduced into food matrices to promote desirable biochemical transformations through their metabolic or enzymatic activities. They are used in several industrial applications (Holzapfel, 1997). A traditional and well-established example is the use of mesophilic and/or thermophilic lactic acid bacteria (LAB) starters, such as *Streptococcus thermophilus* and *Lactobacillus bulgaricus*, which convert lactose into lactic acid in dairy fermentations. This acidification kinetics causes milk proteins to coagulate, resulting in the thick, creamy consistency characteristic of yogurt with desired flavor profiles.

Plant-based fermented foods have gained increasing attention due to their sustainability, perceived health benefits, and compatibility with vegan and lactose-free diets. Soy yogurts, oat-based yogurts, plant-protein "cheese" analogs, vegetable ferments, and novel beverages are some examples of fermented plant-based products that have been investigated (Agarbati et al., 2024; Buatong et al., 2022; Grasso et al., 2021; Gungor et al., 2024; Liu et al., 2023; Okafor & Okoro, 2025; Resende et al., 2025; Wang et al., 2020). However, their development remains challenging because plant matrices, such as soy, oat, pea, and nut milks, as well as vegetable and fruit juices, are variable and may exhibit different carbohydrate and protein compositions, fat contents, and microstructures, which affect microbial communities. These differences often result in a variety of products with distinctive characteristics (Bouakkaz et al., 2024; Malongane & Berejena, 2024; Okafor & Okoro, 2025; Ramos & Schwan, 2017) and sometimes with undesired texture, limited flavor, and reduced stability. In addition to containing antinutritional factors (e.g., phytate) and off-flavors (e.g., beany or grassy notes) that compromise product quality.

Therefore, the selection and design of novel starter cultures and probiotic strains represent a key strategy to minimize these limitations. Yeast and lactic acid bacteria are the main microbial groups investigated, and their application has been carried out as single or mixed cultures (Gänzle et al., 2024; Pereira et al., 2023).

It is crucial that these strains are safe, to adapt to plant matrices, metabolize substrate, and produce desirable sensory and functional attributes, including texture, flavor, aroma, appearance, as well as enhanced nutrient bioavailability, increased bioactive compounds, and improved safety and stability of the product (Agarbati et al., 2024; Buatong et al., 2022; Ferreira et al., 2022; Fonseca et al., 2021b, 2022; Macêdo et al., 2023), as shown in Fig. 11.1.

According to Gänzle et al. (2024), safety (GRAS status), prior use in food fermentations (traditional fermentations), and phylogenetic linkage to desired metabolic traits (e.g., acidification and enzymatic activities) are essential criteria for selecting starter cultures to produce fermented foods. Characteristics such as acidification capacity, proteolytic and lipolytic activities, exopolysaccharide production, and tolerance to environmental stress (e.g., pH, temperature, salinity, and oxygen levels) are crucial for desirable fermentation performance. These traits contribute to the development of optimal texture, flavor, and aroma while ensuring product safety and stability (Contreras et al., 2025; Lavelle et al., 2025).

It is important to highlight that for plant-based substrates, the ability to utilize diverse plant carbohydrates and to degrade antinutritional compounds (e.g., phytase

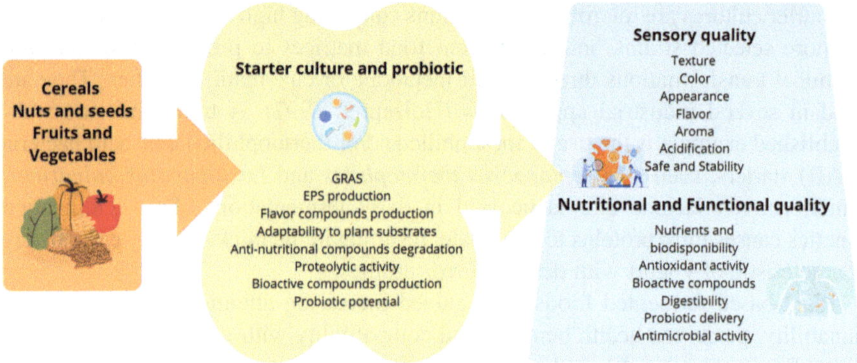

Fig. 11.1 Microbial properties and their impact on sensory, nutritional, and functional quality of plant-based fermentations

activity) to enhance nutritional value and digestibility, as well as the degradation of off-flavor precursors (e.g., aldehydes), are desired microbial properties (Okafor & Okoro, 2025). Furthermore, microbial properties, such as probiotic potential, antioxidant activity, and bioactive compound production, enhance the health benefits of the final product (Agarbati et al., 2024; Kittibunchakul et al., 2021). Therefore, selecting starter cultures with robust technological functionality and health-related benefits is crucial for producing high-quality plant-based fermented foods.

11.2.1 Bacteria and Yeasts as Starter/Probiotic Culture

There are several studies evaluating the use of starter cultures and probiotic strains, bacteria, and yeasts in single and co-culture, to develop a diversity of plant-based fermented foods (Agarbati et al., 2024; Buatong et al., 2022; Gungor et al., 2024; Liu et al., 2023; Resende et al., 2025; Wang et al., 2020). According to the FAO/WHO (2001), Hill et al. (2014), and Marco et al. (2021), probiotics are defined as live microorganisms that, when administered in adequate amounts, confer a health benefit on the host, and therefore represent an important group of functional cultures considered in the design of fermented plant-based products.

The primary challenge in the industrial production of functional fermented foods is selecting a starter or consortium culture that exhibits characteristics like those of spontaneously fermented foods. Some examples of microbial cultures as potential starter cultures and/or probiotic strains, along with their applications in plant-based fermented foods are shown in Table 11.1.

Research has shown that cultures previously evaluated as probiotics can also exhibit desirable sensory, safety, and technological characteristics, adding both technological and functional value to fermented food. For example, the use of *Lactiplantibacillus* (*Lpb.*) *plantarum* PMO-08 as a starter culture for a tomato-

Table 11.1 Microorganisms and their application in different plant-based fermented foods

Microorganisms	Application in fermented foods	Reference
Lachancea thermotolerans Lt101, *Kazachstania unispora* Kum3-B3, *Meyerozyma guilliermondii* Mg112, *Meyerozyma caribbica* Mc58, *Debaryomyces hansenii* Dh36	Non-hopped beer wort and infusion of oxidized rooibos (red rooibos) leaves	Agarbati et al. (2024)
Lactiplantibacillus plantarum L42g	Almond and coix fermented juices	Buatong et al. (2022)
Leuconostoc pseudomesenteroides PD4, *Companilactobacillus alimentarius* PG3	Pistachio-based fermented beverages	Garzón et al. (2025)
Lactiplantibacillus paraplantarum IBVF-10	Chilto (*Solanum betaceum*) fermented juice	Contreras et al. (2025)
Streptococcus thermophilus, *Lactobacillus delburueckii* subsp. *bulgaricus*, *Bifidobacterium* BB-12® (Chr. Hansen), *Propionibacterium* spp. (Ferment Lattici MicroMilk®)	Almond, soy, and coconut fermented beverages	Gungor et al. (2024)
Lactiplantibacillus pentosus 9D3	Brown rice fermented milk	Kittibunchakul et al. (2021)
Lactiplantibacillus plantarum CICC 21796, *Pichia kluyveri* XT110	Bananas, broccoli, and wolfberries fermented juice	Liu et al. (2023)
Leuconostoc lactis Miki-A3	"Miki" (fermented rice saccharified with sweet potato β-amylase)	Ogura et al. (2025)
Pichia kluyveri FrootZen, *Torulaspora delbrueckii* Prelude, *Williopsis saturnus var. mrakii* NCYC2251, *Torulaspora delbrueckii* Biodiva	Green tea fermented slurry	Wang et al. (2020)
Pichia kluyveri CCMA 0615, *Debaryomyces hansenii* CCMA 1761, *Pichia guilliermondii* CCMA 1753, *Lactiplantibacillus plantarum* CCMA 0743	Sunflower seeds, oats, and almonds fermented beverage	Ferreira et al. (2022)
Lactiplantibacillus plantarum CCMA 0743, *Debaryomyces hansenii* CCMA 176	Sunflower seeds, oats, and almonds fermented beverage fortified with Brazilian red propolis (microencapsulated and extracted)	Ferreira et al. (2023)
Lactiplantibacillus plantarum PMO-08	Tomato fermented juice	Oh et al. (2020)
Lactiplantibacillus plantarum CCMA 0743	Passion fruit fermented juice	Fonseca et al. (2022)
Hanseniaspora opuntiae 125, *Issatchenkia terricola* 129, *Hanseniaspora opuntiae* 148	Soursop (*Annona muricata* L.) and umbu-cajá (*Spondias* spp.) fermented juices	Macêdo et al. (2023)
Lacticaseibacillus paracasei subsp. *paracasei* LBC-81, *Saccharomyces cerevisiae* CCMA 0731, *Saccharomyces cerevisiae* CCMA 0732, *Pichia kluyveri* CCMA 0615	Maize-based fermented beverage	Menezes et al. (2018)

(continued)

Table 11.1 (continued)

Microorganisms	Application in fermented foods	Reference
Bifidobacterium longum BL-D5, *Lactobacillus acidophilus* LYO40, *Streptococcus thermophilus* T040, *Lactobacillus delbrueckii* subsp. *bulgaricus* LB340	Soy-based fermented beverage	Pereira et al. (2023)
Leuconostoc lactis CCMA 0415	Fermented yam-based ice cream	Batista et al. (2019a, b)
Lactiplantibacillus plantarum CCMA0743, *Torulaspora delbrueckii* CCMA 0235	Cassava and rice fermented beverage, and rice and corn fermented beverage	Freire et al. (2017a, b)
Lacticaseibacillus rhamnosus LR 32, *Lactobacillus acidophilus* LACA 4, *Lactobacillus delbrueckii* subsp. *bulgaricus* LB 340, *Pediococcus acidilactici* UFLA BFFCX 27.1, *Saccharomyces cerevisiae* UFLA YFFBM 18.03	Peanut-soy fermented beverages	Santos et al. (2014)

juice probiotic beverage resulted in increased total polyphenol and glutamine content, as well as enhanced antioxidant activity and sensory properties (Oh et al., 2020). In another study, the yeast *Pichia kluyveri* XT110 and the bacterium *Lpb. plantarum* CICC 21796 were sequentially inoculated in bananas, broccoli, and wolfberry substrates, resulting in a positive correlation between the production of specific organic acids and sensory quality (Liu et al., 2023). In addition, γ-aminobutyric acid (GABA) concentration and antioxidant activity were also increased, showing positive effects on technological and functional aspects through co-fermentation of *P. kluyveri* and *Lpb. plantarum* (Liu et al., 2023).

Further, the use of different *Lpb. plantarum* strains in single and co-culture with yeasts have been reported to produce a variety of plant-based fermented foods (Buatong et al., 2022; Ferreira et al., 2022, 2023; Fonseca et al., 2022; Freire et al., 2017a, b; Liu et al., 2023; Oh et al., 2020). The strain *Lpb. plantarum* L42g exhibiting high antioxidant activity and antibacterial activity against foodborne pathogens (e.g., *Bacillus cereus*, *Listeria monocytogenes*, *Listeria innocua*, *Staphylococcus aureus*, *Streptococcus agalactiae*, *Escherichia coli*, *Staphylococcus enterica* subsp. *enterica* serovar *Typhimurium*, *Shigella* sp., *Vibrio cholerae*, and *Vibrio parahaemolyticus*) was successfully used to ferment almond and coix juices supplemented with monosodium glutamate (1%), demonstrating its potential as a starter culture for producing functional plant-based fermented beverages (Buatong et al., 2022).

The strain *Leuconostoc lactis* CCMA0415 has demonstrated a promising candidate for yam-based fermentations. This strain decreased pH to below 4.0, produced lactic acid, and maintained its cell viability above 8 log CFU/mL (Batista et al., 2019a). The fermented yam produced by *Leuconostoc lactis* CCMA0415 was successfully used as a suitable substitute for milk in the production of edible ice cream, making it an alternative for lactose-free, plant-based ice cream (Batista et al., 2019b).

In general, LAB are recognized as an important group of probiotic cultures because many species are associated with the gastrointestinal tract and exhibit resistance to harsh conditions, such as low pH, bile salts, and natural growth inhibitors (Yehuala et al., 2024). However, the number of studies searching for probiotic yeasts has increased substantially, including those focused on developing plant-based fermented foods. Unlike probiotic bacteria, yeasts are not affected by antibiotics, allowing their simultaneous use with these drugs. Further, yeasts are generally more heat-, oxygen-, and acidic pH-resistant, which favors their use in food processes and formulations (Czerucka & Rampal, 2019; Hatoum et al., 2012). They have demonstrated potential to be used in co-culture with LAB, for fruits, vegetables, and grains fermentations (Buatong et al., 2022; Ferreira et al., 2022, 2023; Fonseca et al., 2022; Freire et al., 2017a, b; Liu et al., 2023; Menezes et al., 2018) and single cultures (Agarbati et al., 2024; Wang et al., 2020).

The nonconventional probiotic yeasts *Lachancea thermotolerans* Lt101, *Kazachstania unispora* Kum3-B3, *Meyerozyma guilliermondii* Mg112, *Meyerozyma caribbica* Mc58, and *Debaryomyces hansenii* (Dh36) were evaluated to ferment a blend of non-hopped beer wort (60% v/v) with an infusion of oxidized rooibos (red rooibos) leaves (40% v/v). The probiotic yeasts Lt101, Mg112, and Dh36 produced beverages with a balanced nutritional profile, as evidenced by their aromatic and sensory characteristics. On the other hand, the strains Dh36, Mc58, Kum3-B3, and Mg112 exhibited high antioxidant capacity and total phenolic content, supporting the health-promoting effects of the beverages (Agarbati et al., 2024).

The co-cultures of yeast *T. delbrueckii* CCMA 0235 with *Lpb. plantarum* CCMA 0743 and *Lactobacillus* (*L.*) *acidophilus* LAC-04 were used to produce a plant-based beverage similar to the Brazilian indigenous beverage cauim, using a blend of cassava and rice (Freire et al., 2017b). These starter cultures conferred adequate acidification, improved product safety, further enhanced digestibility by reducing starch, and increased antioxidant activity.

In the study performed by Wang et al. (2020), the fermentation performances of non-*Saccharomyces* yeasts (*P. kluyveri* FrootZen, *T. delbrueckii* Prelude, *Williopsis saturnus* var. *mrakii* NCYC2251, and *T. delbrueckii* Biodiva) were evaluated in green tea slurry. According to the authors, the strains *T. delbrueckii* Biodiva, *W. saturnus* NCYC2251, and *P. kluyveri* FrootZen showed potential for green tea fermentation, as evidenced by their desirable antioxidant capacity and flavor profiles. These findings highlight the growing potential of non-*Saccharomyces* yeasts, either alone or in co-culture with LAB, to enhance the nutritional quality, safety, and sensory attributes of plant-based fermented beverages.

Besides LAB and yeasts, *Propionibacterium* spp. has also demonstrated potential as an adjunct starter culture for the production of soy, almond, and coconut fermented beverages (Gungor et al., 2024). The inoculation of the probiotic starter culture *Bifidobacterium* BB-12 in co-culture with *Propionibacterium* spp. in the plant-based substrate maintained their populations at the recommended level (above 7–8 log CFU/g) even during storage (Gungor et al., 2024).

11.2.2 Sensory and Functional Effects of Fermentation

The sensory quality of fermented foods includes texture/viscosity, flavor, aroma, appearance, and color, which are strongly influenced by fermentation and the employed starter culture. One of the primary challenges for plant-based fermented beverages is achieving a creamy and stable texture similar to that of dairy products. In dairy, the casein–whey network supports gel formation and water-holding. In plant substrates, proteins often lack the gelation properties of animal proteins, so fermentation strategies must either supplement or compensate for this deficiency. Thus, the use of starter cultures producing EPS can enhance viscosity and reduce syneresis, contributing to food texture (Hidalgo-Fuentes et al., 2024). There have been reported differences in physical and textural properties, including hardness, adhesiveness, and syneresis, in soy-based yogurt analogs according to starter culture use, demonstrating the importance of selecting EPS-producing strains (Okafor & Okoro, 2025).

Another important issue regarding plant-based fermented foods is their aroma and flavor. It is well known that lipid oxidation pathways or aldehydes/ketones generated during processing can contribute to off-flavor development (e.g., beany, grassy, raw legume). The microbial metabolism can generate a variable profile of volatile compounds during fermentation, producing acids, alcohols, esters, aldehydes, and others, affecting the flavor and aroma of the food (Ferreira et al., 2022; Fonseca et al., 2022; Wang et al., 2020). Off-flavor precursors may be metabolized by selected starter cultures, and desirable volatile compounds (e.g., fruity, floral) can be generated (Hidalgo-Fuentes et al., 2024). Studies have demonstrated that starter culture use can positively impact the flavor intensity in plant-based fermented beverages (Hidalgo-Fuentes et al., 2024; Wang et al., 2020).

Furthermore, microbial activity in fermented foods can promote acidification by lowering pH, which is important for food safety. However, plant matrices often contain pigments (e.g., anthocyanins, carotenoids) that may be changed by some microbial enzymatic activity or pH decrease. Moreover, some EPS or cell-bound polysaccharides may affect beverage turbidity (Cakmak et al., 2025; Hidalgo-Fuentes et al., 2024). Thus, selecting starter cultures for each condition and substrate is crucial for obtaining high-quality fermented foods.

Microbial activity also significantly influences the functional and nutritional properties of plant-based fermented foods. An important consideration is that several plant substrates contain anti-nutritional factors, such as phytate, which chelates minerals (e.g., iron, zinc) and reduces their bioavailability.

Some microbial strains may produce phytase, which reduces phytate levels and increases mineral bioavailability (Gänzle et al., 2024; Menezes et al., 2020b). Additionally, other enzymatic activities, including proteases and amylases, are crucial to improve the digestibility of plant-based fermented foods by metabolizing proteins and polysaccharides into free amino acids and simple carbohydrates, respectively (Cakmak et al., 2025; Freire et al., 2017b; Hidalgo-Fuentes et al., 2024).

According to Pereira et al. (2023), the use of a selected starter culture (*Streptococcus thermophilus* and *Lactobacillus delbrueckii* subsp. *bulgaricus*) in co-culture with probiotic strains (*Bifidobacterium longum* and *Lactobacillus acidophilus*) to produce a fermented soy-based beverage increased the digestibility of soybeans by effectively degrading soy proteins into small peptides and amino acids. For rich-starch substrates, the inoculation of *T. delbrueckii* CCMA 0235 to produce rice-based beverages (similar to cauim) reduced starch content, also improving the product's digestibility (Freire et al., 2017b).

It has been reported that bioactive compounds, including polyphenols, peptides, and vitamins with beneficial biological properties (e.g., antioxidant, anti-inflammatory, and antihypertensive activities), can be enhanced by starter culture activity during fermentation (Agarbati et al., 2024; Cakmak et al., 2025; Ferreira et al., 2022; Freire et al., 2017b; Hidalgo-Fuentes et al., 2024; Menezes et al., 2020b; Wang et al., 2020). Studies by Oh et al. (2020) demonstrated increases in total polyphenol and antioxidant activity in a tomato-based fermented beverage. Meanwhile, Cakmak et al. (2025) argue that starter cultures may contribute to health benefits by producing bacteriocins or metabolites with antimicrobial or immunomodulatory effects.

Nevertheless, although promising, few clinical trials on health outcomes of plant-based fermented foods have been reported, remaining limited compared with in vitro studies (Bruce et al., 2025). A systematic review evaluated the consumption of some fermented plant-based foods, such as sourdough, fermented soy (tempeh, natto, miso), kimchi, kombucha, and sauerkraut, on the "satiety cascade" in humans (Chatonidi et al., 2023). The authors reported that sourdough consumption appears promising for reducing glycemic response, although the mechanisms and relationship with appetite control remain unclear. For the other fermented foods, no or inconsistent effects on satiety and energy consumption were found (Chatonidi et al., 2023).

As previously discussed, many nutritional and sensorial aspects have been evaluated in the development of plant-based fermented foods; however, several practical challenges should be considered when developing commercial plant-based fermented foods and beverages. For example, different substrate compositions associated with physiological microbial characteristics can generate more pronounced acidity or off-flavors, undesired texture or color, and reduced probiotic viability during storage in plant-based products. In addition, selecting starter cultures with probiotic and technological properties adapted to plant substrates remains a challenge, as most strains available on the market are intended for lactic fermentations. Another important aspect is that the formation of structural networks (gelation, emulsification) in plant matrices during fermentation varies with the substrate used, requiring individualized studies, evaluations, and standardization.

This is also important to emphasize some potential risks associated with these products. They are fermented foods, and a lack of hygiene and good manufacturing practices, especially in artisanal production or among small producers without required supervision, can lead to contamination by undesirable or pathogenic microorganisms. Furthermore, undesirable fermentations can produce compounds such as biogenic amines or organic acids at high concentrations, which are undesirable in the final product. It is also worth noting that plant-based products may contain proteins or antinutrients that could trigger allergic or undesirable reactions in certain population groups.

Regulatory requirements require that the production of artisanal and traditional fermented plant-based beverages ensure legal compliance and consumer transparency while maintaining quality, safety, and cultural authenticity. In this sense, there are regulatory barriers worldwide that hinder their commercialization. Many countries maintain lists of microorganisms considered safe for food use (e.g., the "QPS" lists in the European Union or the "GRAS" lists in the USA). New strains isolated from variable matrices require lengthy safety assessment processes. This is an important criterion, since the diversity of substrates requires specific adaptations, especially in the starter culture's adaptability. Additionally, inputs derived from biotechnology or functional ingredients may be classified as "novel foods", which require extensive toxicological documentation and estimated consumption studies. Regulations, such as the European Union's Claims Regulation and the FDA (Food and Drug Administration) rules in the USA, establish rigorous criteria for health claims and probiotic properties, requiring robust scientific evidence. Furthermore, global legislation imposes labeling standards that can be challenging for plant-based fermented products, especially regarding potential allergens derived from raw materials. The absence of internationally applicable regulations among regulatory agencies (e.g., Codex Alimentarius, EFSA, FDA, and MAPA in Brazil) creates barriers to commercialization, as approval requirements in one country may not be recognized in another.

11.3 Microbial Diversity in Spontaneous Fermentation of Plant-Based Substrates: Insights into Potential Starter Cultures

The endogenous microbiota is crucial for understanding the microbial ecology of plant-based fermented foods. Native microbial consortia, naturally associated with raw substrates such as cereals, fruits, and vegetables, play a decisive role in fermentation dynamics. During fermentation, the physicochemical parameters of the matrix, including pH, titratable acidity, sugar consumption, and organic acid production, are modified by fermentation. These changes in composition influence both microbial succession and the development of volatile metabolites.

The interactions among LAB, yeasts, and, in some cases, acetic acid bacteria result in distinctive metabolic activities that influence the sensory, physicochemical, and functional characteristics of fermented products. Furthermore, to better understand the process, studies have investigated on (i) characterizing microbial populations associated with different plant-based fermentations using both culture-dependent and independent of culturing approaches; (ii) monitoring the evolution of physicochemical parameters; and (iii) establishing correlations between metabolic activity and the sensory attributes of the final products.

In this context, the native microbiota of diverse fermented foods and beverages, such as table olives, kimchi, sauerkraut, pickled vegetables, ogi, hausa koko, mawè, boza, pozol, kombucha, and indigenous beverages, including cauim, caxiri, chicha, tarubá, and yakupa, have been increasingly explored as a source of potential starter cultures and probiotic strains, as demonstrated in Table 11.2.

11.3.1 Olive Table

Olive table fermentation involves multiple microbial groups, in which endogenous enzymes interact with microbial metabolic activities, resulting in characteristic flavors and textures (McFeeters, 2004). The microbiota associated with olives and brines is diverse, including members of Enterobacteriaceae, *Clostridium*, *Pseudomonas*, *Staphylococcus*, LAB, yeasts, and sometimes filamentous fungi (Bleve et al., 2014; Heperkan, 2013). However, microbial dynamic change during fermentation. Enterobacteriaceae are generally detected only at the beginning, while *Clostridium* and *Pseudomonas* are absent at later stages due to the inhibitory effect of low pH values (<4.3) (Bleve et al., 2014; Montano et al., 2010). The dominant genera detected in the fermentation include *Celerinatantimonas*, *Lactiplantibacillus*, *Pichia*, *Candida*, and *Wickerhanomyces* (Ricciardi et al., 2025). Other species, such as *Debaryomyces hansenii*, *Wickerhamomyces anomalus*, *Pichia membranifaciens*, and *S. cerevisiae*, have also been reported to contribute to the fermentation process (Arroyo-López et al., 2008; Bleve et al., 2014; Nisiotou et al., 2010).

During fermentation, the yeast population increases from approximately 3 log CFU/mL to 4–5 log CFU/mL by the end of fermentation (Bleve et al., 2014). Yeasts play an important role in bitterness reduction through β-glucosidase activity, which promotes the hydrolyzes of oleuropein, while lipase activity enhances aroma by releasing free fatty acids (Bevilacqua et al., 2012, 2013; Rodríguez-Gómez et al., 2012). Furthermore, the selected strain should not exhibit proteolytic activity, as proteolysis is associated with olive softening and quality loss (Arroyo-López et al., 2012).

After 90 days of fermentation, LAB are predominantly dominated by *Lactiplantibacillus plantarum*, followed by *Leuconostoc* and *Pediococcus*, which contribute to acidification and product stabilization. LAB metabolism results in the accumulation of lactic acid and other organic acids, with acetic acid indicating the presence of heterofermentative pathways (Bleve et al., 2014, 2015).

Microbial succession modulates the volatile profile of fermented olives. The early stages are characterized by higher alcohols and terpenes, whereas later stages (90–180 days) exhibit increases in esters, fatty acids, and hydrocarbons, thus contributing to the aromatic complexity (Bleve et al., 2014, 2015).

Table 11.2 Dominant microbiota in spontaneous fermentation and potential starter cultures for non-dairy fermented foods

Foods/ beverages	Dominant microorganisms	References	Starter cultures	References
Olive table	*Celerinatantimonas*	Ricciardi et al. (2025)	*Lactiplantibacillus pentosus*	Montoro et al. (2016)
	Lactiplantibacillus		*Lacticaseibacillus paracasei CCMA1770*	Simões et al. (2022)
	Pichia		*Lactiplantibacillus pentosus CCMA1768*	
	Candida		*Levilactobacillus brevis CCMA1762*	
	Wickermahomyces		*Saccharomyces cerevisiae CCMA 1746*	Simões et al. (2021)
			Pichia guilliermondii CCMA 1753	
			Candida orthopsilosis CCMA 1748	
			Candida tropicalis CCMA 1751	
			Meyerozyma caribbica CCMA 1758	
			Debaryomyces hansenii CCMA 1761	
			Lactobacillus crispatus MB417	Saeed et al. (2023)
			Lactococcus lactis MB418	
			Carnobacterium divergens MB421	
Kimchi	*Lactobacillus*	Park et al. (2012)	*Lactobacillus* sp. *IJ-K2*	Jung et al. (2012a, b)
	Leuconostoc		*Pediococcus* sp. *IJ-K1*	
	Weissella		*Lacticaseibacillus paracasei DKL121*	Lu et al. (2020)
			Lactiplantibacillus plantarum DKL119	
			Lactobacillus acidophilus DKLLa5	
			Lacticaseibacillus casei DK128	

(continued)

Table 11.2 (continued)

Foods/ beverages	Dominant microorganisms	References	Starter cultures	References
			Lactiplantibacillus plantarum DKL109	
			Lactiplantibacillus plantarum	Lee et al. (2016)
			Leuconostoc mesenteroides	
Sauerkraut	*Lactococcus lactis* subsp. *lactis*	Xiong et al. (2012)	*Lactobacillus*	Touret et al. (2018)
	Enterococcus faecalis			
	Lacticaseibacillus casei		*Lactiplantibacillus plantarum* F42	Han et al. (2025)
	Lactiplantibacillus plantarum			
	Lacticaseibacillus zeae			
	Leuconostoc mesenteroides			
Pickled	*Levilactobacillus namurensis*	Yasir et al. (2022)	*Lactiplantibacillus plantarum*	Guney et al. (2025)
	Lentilactobacillus parafarraginis		*Lactiplantibacillus pentosus*	
	Lactiplantibacillus pentosus		*Lentilactobacillus parabuchneri*	
	Lentilactobacillus buchneri		*Levilactobacillus brevis*	
	Leuconostoc carnosum		*Pediococcus parvulus*	
	Pectobacterium carotovorum			
	Weissella confuse		*Lactiplantibacillus plantarum BH6*	Zhao et al. (2022)
Ogi	*Lactobacillus*	Sanya et al. (2023)	*Lactiplantibacillus plantarum CAM 7*	Adisa et al. (2024)
	Limosilactobacillus		*Levilactobacillus brevis 15F1*	
	Streptococcus		*Lactiplantibacillus plantarum 2.7.17*	
	Weissella		*Pediococcus pentosaceus CP 16*	
			Pediococcus pentosaceus NRIC 0123	
	Saccharomyces		*Lactiplantibacillus plantarum*	Olatunde et al. (2018)

(continued)

Table 11.2 (continued)

Foods/ beverages	Dominant microorganisms	References	Starter cultures	References
			Enterococcus faecalis	
			Enterococcus faecium	
			Lactobacillus brevis	
			Lactobacillus fermentum	
			Lactobacillus reuteri	
			Pediococcus acidilactici	
			Pediococcus pentosaceus	
Hausa koko	*Acetobacter*	Atter et al. (2021)	*Limosilactobacillus fermentum*	Lei and Jakobsen (2004)
	Lactobacillus		*Weisella confusa*	
	Lactococcus			
	Leuconostoc			
	Ligilactobacillus salivarius			
	Limosilactobacillus fermentum			
	Pediococcus			
	Streptococcus			
	Weissella			
	Candida			
	Cyberlindnera			
	Kluyveromyces			
	Nakaseomyces			
	Pichia			
	Saccharomyces			
	Torulaspora			
Mawè	*Limosilactobacillus fermentum*	Houngbédji et al. (2018)	*Kluyveromyces marxianus*	Greppi et al. (2013a)
	Candida glabrata	Greppi et al. (2013b)	*Saccharomyces cerevisiae*	
	Candida krusei			
	Kluyveromyces marxianus			
	Pichia kudrivzevii	Houngbédji et al. (2018)		
Boza	*Lactococcus*	Ucak et al. (2022)	*Levilactobacillus brevis S5*	Zhao et al. (2025)
	Lactobacillus		*Limosilactobacillus fermentum S4*	

(continued)

Table 11.2 (continued)

Foods/ beverages	Dominant microorganisms	References	Starter cultures	References
	Leuconostoc		*Limosilactobacillus fermentum S6*	
	Streptococcus		*Pediococcus acidilactici ST31BZ*	Queiroz et al. (2022)
			Pediococcus pentosaceus ST75BZ	
			Pediococcus pentosaceus ST87BZ	
			Lacticaseibacillus paracasei ST284BZ	Todorov (2009)
			Lactiplantibacillus plantarum ST194BZ	
Pozol	*Streptococcus*	López-Sánchez et al. (2023)	*Bacillus subtilis CS93*	Ray et al. (2000)
Kombucha	*Komagataeibacter*	Arıkan et al. (2020)	*Brettanomyces anomalus*	Areal-Hermida et al. (2024) and Cho et al. (2018)
			Brettanomyces bruxellensis	
	Zygosaccharomyces		*Acetobacter musti LOAAB1*	Wang et al. (2024b)
			Gluconobacter potus GBAAB3	
			Gluconobacter potus LOAAB2	
			Gluconobacter sp	Wang et al. (2014)
			Lactiplantibacillus plantarum	Pei et al. (2020)
			Pichia kudraivzevii GBY1	Wang et al. (2024a)
			Saccharomyces cerevisiae GBY2	
			Pediococcus pentosaceus	Bogdan et al. (2018) and Diguță et al. (2020)
Cauim	*Limosilactobacillus fermentum*	Almeida et al. (2007) and Ramos et al. (2010, 2011)	*L. plantarum* CCMA 0743	Fonseca et al. (2021a, b) and Freire et al. (2017a, b)
	Lactiplantibacillus plantarum		*L. brevis* FFC199	Ramos et al. (2013, 2014)
	Lacticaseibacillus paracasei			
	Levilactobacillus brevis			

(continued)

Table 11.2 (continued)

Foods/ beverages	Dominant microorganisms	References	Starter cultures	References
	Saccharomyces cerevisiae			
	Candida spp.			
	Pichia spp			
	Kluyveromyces lactis			
Caxiri	*Bacillus pumilus*	Santos et al. (2012)	*Pediococcus acidilactici* CCMA 0347	Da Silva et al. (2025), Santos et al. (2014), and Kamda et al. (2014)
	Bacillus cereus group		*Saccharomyces cerevisiae* CCMA 0739	Menezes et al. (2020b)
	Bacillus subtilis		*Saccharomyces cerevisiae* CCMA 0712	Menezes et al. (2020b)
	Saccharomyces cerevisiae		*Saccharomyces cerevisiae* CCMA 0731	Menezes et al. (2018, 2020a, b) and De Almeida Silva Vilela et al. (2020)
			Saccharomyces cerevisiae CCMA 0732	Menezes et al. (2018, 2020a, b)
Chicha	*Bacillus* spp.	Puerari et al. (2015)	*Leuconostoc lactis* CCMA 0415	Batista et al. (2019a, b)
	Lacticaseibacillus			
	Levilactobacillus			
	Lentilactobacillus			
	Lactiplantibacillus			
Taruba	*Lactobacillus plantarum*	Ramos et al. (2015)	*Torulaspora delbrueckii* CCMA 0235	Freire et al. (2017a, b)
	Lactobacillus brevis			
	Leuconostoc mesenteroides			
	Bacillus subtilis			
	Torulaspora delbrueckii			
Yakupa	*Limosilactobacillus fermentum*	Freire et al. (2014)	*Lactobacillus fermentum* CCMA 0215	Freire et al. (2017a)
	Lactiplantibacillus plantarum			
	Weissella cibaria			
	Weissella confusa			
	Saccharomyces cerevisiae			

The interactions between yeasts and LAB determine the chemical composition, sensory attributes, and final quality of fermented olives (Bleve et al., 2014, 2015). Therefore, several strains associated with olive fermentation have been investigated for their potential probiotic properties. These include *Lactiplantibacillus* (*Lpb.*) *pentosus*, *Lactiplantibacillus* (*Lpb.*) *plantarum*, *Lacticaseibacillus* (*Lbs.*) *paracasei* subs. *Paracasei* (Argyri et al., 2013); *Lpb. pentosus* (Montoro et al., 2016); *Lbs. paracasei* CCMA1770, *Lpb. pentosus* CCMA1768, *Levilactobacillus* (*Lvb.*) *brevis* CCMA1762 (Simões et al., 2022); *S. cerevisiae* CCMA 1746, *Meyerozyma guilliermondii* CCMA 1753, *Candida orthopsilosis* CCMA 1748, *Candida tropicalis* CCMA 1751, *Meyerozyma caribbica* CCMA 1758, and *D. hansenii* CCMA 1761 for Simões et al. (2021) and *Lactobacillus* (*Lab.*) *crispatus* MB417, *Lactococcus* (*Lc.*) *lactis* MB418, and *Carnobacterium divergens* MB421 for Saeed et al. (2023) (Table 11.1).

11.3.2 Kimchi

Kimchi is a traditional Korean fermented vegetable characterized by a unique sensory profile and recognized for its health benefits. Initial studies, based on traditional culture methods (morphology and phenotype), frequently identified genera such as *Enterococcus*, *Pediococcus*, and *Streptococcus* as dominant (Jeong et al., 2013a, b). However, molecular technologies (16S rRNA gene analysis) have revealed that these groups are rarely dominant, suggesting that they play secondary roles (Jeong et al., 2013a; Jung et al., 2013). *Leuconostoc*, *Lactobacillus*, and *Weissella* are the dominant microbiota throughout the process and ecology roles of these LAB, as well as their metabolic and functional contributions have been investigated (Cho et al., 2006; Jeong et al., 2013a, b, c; Jung et al., 2012b; Kim & Chun, 2005; Park et al., 2010, 2012).

In the initial stage of fermentation, the genera *Leuconostoc* is detected more frequently. Its dominance initially is due to its lower tolerance to acids. Species such as *Leu. mesenteroides* and *Leu. citreum*, are important in the early stages due to its metabolic characteristics, while *Leu. gasicomitatum* and *Leu. gelidum* can persist into later stages due to its greater acid tolerance (Jeong et al., 2013a, b; Jung et al., 2011). At the end of the process, *Lactobacillus* becomes the most dominant genus (Jeong et al., 2013a, b; Jung et al., 2011). In this stage of fermentation, the environment becomes more acidic and anaerobic.

Weisella becomes dominant in kimchi fermentation under specific conditions related to the substrates used, the fermentation stage, and storage temperature. The addition of red pepper powder substantially increases the proportion of *Weissella* compared to the genera *Leuconostoc* and *Lactobacillus* (Jeong et al., 2013b). This is because pepper appears to slow down initial fermentation and may have a greater inhibitory effect on the other genera than on *Weissella*. Furthermore, this genus was described as dominant young radish kimchi (Cha et al., 2023). Species such as *W. koreensis* tend to become more dominant as fermentation progresses into the middle stages, due to their greater tolerance of acidic and anaerobic environments, and mainly produce lactic acid, acetic acid, carbon dioxide, and ethanol as major end products (Cheigh et al., 1994).

Microbial succession of LAB is a key feature of kimchi fermentation, and this microbial profile is similar to that of other vegetable fermentations, such as sauerkraut and pickles (Breidt et al., 2013). This dominant microbiota frequently reaches 9–10 log CFU/g or mL and is responsible for producing compounds that define the sensory properties of kimchi, such as lactic acid, acetic acid, carbon dioxide, ethanol, mannitol, and amino acids (Cheigh et al., 1994; Jeong et al., 2013a, b; Jung et al., 2011, 2012a).

Yeasts also play a significant role in kimchi fermentation. They act primarily in the final stages of fermentation. Species such as *Saccharomyces*, *Candida*, *Pichia*, and *Kluyveromyces* are frequently isolated and tolerate low pH conditions (Chang et al., 2008; Jeong et al., 2013a, c).

Saccharomyces and *Candida* are dominant genera. *Saccharomyces* shows a significant increase in population after 15 days of fermentation, reaching a maximum value around 40 days (Jeong et al., 2013a). The *Candida* population begins to increase after the decline in the *Saccharomyces* population, reaching its peak around 90 days of fermentation. The yeast contributes to the metabolite profile, however excessive growth of these microorganisms in the later stages is often associated by the reduction of free sugars, and amino acids including glutamine, glutamic acid, glycine, alanine, arginine, asparagine, lysine, leucine, and tyrosine, which decrease gradually after 30 or 60 days of fermentation characterizing the end of the fermentation process (Chang et al., 2008; Jeong et al., 2013a).

Organic acids (lactic acid and acetic acid), amino acids, mannitol, and carbohydrates (glucose and fructose) are determinants in the formation of kimchi flavor. The mannitol is synthesized by *Leuconostoc* species through fructose reduction with the consumption of NADH by mannitol dehydrogenase (Wisselink et al., 2002). The ability to utilize fructose as an electron acceptor provides them with a significant selective advantage because it increases their energy yield during fermentation. This process allows for the production of an additional mole of ATP for every two moles of mannitol produced, enhancing their competitiveness and supporting their dominance during the early and middle stages of fermentation.

While lactic and acetic acids are the main organic acids produced, propionic, butanoic, succinic, fumaric, and tartaric acids are present in small quantities. The organic acids contribute to flavor complexity, sourness, and preservation (Shim et al., 2012). Additionally, amino acid accumulation is a critical factor in sensory quality, particularly in defining the umami profile. These compounds accumulate through the proteolytic activity of both the raw substrate and the metabolic activity of LAB (Jeong et al., 2013a, c).

From a functional and nutritional perspective, LAB in kimchi can produce γ-aminobutyric acid (GABA), vitamins, and volatile compounds (Kim & Kim, 2012; Kwon et al., 2025). GABA functions as an inhibitory neurotransmitter and is produced from glutamate by the enzyme glutamate decarboxylase. While various LAB can produce it, *Lactobacillus* species (such as *Lb. sakei* and *Lb. buchneri*) are identified as the major GABA producers during kimchi fermentation, adding a functional health dimension to the food's metabolic profile. Indeed, microorganisms from kimchi can survive gastrointestinal passage and remain metabolically active in the gut, supporting host health and microbial balance (Cha et al., 2023).

The microbial ecology of kimchi fermentation is modulated by dynamic interactions among LAB and yeasts, environmental parameters, and substrate composition. The microbial succession from *Leuconostoc* followed by *Lactobacillus* and *Weissella*, coupled with yeast activity, ensures both product stability and the development of kimchi's complex sensory and functional attributes. Furthermore, several microorganisms have been evaluated as potential functional microbial resources, such as *Lactobacillus* sp. IJ-K2, *Pediococcus* sp. IJ-K1, *Lbs. paracasei* DKL121, *Lpb. plantarum* DKL11924, *Lactobacillus acidophilus* DKLLa5, *Lacticaseibacillus* (*Lbs.*) *casei* DK128,24, *Lpb. plantarum* DKL109, *Lpb. plantarum* and *Leu. mesenteroides* (Jung et al., 2012a; Lu et al., 2020; Lee et al., 2016).

11.3.3 Sauerkraut

Chinese sauerkraut (or pao cai) is the spontaneous fermentation of cabbage-based foods. Microbial diversity modulates the metabolic and sensory profiles of the final product. This microbiota is influenced by substrate composition, fermentative conditions, and geographic location (Yang et al., 2020).

A wide array of microorganisms characterizes this process. Predominantly composed of *Lactobacillus*, *Lactococcus*, *Leuconostoc*, *Weissella*, *Enterobacteriaceae*, and occasionally *Pseudomonas* and *Clostridium* (Wouters et al., 2013; Xiong et al., 2012). *Leuconostoc mesenteroides* subsp. *mesenteroides* dominated the beginning of the process. This bacterium is heterofermentative and produces CO_2, creating the anaerobic conditions necessary for subsequent species (Xiong et al., 2012). Subsequently, *Enterococcus faecalis* and *Lactococcus lactis* produce bacteriocins that inhibit foodborne pathogens and undesirable Gram-negative organisms (Xiong et al., 2012). *Lactobacillus plantarum* and *Lbs. casei* emerge and dominate the final stage of fermentation when the environment becomes acidic (Cheigh et al., 1994; Pederson & Albury, 1969; Zhang et al., 2022).

Firmicutes and Proteobacteria are the predominant bacterial phyla detected during the spontaneous fermentation of sauerkraut. *Weissella* is the dominant genus due to its resistance under acidic conditions. *Lactobacillus*, *Clostridium*, and *Leuconostoc* are detected at low levels (Guan et al., 2020; Yang et al., 2020). *Clostridium* and *Enterobacter* can be detected in some household fermentations (Jung et al., 2012b).

The LAB are essential to promoting the accumulation of lactic and acetic acids. This group of microorganisms not only produces organic acids but also bacteriocins that inhibit nitrate-reducing bacteria such as *Enterobacter*. This results in a beverage with suppression of spoilage microorganisms and lower, safer nitrite levels (below

20 mg/kg), in addition to influencing the aroma and texture of the final product (Bell et al., 2017; Pederson & Albury, 1969; Zhang et al., 2019; Zhou et al., 2018).

Beyond acidification, proteolysis and amino acid metabolism are essential to the development of the sensory and functional characteristics of fermented cabbage products like sauerkraut and Pao Cai. The breakdown of cabbage proteins leads to the accumulation of free amino acids such as glutamate, aspartate, and alanine, which contribute to umami and sweet notes, as well as γ-aminobutyric acid (GABA), which is produced through the microbial decarboxylation of glutamate and may increase by as much as 30- to 50-fold compared to the initial raw cabbage (Rizzello et al., 2018; Zhao et al., 2016). Furthermore, the esterification of lactic acid with alcohol enhances aromatic complexity, producing fruity and floral notes such as ethyl lactate and ethyl butanoate (strawberry-like aroma) (Ricci et al., 2018; Wu et al., 2015). Terpenes represent another class of volatile compounds that confer antioxidant and antimicrobial properties and are produced by the action of microbial enzymes, such as glycosylases, which break the bond between terpenes and sugars, or by their possible ex novo production (Minervini et al., 2020; Ricci et al., 2018).

Esters, alcohols, aldehydes, terpenes, and fatty acids are produced by microbial metabolism of carbohydrates, amino acids, and lipids. Distinct household fermentations have exhibited divergent volatile compound profiles. For example, one household fermentation was dominated by acids and lactones, another by alcohols, aldehydes, and esters, and a third enriched in terpenes and nitriles (Yang et al., 2020).

LAB, particularly *Lactobacillus* and *Weissella*, have been associated with increased formation of esters, higher alcohols, and other volatile compounds, suggesting a major role in flavor development in fermented plant-based foods (Lee et al., 2018; Walsh et al., 2016). *Weissella* correlated positively with lactones, acids, and terpenes. *Leuconostoc* contributed to the formation of alcohols, aldehydes, and nitriles. Furthermore, the *Lactobacillus* strain has demonstrated a significant positive correlation with the production of acids, such as octanoic and 3-methylbutanoic acids, esters, and terpenes. In addition, *Lactobacillus* strains demonstrate functional activity by producing antimicrobial substances, such as bacteriocins, and beneficial compounds like GABA, which has hypotensive and nervous system regulatory effects (Touret et al., 2018; Han et al., 2025; Verni et al., 2019; Yang et al., 2020). *Clostridium* and *Enterobacter* showed positive correlations with the production of terpenes, alcohols, and esters (He et al., 2020).

The metabolic activity of the microbial community modulates the biochemical and sensory complexity of traditional fermented vegetables. These fermentations are carried out in environmental heterogeneity, with diverse raw materials and microbial adaptation to acidic stress. However, uncontrolled fermentations may compromise food safety due to the presence of contaminants such as *Pseudomonas*, *Sphingomonas*, *Rhizobium*, *Aureimonas*, *Serratia*, and *Stenotrophomonas*, which are often associated with spoilage and reduced product quality (Cao et al., 2017; Pérez-Díaz et al., 2019). For this reason, it is recommended that spontaneously fermented sauerkraut not be consumed for at least 4 weeks to ensure that the LAB has stabilized the product and eliminated competing contaminants. Understanding these dynamics is essential not only for preserving traditional practices but also for the selection of starter cultures with probiotic and technological potential for controlled fermentation systems (Abriouel et al., 2012; Bleve et al., 2015).

11.3.4 Pickled Vegetables

Fermented pickled vegetables are a diverse and ancient category of plant-based foods, produced primarily through spontaneous fermentation. This process relies on the activity of autochthonous microorganisms present in raw materials and the processing environment. At the start of the process, raw vegetables are often dominated by Proteobacteria and spoilage-related genera such as *Pseudomonas*, *Flavobacterium*, and *Arthrobacter*. As fermentation progresses, LAB, including the genera *Lactobacillus*, *Weissella*, *Pediococcus*, and *Leuconostoc*, become the dominant microbial group. Halophilic bacteria, such as *Halomonas* and *Idiomarina*, as well as halophilic yeasts, such as *Debaryomyces hansenii*, are also present (Zhang et al., 2021).

The quantitative microbiome profiling (QMP) and high-throughput sequencing (HTS) techniques have evaluated traditional pickled vegetables, including Taiwanese pickled cabbage, Chinese pickled cowpea (*Vigna unguiculata*), and Zhacai (pickled mustard tuber) (Guo et al., 2021; Hu et al., 2024; Zhang et al., 2021). In Taiwanese pickled cabbage, real-time qPCR combined with QMP revealed a unique, detailed microbial succession. During the first spontaneous fermentation round, the total bacterial load increased from 6.1 to 10 log CFU/mL, dominated sequentially by *Lactococcus*, *Leuconostoc*, and *Lactiplantibacillus* species (Hu et al., 2024). The application of a backslopping starter and vinegar during the second fermentation is essential for stabilization of the community, resulting in a homogeneous population (~90%) dominated by *Lactiplantibacillus*. When vinegar was omitted from the backslopping process, *Leuconostoc* consistently dominated the fermentation (Hu et al., 2024).

The microbiota of pickled cowpea was dominated by the phylum Firmicutes. The main family identified includes Lactobacillaceae (*Lactiplantibacillus*, *Levilactobacillus*, *Companilactobacillus*), Leuconostocaceae (*Weissella*), and Pediococcaceae (*Pediococcus*) (Guo et al., 2021; Liu & Tong, 2017). Culture-dependent methods confirmed the predominance of the *Lpb. plantarum* group followed by *Limosilactobacillus* (*Lmb.*) *fermentum* and *Lpb. rhamnosus*, reinforcing the probiotic potential ability to inhibit pathogens (Guo et al., 2021). *Levilactobacillus namurensis*, *Lentilactobacillus* (*Llb.*) *buchneri*, *Lentilactobacillus parafarraginis*, *Lpb. Pectobacterium carotovorum*, *Pectobacterium pentosus*, *Leuconostoc carnosum*, and *Weissella confuse* were found in different populations in most of the 18 samples from six pickled vegetables (Yasir et al., 2022).

Organic acids, amino acids, esters, and other compounds produced by LAB define the flavor and safety of food (Di Cagno et al., 2013). On the other hand, the occasional detection of *Klebsiella* and *Pseudomonas* underscores the need for strict hygienic control during spontaneous fermentation (Guo et al., 2021; Jung et al., 2018).

Zhacai is a traditional, spontaneous fermentation of pickled mustard tubers, made using the dry-salting method, containing 12–14% (w/w) sodium chloride. This high-salt environment creates conditions that favor the interaction between halophilic bacteria (HAB) and LAB (Zhang et al., 2021). During 90 days of fermentation, microbial succession led to the dominance of *Halomonas*, *Idiomarina*, *Companilactobacillus* (*Cob.*) *alimentarius*, *Companilactobacillus versmoldensis*, and *Weissella* spp., along with the yeasts *D. hansenii*, *Sterigmatomyces halophilus*, and *Sporidiobolus* spp. (Liang et al., 2020; Zhang et al., 2021). These interactions among HAB, LAB, and yeasts drive the biochemical transformations that define the quality of mature Zhacai. The main metabolites include lactic and acetic acids, free amino acids (e.g., glutamic acid, alanine, and lysine), and volatile compounds (e.g., alcohols, esters, aldehydes, and ketones) (Guan et al., 2020; Zhao et al., 2016). Halophilic species, such as *Halomonas*, are oxidase- and catalase-positive and contribute significantly to acetate production (Oguntoyinbo et al., 2017; Zhang et al., 2021). *Idiomarina loihiensis* and *Halomonas* spp. are strongly correlated with the accumulation of free amino acids (FAAs) contributing to the levels of glutamic acid (umami), as well as alanine and lysine (sweet tastes). The yeasts, such as *D. hansenii*, which can reach relative abundances of up to 60% and play essential roles in flavor formation through ester synthesis and proteolysis, along with LAB, and yeasts, which enrich the aromatic profile with floral and fruity notes (Zhang et al., 2021).

Therefore, the autochthonous microbial consortia of vegetables are responsible for the biochemical transformations that establish the functional, sensory, and safety characteristics of traditional fermented vegetables. *Lpb. plantarum*, *Lpb. pentosus*, *Llb. parabuchneri*, *Lvb. brevis*, *Pediococcus parvulus*, and *Lpb. plantarum* BH6, offering biotechnological potential for the design of controlled and functional plant-based fermentations (Guney et al., 2025; Zhao et al., 2022).

11.3.5 Ogi

Ogi is a traditional fermented cereal gruel mainly consumed in Nigeria, Benin, and Ghana. This beverage plays a key role in the nutrition and economy of these countries. It is generally produced from maize (Zea mays: white and yellow), although sorghum or millet may also be used. The process is spontaneous and occurs naturally due to the microbiota present on grains, utensils, and in the processing environment (Adisa & Enujiugha, 2020; Nago et al., 1998).

LAB and yeasts dominate the microbial community in Ogi fermentation, with LAB accounting for approximately 60–65% of total isolates (Oyedeji et al., 2013; Okeke et al., 2015; Sanya et al., 2023). Firmicutes was the predominant bacterial phylum identified, including members of the genera *Lactobacillus*, *Limosilactobacillus*, *Streptococcus*, and *Weissella* (Sanya et al., 2023).

The fermentation process begins with the predominance of facultatively heterofermentative LAB. *Leuconostoc mesenteroides* and *Weissella confusa* are commonly detected at this stage and initiate acidification by producing lactic acid and carbon dioxide (CO_2). Members of the family Enterobacteriaceae, along with other environmental bacteria such as *Bacillus*, may persist during the first 48 h of fermentation.

As fermentation progresses, the production of organic acids lowers the pH, creating a selective environment for homofermentative bacteria, including *Lpb. plantarum*, *Lmb. fermentum*, *Lactobacillus cellobiosus*, and *Pediococcus acidilactici* (Nago et al., 1998; Oyedeji et al., 2013). This microbial succession is similar to that of other West African fermented cereals, such as mawè and Hausa koko, in which heterofermentative LAB initiate fermentation, followed by acid-tolerant lactobacilli that dominate the later stages.

The dominant LAB species not only promote acidification but also contribute to the biochemical transformation of substrates through carbohydrate fermentation, proteolysis, and detoxification of undesirable compounds. *Lpb. plantarum* and *P. acidilactici* are frequently associated with strong acidification capacity and the production of antimicrobial metabolites, which confer a selective advantage over other bacterial groups (Adisa & Enujiugha, 2020; Sanya et al., 2023).

During Ogi fermentation, the most common yeast genera include *Candida*, *Saccharomyces*, *Pichia*, *Geotrichum*, and *Kluyveromyces* (Adisa & Enujiugha, 2020; Nago et al., 1998; Oyedeji et al., 2013; Sanya et al., 2023). Among them, *C. krusei* (formerly *P. kudriavzevii*), *C. tropicalis*, and *S. cerevisiae* are commonly detected as predominant species during the fermentation process (Adisa & Enujiugha, 2020; Nago et al., 1998; Oyedeji et al., 2013; Sanya et al., 2023). These yeasts contribute to the sensory complexity of Ogi by producing volatile flavor compounds (ethanol, esters, aldehydes) and releasing CO_2. Some species, such as *C. krusei* and *S. cerevisiae*, also help reduce anti-nutritional factors and may exert probiotic effects by competitively excluding spoilage microorganisms.

Synergistic interactions between LAB and yeasts result in the production of organic acids by LAB that lower the pH, creating an environment favorable for acid-tolerant yeasts, while yeasts provide growth factors, such as vitamins and amino acids, that enhance LAB proliferation (Adisa & Enujiugha, 2020; Itaman & Nwachukwu, 2021; Nago et al., 1998). This cooperative dynamic stabilizes the fermentation ecosystem, ensuring that beneficial microbial groups dominate over potential contaminants.

The microbiota of Ogi reflects a well-adapted and self-regulating ecosystem whose metabolic activities modulate acidification and flavor formation and improve safety. A potential starter culture can be selected from this substrate. Adisa et al. (2023) described strains such as *Lpb. plantarum* CAM7, *Lvb. brevis* 15F1, *Lpb. plantarum* 2.7.17, *P. pentosaceus* CP16, and *P. pentosaceus* NRIC 0123 as promising starter cultures, with *Lpb. plantarum* CAM7 and 2.7.17 showing probiotic characteristics. According to Olatunde et al. (2018), strains such as *Lpb. plantarum*, *E. faecalis*, *E. faecium*, *Lvb. brevis*, *Lmb. fermentum*, *L. reuteri*, *P. acidilactici*, and *P. pentosaceus* exhibited probiotic potential.

11.3.6 Hausa Koko

Hausa koko is a spontaneously fermented millet-based porridge widely consumed in Ghana and other parts of West Africa. The traditional process involves steeping millet grains in water for 12–24 h, followed by wet milling together with various spices. The mixture is homogenized with water to form a slurry. The slurry is then sieved and allowed to ferment (8–12 h). The liquid top layer is boiled for 1–2 h, and finally, the sedimented bottom layer is gradually added back until the desired consistency is achieved (Lei & Jakobsen, 2004).

The microbiota that develops during processing is composed of LAB and yeasts. The synergy between them promotes biochemical and sensory changes that determine the quality, safety, and acceptability of Hausa koko (Atter et al., 2021, 2024; Lei & Jakobsen, 2004). At the beginning of fermentation, the microbial community is diverse, including environmental and plant-associated bacteria. Dried grains show a high abundance of Gram-negative genera, such as *Pantoea*, *Pseudomonas*, *Staphylococcus*, *Bacteroides*, and members of the Enterobacteriaceae family (including *Escherichia-Shigella*).

However, as fermentation progresses, LAB dominate the process, promoting the pH reduction from approximately 6.0–6.5 in the steeped grains to 3.2–3.7. After 24–48 h of fermentation, leading to a rapid decline of Gram-negative species (Lei & Jakobsen, 2004). This acidification is primarily mediated by homo- and heterofermentative LAB, which can reach populations of 10^8–10^9 CFU/g and metabolize cereal carbohydrates into lactic and acetic acids (Atter et al., 2024; Lei & Jakobsen, 2004).

Microbiota research shows that the bacterial community becomes dominated by *Acetobacter*, *Lactobacillus*, *Lactococcus*, *Leuconostoc*, *Ligilactobacillus salivarius*, *Lmb. fermentum*, *Pediococcus*, *Streptococcus*, and *Weissella* (Atter et al., 2021). *Weissella confusa* and *Lmb. fermentum* are described as pioneer species in the liquid fraction known as koko sour water, indicating their crucial role in initiating fermentation and establishing the acidic environment (Lei & Jakobsen, 2004).

These LAB species contribute to several essential biochemical transformations that determine the nutritional and sensory characteristics of Hausa koko. Lactic and acetic acids produced lower pH, enhancing microbial safety by suppressing pathogens and spoilage organisms. In addition, many of the dominant LAB, particularly *Lpb. plantarum* and *P. acidilactici*, possess strong proteolytic and amylolytic activities that facilitate the release of free amino acids, peptides, and reduce sugars from millet substrates (Lim et al., 2019; Oguntoyinbo & Narbad, 2015). These compounds not only improve digestibility and nutritional quality but also serve as precursors to flavor formation.

Yeasts coexist with LAB throughout the fermentation and are indispensable for flavor development. *Saccharomyces* followed by *Candida*, *Cyberlindnera*, *Kluyveromyces*, *Nakaseomyces*, *Pichia*, *Saccharomyces*, and *Torulaspora* are the dominant yeasts in Hausa koko (Atter et al., 2021). These yeasts metabolize sugars and organic acids to produce ethanol, esters, aldehydes, and other volatile compounds responsible for the complex aromatic profile of this porridge. Some yeast species, particularly *P. kudriavzevii* and *S. cerevisiae*, have also been associated with the detoxification of phenolic and anti-nutritional compounds and are proposed to possess probiotic potential through competitive exclusion of undesirable microbes (Atter et al., 2024; Ganapathiwar, 2023; Menezes et al., 2020b).

The microbial interactions perform the production of organic acids (mainly lactic, acetic, and succinic acids) and amino acids (including alanine, leucine, and glutamic acid), which enhance some taste attributes, alongside the synthesis of esters and aldehydes, contributing fruity and floral notes (Atter et al., 2021, 2024; Ponomarova et al., 2017). The balance between acidification, proteolysis, and volatile formation defines the characteristic flavor complexity, nutritional enhancement, and microbial safety of Hausa koko.

Hausa koko demonstrates how complex microbial consortia transform raw millet into a nutritionally enriched and sensorially complex food. The interactions among LAB and yeasts not only ensure acidification and flavor development but also contribute to safety, shelf life, and functional potential. For this reason, the indigenous microbiota of Hausa koko, particularly strains of *Lmb. fermentum* and *W. confuse*, have been proposed as promising candidates for the development of controlled starter cultures and functional probiotics applicable to plant-based fermented foods.

11.3.7 Mawè

Mawè is a traditional spontaneously fermented cereal dough widely consumed in Benin and Togo, typically produced from dehulled or underhulled maize, sorghum, millet, or rice. Fermentation proceeds for 24–48 h, yielding an uncooked dough used for a variety of dishes, including *makumè*, *akassa*, and *akpan* (Houngbédji et al., 2018; Hounhouigan et al., 1994). The process is carried out by the indigenous microbiota of the raw materials and processing environment. The fermentation involves both LAB and yeasts as the predominant microbial groups, including *Lmb. fermentum*, *Candida glabrata*, *C. krusei*, and *Kluyveromyces marxianus* (Greppi et al., 2013a, b; Houngbédji et al., 2018; Hounhouigan et al., 1994).

The LAB population dominates fermentation, ranging from approximately 7.5 to 9.2 log CFU/g, while yeast counts increase from 4.8 to about 7.4 log CFU/g (Houngbédji et al., 2018). The LAB community is characterized by *Lmb fermentum* and *Lpb. plantarum*, accompanied by *W. confusa*, *Lvb. brevis*, *L. reuteri*, and *P. acidilactici* (Houngbédji et al., 2018; Hounhouigan et al., 1993, 1994). These species exhibit complementary functional traits, including rapid acidification and tolerance to low pH, which, together, suppress spoilage organisms and create a stable, acidic matrix.

The yeast community includes *P. kudriavzevii* (syn. *Issatchenkia orientalis*), *S. cerevisiae*, *K. marxianus*, and *Candida glabrata* (Greppi et al., 2013a; Houngbédji et al., 2018; Hounhouigan et al., 1993, 1994). *Pichia kudriavzevii* has been reported as the dominant yeast species in mawè fermentations, exhibiting remarkable persistence under acidic conditions (Houngbédji et al., 2018). *S. cerevisiae* and *K. marxianus*, although less abundant, are recurrently detected in similar cereal fermentations and are known to produce CO_2 and ethanol, contributing to the development of texture and flavor (Greppi et al., 2013b). In contrast, *C. glabrata* occasionally appears as a transient species with limited technological importance and potential safety concerns (Fidel et al., 1999; Greppi et al., 2013a).

Stable symbiotic association between LAB and yeasts produces organic acids (lactic and acetic acids) that lower the pH, favoring the growth of acid-tolerant yeasts, while yeasts, in turn, release amino acids, peptides, and vitamins that stimulate bacterial growth (Greppi et al., 2013a; Oguntoyinbo, 2008). Other yeast metabolites, including CO_2, pyruvate, acetate, and succinate, have also been shown to stimulate the development of lactobacilli (Pswarayi & Gänzle, 2019).

These microbial transformations, including acidification, proteolysis, and volatile formation, contribute to flavor complexity, nutritional enhancement, and microbial safety. *Lmb. fermentum*, *P. acidilactici*, and *P. kudriavzevii* demonstrated technological potential as candidates for the development of controlled starter cultures, whereas *K. marxianus* and *S. cerevisiae* have been classified as microorganisms with probiotic potential.

11.3.8 Boza

Boza is a traditional low-alcohol fermented cereal beverage widely consumed in the Balkan Peninsula, Turkey, and parts of the Middle East. Its production typically involves spontaneous or back-slopped fermentation of a cooked cereal base (e.g., millet, maize, wheat, or rice) (Kabak & Dobson, 2011). The fermentation process is carried out primarily by LAB and yeasts, which interact synergistically to modulate the microbial ecology, biochemical composition, and sensory characteristics of the product (Hancioğlu & Karapinar, 1997; Osimani et al., 2015; Queiroz et al., 2022).

The LAB community of Boza is dominated by *Lactococcus*, *Lactobacillus*, *Leuconostoc*, and *Streptococcus*. Other species commonly detected include *Lb. acidophilus*, *Lvb. brevis*, *Loigolactobacillus coryniformis*, *Lmb. fermentum*, *Lbs. paracasei*, *Lpb. plantarum*, *Lc. lactis* subsp. *lactis*, *Leu. mesenteroides*, *Leuconostoc lactis*, and *Llb. parabuchneri* (Borcakli et al., 2018; Gotcheva et al., 2001; Osimani et al., 2015; Queiroz et al., 2022; Aila et al., 2020). These bacteria are responsible for acidification by producing lactic and acetic acids, lowering the pH to approximately 3.8–4.2. The acidic conditions inhibit the growth of spoilage and pathogenic microorganisms (Gül et al., 2025; Hancioğlu & Karapinar, 1997).

Among LAB, *Pediococcus* spp. and *Lactiplantibacillus* spp. play important roles during fermentation. Genomic investigations have revealed that *P. pentosaceus* and *P. acidilactici* strains isolated from Boza possess operons encoding bacteriocins such as pediocin PA-1 and penocin A, conferring vigorous antimicrobial activity against *L. monocytogenes*, *E. faecium*, and other Gram-positive bacteria (Queiroz et al., 2022). These bacteriocinogenic LABs contribute to Boza's biopreservation capacity, extending its shelf life and enhancing its safety. Additionally, these strains harbor genes for sugar transport and amino acid biosynthesis (arginine, serine, alanine, and glutamate), which contribute to metabolic versatility and flavor development through carbohydrate fermentation and proteolysis (Queiroz et al., 2022). Furthermore, some *Pediococcus* strains contribute to safety through phytase activity, which degrades phytates, antinutrients found in cereals that interfere with mineral absorption.

The yeast community of Boza is composed of *S. cerevisiae*, *S. uvarum*, *C. tropicalis*, *C. glabrata*, and *P. kudriavzevii*. Other genera, such as *Pichia*, *Torulaspora*, and *Geotrichum*, are often present. Yeasts and LAB undergo concurrent but distinct fermentation processes that define Boza's profile. While LAB are primarily responsible for lactic acid production and the acidic character of the food, yeasts contribute carbon dioxide, ethanol, and aromatic metabolites. The main aromatic metabolites produced are esters and geraniol, which provide a palatable, sweet, fruity character. Species like *Pichia fermentans* are specifically noted for their contribution to these aromatic compounds (Caputo et al., 2012; Osorio-Cadavid et al., 2008; Rojas et al., 2001).

Boza is characterized by the accumulation of lactic and acetic acids, ethanol, and free amino acids, along with moderate reductions in complex carbohydrates. Genomic and functional studies have demonstrated that strains isolated from Boza possess genes associated with carbohydrate and amino acid metabolism. These metabolic activities contribute to the liberation of peptides and reducing sugars, enriching the substrate available for microbial growth and flavor development (Osimani et al., 2015; Queiroz et al., 2022). The resulting metabolite profile enhances mouthfeel and acidity perception, yielding a balanced sour-sweet flavor and a smooth, viscous consistency (Hancioğlu & Karapinar, 1997; Osimani et al., 2015; Sutherland, 1998; Werning et al., 2006). The production of exopolysaccharides by some LAB species also contributes to Boza's characteristic rheological properties, increasing viscosity and stabilizing the suspension of starch and proteins (Osimani et al., 2015).

In addition, *Lvb. brevis* S5, *Lmb. fermentum* S4 and *Lmb. fermentum* S6 were classified as probiotic strains, and *P. acidilactici* ST31BZ, *P. pentosaceus* ST75BZ, *P. pentosaceus* ST87BZ, and *Lpb. plantarum* ST194BZ were classified as bacteriocin-producing strains (Gül et al., 2025; Queiroz et al., 2022). Modern processing innovations, such as spray-drying and fluidized-bed agglomeration, aim to preserve LAB viability, enabling the traditional beverage to be consumed year-round without loss of sensory or nutritional quality (Gül et al., 2025).

11.3.9 Pozol

Pozol is a traditional fermented beverage regarded as one of the most microbiologically complex naturally fermented cereal products. Its fermentation involves interactions among LAB, yeasts, and filamentous fungi within the nixtamalized maize matrix (Cañas-Urbina et al., 1993; Ulloa et al., 1983). The fermentation process begins with a microbial community originating from nixtamalized dough, banana leaves, water, and utensils, each of which contributes specific microorganisms that become active once the dough is molded and wrapped (Nuraida et al., 1995; Wacher et al., 1993, 2000).

Firmicutes (70.61%) and *Proteobacteria* (22.62%) represent the main phyla detected during fermentation. LAB predominates in fermentation, accounting for 64.63%. They are fundamental to product stability, inhibiting unwanted microorganisms by producing organic acids and lowering pH. *Streptococcus* was the predominant genus in this community, accounting for 47.87%, mainly due to *Streptococcus infantarius* subsp. *infantarius* (Sii). This dominance is due to its ability to withstand extreme pH conditions (both alkaline and acidic) and to hydrolyze starch, the main carbohydrate in corn (Domínguez-Ramírez et al., 2020). *Lactobacillus* genus (7.1%), *Leuconostoc* (4.69%), *Enterococcus* (3.41%), and *Lactococcus* (1.56%) were identified but in minor abundance (Omar & Ampe, 2000; López-Sánchez et al., 2023; Rizo et al., 2021).

The microbial succession during pozol fermentation is a highly dynamic process, characterized by the transition from an initially heterogeneous microbiota to a community dominated by LAB and specific fungi. Between 9 and 24 h, the genus *Streptococcus* (especially *S. infantarius*) is the most abundant, accounting for 45–48% of the community. Other genera present in this initial phase include *Leuconostoc* (11% at 9 am), *Citrobacter* (5%), *Acinetobacter* (3%), and *Weissella*. As fermentation progresses (24–48 h), there is a increase in facultatively heterofermentative LAB, such as *Lmb. fermentum* and *Leuconostoc* species. This period coincides with the production of high levels of ethanol and a significant drop in pH. After 48 h, homofermentative species that are more tolerant of acid, such as *Lpb. plantarum*, *L. casei*, and *L. delbrueckii*, dominate the process. These bacteria can continue acidification even when the pH drops to values close to or below 4.0 (López-Sánchez et al., 2023; Omar & Ampe, 2000; Rizo et al., 2021). The fungal community is initially dominated by *Schizosaccharomyces*, with subsequent participation from *Candida* and *Saccharomyces*. After 48 h of fermentation, *Neurospora* became the most abundant genus, while *Schizosaccharomyces* and *Saccharomyces* reduced their populations (Rizo et al., 2021).

Amylolytic strains, such as *Streptococcus infantarius* subsp. *infantarius*, are essential in starch degradation within the nixtamalized maize matrix. The production of intracellular α-amylase and extracellular amylopullulanase, releasing fermentable sugars that contribute to microbial succession. This species exhibits physiological adaptability, with pH growth limits ranging from 4.0 to 11.0 and an optimum between 6.6 and 8.0. It maintains metabolic activity under both alkaline and acidic stress, showing induction of acid- and alkali-tolerance responses and cross-protection mechanisms (upregulation of dnaK, groEL, ptsG, and atpB). Such versatility likely explains its persistence from the highly alkaline post-nixtamalization environment to the acidic stages of fermentation (Bolaños-Núñez et al., 2021; Domínguez-Ramírez et al., 2020; Rizo et al., 2021).

Beyond starch hydrolysis, LAB further modulates the biochemical and sensory profile of pozol through proteolytic activity and amino acid fermentation. During fermentation, free amino acids, such as alanine, leucine, and glutamic acid, are accumulated. Furthermore, there is an increase in the concentration of essential amino acids, including lysine, isoleucine, threonine, and tryptophan, which improve the biological value of the beverage.

Concurrently, the production of acetoin and 2,3-butanediol, mainly by specific *Lactobacillus* and *Weissella* strains, contributes mild buttery and malty notes that characterize the mature product (López-Sánchez et al., 2023; Omar & Ampe, 2000; Rizo et al., 2021). The metabolite profile obtained, dominated by lactic and acetic acids, glycerol, and ethanol, contributes to the characteristic acidity and aromatic complexity of pozol while contributing to its microbial stability and safety (López-Sánchez et al., 2023; Omar & Ampe, 2000; Wacher et al., 2000; Rizo et al., 2021).

Yeasts and filamentous fungi coexist with LAB during pozol fermentation. Bacterial action promotes starch degradation and nutrient release, increasing the availability of amino acids and fermentable sugars that sustain both microbial groups. The yeast contributes to the production of CO_2 and ethanol through carbohydrate fermentation (López-Sánchez et al., 2023; Omar & Ampe, 2000; Rizo et al., 2021; Wacher et al., 2000). *Schizosaccharomyces* dominates in the first hours of fermentation, followed by *Saccharomyces*, *Candida*. Other genera detected include *Rhodotorula* (*R. minuta* and *R. mucilaginosa*), *Debaryomyces* (*D. hansenii*), *Kluyveromyces* (*K. lactis*), *Hanseniaspora*, and *Pichia.* Filamentous fungi develop mainly on the surface of dough balls due to oxygen availability and low pH. *Neurospora* is the most abundant genus after 48 h of fermentation. Other genera, such as *Geotrichum* (*Geotrichum candidum*), *Cladosporium* (*C. cladosporioides*), *Phoma* (*P. glomerata*), *Penicillium* (*P. fellutanum*), and *Aspergillus* can be found (Rizo et al., 2021).

This microbial diversity not only ensures ecological stability but also provides strains with technological and functional potential. Several isolates from pozol have been proposed as potential starter cultures or probiotic candidates due to their acid tolerance, amylolytic capacity, and antimicrobial activity. Strains of *Lpb. plantarum*, *Lmb. fermentum*, *P. acidilactici*, and *S. infantarius* exhibit strong acidification ability, survive under simulated gastrointestinal conditions, and display antagonism toward foodborne pathogens such as *Escherichia coli* and *S. aureus* (López-Sánchez et al., 2023; Omar & Ampe, 2000; Olivares-Illana et al., 2002; Rizo et al., 2021). *Streptococcus infantarius* showed starter culture potential by demonstrating resistance to stress and the ability to synthesize vitamins (Domínguez-Ramírez et al., 2020; Rizo et al., 2021). Furthermore, other strains, such as *Bacillus subtilis* CS93, have been proposed as potential starter cultures or probiotic candidates due to their acid tolerance, amylolytic capacity, and antimicrobial activity (Ray et al., 2000). In addition to the strains of *Lpb. plantarum*, *Lmb. fermentum*, *P. acidilactici*, and *Streptococcus infantarius* exhibit strong acidification ability, survive under simulated gastrointestinal conditions, and display antagonism toward foodborne pathogens.

11.3.10 Kombucha

Kombucha is a traditional fermented tea beverage obtained through the symbiotic association of bacteria and yeasts embedded within a cellulose-based biofilm known as SCOBY (*Symbiotic Culture of Bacteria and Yeast*) (Kruk et al., 2021). The microbial community comprises acetic acid bacteria (AAB), LAB, and yeasts. This consortium is dynamic and drives the transformation of sweetened tea into an acidic, effervescent, and bioactive (Kruk et al., 2021).

Komagataeibacter spp. (formerly *Gluconacetobacter*), including *K. xylinus*, *K. rhaeticus*, *K. intermedius*, and *K. saccharivorans*, are bacteria responsible for the synthesis of bacterial cellulose and organic acids (Tran et al., 2020; Villarreal-Soto et al., 2018; Yamada et al., 2012). These bacteria oxidize ethanol and glucose to acetic, gluconic, and glucuronic acids, contributing to acidification (pH 2.5–3.5) and the generation of the cellulose pellicle that structures the microbial consortium. They polymerize glucose residues into β-1,4-glucan chains, which organize themselves into a three-dimensional matrix on top of the liquid (air-liquid interface), facilitating oxygen access for aerobic microorganisms. Other bacterial genera include *Acetobacter*, *Gluconobacter*, *and Tanticharoenia* (Coton et al., 2017; Tran et al., 2020; Villarreal-Soto et al., 2018; Yamada et al., 1997, 2012). Although AABs can form biofilms independently, the presence of yeasts increases the robustness and resilience of this structure.

The yeast population is more diverse, including *B. bruxellensis*, *Zygosaccharomyces bailii*, *Schizosaccharomyces pombe*, *S. cerevisiae*, *T. delbrueckii*, and *P. kudriavzevii*, whose population and composition vary according to geographic origin and tea substrate (Chen & Liu, 2000; Marsh et al., 2014; Mayser et al., 1995; Teoh et al., 2004; Villarreal-Soto et al., 2020). However, *Zygosaccharomyces* sp. is the dominant genus detected in the pellicle and during fermentation (Arıkan et al., 2020; Marsh et al., 2014).

Yeasts initiate the fermentative phase by converting sucrose into ethanol and CO_2 through invertase and zymase activity, while AAB subsequently oxidize ethanol to acetic acid, completing the biphasic fermentation cycle characteristic of kombucha (Jayabalan et al., 2014; Sievers et al., 1995; Teoh et al., 2004; Tran et al., 2020). This metabolic cooperation not only ensures a continuous supply of substrates but also shapes the beverage's sensorial profile. Yeast-derived alcohols and esters contribute fruity and floral notes, whereas bacterial acids impart sour and vinegar-like attributes (Kapp & Sumner, 2019; Villarreal-Soto et al., 2018).

Although LAB occur at lower abundance than AAB, isolates have been shown to tolerate the acidic environment of kombucha and to synthesize bacteriocins and antioxidant peptides, highlighting their potential functional contributions (Bogdan et al., 2018; Diguță et al., 2020; Pei et al., 2020).

The complex metabolome of kombucha includes lactic, acetic, gluconic, and glucuronic acids, glycerol, ethanol, fructose, glucose, amino acids, and a wide spectrum of polyphenols, flavonoids, vitamins (B_1, B_2, B_6, B_{12}, C), and minerals, conferring both nutritional and antioxidant properties (Villarreal-Soto et al., 2018).

Polyphenolic compounds derived from tea, including catechins, epicatechins, and theaflavins, undergo biotransformation by microbial enzymes into bioavailable phenolic acids with enhanced antioxidant potential (Cardoso et al., 2020; Jakubczyk et al., 2020, 2022; Villarreal-Soto et al., 2018; Zhenjun et al., 2018).

Concurrently, bacterial cellulose exopolymers contribute to the structural integrity of the SCOBY and may function as dietary fiber with prebiotic-like effects (Vargas et al., 2021). Several kombucha isolates, including *Brettanomyces* strains UVI55 and UVI56, have been characterized as probiotic candidates, exhibiting survival under simulated gastrointestinal conditions, bile salt tolerance, and antimicrobial activity against foodborne pathogens, including *E. coli*, *L. monocytogenes*, *S. aureus*, and *Salmonella enterica* (Areal-Hermida et al., 2024).

Brettanomyces bruxellensis and *B. anomalus* strains demonstrated high adhesion to intestinal epithelial cells, with antioxidant capacities exceeding 80%, and suppression of pathogenic bacteria, while prolonging the lifespan of *Caenorhabditis elegans* "in vivo" (Areal-Hermida et al., 2024; Cho et al., 2018). *Pediococcus. pentosaceus* strains exhibited tolerance to gastrointestinal conditions, the ability to adhere to intestinal epithelial cells, and the production of bacteriocins with antioxidant and antimicrobial activities (Bogdan et al., 2018; Diguță et al., 2020). Similarly, *Lpb. plantarum* isolated from kombucha was shown to produce a Class IId bacteriocin that is active against both Gram-positive (G+) and Gram-negative (G–) bacteria and can increase cell membrane permeability, leading to bacterial cell death (Pei et al., 2020).

Potential probiotic features have also been identified among acetic acid bacteria. *Gluconacetobacter* sp. strains have been reported to produce D-saccharic acid 1,4-lactone (DSL), a compound with hepatoprotective effects (Wang et al., 2014). *Acetobacter musti* LOAAB1, *Gluconobacter potus* GBAAB3, and *Gluconobacter potus* LOAAB2 showed excellent antioxidant activity, besides auto-aggregation and co-aggregation with different pathogens (Wang et al., 2024b). *Pichia kudriavzevii* GBY1 and *S. cerevisiae* GBY2 exhibited high surface hydrophobicity, strong auto-aggregation, and antioxidant activity, as well as resistance to multiple antibiotics, indicating overall promising probiotic potential for functional food applications (Wang et al., 2024a).

In general, kombucha is a functionally enriched matrix in which synergistic microbial interactions transform sweetened tea into a beverage with significant chemical and biological complexity. Kombucha remains a valuable system for designing new fermentation processes and understanding the importance of microbial biodiversity in functional food development (Içen et al., 2023).

11.3.11 Cauim

Cauim (also known as kawi) is a traditional non-alcoholic fermented beverage produced by several Brazilian Indigenous groups, such as the Tapirapé, Araweté, Kayapó, Karajá, and Javaé. Although alcoholic versions are produced for ritual festivals, Cauim consumed daily is generally a non-alcoholic beverage that serves as a staple food for adults and, crucially, as a primary source of nutrition for infants under 2 years of age (Almeida et al., 2007). Cauim can be produced from a variety of raw materials, including cassava, rice, maize, peanuts, cottonseed, banana, and pumpkin (Almeida et al., 2007).

The fermentation of Cauim is characterized by complex microbial interactions between LAB and yeasts, originating from the raw materials and the traditional inoculum prepared by female tribe members through the mastication of sweet potato. During fermentation, microbial populations typically range from 7.4 to 8.4 log CFU/mL for LAB and from 4.0 to 6.6 log CFU/mL for yeasts. The production of organic acids promotes acidification, with pH decreasing from initial values of approximately 6.9–5.5 to around 4.0–3.4 within 24–48 h (Almeida et al., 2007; Ramos et al., 2010, 2011).

The LAB are the dominant group (Almeida et al., 2007; Ramos et al., 2010) and include *Lmb. fermentum*, *Lpb. plantarum*, *Lbs. paracasei*, and *Levilactobacillus brevis* (Almeida et al., 2007; Ramos et al., 2010, 2011). *S. cerevisiae*, *Candida* spp., *Pichia* spp., and *Kluyveromyces lactis* are commonly detected.

Both microbial groups contribute to aroma development and nutrient release through enzymatic activity, including amylases and proteases (Ramos et al., 2010; Schwan et al., 2007). Yeasts produce complex, long-chain alcohols and derived esters (such as phenylethyl alcohol, which has a rose-like odor) that are critical to the final product's flavor profile. Furthermore, lactic acid is the predominant metabolite, reaching approximately 24 g/L, while smaller amounts of ethanol and acetic acid are also produced (Almeida et al., 2007; Ramos et al., 2010, 2011).

Potential probiotic microorganisms have been isolated from Cauim and exhibit desirable physiological characteristics, including tolerance to acidic pH and bile salts, autoaggregation, adhesion to intestinal epithelial cells, and antagonism against foodborne pathogens. Promising strains from Cauim include *Pediococcus acidilactici* CCMA 0347, *Lpb. plantarum* CCMA 0743, and *L. brevis* FFC199 that have shown inhibition of pathogen adhesion and improvement of intestinal barrier integrity, highlighting Cauim as a relevant source of novel starter cultures and probiotic candidates (Fonseca et al., 2021a, b; Freire et al., 2017a, b; Ramos et al., 2013, 2014).

11.3.12 Caxiri

Caxiri is recognized in the sources as a traditional indigenous beverage and a significant reservoir of potential probiotic microorganisms. This fermented alcoholic beverage is traditionally produced from cassava and sweet potatoes by the indigenous Juruna or Yudjá people in Brazil and is consumed in religious rituals, social celebrations, and collective activities such as land clearing for cultivation (Lima, 2005).

The traditional preparation involves submerging cassava roots in running water for 2 days to soften the skin, promote detoxification of cyanogenic glycosides, and allow natural acidification by LAB. After pressing and roasting into flour, the mixture is rehydrated and left to ferment for up to 120 h under spontaneous fermentation conditions. Microorganisms originate from raw materials, utensils, and the surrounding environment, characterizing caxiri as a complex and dynamic microbial ecosystem (Santos et al., 2012).

During fermentation, the bacterial population ranged from 3.05 to 5.33 log CFU/mL, while yeasts became dominant after 48 h, reaching 7.34 log CFU/mL. Dominant yeast and bacteria include *S. cerevisiae* and *Bacillus* (with *Bacillus pumilus*, *Bacillus cereus* group, and *B. subtilis*) (Santos et al., 2012).

As fermentation progresses, the microbiota consumes reducing sugars and several organic acids, including malic, tartaric, succinic, oxalic, and propionic acids. These metabolic conversions lead to the accumulation of ethanol (approximately 83.9 g/L) and lactic acid (~27.9 g/L), resulting in a marked acidification of the beverage, with the pH dropping from 4.76 to around 3.15 by the end of fermentation (Santos et al., 2012).

Caxiri serves as a vital source of functional microorganisms, such as *Pediococcus acidilactici* CCMA 0347 and *S. cerevisiae* (CCMA 0739, CCMA 0712, CCMA 0731, CCMA 0732) have demonstrated key probiotic traits, including tolerance to low pH and bile salts, adhesion to epithelial cells, production of antimicrobial compounds, and phytase activity, which increases mineral bioavailability (Da Silva et al., 2025; De Almeida Silva Vilela et al., 2020; Kamda et al., 2014; Menezes et al., 2018, 2020a, b).

11.3.13 Chicha

Chicha is a fermented rice beverage produced by the indigenous Umutina people in Brazil. It serves as a vital component of their daily diet and plays a central role in rituals and collective celebrations, where it is consumed by individuals of all ages (Puerari et al., 2015).

The beverage is prepared by soaking whole rice grains in water for several hours, then grinding to obtain rice flour, which is cooked in boiling water. A portion of the flour is separately toasted and chewed, a traditional step that introduces salivary amylases and the native oral microbiota, which initiate starch hydrolysis and support fermentation. After cooking, sugar and additional water are added, and the mixture is fermented at room temperature (approximately 30–35 °C) for approximately 36 h. The microbiota originates from raw materials, utensils, handlers' hands, and the surrounding environment (Puerari et al., 2015).

During fermentation, bacteria dominate, with populations ranging from 0.1 to 6.83 log CFU/mL. LAB and *Bacillus* spp. are the main groups involved, including species formerly classified within the *Lactobacillus* genus such as *Lacticaseibacillus*, *Levilactobacillus*, *Lentilactobacillus*, and *Lactiplantibacillus*, as well as *Leuconostoc*, *Enterococcus*, *Streptomyces*, *Enterobacter*, *Acinetobacter*, *Escherichia*, *Cronobacter*, and *Klebsiella*. *Bifidobacterium* and *Propionibacterium* have also been detected throughout fermentation (Puerari et al., 2015).

Bacillus plays a crucial role in producing amylase enzymes that break down rice starch into fermentable sugars (such as glucose and maltose), which serve as a substrate for lactic acid bacteria. Lactic acid is the major metabolite produced by LAB, leading to a pH reduction from 5.2 to ~3.9 within the first 36 h, which improves microbiological safety and contributes to the beverage's characteristic flavor (Puerari et al., 2015).

In addition, Chicha is essential not only for cultural preservation but also for identifying functional microorganisms for industrial applications. *Leuconostoc lactis* CCMA 0415 has demonstrated rapid carbohydrate consumption, high lactic acid production, phytase activity, and viability above 8 log CFU/mL during fermentation of starchy substrates, supporting its application as a starter culture in root-, tuber-, and cereal-based functional beverages (Batista et al., 2019a, b).

11.3.14 Tarubá

Tarubá is a beverage produced by Amerindian communities in the state of Amazonas, Brazil. Its production via solid-state fermentation distinguishes it from other traditional cassava-derived products, such as cauim, caxiri, and yakupa, which typically involve submerged fermentation (Ramos et al., 2015).

The microbiota involved in tarubá fermentation is characterized by a diverse association of yeasts and bacteria. *T. delbrueckii* is a dominant yeast species throughout the process. Other yeast species detected include *Pichia exigua*, *Candida rugosa*, *T. delbrueckii*, *Candida tropicalis*, *Pichia kudriavzevii*, *Wickerhamomyces anomalus*, and *Candida ethanolica* (Ramos et al., 2015). The bacterial community is primarily composed of LAB, such as *Lpb. plantarum*, *Lactobacillus brevis*, and *Leuconostoc mesenteroides*, as well as *Bacillus subtilis* (Ramos et al., 2015).

The technological capabilities of the microorganisms involved in the fermentation of tarubá are critical for transforming the solid cassava substrate into the final beverage. These abilities include amylolytic, proteolytic, and thermotolerant activities. *Lpb. plantarum*, *Bacillus subtilis*, *B. amyloliquefaciens*, *B. licheniformis*, *Bacillus* sp., and *Chitinophaga terrae*, which exhibit amylolytic activity. Additionally, several yeast species, including *T. delbrueckii*, *P. exigua*, *P. rugosa*, *Candida tropicalis*, *Pichia kudriavzevii*, *Wickerhamomyces anomalus*, and *Candida ethanolica*, also demonstrate amylolytic capabilities. Proteolytic activity is crucial for flavor development and can contribute to the initial increase in pH observed during fermentation by releasing ammonia, and was detected in *Bacillus* spp., as well as in the yeasts *P. exigua* and *T. delbrueckii.* Furthermore, the strains have also demonstrated thermotolerance at 38 °C and at 42 °C, except *Hanseniaspora uvarum* (Ramos et al., 2015).

The enzymatic activities, thermotolerance, and fermentative performance of *T. delbrueckii* CCMA 0235 and *Lmb. fermentum* CCMA 0215 are considered ideal for standardizing production and creating innovative, high-quality, and safe functional non-dairy beverages (Freire et al., 2015, 2017a, b; Kamda et al., 2014).

11.3.15 Yakupa

Yakupa is an acid-cassava fermented product traditionally prepared by women within Brazilian Amerindian tribes. It is widely consumed both in daily activities and for its refreshing properties (Freire et al., 2014).

The microbial community, composed of yeasts and bacteria, reached 7 log CFU/mL after 36 h of fermentation (Freire et al., 2014). The main bacteria and yeast detected were *Lmb. fermentum*, *Lpb. plantarum*, *W. cibaria*, *W. confusa*, *S. cerevisiae*, and *P. kudriavzevii*.

The interaction between these specific bacteria and yeasts is what defines the beverage's profile (Freire et al., 2014). Maltose (5.13 g/L), ethanol (2.3 g/L), and lactic acid (4.5 g/L) are the main chemical compounds produced. Aldehydes, acids, alcohols, and esters were the most representative volatile compounds, responsible for the fruity and acidic aroma profile of the final beverage (Freire et al., 2014).

Limosilactobacillus fermentum CCMA 0215, originally isolated from yakupa, exhibits desirable technological and functional traits. It includes rapid acidification, good growth in cassava substrates, and co-culturing compatibility with yeasts. This strain is considered a promising starter culture for cassava-based beverages and has demonstrated the ability to enhance safety, sensory profile, and pH reduction under controlled fermentation conditions (Freire et al., 2015).

In this way, spontaneous plant fermentations offer valuable microbial and metabolic diversity for food technology, including candidate starter cultures and functional criteria (acidification, degradation of anti-nutrients, and volatile profile). The controlled processes require integration between metagenomic characterization, functional evaluation of isolated microorganisms, and sensory analyses. However, applying this knowledge to industrial practice remains challenging, as plant substrates differ substantially in composition, fermentation behavior, and processing constraints. Therefore, understanding the technological barriers and strategies to overcome them is essential for developing safe, stable, and consumer-acceptable plant-based fermented foods and beverages.

11.4 Challenges and Technology in the Development of Plant-Based Foods and Beverages

Research in science and technology contributed to fermented food processes, increasing product safety and quality, and achieving greater consistency and standardization in production. Traditional fermented products such as beer, wine, bread, and dairy exemplify this evolution. Despite industrialization, small-scale fermentation remains a common practice in many regions (Cuamatzin-García et al., 2022; Singh & Kumar, 2025).

Consumer behavior has undergone evolution, driving innovation in the food sector. The growing interest in plant-based products reflects societal shifts driven by health awareness, ethical concerns, and environmental sustainability. This shift has encouraged industries and researchers to develop novel formulations and technologies that revalorize traditional fermentation through modern scientific approaches (Sotelo-Lara et al., 2024; Tachie et al., 2023).

The production of plant-based foods and beverages has expanded rapidly due to technological innovations that improve their nutritional, sensory, and environmental performance. The adoption of sustainable processing methods, combined with increased awareness of dietary restrictions and their impact on climate change, has positioned plant-based foods as viable alternatives to traditional products (Tachie et al., 2023). Within this context, fermentation using selected microorganisms plays a fundamental role in enhancing flavor, stability, digestibility, and nutrient bioavailability in plant-based matrices (Pizarro-Oteíza et al., 2025).

Fermented plant beverages are often pasteurized, ultra-high-temperature (UHT) treated, or subjected to other heat-based preservation processes to ensure microbiological safety. However, these thermal methods may alter the structure of proteins, polysaccharides, and starch, leading to changes in viscosity and potential nutritional losses (Galván-D'Alessandro & Carciochi, 2018).

To minimize such limitations, non-thermal technologies have emerged as promising alternatives for extending shelf life and maintaining quality. Techniques such as ultrasound, high hydrostatic pressure processing (HHP), PEF, and ultra-high-pressure homogenization (UHPH) can preserve the sensory and nutritional attributes of plant-based fermented food and beverages (Pizarro-Oteíza et al., 2025). Moreover, precision fermentation has also gained attention by employing genetically modified microorganisms to produce specific compounds, enhance flavor complexity, and improve the safety and functionality of plant-based products (Singh & Kumar, 2025).

Despite the remarkable progress, there are still challenges that must be prioritized according to their impact on the implementation of these techniques: (i) economic feasibility and capital investment represent the primary barrier, with equipment costs for technologies such as HHP and PEF exceeding conventional thermal systems by approximately 30–200%, depending on processing capacity (Aganovic et al., 2021; Nawawi et al., 2023); (ii) process scalability and operational integration constitute the second most critical constraint, particularly for ultrasound and precision fermentation, which require matrix-specific optimization and process control; and (iii) regulatory definition and the lack of standardized clinical or functional validation metrics represent emerging constraints, especially for precision fermentation.

From a technological readiness perspective, HHP and PEF currently exhibit the highest industrial maturity, while ultrasound shows intermediate maturity due to variability in scale-up efficiency (Souza et al., 2025).

Therefore, the selection and prioritization of these technologies should be aligned with the intended processing objective. HHP is particularly suitable for microbial inactivation and shelf-life extension, while PEF are more appropriate for liquid plant-based beverages that require preservation of sensory and nutritional attributes. Ultrasound, in turn, has shown strong potential to enhance fermentation performance and facilitate the extraction of functional compounds. In this context, the combined application of these technologies may provide complementary effects, contributing to improved process performance and overall product quality.

The following sections discuss the main emerging technologies used in the formulation and improvement of plant-based foods and beverages, focusing on their technological potential, limitations, and prospects for industrial implementation.

11.4.1 Non-thermal Processes for Plant-Based Fermented Foods and Beverages

11.4.1.1 Ultrasound

Ultrasound is a technology to produce plant-based foods and beverages, particularly in fermented products such as olives, kimchi, sauerkraut, and pickled vegetables. Its applications range from the optimization of fermentation, the improvement of sensory and nutritional properties, to the control of microbial activity. Then, the technique holds great potential for high-quality fermented foods (Sotelo-Lara et al., 2024; Yu et al., 2021).

Ultrasound can be classified into three main frequency ranges: high-power and low-frequency (20–100 kHz), intermediate-power and frequency (100 kHz–1 MHz), and low-power and high-frequency (1–100 MHz) (Kh et al., 2025). At low frequencies, this technology increases mass transfer and cell membrane permeability, accelerating fermentation and achieving higher yields and shorter total production time. Therefore, they are most used in food processing and are responsible for several beneficial effects in the food matrix (Kh et al., 2025; Sotelo-Lara et al., 2024; Tan et al., 2024).

Furthermore, the application of this technology can stimulate the growth of beneficial microorganisms and enhance enzymatic activity, thereby contributing to the formation of desirable compounds and the development of the sensory profile of fermented products. Ultrasound can accelerate fermentation time, improve sensory characteristics such as texture, color, and flavor, and enhance the nutritional value of the products. Additionally, technology enables the reduction or even elimination of thermal steps, thereby preserving heat-sensitive compounds (Yu et al., 2021).

Higher-intensity ultrasonics can help improve food safety (Yu et al., 2021) by promoting microbial inactivation through mechanical damage, free radical formation, and degradation of essential cellular components, including DNA, lipids, and proteins (Kh et al., 2025; Andressa et al., 2024; Mustapha et al., 2024; Sotelo-Lara et al., 2024; Tan et al., 2024).

The application of ultrasonic waves for sterilization has demonstrated effectiveness against various microorganisms, including bacteria, fungi, and viruses, such as *E. coli*, *S. aureus*, and hepatitis A virus. The degree of inactivation depends on the species and its structural characteristics, with Gram-positive bacteria and yeasts being more resistant than Gram-negative bacteria due to their thicker cell walls. Similarly, cocci show greater tolerance compared to bacilli due to their smaller surface-to-volume ratio (Kh et al., 2025; Tan et al., 2024).

However, this treatment was not significantly effective in reducing high levels of microbial contamination in olives, despite being safe and environmentally friendly. Habibi et al. (2015, 2016) reported that ultrasound treatment accelerated and promoted the debittering of olives without NaOH. Thus, the number of washing cycles required to complete the de-bittering process was reduced, and simultaneously, water consumption was reduced (Campus et al., 2018).

In recent years, ultrasound has emerged as a promising tool to enhance the activity of probiotic cells, offering broad application possibilities in the functional food industry (Kh et al., 2025).

According to Manyatsi et al. (2024), this technology can significantly improve fermentation processes, especially when using probiotic cultures such as *Bifidobacterium*, *Lvb. brevis*, *Lc. lactis*, *Lbs. casei*, *Leu. mesenteroides*, and *Lpb. plantarum* in food formulations. For example, Tan et al. (2024) reported that treatment with 20 kHz ultrasound resulted in a twofold increase in β-galactosidase activity in *Lab. acidophilus* BCRC 1069. Bacterial growth is stimulated during the logarithmic phase, with an increase in permeability of LAB cell membranes and the synthesis of enzymes, such as β-galactosidase, which facilitates the release of bioactive compounds, including anthocyanins, flavonoids, and phenolics (Kh et al., 2025; Tan et al., 2024).

The β-glucosidase enzyme plays an essential role in plant-based fermentation processes by hydrolyzing glycosylated flavonoids to their aglycone forms (Tan et al., 2024). Thus, optimizing ultrasonic conditions in lactic fermentation can contribute to increased viability of probiotic microorganisms, improved sustainable industrial processes, and the development of high-value functional foods enriched with bioactive components (Kh et al., 2025).

Promising results with ultrasound have also been observed in kombucha, leading to increased production of organic acids and bioactive compounds. In beverages based on legumes and cereals, this technology has demonstrated the ability to improve protein digestibility and reduce fermentation time without compromising the sensory properties or nutritional value of the final product (Yusoff et al., 2022). Additionally, ultrasonic treatment has been shown to effectively preserve the firmness and texture of fermented vegetables, such as cabbage, thereby enhancing their palatability (Tan et al., 2024).

Despite recent advances, the application of ultrasound on an industrial scale still presents technical limitations, including the need for more robust equipment and the precise optimization of operational parameters, such as exposure time, intensity, and frequency, which must be adjusted according to the food matrix to avoid nutritional losses or impacts on microbial viability (Tan et al., 2024).

Even so, this technology has high potential as a versatile, sustainable tool for the plant-based food and beverage industry, contributing to improved process efficiency, product quality, and microbiological safety.

11.4.1.2 Pulsed Electric Fields

PEF is a non-thermal technology used for microbial inactivation in different food products, with low impact on sensorial characteristics and nutritional value. In recent years, PEF technology has advanced significantly, evolving from laboratory to large-scale industrial applications (Yan et al., 2024).

This technique uses high-intensity, short-duration pulses (10–40 kV/cm) and moderate temperatures of 30–40 °C (Munekata et al., 2020; Pizarro-Oteíza et al., 2025; Xie et al., 2023). However, in practice, intensities between 0.1 and 5 kV/cm and pulse durations between 50 and 1000 μs are more common (Galván-D'Alessandro & Carciochi, 2018).

The cell membrane of microorganisms can be compared to a capacitor filled with a material of low dielectric constant. When cells are exposed to an electric field, intracellular ions move in the direction of the field, accumulating on the inner and outer membrane surfaces. This free charge accumulation affects cellular signaling and transduction processes (Galván-D'Alessandro & Carciochi, 2018). Loss of membrane semipermeability occurs when the transmembrane potential exceeds a critical value, generally between 0.2 and 1 V (Galván-D'Alessandro & Carciochi, 2018; Pizarro-Oteíza et al., 2025). Although PEF is not entirely effective against microbial spores, its efficiency can be increased by adjusting the pH with the addition of organic acids. However, its industrial use remains limited due to the high initial cost and the availability of large-scale equipment (Pataro & Ferrari, 2020).

The PEF system structure is a chamber that can operate in both static and continuous modes. When operated in static mode, the feed remains between two parallel electrodes during processing. In continuous mode, the product flows between the electrodes via a pumping system, enabling continuous processing and greater operational efficiency (Galván-D'Alessandro & Carciochi, 2018).

In addition to the mode of operation, the process efficiency depends on the optimization of operational parameters, such as time, intensity, frequency, and pulse width. As with ultrasound techniques, vegetative cells of Gram-positive bacteria are more resistant to PEF treatment than Gram-negative cells. Larger cells, such as yeasts, are more sensitive to low electric field intensities than smaller cells, due to the larger surface area in contact with PEF (Galván-D'Alessandro & Carciochi, 2018).

PEF has been studied mainly to increase the shelf life of beverages with low conductivity and low viscosity, such as plant-based beverages. In soy beverages, for example, PEF has been shown to inactivate *E. coli* and *S. aureus* without altering sensory attributes, making it an interesting alternative to thermal pasteurization (Li et al., 2013). Shi et al. (2017) also studied the application of PEF in soy beverages and found that, at 42 kV/cm, 1036 μs, 400 Hz, 2 μs, and 25 °C, the activity of soy lipoxygenase was reduced by 88%.

Cortés et al. (2004) evaluated the effect of PEF with different parameters on the quality of horchata (a traditional Spanish plant-based beverage). The authors found that PEF attenuated the reduction in pH during storage, with longer times associated with higher values. The treatment also partially inhibited peroxidase and, despite some enzymatic regeneration, the final activity remained lower than that of the control samples.

Depending on the treatment conditions, PEF can also be used to stimulate the metabolism of LAB, such as *Lbs. casei*, *Lpb. plantarum*, and *Lc. lactis*. Increased cell growth, lactic acid production, and bioactive compound synthesis have been observed (Kh et al., 2025; Xie et al., 2023).

Studies compiled by Galván-D'Alessandro and Carciochi (2018) confirmed that PEF treatment can also exert a stimulating effect on the growth of *S. cerevisiae*, especially when low frequencies and intensities of up to 1625 kV/cm are applied.

Although the exact mechanism of electrostimulation is not yet known, several hypotheses have been proposed to explain this phenomenon. One possibility is that PEF acts on enzymatic systems and metabolic pathways associated with the cell membrane, promoting electrical activation and triggering changes in the cell cycle. This can happen, possibly due to the acceleration of signal transduction pathways that bring the cell to the starting point of the cycle more quickly, or to the shortening of the G1 phase, which anticipates the transition to the S phase and thus results in shorter cell cycles (Galván-D'Alessandro & Carciochi, 2018).

Optimizing treatment conditions also enhances the viability of probiotics during fermentation, thereby improving the functional and nutritional properties of fermented foods (Kh et al., 2025; Xie et al., 2023).

11.4.1.3 High-Pressure Processing or High-Hydrostatic Pressure

HPP or HHP is also a non-thermal technology used to preserve beverages and foods. In this technique, products are packaged in flexible packaging and placed in a chamber filled with water, usually at room temperature, under pressure ranging from 100 to 800 MPa. HHP can be conducted in either batch or semi-continuous mode, depending on the product's characteristics and the operating conditions follow the principle of Pascal's Law: the pressure is distributed uniformly throughout the volume of the product, avoiding structural deformations, excessive heating, and loss of quality (Kh et al., 2025; Argyri et al., 2014; Bao et al., 2016; Campus et al., 2018).

Although it operates differently from the other techniques presented previously, HHP also stands out for its efficiency in inactivating spoilage microorganisms and enzymes, thereby extending product shelf life without compromising sensory or nutritional properties (Andressa et al., 2024). One point of attention is that microbial cells can suffer sublethal damage or destruction, depending on the conditions applied (Kh et al., 2025).

This technique also reduces the size of micelles and lipid droplets, preventing foam formation and coalescence of the lipid fraction (Andressa et al., 2024). HHP preserves the primary structure of molecules such as proteins, but can induce changes in secondary, tertiary, and quaternary structures. These changes can be positive, as they improve digestibility, bioavailability, and the functional properties of nutrients. However, they directly influence texture, sensory properties, and nutritional value (Munekata et al., 2020). With recent advances in industrial equipment, the use of HHP has expanded to various food matrices, enabling the production of safe, stable, and high-quality foods (Kh et al., 2025; Argyri et al., 2014).

During processing, low-molecular-weight compounds, such as vitamins, amino acids, and volatile compounds, tend to be less susceptible to structural changes. HHP acts on cell membranes by compressing the phospholipid bilayer and altering its fluidity, thereby compromising cell integrity and microbial viability (Kh et al., 2025).

High-pressure treatment has also been studied in the preservation of fermented and vegetable products. In the case of olives and their brines, a reduction in microbial populations to levels below the detection limit was observed immediately after treatment, except for the 400 MPa for 15 min condition, in which yeast recovery occurred during storage after 7 days at 20 °C, LAB and yeasts were again detected in all treatments, except in the 500 MPa for 30 min condition, where no detectable growth occurred. LAB reappeared in samples treated at 400 MPa (15–30 min) and 450 MPa (15 min), while yeasts showed greater resistance, being recovered in almost all conditions, except at 500 MPa for 30 min (Argyri et al., 2014).

In a review written by Campus et al. (2018), the authors listed the main advantages of strategically applying the technique to improve the sensory characteristics of food products and extend shelf life. The authors also highlighted the use of HHP in the preservation and quality improvement of table olives, ensuring that the product presented microbiological safety and physicochemical stability.

Argyri et al. (2014) showed that pressure levels of 250–600 MPa for 5–30 min are more effective than thermal pasteurization (80 °C for 20 min) in reducing yeasts, fungi, and some mycotoxins.

Similar results were obtained by Pradas et al. (2012) in Cornezuelo olives treated with 400 and 600 MPa, which showed greater pH stability, absence of gas formation, and better sensory performance. HPP combined with the addition of ascorbic acid (15 g/L) or purging with nitrogen gas has proven effective in preserving and intensifying the intense green color of olives, an important characteristic for the commercial acceptance of this type of product (Argyri et al., 2014; Campus et al., 2018; Pradas et al., 2012).

In addition, HHP has been investigated for its potential in reducing the allergenicity of plant-based beverages. Dhakal et al. (2014) evaluated the impact of the treatment on allergenic almond proteins, especially amandin (the main protein responsible for allergic reactions in raw almond extract). Samples treated under 450 and 600 MPa for up to 600 s at 30 °C showed a significant reduction in amandin levels.

HHP has demonstrated potential for preserving the quality of fermented foods during refrigerated storage. Peñas et al. (2010) reported that HHP-treated sauerkraut showed a microbial reduction of approximately 4–5 log CFU/g after 3 months under refrigeration. Comparable findings were described by Li et al. (2010) for fermented Chinese cabbage. In addition, pressures exceeding 400 MPa were found to limit excessive acidification and texture softening in kimchi, contributing to improved product stability (Bao et al., 2016).

In another study, Bao et al. (2016) investigated the effects of HHP on microbial inactivation and the maintenance of quality in pickled radish during 90 days of storage at 4 °C. The treatment eliminated yeasts and fungi. Compared with heat treatment, HPP (5 min at 550 MPa) showed better preservation of texture and flavor. This improvement was attributed to the action of pectin methyl esterase, which, under high pressure, catalyzes the demethylation of pectin, forming gel networks with divalent ions and thus contributing to the firmness of the product.

Given the advantages presented above, HPP reinforces its potential as a viable alternative to conventional thermal methods. As technological advancements make the process more accessible and energy-efficient, HPP is expanding its application in the production of safe, stable, and high-value-added foods.

Non-thermal technologies have emerged as effective strategies for improving the quality, safety, and nutritional value of plant-based fermented foods and beverages. Although they operate through different physical mechanisms, these approaches offer advantages with reduced undesirable effects associated with heating. Nevertheless, factors such as implementation costs, large-scale feasibility, and the need for optimization across different matrices remain decisive for their industrial adoption. Table 11.3 compares the mechanisms, applications, advantages, and limitations of non-thermal techniques.

11.4.2 Precision Fermentation Technology

Precision fermentation integrates biotechnology, genetic engineering, and biology to direct microorganisms to synthesize desired compounds. Strains engineered are used to express specific pathways that produce different metabolites, such as proteins, enzymes, and lipids with high purity, reproducibility, and safety, favoring economically competitive processes (Augustin et al., 2024; Mirsalami & Mirsalami, 2025; Verma et al., 2025).

The principle of precision fermentation is based on introducing genes encoding the target molecules into well-characterized hosts, which are cultivated in bioreactors under strictly controlled conditions (pH, temperature, and aeration). The engineering strategies and cultivation condition optimization determine yield and production stability. These metabolites and biomolecules produced are subsequently recovered and purified through downstream processing steps (Augustin et al., 2024; Verma et al., 2025).

Table 11.3 Comparison of non-thermal technologies applied to plant-based fermented foods and beverages

Technology	Working principle	Products studied	Main positive effects	Limitations	References
Ultrasound (US)	High-energy sound waves (20–100 kHz) induce cavitation, enhancing mass transfer, stimulating microbial activity, and releasing bioactive compounds	Kombucha, legume-based beverages, olives, kimchi, sauerkraut, and other fermented vegetables	Accelerates fermentation, improves texture, color, and flavor, enhances phenolic bioaccessibility and probiotic viability, and reduces processing time	Parameter standardization challenges: requires precise control of intensity and frequency; high cost for industrial scaling	Kh et al. (2025), Tan et al. (2024), and Yu et al. (2021)
Pulsed Electric Fields (PEF)	High-voltage pulses (10–40 kV/cm) induce electroporation in microbial membranes, leading to enzyme and cell inactivation	Plant-based beverages (soy, horchata), fermentations with yeasts and LAB	Maintains color, flavor, and nutrients; inactivates microorganisms; may stimulate LAB metabolism and production of bioactive compounds	High equipment cost; limited effectiveness against spores; best suited for low-conductivity liquids	Galván-D'Alessandro and Carciochi (2018), Li et al. (2013), and Xie et al. (2023)
High Hydrostatic Pressure (HHP)	Uniform application of 100–800 MPa, preserving molecular structure and heat-sensitive compounds	Olives, sauerkraut, kimchi, plant-based beverages, and almond milk	Inactivates microorganisms and enzymes; maintains texture, color, and flavor; reduces allergenicity; extends shelf life; preserves bioactive compounds	High initial cost; possible color alterations; microbial regrowth under certain conditions	Argyri et al. (2014), Bao et al. (2016), and Dhakal et al. (2014)

The choice of the host microorganism represents a critical step in development, and should be GRAS strains. Among the most widely used hosts in food applications are bacteria (e.g., *Bacillus* spp.), industrial yeasts (e.g., *S. cerevisiae*, *K. phaffii*, *Kluyveromyces* spp.), and filamentous fungi such as *Trichoderma reesei* (Verma et al., 2025).

Technical innovation in the field stems from both the exploration of new species and the improvement of metabolic pathways that enhance the expression and conversion of substrates, increasing product yield. Microbial biodiversity still offers a vast range of potential hosts, and, with the support of omics tools, it is possible to identify and design more efficient and safer microorganisms for industrial applications (Verma et al., 2025).

Precision fermentation has already enabled the production of ingredients previously restricted to animal sources, such as dairy proteins (casein and beta-lactoglobulin) and egg albumin, with sensory and nutritional properties comparable to conventional versions. The technology is also used in the synthesis of industrial enzymes, organic acids, and bioactive compounds, as well as in the formulation of functional products, such as kombucha enriched with gamma-aminobutyric acid (GABA) using combined starter cultures of *Lpb. plantarum*, *Acetobacter pasteurianus*, and *S. cerevisiae* (Kim et al., 2025; Mirsalami & Mirsalami, 2025; Singh & Kumar, 2025).

Advances in automation, artificial intelligence, and biology are enabling the design and optimization of microbial strains with improved capacity to synthesize aroma molecules, pigments, bioactive compounds, and functional proteins. Therefore, precision fermentation offers new opportunities to produce ingredients with standardized quality and tailored functional or nutritional characteristics, while also supporting more efficient and scalable production systems. Despite this potential, its large-scale implementation warrants attention due to regulatory complexity, persistent concerns about consumer perceptions, and transparency in processes (Augustin et al., 2024; Verma et al., 2025).

Regarding plant-based foods, precision fermentation has been successfully used to enhance the flavor and aroma of these products. Yeasts and LAB have been genetically optimized to produce aromatic volatile compounds. For example, strains of *S. cerevisiae* have been modified to increase volatile thiols that enhance fruity notes in wines and fermented beverages, and to synthesize compounds that reproduce the aromatic profile of hops, allowing the production of beers with a hoppy character without using hops themselves. In parallel, the use of genetically modified microorganisms that produce natural antimicrobial agents (bacteriocins and endolysins) is helping improve the microbiological safety of plant-based foods and extend their shelf life (Singh & Kumar, 2025).

Despite these advances, industrial scaling faces significant barriers: the transition from the laboratory to commercial bioreactors presents uncertainties about microbial adaptation to the new environment, involving unpredictable physiological, ecological, and evolutionary interactions (Augustin et al., 2024; De Lorenzo & Couto, 2019; Delvigne et al., 2017). Efficient large-scale production requires integration of strain development, fermentation process control, and downstream purification operations, while accounting for the effects of metabolic engineering on byproducts and separation costs (Augustin et al., 2024; Rangel et al., 2020).

The CRISPR-Cas9 system can be a very useful tool for enhancing microorganisms used in the fermentation of plant-based foods. One example is *Lc. lactis*, traditionally used in lactic fermentation, has been genetically modified to increase the biosynthesis of vitamin B9, an essential micronutrient involved in DNA synthesis, cell division, and overall health (Dhiman et al., 2025).

Vitamin B12 is essential for the functioning of the nervous system and for the formation of red blood cells, but its absence is frequent in plant-based diets. Starting from this point, CRISPR was used to insert a gene related to vitamin B12 biosynthesis into *E. coli*, allowing this microorganism to produce the compound (Dhiman et al., 2025).

In addition to micronutrients, CRISPR has also been explored as a means to enhance the generation of bioactive peptides during the fermentation of plant-based foods. These protein fragments, resulting from the biotransformation of food proteins, have been recognized for their beneficial physiological effects and can enhance both the nutritional value and functional properties of fermented products. Thus, CRISPR-mediated editing of microorganisms emerges as a strategy for developing more nutritious, functional foods that meet the needs of consumers adopting plant-based diets (Dhiman et al., 2025).

In this way, it becomes evident that recent advances indicate precision fermentation is an interesting strategy for developing plant-based fermented foods characterized by more complex flavors and aromas, greater functionality, and enhanced microbiological safety (Singh & Kumar, 2025).

11.5 Conclusion and Perspectives

The potential for innovation in the plant-based fermented foods market is considerable, driven by advances in microbial biotechnology, rising consumer demand for health-oriented, sustainable products, and the valorization of traditional fermentation knowledge. A key priority is the identification and characterization of microbial strains that can improve nutritional, functional, and sensory attributes, as well as the definition of clear regulatory and clinical metrics.

The search for novel starter cultures and probiotic strains is a promising strategy for developing plant-based fermented foods, including vegan and non-dairy products. Studies on the microbial succession of traditional spontaneous fermented foods, including kimchi, sauerkraut, kombucha, pozol, and cereal- and cassava-based beverages, have demonstrated that these foods are potential sources of strains with probiotic and technological properties for plant-based matrices.

Despite this, significant gaps remain regarding the consistency, safety, and functional performance of these strains when applied to controlled fermentations. In particular, the impact of selected cultures on sensory acceptance, metabolite production, and functional properties remains unpredictable. While lactic acid bacteria and non-*Saccharomyces* yeasts can enhance flavor complexity, synthesize bioactive compounds, and degrade anti-nutritional factors, strain-specific variability represents a critical challenge for industrial application.

Although the use of appropriate microbial cultures is crucial for developing novel plant-based fermented foods and beverages, the choice and treatment of the substrate to ensure microbiological safety are also important for obtaining high-quality products. As replacements for thermal methods, non-thermal technologies such as ultrasound, high HHP, HPP, and PEF have been proposed as alternatives to extend shelf life and maintain product quality. Despite presenting implementation challenges on an industrial scale, these methods have demonstrated effectiveness in preserving probiotic viability, maintaining nutritional integrity, and extending shelf life without compromising the nutritional and sensory characteristics of plant-based fermented food and beverages.

Therefore, studies that integrate microbiological, metabolomics, and sensory parameters are crucial for a better understanding of microbe–matrix interactions and for enabling precision fermentation approaches. Collaborative efforts among microbiologists, food technologists, sensory scientists, and regulatory bodies will be essential to scale up from laboratory to industrial applications. Furthermore, the use of omics-based tools to identify and characterize novel bacteria and yeasts with technological and health-promoting potential is an important strategy for bioprospecting microbial cultures with desired characteristics for plant-based fermentations. Future research should therefore focus on elucidating microbe–matrix interactions to enable the design of optimized starter cultures tailored to specific plant substrates.

References

Abriouel, H., Benomar, N., Cobo, A., Caballero, N., Fernández Fuentes, M. A., & Pérez-Pulido, R. (2012). Characterization of lactic acid bacteria from naturally fermented Manzanilla Aloreña green table olives. *Food Microbiology, 32*(2), 308–316. https://doi.org/10.1016/j.fm.2012.07.006

Adisa, A. M., & Enujiugha, V. N. (2020). Microbiology and safety of ogi fermentation: A review. *European Journal of Nutrition & Food Safety, 12*(5), 90–100. https://doi.org/10.9734/ejnfs/2020/v12i530231

Adisa, A. M., Badejo, A. A., Ifesan, B. O., & Enujiugha, V. N. (2023). Phenotypic and 16S rRNA gene sequence differentiation of lactic acid bacteria starter cultures for fonio millet ogi fermentation. *SSRN Electronic Journal*, 1–24. https://doi.org/10.2139/ssrn.4600073

Adisa, A. M., Badejo, A. A., Ifesan, B. O. T., & Enujiugha, V. N. (2024). Phenotypic and molecular differentiation of lactic acid bacteria in fonio millet ogi fermentation and their potential as starter cultures. *Food and Humanity, 2*, 100230. https://doi.org/10.1016/j.foohum.2024.100230

Aganovic, K., Hertel, C., Vogel, R. F., et al. (2021). Aspects of high hydrostatic pressure food processing: Perspectives on technology and food safety. *Comprehensive Reviews in Food Science and Food Safety, 20*, 3225–3266. https://doi.org/10.1111/1541-4337.12763

Agarbati, A., Canonico, L., Ciani, M., Morresi, C., Damiani, E., Bacchetti, T., & Comitini, F. (2024). Functional potential of a new plant-based fermented beverage: Benefits through non-conventional probiotic yeasts and antioxidant properties. *International Journal of Food Microbiology, 424*, 110857. https://doi.org/10.1016/j.ijfoodmicro.2024.110857

Aila, R., Saburjiang, A., Alim, A., Mahemuti, A., & Kelimu, A. (2020). Screening and fermentation characteristics of excellent Boza strain. *Journal of Food and Nutrition Research, 8*(4), 160–167. https://doi.org/10.12691/jfnr-8-4-1

Almeida, E. G., Rachid, C. C. T. C., & Schwan, R. F. (2007). Microbial population present in fermented beverage 'cauim' produced by Brazilian Amerindians. *International Journal of Food Microbiology, 120*, 146–151. https://doi.org/10.1016/j.ijfoodmicro.2007.06.020

Andressa, I., do Nascimento, G. K. S., dos Santos, T. M., Rodrigues, R. D. S., de Oliveira Teotônio, D., Paucar-Menacho, L. M., Machado Benassi, V., & Schmiele, M. (2024). Technological and health properties and main challenges in the production of vegetable beverages and dairy analogs. *Food & Function, 15*(2), 460–480. https://doi.org/10.1039/d3fo04199a

Anjos, O., Pires, P. C. P., Gonçalves, J., Estevinho, L. M., Mendonça, A. G., & Guiné, R. P. F. (2024). Plant-based beverages: Consumption habits, perception, and knowledge on a sample of Portuguese citizens. *Foods, 13*(20), 3235. https://doi.org/10.3390/foods13203235

Areal-Hermida, L., Coelho, P., Pichardo-Gallardo, C., Prudêncio, C., & Sieiro, C. (2024). Potential probiotic and functional properties of *Brettanomyces* strains isolated from kombucha tea. *Frontiers in Microbiology, 15*, 1415616. https://doi.org/10.3389/fmicb.2024.1415616

Argyri, A. A., Zoumpopoulou, G., Karatzas, K. A. G., Tsakalidou, E., Nychas, G. J. E., Panagou, E. Z., & Tassou, C. C. (2013). Selection of potential probiotic lactic acid bacteria from fermented olives by in vitro tests. *Food Microbiology, 33*(2), 282–291. https://doi.org/10.1016/j.fm.2012.10.005

Argyri, A. A., Panagou, E. Z., Nychas, G. J. E., & Tassou, C. C. (2014). Nonthermal pasteurization of fermented green table olives by means of high hydrostatic pressure processing. *BioMed Research International*, 515623. https://doi.org/10.1155/2014/515623

Arıkan, M., Mitchell, A. L., Finn, R. D., & Gürel, F. (2020). Microbial composition of Kombucha determined using amplicon sequencing and shotgun metagenomics. *Journal of Food Science, 85*(2), 455–464. https://doi.org/10.1111/1750-3841.14992

Arroyo-López, F. N., Querol, A., Bautista-Gallego, J., & Garrido-Fernández, A. (2008). Role of yeasts in table olive production. *International Journal of Food Microbiology, 128*(2), 189–196. https://doi.org/10.1016/j.ijfoodmicro.2008.08.018

Arroyo-López, F. N., Romero-Gil, V., Bautista-Gallego, J., Rodriguez-Gómez, F., Jiménez-Díaz, R., & García-García, P. (2012). Yeasts in table olive processing: Desirable or spoilage microorganisms? *International Journal of Food Microbiology, 160*(1), 42–49. https://doi.org/10.1016/j.ijfoodmicro.2012.08.003

Atter, A., Diaz, M., Tano-Debrah, K., Kunadu, A. P., Mayer, M. J., Colquhoun, I. J., Nielsen, D. S., Baker, D., Narbad, A., & Amoa-Awua, W. (2021). Microbial diversity and metabolite profile of fermenting millet in the production of *Hausa koko*, a Ghanaian fermented cereal porridge. *Frontiers in Microbiology, 12*, 681983. https://doi.org/10.3389/fmicb.2021.681983

Atter, A., Diaz, M., Tano-Debrah, K., Kunadu, A. P., Mayer, M. J., Sayavedra, L., Misita, C., Amoa-Awua, W., & Narbad, A. (2024). The predominant lactic acid bacteria and yeasts involved in the spontaneous fermentation of millet during the production of the traditional porridge *Hausa koko* in Ghana. *BMC Microbiology, 24*(1), 33. https://doi.org/10.1186/s12866-024-03317-1

Augustin, M. A., Hartley, C. J., Maloney, G., & Tyndall, S. (2024). Innovation in precision fermentation for food ingredients. *Critical Reviews in Food Science and Nutrition, 64*(18), 6218–6238. https://doi.org/10.1080/10408398.2023.2166014

Bambridge-Sutton, A. (2025, August 1). *Plant-based food market could triple by 2035.* FoodNavigator.com. https://www.foodnavigator.com/Article/2025/08/01/plant-based-food-market-to-triple-by-2035/#:~:text=Plant%2Dbased%20food%20market%20to%20triple%20by%202035

Bao, R., Fan, A., Hu, X., Liao, X., & Chen, F. (2016). Effects of high-pressure processing on the quality of pickled radish during refrigerated storage. *Innovative Food Science & Emerging Technologies, 38*, 206–212. https://doi.org/10.1016/j.ifset.2016.10.009

Batista, N. N., Ramos, C. L., De Figueiredo Vilela, L., Dias, D. R., & Schwan, R. F. (2019a). Fermentation of yam (*Dioscorea* spp. L.) by indigenous phytase-producing lactic acid bacteria strains. *Brazilian Journal of Microbiology, 50*(2), 507–514. https://doi.org/10.1007/s42770-019-00059-5

Batista, N. N., Ramos, C. L., Pires, J. F., Moreira, S. I., Alves, E., Dias, D. R., & Schwan, R. F. (2019b). Nondairy ice cream based on fermented yam (*Dioscorea* sp.). *Food Science & Nutrition, 7*(5), 1899–1907. https://doi.org/10.1002/fsn3.1051

Bell, V., Ferrão, J., & Fernandes, T. (2017). Nutritional guidelines and fermented food frameworks. *Foods, 6*(8), 65. https://doi.org/10.3390/foods6080065

Bevilacqua, A., Corbo, M. R., & Sinigaglia, M. (2012). Selection of yeasts as starter cultures for table olives: A step-by-step procedure. *Frontiers in Microbiology, 3*, 194. https://doi.org/10.3389/fmicb.2012.00194

Bevilacqua, A., Beneduce, L., Sinigaglia, M., & Corbo, M. R. (2013). Selection of yeasts as starter cultures for table olives. *Journal of Food Science, 78*(5), 742–751. https://doi.org/10.1111/1750-3841.12117

Bleve, G., Tufariello, M., Durante, M., Perbellini, E., Ramires, F. A., Grieco, F., Cappello, M. S., De Domenico, S., Mita, G., Tasioula-Margari, M., & Logrieco, A. F. (2014). Physico-chemical and microbiological characterization of spontaneous fermentation of *Cellina di Nardò* and *Leccino* table olives. *Frontiers in Microbiology, 5*, 570. https://doi.org/10.3389/fmicb.2014.00570

Bleve, G., Tufariello, M., Durante, M., Grieco, F., Ramires, F. A., & Mita, G. (2015). Physico-chemical characterization of the natural fermentation process of *Conservolea* and *Kalamata* table olives and development of a protocol for the pre-selection of fermentation starters. *Food Microbiology, 46*, 368–382. https://doi.org/10.1016/j.fm.2014.08.021

Bogdan, M., Justine, S., Filofteia, D. C., Petruța, C. C., Gabriela, L., Roxana, U. E., & Florentina, M. (2018). Lactic acid bacteria strains isolated from kombucha with potential probiotic effect. *Romanian Biotechnological Letters, 23*(3), 13592–13598.

Bolaños-Núñez, S., Santiago-Urbina, J. A., Guyot, J., Díaz-Ruiz, G., & Wacher, C. (2021). Microbial interactions between amylolytic and non-amylolytic lactic acid bacteria strains isolated during the fermentation of *Pozol. Foods, 10*(11), 2607. https://doi.org/10.3390/foods10112607

Borcakli, M., Öztürk, T., & YeşiLada, E. (2018). Cereal source and microbial consortia of the starter culture influence the chemical composition and physicochemical characteristics of boza. *Turkish Journal of Agriculture and Forestry, 42*(6), 412–422. https://doi.org/10.3906/tar-1802-3

Bouakkaz, S., Zerizer, H., Rachedi, K., Accettulli, A., Racioppo, A., & Bevilacqua, A. (2024). African cereal-based fermented foods: Microbiota, functional microorganisms, starter cultures and nutritional properties. *Food Bioscience, 62*, 105212. https://doi.org/10.1016/j.fbio.2024.105212

Breidt, F., McFeeters, R. F., Perez-Diaz, I., & Lee, C. H. (2013). Fermented vegetables. In M. P. Doyle & L. R. Beuchat (Eds.), *Food microbiology: Fundamentals and frontiers* (4th ed., pp. 783–793). ASM Press.

Bruce, B. B., Boateng, I. D., & Boateng, C. (2025). Recent advances in bioactive peptides from fermented plant-based foods and their bioactivities. *Food Chemistry: X, 32*, 103291. https://doi.org/10.1016/j.fochx.2024.103291

Buatong, A., Meidong, R., Trongpanich, Y., & Tongpim, S. (2022). Production of plant-based fermented beverages possessing functional ingredients, antioxidants, γ-aminobutyric acid, and antimicrobials using a probiotic *Lactiplantibacillus plantarum* strain L42g as an efficient starter culture. *Journal of Bioscience and Bioengineering, 134*(3), 226–232. https://doi.org/10.1016/j.jbiosc.2022.05.008

Cakmak, H., Düven, G., Subasi, B. G., Buyukkestelli, H. I., Bravo-Núñez, A., Ozyurt, V. H., Sogut, E., & Simsek, S. (2025). Plant-based fermented foods: General concepts, improvement of the quality, metabolites, and health-related functions. *Food Bioscience, 69*, 106960. https://doi.org/10.1016/j.fbio.2025.106960

Campus, M., Degirmencioglu, N., & Comunian, R. (2018). Technologies and trends to improve table olive quality and safety. *Frontiers in Microbiology, 9*, 617. https://doi.org/10.3389/fmicb.2018.00617

Cañas-Urbina, A. O., Bgrzana-Garcia, E., Owens, J. D., & Wacher-Rodarte, M. C. (1993). La elaboraci6n de *pozol* en los Altos de Chiapas. *Ciencia (México), 44*, 219–229.

Cao, J., Yang, J., Hou, Q., Xu, H., Zheng, Y., & Zhang, H. (2017). Assessment of bacterial profiles in aged, home-made Sichuan *paocai* brine with varying titratable acidity by PacBio SMRT sequencing technology. *Food Control, 78*, 14–23. https://doi.org/10.1016/j.foodcont.2017.02.006

Caputo, L., Quintieri, L., Baruzzi, F., Borcakli, M., & Morea, M. (2012). Molecular and phenotypic characterization of Pichia fermentans strains found among Boza yeasts. *Food Research International, 48*(2), 755–762. https://doi.org/10.1016/j.foodres.2012.06.022

Cardoso, R. R., Neto, R. O., Dos Santos D'Almeida, C. T., do Nascimento, T. P., Pressete, C. G., Azevedo, L., Martino, H. S. D., Cameron, L. C., Ferreira, M. S. L., & de Barros, F. A. R. (2020). Kombuchas from green and black teas have different phenolic profiles, which impact their antioxidant capacities, antibacterial and antiproliferative activities. *Food Research International, 128*, 108782. https://doi.org/10.1016/j.foodres.2020.108782

Cha, J., Kim, Y. B., Park, S. E., Lee, S. H., Roh, S. W., Son, H. S., & Whon, T. W. (2023). Does kimchi deserve the status of a probiotic food? *Critical Reviews in Food Science and Nutrition, 64*(19), 6512–6525. https://doi.org/10.1080/10408398.2023.2170319

Chang, H. W., Kim, K. H., Nam, Y. D., Roh, S. W., Kim, M. S., Jeon, C. O., Oh, H. M., & Bae, J. W. (2008). Analysis of yeast and archaeal population dynamics in kimchi using denaturing gradient gel electrophoresis. *International Journal of Food Microbiology, 126*(1–2), 159–166. https://doi.org/10.1016/j.ijfoodmicro.2008.05.027

Chatonidi, G., Poppe, J., & Verbeke, K. (2023). Plant-based fermented foods and the satiety cascade: A systematic review of randomized controlled trials. *Trends in Food Science & Technology, 133*, 127–137. https://doi.org/10.1016/j.tifs.2023.01.018

Cheigh, H. S., Park, K. Y., & Lee, C. Y. (1994). Biochemical, microbiological, and nutritional aspects of kimchi (Korean fermented vegetable products). *Critical Reviews in Food Science and Nutrition, 34*(2), 175–203. https://doi.org/10.1080/10408399409527668

Chen, C., & Liu, B. Y. (2000). Changes in major components of tea fungus metabolites during prolonged fermentation. *Journal of Applied Microbiology, 89*(5), 834–839. https://doi.org/10.1046/j.1365-2672.2000.01188.x

Cho, J., Lee, D., Yang, C., Jeon, J., Kim, J., & Han, H. (2006). Microbial population dynamics of kimchi, a fermented cabbage product. *FEMS Microbiology Letters, 257*(2), 262–267. https://doi.org/10.1111/j.1574-6968.2006.00179.x

Cho, Y. J., Kim, D. H., Jeong, D., Seo, K. H., Jeong, H. S., Lee, H. G., & Kim, H. (2018). Characterization of yeasts isolated from kefir as probiotics and their synergistic interaction with the wine byproduct grape seed flour/extract. *LWT – Food Science and Technology, 90*, 535–539. https://doi.org/10.1016/j.lwt.2018.01.010

Contreras, L., Isla, M. I., & Torres, S. (2025). Selection of novel starter lactic acid bacteria with probiotic potential native to *chilto* (*Solanum betaceum*) fruits. *Food Bioscience, 71*, 107200. https://doi.org/10.1016/j.fbio.2025.107200

Cortés, C., Esteve, M., FnGola, A., & Torregrosa, F. (2004). Quality characteristics of horchata (a Spanish vegetable beverage) treated with pulsed electric fields during shelf-life. *Food Chemistry, 91*(2), 319–325. https://doi.org/10.1016/j.foodchem.2004.06.014

Coton, M., Pawtowski, A., Taminiau, B., Burgaud, G., Deniel, F., Coulloumme-Labarthe, L., Fall, A., Daube, G., & Coton, E. (2017). Unraveling microbial ecology of industrial-scale kombucha fermentations by metabarcoding and culture-based methods. *FEMS Microbiology Ecology, 93*(5), fix048. https://doi.org/10.1093/femsec/fix048

Cuamatzin-García, L., Rodríguez-Rugarcía, P., El-Kassis, E. G., Galicia, G., Meza-Jiménez, M. L., Baños-Lara, M. D. R., Zaragoza-Maldonado, D. S., & Pérez-Armendáriz, B. (2022). Traditional fermented foods and beverages from around the world and their health benefits. *Microorganisms, 10*(6), 1151. https://doi.org/10.3390/microorganisms10061151

Czerucka, D., & Rampal, P. (2019). Diversity of *Saccharomyces boulardii* CNCM I-745 mechanisms of action against intestinal infections. *World Journal of Gastroenterology, 25*(18), 2188–2203. https://doi.org/10.3748/wjg.v25.i18.2188

Da Silva, L. L., Santos, C. C. A. D. A., Da Silva Libeck, B., De Sousa Melo, D., Dias, D. R., & Schwan, R. F. (2025). Indigenous-based probiotic beverage from peanuts and soybean: Development, optimization, and characterization. *FEMS Microbes, 6*, xtaf006. https://doi.org/10.1093/femsmc/xtaf006

De Almeida Silva Vilela, J., De Figueiredo Vilela, L., Ramos, C. L., & Schwan, R. F. (2020). Physiological and genetic characterization of indigenous *Saccharomyces cerevisiae* for potential use in productions of fermented maize-based-beverages. *Brazilian Journal of Microbiology, 51*(3), 1297–1307. https://doi.org/10.1007/s42770-020-00271-8

De Lorenzo, V., & Couto, J. (2019). The important versus the exciting: Reining contradictions in contemporary biotechnology. *Microbial Biotechnology, 12*(1), 32–34. https://doi.org/10.1111/1751-7915.13348

Delvigne, F., Takors, R., Mudde, R., van Gulik, W., & Noorman, H. (2017). Bioprocess scale-up/down as integrative enabling technology: From fluid mechanics to systems biology and beyond. *Microbial Biotechnology, 10*(5), 1267–1274. https://doi.org/10.1111/1751-7915.12803

Dhakal, S., Liu, C., Zhang, Y., Roux, K. H., Sathe, S. K., & Balasubramaniam, V. M. (2014). Effect of high-pressure processing on the immunoreactivity of almond milk. *Food Research International, 62*, 215–222. https://doi.org/10.1016/j.foodres.2014.02.024

Dhiman, S., Kaur, S., Thakur, B., Singh, P., & Tripathi, M. (2025). Nutritional enhancement of plant-based fermented foods: Microbial innovations for a sustainable future. *Fermentation, 11*(6), 346. https://doi.org/10.3390/fermentation11060346

Di Cagno, R., Coda, R., De Angelis, M., & Gobbetti, M. (2013). Exploitation of vegetables and fruits through lactic acid fermentation. *Food Microbiology, 33*(1), 1–10. https://doi.org/10.1016/j.fm.2012.09.003

Diguță, C. F., Nițoi, G. D., Matei, F., Luță, G., & Cornea, C. P. (2020). The biotechnological potential of *Pediococcus* spp. isolated from Kombucha Microbial Consortium. *Foods, 9*(12), 1780. https://doi.org/10.3390/foods9121780

Domínguez-Ramírez, L. L., Rodríguez-Sanoja, R., Tecante, A., García-Garibay, M., Sainz, T., & Wacher, C. (2020). Tolerance to acid and alkali by *streptococcus infantarius* subsp. *infantarius* strain 25124 isolated from fermented nixtamal dough (*pozol*): Studies in APT broth. *Food Microbiology, 90*, 103458. https://doi.org/10.1016/j.fm.2020.103458

FAO, F. and A. O. of the U. N., & WHO, W. H. O. (2001). Health and nutritional properties of probiotics in food including powder milk with live lactic acid bacteria. *Prevention, 5*(1), 1–34.

Ferreira, I., Melo, D. S., Menezes, A. G. T., Fonseca, H. C., de Assis, B. B. T., Ramos, C. L., Magnani, M., Dias, D. R., & Schwan, R. F. (2022). Evaluation of potentially probiotic yeasts and *Lactiplantibacillus plantarum* in co-culture for the elaboration of a functional plant-based fermented beverage. *Food Research International, 160*, 111697. https://doi.org/10.1016/j.foodres.2022.111697

Ferreira, I., Melo, D. S., Santos, M. S., Dias, D. R., Souza, C. O., Favaro-Trindade, C. S., Pinho, L. S., Almeida, R. C. C., Magalhães, K. T., & Schwan, R. F. (2023). Non-lactic probiotic beverage enriched with microencapsulated red propolis: Microorganism viability, physicochemical characteristics, and sensory perception. *Fermentation-Basel, 9*(3), 234.

Fidel, P. L., Vazquez, J. A., & Sobel, J. D. (1999). *Candida glabrata*: Review of epidemiology, pathogenesis, and clinical disease with comparison to *C. albicans*. *Clinical Microbiology Reviews, 12*(1), 80–96. https://doi.org/10.1128/CMR.12.1.80

Fonseca, H. C., de Sousa Melo, D., Ramos, C. L., Menezes, A. G. T., & Schwan, R. F. (2021a). Probiotic properties of *Lactobacilli* and their ability to inhibit the adhesion of enteropathogenic bacteria to Caco-2 and HT-29 cells. *Probiotics and Antimicrobial Proteins, 13*, 102–112. https://doi.org/10.1007/s12602-020-09659-2

Fonseca, H. C., Melo, D. S., Ramos, C. L., Menezes, A. G. T., Dias, D. R., & Schwan, R. F. (2021b). *Lactiplantibacillus plantarum* CCMA 0743 and *Lacticaseibacillus paracasei* subsp.

paracasei LBC-81 metabolism during the single and mixed fermentation of tropical fruit juices. *Brazilian Journal of Microbiology, 52*(4), 2307–2317. https://doi.org/10.1007/s42770-021-00628-7

Fonseca, H. C., Melo, D. S., Ramos, C. L., Menezes, A. G. T., & Dias, D. R. (2022). Sensory and flavor-aroma profiles of passion fruit juice fermented by potentially probiotic *Lactiplantibacillus plantarum* CCMA 0743 strain. *Food Research International, 152*, 110710. https://doi.org/10.1016/j.foodres.2021.110710

Freire, A. L., Ramos, C. L., De Almeida, E. G., Duarte, W. F., & Schwan, R. F. (2014). Study of the physicochemical parameters and spontaneous fermentation during the traditional production of yakupa, an indigenous beverage produced by Brazilian Amerindians. *World Journal of Microbiology and Biotechnology, 30*(2), 567–577. https://doi.org/10.1007/s11274-013-1476-0

Freire, A. L., Ramos, C. L., & Schwan, R. F. (2015). Microbiological and chemical parameters during cassava based-substrate fermentation using potential starter cultures of lactic acid bacteria and yeast. *Food Research International, 76*, 787–795. https://doi.org/10.1016/j.foodres.2015.07.041

Freire, A. L., Ramos, C. L., & Schwan, R. F. (2017a). Effect of symbiotic interaction between a fructooligosaccharide and probiotic on the kinetic fermentation and chemical profile of maize blended rice beverages. *Food Research International, 100*, 698–707. https://doi.org/10.1016/j.foodres.2017.07.070

Freire, A. L., Ramos, C. L., Da Costa Souza, P. N., Cardoso, M. G. B., & Schwan, R. F. (2017b). Nondairy beverage produced by controlled fermentation with potential probiotic starter cultures of lactic acid bacteria and yeast. *International Journal of Food Microbiology, 248*, 39–46. https://doi.org/10.1016/j.ijfoodmicro.2017.02.011

Galván-D'Alessandro, L., & Carciochi, R. A. (2018). Fermentation assisted by pulsed electric field and ultrasound: A review. *Fermentation, 4*(1), 1. https://doi.org/10.3390/fermentation4010001

Ganapathiwar, S. (2023). *In vitro* assessment for the probiotic potential of *Pichia kudriavzevii*. *Bioinformation, 19*(4), 441–444. https://doi.org/10.6026/97320630019441

Gänzle, M. G., Monnin, L., Zheng, J., Zhang, L., Coton, M., Sicard, D., & Walter, J. (2024). Starter culture development and innovation for novel fermented foods. *Annual Review of Food Science and Technology, 15*, 211–239. https://doi.org/10.1146/annurev-food-072023-034207

Garzón, A. G., Puppo, M. C., & Di Renzo, T. (2025). Mineral bioaccessibility of pistachio-based beverages: The effect of lactic acid bacteria fermentation. *Plant Foods for Human Nutrition, 80*(4). https://doi.org/10.1007/s11130-024-01239-x

Gotcheva, V., Pandiella, S. S., Angelov, A., Roshkova, Z., & Webb, C. (2001). Monitoring the fermentation of the traditional Bulgarian; beverage boza. *International Journal of Food Science & Technology, 36*(2), 129–134. https://doi.org/10.1046/j.1365-2621.2001.00429.x

Grasso, N., Roos, Y. H., Crowley, S. V., Arendt, E. K., & O'Mahony, J. A. (2021). Composition and physicochemical properties of commercial plant-based block-style products as alternatives to cheese. *Future Foods, 4*, 100048. https://doi.org/10.1016/j.fufo.2021.100048

Greppi, A., Rantsiou, K., Padonou, W., Hounhouigan, J., Jespersen, L., Jakobsen, M., & Cocolin, L. (2013a). Yeast dynamics during spontaneous fermentation of *mawè* and *tchoukoutou*, two traditional products from Benin. *International Journal of Food Microbiology, 165*(2), 200–207. https://doi.org/10.1016/j.ijfoodmicro.2013.05.004

Greppi, A., Rantsiou, K., Padonou, W., Hounhouigan, J., Jespersen, L., Jakobsen, M., & Cocolin, L. (2013b). Determination of yeast diversity in *ogi*, *mawè*, *gowé*, and *tchoukoutou* by using culture-dependent and -independent methods. *International Journal of Food Microbiology, 165*(2), 84–88. https://doi.org/10.1016/j.ijfoodmicro.2013.05.005

Guan, Q., Zheng, W., Huang, T., Xiao, Y., & Xiong, T. (2020). Comparison of microbial communities and physicochemical characteristics of two traditionally fermented vegetables. *Food Research International, 128*, 108755. https://doi.org/10.1016/j.foodres.2019.108755

Gül, O., Atalar, I., Özgeçen, A. B., Gül, L. B., Törnük, F., Sert, D., & Yazıcı, F. (2025). Instant cereal-based fermented beverage (*Boza*) powder: Characterization of physical, rheological, and structural properties. *Journal of Food Measurement and Characterization*. https://doi.org/10.1007/s11694-025-03463-8

Guney, D., Başdoğan, M. G. B., & Sengun, I. (2025). Probiotic characterisation of lactic acid bacteria isolated from pickles and their potential application as presumptive probiotic starter culture in cucumber pickles. *Journal of Food Measurement & Characterization, 19*(3), 2077–2097. https://doi.org/10.1007/s11694-025-03101-3

Gungor, G., Akpinar, A., & Yerlikaya, O. (2024). Production of plant-based fermented beverages using probiotic starter cultures and *Propionibacterium* spp. *Food Bioscience, 59*, 103840. https://doi.org/10.1016/j.fbio.2024.103840

Guo, Z., Wang, Y., Xiang, F., Zhao, Y., & Liu, J. (2021). Bacterial diversity in pickled cowpea (*Vigna unguiculata* [Linn.] Walp) as determined by Illumina MiSeq sequencing and culture-dependent methods. *Current Microbiology, 78*(4), 1286–1297. https://doi.org/10.1007/s00284-021-02382-3

Habibi, M., Golmakani, M. T., Mesbahi, G., Majzoobi, M., & Farahnaky, A. (2015). Ultrasound-accelerated debittering of olive fruits. *Innovative Food Science & Emerging Technologies, 31*, 105–115. https://doi.org/10.1016/j.ifset.2015.06.014

Habibi, M., Golmakani, M. T., Farahnaky, A., Mesbahi, G., & Majzoobi, M. (2016). NaOH-free debittering of table olives using power ultrasound. *Food Chemistry, 192*, 775–781. https://doi.org/10.1016/j.foodchem.2015.07.086

Han, M., Lou, W., & Li, M. (2025). Lactic acid bacteria in Chinese sauerkraut: Its isolation and in vitro probiotic properties. *Foods, 14*(15), 2690. https://doi.org/10.3390/foods14152690

Hancioğlu, Ö., & Karapinar, M. (1997). Microflora of *boza*, a traditional fermented Turkish beverage. *International Journal of Food Microbiology, 35*(3), 271–274. https://doi.org/10.1016/s0168-1605(96)01230-5

Hatoum, R., Labrie, S., & Fliss, I. (2012). Antimicrobial and probiotic properties of yeasts: From fundamental to novel applications. *Frontiers in Microbiology, 3*, 421. https://doi.org/10.3389/fmicb.2012.00421

He, Z., Chen, H., Wang, X., Lin, X., Ji, C., Li, S., & Xiong, T. (2020). Effects of different temperatures on bacterial diversity and volatile flavor compounds during the fermentation of *suancai*, a traditional fermented vegetable food from northeastern China. *LWT – Food Science and Technology, 118*, 108773. https://doi.org/10.1016/j.lwt.2019.108773

Heperkan, D. (2013). Microbiota of table olive fermentations and criteria of selection for their use as starters. *Frontiers in Microbiology, 4*, 143. https://doi.org/10.3389/fmicb.2013.00143

Hidalgo-Fuentes, B., de Jesús-José, E., Cabrera-Hidalgo, A. J., Sandoval-Castilla, O., Espinosa-Solares, T., González-Reza, R. M., Zambrano-Zaragoza, M. L., Liceaga, A. M., & Aguilar-Toalá, J. E. (2024). Plant-based fermented beverages: Nutritional composition, sensory properties, and health benefits. *Foods, 13*(6), 844. https://doi.org/10.3390/foods13060844

Hill, C., Guarner, F., Reid, G., Gibson, G. R., Merenstein, D. J., Pot, B., Morelli, L., Canani, R. B., Flint, H. J., Salminen, S., Calder, P. C., & Sanders, M. E. (2014). The International Scientific Association for Probiotics and Prebiotics consensus statement on the scope and appropriate use of the term probiotic. *Nature Reviews Gastroenterology & Hepatology, 11*(8), 506–514. https://doi.org/10.1038/nrgastro.2014.66

Holzapfel, W. (1997). Use of starter cultures in fermentation on a household scale. *Food Control, 8*(5–6), 241–258. https://doi.org/10.1016/s0956-7135(97)00017-0

Houngbédji, M., Johansen, P., Padonou, S. W., Akissoé, N., Arneborg, N., Nielsen, D. S., Hounhouigan, D. J., & Jespersen, L. (2018). Occurrence of lactic acid bacteria and yeasts at species and strain level during spontaneous fermentation of *mawè*, a cereal dough produced in West Africa. *Food Microbiology, 76*, 267–278. https://doi.org/10.1016/j.fm.2018.05.018

Hounhouigan, D. J., Nout, M. R., Nago, C. M., Houben, J. H., & Rombouts, F. M. (1993). Composition and microbiological and physical attributes of *mawè*, a fermented maize dough from Benin. *International Journal of Food Science & Technology, 28*(5), 513–517. https://doi.org/10.1111/j.1365-2621.1993.tb01300.x

Hounhouigan, D. J., Nout, M. J. R., Nago, C. M., Houben, J. H., & Rombouts, F. M. (1994). Microbiological changes in *mawè* during natural fermentation. *World Journal of Microbiology and Biotechnology, 10*(4), 410–413. https://doi.org/10.1007/bf00144462

Hu, Y., Lo, I., Hsiao, J., & Sheu, F. (2024). Real-time PCR-based quantitative microbiome profiling elucidates the microbial dynamic succession in backslopping fermentation of Taiwanese pickled cabbage. *Journal of the Science of Food and Agriculture, 104*(14), 8604–8612. https://doi.org/10.1002/jsfa.13688

Içen, H., Corbo, M. R., Sinigaglia, M., Korkmaz, B. I. O., & Bevilacqua, A. (2023). Microbiology and antimicrobial effects of kombucha: A short overview. *Food Bioscience, 56*, 103270. https://doi.org/10.1016/j.fbio.2023.103270

Itaman, V. O., & Nwachukwu, E. (2021). Microbial profile of fermented maize flour (*ogi*) and African oil bean (*Pentaclethra macrophylla*) seeds. *Journal of Advances in Microbiology, 21*(8), 16–27. https://doi.org/10.9734/jamb/2021/v21i830373

Jakubczyk, K., Kałduńska, J., Kochman, J., & Janda, K. (2020). Chemical profile and antioxidant activity of kombucha beverage derived from white, green, black, and red tea. *Antioxidants, 9*(5), 447. https://doi.org/10.3390/antiox9050447

Jakubczyk, K., Kupnicka, P., Melkis, K., Mielczarek, O., Walczyńska, J., Chlubek, D., & Janda-Milczarek, K. (2022). Effects of fermentation time and type of tea on the content of micronutrients in kombucha fermented tea. *Nutrients, 14*(22), 4828. https://doi.org/10.3390/nu14224828

Jayabalan, R., Malbaša, R. V., Lončar, E. S., Vitas, J. S., & Sathishkumar, M. (2014). A review on kombucha tea—Microbiology, composition, fermentation, beneficial effects, toxicity, and tea fungus. *Comprehensive Reviews in Food Science and Food Safety, 13*(4), 538–550. https://doi.org/10.1111/1541-4337.12073

Jeong, S. H., Jung, J. Y., Lee, S. H., Jin, H. M., & Jeon, C. O. (2013a). Microbial succession and metabolite changes during fermentation of *dongchimi*, traditional Korean watery kimchi. *International Journal of Food Microbiology, 164*(1), 46–53. https://doi.org/10.1016/j.ijfoodmicro.2013.03.010

Jeong, S. H., Lee, H. J., Jung, J. Y., Lee, S. H., Seo, H. Y., Park, W. S., & Jeon, C. O. (2013b). Effects of red pepper powder on microbial communities and metabolites during kimchi fermentation. *International Journal of Food Microbiology, 160*(3), 252–259. https://doi.org/10.1016/j.ijfoodmicro.2012.11.003

Jeong, S. H., Lee, S. H., Jung, J. Y., Choi, E. J., & Jeon, C. O. (2013c). Microbial succession and metabolite changes during long-term storage of kimchi. *Journal of Food Science, 78*(5), M763–M769. https://doi.org/10.1111/1750-3841.12108

Jung, J. Y., Lee, S. H., Kim, J. M., Park, M. S., Bae, J., Hahn, Y., Madsen, E. L., & Jeon, C. O. (2011). Metagenomic Analysis of Kimchi a Traditional Korean Fermented Food. *ABSTRACT Applied and Environmental Microbiology, 77*(7), 2264–2274. https://doi.org/10.1128/AEM.02157-10

Jung, J. Y., Lee, S. H., Lee, H. J., Seo, H. Y., Park, W. S., & Jeon, C. O. (2012a). Effects of *Leuconostoc mesenteroides* starter cultures on microbial communities and metabolites during kimchi fermentation. *International Journal of Food Microbiology, 153*(3), 378–387. https://doi.org/10.1016/j.ijfoodmicro.2011.11.014

Jung, S., Park, J., Cho, I., Lee, N., Yeo, I., Kim, B., Kim, H., & Hahm, Y. (2012b). Characterization of lactic acid bacteria isolated from sauce-type kimchi. *Preventive Nutrition and Food Science, 17*(3), 217–222. https://doi.org/10.3746/pnf.2012.17.3.217

Jung, J. Y., Lee, S. H., Jin, H. M., Hahn, Y., Madsen, E. L., & Jeon, C. O. (2013). Metatranscriptomic analysis of lactic acid bacterial gene expression during kimchi fermentation. *International Journal of Food Microbiology, 163*(2–3), 171–179. https://doi.org/10.1016/j.ijfoodmicro.2013.02.022

Jung, M. J., Kim, M. S., Yun, J. H., Lee, J. Y., Kim, P. S., Lee, H. W., Ha, J. H., Roh, S. W., & Bae, J. W. (2018). Viral community predicts the geographical origin of fermented vegetable foods more precisely than bacterial community. *Food Microbiology, 76*, 319–327. https://doi.org/10.1016/j.fm.2018.06.010

Kabak, B., & Dobson, A. D. W. (2011). An introduction to the traditional fermented foods and beverages of Turkey. *Critical Reviews in Food Science and Nutrition, 51*(3), 248–260. https://doi.org/10.1080/10408390903569640

Kamda, A. G. S., Ramos, C. L., Fokou, E., Duarte, W. F., Mercy, A., Germain, K., Dias, D. R., & Schwan, R. F. (2014). In vitro determination of volatile compound development during starter culture-controlled fermentation of Cucurbitaceae cotyledons. *International Journal of Food Microbiology, 192*, 58–65. https://doi.org/10.1016/j.ijfoodmicro.2014.09.030

Kapp, J. M., & Sumner, W. (2019). Kombucha: A systematic review of the empirical evidence of human health benefit. *Annals of Epidemiology, 30*, 66–70. https://doi.org/10.1016/j.annepidem.2018.11.001

Kh, A. A., Abdo, A. A., Khan, S., Aleryani, H., Mi, S., & Wang, X. (2025). Advancing pickling techniques to enhance bioactive compounds and probiotic content in pickled vegetables. *Food Reviews International, 42*(1), 31–57. https://doi.org/10.1080/87559129.2025.2473009

Kim, M., & Chun, J. (2005). Bacterial community structure in kimchi, a Korean fermented vegetable food, as revealed by 16S rRNA gene analysis. *International Journal of Food Microbiology, 103*(1), 91–96. https://doi.org/10.1016/j.ijfoodmicro.2004.11.030

Kim, M., & Kim, K. (2012). Isolation and identification of γ-aminobutyric acid (GABA)-producing lactic acid bacteria from kimchi. *Journal of the Korean Society for Applied Biological Chemistry, 55*(6), 777–785. https://doi.org/10.1007/s13765-012-2174-6

Kim, G.-H., Baek, K.-R., Lee, G.-E., Lee, J.-H., Moon, J.-H., & Seo, S.-O. (2025). Development of starter cultures for precision fermentation of kombucha with enriched gamma-aminobutyric acid (GABA) content. *Fermentation, 11*(1), 17. https://doi.org/10.3390/fermentation11010017

Kittibunchakul, S., Yuthaworawit, N., Whanmek, K., Suttisansanee, U., & Santivarangkna, C. (2021). Health beneficial properties of a novel plant-based probiotic drink produced by fermentation of brown rice milk with GABA-producing *Lactobacillus pentosus* isolated from Thai pickled weed. *Journal of Functional Foods, 86*, 104710. https://doi.org/10.1016/j.jff.2021.104710

Kruk, M., Trząskowska, M., Ścibisz, I., & Pokorski, P. (2021). Application of the "SCOBY" and kombucha tea for the production of fermented milk drinks. *Microorganisms, 9*(1), 123. https://doi.org/10.3390/microorganisms9010123

Kwon, M.-J., Kim, J.-E., & Lee, S.-P. (2025). Functional kimchi beverage enhanced with γ-aminobutyric acid (GABA) through serial co-fermentation using *Leuconostoc citreum* S5 and *Lactiplantibacillus plantarum* KS2020. *Fermentation, 11*(1), 44. https://doi.org/10.3390/fermentation11010044

Lavelle, K., Mahony, J., & van Sinderen, D. (2025). Adaptation of bacterial starter cultures from dairy to plant-based substrates: Challenges and opportunities. *International Journal of Food Microbiology, 441*, 111304. https://doi.org/10.1016/j.ijfoodmicro.2025.111304

Lee, K. W., Shim, J. M., Park, S., Heo, H., Kim, H., Ham, K., & Kim, J. H. (2016). Isolation of lactic acid bacteria with probiotic potentials from kimchi, traditional Korean fermented vegetable. *LWT – Food Science and Technology, 71*, 130–137. https://doi.org/10.1016/j.lwt.2016.03.029

Lee, S. M., Lim, H. J., Chang, J. W., Hurh, B. S., & Kim, Y. S. (2018). Investigation on the formation of volatile compounds, fatty acids, and γ-lactones in white and brown rice during fermentation. *Food Chemistry, 269*, 347–354. https://doi.org/10.1016/j.foodchem.2018.07.037

Lei, V., & Jakobsen, M. (2004). Microbiological characterization and probiotic potential of *koko* and *koko sour water*, African spontaneously fermented millet porridge and drink. *Journal of Applied Microbiology, 96*(2), 384–397. https://doi.org/10.1046/j.1365-2672.2004.02162.x

Li, L., Feng, L., Yi, J., Hua, C., Chen, F., Liao, X., & Hu, X. (2010). High hydrostatic pressure inactivation of total aerobic bacteria, lactic acid bacteria, and yeasts in sour Chinese cabbage. *International Journal of Food Microbiology, 142*(1–2), 180–184. https://doi.org/10.1016/j.ijfoodmicro.2010.06.021

Li, Y. Q., Tian, W. L., Mo, H. Z., Zhang, Y. L., & Zhao, X. Z. (2013). Effects of pulsed electric field processing on quality characteristics and microbial inactivation of soymilk. *Food and Bioprocess Technology, 6*(7), 1907–1916. https://doi.org/10.1007/s11947-012-0844-z

Liang, H., He, Z., Wang, X., Song, G., Chen, H., Lin, X., Ji, C., & Li, S. (2020). Effects of salt concentration on microbial diversity and volatile compounds during *suancai* fermentation. *Food Microbiology, 91*, 103537. https://doi.org/10.1016/j.fm.2020.103537

Lim, Y. H., Foo, H. L., Loh, T. C., Mohamad, R., & Abdullah, N. (2019). Comparative studies of versatile extracellular proteolytic activities of lactic acid bacteria and their potential for extracellular amino acid production as feed supplements. *Journal of Animal Science and Biotechnology, 10*(1), 1–10. https://doi.org/10.1186/s40104-019-0323-z

Lima, T. S. (2005). *Um peixe olhou para mim: o povo Yudjá e a perspectiva*, NUTI (ed), Rio de Janeiro.

Liu, D., & Tong, C. (2017). Bacterial community diversity of traditional fermented vegetables in China. *LWT – Food Science and Technology, 86*, 40–48. https://doi.org/10.1016/j.lwt.2017.07.040

Liu, H., Ni, Y., Yu, Q., & Fan, L. (2023). Evaluation of co-fermentation of *Lactiplantibacillus plantarum* and *Pichia kluyveri* for a plant-based fermented beverage: Physicochemical, functional, and sensory properties. *Food Research International, 172*, 113060. https://doi.org/10.1016/j.foodres.2023.113060

López-Sánchez, R., Hernández-Oaxaca, D., Escobar-Zepeda, A., Cerrillo, B. R., López-Munguía, A., & Segovia, L. (2023). Analysing the dynamics of the bacterial community in *pozol*, a Mexican fermented corn dough. *Microbiology, 169*(7), 1355. https://doi.org/10.1099/mic.0.001355

Lu, R., Shang, M., Zhang, Y., Jiao, Y., Xia, Y., Garrett, S., Bakke, D., Bäuerl, C., Martinez, G. P., Kim, C., Kang, S., & Sun, J. (2020). Lactic acid bacteria isolated from Korean kimchi activate the vitamin D receptor–autophagy signaling pathways. *Inflammatory Bowel Diseases, 26*(8), 1199–1211. https://doi.org/10.1093/ibd/izaa049

Macêdo, E. L. C., Pimentel, T. C., Melo, D. S., de Souza, A. C., de Morais, J. S., Lima, M. S., Dias, D. R., Schwan, R. F., & Magnani, M. (2023). Yeasts from fermented Brazilian fruits as biotechnological tools for increasing phenolics bioaccessibility and improving the volatile profile in derived pulps. *Food Chemistry, 401*, 134200. https://doi.org/10.1016/j.foodchem.2022.134200

Malongane, F., & Berejena, T. (2024). Exploring the microbiome present in fermented indigenous African foods and their potential impact on human health. *Journal of Agriculture and Food Research, 16*, 101101. https://doi.org/10.1016/j.jafr.2024.101101

Manyatsi, T. S., Khaneghah, A. M., & Gavahian, M. (2024). The effects of ultrasound on probiotic functionality: An updated review. *Critical Reviews in Food Science and Nutrition*, 11643–11660. https://doi.org/10.1080/10408398.2023.2242490

Marco, M. L., Sanders, M. E., Gänzle, M., Arrieta, M. C., Cotter, P. D., De Vuyst, L., Hill, C., Holzapfel, W., Lebeer, S., Merenstein, D., Reid, G., Wolfe, B. E., & Hutkins, R. (2021). The International Scientific Association for Probiotics and Prebiotics (ISAPP) consensus statement on fermented foods. *Nature Reviews Gastroenterology & Hepatology, 18*, 196–208. https://doi.org/10.1038/s41575-020-00390-5

Marsh, A. J., O'Sullivan, O., Hill, C., Ross, R. P., & Cotter, P. D. (2014). Sequence-based analysis of the bacterial and fungal compositions of multiple kombucha (tea fungus) samples. *Food Microbiology, 38*, 171–178. https://doi.org/10.1016/j.fm.2013.09.003

Mayser, P., Fromme, S., Leitzmann, G., & Gründer, K. (1995). The yeast spectrum of the 'tea fungus kombucha'. *Mycoses, 38*(7–8), 289–295. https://doi.org/10.1111/j.1439-0507.1995.tb00410.x

McFeeters, R. F. (2004). Fermentation, microorganisms, and flavor changes in fermented foods. *Journal of Food Science, 69*(1), 35–36. https://doi.org/10.1111/j.1365-2621.2004.tb17876.x

Menezes, A. G. T., Ramos, C. L., Dias, D. R., & Schwan, R. F. (2018). Combination of probiotic yeast and lactic acid bacteria as starter culture to produce maize-based beverages. *Food Research International, 111*, 187–197. https://doi.org/10.1016/j.foodres.2018.04.065

Menezes, A. G. T., De Sousa Melo, D., Ramos, C. L., Moreira, S. I., Alves, E., & Schwan, R. F. (2020a). Yeasts isolated from Brazilian fermented foods in the protection against infection by pathogenic food bacteria. *Microbial Pathogenesis, 140*, 103969. https://doi.org/10.1016/j.micpath.2020.103969

Menezes, A. G. T., Ramos, C. L., Cenzi, G., Melo, D. S., Dias, D. R., & Schwan, R. F. (2020b). Probiotic potential, antioxidant activity, and phytase production of indigenous yeasts isolated from traditional fermented foods. *Probiotics and Antimicrobial Proteins, 12*(1), 280–288. https://doi.org/10.1007/s12602-019-9518-z

Minervini, F., Missaoui, J., Celano, G., Calasso, M., Achour, L., Saidane, D., et al. (2020). Use of autochthonous *Lactobacilli* to increase the safety of *Zgougou*. *Microorganisms, 8*(1), 29. https://doi.org/10.3390/microorganisms8010029

Mirsalami, S. M., & Mirsalami, M. (2025). Advances in genetically engineered microorganisms: Transforming food production through precision fermentation and synthetic biology. *Future Foods, 11*, 100601. https://doi.org/10.1016/j.fufo.2025.100601

Montano, A., Sanchez, A. H., Lopez-Lopez, A., De Castro, A., & Rejano, L. (2010). Chemical composition of fermented green olives: Acidity, salt, moisture, fat, protein, ash, fiber, sugar, and polyphenol. In V. R. Preedy & R. R. Watson (Eds.), *Olives and olive oil in health and disease prevention* (pp. 291–296). Academic.

Montoro, B. P., Benomar, N., Lerma, L. L., Gutiérrez, S. C., Gálvez, A., & Abriouel, H. (2016). Fermented Aloreña table olives as a source of potential probiotic *Lactobacillus pentosus* strains. *Frontiers in Microbiology, 7*, 1583. https://doi.org/10.3389/fmicb.2016.01583

Munekata, P. E. S., Domínguez, R., Budaraju, S., Roselló-Soto, E., Barba, F. J., Mallikarjunan, K., Roohinejad, S., & Lorenzo, J. M. (2020). Effect of innovative food processing technologies on the physicochemical and nutritional properties and quality of non-dairy plant-based beverages. *Foods, 9*(3), 288. https://doi.org/10.3390/foods9030288

Mustapha, A. T., Wahia, H., Ji, Q., Fakayode, O. A., Zhang, L., & Zhou, C. (2024). Multiple-frequency ultrasound for the inactivation of microorganisms on food: A review. *Journal of Food Process Engineering, 47*(4), e14587. https://doi.org/10.1111/jfpe.14587

Nago, M. C., Hounhouigan, J. D., Akissoe, N., Zanou, E., & Mestres, C. (1998). Characterization of the Beninese traditional *ogi*, a fermented maize slurry: Physicochemical and microbiological aspects. *International Journal of Food Science & Technology, 33*(3), 307–315. https://doi.org/10.1046/j.1365-2621.1998.00169.x

Nawawi, N. I. M., Ijod, G., Senevirathna, S. S. J., Aadil, R. M., Yusof, N. L., Yusoff, M. M., Adzahan, N. M., & Azman, E. M. (2023). Comparison of high pressure and thermal pasteurization on the quality parameters of strawberry products: a review. *Food Science and Biotechnology, 32*(6), 729–747. https://doi.org/10.1007/s10068-023-01276-3

Nisiotou, A. A., Chorianopoulos, N., Nychas, G. J. E., & Panagou, E. Z. (2010). Yeast heterogeneity during spontaneous fermentation of black *Conservolea* olives in different brine solutions. *Journal of Applied Microbiology, 108*(1), 396–405. https://doi.org/10.1111/j.1365-2672.2009.04424.x

Nuraida, L., Wacher, M. C., & Owens, J. D. (1995). Microbiology of *pozol*, a Mexican fermented maize dough. *World Journal of Microbiology and Biotechnology, 11*(5), 567–571. https://doi.org/10.1007/bf00286375

Oguntoyinbo, F. A. (2008). Functional properties of microorganisms in fermented foods and beverages. *African Journal of Biotechnology, 7*(25), 4754–4760.

Oguntoyinbo, F. A., & Narbad, A. (2015). Multifunctional properties of *Lactobacillus plantarum* strains isolated from fermented cereal foods. *Journal of Functional Foods, 17*, 621–631. https://doi.org/10.1016/j.jff.2015.06.022

Oguntoyinbo, F. A., Cnockaert, M., Cho, G., Kabisch, J., Neve, H., Bockelmann, W., Wenning, M., Franz, C. M. A. P., & Vandamme, P. (2017). *Halomonas nigrificans* sp. nov., isolated from cheese. *International Journal of Systematic and Evolutionary Microbiology, 68*(1), 371–376. https://doi.org/10.1099/ijsem.0.002515

Ogura, O., Okamoto, N., Nakamura, A., Takahashi, H., Kamata, R., & Kuda, T. (2025). Properties of *Leuconostoc lactis* isolated from traditional fermented rice beverages as a starter. *Process Biochemistry, 157*, 84–92. https://doi.org/10.1016/j.procbio.2025.06.019

Oh, Y. J., Kim, T. S., Moon, H. W., Lee, S. Y., Ji, G. E., & Hwang, K. T. (2020). *Lactobacillus plantarum* PMO 08 as a probiotic starter culture for plant-based fermented beverages. *Molecules, 25*(21), 5056. https://doi.org/10.3390/molecules25215056

Okafor, C. A., & Okoro, M. C. (2025). Fermentation profiles and probiotic viability in plant-based yogurt alternatives compared to traditional dairy yogurts: A critical review. *RKGP Journal of Food Science, 1*(1), 29–44. https://rkgpjournals.org/index.php/RJFS/article/view/18

Okeke, C. A., Ezekiel, C. N., Nwangburuka, C. C., Sulyok, M., Ezeamagu, C. O., Adeleke, R. A., Dike, S. K., & Krska, R. (2015). Bacterial diversity and mycotoxin reduction during maize fermentation (steeping) for *ogi* production. *Frontiers in Microbiology, 6*, 1402. https://doi.org/10.3389/fmicb.2015.01402

Olatunde, O. O., Obadina, A. O., Omemu, A. M., Oyewole, O. B., Olugbile, A., & Olukomaiya, O. O. (2018). Screening and molecular identification of potential probiotic lactic acid bacteria in effluents generated during ogi production. *Annals of Microbiology, 68*(7), 433–443. https://doi.org/10.1007/s13213-018-1348-9

Olivares-Illana, V., Wacher-Rodarte, C., Borgne, S. L., & López-Munguía, A. (2002). Characterization of a cell-associated inulosucrase from a novel source: A *Leuconostoc citreum* strain isolated from *pozol*, a fermented corn beverage of Mayan origin. *Journal of Industrial Microbiology & Biotechnology, 28*(2), 112–117. https://doi.org/10.1038/sj/jim/7000224

Omar, N. B., & Ampe, F. (2000). Microbial community dynamics during production of the Mexican fermented maize dough *pozol*. *Applied and Environmental Microbiology, 66*(9), 3664–3673. https://doi.org/10.1128/aem.66.9.3664-3673.2000

Osimani, A., Garofalo, C., Aquilanti, L., Milanović, V., & Clementi, F. (2015). Unpasteurised commercial *boza* as a source of microbial diversity. *International Journal of Food Microbiology, 194*, 62–70. https://doi.org/10.1016/j.ijfoodmicro.2014.11.011

Osorio-Cadavid, E., Chaveslopez, C., Tofalo, R., Paparella, A., & Suzzi, G.(2008). Detection and identification of wild yeasts in Champús a fermented Colombian maize beverage. *Food Microbiology, 25*(6), 771–777. https://doi.org/10.1016/j.fm.2008.04.014

Oyedeji, O., Ogunbanwo, S. T., & Onilude, A. A. (2013). Predominant lactic acid bacteria involved in the traditional fermentation of *fufu* and *ogi*, two Nigerian fermented food products. *Food and Nutrition Sciences, 4*(11), 40–46. https://doi.org/10.4236/fns.2013.411a006

Park, J. M., Shin, J. H., Lee, D. W., Song, J. C., Suh, H. J., Chang, U. J., & Kim, J. M. (2010). Identification of lactic acid bacteria in kimchi according to fermentation stages using PCR and 16S rRNA gene sequence analysis. *Food Science and Biotechnology, 19*(2), 541–546. https://doi.org/10.1007/s10068-010-0075-7

Park, E. J., Chun, J., Cha, C. J., Park, W. S., Jeon, C. O., & Bae, J. W. (2012). Bacterial community analysis during fermentation of ten representative kinds of kimchi with barcoded pyrosequencing. *Food Microbiology, 30*(1), 197–204. https://doi.org/10.1016/j.fm.2011.10.011

Pataro, G., & Ferrari, G. (2020). Limitations of pulsed electric field utilization in the food industry. In G. Ferrari & G. Pataro (Eds.), *Pulsed electric fields to obtain healthier and sustainable food for tomorrow* (pp. 283–310). Elsevier. https://doi.org/10.1016/B978-0-12-816402-3.00013-4

Pederson, C. S., & Albury, M. N. (1969). *The sauerkraut fermentation* (Technical Bulletin 824). New York State Agricultural Experiment Station.

Pei, J., Jin, W., Abd El-Aty, A. M., Baranenko, D. A., Gou, X., Zhang, H., Geng, J., Jiang, L., Chen, D., & Yue, T. (2020). Isolation, purification, and structural identification of a new bacteriocin made by *Lactobacillus plantarum* found in conventional kombucha. *Food Control, 110*, 106923. https://doi.org/10.1016/j.foodcont.2019.106923

Peñas, E., Frias, J., Gomez, R., & Vidal-Valverde, C. (2010). High hydrostatic pressure can improve the microbial quality of sauerkraut during storage. *Food Control, 21*(4), 524–528. https://doi.org/10.1016/j.foodcont.2009.08.010

Pereira, W. A., Piazentin, A. C. M., da Silva, T. M. S., Mendonça, C. M. N., Figueroa Villalobos, E., Converti, A., & Oliveira, R. P. S. (2023). Alternative fermented soy-based beverage: Impact of inulin on the growth of probiotic strains and starter culture. *Fermentation, 9*(11), 961. https://doi.org/10.3390/fermentation9110961

Pérez-Díaz, I. M., Hayes, J. S., Medina, E., Webber, A. M., Butz, N., Dickey, A. N., et al. (2019). Assessment of the non-lactic acid bacteria microbiota in fresh cucumbers and commercially fermented cucumber pickles brined with 6% NaCl. *Food Microbiology, 77*, 10–20. https://doi.org/10.1016/j.fm.2018.08.003

Pizarro-Oteíza, S., Salazar, F., Cea, R., Cavieres, O., & Meenu, M. (2025). Impact of high hydrostatic pressure, ultrasound, and pulsed electric field in beverage fermentation: A review of nutritional, functional, and sensory aspects and the future. *Foods, 14*(20), 3576. https://doi.org/10.3390/foods14203576

Ponomarova, O., Gabrielli, N., Sévin, D. C., Mülleder, M., Zirngibl, K., Bulyha, K., Andrejev, S., Kafkia, E., Typas, A., Sauer, U., Ralser, M., & Patil, K. R. (2017). Yeast creates a niche for symbiotic lactic acid bacteria through nitrogen overflow. *Cell Systems, 5*(4), 345–357. https://doi.org/10.1016/j.cels.2017.09.002

Pradas, I., Del Pino, B., Peña, F., Ortiz, V., Moreno-Rojas, J. M., Fernández-Hernández, A., & Moreno, J. (2012). The use of high hydrostatic pressure (HHP) treatments for table olive preservation. *Innovative Food Science & Emerging Technologies, 13*, 64–68. https://doi.org/10.1016/j.ifset.2011.10.011

Pswarayi, F., & Gänzle, M. G. (2019). Composition and origin of the fermentation microbiota of mahewu, a Zimbabwean fermented cereal beverage. *Applied and Environmental Microbiology, 85*(11), e03130-18. https://doi.org/10.1128/aem.03130-18

Puerari, C., Magalhães-Guedes, K. T., & Schwan, R. F. (2015). Physicochemical and microbiological characterization of chicha, a rice-based fermented beverage produced by Umutina Brazilian Amerindians. *Food Microbiology, 46*, 210–217. https://doi.org/10.1016/j.fm.2014.08.009

Queiroz, L. L., Hoffmann, C., Lacorte, G. A., de Melo Franco, B. D. G., & Todorov, S. D. (2022). Genomic and functional characterization of bacteriocinogenic lactic acid bacteria isolated from boza, a traditional cereal-based beverage. *Scientific Reports, 12*(1), 5086. https://doi.org/10.1038/s41598-022-05086-1

Ramos, C. L., & Schwan, R. F. (2017). Technological and nutritional aspects of indigenous Latin America fermented foods. *Current Opinion in Food Science, 13*, 97–102. https://doi.org/10.1016/j.cofs.2017.07.001

Ramos, C. L., de Almeida, E. G., Pereira, G. V., Cardoso, P. G., Dias, E. S., & Schwan, R. F. (2010). Determination of dynamic characteristics of microbiota in a fermented beverage produced by Brazilian amerindians using culture-dependent and culture-independent methods. *International Journal of Food Microbiology, 140*(2–3), 225–231. https://doi.org/10.1016/j.ijfoodmicro.2010.03.029

Ramos, C. L., de Almeida, E. G., Freire, A. L., & Freitas Schwan, R. (2011). Diversity of bacteria and yeast in the naturally fermented cotton seed and rice beverage produced by Brazilian Amerindians. *Food Microbiology, 28*(7), 1380–1386. https://doi.org/10.1016/j.fm.2011.06.012

Ramos, C. L., Thorsen, L., Schwan, R. F., & Jespersen, L. (2013). Strain-specific probiotics properties of *Lactobacillus fermentum*, *Lactobacillus plantarum* and *Lactobacillus brevis* isolates from Brazilian food products. *Food Microbiology, 36*(1), 22–29. https://doi.org/10.1016/j.fm.2013.03.010

Ramos, C. L., Thorsen, L., Ryssel, M., Nielsen, D. S., Siegumfeldt, H., Schwan, R. F., & Jespersen, L. (2014). Effect of the gastrointestinal environment on pH homeostasis of *Lactobacillus plantarum* and *Lactobacillus brevis* cells as measured by real-time fluorescence ratio-imaging microscopy. *Research in Microbiology, 165*(3), 215–225. https://doi.org/10.1016/j.resmic.2014.02.005

Ramos, C. L., De Sousa, E. S. O., Ribeiro, J., Almeida, T. M. M., Santos, C. C. D. A., Abegg, M. A., & Schwan, R. F. (2015). Microbiological and chemical characteristics of tarubá, an indigenous beverage produced from solid cassava fermentation. *Food Microbiology, 49*, 182–188. https://doi.org/10.1016/j.fm.2015.02.005

Rangel, A. E. T., Gómez Ramírez, J. M., & González Barrios, A. F. (2020). From industrial by-products to value-added compounds: The design of efficient microbial cell factories by coupling systems metabolic engineering and bioprocesses. *Biofuels, Bioproducts and Biorefining, 14*(6), 1228–1238. https://doi.org/10.1002/bbb.2127

Ray, P., Sanchez, C., O'Sullivan, D., & McKay, L. (2000). Classification of a bacterial isolate from pozol exhibiting antimicrobial activity against several Gram-positive and Gram-negative bacteria, yeasts, and molds. *Journal of Food Protection, 63*(8), 1123–1132. https://doi.org/10.4315/0362-028x-63.8.1123

Resende, I. F., Martins, P. M. M., Melo, D. S., Magnani, M., Dias, D. R., & Schwan, R. F. (2025). Development and characterization of microencapsulated *Pichia kluyveri* CCMA 0615 with probiotic properties and its application in fermented beverages. *International Journal of Food Microbiology, 427*, 110967. https://doi.org/10.1016/j.ijfoodmicro.2024.110967

Ricci, A., Cirlini, M., Levante, A., Dall'Asta, C., Galaverna, G., & Lazzi, C. (2018). Volatile profile of elderberry juice: Effect of lactic acid fermentation using *Lactobacillus plantarum*, *L. rhamnosus*, and *L. casei* strains. *Food Research International, 105*, 412–422. https://doi.org/10.1016/j.foodres.2017.11.042

Ricciardi, A., López, F. N. A., Giavalisco, M., Pietrafesa, R., & Parente, E. (2025). Determining the core bacterial and fungal genera in table olive fermentations. *International Journal of Food Microbiology, 442*, 111344. https://doi.org/10.1016/j.ijfoodmicro.2025.111344

Rizo, J., Guillén, D., Díaz-Ruiz, G., Wacher, C., Encarnación, S., Sánchez, S., & Rodríguez-Sanoja, R. (2021). Metaproteomic insights into the microbial community in pozol. *Frontiers in Nutrition, 8*, 714814. https://doi.org/10.3389/fnut.2021.714814

Rizzello, C. G., Coda, R., Wang, Y., Verni, M., Kajala, I., Katina, K., et al. (2018). Characterization of indigenous *Pediococcus pentosaceus*, *Leuconostoc kimchii*, *Weissella cibaria* and *Weissella confusa* for faba bean bioprocessing. *International Journal of Food Microbiology, 302*, 24–34. https://doi.org/10.1016/j.ijfoodmicro.2018.08.014

Rodríguez-Gómez, F., Romero-Gil, V., Bautista-Gallego, J., Garrido-Fernández, A., & Arroyo-López, F. N. (2012). Multivariate analysis to discriminate yeast strains with technological applications in table olive processing. *World Journal of Microbiology and Biotechnology, 28*(5), 1761–1770. https://doi.org/10.1007/s11274-011-0990-1

Rojas, V., Gil, J. V., Piñaga, F., & Manzanares, P. (2001). Studies on acetate ester production by non-Saccharomyces wine yeasts. *International Journal of Food Microbiology, 70*(3), 283–289. https://doi.org/10.1016/S0168-1605(01)00552-9

Saeed, A., Yasmin, A., Baig, M., Khan, K., Heyat, M. B. B., Akhtar, F., Batool, Z., Kazmi, A., Wahab, A., Shahid, M., Ahmed, M. A., Abbas, S., Muaad, A. Y., Shahzad, A., & Ahmad, I. (2023). Isolation and characterization of *Lactobacillus crispatus, Lactococcus lactis,* and *Carnobacterium* divergens as potential probiotic bacteria from fermented black and green olives (*Olea europaea*): An exploratory study. *BioMed Research International, 2023*(1), 8726320. https://doi.org/10.1155/2023/8726320

Santos, C. C. A. A., De Almeida, E. G., De Melo, G. V. P., & Schwan, R. F. (2012). Microbiological and physicochemical characterisation of caxiri, an alcoholic beverage produced by the indigenous Juruna people of Brazil. *International Journal of Food Microbiology, 156*(2), 112–121. https://doi.org/10.1016/j.ijfoodmicro.2012.03.010

Santos, C. C. A. A., Libeck, B. S., & Schwan, R. F. (2014). Co-culture fermentation of peanut-soy milk for the development of a novel functional beverage. *International Journal of Food Microbiology, 186*, 32–41. https://doi.org/10.1016/j.ijfoodmicro.2014.06.011

Sanya, A. C., Madode, Y. E., Schoustra, S. E., Smid, E. J., & Linnemann, A. R. (2023). Technological variations, microbial diversity and quality characteristics of maize ogi used for akpan production in Benin. *Food Research International, 170*, 113038. https://doi.org/10.1016/j.foodres.2023.113038

Schwan, R. F., Almeida, E. G., Souza-Dias, M. A., & Jespersen, L. (2007). Yeast diversity in rice-cassava fermentations produced by the indigenous Tapirapé people of Brazil. *FEMS Yeast Research, 7*, 966–972. https://doi.org/10.1111/j.1567-1364.2007.00241.x

Shi, L., Mu, K., Arntfield, S. D., & Nickerson, M. T. (2017). Changes in levels of enzyme inhibitors during soaking and cooking for pulses available in Canada. *Journal of Food Science and Technology, 54*(4), 1014–1022. https://doi.org/10.1007/s13197-017-2523-3

Shim, S. M., Kim, J. Y., Lee, S. M., Park, J. B., Oh, S. K., & Kim, Y. S. (2012). Profiling of fermentative metabolites in kimchi: Volatile and non-volatile organic acids. *Journal of the Korean Society for Applied Biological Chemistry, 55*(4), 463–469. https://doi.org/10.1007/s13765-012-2074-9

Sievers, M., Lanini, C., Weber, A., Schuler-Schmid, U., & Teuber, M. (1995). Microbiology and fermentation balance in a kombucha beverage obtained from a tea fungus fermentation. *Systematic and Applied Microbiology, 18*(4), 590–594. https://doi.org/10.1016/S0723-2020(11)80420-0

Simões, L., De Souza, A. C., Ferreira, I., Melo, D., Lopes, L., Magnani, M., Schwan, R., & Dias. (2021). Probiotic properties of yeasts isolated from Brazilian fermented table olives. *Journal of Applied Microbiology, 131*(4), 1983–1997. https://doi.org/10.1111/jam.15065

Simões, L., Fernandes, N., De Souza, A., Santos, L. D., Magnani, M., Abrunhosa, L., Teixeira, J., Schwan, R. F., & Dias, D. R. (2022). Probiotic and antifungal attributes of lactic acid bacteria isolates from naturally fermented Brazilian table olives. *Fermentation, 8*(6), 277. https://doi.org/10.3390/fermentation8060277

Singh, A., & Kumar, S. (2025). Exploring the functionality of microbes in fermented foods: Technological advancements and future directions. *Fermentation, 11*(6), 300. https://doi.org/10.3390/fermentation11060300

Sotelo-Lara, D. M., Amador-Espejo, G. G., Álvarez-Araiza, D. F., Cordero-Rivera, A. K., Millán-Quintero, K. G., Campos-Vega, R., & Velázquez-Estrada, R. M. (2024). Ultrasound and thermosonication as promising technologies for processing plant-based beverages: A review. *Food Technology and Biotechnology, 62*(4), 538–552. https://doi.org/10.17113/ftb.62.04.24.8624

Souza, F. E. B., Rodrigues, S., & Fonteles, T. V. (2025). Non-thermal technologies in food fermentation: Mechanisms, benefits, and industrial perspectives for sustainable development. *Processes, 13*(9), 2988. https://doi.org/10.3390/pr13092988

Sutherland, I. W. (1998). Novel and established applications of microbial polysaccharides. *Trends in Biotechnology, 16*(1), 41–46. https://doi.org/10.1016/s0167-7799(97)01139-6

Tachie, C., Nwachukwu, I. D., & Aryee, A. N. A. (2023). Trends and innovations in the formulation of plant-based foods. *Food Production, Processing and Nutrition, 5*(1), 16. https://doi.org/10.1186/s43014-023-00129-0

Tan, X., Cui, F., Wang, D., Lv, X., Li, X., & Li, J. (2024). Fermented vegetables: Health benefits, defects, and current technological solutions. *Foods, 13*(1), 38. https://doi.org/10.3390/foods13010038

Teoh, A. L., Heard, G., & Cox, J. (2004). Yeast ecology of kombucha fermentation. *International Journal of Food Microbiology, 95*(2), 119–126. https://doi.org/10.1016/j.ijfoodmicro.2003.12.020

Todorov, S. D. (2009). Bacteriocins from *Lactobacillus plantarum* production, genetic organization and mode of action: Production, genetic organization and mode of action. *Brazilian Journal of Microbiology, 40*(2), 209–221. https://doi.org/10.1590/s1517-83822009000200001

Touret, T., Oliveira, M., & Semedo-Lemsaddek, T. (2018). Putative probiotic lactic acid bacteria isolated from sauerkraut fermentations. *PLoS One, 13*(9), e0203501. https://doi.org/10.1371/journal.pone.0203501

Tran, T., Grandvalet, C., Verdier, F., Martin, A., Alexandre, H., & Tourdot-Maréchal, R. (2020). Microbial dynamics between yeasts and acetic acid bacteria in kombucha: Impacts on the chemical composition of the beverage. *Foods, 9*(7), 963. https://doi.org/10.3390/foods9070963

Ucak, S., Yurt, M. N. Z., Tasbasi, B. B., Acar, E. E., Altunbas, O., Soyucok, A., Aydin, A., Ozalp, V. C., & Sudagidan, M. (2022). Identification of bacterial communities of fermented cereal beverage Boza by metagenomic analysis. *LWT, 153*, 112465. https://doi.org/10.1016/j.lwt.2021.112465

Ulloa, M., Herrera, T., & Taboada, J. (1983). Pozol, a fermented maize dough consumed in southern Mexico. In K. H. Steinkraus et al. (Eds.), *Handbook of indigenous fermented foods* (pp. 226–233). Marcel Dekker.

Vargas, B. K., Fabricio, M. F., & Záchia Ayub, M. A. (2021). Health effects and probiotic and prebiotic potential of kombucha: A bibliometric and systematic review. *Food Bioscience, 44*, 101332. https://doi.org/10.1016/j.fbio.2021.101332

Verma, K., Duhan, P., Pal, D., Verma, P., & Bansal, P. (2025). Precision fermentation for the next generation of food ingredients: Opportunities and challenges. *Future Foods, 12*, 100750. https://doi.org/10.1016/j.fufo.2025.100750

Verni, M., De Mastro, G., De Cillis, F., Gobbetti, M., & Rizzello, C. G. (2019). Lactic acid bacteria fermentation to exploit the nutritional potential of Mediterranean faba bean local biotypes. *Food Research International, 125*, 108571. https://doi.org/10.1016/j.foodres.2019.108571

Villarreal-Soto, S. A., Beaufort, S., Bouajila, J., Souchard, J. P., & Taillandier, P. (2018). Understanding kombucha tea fermentation: A review. *Journal of Food Science, 83*(3), 580–588. https://doi.org/10.1111/1750-3841.14068

Villarreal-Soto, S. A., Bouajila, J., Pace, M., Leech, J., Cotter, P. D., Souchard, J., Taillandier, P., & Beaufort, S. (2020). Metabolome-microbiome signatures in the fermented beverage, Kombucha. *International Journal of Food Microbiology, 333*, 108778. https://doi.org/10.1016/j.ijfoodmicro.2020.108778

Wacher, C., Cañas, A., Cook, P. E., Barzana, E., & Owens, J. D. (1993). Sources of microorganisms in pozol, a traditional Mexican fermented maize dough. *World Journal of Microbiology and Biotechnology, 9*(2), 269–274. https://doi.org/10.1007/bf00327853

Wacher, C., Cañas, A., Bárzana, E., Lappe, P., Ulloa, M., & Owens, J. D. (2000). Microbiology of Indian and Mestizo pozol fermentations. *Food Microbiology, 17*(3), 251–256. https://doi.org/10.1006/fmic.1999.0310

Walsh, A. M., Crispie, F., Kilcawley, K., O'Sullivan, O., O'Sullivan, M. G., Claesson, M. J., et al. (2016). Microbial succession and flavor production in the fermented dairy beverage kefir. *mSystems, 1*(5), e00052-16. https://doi.org/10.1128/mSystems.00052-16

Wang, Y., Ji, B., Wu, W., Wang, R., Yang, Z., Zhang, D., & Tian, W. (2014). Hepatoprotective effects of kombucha tea: Identification of functional strains and quantification of functional components. *Journal of the Science of Food and Agriculture, 94*(2), 265–272. https://doi.org/10.1002/jsfa.6245

Wang, R., Sun, J., Lassabliere, B., Yu, B., & Liu, S. Q. (2020). Fermentation characteristics of four non-*Saccharomyces* yeasts in green tea slurry. *Food Microbiology, 92*, 103609. https://doi.org/10.1016/j.fm.2020.103609

Wang, B., Rutherfurd-Markwick, K., Liu, N., Zhang, X., & Mutukumira, A. N. (2024a). Evaluation of the probiotic potential of yeast isolated from kombucha in New Zealand. *Current Research in Food Science, 8*, 100711. https://doi.org/10.1016/j.crfs.2024.100711

Wang, B., Rutherfurd-Markwick, K., Liu, N., Zhang, X., & Mutukumira, A. N. (2024b). Probiotic potential of acetic acid bacteria isolated from kombucha in New Zealand in vitro. *The Microbe, 4*, 100130. https://doi.org/10.1016/j.microb.2024.100130

Werning, M. L., Ibarburu, I., Dueñas, M. T., Irastorza, A., Navas, J., & López, P. (2006). *Pediococcus parvulus gtf* gene encoding the GTF glycosyltransferase and its application for specific PCR detection of α-D-glucan–producing bacteria in foods and beverages. *Journal of Food Protection, 69*(1), 161–169. https://doi.org/10.4315/0362-028x-69.1.161

Wisselink, H. W., Weusthuis, R. A., Eggink, G., Hugenholtz, J., & Grobben, G. J. (2002). Mannitol production by lactic acid bacteria: A review. *International Dairy Journal, 12*(2–3), 151–161. https://doi.org/10.1016/S0958-6946(01)00155-3

Wouters, D., Bernaert, N., Conjaerts, W., Van Droogenbroeck, B., De Loose, M., & De Vuyst, L. (2013). Species diversity, community dynamics, and metabolite kinetics of spontaneous leek fermentations. *Food Microbiology, 33*(2), 185–196. https://doi.org/10.1016/j.fm.2012.09.016

Wu, R., Yu, M., Liu, X., Meng, L., Wang, Q., Xue, Y., & Zhang, W. (2015). Changes in flavour and microbial diversity during natural fermentation of suan-cai, a traditional food made in Northeast China. *International Journal of Food Microbiology, 211*, 23–31. https://doi.org/10.1016/j.ijfoodmicro.2015.06.028

Xie, A., Dong, Y., Liu, Z., Li, Z., Shao, J., Li, M., & Yue, X. (2023). A review of plant-based drinks addressing nutrients, flavor, and processing technologies. *Foods, 12*(21), 3952. https://doi.org/10.3390/foods12213952

Xiong, T., Guan, Q., Song, S., Hao, M., & Xie, M. (2012). Dynamic changes of lactic acid bacteria flora during Chinese sauerkraut fermentation. *Food Control, 26*(1), 178–181. https://doi.org/10.1016/j.foodcont.2012.01.027

Yamada, Y., Hoshino, K., & Ishikawa, T. (1997). The phylogeny of acetic acid bacteria based on partial sequences of 16S rRNA: Elevation of the subgenus *Gluconoacetobacter* to generic level. *Bioscience, Biotechnology and Biochemistry, 61*(8), 1244–1251. https://doi.org/10.1271/bbb.61.1244

Yamada, Y., Yukphan, P., Vu, H. T. L., Muramatsu, Y., Ochaikul, D., Tanasupawat, S., & Nakagawa, Y. (2012). Description of *Komagataeibacter* gen. nov., with proposals of new combinations (Acetobacteraceae). *Journal of General and Applied Microbiology, 58*(5), 397–404. https://doi.org/10.2323/jgam.58.397

Yan, B., Li, J., Liang, Q., Huang, Y., Cao, S., Wang, L., & Zeng, X. (2024). From laboratory to industry: The evolution and impact of pulsed electric field technology in food processing. *Food Reviews International, 41*(2), 373–398. https://doi.org/10.1080/87559129.2024.2401433

Yang, X., Hu, W., Xiu, Z., Jiang, A., Saren, G., Ji, Y., Guan, Y., & Feng, K. (2020). Microbial community dynamics and metabolome changes during spontaneous fermentation of northeast sauerkraut from different households. *Frontiers in Microbiology, 11*, 1878. https://doi.org/10.3389/fmicb.2020.01878

Yasir, M., Al-Zahrani, I. A., Bibi, F., Ghany, M. A. E., & Azhar, E. I. (2022). New insights of bacterial communities in fermented vegetables from shotgun metagenomics and identification of antibiotic resistance genes and probiotic bacteria. *Food Research International, 157*, 111190. https://doi.org/10.1016/j.foodres.2022.111190

Yehuala, G. A., Shibeshi, N. T., Kim, S. H., & Park, M. K. (2024). Characterization of autochthonous lactic acid bacteria isolated from a traditional Ethiopian beverage, tella. *Foods, 13*(4), 575. https://doi.org/10.3390/foods13040575

Yu, Z., Su, Y., Zhang, Y., Zhu, P., Mei, Z., Zhou, X., & Yu, H. (2021). Potential use of ultrasound to promote fermentation, maturation, and properties of fermented foods: A review. *Food Chemistry, 357*, 129805. https://doi.org/10.1016/j.foodchem

Yusoff, I. M., Mat Taher, Z., Rahmat, Z., & Chua, L. S. (2022). A review of ultrasound-assisted extraction for plant bioactive compounds: Phenolics, flavonoids, thymols, saponins, and proteins. *Food Research International, 157*, 111268. https://doi.org/10.1016/j.foodres.2022.111268

Zhang, P., Wu, J., Tao, D., & Wu, R. (2019). Effects of *Leuconostoc mesenteroides* on physicochemical and microbial succession characterization of soybean paste (Da-jiang). *LWT – Food Science and Technology, 115*, 108028. https://doi.org/10.1016/j.lwt.2019.04.029

Zhang, C., Zhang, J., & Liu, D. (2021). Biochemical changes and microbial community dynamics during spontaneous fermentation of Zhacai, a traditional pickled mustard tuber from China. *International Journal of Food Microbiology, 347*, 109199. https://doi.org/10.1016/j.ijfoodmicro.2021.109199

Zhang, S., Zhang, Y., Wu, L., Zhang, L., & Wang, S. (2022). Characterization of microbiota of naturally fermented sauerkraut by high-throughput sequencing. *Food Science and Biotechnology, 32*(6), 855–862. https://doi.org/10.1007/s10068-022-01221-w

Zhao, C. J., Schieber, A., & Gänzle, M. G. (2016). Formation of taste-active amino acids, amino acid derivatives, and peptides in food fermentations: A review. *Food Research International, 89*, 39–47. https://doi.org/10.1016/j.foodres.2016.08.042

Zhao, X., Hu, R., He, Y., Li, S., Yang, J., Zhang, J., Zhou, J., & Xue, T. (2022). Screening of isolated potential probiotic lactic acid bacteria from Sichuan pickle for cholesterol lowering property and triglycerides lowering activity. *Food Science and Technology, 42*. https://doi.org/10.1590/fst.09122

Zhao, X., Pei, L., Wang, X., Luo, M., Hou, S., Ye, X., Liu, W., & Zhou, Y. (2025). Screening and phenotyping of lactic acid bacteria in Boza. *Microorganisms, 13*(8), 1767. https://doi.org/10.3390/microorganisms13081767

Zhenjun, Z., Yucheng, S., Huawei, W., Caibi, Z., Xianchun, H., Jian, Z. (2018). Flavour chemical dynamics during fermentation of kombucha tea. *Emirates Journal of Food and Agriculture, 732*. https://doi.org/10.9755/ejfa.2018.v30.i9.1794

Zhou, Q., Zang, S., Zhao, Z., & Li, X. (2018). Dynamic changes of bacterial communities and nitrite character during northeastern Chinese sauerkraut fermentation. *Food Science and Biotechnology, 27*(1), 79–85. https://doi.org/10.1007/s10068-017-0279-8

Chapter 12
Plant-Based Fermented Foods in Climate Resilience, Food Security, and Sustainable Agriculture

Ravneet Kaur, Nishtha Paul, Vijay Kumar, Gagan Preet Kour Bali, and Tamilarasan Kamalanathan

Abstract Fermentation plays an important role in ancient food processing technology that extends shelf life, safety, and nutritional quality. In the modern age of high-speed change of the climate, fermentation takes a central role in the climate-resistant food system. Fermentation enhances food security by transforming underused crops and climate-resistant plant-based resources into stable, nutrient-rich foods, reducing post-harvest loss, and thus increasing food security. The same action promotes protein digestibility and bioavailability of micronutrients, which provide a solution to the prevalent nutritional deficiencies. Fermentation enterprises further support local circular economies by converting surplus produce and agricultural by-products into value-added products, thereby creating income and reducing waste. Fermentation contributes to climate adaptation in two key ways: first, by preserving seasonal food surpluses and diversifying food supplies through culturally acceptable products; and second, by creating economic incentives for the cultivation of sustainable crop varieties through the establishment of reliable processing and consumption pathways. The integration of traditional fermentation practices with modern scientific advances, including detailed microbial characterization and state-of-the-art bioprocess engineering, has unlocked substantial potential for amplifying the benefits of fermentation. Furthermore, fermented products also carry useful microbiota, which improves the well-being and health of the gastrointestinal system and resistance to the microbiota at the community scale. Subsequently, the dual effects of fermentation in climate adaptation include food resource preservation and diversifi-

R. Kaur · N. Paul (✉) · G. P. K. Bali
Department of Biosciences, University Institute of Biotechnology, Chandigarh University, Gharuan, Mohali, Punjab, India

V. Kumar
Patanjali Herbal Research Department, Patanjali Research Institute, Haridwar, Uttarakhand, India

T. Kamalanathan
Eagle Pest Control, Al Ain, Abu Dhabi, United Arab Emirates

S. Dhiman, S. Kaur (eds.), *Next-Gen Plant Based Fermented Foods: Innovations for a Sustainable Future*, https://doi.org/10.1007/978-3-032-29927-7_12

cation as well as promoting sustainable agronomic use. The intersection of ancient wisdom and contemporary technology is potentially fruitful in broadening the scope of fermentation, making it a key pilot point between food security and sustainable agriculture in the climate-resilient future.

Keywords Traditional fermentation · Climate resilience · Food security · Sustainable agriculture · Plant-based foods · Nutrition

Abbreviations

AI	Artificial intelligence
Cas9	Crispr associated protein 9
CRISPR	Clustered Regularly interspaced short palindromic sequences
LAB	Lactic acid bacteria
SCOBY	symbiotic culture of bacteria and yeast

12.1 Introduction

Climate changes pose an inevitable risk to global food security, which directly impacts the agricultural productivity and leads to disastrous effects on existing food system. The challenges posed by conventional agriculture practices leads to ethical and environmental problems, therefore a transition from sustainable food systems is essential (Abbaspour, 2024). In this context, plant-based foods especially those enhanced using ancient biotechnology fermentation are catching public attention. Fermentation technologies are transforming the plant-based food sector by facilitating the development of plant-based meat analogues, dairy alternatives, and protein ingredients with sensory and nutritional properties that closely resemble those of conventional animal-derived products (Boukid et al., 2023). Plant-based foods offer significant advantages over animal-based products by simultaneously addressing ethical concerns, reducing environmental footprints, and improving sensory attributes (Zhang et al., 2024). There are many environmental advantages of adopting fermentation in plant-based food production. They significantly reduce the greenhouse gas emission leading to mitigate the deforestation and water scarcity (Agrawal et al., 2023). From the health perspective, fermentation helps to infuse foods with probiotic and prebiotic aspects which helps in healthy gut and improved immunity which becomes a viable option for people with specific food allergies (Ibrahim et al., 2023). Plant based fermented foods have gained attention not only because the probiotic and industrial potential of their microbial communities but also for their emerging role as non-dairy alternatives for probiotic delivery (D'Almeida et al., 2024). Plant based foods are known to provide essential nutrients, vitamins, minerals, antioxidants, and dietary fibre, while typically containing low levels of sugar. Novel strategies have enabled the development of functional foods that integrate plant-based ingredients with resilient probiotic strains able to thrive within plant matrices (De Bellis et al., 2021).

A major gap remains in understanding how specific processing technologies can transform these dietary approaches into viable and scalable strategies for climate change adaptation and mitigation (Ashoka et al., 2023). This chapter addresses the nutritional, socio-economic, policy, and technological gaps associated with plant-based foods (Abe-Inge et al., 2024). The chapter highlights how plant-based fermentation can serve as a strategic link between improved post-harvest management, increased nutrient bioavailability, and the promotion of decentralized and resilient local economies (Abbaspour, 2024). Finally, the chapter emphasizes on the need to legitimize the introduction of both conventional and innovative modalities of fermentation into the very core of food policy and system in the future.

12.2 Evolution of Fermentation: From Traditional Practices to Modern Technologies

Fermentation is known to occupy a very significant place in the entire history of human discovery in terms of written and archaeological evidence tracing its applications dating back to ancient Egypt, China and America at least several millennia ago (Huang, 2017). The process of fermentation was developed as the ancient people started to discover the microbes-driven transformation of raw materials could lead to preservation of food, improve its nutritional quality and enhance its flavour (Singh et al., 2023). Traditional form of fermentation foods such as sauerkraut, kimchi and tempeh are primarily made out of natural fermentation in which microbes that are found in raw materials or the surrounding environment are used in making the fermented product (Sionek et al., 2023).

The chapter aims to draw a clear distinction between the traditional as well as modern practices involved in fermentation. Traditional fermentation aims to provide cultural specificity, complex flavour development which arises from diverse microbial communities along with the requirement of low resource conditions (Contreras-Negrete et al., 2024). On the other hand, it is reported that contemporary modern fermentation involves industrial approaches that are specific to scalability, food safety and standardisation through specific use of controlled starer cultures (Rai & Appaiah, 2021).

However, modern fermentation provides a significant edge in terms of development of functional properties, large-scale production of active constituents in bioactive compounds and its precision abilities (Rai & Appaiah, 2021). This contrast critically highlights a central challenge in the prevailing food system development which points out on balancing the safety and efficacy of industrial processes along with cultural richness, local adaptability found and inherited in traditional form of fermentation practice along with biodiversity.

12.3 Conventional Fermentation Processes

Conventional fermentation practices play a significant role in enhancing the safety, nutritional quality and functional properties of plant-based foods (Abbaspour, 2024). These traditional processes depend upon the natural microbes, particularly the lactic acid bacteria (LAB), Yeast and molds, that metabolise plant substrate into value-added products (Sionek et al., 2023). The traditional fermentation plays a key role is food preservation as the when the microbial production of organic acids such as lactic acid lowers pH and further inhibits spoilage pathogens thereby extending the shelf life (Lys, 2025). Conventional fermentation also improves the nutritional quality by degrading anti-nutritional factors such as phytates and tannins which enhances protein digestibility and mineral bioavailability (Thorat et al., 2017). Moreover, conventional fermentation promotes numerous functional and health benefits, as microbial activity can enhance the production of bioactive compounds, vitamins, and probiotics (Gormaz et al., 2010) (Fig. 12.1). These methods are considered cost-effective and sustainable, requiring minimal technological input. Conventional fermentation also supports local food systems and helps preserve cultural heritage associated with traditional food practices (Abbaspour, 2024). Therefore, traditional fermentation practices play a crucial role in improving the quality, nutritional value, functionality, and sustainability of plant-based foods (Dhiman et al., 2025).

Lactic acid fermentation quantitatively causes acidification of substrates and further leads to maintain vegetables such as kimchi in Korea and sauerkraut in Europe and aids in production of probiotics (Das et al., 2016). In fungal fermentation, e.g., in Indonesian tempeh or Japanese miso, mycelial growth raises the

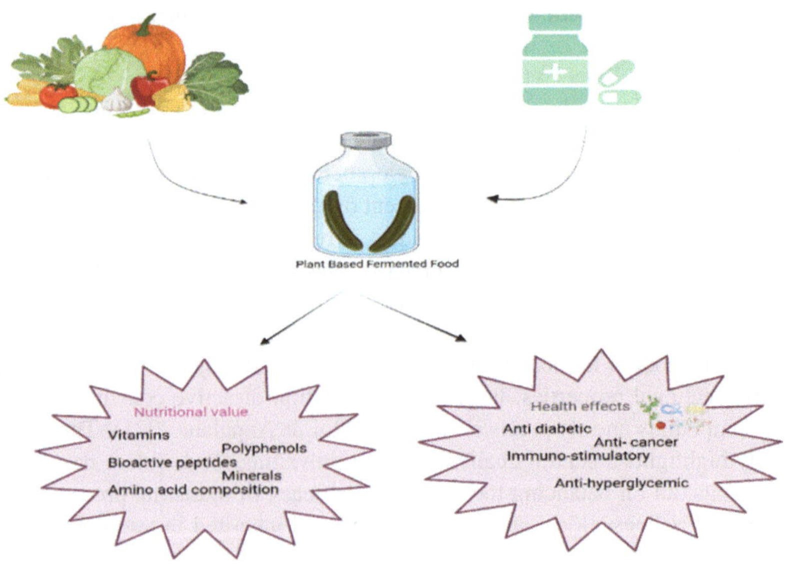

Fig. 12.1 Utilities of plant based fermented food and its components

digestibility of proteins and vitamin levels by enzymatic activity (Azri et al., 2026). An example of alkaline fermentation such as the African condiment soumbala, utilize Bacillus spp. to degrade proteins, thereby enhancing flavor development and improving nutrient bioavailability (Dabiré et al., 2022). These processes are typically low in energy demand and often rely on ambient temperatures and naturally occurring microorganisms, contributing to their sustainability (Abbaspour, 2024). Furthermore, such traditional fermentation practices promote the utilization of locally grown and underutilized crops, thereby supporting regional food sovereignty, agrobiodiversity, and dietary diversity (Bisht et al., 2018). According to quantitative research, these types of fermentation can lead to a reduction of up to 30% post-harvest losses of some perishable products and, as a result, food waste can be directly mitigated (Ortiz-Sanchez et al., 2023). According to meta-analysis, intake of fermented kimchi significantly lowered fasting blood glucose levels relative to the control group (WMD: −1.93 mg/dL; 95% CI: −3.82, −0.03; $I^2 = 17.4\%$) (Ahn et al., 2025).

These processes make the local, climate-resilient crops, including millets, cassava, or native legumes, into the stable, nutrient-dense foods, thus enhancing the resilience of the decentralized food systems to the climatic shocks (Kumar et al., 2022). Traditional fermentation is therefore known to be a source of biocultural knowledge as it provides an adaptive model for developing resilient and sustainable food systems that are based on the microbial exploitation of plant-based resources (Sharma et al., 2026).

12.4 Current Advances in Plant-Based Fermentation

Plant-based fermentation is recognized for their nutritional and functional benefits but also known to face several safety risks, important limitations, and regulatory barriers that somehow affect both its wider industrial adoption and commercialization (Dhiman et al., 2025). There are certain shortcomings in the process of traditional fermentation systems such as lack of process standardization (Skowron et al., 2022). These plant-based fermentation leads to variability in inconsistent production of bioactive compounds, organic compounds and flavour metabolites, texture profiles causing changes in product quality and reproducibility and safety risks (Van Wyk & Prinsloo, 2020). It has been reported that insufficient acidification resulting from microbial inconsistency may fail to reduce the pH below the critical safety threshold of approximately 4.5, thereby allowing the growth of pathogenic microorganisms such as *Listeria monocytogenes* and *Bacillus cereus* (Mohammad et al., 2020). This type of variability can cause a significant challenge to industrial scalability where regulatory compliance, predictability and uniformity become an essential part to be considered in the fermentation process cycle (Olsson et al., 2022). Microbial safety emerges as another concern when it comes to plant-based fermentation. It is reported that although the process of fermentation usually inhibits pathogens through acid production and microbial competitions but poor hygiene and improper fermentation conditions or inadequate monitoring can result in formation of harmful substances like biogenic amines like histamines and tyramine and leads to contamination (Roy et al., 2023). The formation of these compounds (Histamines and tyramines) occurs through microbial decarboxylation of amino acids which causes adverse effects such as headaches, allergic reactions and hypertension when consumed in high doses (Saha Turna et al., 2024). The concentration of these toxic substances is influenced by temperature, microbial composition and fermentation duration which highlights the need to control fermentation and ensure strict quality control measures (Zhao et al., 2024).

Additionally, the incomplete degradation of natural toxins and anti-nutritional factors further pose a risk to the environment. While fermentation is known to reduce substances like tannins, phytates and cyanogenic glycosides, the insufficient fermentation and improper removal may leave residual levels that impair nutrient absorption and pose toxicity risks to environment (Arsov et al., 2024). For instance, it is reported that insufficient fermentation of cassava may cause incomplete removal of cyanogenic substances which could release hydrogen cyanide and cause serious neurological disorders (Sambala et al., 2019). It is demonstrated that the efficiency and effectiveness of fermentation in detoxification is majorly dependent on optimum process conditions and microbial efficiency (Gomes et al., 2022).

Technological and economic constraints continue to limit the large-scale implementation of plant-based fermentation. Successful fermentation processes require precise control of key parameters, including temperature, pH, oxygen levels, and microbial population dynamics (Abbaspour, 2024). Studies have shown that maintaining precise control over fermentation conditions necessitates the use of advanced bioreactor systems, optimised starter cultures and real-time monitoring technologies, thereby increasing both operational costs and technical complexity (Mitra & Murthy, 2022). In addition, the identification, development, and validation of substrate-specific microbial starter cultures demand substantial research efforts, which may hinder the pace of industrial innovation and commercialization (Muhammed et al., 2025).

According to Abe-Inge et al. (2024), regulatory barriers represent a significant challenge to the the commercialization of fermented plant-based fermentation. It is reported that regulatory authorities such as European Food Safety Authority and Food and Agricultural Organization require scientific evidence demonstrating product stability, absence of toxic compounds and microbial safety before approval for market entry (Ackerl et al., 2023). Fermented foods that contain novel microbial strains are often regulated under novel food frameworks and must undergo comprehensive safety assessments before market approval. These regulatory procedures are frequently lengthy, costly, and resource-intensive, thereby creating barriers to commercialization (Ağagündüz et al., 2022). Studies have highlighted that stringent labeling requirements and restrictions on health claims pose significant challenges to manufacturers. Moreover, the absence of harmonized global regulatory frameworks creates obstacles in product approval processes across different countries, thereby limiting international trade, market penetration, and the global expansion of fermented food products.

Dhiman et al. (2025) reported that although plant-based fermentation offers certain benefits for improving nutrition, sustainability and food quality, its broader application is hindered by microbial variability, safety risks and technological loopholes. Evidence suggests that overcoming these barriers will require the development of standardized and well-characterized starter cultures, the implementation of advanced process-control technologies, and the establishment of clear and harmonized regulatory framework. Such measures are essential to ensure product consistency, safety, consumer confidence, and global market acceptance.

12.5 Climate Change and Food Security

Mukherjee (2021) reported that climate change poses a serious threat to global food security and further exacerbates the vulnerability of existing weak agricultural systems through both direct and compound effects. Knox et al. (2012) reported that each 1 °C increase in global mean temperature is associated with a decline of approximately 6–10% in the average yields of major staple crops, including wheat, maize, and rice.

It was further demonstrated that these impacts are not uniformly distributed across the globe but instead they reflect a more pronounced regional inequality (Ma et al., 2024). A study reported that increasing temperatures and increasingly unpredictable precipitation patterns are projected to reduce the yields of rain-fed crops, particularly in arid and semi-arid regions of Sub-Saharan Africa and South Asia, by up to 30% by the mid-twenty-first century (Knox et al., 2012). In contrast, modest yield increases have been projected for some temperate and high-latitude regions, including parts of Northern Europe and Canada, owing to extended growing seasons (Vogel et al., 2019). Ullah et al. (2022) noted that these potential gains are increasingly being offset by heightened climatic variability and the emergence of new biotic stresses, such as pests and diseases.

Zenda et al. (2022) reported that rising temperatures accelerate the plant phenological development and thereby shorten the critical growth stages specifically in grain-filled period that ultimately reduces the yield potential and biomass accumulation. It is reported that heat stress during key reproductive stages can cause severe reproductive failure and pollen sterility in cereal crops during flowering (Ullah et al., 2022). Zhang et al. (2021) reported that increasing instability in the global hydrological cycle is contributing to greater variability in precipitation patterns. The study further projected that the frequency and intensity of drought events in major global breadbasket regions could increase by more than 50% during the twenty-first century, posing a significant threat to rain-fed agricultural systems that support large populations worldwide.

Yan et al. (2022) reported that rising temperature facilitate the expansion and increase the reproductive potential of agricultural pests and pathogen. For example, it was reported that the geographical range of major crop pest *Spodoptera frugiperda* (fall armyworm) is said to expand by 20% in tropical and subtropical regions posing a serious threat to sorghum and maize production further increasing the problem of pest management. The interconnected heat-water-biotic stress nexus further highlights the global challenge to agricultural yield and productivity and underscores the urgent need for transformative adaptation strategies to enhance resilience in agro-ecological and socio-economic systems as reported by Magliocca & Gonzalez-Jimenez, 2020.

12.6 Climate-Induced Vulnerabilities in Global Food Systems

Gomez-Zavaglia et al. (2020) reported that climate -induced vulnerabilities in existing food systems pose a profound threat to plant based fermented foods such as tempeh, kimchi and sauerkraut primarily through escalated drought, flooding and extreme heatwaves and rising CO_2 levels. Escalating drought events, flooding, extreme heatwaves, and rising atmospheric CO_2 concentrations can disrupt agricultural production systems, thereby affecting the quality, yield, and availability of essential feedstocks, including soybean, lentils, and chickpeas. For example, Ackerl et al. (2023) demonstrated that elevated atmospheric CO_2 concentrations, projected to reach approximately 550 ppm by 2050, in combination with higher temperatures, can reduce starch content in cereals and legumes by 10–20%. These changes may also alter the levels of anti-nutritional compounds, including certain phenolics and flavonoids, potentially reducing microbial fermentation efficiency and adversely affecting product texture reported that synergistic drought and heat stress during critical growth period can slash the soyabean photosynthetic rates, protein accumulation and pod filling by 40%–50% directly limiting high-quality substrate for *Rhizopus oligosporus* in tempeh production leading to global supply shortage in subtropical and tropical areas.

These risks are further exacerbated by deep- rooted socioeconomic inequalities where climate change affects the low-income populations despite the fact that they have relatively low carbon footprints as reported by Dwarkasing (2023). It was found that small land holders, farmers, pastoralists and fishing communities are specifically vulnerable as they depend directly upon climate- sensitive natural resources and often have no access to advanced technology, infrastructure and financial support (Cohn et al., 2017). The Food and Agriculture Organization estimates that about 60% or more of the population in sub-Saharan region relies upon climate-sensitive livelihood thereby highlighting the extent of vulnerability and exposure (Ibe & Amikuzuno, 2019).

Gomez-Zavaglia et al. (2020) reported that another loophole is the dependence on long and centralized supply chain which are highly susceptible to climatic disruptions. It was reported that extreme weather events such and floods that cause damage to transportation infrastructure and heatwaves disrupt logistics systems can further trigger cascading effects like price volatility and post-harvest losses (Zhao et al., 2024).

These disruptions are found to increase the post- harvest losses by 30%–50% under high- emission scenarios causing further destabilization of food accessibility and availability (Hasegawa et al., 2021).

Studies have indicated that biological homogenization, socioeconomic inequality, supply chain fragility, and climate-induced nutrient dilution collectively contribute to the vulnerability of the global food system under increasing climatic uncertainty. Knox et al. (2012) highlighted that overcoming these interconnected challenges requires a paradigm shift toward diversified and resilient agricultural systems, equitable access to adaptive resources, and the integration of nutrition-sensitive and climate-resilient food production and processing strategies. Collectively, these approaches can enhance the long-term resilience, sustainability, and adaptive capacity of global food systems (Dhiman et al., 2025).

12.7 Plant-Based Fermented Foods: Types, Benefits, and Preservation

The vegetable fermented foods represent a diverse and nutritionally interesting group which are to known to be produced by the microbial transformation of plant- based materials that can be used to supplement their sensory qualities, increase shelf life, and enhance health-promoting functionality as reported by Das et al., 2016.

12.7.1 Types of Fermented Foods

According to Zhou et al. (2016), solid foods include tempeh, a fermented soybean food that has its origins in Indonesia and has been known to contain high levels of protein and a firm texture and kimchi, a fermented dish of vegetables, spiced up by various ingredients and celebrated due to its high complexity of microbial consortia and probiotic properties. Kombucha is one of the most interesting beverages; it is a fermented tea that is created by a symbiotic culture of bacteria and yeast (SCOBY) with a long-standing reputation of the generation of organic acids and the possible health-promoting effects. Plant based yogurts (fermented milk, made of coconut, soy, or oat milk, and the presence of probiotic bacteria) have grown at a very fast rate, providing dairy-free options with the characteristic acidity and texture. Even grain-based fermentations are based on fermentation, as is the case with sourdough bread, which uses lactic acid bacteria and wild yeasts to make dough light and therefore tastier and possibly healthier in nutrient content (Dhiman et al., 2025).

12.7.2 Nutritional Benefits

Fermentation is a major contributor to the nutritional value of plant-based foods. One of the major advantages of fermentation is the improved bioavailability of micronutrients such as iron and zinc, which are made more accessible by the degradation of antinutrients such as phytates (Fig. 12.2) (Dhiman et al., 2025). In

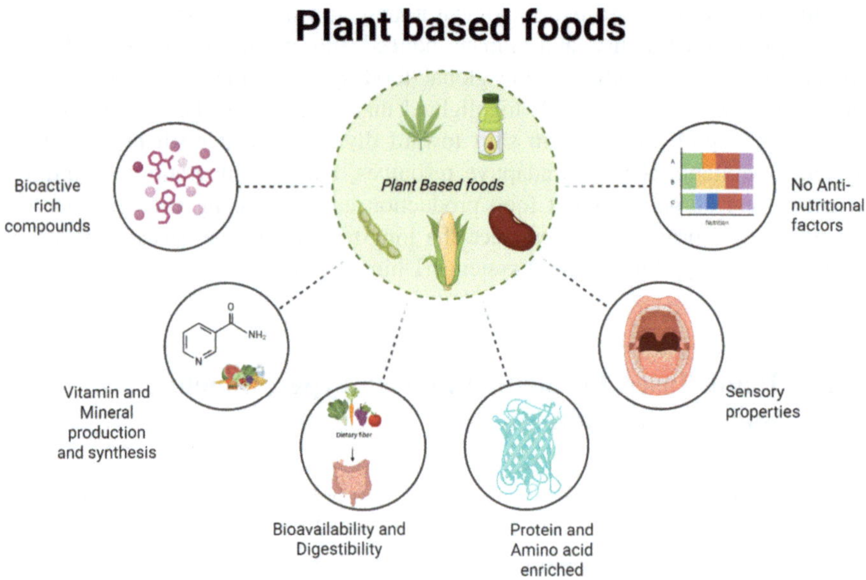

Fig. 12.2 Use of plant based-fermented foods and their significance

addition, fermentation may lead to an increase in the levels of certain vitamins B, such as folate and riboflavin, which are produced through microbial fermentation (Abbaspour, 2024). It also leads to improved protein digestibility, which is achieved through the conversion of complex proteins to peptides and free amino acids (Zhang et al., 2021). Microbial fermentation also leads to the production of other bioactive compounds such as antioxidants and short-chain fatty acids, which may have health-promoting properties in addition to their nutritional value (Dhiman et al., 2025).

12.7.3 Environmental Benefits

Plant-based fermentation practices are in line with the principles of sustainable food systems. Life cycle assessment studies have shown that plant-based production systems, including fermentation, tend to have lower land, water, and greenhouse gas emissions than their animal-based counterparts. Fermentation, in particular, can add value to local crops that are not well utilized, and this process can help promote diverse and low-input agricultural systems, which have lower environmental impacts and higher ecosystem resilience.

12.7.4 Post-Harvest Preservation

Fermentation is a low-cost and efficient post-harvest preservation technique that is essential for minimizing food loss and improving supply chain stability, especially in areas with limited refrigeration capacity (Sharma et al., 2026). By the selective formation of a microenvironment rich in organic acids and other antimicrobial substances, fermentation is an effective way to suppress the growth of the spoilage and pathogenic microorganisms. Because of this, perishable sources like cabbage are transformed into steadier items, like sauerkraut as well as kimchi (Dhiman et al., 2025). This will be necessary in order to prolong shelf life, minimize waste and have a stable food supply channel, based on seasonal harvests (Chang & Chang, 2011).

Plant-based fermentation increases food diversity, improves nutritional quality through biochemical conversion, reduces the environmental impact of food production, and offers a reliable, low-tech preservation strategy for agricultural harvests (Abbaspour, 2024). Collectively, these factors make it an essential tool in the construction of sustainable and resilient global food systems.

12.8 Research Gaps and Future Directions

Although the promise is great, the actualization of plant fermentation in the sustainable food system agenda must fill important interdisciplinary research gaps (Dhiman et al., 2025).

12.8.1 Scientific Gaps

There are important knowledge gaps in the mechanistic foundations and long-term consequences of fermentation (Dhiman et al., 2025). It was demonstrated that although the general ability of fermentation to improve the bioavailability of nutrients is recognized, the metabolic mechanisms and the particular interactions between microbial communities and plant materials are not well understood (Yang et al., 2020). There is a large gap in the knowledge of the long-term health consequences of the consumption of new fermented foods, especially those produced by microbial strains engineered for this purpose or those produced from non-traditional plant materials (Dhiman et al., 2025). In addition, there are few life cycle assessments of the environmental impacts of fermentation processes, from plant growth to waste streams, that would allow a precise assessment of the sustainability of the process (Meena et al., 2025).

12.8.2 Technological Gaps

Despite the growing potential of plant-based fermented foods to support climate resilience, food security, and sustainable agriculture, several technological gaps continue to limit their large-scale adoption. According to the transition of the artisanal, ad hoc fermentations to the systematic and large-scale manufacturing paradigms also brings with it some significant technological challenges as can be found in the literature (Bisht et al., 2018). A key technological gap in plant-based fermentation is the reliance spontaneous fermentations processes, which are inherently variable and difficult to control. Such variability can affect product quality and safety and may lead to the formation of undesirable metabolites, including biogenic amines, thereby limiting process standardization and industrial scalability (Ruane et al., 2024). Many traditional fermentation processes rely on spontaneous microbial communities, resulting in variability in product quality, safety, and functional attributes. The lack of standardized starter cultures, real-time process monitoring systems, and optimized fermentation technologies further constrains industrial scalability and commercial viability. In addition, limited understanding of microbe–substrate interactions across diverse plant materials hinders process optimization and product innovation. Bioprocess engineering should be aiming at improving the economic viability as well as the energy profile of fermentation processes especially in the solid-state systems so that they can be able to compete with traditional staples at competitive prices (Yee et al., 2025). Addressing these technological challenges through advances in microbial biotechnology, bioprocess engineering, digital technologies, like artificial intelligence and digital process control will be essential for ensuring consistent, safe, and sustainable production of plant-based fermented foods (Dhiman et al., 2025).

12.8.3 Socioeconomic Gaps

Despite their potential to enhance food security and sustainability, the widespread adoption of plant-based fermented foods is constrained by several socioeconomic challenges. The transition from artisanal and small scale fermentation practices to systematic and industrial-scale manufacturing represents a critical step toward commercialization; however, the literature highlights several technological challenges that must be overcome to achieve consistent product quality, process reliability, and economic viability (Knox et al., 2012). Natural fluctuations of spontaneous fermentations, accompanied by the corresponding safety risks, such as the potential production of biogenic amines, are still a major challenge (Ruane et al., 2024). According to Cohn et al. (2017), the development of robust, locally adapted starter cultures tailored to regional crop conditions is essential for ensuring product quality and safety; however, achieving this objective remains inherently challenging. In addition, bioprocess engineering efforts should focus on improving both the economic viability and energy efficiency of fermentation processes, particularly in solid-state fermentation systems, to enhance their competitiveness with conventional food products and staple foods (Taiwo et al., 2023). The next significant technological frontier is the amalgamation of digital technologies, like artificial intelligence to optimise processes, and blockchain to track the supply chain, into these historically low-tech practices (Roy et al., 2025). Limited consumer awareness, cultural preferences, and misconceptions regarding the taste, safety, and nutritional value of fermented products can reduce market acceptance. In many low- and middle-income regions, inadequate access to infrastructure, processing facilities, financing, and technical expertise further restricts the commercialization and scaling of fermentation-based enterprises. Additionally, unequal access to resources, markets, and value chains may prevent smallholder farmers and local producers from fully benefiting from fermentation technologies. Addressing these socioeconomic gaps is essential to ensure the equitable adoption and long-term sustainability of plant-based fermented food systems. The development of new economic models that promote small-scale, localized fermented food producers, especially women-owned enterprises in developing countries, is essential for ensuring equitable benefits and community resilience.

12.8.4 Policy and Regulatory Gaps

It was reported that the current policy environment is not suited for fermentation innovation, especially for genetically modified microorganisms or new food ingredients (Nascimento & Barros, 2025). Knox et al. (2012) reported that current policies are often disjointed, process-oriented rather than product-oriented, and time-consuming, making it difficult for investors and producers to navigate. There is a pressing need for harmonized, science-based, and internationally aligned policies that ensure safety without inhibiting innovation. It was reported that policies should promote sustainable practices such as the upcycling of agricultural by-products and research on traditional fermented foods as a form of biocultural heritage (Zhang et al., 2024). Dhiman et al. (2025) reported that future research should focus on community-based research collaborations to preserve indigenous knowledge while applying modern microbiology. Public-private partnerships to mitigate the risks of and scale up affordable technologies and the development of policy frameworks that are participatory and address trade-offs of safety, innovation, and equity (Nascimento & Barros, 2025). By systematically closing these gaps, plant-based fermentation can move from a promising niche to a foundation of resilient, healthy, and sustainable global food systems (Abbaspour, 2024).

12.9 Conclusion

Plant-based fermented foods offer a holistic solution to the challenges of food security and sustainable agriculture by integrating traditional knowledge with modern technological innovations (Abbaspour, 2024). These foods enhance nutrient bioavailability and reduce post-harvest losses, and facilitate the valorization of agricultural waste into useful resources, thereby contributing to more resilient and sustainable food systems. In addition, plant-based fermented foods can support the utilization of climate-resilient crops and strengthen food system sustainability.

Owing to their relatively low energy requirements and reduced environmental footprint compared with conventional animal based and highly processed foods, they also offer potential benefits for climate change mitigation (Ruane et al., 2024). The accumulated evidence indicates an urgent need for coordinated action by researchers, policymakers, and practitioners. Future research should focus on three key areas: evaluating the long-term health effects of fermented foods, developing fermentation technologies suitable for resource-constrained settings, and comprehensively assessing their environmental impacts (Rastogi et al., 2022). The policymakers must develop safety-focused regulation frameworks that facilitate the development of innovation, provide financial assistance program aimed at enabling farmers and producers to engage in sustainable practices at scale (Zhang et al., 2024). Farmers, food processors, nutritionists, and other practitioners must work collaboratively to develop plant-based fermented food products that meet diverse cultural preferences while ensuring an equitable and inclusive transition toward sustainable food systems (Abbaspour, 2024). The advancement of Plant-based fermentation processes requires close collaboration among farmers, food processors, nutritionists, and other stakeholders to develop culturally appropriate products that reflect diverse food traditions, consumer preferences, and dietary patterns. Such efforts are essential for enhancing consumer acceptance and facilitating the widespread adoption of fermented plant-based food (Abbaspour, 2024). Ultimately, the full potential of the plant-based fermentation can be unlocked only through coordinated and interdisciplinary efforts involving researchers, policymakers, industry stakeholders, farmers, and food-system practitioners. By integrating scientific innovation, supportive policies, and sustainable production practices, plant-based fermentation has the potential to become a cornerstone of resilient, sustainable, and nutritionally secure food systems for the future.

References

Abbaspour, N. (2024). Fermentation's pivotal role in shaping the future of plant-based foods: An integrative review of fermentation processes and their impact on sensory and health benefits. *Applied Food Research, 4*(2), 100468. https://doi.org/10.1016/j.afres.2024.100468

Abe-Inge, V., Aidoo, R., Moncada De La Fuente, M., & Kwofie, E. M. (2024). Plant-based dietary shift: Current trends, barriers, and carriers. *Trends in Food Science & Technology, 143*, 104292. https://doi.org/10.1016/j.tifs.2023.104292

Ackerl, R., Mendes, V., Germini, A., McArdle, H. J., Neuhäuser-Berthold, M., Roldán-Torres, R., & Ververis, E. (2023). Assessment of protein quality in novel foods by the European food safety authority: Methodology and challenges. In *The 14th European Nutrition Conference FENS* (p. 44). MDPI. https://doi.org/10.3390/proceedings2023091044

Ağagündüz, D., Yılmaz, B., Koçak, T., Altıntaş Başar, H., Rocha, J., & Özoğul, F. (2022). Novel candidate microorganisms for fermentation technology: From potential benefits to safety issues. *Food, 11*(19), 3074. https://doi.org/10.3390/foods11193074

Agrawal, D., Awani, K., Nabavi, S. A., Balan, V., Jin, M., Aminabhavi, T. M., Dubey, K. K., & Kumar, V. (2023). Carbon emissions and decarbonisation: The role and relevance of fermentation industry in chemical sector. *Chemical Engineering Journal, 475*, 146308. https://doi.org/10.1016/j.cej.2023.146308

Ahn, S., Darooghegi Mofrad, M., Nosal, B. M., Chun, O. K., & Joung, H. (2025). Effects of fermented kimchi consumption on anthropometric and blood cardiometabolic indicators: A systematic review and meta-analysis of intervention studies and prospective cohort studies. *Nutrition Reviews, 83*(7), e1441–e1457. https://doi.org/10.1093/nutrit/nuae167

Arsov, A., Tsigoriyna, L., Batovska, D., Armenova, N., Mu, W., Zhang, W., Petrov, K., & Petrova, P. (2024). Bacterial degradation of antinutrients in foods: The genomic insight. *Food, 13*(15), 2408. https://doi.org/10.3390/foods13152408

Ashoka, P., Raut, D., Sudeepthi, B., Gawande, K. N., Reddy, G. S. V., Padhan, S. R., & Panigrahi, C. K. (2023). Millet's role as a climate resilient staple for future food security: A review. *International journal of environment and climate change, 13*(11), 4542–4552. https://doi.org/10.9734/IJECC/2023/v13i113634

Azri, R. N., Panggabean, A., Dwi, C. P., Fadilla, A., Balqis, Z. S., & Pakpahan, E. H. (2026). The role of rhizopus oligosporus in enhancing protein and mineral bioavailability in tempeh: A literature review. *Multidisciplinary Scientific Journal, 5*(01), 51–56. https://doi.org/10.56127/jukim.v5i01.2506

Bisht, I., Mehta, P., Verma, S., & Negi, K. (2018). Traditional land and food systems: A case of Uttarakhand state in north-western indian himalayas. *Indian Journal of Plant Genetic Resources, 31*(3), 215. https://doi.org/10.5958/0976-1926.2018.00026.8

Boukid, F., Hassoun, A., Zouari, A., Tülbek, M., Mefleh, M., Aït-Kaddour, A., & Castellari, M. (2023). Fermentation for designing innovative plant-based meat and dairy alternatives. *Food, 12*(5), 1005. https://doi.org/10.3390/foods12051005

Chang, J. Y., & Chang, H. C. (2011). Growth inhibition of foodborne pathogens by kimchi prepared with bacteriocin-producing starter culture. *Journal of Food Science, 76*(1), M72–M78. https://doi.org/10.1111/j.1750-3841.2010.01965.x

Cohn, A. S., Newton, P., Gil, J. D. B., Kuhl, L., Samberg, L., Ricciardi, V., Manly, J. R., & Northrop, S. (2017). Smallholder agriculture and climate change. *Annual Review of Environment and Resources, 42*(1), 347–375. https://doi.org/10.1146/annurev-environ-102016-060946

Contreras-Negrete, G., Valiente-Banuet, A., Molina-Freaner, F., Partida-Martínez, L. P., & Hernández-López, A. (2024). Agricultural practices and environmental factors drive microbial communities in the mezcal-producing agave angustifolia haw. *Microbial Ecology, 87*(1), 181. https://doi.org/10.1007/s00248-025-02496-2

D'Almeida, A. P., Neta, A. A. I., De Andrade-Lima, M., & De Albuquerque, T. L. (2024). Plant-based probiotic foods: Current state and future trends. *Food Science and Biotechnology, 33*(15), 3401–3422. https://doi.org/10.1007/s10068-024-01674-1

Dabiré, Y., Somda, N. S., Somda, M. K., Mogmenga, I., Traoré, A. K., Ezeogu, L. I., Traoré, A. S., Ugwuanyi, J. O., & Dicko, M. H. (2022). Molecular identification and safety assessment of Bacillus strains isolated from Burkinabe traditional condiment "soumbala". *Annals of Microbiology, 72*(1), 10. https://doi.org/10.1186/s13213-022-01664-w

Das, R., Pandey, H., Das, B., & Sarkar, S. (2016). Fermentation and its application in vegetable preservation: A review. *International Journal of Food and Fermentation Technology, 6*(2), 207. https://doi.org/10.5958/2277-9396.2016.00044.1

De Bellis, P., Sisto, A., & Lavermicocca, P. (2021). Probiotic bacteria and plant-based matrices: An association with improved health-promoting features. *Journal of Functional Foods, 87*, 104821. https://doi.org/10.1016/j.jff.2021.104821

Dhiman, S., Kaur, S., Thakur, B., Singh, P., & Tripathi, M. (2025). Nutritional enhancement of plant-based fermented foods: Microbial innovations for a sustainable future. *Fermentation, 11*(6), 346. https://doi.org/10.3390/fermentation11060346

Dwarkasing, C. (2023). Inequality determined social outcomes of low-carbon transition policies: A conceptual meta-review of justice impacts. *Energy Research & Social Science, 97*, 102974. https://doi.org/10.1016/j.erss.2023.102974

Gomes, M. A., Santos-Rocha, M. S. R. D., Barbosa, K. L., Vieira, R. C., Antunes, D. P. C., Silva, C. E. D. F., Almeida, R. M. R. G., & Albuquerque, E. C. D. M. C. (2022). The fermentation efficiency exhibited by *Saccharomyces cerevisiae* on sugarcane bagasse hydrolysate, by analyzing the effects of pre-treatment and detoxification. *Semina: Ciências Agrárias, 43*(5), 2155–2170. https://doi.org/10.5433/1679-0359.2022v43n5p2155

Gomez-Zavaglia, A., Mejuto, J. C., & Simal-Gandara, J. (2020). Mitigation of emerging implications of climate change on food production systems. *Food Research International, 134*, 109256. https://doi.org/10.1016/j.foodres.2020.109256

Gormaz, J. G., Rodrigo, R., Videla, L. A., & Beems, M. (2010). Biosynthesis and bioavailability of long-chain polyunsaturated fatty acids in non-alcoholic fatty liver disease. *Progress in Lipid Research, 49*(4), 407–419. https://doi.org/10.1016/j.plipres.2010.05.003

Hasegawa, T., Sakurai, G., Fujimori, S., Takahashi, K., Hijioka, Y., & Masui, T. (2021). Extreme climate events increase risk of global food insecurity and adaptation needs. *Nature Food, 2*(8), 587–595. https://doi.org/10.1038/s43016-021-00335-4

Huang, Y. (2017). Fermented food and ancient civilization. *DEStech Transactions on Social Science, Education and Human Science, (ICSS)*. https://doi.org/10.12783/dtssehs/icss2016/9181

Ibe, G. O., & Amikuzuno, J. (2019). Climate change in sub-Saharan Africa: A menace to agricultural productivity and ecological protection. *Journal of Applied Sciences and Environmental Management, 23*(2), 329. https://doi.org/10.4314/jasem.v23i2.20

Ibrahim, S. A., Yeboah, P. J., Ayivi, R. D., Eddin, A. S., Wijemanna, N. D., Paidari, S., & Bakhshayesh, R. V. (2023). A review and comparative perspective on health benefits of probiotic and fermented foods. *International Journal of Food Science & Technology, 58*(10), 4948–4964. https://doi.org/10.1111/ijfs.16619

Knox, J., Hess, T., Daccache, A., & Wheeler, T. (2012). Climate change impacts on crop productivity in Africa and South Asia. *Environmental Research Letters, 7*(3), 034032. https://doi.org/10.1088/1748-9326/7/3/034032

Kumar, L., Naresh, R. K., Tiwari, H., Kataria, S. K., Saharan, S., Reddy, B. R., Singh, O., Qidwai, S., & Singh, R. P. (2022). Millets for food and nutritional security in the context of climate resilient agriculture: A review. *International Journal of Plant & Soil Science, 34*(23), 939–953. https://doi.org/10.9734/ijpss/2022/v34i232504

Lys, I. M. (2025). The role of lactic fermentation in ensuring the safety and extending the shelf life of African indigenous vegetables and its economic potential. *Applied Research, 4*(1), e202400131. https://doi.org/10.1002/appl.202400131

Ma, L., Wei, T., Nordling, K., & Aaheim, A. (2024). Impacts of global warming on regional energy and economy: Uneven consequences arising from global warming-induced heating and cooling demand of households. *Climate Change Economics, 15*(02), 2450005. https://doi.org/10.1142/S2010007824500052

Magliocca, N. R., & Gonzalez-Jimenez, E. (2020). Costa Rica's water paradox: Linking rainforests and droughts through the water-energy-food-environment nexus in Guana caste province. *Case Studies in the Environment, 4*(1), 1–9. https://doi.org/10.1525/cse.2019.002253

Meena, P. K., Kumar, D., Sharma, S., Didwania, M., & Singh, L. (2025). Advancing sustainable energy: Biohydrogen, biogas, and biohythane production from waste materials through life cycle analysis. *Biofuels, 16*(7), 789–805. https://doi.org/10.1080/17597269.2025.2455813

Mitra, S., & Murthy, G. S. (2022). Bioreactor control systems in the biopharmaceutical industry: A critical perspective. *Systems Microbiology and Biomanufacturing, 2*(1), 91–112. https://doi.org/10.1007/s43393-021-00048-6

Mohammad, Z. H., Payton, L., & Sirsat, S. A. (2020). Efficacy of sushi rice acidification: Quantification of *Bacillus cereus* and *Clostridium perfringens* during simulation of retail practices. *LWT - Food Science and Technology, 131*, 109884. https://doi.org/10.1016/j.lwt.2020.109884

Muhammed, Y. M. R., Minervini, F., & Cavoski, I. (2025). From ancient fermentations to modern biotechnology: Historical evolution, microbial mechanisms, and the role of natural and commercial starter cultures in shaping organic and sustainable food systems. *Food, 14*(24), 4240. https://doi.org/10.3390/foods14244240

Mukherjee, D. (2021). Food security under the era of climate change threat. *Journal of Advanced Agriculture & Horticulture Research, 1*(1), 1–4. https://doi.org/10.55124/jahr.v1i1.78

Nascimento, A. P. S., & Barros, A. N. (2025). Sustainable innovations in food microbiology: Fermentation, biocontrol, and functional foods. *Food, 14*(13), 2320. https://doi.org/10.3390/foods14132320

Olsson, L., Rugbjerg, P., Torello Pianale, L., & Trivellin, C. (2022). Robustness: Linking strain design to viable bioprocesses. *Trends in Biotechnology, 40*(8), 918–931. https://doi.org/10.1016/j.tibtech.2022.01.004

Ortiz-Sanchez, M., Inocencio-García, P.-J., Alzate-Ramírez, A. F., & Alzate, C. A. C. (2023). Potential and restrictions of food-waste valorization through fermentation processes. *Fermentation, 9*(3), 274. https://doi.org/10.3390/fermentation9030274

Rai, A. K., & Appaiah, K. A. (2021). *Bioactive compounds in fermented foods: Health aspects* (1st ed.). CRC Press. https://doi.org/10.1201/9780429027413

Rastogi, Y. R., Thakur, R., Thakur, P., Mittal, A., Chakrabarti, S., Siwal, S. S., Thakur, V. K., Saini, R. V., & Saini, A. K. (2022). Food fermentation–Significance to public health and sustainability challenges of modern diet and food systems. *International Journal of Food Microbiology, 371*, 109666. https://doi.org/10.1016/j.ijfoodmicro.2022.109666

Roy, S., Shaw, D., Sarkar, T., & Choudhury, L. (2023). Mycotoxins in fermented foods: A comprehensive review. *Novel Research in Microbiology Journal, 7*(2), 1897–1917. https://doi.org/10.21608/nrmj.2023.293196

Roy, P., Kumar, D., & Kumar, K. (2025). Harnessing AI and emerging technologies for sustainable food systems: Innovations in automation and intelligent production. *Asian Journal of Research in Computer Science, 18*(4), 115–135. https://doi.org/10.9734/ajrcos/2025/v18i4611

Ruane, A. C., Phillips, M., Jägermeyr, J., & Müller, C. (2024). Non-linear climate change impacts on crop yields may mislead stakeholders. *Earth's Future, 12*(4), e2023EF003842. https://doi.org/10.1029/2023EF003842

Saha Turna, N., Chung, R., & McIntyre, L. (2024). A review of biogenic amines in fermented foods: Occurrence and health effects. *Heliyon, 10*(2), e24501. https://doi.org/10.1016/j.heliyon.2024.e24501

Sambala, E. Z., Ndwandwe, D. E., Imaan, L. M., & Wiysonge, C. S. (2019). Evaluation of influenza surveillance systems in sub-Saharan Africa: A systematic review protocol. *BMJ Open, 9*(1), e023335. https://doi.org/10.1136/bmjopen-2018-023335

Sharma, R. S., Manjkhola, I., Bharati, V., Sharma, A., Sukhralia, M., Chishty, S., & Mishra, V. (2026). Traditional socio-cultural practices in Himalayan food systems for climate adaptation. *Frontiers in Sustainable Food Systems, 9*, 1718684. https://doi.org/10.3389/fsufs.2025.1718684

Singh, K. A., Saroch, A., & Rajput, R. (2023). Microbial diversity of fermented foods. *Current Journal of Applied Science and Technology, 42*(47), 20–31. https://doi.org/10.9734/cjast/2023/v42i474313

Sionek, B., Szydłowska, A., Küçükgöz, K., & Kołożyn-Krajewska, D. (2023). Traditional and new microorganisms in lactic acid fermentation of food. *Fermentation, 9*(12), 1019. https://doi.org/10.3390/fermentation9121019

Skowron, K., Budzyńska, A., Grudlewska-Buda, K., Wiktorczyk-Kapischke, N., Andrzejewska, M., Wałecka-Zacharska, E., & Gospodarek-Komkowska, E. (2022). Two faces of fermented foods—The benefits and threats of its consumption. *Frontiers in Microbiology, 13*, 845166. https://doi.org/10.3389/fmicb.2022.845166

Taiwo, A. E., Tom-James, A., & Musonge, P. (2023). Economic assessment of cellulase production in batch and semi-batch solid-state fermentation processes. *International Journal of Low-Carbon Technologies, 18*, 204–211. https://doi.org/10.1093/ijlct/ctad009

Thorat, P. P., Sadawarte, S. K., Sawate, A. R., & Machewad, G. M. (2017). Studies on effect of fermentation on physicochemical properties of vegetables and preparation of sauce. *International Journal of Current Microbiology and Applied Sciences, 6*(8), 3537–3545. https://doi.org/10.20546/ijcmas.2017.608.423

Ullah, A., Nadeem, F., Nawaz, A., Siddique, K. H. M., & Farooq, M. (2022). Heat stress effects on the reproductive physiology and yield of wheat. *Journal of Agronomy and Crop Science, 208*(1), 1–17. https://doi.org/10.1111/jac.12572

Van Wyk, A. S., & Prinsloo, G. (2020). Health, safety and quality concerns of plant-based traditional medicines and herbal remedies. *South African Journal of Botany, 133*, 54–62. https://doi.org/10.1016/j.sajb.2020.06.031

Vogel, E., Donat, M. G., Alexander, L. V., Meinshausen, M., Ray, D. K., Karoly, D., Meinshausen, N., & Frieler, K. (2019). The effects of climate extremes on global agricultural yields. *Environmental Research Letters, 14*(5), 054010. https://doi.org/10.1088/1748-9326/ab154b

Yan, X.-R., Wang, Z.-Y., Feng, S.-Q., Zhao, Z.-H., & Li, Z.-H. (2022). Impact of temperature change on the fall armyworm, *Spodoptera frugiperda* under global climate change. *Insects, 13*(11), 981. https://doi.org/10.3390/insects13110981

Yang, B., Dong, Y., Wang, F., & Zhang, Y. (2020). Nanoformulations to enhance the bioavailability and physiological functions of polyphenols. *Molecules, 25*(20), 4613. https://doi.org/10.3390/molecules25204613

Yee, C. S., Zahia-Azizan, N. A., Abd Rahim, M. H., Mohd Zaini, N. A., Raja-Razali, R. B., Ushidee-Radzi, M. A., & Wan-Mohtar, W. A. A. Q. I. (2025). Smart fermentation technologies: Microbial process control in traditional fermented foods. *Fermentation, 11*(6), 323. https://doi.org/10.3390/fermentation11060323

Zenda, T., Wang, N., Dong, A., Zhou, Y., & Duan, H. (2022). Reproductive-stage heat stress in cereals: Impact, plant responses and strategies for tolerance improvement. *International Journal of Molecular Sciences, 23*(13), 6929. https://doi.org/10.3390/ijms23136929

Zhang, W., Furtado, K., Wu, P., Zhou, T., Chadwick, R., Marzin, C., Rostron, J., & Sexton, D. (2021). Increasing precipitation variability on daily-to-multiyear time scales in a warmer world. *Science Advances, 7*(31), eabf8021. https://doi.org/10.1126/sciadv.abf8021

Zhang, C., Wu, X., Chen, J., & Zhou, J. (2024). Novel fungal alternative proteins from penicillium limosum for enhancing structural and functional properties of plant-based meat analogues. *Food Chemistry, 444*, 138627. https://doi.org/10.1016/j.foodchem.2024.138627

Zhao, A. P., Li, S., Xie, D., Hu, P. J.-H., Wu, C., Fei, F. X., Li, T. T., Xiang, Y., Gu, C., & Cao, Z. (2024). Extreme events threat water-energy-carbon nexus through cascading effects. *Next Energy, 5*, 100151. https://doi.org/10.1016/j.nxener.2024.100151

Zhou, S., Xu, R., He, F., Zhou, J., Wang, Y., Zhou, J., Wang, M., & Zhou, W. (2016). Diversity of gut microbiota metabolic pathways in 10 pairs of Chinese infant twins. *PLoS One, 11*(9), e0161627. https://doi.org/10.1371/journal.pone.0161627

Chapter 13
Zero-Waste Fermentation—Up-Cycling Food Waste into High-Value Products

Hulya Cakmak, Ece Sogut, and Gamze Duven

Abstract One third of the total food produced globally —approximately 1.3 billion tons per year is lost or wasted, with nearly half of all fruit and vegetables being squandered. The loss of still-edible food is only the tip of the iceberg, as vital resources such as water, land, energy, labour, and capital are also needlessly consumed. Food waste, rich in carbon and biological value has conventionally been disposed of in landfills, but emerging technologies now enable up-cycling these wastes into valuable bio-based materials. Among other biotransformation methods, fermentation stands out by allowing precise control of the process and unlocking the potential of producing a wide range of economically viable end products. Drawing from recent advances, including combined bioprocessing, solid-state fermentation, submerged fermentation and integrated biorefinery approaches, this chapter reviews how microbial pathways can convert food waste into enzymes, organic acids, biofuels, bioplastics, and nutraceuticals. However, there are several bottlenecks which need to be eliminated including optimization and scaling-up of the processes, mitigation of possible contaminants or risks associated with the wastes, and enhancement of the product yield and purity. Zero-waste fermentation plays a key role in advancing circular economy principles by exploring the classification and characteristics of different food waste types, alongside various

H. Cakmak (✉)
Izmir Katip Celebi University, Faculty of Tourism, Department of Tourism and Gastronomy Management Programs, Izmir, Turkey

Hitit University, Faculty of Engineering and Natural Sciences, Department of Food Engineering, Corum, Turkey
e-mail: hulya.cakmak@ikc.edu.tr

E. Sogut
Suleyman Demirel University, Faculty of Engineering and Natural Sciences, Department of Food Engineering, Isparta, Turkey
e-mail: ececagdas@sdu.edu.tr

G. Duven
Bursa Uludag University, Karacabey Vocational School, Milk and Dairy Technology Program, Bursa, Turkey
e-mail: gamzeduven@uludag.edu.tr

S. Dhiman, S. Kaur (eds.), *Next-Gen Plant Based Fermented Foods: Innovations for a Sustainable Future*, https://doi.org/10.1007/978-3-032-29927-7_13

fermentation techniques used in waste valorisation presenting a transformative opportunity to shift food waste management from a cost-driven environmental liability into a value-generating pillar of sustainable bioproduction.

Keywords Zero-waste fermentation · Food waste valorisation · Biorefinery · Circular bioeconomy · Solid-state fermentation · Bioplastics · Organic acids

13.1 Introduction

Food losses can occur during harvesting and post-harvesting, transporting, processing, retailing, and household consumption, whereas inappropriate storage conditions or overbuying/impulse purchases may lead to insurmountable waste (Bilal et al., 2022). This is not only related to the still edible food being lost, but the water, land, energy, labour and capital to produce this food are also eliminated (UN., 2025). To tackle these wastes, additional burdens on global warming and climate change occur with increased greenhouse gas emissions. Every year 1.3 billion tons of food is lost or wasted, corresponding to nearly 30% of the produce for human consumption (FAO, 2015). Thus, every actor in the food supply chain should take responsibility and improve their responses individually or collectively to minimize food waste and provide food security.

Landfilling, composting, utilization as animal feed, and incineration are conventional methods of food and agricultural waste management. However, agro-industrial and food wastes have significant potential as promising biomass with its carbon and organic material rich composition to be converted into economically viable products—such as biofuels, enzymes, vitamin supplements, antioxidants, livestock feed, antibiotics, biofertilizers, and other compounds—through solid-state fermentation (SSF) (Bibi et al., 2023).

Food and agricultural wastes or by-products that are utilized as biomass can be collected from different sources including households, restaurants, hotels, university canteens, and industrial facilities. However, its composition may vary greatly, which might alter proliferation of microorganisms and reduce product yield (Bilal et al., 2022; Pan et al., 2025). Synthetic food waste models that provide a homogeneous distribution of nutrients have been developed to overcome the heterogeneous structure of naturally occurring waste, thereby boosting microbial activity during fermentation (Braguglia et al., 2018; Bühlmann et al., 2022).

Biorefineries converting biomass for energy harvesting reduce energy costs and greenhouse gas emissions (GHG) compared to traditional energy production plants. Bioprocesses that employ food waste as a substrate within the biorefinery framework are generally classified into four categories based on the final product: (i) bioenergy production (e.g. bioethanol, biodiesel or biohydrogen), (ii) high value-added bioproducts (i.e. bio-oil, wax, bioplastics, lactic acid, tri- and di-carboxylic acids, biosurfactants, or biolubricants); (iii) isolated or extracted compounds with significant applications in pharmaceutical and food industries, and (iv) other products, including those used in enzymatic bioremediation, nanoparticles or foam production (Carmona-Cabello et al., 2018).

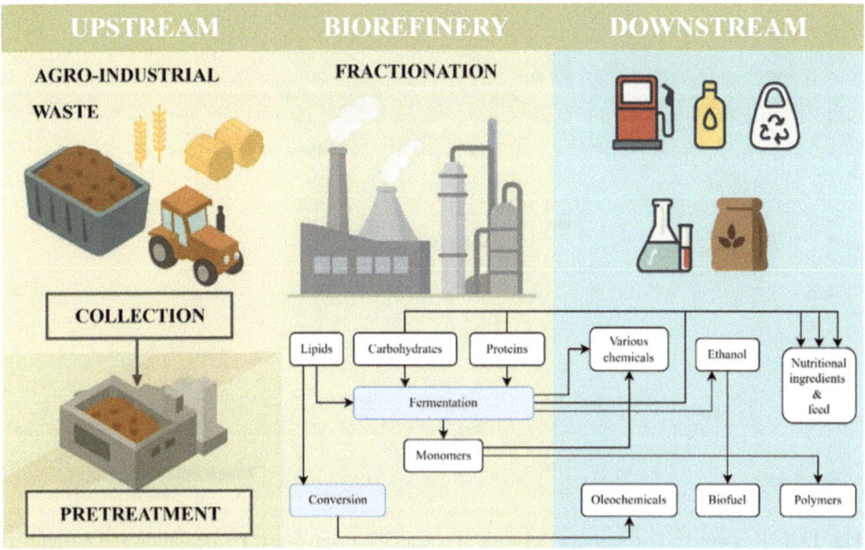

Fig. 13.1 A schematic representation of a zero-waste biorefinery concept

Waste and by-product streams from existing industrial sectors (e.g., food industry, pulp and paper industry, biodiesel and bioethanol production) can serve as renewable resources for both biorefinery development and the production of nutrient-complete fermentation feedstocks. In a zero-waste biorefinery, the entire processing chain, from upstream pre-treatment to microbial fermentation and downstream recovery, is designed to achieve complete resource valorisation and circular material flow (Fig. 13.1). However, this term should be used with caution; the phrase "towards zero-waste valorisation" is often more accurate than "zero-waste valorisation", as many studies claim zero-waste approach still generates some residual waste streams (Taheri & Hosseini, 2025).

The upstream operations are the first steps including physical, chemical, or enzymatic hydrolysis treatment of agricultural and food residues to produce fermentable sugars and other essential nutrients. These produced intermediates can act as substrates in the fermentation stage, which is the metabolic center of the system. These substrates are converted into high-value bioproducts including organic acids, solvents, enzymes, and biodegradable polymers by various microorganisms via optimized microbial pathways (Moodley & Trois, 2022). The inclusion of feedback loop links in each stage provides useful feedstock for subsequent cycles, while reducing landfill accumulation, energy losses, and emissions (Moodley & Trois, 2022). Zero-waste fermentation can stand as a self-sustaining biorefinery model strengthened by metabolic engineering, co-culture systems, and process integration of waste reduction, renewable resource utilization, and sustainable product synthesis within a single, closed-loop industrial ecosystem.

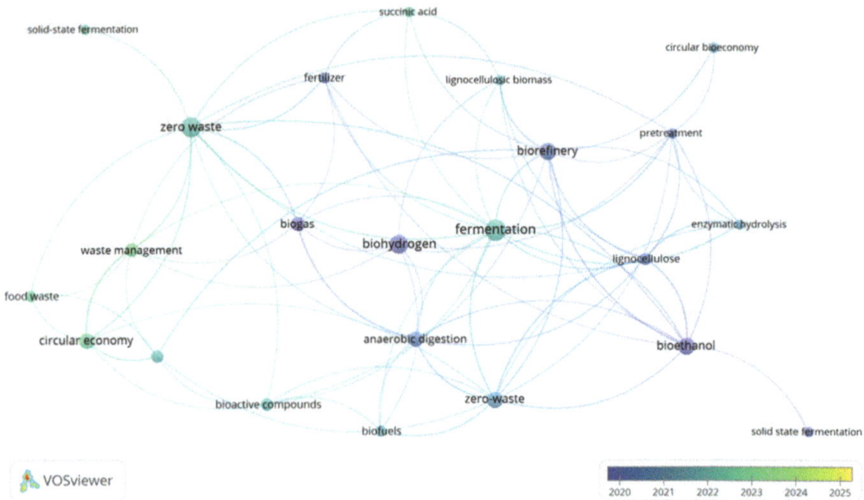

Fig. 13.2 Keyword co-occurrence network of research on zero-waste fermentation and biorefinery (2020–2025). (This visual map represents the co-occurrence relationships among frequently used keywords in the field of zero-waste fermentation based on bibliometric analysis of Web of Science Core Collection publications from 2020 to 2025. The network was constructed using VOSviewer software)

Figure 13.2 illustrates the bibliometric network associated with zero-waste biorefinery and fermentation-based valorization, capturing the field's shifting focus. The tight connection between terms likes "zero-waste," "fermentation," "biorefinery," and "circular economy" points to the growing role of microbial conversion in sustainable production frameworks. While established concepts such as anaerobic digestion and bioethanol continues to connect energy recovery with waste reduction, newer nodes including bioactive compounds and solid-state fermentation that is a reflection of a move toward the diversification of value-added products. A clear trend between 2020 and 2025, demonstrates a shift from traditional biofuels to improved circular bioeconomy and multi-output fermentation methods. Consequently, future research efforts must focus on bridging the gap between mature energy recovery protocols and emerging high-value product extraction to achieve a true circular zero-waste model.

13.2 Bioproducts Derived from Fermentation of Food Waste

13.2.1 Enzymes

Food and agricultural wastes and by-products are valuable substrates for producing bio-based materials such as enzymes, organic acids, single-cell proteins and microbial biomass, functional bioactive compounds, bioplastics, and biofuels (Ray et al.,

2025). These materials have potential applications across various industries including food, chemical, textile, paper and pulp, pharmaceutical and energy sectors. Among the biomaterials produced through microbial fermentation, enzymes such as cellulase, xylanase, amylase, lipase, protease, and pectinase are the most commercially significant examples (Uçkun Kiran et al., 2014; Sharma et al., 2022). A wide range of food wastes, including fruit and vegetable peels and pomace, oil seed cakes, bread and coffee waste have been utilised for enzyme production, and some recent studies are summarized in Table 13.1 (Uçkun Kiran et al., 2014; Costa et al., 2022; Gooruee et al., 2024).

In addition to acting as substrates for biomaterial production through microbial fermentation, wastes can also be employed as support material for enzyme immobilization in various applications, including bioactive compound and biodiesel production, as well as food processing. Their high surface area, high porosity and the presence of functional groups such as amino, carboxyl, hydroxyl, phosphate and thiol contribute to their effectiveness (Sharma et al., 2022). Moreover, these materials are low-cost, biodegradable and non-toxic materials and they enhance enzyme stability under harsh environmental conditions (Ray et al., 2025).

13.2.2 Organic Acids

According to recent reports, approximately 1.05 billion tons of food waste had been generated worldwide in 2022 (Forbes et al., 2024), while in Europe, the annual amount exceeded 58 million tons (Eurostat, 2025). This situation indicates that one-fifth (19%) of the food made available to consumers globally is wasted at the retail, food service, and household levels, and in Europe, it corresponds to an annual economic loss of 132 billion euros. Therefore, it is accepted as a rapidly growing global issue. Despite this, food waste could be considered as a potential raw material for producing value-added products (Wang et al., 2020a). Food waste rich in carbohydrates, proteins, and lipids provides suitable substrates for producing high-value compounds through biotechnological processes (Yang et al., 2022a). Among these, the production of organic acids has gained significant attention in recent years. Organic acids are essential to various industries including textile, cosmetic, food, pharmaceutical, chemical, and polymer (Lu et al., 2020). The global organic acids market reached 16.6 billion dollar in 2022, and is expected to reach 29.4 billion dollar by 2030, with an annual growth rate of 7.5% (KBV Research, 2024). Acetic acid, butyric acid, citric acid, formic acid, lactic acid, propionic acid, succinic acid, as well as volatile fatty acids (VFAs) could be converted into high-value products, and further be used in applications like bioplastic production, food and pharmaceutical formulations, and renewable energy (Yu et al., 2021; Sapmaz et al., 2022; Gottardo et al., 2024; Song et al., 2024). Thus, converting food waste into organic acids represents a strategic pathway for waste valorisation and advancing circular bioeconomy (LaTurner et al., 2020). This section discusses recent approaches to producing various organic acids from food waste, microbial fermentation processes,

Table 13.1 Enzymes produced by microbial fermentation from food and agricultural wastes

Waste type	Microorganism	Pre-treatment	Fermentation type	Enzyme produced	Reference
Apple pomace, orange peel, hazelnut shell	*Bacillus subtilis*	Acid hydrolysis (3% H_2SO_4)	Submerged fermentation	Cellulase, pectinase	Ozzeybek and Cekmecelioglu (2024)
Grape pomace, soybean hull	*Aspergillus niger* *A. oryzae*	–	Solid-state fermentation	Cellulase, pectinase, tannase, α-amylase	Cabezudo et al. (2023)
Lemon peel	*Trichoderma aureoviride NAS106* *T. afroharzianum NAS107* *T. ghanense NAS108* *T. pleuroticola NAS109* *T. harzianum NAS110* *T. lixii NAS114*	–	Submerged fermentation	Cellulase, pectinase, xylanase	Gooruee et al. (2024)
Olive pomace	*A. niger* *A. ibericus* *A. uvarum* *A. carbonarius* *Trametes versicolor*	Ultrasound (750 W, 20 kHz/ 5–-15 min) + acid hydrolysis (3% H_2SO_4)	Solid-state fermentation	Cellulase, xylanase	Leite et al. (2016)
Olive pomace	*Crypthecodinium cohnii*	Organosolv process (ethanol/water; 50% v/v and 0.5% (w/v) H_2SO_4)	Submerged fermentation	Polygalacturonase	Paz et al. (2021)
Olive pomace +(wheat bran)	*A. ibericus* *A. niger* *A. tubingensis*	–	Solid-state fermentation	Lipase	Oliveira et al. (2016)
Orange peel, crab shell	*B. cereus* spH1	–	Submerged fermentation	α-Amylase	Ben Hadj Hmida et al. (2024)
Sorghum grass waste, cottonseed oil cake	*Trichoderma reesei MTCC 4876*	–	Solid-state fermentation	Cellulase	Bhattacharya et al. (2024)

pre-treatment methods, and process optimization strategies. Additionally, the contributions of these production technologies to environmental and economic sustainability will be assessed, along with future research trends.

13.2.2.1 Citric Acid

Citric acid ($C_6H_8O_7$) is a water-soluble and colourless weak organic acid. It can be found naturally in all citrus fruits including lemon, orange, mandarin, lime, bergamot etc. It is widely used in the food industry as a preservative, colourant, antioxidant, acidity regulator, and flavouring agent (Lende et al., 2021). Citric acid is the most widely produced organic acid globally and is used primarily in food industry followed by pharmaceutical, cosmetic, and metal industries (Max et al., 2010). Citric acid can be produced via extraction, chemical synthesis, or fermentation. Over 90% of global production is achieved via fermentation, primarily by *A. niger* under batch submerged fermentation (Odu et al., 2020; Mores et al., 2021). Common substrates for citric acid production are hydrocarbons, molasses, and starch involving food or agricultural materials (Dienye et al., 2018).

In line with the sustainability and zero-waste goals, recent studies have explored industrial and household food waste as substrates for the production of citric acid. Although submerged fermentation offers several advantages, it generates substantial waste. For this reason, researchers are seeking eco-friendly processes that utilise less water and energy, while simplifying the downstream purification step. For example, orange, banana, pineapple, and sweet lime peels (Satheeshkumar et al., 2019), cassava peels (Ajala et al., 2020), waste cooking oil, and glycerol (Kaur & Chavan, 2022), had been employed for citric acid production using solid-state, submerged, and surface fermentation methods. Using *A. niger* NRRL 567 and the solid-state fermentation method, 65.6 g/dm^3 of citric acid was produced from apple waste, while 8.3 g/dm^3 was obtained through surface fermentation (Dhillon et al., 2011). From date waste and whey, *A. niger* ATCC 6275 produced 32.4 g/dm^3 of citric acid via surface fermentation (Mehyar et al., 2005). Using *A. niger* von Tiegh 1867. 64 g/kg of citric acid was produced from orange waste and cane molasses through surface fermentation (Hamdy, 2013). Similarly, *A. niger* A20 produced 29.7 g/dm^3, and *A. niger* W78B produced 110 g/dm^3 of citric acid from beet molasses using the same method (Adham, 2002). From pineapple waste, *A. niger* DS1 produced citric acid with a 54.2% yield via solid-state fermentation (Kumar et al., 2003). Among these methods, solid-state fermentation has been found to generate less waste, consume less water, and energy compared to others (Roukas & Kotzekidou, 2020; Lende et al., 2021; Książek, 2024).

In submerged fermentation, the final purification step involves precipitation with calcium salts, and recovery steps with sulfuric acid. Although food or agricultural wastes could be used as substrates, considerable amount of residues are ejected in addition to numerous chemical required which conflicts with zero-waste principles (Mores et al., 2021). Major challenge in citric acid recovery from food waste is achieving high product purity. Therefore, downstream processes such as membrane separation, adsorption, and solvent extraction are considered potential alternatives to the conventional methods (Dienye et al., 2018; Vidya et al., 2018; Tornuk & Akman, 2025). These innovative techniques not only reduce chemical use and waste generation but also enable the production of a pure product with shorter processing periods (Mores et al., 2021; Książek, 2024).

13.2.2.2 Lactic Acid

Lactic acid ($C_3H_6O_3$) is an organic acid and is widely used in food, chemical, cosmetic, packaging, and textile industries (Tang et al., 2024). Approximately 80–90% of global production is achieved via fermentation primarily by the homofermentative *Lactobacillus* spp. (Costa et al., 2024; Villanueva-Galindo et al., 2025). In recent years, recombinant *Escherichia coli*, *Bacillus coagulans*, *Corynebacterium glutamicum*, *B. licheniformis*, and metabolically engineered yeasts have also been employed for the production of lactic acid (Villanueva-Galindo et al., 2025). Traditionally, sugars from sugarcane, cassava, or corn have served as carbon sources; however, utilization of these substrates increases production costs (Costa et al., 2024). Therefore, food waste is emerged as a cost-effective and sustainable alternative to the conventional raw materials. Fermentation from food and dairy wastes has been extensively studied, with several processes already implemented industrially (Bühlmann et al., 2023). However, the complex carbohydrate composition of food waste may inhibit microbial activity, and often requires multiple culture systems or pre-treatment, depending on the type of waste and its nutrient content (Rahman et al., 2025).

Lactic acid production typically involves three steps in agreement with the zero-waste principles. The first is the pre-treatment step, which involves removing undesirable substances, decreasing particle size, and setting up the substrate for hydrolysis (Wang, Li, et al., 2020b; Tang et al., 2024). Depending on the hydrolysation method, various acids, alkali, or enzymes are used to break down complex carbohydrates of the waste into monosaccharides. This thereby improves process stability and product yield. However, the composition of food waste may increase the production costs by demanding more complex and labour-intensive pre-treatment (Pau et al., 2024).

The second step of the production is fermentation, which is carried out using lactic acid bacteria, or fungi (Huang et al., 2023; Pau et al., 2024). Using starter cultures can be expensive on an industrial scale, so the natural microbiota of food waste is often utilized. Spontaneous fermentation reduces the production costs; however, the diversity of microorganisms available in waste can affect the final yield and optical purity of lactic acid. The initial pH and the content of fermentable sugars in substrate are critical parameters that affect fermentation efficiency (Lobeda et al., 2022; Tang et al., 2024). Studies have shown that maintaining an initial pH of 5–6 during fermentation increases lactic acid production. For example, lactic acid yields are found as 50 g/L at pH 6.5 from cassava, 102 g/L at pH 6.0 from rice noodle waste, 14–40 g/L at pH 6.0–6.5 from household food waste, and 20.7 g/L at pH 7.29 from cafeteria waste, respectively (Lobeda et al., 2022; Villanueva-Galindo et al., 2025).

The final step of the process is downstream processing, which involves separating and purifying lactic acid from the fermentation broth. This step is rather important as it affects cost and waste management. For improving effectiveness, solvent extraction, membrane filtration, electrodialysis, precipitation, adsorption, molecular distillation, and esterification can be utilized (Costa et al., 2024). The purity of the product and the composition of residual waste depend on the purification method or their combination. During lactic acid fermentation, microorganisms primarily consume soluble carbohydrates, leaving organic residues with a high risk of environmental pollution after separation and purification (Yang et al., 2022a).

Studies have shown that these residues can be reused for methane production without reducing biogas yield (Chenebault et al., 2022). In some cases, the higher protein and lipid content of the residues even improves methane yield (Demichelis et al., 2017). Additionally, fermentation residues or digestate can be converted into biochar, providing another valuable up-cycling option. Biochar can further be used in agriculture, industry, and the production new materials (Tang et al., 2024).

13.2.2.3 Succinic Acid

Succinic acid ($C_4H_6O_4$) is an organic acid with two carboxyl groups on the terminal carbon atoms (Kumar et al., 2024). As an intermediate of the tricarboxylic acid cycle, it is also recognized as an important compound used in the production of high-value-added chemicals such as 1,4-butanediol, γ-butyrolactone, adipic acid, and others. Succinic acid has extensive applications in the food and beverage industry, cosmetics, and pharmaceuticals, polymer and paint production (Lin et al., 2024). Due to its significant contribution to the circular economy, researchers have focused on finding low-cost raw materials for sustainable succinic acid production, which has led to an interest in food waste. Food waste used for succinic acid production originates from household, industry and commercial enterprises including cafeterias, restaurants, and school canteens (Wang et al., 2022; Kuglarz & Angelidaki, 2023). In a study conducted in Hong Kong, bakery waste (pastries and cakes) from a global coffee company was used as a substrate for succinic acid production with *Actinobacillus succinogenes*, resulted in a yield of 24.8–31.7 g/L. The product was purified from the fermentation bulk using a cation exchange resin-based process with 96%–97.7% purity (Zhang et al., 2013). In another study, fruit and vegetable waste as a substrate was hydrolysed with the crude enzyme mixtures produced by *A. niger* and *Rhizopus oryzae*, generating a sugar-rich medium. From this medium, *Actinobacillus succinogenes* produced 27.03 g/L of succinic acid. It demonstrates high-yield and environmentally friendly bioproduction from this waste (Dessie et al., 2018). Household kitchen waste is also another promising substrate for succinic acid production. Succinic acid is produced from these wastes containing 81–85 g/L sugars by fermentation with *A. succinogenes* after enzymatic hydrolysis with a high yield of 74%–75%. Carbon dioxide and magnesium carbonate supplementation further increased the yield of succinic acid, while the methane content of the biogas exceeded 91% (Kuglarz & Angelidaki, 2023). Similarly, 32.2 g/L of succinic acid was produced from potato waste using *A. succinogenes*, achieving a 37% higher yield compared to pure glucose (Escanciano et al., 2023). Fermentation of acid whey and oat pomace produced approximately 20 g/L succinic acid in 72 hours. After a three-stage purification process, organic acids and salts were removed, and 53% of the succinic acid was recovered at purity greater than 99% (Kleps et al., 2024).

13.2.2.4 Acetic Acid

Acetic acid (CH_3COOH) is known as "vinegar acid" due to its sour taste and strong characteristic vinegar odour (Li et al., 2015). It is a water-soluble, colourless and weak organic acid. It has been due to its sour taste and strong odour of vinegar. In the food industry, it is recognized as an additive under the code E260. It can be used as a food preservative and acidity regulator in various food formulations. Owing to its chemical structure, acetic acid can also serve as a precursor in the synthesis of various chemicals, and is extensively used in the production of PET plastic (Wang et al., 2025). Global acetic acid production is mainly carried out chemically (approx. 78%) through the carbonylation of methanol. In contrast, only 15% of the acetic acid is generated through fermentation which the majority is used for vinegar production (Chozhavendan et al., 2021).

Fermentation has several advantages over chemical synthesis in the production of acetic acid. These advntages include the utilization of food waste as a substrate, lower chemical inputs, and more environmentally and economically sustainable process (Mizzi et al., 2022; Saha & Kumar Das, 2024). Fermentation enables the efficient conversion of waste from fruit and vegetable industry, dairy industry (whey), and kitchen waste into acetic acid (Wang et al., 2020b, 2025). Such waste materials are generally preferred because they are renewable, reduce production costs, minimize environmental pollution, and allow for high-yield acetic acid production (Chozhavendan et al., 2021; Upadhyay et al., 2023). In one study, acetic acid was produced through an ethanol-type fermentation using yeast and acetic acid bacteria from food waste collected at a university cafeteria. As a result, 25.88 g/L of acetic acid was obtained, and high-purity acetic acid was recovered from fermentation broth by distillation (Li et al., 2015). Researchers investigating acetic acid production from kitchen waste via fermentation have reported that the resulting acetic acid can serve as a low-carbon substrate for microbial lipid synthesis. It has been demonstrated that oleaginous yeasts can convert acetic acid into short-chain fatty acids (SCFA). Thus, the conversion of kitchen waste into high-value-added products by zero-waste principle proceeds into two stages: first fermentation step which transforms waste into acetic acid, and then bioconversion into SCFA (Wang et al., 2025).

Recent advances in genetic and metabolic engineering techniques have been employed to overcome challenges such as acid stress, flux distribution, dependence on a single carbon source, and low production yields. Studies have shown that genetically modified strains can significantly enhance acetic acid production yields (Chozhavendan et al., 2021; Saha & Kumar Das, 2024).

Downstream processes following fermentation such as condensation, evaporation, and, in some cases, removal of secondary metabolites are required to obtain pure acetic acid. Moreover, after separation of acetic acid from the fermentation broth, the nutrient-rich waste can be used as a soil enrichment medium. In this manner, the zero-waste principles can be effectively implemented in acetic acid production aligned with the Sustainable Development Goals (SDGs), (Upadhyay et al., 2023).

13.2.3 Nutraceuticals and Functional Bioactives

Biologicially active compounds possessing health benefits can be obtained by solid-state and submerged fermentation processes utilizing food and agricultural wastes as potential substrates (Sadh et al., 2018). For example, bioactive peptides can be produced by fermentation with filamentous fungi; however, the composition of bioactive peptides may vary depending on the protease specificity and the composition of proteins in the waste substrate (Rousta et al., 2024). Animal waste proteins could be used as a potential substrate for fermentation, and the proteolytic enzymes secreted by native or additional microorganisms during their metabolism break down proteins into smaller molecules, and release peptides and amino acids (Cakmak et al., 2025; Dhiman et al., 2025).

Mucor spp. strains are able to produce gamma-linolenic acid and β-carotene together in submerged fermentation of brans, spent malt grains, distiller grains, etc. (Martí-Quijal et al., 2021). In addition, *Blakeslea trispora* utilizing carrot, orange, and papaya peels as substrate produces β-carotene via solid-state fermentation. The process yield was dependent on the pH of the medium and the fermentation temperature, and pH 6.2 and 28 °C were determined as the optimum conditions for maximization of the carotenoid content.

Enhancing the antioxidant activity of phenolic compounds and producing new metabolites from food wastes through microbial fermentation is a promising approach for the efficient utilization of natural resources (Gulsunoglu-Konuskan & Kilic-Akyilmaz, 2022). After fermentation, the concentration of some bioactive compounds can increase, and new metabolites may be formed that are not found in the substrate. In the study of Abd Razak et al. (2015), rice bran fermented with a mixed culture of *R. oligosporus* and *Monascus purpureus* showed significantly higher total phenolic content and antioxidant activity compared to the non-fermented rice bran or rice bran fermented with only one culture (Abd Razak et al., 2015). Moreover, vanillic, caffeic, and 4-hydroxybenzoic acid which were not found in the substrate, were detected in the fermented samples. Similarly, the ferulic acid content of rice bran was significantly increased through solid-state fermentation with *R. oryzae* (Sadh et al., 2018). However, enzymatic hydrolysis with esterase and glucosidases during fermentation of the wastes is essential for releasing phenolic compounds that are present conjugated forms (Martinez-Avila et al., 2014).

In another study, the concentrations of pelargonidin 3-glucoside and *p*-coumaric acid increased significantly from day 0 to day 1 of fermentation, indicating that fermentation at pH 4 was able to effectively release these compounds from the strawberry waste (Serrano et al., 2020). However, the overall anthocyanin content decreased following fermentation. Similarly, total phenolic content of acetic beverage produced from fermented olive mill wastewater by two-step fermentation process—initial alcoholic fermentation with *Saccharomyces cerevisiae*, followed by acetic acid fermentation with *Acetobacter pasteurianus*— was improved (Signorello et al., 2025). This finding was attributed to the hydrolysis of oligomeric phenols during acetic fermentation.

Polyphenols found in okara biomass could be converted into phenyllactic, p-hydroxyphenyllactic and indole-3-lactic acids at increased amounts by co-culturing with *Lactobacillus acidophilus*, *Lacticaseibacillus rhamnosus* and *Pediococcus acidilacti* (Khubber et al., 2022). Likewise, the production yield of polyphenol from pomegranate wastes could be highly dependent on the microorganisms involved in fermentation, as *S. cerevisiae* caused the production of five-fold more ellagic acid content than *A. niger* (Moccia et al., 2019).

Furthermore, fermentation conditions may have a noticeable effect on the release of bioactive compounds. The pistachio green hull (PGH), as a non-edible fraction and regarded as agro-industrial waste, could be utilized for its phenolic compounds, which were improved by fermentation. Ordoñez-Cano et al. (2025) investigated the potential impacts of the fermentation conditions, namely, moisture, inoculum concentration, and aeration rate on the total phenolic content and antioxidant activity of PGH fermented with *A. niger* GH1. They concluded that moisture and aeration rates had a significant impact on bioactives, and the maximum total phenolics content and antioxidant activity were reached under the conditions of 60% moisture, 5×10^6 spore/mL, and 1 L/kgwm min aeration rates, which were increased by 129% and 1039%, respectively, through fermentation.

The brewing process generates around 30% of waste in different forms and spent grains alone account for 80%–85% of that waste (Ahuja et al., 2024). BSG typically contains a considerable amount of dietary fibre (30%–50%), 19%–30% of proteins, 10% of lipids, and 2%–5% of ash. Barley proteins are composed of hordeins or barley prolamins, glutelins, globulins, and albumins. Even after brewing, a significant amount of protein is left insoluble and unutilized. Therefore, BSG has great potential to be utilized as a substrate. BSG as substrate can be used for enzyme production (mainly xylanase and cellulase), and phenolic compounds due to its rich nutrient composition. Besides, hydroxycinnamic acid and caffeic acid are responsible for the total antioxidant capacity of BSG (Leite et al., 2019). The amount of protein, amino acids, and other bioactive compounds increases after fermentation, improving the nutritional profile of BSG (brewery spent grain) and indirectly providing health benefits (Ahuja et al., 2024). The improvements in antioxidant capacity of BSG via fermentation are associated with the amino acids involving bioactive peptides released via fermentation (Chai et al., 2020). Similarly, fermenting okara results in breakdown of proteins into free amino acids, which improves digestibility as well as providing umami flavour (Vong et al., 2016).

The improvements in bioactive materials in the fermented wastes could be attributed to microbial hydrolysis, which then increases the quantity of phenolic compounds and flavonoids (Erskine et al., 2023). Besides, proteolysis of proteins may result in increased contents of bioactive peptides that enhance antioxidant activity (Vong et al., 2016; Vicenssuto & de Castro, 2020).

13.2.3.1 Single-Cell Proteins and Microbial Biomass

Globally, food waste has reached alarming levels, while a significant portion of the population still lacks access to sufficient and safe food. Hunger and protein deficiency are becoming increasingly severe problems, particularly in underdeveloped and developing countries (Forbes et al., 2024). Consequently, in recent years, the biotechnological conversion of food waste into single-cell protein (SCP) and microbial biomass has emerged as a sustainable strategy for both waste management and meeting the growing demand for protein (Khan et al., 2024).

The dead and dried cellular biomass of microorganisms such as bacteria, yeast, fungi, and algae is regarded as SCP (Kaur & Chavan, 2022; Sun et al., 2024). It serves as an alternative high-protein biomass and renewable protein source for the food and feed industries (Li et al., 2024; Aamir Shahzad et al., 2025). In addition to being a protein souce, microbial biomass derived from food waste can also be a good source for the production of enzymes, pigments, lipids, and various bioactive compounds (Salazar-López et al., 2022). Therefore, the biotechnological conversion of food waste into SCPs not only contributes to meeting global protein demand, but also generates multifaceted value that extends beyond the single-stage protein recovery.

Fruit and vegetable peels, by-products from the brewing and dairy industries, and household kitchen wastes are commonly used as substrates for SCP production (Thiviya et al., 2022; Sun et al., 2024). Firstly, appropriate pre-treatments, such as hydrolysis, pH adjustment, or enzymatic degradation are required to enhance nutrient availability for microorganisms that are responsible for fermentation. In SCP production, the selection of starter cultures and the optimization of fermentation conditions play crucial roles in determining the yield and quality of the final product (Rajput et al., 2024). Despite the advantages of zero-waste approach, various technical and economic challenges remain in the production of SCPs including heterogeneity of waste, high pre-treatment costs, and concerns regarding product safety. Nevertheless, the production of SCP and microbial biomass from food waste is gaining importance as an innovative biotechnological method that supports the circular bioeconomy, mitigates environmental impact, and creates economic and social value (Onyeaka et al., 2022; Sekoai et al., 2024).

In this section, current scientific progress in the production of microbial biomass and SCP from food waste is discussed in detail. The types of microorganisms used, food waste as feedstocks employed, process optimization strategies, production challenges, and the technological and sustainability aspects of this novel method are categorically evaluated.

13.2.3.1.1 Zero-Waste Fermentation for Up-Cycling Food Waste into Single-Cell Protein

SCPs, also called microbial proteins, are derived from the cellular biomass of microorganisms such as bacteria, yeasts, and algae grown on various carbon-containing substrates (Najafpour, 2007). SCP is commercially available product that serves as a protein source in animal feed, primarily derived from heterotrophic bacteria and algae (Hülsen et al., 2018). Although SCP is widely used as a feed particularly for aquaculture, its direct consumption in human nutrition is restricted by many authorities due to its relatively high nucleic acid content (Yadav et al., 2016; Hülsen et al., 2018). Therefore, various downstream processes and methods must be employed to make them suitable for human consumption (Abdullahi et al., 2021; Fernández-López et al., 2024).

In addition to its protein content, SCP is a highly nutritious bioproduct that contains essential vitamins, minerals, bioactive peptides and other nutritious compounds (Yadav et al., 2016; Sekoai et al., 2024). The food waste used as a substrate for SCP production should provide specific characteristics, including being non-toxic, abundant, renewable, inexpensive, easy to process, nutritionally rich enough to support microbial growth, and be sustainable (Reihani & Khosravi-Darani, 2019; Yang et al., 2022b).

The substances that can be used in SCP production include waste materials from the fruit, vegetable, and cereal processing industry (e.g., orange peel, rice husk, cassava waste, sugarcane bagasse, wheat straw, corn cobs, sugar beet pulp, coconut residues, grape pomace, mango waste), by-products from the dairy industry (e.g., whey), and wastewater from meat, poultry, dairy, and sugar processing plants (Hülsen et al., 2018; Reihani & Khosravi-Darani, 2019; Li et al., 2024).

Selecting a microorganism that can adapt well to the substrate is essential for improving the yield and quality of the SCP product. *Kluyveromyces, Candida, Saccharomyces, Pichia, Aspergillus, Fusarium*, and *Trichoderma* are commonly utilized in SCP production due to their biochemical composition, high content of B-complex vitamins and folic acid, and essential amino acids (notably lysine and threonine) consistent with the nutritional recommendations of FAO (Khan et al., 2024; Tornuk & Akman, 2025). Additionally, *Yarrowia lipolytica* which is a yeast species with GRAS status has a unique metabolic pathway that enables to utilize a wide range of carbon sources. This species demonstrates significant biotechnological potential for converting organic food wastes into SCP (Groenewald et al., 2014; Dobrowolski et al., 2020). For instance, *Y. lipolytica* has been employed to produce SCP from organic food wastes in a two-stage fermentation process, resulted in achieving a chemical oxygen demand (COD) removal efficiency of 85.5%, and a biomass protein content of 38.8% (Yang et al., 2022b). Similarly, *S. cerevisiae*, also known as baker's yeast, has been used to produce SCPs from date juice using an under batch fermentation system, and it has been yielding a protein content of 67.8% (Sorour et al., 2021). From fruit waste substrates such as mango, soursop, papaya, pineapple, banana, mangosteen, cashew apple, cacao, jackfruit and pomegranate, using *S. cerevisiae* reached protein amounts up to 48.3% under batch process (Dunuweera et al., 2021; Koukoumaki et al., 2024).

Fermentation mode, fermentation conditions (e.g., temperature, pH, oxygen concentration, inoculum cell density, and the age of the microorganism, and the carbon to nitrogen ratio of the substrate are crucial parameters that significantly influence both the quality and quantity of the final product. These parameters must be optimised for each specific feedstock to enable efficient and economically viable industrial-scale production (Reihani & Khosravi-Darani, 2019; Li et al., 2024).

Nevertheless, concerns remain regarding the potential risks associated with SCPs, including high RNA content, toxins and allergens which are related to the producer strains, process conditions and product formulations (Ye et al., 2024). Therefore, incorporation of SCPs into human foods requires detailed observations and implementation of regulations by relevant authorities.

13.2.4 Bioplastics

Microbial fermentation processes use microorganisms to make polymers that are very pure and have a high yield with more eco-friendly way (Thakur et al., 2022). Bacteria, fungi, and yeast can be engineered to synthesize polymers using renewable carbon sources (Beena Unni & Muringayil Joseph, 2024). Bio-based polyesters, particularly polyhydroxyalkanoates (PHAs), polylactic acid (PLA), and polybutylene succinate (PBS), are synthesized through the fermentation of carbohydrates or lipids, such as sugars, starch, and plant oils (Mandal et al., 2024).

In contrast, bio-based polyolefins, such as bio-polyethylene and bio-polypropylene, are derived from biomass-based intermediates including bioethanol or biopropane (Mandal et al., 2024). Bio-based polyamides (e.g., PA 6, PA 11, PA 10.10) are made from monomers derived from microbes or plants, such as amino acids and sebacic acid (Mandal et al., 2024). Another important group includes polysaccharide-based polymers, such as exopolysaccharides (EPS) (Sharma et al., 2024), bacterial cellulose (BC), kefiran, pullulan, levan, gellan (Vijayendra & Shamala, 2014), and thermoplastic starch.

Emerging bio-based polyacrylates (e.g., bio-PMMA) are produced from ethanol-derived monomers, offering bio-origin combined with the optical clarity and rigidity characteristics of conventional plastics. Additionally, microbial and hybrid polymers, such as bio-polyurethanes, bio-PVC, polythioesters, and furan-based polymers, represent an expanding frontier in bio-based chemistry. A broad range of sustainable and high-value products have been extracted from agricultural residues and other renewable feedstocks, thus emphasizing the adaptability of white biotechnology in these systems. Fermentation allows the combination of waste valorisation with sustainable material development which makes them suitable candidates for zero-waste initiatives (Fatima, 2024).

An example of a zero-waste fermentation model can be found in a bioplastic production unit using cultivation of EPS-producing microorganisms to convert solid fractions of food waste into PHAs or BC as well as reintroducing the biomass into the process as fertilizers or feed additives (Koutinas et al., 2014). In these types of

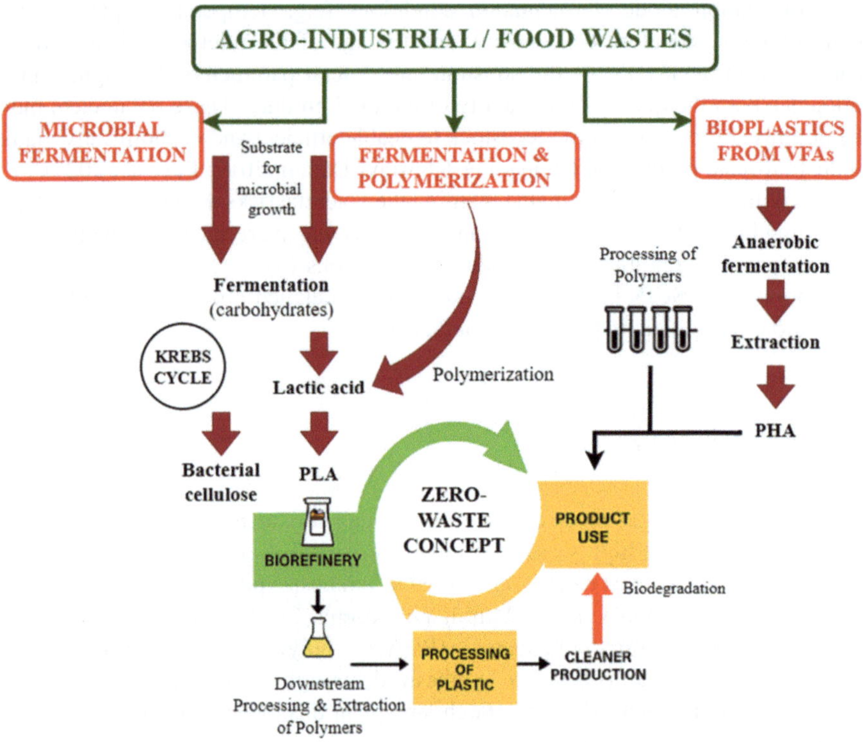

Fig. 13.3 A zero-waste biorefinery approach for bioplastic (PHA, PLA, and PHBV) production from agricultural food waste, product use, biodegradation, and recycling via three main pathways: (1) microbial fermentation, (2) fermentation followed by polymerization, and (3) VFA/SCFA-based bioplastic production

closed-loop systems, biodegradable bioplastics are capable of replacing conventional synthetic plastics (Fig. 13.3). These approaches are well aligned with United Nations Sustainable Development Goals (SDGs), particularly responsible consumption and production (SDG 12) and climate action (SDG 13) (Mandal et al., 2024).

Various carbon and nitrogen sources found in wastes are utilized as nutrients in fermentation to produce microbial biopolymers within a zero-waste fermentation strategy. Biopolymers synthesized through different fermentation techniques have been extensively studied by researchers (Koutinas et al., 2014; Talan et al., 2022; Khan et al., 2023; Swetha et al., 2023; Artola et al., 2024; Kusuma et al., 2024; Mandal et al., 2024; Marotta et al., 2025). Bioplastic production via zero-waste methodology focusing on multi-product utilization is summarized in Table 13.2.

Significant progress has been achieved; however, scaling up zero-waste fermentation processes is still a challenge due to high costs, complex downstream processes difficulties in bioplastic recovery and purification, as well as, variability in the composition of food waste affecting the fermentation yields (Sharma et al., 2024). Consequently, future research should focus on the genetic engineering of microbial

Table 13.2 Studies on multi-product utilization including bioplastics from agro-industrial residues via fermentation

Source	Processing strategy	Obtained bioproducts	Zero-waste integration	Key findings	References
Food waste and agro-industrial residues	(1) Sequential anaerobic digestion (2) Fermentation (3) Biopolymer recovery	Biohydrogen, VFAs, PHAs	Integration of multiple fermentative routes	Demonstrated feasibility of recovering multiple energy and material streams from food waste biorefineries	Uçkun Kiran and Liu (2015)
Agro-industrial residues	Multi-stage biorefinery simulation (fermentation + recovery)	Biopolymers, organic acids	Techno-economic modeling of integrated valorisation chain	Identified fermentation yield and energy integration as economic bottlenecks	Helenas Perin et al. (2020)
Organic fraction of municipal solid waste (OF-MSW)	(1) Ethanol organosolv pre-treatment (2) Enzymatic hydrolysis (3) Fermentation (*Enterobacter aerogenes*)	Ethanol, 2,3-butanediol, hydrogen, PHAs, methane	Valorisation of all fractions: Fermentation + anaerobic digestion for residue; no waste disposal	Achieved 8236.9 kJ energy recovery; produced multiple fuels and bioplastics within an integrated process	Ebrahimian et al. (2020)
Beverage industry residues (spent coffee grounds, apple pomace)	(1) Sequential extraction (2) Oil valorisation (3) Microbial conversion (4) Biopolymer formation	Bioplastics, colorants, prebiotics	Multi-valorisation cascade recovering lipids, caffeine, polysaccharides, and pectin for polymer and food ingredient production	Developed a complete waste-free pipeline converting both lipidic and carbohydrate fractions into value-added biopolymers and nutraceuticals	Williamson (2021)
Almond hulls	(1) Acidic extraction (pectin, phenolics) (2) Enzymatic hydrolysis (3) microbial fermentation (*Aureobasidium pullulans*)	Pectin, phenolics, pullulan, single-cell protein	Multi-product valorisation; residues used for microbial fermentation	Recovered pectin (26.3%), phenolics (6.97%), pullulan (34.3 g/L), and SCP (19.3%) with full residue utilization	Najari et al. (2022)
Mandarin peel waste	Integrated hydrolysis + fermentation + polymerization	Lactic acid, PLA	Full conversion of citrus waste; solvent recycling	Demonstrated sustainable PLA synthesis directly from food waste hydrolysate	Jang et al. (2022)

(continued)

Table 13.2 (continued)

Source	Processing strategy	Obtained bioproducts	Zero-waste integration	Key findings	References
Microbial biomass (*Phaffia rhodozyma*)	(1) Fermentative production of carotenoid-rich biomass (2) Solvent-based extraction (ethanol + ethyl acetate) (3) Sequential recovery of astaxanthin, β-carotene, lipids, and proteins (4) Integration with BC matrix	Astaxanthin, β-carotene, lipids, proteins, antioxidant bioplastics	Multi-product valorisation with complete utilization of microbial biomass; all fractions recovered and reused; coupling fermentation with downstream material fabrication	Recovered over 70% astaxanthin and 90% β-carotene; created antioxidant BC-based bioplastics, demonstrated a zero-residue microbial biorefinery platform	Mussagy et al. (2022)
Mixed food waste	Fermentation (lactic acid bacteria) + chemical synthesis (PLA) + biodiesel co-production	PLA, biodiesel	Integrated system using both carbohydrate and lipid fractions	Minimum selling price of PLA reduced to 4.29 USD/kg, energy integration crucial for economics	Rajendran and Han (2022)
Apple pomace	Microwave-assisted co-extraction + microbial fermentation	Pectin, phenolics, pullulan, SCP	Zero-waste system: Co-extraction residues used as carbon source for microbial products	Pullulan (43.7 g/L) and SCP (33.5%), pectin and pullulan used as edible coating for strawberries	Salimi et al. (2024)
Municipal solid waste (organic fraction)	Enzymatic hydrolysis + anaerobic digestion / solid-state fermentation	Enzymes, biopesticides, bioenergy	Enzyme self-production + dual valorisation routes (energy + bioproducts)	On-site enzyme production + SSF increased profit margin 74%, IRR 13%, and 6 years payback	Molina-Peñate et al. (2024)
Fruit and vegetable processing wastes	Integrated biorefinery employing enzymatic hydrolysis + mixed microbial fermentation + downstream valorisation	PHAs, organic acids, bioenergy	Sequential fermentation stages using consortia; recovery of solid residues for fertilizer and biogas	Achieved 75–80% carbon recovery, demonstrated operational circularity via energy and nutrient recirculation	Rodriguez-Perez et al. (2022)

BC bacterial cellulose, *IRR* internal rate of return, *PHA* polyhydroxyalkanoates, *PLA* polylactic acid, *SCP* single-cell protein, *SSF* solid-state fermentation, *VFA* volatile fatty acid strains to enhance the product yield and tolerance to heterogeneous waste feedstocks; the optimization of processes via continuous fermentation and integrated bioreactors. Besides, development of polymer blends and composites will improve the mechanical and barrier properties of biopolymer films; and life cycle assessments will ensure authentic environmental benefits across production, utilization, and end-of-life phases (Koutinas et al., 2014; Khaleel et al., 2025).

13.2.4.1 Polyesters

Figure 13.4 presents different pathways to produce microbial polyesters, including PHAs, PLA, and polybutylene succinate (PBS) (Vijayendra & Shamala, 2014; Mohanta et al., 2025).

PHAs constitute a family of biopolyesters that include poly-3-hydroxybutyrate (P3HB), poly-3- hydroxyvalerate (P3V), poly-3-hydroxyhexanoate (P3H), poly-3-hydroxyoctanoate (PHO), and poly-4-hydroxybutyrate (P4HB). Limited nutrients combined with an excess of carbon sources facilitate the production of these bioplastics within bacterial cells (Ali et al., 2022). Polyhydroxyalkanoates (PHAs), as energy stocks in microbial cells, have complete biodegradability in natural environments (Bhatia, 2023). PHB, the most common form among PHAs, is produced by *Ralstonia eutropha*, *Pseudomonas* spp., and *Bacillus* spp. (Vijayendra & Shamala, 2014). Starch-rich residues, including potato peels and cassava, along with whey, molasses, and lipid-rich streams, are utilized as economical carbon sources for

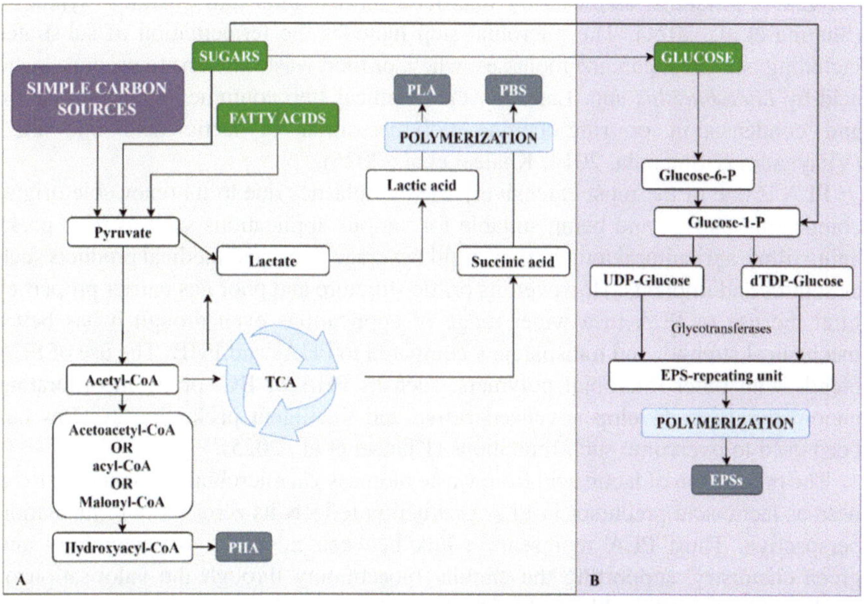

Fig. 13.4 Schematic representation of simple production pathways for polyesters (**a**), and EPSs (**b**)

the biosynthesis of PHA and PHB. Biopolymers possess the potential to substitute traditional petrochemical plastics in food packaging, agricultural films, and medical devices (Khaleel et al., 2025).

Integrating PHA production into biorefineries for bioethanol, biodiesel, or food processing improves zero-waste processes (Andhalkar et al., 2022; Jayalath & de Alwis, 2025). For instance, fermentable sugars found in lignocellulosic residues of bioethanol refineries (e.g., sugarcane bagasse), crude glycerol from biodiesel production, and organic wastes of foods (e.g., fruit peels, starch residues, and spent coffee grounds) can be utilized as feedstock for VFA production and subsequent PHA biosynthesis to reduce waste disposal costs, and enhance overall feasibility of biorefineries (Zhou et al., 2023; Zhang et al., 2024; Ilhami et al., 2025).

Algae-based biorefineries have potential for improving economic sustainability via co-production of biodiesel, astaxanthin, and PHAs. Besides, wastewater treatment and PHA recovery from fermentation by-products, simultaneously within a zero-waste system can further promote carbon reuse, nutrient recovery, and biogas generation (Jayalath & de Alwis, 2025). PHA synthesis together with the production of other products within lignocellulosic biorefineries supports valorising waste streams through scalable industrial processes (Dietrich et al., 2019). These types of co-production strategies have potential to lower downstream processing costs and to enable efficient utilization of residual biomass from both economic and environmental perspectives. The co-synthesis potential of pigments, carotenoids, amino acids, hydrogen, surfactants, and PHA was reported by Kumar and Kim (2018), while Koller et al. (2017) demonstrated the simultaneous production of bioethanol as a coproduct with PHA.

PLA is produced stepwise by microbial fermentation and chemical synthesis (Sharma et al., 2024). The microbial step includes the fermentation of substrates including, starch, sugarcane molasses, whey, or food waste streams to produce lactic acid by *Lactobacillus* spp. Later on, the chemical step continues with purification and condensation or ring-opening polymerization of lactic acid into PLA (Vijayendra & Shamala, 2014; Khaleel et al., 2025).

PLA is one of the most extensively used bioplastics due to its renewable origin, composting ability, and being suitable for various applications such as food packaging films, agricultural mulch films, rigid beverage cups, and medical products such as sutures and implants. However, its brittle structure and poor gas barrier properties limit the use of PLA in a wider range of applications even though it has better mechanical strength and transparency compared to PHAs and PHB. The use of PLA blends with other microbial polymers, such as PHA or BC, or by incorporating nanomaterials to develop advanced active and intelligent packaging systems has been used to overcome such limitations (Khaleel et al., 2025).

The production of lactic acid from waste biomass via microbial fermentation to be used as lactic acid precursor in PLA production reflects its zero-waste fermentation perspective. Thus, PLA represents a link between microbial biotechnology and green chemistry, supporting the circular bioeconomy through the valorisation of food waste into sustainable packaging materials.

13.2.5 Exopolysaccharides (EPS)

Exopolysaccharides are biopolymers secreted by microorganisms into the extracellular environment during fermentation (Fig. 13.4). Common examples include xanthan gum, gellan gum, kefiran, levan, and pullulan which can be used as thickening agents, stabilizers, film-forming materials in edible coatings and active packaging (Vijayendra & Shamala, 2014). Using waste-derived substrates enhances the cost-effectiveness and sustainability of EPS production. Agro-industrial by-products such as molasses, rice bran, and sugarcane bagasse have been successfully used to obtain high EPS yields while simultaneously addressing waste management challenges. Moreover, incorporating antimicrobial agents or nanoparticles into EPS-based films can produce active and intelligent packaging systems that extend the shelf life of food products (Khaleel et al., 2025).

Sugar monomers (glucose, mannose, galactose, and rhamnose, etc.) with non-carbohydrate moieties like uronic acids, acetyl, pyruvyl, or phosphate groups are primary components of EPSs (Vijayendra & Shamala, 2014). EPSs, including pullulan, kefiran, levan, gellan, and BC, with their gelling, viscosity-modifying, emulsifying, and film-forming properties have shown their unique industrial application potential. Pullulan, a linear, neutral polysaccharide composed of maltotriose units linked via α-1,6-glycosidic bonds, is produced predominantly by *Aureobasidium pullulans*. It forms transparent and oxygen-impermeable films. Kefiran, synthesised by *Lactobacillus kefiranofaciens* in kefir grains, is a water-soluble glucose–galactose polymer and gelling ability in addition to its antibacterial, antifungal, and antihypertensive properties.

Levan is synthesized by *Bacillus*, *Pseudomonas*, and *Zymomonas* species with a highly branched β-2,6-linked fructan structure and displays emulsifying, antioxidant, and moisture-retention properties suitable for films and coatings. Furthermore, some unique EPSs adapted to extreme environments, can be produced via *Halophilic* archaea such as *Haloferax mediterranei*, and show bioactive properties with emulsifying ability. Likewise, gellan gum, secreted by *Sphingomonas paucimobilis*, forms strong, thermally stable gels widely utilised in food and biomedical formulations (Vijayendra & Shamala, 2014). Meanwhile, bacterial cellulose, primarily produced by *Komagataeibacter* spp., *Acetobacter* spp., *Sarcina ventriculi*, and *Pseudomonas* spp., is a highly pure form of cellulose synthesised via microbial fermentation using various waste-derived substrates, including agricultural residues (rice husk, corn stalk, oat hull), industrial effluents (biodiesel, confectionery, fermentation wastewater), food wastes (fruit juices, sugarcane molasses, jaggery waste) and cotton-textile waste (Tsouko et al., 2015; Cheng et al., 2017; Skiba et al., 2020; Sharma et al., 2024). It exhibits better mechanical strength, water-holding capacity, and can form films, making it useful for biodegradable packaging, wound dressings, and tissue engineering scaffolds due to its nanofibrillar structure of β-1,4-linked glucose chains and extensive hydrogen bonding (Wang et al., 2023). Although agitated fermentation provides higher yields than static methods, scaling up BC production is challenged by high costs and low yields (Zhong, 2020).

Within a zero-waste context, the carbohydrate-rich biomass residues and solid food waste streams appear as the most promising feedstocks, as they allow the simultaneous recovery of bioplastics, bioenergy, and fertilizers in closed-loop systems. However, high costs of downstream purification and the variable nature of waste composition are some of the primary technical bottlenecks, resulting in reduced fermentation yield and product consistency. Therefore, critical research gaps exist in genetically engineering microbial strains for enhanced tolerance to heterogeneous substrates and performing comprehensive life cycle assessments to validate the environmental viability of these integrated biorefinery models.

13.2.6 Biofuels

Food waste can be converted to bio-methane, bio-hydrogen, bioethanol, bio-butanol, and biodiesel (Garita-Cambronero et al., 2021; Kazimierowicz et al., 2021; Zeng et al., 2022; Bravo-Venegas et al., 2023). Protein, fat, starch, sugar, and cellulose that are found in food wastes are used for biofuel production. Especially carbohydrate-rich substrates such as sugar beet pulp, corn silage, and fruit and vegetable waste are potential and sustainable alternatives for biohydrogen production (Cieciura-Włoch et al., 2020). Sugar beet and black strap molasses, which are wastes from the sugar processing industry, are promising alternatives for photofermentative biohydrogen production with *Rhodobacter capsulatus* JP91 (Keskin & Hallenbeck, 2012). However, the sugar concentration is found as a limiting factor since over 2 g/L concentrations were stated to have a toxic effect, either through causing excessive acidification or through the increased concentration of inhibitors.

Sugar-based waste materials are also alternative substrates for bio-butanol production with a zero-waste approach. In the study of Bravo-Venegas et al. (2023), apple pomace was used as a substrate for the production of bio-butanol by acetone-butanol-ethanol fermentation with *Clostridium beijerinckii* CECT 508. A major challenge during butanol production is the problem of pH control (acid crash); which stops cell growth, consumption of substrate and solvent fermentation thus inhibiting further butanol production. However, the authors concluded that butanol production could be optimized by controlling the fermentation conditions, that is, pH, temperature, and calcium carbonate concentration.

The bacteria consortium converts food waste into biofuel in a consecutive process involving hydrolysis, acidogenesis, acetogenesis, and methanogenesis (Ortiz-Sanchez et al., 2023). In the hydrolysis step complex organic molecules are converted into smaller molecules like glycerol, short-chain fatty acids, cellobiose, and glucose (Ortiz-Sanchez et al., 2023; Wu et al., 2025). Following hydrolysis, soluble organic materials are converted into VFAs by acidogenic microorganisms (Beschkov & Angelov, 2025). VFA is an intermediate product of this process and could be utilized as a precursor in the chemical industry or biodegradable plastic production like polyhydroxyalkanoates, etc. (Patel et al., 2021; Bhatia et al., 2024). In addition to utilization in different applications, VFAs are further fermented by acetogenic microorganisms (*Acetobacter* and *Clostridium*) leading to acetic acid, hydrogen, and carbon dioxide (Beschkov & Angelov, 2025). Eventually, methanogenic microorganisms transform these generated products into methane and carbon dioxide (Ortiz-Sanchez et al., 2023; Beschkov & Angelov, 2025). VFAs release during methane fermentation may limit the biogas production yield (Kazimierowicz et al., 2021). Food waste feedstock, rich in biodegradable organics, can quickly lead to instability in the production/consumption of VFAs, and elate the methanogenesis. This leads to the failure of the process due to VFA accumulation and a decrease in the pH, which then causes microbial stress and failure of the digester (Braguglia et al., 2018; Al-Sulaimi et al., 2022; Beschkov & Angelov, 2025). Therefore, the addition of methanogenesis inhibitors, such as biomass tar, is beneficial for the accumulation of VFAs (Liu et al., 2024). Integrating biomass tar from pyrolysis with anaerobic fermentation can efficiently meet biorefinery approaches by fully utilizing diverse organic wastes for sustainable energy production. Spent coffee grounds, meat industry waste, palm oil effluents, fruit and vegetable waste, municipal food waste or simulated food waste can effectively be used as substrate for biomethane production (Ortiz-Sanchez et al., 2023; Liu et al., 2024; Dhiman et al., 2025; Lawrence et al., 2025).

Lignocellulosic biomass from agricultural wastes such as rice straw, sweet potato waste, sawdust, potato waste, corn stalks, sugarcane bagasse, sugar beet waste, and forestry residues- is used in second-generation bioethanol production. This emerging and sustainable method has potential to replace corn and sugarcane as the primary feedstocks used for first-generation bioethanol production (Chatterjee & Venkata Mohan, 2021; Salafia et al., 2022; Faria et al., 2023). However, enzymatic hydrolysis is recommended prior to fermentation for releasing the fermentable sugars and amino nitrogen which could be used as substrates (Chen et al., 2022; Dhiman et al., 2025). Basically, bioethanol production from food wastes includes five steps: (i) feedstock preparation, (ii) pre-treatment, (iii) release of free fermentable sugars, (iv) fermentation, and (v) purification (Pour & Makkawi, 2021). For bioethanol production from agri-food waste, *Bacillus* spp., *S. cerevisiae*, *Kluyveromyces marxianus* and *Lactococcus lactis* are utilised depending on the source of waste biomass (Williams et al., 2019; Chatterjee & Venkata Mohan, 2021; Salafia et al., 2022; Mahish et al., 2024). However, hydrolysis by-products and fermentation inhibitors produced from the fermentation of lignocellulosic-based feedstock appear as a major disadvantage (Ahuja et al., 2024). Besides, some bioactive compounds having antimicrobial activity may inhibit the yeast fermentation. D-limonene found in citrus fruit wastes has such an effect, thus inhibiting ethanol production (Sabater et al., 2020).

13.3 Technical Challenges and Process Related Limitations

Feedstock cost remains a major determinant of the economic viability of fermentation-based products in industrial scale applications (Kumar et al., 2024). The valorisation of food waste streams can considerably reduce these costs due to the abundant and minimal-cost nature of such substrates. However, the efficient utilization of food waste through fermentation still faces challenges, particularly in scaling-up to larger production volumes (Martí-Quijal et al., 2021). These challenges arise from the heterogeneity of food waste collected in different batches, as well as process-related issues such as excess heat generation and high operating costs (Uçkun Kiran et al., 2014; Paz et al., 2021; Ray et al., 2025). Also, downstream operations, including product recovery and purification, are often complex and expensive (Wang et al., 2020b; Bibi et al., 2023; Ray et al., 2025). Furthermore, the generation of single product from food waste fermentation may limit the overall utilization potential and fail to achieve true zero-waste goals (Taheri & Hosseini, 2025).

Microbial inocula that are employed in food waste fermentation are diverse, either single pure cultures, mixed indigenous consortium of cultures or genetically modified microorganisms might be involved (Ray et al., 2025). Mixed cultures are generally preferred due to chemically diverse and variable composition of food wastes (Bhatia et al., 2024; Zhao et al., 2025). In anaerobic digestion systems, microorganism consortium work synergistically to degrade complex organic materials. Therefore, it is essential to evaluate the metabolic pathways of these consortia and identify the possible bottlenecks that are limiting the efficiency of conversion based on the chemical composition of waste (Ortiz-Sanchez et al., 2023).

The quality, nutrient composition and stability of food waste are also affected by storage conditions. Prolonged exposure to heat or moisture accelerates decomposition, nurture microbial competition, and provokes acidification, which then reduces the nutritional bioavailability and produces inhibitory compounds such as VFAs (Wu et al., 2025).

The fermentation efficiency is linked with the pre-treatment applications, particularly for improving the yield when lignocellulosic feedstocks are used as substrates. For example, thermal treatment can improve anaerobic fermentation; however, high temperature applications (> 100 °C) may decrease the bioavailability of phenolic compounds. Besides, the formation of Maillard reaction compounds such as acrylamide are promoted, which poses toxicity risks (Braguglia et al., 2018). Similarly, furfural and hydroxymethylfurfural from the waste may be released during acid hydrolysis used as pre-treatment, and both of these substances inhibit the microbial activity and increase the toxicity (Wu et al., 2025).

Beyond technical limitations, the lack of consumer acceptance of upcycled materials by fermentation from agricultural or food waste-based substrates remains a major obstacle to the market adoption of fermented products (Javed et al., 2025). Additional challenges include ensuring product safety, traceability, and certification (Pinela et al., 2024).

Nowadays, the commercial animal feed is substituted or partially replaced with fermented food waste. However, heavy metal accumulation in fish feed can cause potential health risks to humans following consumption. In the study of Mo et al. (2019), a risk assessment study was performed using the fermented food waste collected from restaurants and food processing facilities that was used as fish feed formulations to determine if any heavy metals exceeded the maximum permissible limit for human consumption. Their results showed no significant differences in metal/loid concentrations between the fermented food waste-based diets and control (commercial feeds) diets of cultured fish. Moreover, the growth performance of the fishes fed was unaffected, and they were considered safe for human consumption.

Rich organic material composition of food wastes makes them potential candidates to be valorized as biofertilizers after their bioconversion. Therefore, it is important to assess the microbial safety of fermented food waste regarded as a biofertilizer. Södergren et al. (2022) evaluated the risk of contamination by foodborne pathogens during hydroponic vegetable cultivation using fermented food waste-based biofertilizer. The upcycled biofertilizer was found to be microbiologically safe, though continuous risk assessment was highly recommended due to the compositional changes in wastes.

13.4 Conclusions

This chapter focuses on how zero-waste fermentation can transform food and agricultural waste into multiple upcycled products simultaneously, including organic acids, enzymes, nutraceuticals, bioplastics, and biofuels with appreciable amounts of economic value.

Shifting from conventional waste management to zero-waste fermentation systems represents a fundamental move in how waste is perceived, and transforms it from an environmental problem into a renewable source. Despite significant enhancements in laboratory-scale research, achieving a completely circular biorefinery with no residual waste streams remains challenging. The compositional variability and heterogeneity of feedstock, difficulties in process scale up, high purification costs, and regulatory limitations to the upcycled bio-based products continue to hinder industrial implementation. A comprehensive strategy that integrates microbial strain engineering, continuous process optimization and intelligent valorization pathways to enable the complete reutilization of waste streams is required to tackle those challenges. Furthermore, introducing vigorous standards for safety, traceability, and consumer acceptance is critical to ensuring the market success of products derived from zero-waste fermentation. It is also important to establish biorefinery systems close to local waste collection centres and improve the knowledge of consumers for better waste classification of organic wastes. Accordingly, this will prevent the contamination of still valuable waste which later used as substrates.

Future research should focus on developing hybrid bioprocesses that integrate fermentation with other sustainable conversion technologies, such as enzymatic hydrolysis, pyrolysis, and anaerobic digestion. The incorporation of advanced digital monitoring tools like IoT, generative AI for hands-on predicting and adapting fermentation conditions, and comprehensive life-cycle assessment methodologies will be essential for accurately evaluating both environmental impact and techno-economic performance.

References

Aamir Shahzad, H. M., Asim, Z., Elkahlout, Z. I., Mahmoud, K. A., Almomani, F., & Rasool, K. (2025). Integrated bioprocess for single-cell protein production: Converting waste-derived biogas into sustainable feed protein. *Energy Conversion and Management: X, 26*, 101036. https://doi.org/10.1016/j.ecmx.2025.101036

Abd Razak, D. L., Abd Rashid, N. Y., Jamaluddin, A., Sharifudin, S. A., & Long, K. (2015). Enhancement of phenolic acid content and antioxidant activity of rice bran fermented with *Rhizopus oligosporus* and *Monascus purpureus*. *Biocatalysis and Agricultural Biotechnology, 4*(1), 33–38. https://doi.org/10.1016/j.bcab.2014.11.003

Abdullahi, N., Abba Dandago, M., & Kabiru Yunusa, A. (2021). Review on production of single-cell protein from food wastes. *Turkish Journal of Agriculture - Food Science and Technology, 9*(6), 968–974. https://doi.org/10.24925/turjaf.v9i6.968-974.3758

Adham, N. (2002). Attempts at improving citric acid fermentation by *Aspergillus niger* in beet-molasses medium. *Bioresource Technology, 84*(1), 97–100. https://doi.org/10.1016/S0960-8524(02)00007-X

Ahuja, V., Chauhan, S., Purewal, S. S., Mehariya, S., Patel, A. K., Kumar, G., Megharaj, M., Yang, Y. H., & Bhatia, S. K. (2024). Microbial alchemy: Upcycling of brewery spent grains into high-value products through fermentation. *Critical Reviews in Biotechnology, 44*(7), 1367–1385. https://doi.org/10.1080/07388551.2023.2286430

Ajala, A. S., Adeoye, A. O., Olaniyan, S. A., & Fasonyin, O. T. (2020). A study on effect of fermentation conditions on citric acid production from cassava peels. *Scientific African, 8*, e00396. https://doi.org/10.1016/j.sciaf.2020.e00396

Ali, S. S., Elsamahy, T., Abdelkarim, E. A., Al-Tohamy, R., Kornaros, M., Ruiz, H. A., Zhao, T., Li, F., & Sun, J. (2022). Biowastes for biodegradable bioplastics production and end-of-life scenarios in circular bioeconomy and biorefinery concept. *Bioresource Technology, 363*, 127869. https://doi.org/10.1016/j.biortech.2022.127869

Al-Sulaimi, I. N., Nayak, J. K., Alhimali, H., Sana, A., & Al-Mamun, A. (2022). Effect of volatile fatty acids accumulation on biogas production by sludge-feeding thermophilic anaerobic digester and predicting process parameters. *Fermentation, 8*(4), 184. https://doi.org/10.3390/fermentation8040184

Andhalkar, V. V., Ahorsu, R., Domínguez de María, P., Winterburn, J., Medina, F., & Constantí, M. (2022). Valorization of lignocellulose by producing polyhydroxyalkanoates under circular bioeconomy premises: Facts and challenges. *ACS Sustainable Chemistry & Engineering, 10*(50), 16459–16475. https://doi.org/10.1021/acssuschemeng.2c04925

Artola, A., Font, X., Moral-Vico, J., & Sánchez, A. (2024). The role of solid-state fermentation to transform existing waste treatment plants based on composting and anaerobic digestion into modern organic waste-based biorefineries, in the framework of circular bioeconomy. *Frontiers in Chemical Engineering, 6*. https://doi.org/10.3389/fceng.2024.1463785

Beena Unni, A., & Muringayil Joseph, T. (2024). Enhancing polymer sustainability: Eco-conscious strategies. *Polymers, 16*(13), 1769. https://doi.org/10.3390/polym16131769

Ben Hadj Hmida, B., Ben Mabrouk, S., Fendri, A., Hmida-Sayari, A., & Sayari, A. (2024). Optimization of newly isolated *Bacillus cereus* α-amylase production using orange peels and crab shells and application in wastewater treatment. *3 Biotech, 14*(4), 1–19. https://doi.org/10.1007/s13205-024-03962-3

Beschkov, V. N., & Angelov, I. K. (2025). Volatile fatty acid production vs. methane and hydrogen in anaerobic digestion. *Fermentation, 11*(4), 172. https://doi.org/10.3390/fermentation11040172

Bhatia, S. K. (2023). Microbial biopolymers: Trends in synthesis, modification, and applications. *Polymers, 15*(6), 1364. https://doi.org/10.3390/polym15061364

Bhatia, S. K., Patel, A. K., & Yang, Y. H. (2024). The green revolution of food waste upcycling to produce polyhydroxyalkanoates. *Trends in Biotechnology, 42*(10), 1273–1287. https://doi.org/10.1016/j.tibtech.2024.03.002

Bhattacharya, R., Arora, S., & Ghosh, S. (2024). Bioprocess optimization for food-grade cellulolytic enzyme production from sorghum waste in a novel solid-state fermentation bioreactor for enhanced apple juice clarification. *Journal of Environmental Management, 358*, 120781. https://doi.org/10.1016/j.jenvman.2024.120781

Bibi, F., Ilyas, N., Saeed, M., Shabir, S., Shati, A. A., Alfaifi, M. Y., Amesho, K. T. T., Chowdhury, S., & Sayyed, R. Z. (2023). Innovative production of value-added products using agro-industrial wastes via solid-state fermentation. *Environmental Science and Pollution Research, 30*(60), 125197–125213. https://doi.org/10.1007/s11356-023-28765-6

Bilal, M., Mehmood, T., Nadeem, F., Barbosa, A. M., de Souza, R. L., Pompeu, G. B., Meer, B., Ferreira, L. F. R., & Iqbal, H. M. N. (2022). Enzyme-assisted transformation of lignin-based food bio-residues into high-value products with a zero-waste theme: A review. *Waste and Biomass Valorization, 13*(4), 1807–1824. https://doi.org/10.1007/s12649-021-01618-4

Braguglia, C. M., Gallipoli, A., Gianico, A., & Pagliaccia, P. (2018). Anaerobic bioconversion of food waste into energy: A critical review. *Bioresource Technology, 248*, 37–56. https://doi.org/10.1016/j.biortech.2017.06.145

Bravo-Venegas, J., Prado-Acebo, I., Gullón, B., Lú-Chau, T. A., & Eibes, G. (2023). Avoiding acid crash: From apple pomace hydrolysate to butanol through acetone-butanol-ethanol fermentation in a zero-waste approach. *Waste Management, 164*, 47–56. https://doi.org/10.1016/j.wasman.2023.03.039

Bühlmann, C. H., Mickan, B. S., Tait, S., Batstone, D. J., Mercer, G. D., & Bahri, P. A. (2022). Lactic acid from mixed food waste fermentation using an adapted inoculum: Influence of pH and temperature regulation on yield and product spectrum. *Journal of Cleaner Production, 373*, 133716. https://doi.org/10.1016/j.jclepro.2022.133716

Bühlmann, C. H., Mickan, B. S., Tait, S., Batstone, D. J., & Bahri, P. A. (2023). Lactic acid production from food waste at an anaerobic digestion biorefinery: Effect of digestate recirculation and sucrose supplementation. *Frontiers in Bioengineering and Biotechnology, 11*, 1–12. https://doi.org/10.3389/fbioe.2023.1177739

Cabezudo, I., Galetto, C. S., Romanini, D., Furlán, R. L. E., & Meini, M. R. (2023). Production of gallic acid and relevant enzymes by *Aspergillus niger* and *aspergillus oryzae* in solid-state fermentation of soybean hull and grape pomace. *Biomass Conversion and Biorefinery, 13*(16), 14939–14947. https://doi.org/10.1007/s13399-022-03435-8

Cakmak, H., Düven, G., Gultekin Subasi, B., Ilyasoglu Buyukkestelli, H., Bravo-Núñez, Á., Ozyurt, V. H., Sogut, E., & Simsek, S. (2025). Plant-based fermented foods: General concepts, improvement of the quality, metabolites, and health-related functions. *Food Bioscience, 69*, 106960. https://doi.org/10.1016/j.fbio.2025.106960

Carmona-Cabello, M., Garcia, I. L., Leiva-Candia, D., & Dorado, M. P. (2018). Valorization of food waste based on its composition through the concept of biorefinery. *Current Opinion in Green and Sustainable Chemistry, 14*, 67–79. https://doi.org/10.1016/j.cogsc.2018.06.011

Chai, K. F., Voo, A. Y. H., & Chen, W. N. (2020). Bioactive peptides from food fermentation: A comprehensive review of their sources, bioactivities, applications, and future development. *Comprehensive Reviews in Food Science and Food Safety, 19*(6), 3825–3885. https://doi.org/10.1111/1541-4337.12651

Chatterjee, S., & Venkata Mohan, S. (2021). Refining of vegetable waste to renewable sugars for ethanol production: Depolymerization and fermentation optimization. *Bioresource Technology, 340*, 125650. https://doi.org/10.1016/j.biortech.2021.125650

Chen, X., Zheng, X., Pei, Y., Chen, W., Lin, Q., Huang, J., Hou, P., Tang, J., & Han, W. (2022). Process design and techno-economic analysis of fuel ethanol production from food waste by enzymatic hydrolysis and fermentation. *Bioresource Technology, 363*, 127882. https://doi.org/10.1016/j.biortech.2022.127882

Chenebault, C., Moscoviz, R., Trably, E., Escudié, R., & Percheron, B. (2022). Lactic acid production from food waste using a microbial consortium: Focus on key parameters for process upscaling and fermentation residues valorization. *Bioresource Technology, 354*, 127230. https://doi.org/10.1016/j.biortech.2022.127230

Cheng, Z., Yang, R., Liu, X., Liu, X., & Chen, H. (2017). Green synthesis of bacterial cellulose via acetic acid pre-hydrolysis liquor of agricultural corn stalk used as carbon source. *Bioresource Technology, 234*, 8–14. https://doi.org/10.1016/j.biortech.2017.02.131

Chozhavendan, S., Aniskumar, M., Pradeepa, S., Karthika Devi, S., & Mathumitha, N. (2021). Acetic acid production by wastes and strains. *Journal of Emerging Technologies and Innovative Research, 8*(12), 335–343.

Cieciura-Włoch, W., Borowski, S., & Otlewska, A. (2020). Biohydrogen production from fruit and vegetable waste, sugar beet pulp and corn silage via dark fermentation. *Renewable Energy, 153*, 1226–1237. https://doi.org/10.1016/j.renene.2020.02.085

Costa, A. R., Salgado, J. M., Lopes, M., & Belo, I. (2022). Valorization of by-products from vegetable oil industries: Enzymes production by *Yarrowia lipolytica* through solid state fermentation. *Frontiers in Sustainable Food Systems, 6*, 1006467. https://doi.org/10.3389/fsufs.2022.1006467

Costa, S., Summa, D., Radice, M., Vertuani, S., Manfredini, S., & Tamburini, E. (2024). Lactic acid production by *Lactobacillus casei* using a sequence of seasonally available fruit wastes as sustainable carbon sources. *Frontiers in Bioengineering and Biotechnology, 12*, 1–12. https://doi.org/10.3389/fbioe.2024.1447278

Demichelis, F., Pleissner, D., Fiore, S., Mariano, S., Navarro Gutiérrez, I. M., Schneider, R., & Venus, J. (2017). Investigation of food waste valorization through sequential lactic acid fermentative production and anaerobic digestion of fermentation residues. *Bioresource Technology, 241*, 508–516. https://doi.org/10.1016/j.biortech.2017.05.174

Dessie, W., Zhang, W., Xin, F., Dong, W., Zhang, M., Ma, J., & Jiang, M. (2018). Succinic acid production from fruit and vegetable wastes hydrolyzed by on-site enzyme mixtures through solid state fermentation. *Bioresource Technology, 247*, 1177–1180. https://doi.org/10.1016/j.biortech.2017.08.171

Dhillon, G. S., Brar, S. K., Verma, M., & Tyagi, R. D. (2011). Utilization of different agro-industrial wastes for sustainable bioproduction of citric acid by *Aspergillus niger*. *Biochemical Engineering Journal, 54*(2), 83–92. https://doi.org/10.1016/j.bej.2011.02.002

Dhiman, S., Thakur, B., Kaur, S., Ahuja, M., Gantayat, S., Sarkar, S., Singh, R., & Tripathi, M. (2025). Closing the loop: Technological innovations in food waste valorisation for global sustainability. *Discover Sustainability, 6*(1), 258. https://doi.org/10.1007/s43621-025-01073-4

Dienye, B. N., Ahaotu, I., Agwa, O. K., & Odu, N. N. (2018). Citric acid production potential of *Aspergillus niger* using *Chrysophyllum albidum* peel. *Advances in Bioscience and Biotechnology, 9*, 190–203. https://doi.org/10.4236/abb.2018.94013

Dietrich, K., Dumont, M.-J., Del Rio, L. F., & Orsat, V. (2019). Sustainable PHA production in integrated lignocellulose biorefineries. *New Biotechnology, 49*, 161–168. https://doi.org/10.1016/j.nbt.2018.11.004

Dobrowolski, A., Drzymała, K., Mituła, P., & Mirończuk, A. M. (2020). Production of tailor-made fatty acids from crude glycerol at low pH by *Yarrowia lipolytica*. *Bioresource Technology, 314*, 123746. https://doi.org/10.1016/j.biortech.2020.123746

Dunuweera, A. N., Nikagolla, D. N., & Ranganathan, K. (2021). Fruit waste substrates to produce single-cell proteins as alternative human food supplements and animal feeds using baker's yeast (*Saccharomyces cerevisiae*). *Journal of Food Quality, 2021*(1). https://doi.org/10.1155/2021/9932762

Ebrahimian, F., Karimi, K., & Kumar, R. (2020). Sustainable biofuels and bioplastic production from the organic fraction of municipal solid waste. *Waste Management, 116*, 40–48. https://doi.org/10.1016/j.wasman.2020.07.049

Erskine, E., Ozkan, G., Lu, B., & Capanoglu, E. (2023). Effects of fermentation process on the antioxidant capacity of fruit byproducts. *ACS Omega, 8*(5), 4543–4553. https://doi.org/10.1021/acsomega.2c07602

Escanciano, I. A., Santos, V. E., Blanco, Á., & Ladero, M. (2023). Bioproduction of succinic acid from potato waste. Kinetic modeling. *Industrial Crops and Products, 203*, 117124. https://doi.org/10.1016/j.indcrop.2023.117124

Eurostat. (2025). *130 kg of food wasted per person annually in the EU*. Eurostat. https://ec.europa.eu/eurostat/web/products-eurostat-news/w/ddn-20251016-2

FAO. (2015). *Lost or wasted* (p. 4807). FAO. http://www.fao.org/3/i4807e/i4807e.pdf

Faria, D. J., de Carvalho, A. P. A., & Conte-Junior, C. A. (2023). Valorization of fermented food wastes and byproducts: Bioactive and valuable compounds, bioproduct synthesis, and applications. *Fermentation, 9*(10), 920. https://doi.org/10.3390/fermentation9100920

Fatima, S. N. (2024). Microbial biopolymers: A sustainable alternative to traditional petroleum-based polymers for a greener future. *Materials Today Communications, 40*, 109846. https://doi.org/10.1016/j.mtcomm.2024.109846

Fernández-López, L., González-García, P., Fernández-Ríos, A., Aldaco, R., Laso, J., Martínez-Ibáñez, E., Gutiérrez-Fernández, D., Pérez-Martínez, M. M., Marchisio, V., Figueroa, M., de Sousa, D. B., Méndez, D., & Margallo, M. (2024). Life cycle assessment of single cell protein production–A review of current technologies and emerging challenges. *Cleaner and Circular Bioeconomy, 8*, 100079. https://doi.org/10.1016/j.clcb.2024.100079

Forbes, H., Peacock, E., Abbot, N., & Jones, M. (2024). *Food waste index report*. WeDocs. https://wedocs.unep.org/handle/20.500.11822/45230

Garita-Cambronero, J., Paniagua-García, A. I., Hijosa-Valsero, M., & Díez-Antolínez, R. (2021). Biobutanol production from pruned vine shoots. *Renewable Energy, 177*, 124–133. https://doi.org/10.1016/j.renene.2021.05.093

Gooruee, R., Hojjati, M., Behbahani, B. A., Shahbazi, S., & Askari, H. (2024). Extracellular enzyme production by different species of Trichoderma fungus for lemon peel waste bioconversion. *Biomass Conversion and Biorefinery, 14*(2), 2777–2786. https://doi.org/10.1007/s13399-022-02626-7

Gottardo, M., Adele Tuci, G., Pavan, P., Dosta, J., & Valentino, F. (2024). Short and medium chain organic acids production from hydrolyzed food waste: Technical–economic evaluation and insight into the product's quality. *Chemical Engineering Science, 284*, 119539. https://doi.org/10.1016/j.ces.2023.119539

Groenewald, M., Boekhout, T., Neuvéglise, C., Gaillardin, C., van Dijck, P. W. M., & Wyss, M. (2014). *Yarrowia lipolytica* : Safety assessment of an oleaginous yeast with a great industrial potential. *Critical Reviews in Microbiology, 40*(3), 187–206. https://doi.org/10.3109/1040841X.2013.770386

Gulsunoglu-Konuskan, Z., & Kilic-Akyilmaz, M. (2022). Microbial bioconversion of phenolic compounds in agro-industrial wastes: A review of mechanisms and effective factors. *Journal of Agricultural and Food Chemistry, 70*(23), 6901–6910. https://doi.org/10.1021/acs.jafc.1c06888

Hamdy, H. S. (2013). Citric acid production by *Aspergillus niger* grown on orange peel medium fortified with cane molasses. *Annals of Microbiology, 63*(1), 267–278. https://doi.org/10.1007/s13213-012-0470-3

Helenas Perin, J. K., Biesdorf Borth, P. L., Torrecilhas, A. R., Santana da Cunha, L., Kuroda, E. K., & Fernandes, F. (2020). Optimization of methane production parameters during anaerobic co-digestion of food waste and garden waste. *Journal of Cleaner Production, 272*, 123130. https://doi.org/10.1016/j.jclepro.2020.123130

Huang, Y., Wang, Y., Shang, N., & Li, P. (2023). Microbial fermentation processes of lactic acid: Challenges, solutions, and future prospects. *Food, 12*(12), 2311. https://doi.org/10.3390/foods12122311

Hülsen, T., Hsieh, K., Lu, Y., Tait, S., & Batstone, D. J. (2018). Simultaneous treatment and single cell protein production from agri-industrial wastewaters using purple phototrophic bacteria or microalgae—A comparison. *Bioresource Technology, 254*, 214–223. https://doi.org/10.1016/j.biortech.2018.01.032

Ilhami, S., Rahman, S. N. S. A., Iqhrammullah, M., Hamid, Z., Chai, Y. H., & Lam, M. K. (2025). Polyhydroxyalkanoates production from microalgae for sustainable bioplastics: A review. *Biotechnology Advances, 79*, 108529. https://doi.org/10.1016/j.biotechadv.2025.108529

Jang, S.-K., Jung, C.-D., Seong, H., Myung, S., & Kim, H. (2022). An integrated biorefinery process for mandarin peel waste elimination. *Journal of Cleaner Production, 371*, 133594. https://doi.org/10.1016/j.jclepro.2022.133594

Javed, A., Hashmi, M. S., Nadeem, U., Javaid, U., & Rehman, T. (2025). Valorization of vegetable waste for developing high-value food products: An upcycling approach. *International Journal of Vegetable Science, 31*(5), 711–767. https://doi.org/10.1080/19315260.2025.2515427

Jayalath, S. U., & de Alwis, A. P. (2025). PHA, the greenest plastic so far: Advancing microbial synthesis, recovery, and sustainable applications for circularity. *ACS Omega, 10*(30), 32564–32586. https://doi.org/10.1021/acsomega.5c00684

Kaur, N., & Chavan, A. (2022). Production of single-cell protein from fruit and vegetable waste using an isolated strain of *Lactobacillus* sp. *Industrial Biotechnology, 18*(5), 304–313. https://doi.org/10.1089/ind.2022.0023

Kazimierowicz, J., Dzienis, L., Dębowski, M., & Zieliński, M. (2021). Optimisation of methane fermentation as a valorisation method for food waste products. *Biomass and Bioenergy, 144*. https://doi.org/10.1016/j.biombioe.2020.105913

KBV Research. (2024). *Organic acids market*. KBV Research. https://www.kbvresearch.com/organic-acids-market/

Keskin, T., & Hallenbeck, P. C. (2012). Hydrogen production from sugar industry wastes using single-stage photofermentation. *Bioresource Technology, 112*, 131–136. https://doi.org/10.1016/j.biortech.2012.02.077

Khaleel, G., Sharanagat, V. S., Upadhyay, S., Desai, S., Kumar, K., Dhiman, A., & Suhag, R. (2025). Sustainable approach toward biodegradable packaging through naturally derived biopolymers: An overview. *Journal of Packaging Technology and Research, 9*(1), 19–46. https://doi.org/10.1007/s41783-024-00180-3

Khan, S., Ul-Islam, M., Fatima, A., Manan, S., Khattak, W. A., Ullah, M. W., & Yang, G. (2023). Potential of food and agro-industrial wastes for cost-effective bacterial cellulose production: An updated review of literature. *ES Food & Agroforestry, 13*, 905. https://doi.org/10.30919/esfaf905

Khan, A. R., Ahmad, B., Khan, M., Khan, M. A., Sultan, S., Sultana, K., & Hassan, S. (2024). Production of single cell protein (SCP) from the peel waste of pea, potato, and banana by *Aspergillus flavus* NRRL 21882 as an efficient organic poultry supplement. *ACS Omega, 9*(36), 37763–37770. https://doi.org/10.1021/acsomega.4c03317

Khubber, S., Marti-Quijal, F. J., Tomasevic, I., Remize, F., & Barba, F. J. (2022). Lactic acid fermentation as a useful strategy to recover antimicrobial and antioxidant compounds from food and by-products. *Current Opinion in Food Science, 43*, 189–198. https://doi.org/10.1016/j.cofs.2021.11.013

Kleps, C., Schneider, R., Venus, J., & Pleissner, D. (2024). Investigation of the utilization of oat pomace and acid whey in technical scale succinic acid fermentation including downstream processing. *Sustainable Chemistry One World, 4*, 100028. https://doi.org/10.1016/j.scowo.2024.100028

Koller, M., Maršálek, L., de Sousa Dias, M. M., & Braunegg, G. (2017). Producing microbial polyhydroxyalkanoate (PHA) biopolyesters in a sustainable manner. *New Biotechnology, 37*, 24–38. https://doi.org/10.1016/j.nbt.2016.05.001

Koukoumaki, D. I., Tsouko, E., Papanikolaou, S., Ioannou, Z., Diamantopoulou, P., & Sarris, D. (2024). Recent advances in the production of single cell protein from renewable resources and applications. *Carbon Resources Conversion, 7*(2), 100195. https://doi.org/10.1016/j.crcon.2023.07.004

Koutinas, A. A., Vlysidis, A., Pleissner, D., Kopsahelis, N., Lopez Garcia, I., Kookos, I. K., Papanikolaou, S., Kwan, T. H., & Lin, C. S. K. (2014). Valorization of industrial waste and by-product streams via fermentation for the production of chemicals and biopolymers. *Chemical Society Reviews, 43*(8), 2587. https://doi.org/10.1039/c3cs60293a

Książek, E. (2024). Citric acid: Properties, microbial production, and applications in industries. *Molecules, 29*(1), 22. https://doi.org/10.3390/molecules29010022

Kuglarz, M., & Angelidaki, I. (2023). Succinic production from source-separated kitchen biowaste in a biorefinery concept: Focusing on alternative carbon dioxide source for fermentation processes. *Fermentation, 9*(3), 259. https://doi.org/10.3390/fermentation9030259

Kumar, P., & Kim, B. S. (2018). Valorization of polyhydroxyalkanoates production process by co-synthesis of value-added products. *Bioresource Technology, 269*, 544–556. https://doi.org/10.1016/j.biortech.2018.08.120

Kumar, D., Jain, V. K., Shanker, G., & Srivastava, A. (2003). Utilisation of fruits waste for citric acid production by solid state fermentation. *Process Biochemistry, 38*(12), 1725–1729. https://doi.org/10.1016/S0032-9592(02)00253-4

Kumar, V., Kumar, P., Maity, S. K., Agrawal, D., Narisetty, V., Jacob, S., Kumar, G., Bhatia, S. K., Kumar, D., & Vivekanand, V. (2024). Recent advances in bio-based production of top platform chemical, succinic acid: An alternative to conventional chemistry. *Biotechnology for Biofuels and Bioproducts, 17*(1), 72. https://doi.org/10.1186/s13068-024-02508-2

Kusuma, H. S., Sabita, A., Putri, N. A., Azliza, N., Illiyanasafa, N., Darmokoesoemo, H., Amenaghawon, A. N., & Kurniawan, T. A. (2024). Waste to wealth: Polyhydroxyalkanoates (PHA) production from food waste for a sustainable packaging paradigm. *Food Chemistry: Molecular Sciences, 9*, 100225. https://doi.org/10.1016/j.fochms.2024.100225

LaTurner, Z. W., Bennett, G. N., San, K.-Y., & Stadler, L. B. (2020). Single cell protein production from food waste using purple non-sulfur bacteria shows economically viable protein products have higher environmental impacts. *Journal of Cleaner Production, 276*, 123114. https://doi.org/10.1016/j.jclepro.2020.123114

Lawrence, J., Oliva, A., Papirio, S., Murphy, J. D., & Lens, P. N. L. (2025). Biofuel and volatile fatty acids production from raw and pretreated spent coffee grounds in semi-continuous bioreactors. *Journal of Water Process Engineering, 71*, 107403. https://doi.org/10.1016/j.jwpe.2025.107403

Leite, P., Salgado, J. M., Venâncio, A., Domínguez, J. M., & Belo, I. (2016). Ultrasounds pretreatment of olive pomace to improve xylanase and cellulase production by solid-state fermentation. *Bioresource Technology, 214*, 737–746. https://doi.org/10.1016/j.biortech.2016.05.028

Leite, P., Silva, C., Salgado, J. M., & Belo, I. (2019). Simultaneous production of lignocellulolytic enzymes and extraction of antioxidant compounds by solid-state fermentation of agro-industrial wastes. *Industrial Crops and Products, 137*, 315–322. https://doi.org/10.1016/j.indcrop.2019.04.044

Lende, S. V., Karemore, H., & Umekar, M. J. (2021). Review on production of citric acid by fermentation technology. *GSC Biological and Pharmaceutical Sciences, 17*(3), 085–093. https://doi.org/10.30574/gscbps.2021.17.3.0313

Li, Y., He, D., Niu, D., & Zhao, Y. (2015). Acetic acid production from food wastes using yeast and acetic acid bacteria micro-aerobic fermentation. *Bioprocess and Biosystems Engineering, 38*(5), 863–869. https://doi.org/10.1007/s00449-014-1329-8

Li, Y. P., Ahmadi, F., Kariman, K., & Lackner, M. (2024). Recent advances and challenges in single cell protein (SCP) technologies for food and feed production. *npj Science of Food, 8*(1), 66. https://doi.org/10.1038/s41538-024-00299-2

Lin, F., Li, W., Wang, D., Hu, G., Qin, Z., Xia, X., Hu, L., Liu, X., & Luo, R. (2024). Advances in succinic acid production: The enhancement of CO2 fixation for the carbon sequestration benefits. *Frontiers in Bioengineering and Biotechnology, 12*, 1–13. https://doi.org/10.3389/fbioe.2024.1392414

Liu, Q., Ji, X., Zhu, K., Wang, W., Bao, Z., & Zhang, L. (2024). Biorefinery-oriented pyrolysis and anaerobic fermentation synergies: Leveraging biomass tar for enhanced chemical production from food waste. *Journal of Cleaner Production, 477*, 143830. https://doi.org/10.1016/j.jclepro.2024.143830

Lobeda, K., Jin, Q., Wu, J., Zhang, W., & Huang, H. (2022). Lactic acid production from food waste hydrolysate by lactobacillus pentosus : Focus on nitrogen supplementation, initial sugar concentration, pH, and fed-batch fermentation. *Journal of Food Science, 87*(7), 3071–3083. https://doi.org/10.1111/1750-3841.16205

Lu, J., Lv, Y., Qian, X., Jiang, Y., Wu, M., Zhang, W., Zhou, J., Dong, W., Xin, F., & Jiang, M. (2020). Current advances in organic acid production from organic wastes by using microbial co-cultivation systems. *Biofuels, Bioproducts and Biorefining, 14*(2), 481–492. https://doi.org/10.1002/bbb.2075

Mahish, P. K., Verma, D. K., Ghritlahare, A., Arora, C., & Otero, P. (2024). Microbial bioconversion of food waste to bio-fertilizers. *Sustainable Food Technology, 2*(3), 689–708. https://doi.org/10.1039/d3fb00041a

Mandal, M., Roy, A., Mitra, D., & Sarkar, A. (2024). Possibilities and prospects of bioplastics production from Agri-waste using bacterial communities: Finding a silver-lining in waste management. *Current Research in Microbial Sciences, 7*, 100274. https://doi.org/10.1016/j.crmicr.2024.100274

Marotta, A., Borriello, A., Khan, M. R., Cavella, S., Ambrogi, V., & Torrieri, E. (2025). Boosting food packaging sustainability through the valorization of agri-food waste and by-products. *Polymers, 17*(6), 735. https://doi.org/10.3390/polym17060735

Martinez-Avila, G. C. G., Aguilera, A. F., Saucedo, S., Rojas, R., Rodriguez, R., & Aguilar, C. N. (2014). Fruit wastes fermentation for phenolic antioxidants production and their application in manufacture of edible coatings and films. *Critical Reviews in Food Science and Nutrition, 54*(3), 303–311. https://doi.org/10.1080/10408398.2011.584135

Martí-Quijal, F. J., Khubber, S., Remize, F., Tomasevic, I., Roselló-Soto, E., & Barba, F. J. (2021). Obtaining antioxidants and natural preservatives from food by-products through fermentation: A review. *Fermentation, 7*(3), 1–11. https://doi.org/10.3390/fermentation7030106

Max, B., Salgado, J. M., Rodríguez, N., Cortés, S., Converti, A., & Domínguez, J. M. (2010). Biotechnological production of citric acid. *Brazilian Journal of Microbiology, 41*(4), 862–875. https://doi.org/10.1590/S1517-83822010000400005

Mehyar, G. F., Delaimy, K. S., & Ibrahim, S. A. (2005). Citric acid production by *Aspergillus niger* using date-based medium fortified with whey and additives. *Food Biotechnology, 19*(2), 137–144. https://doi.org/10.1081/FBT-200063458

Mizzi, J., Gaggìa, F., Bozzi Cionci, N., Di Gioia, D., & Attard, E. (2022). Selection of acetic acid bacterial strains and vinegar production from local Maltese food sources. *Frontiers in Microbiology, 13*, 1–13. https://doi.org/10.3389/fmicb.2022.897825

Mo, W. Y., Man, Y. B., Zhang, F., & Wong, M. H. (2019). Fermented food waste for culturing jade perch and Nile tilapia: Growth performance and health risk assessment based on metal/loids. *Journal of Environmental Management, 236*, 236–244. https://doi.org/10.1016/j.jenvman.2019.01.102

Moccia, F., Flores-Gallegos, A. C., Chávez-González, M. L., Sepúlveda, L., Marzorati, S., Verotta, L., Panzella, L., Ascacio-Valdes, J. A., Aguilar, C. N., & Napolitano, A. (2019). Ellagic acid recovery by solid state fermentation of pomegranate wastes by *Aspergillus niger* and *Saccharomyces cerevisiae*: A comparison. *Molecules, 24*(20), 3689. https://doi.org/10.3390/molecules24203689

Mohanta, Y. K., Mishra, A. K., Lakshmayya, N. S. V., Panda, J., Thatoi, H., Sarma, H., Rustagi, S., Baek, K.-H., & Mishra, B. (2025). Agro-waste-derived bioplastics: Sustainable innovations for a circular economy. *Waste and Biomass Valorization, 16*(7), 3331–3355. https://doi.org/10.1007/s12649-025-03070-0

Molina-Peñate, E., Artola, A., & Sánchez, A. (2024). Exploring biorefinery alternatives for biowaste valorization: A techno-economic assessment of enzymatic hydrolysis coupled with anaerobic digestion or solid-state fermentation for high-value bioproducts. *Bioengineered, 15*(1), 2307668. https://doi.org/10.1080/21655979.2024.2307668

Moodley, P., & Trois, C. (2022). Circular closed-loop waste biorefineries: Organic waste as an innovative feedstock for the production of bioplastic in South Africa. *South African Journal of Science, 118*(1), 1–6. https://www.ajol.info/index.php/sajsci/article/view/231417

Mores, S., Vandenberghe, L. P. D. S., Magalhães Júnior, A. I., de Carvalho, J. C., de Mello, A. F. M., Pandey, A., & Soccol, C. R. (2021). Citric acid bioproduction and downstream processing: Status, opportunities, and challenges. *Bioresource Technology, 320*, 124426. https://doi.org/10.1016/j.biortech.2020.124426

Mussagy, C. U., Remonatto, D., Picheli, F. P., Paula, A. V., Herculano, R. D., Santos-Ebinuma, V. C., Farias, R. L., Onishi, S. D., Ribeiro, S. J. L., Pereira, F. B. J., & Pessoa, A., Jr. (2022). A look into Phaffia rhodozyma biorefinery: From the recovery and fractionation of carotenoids, lipids and proteins to the sustainable manufacturing of biologically active bioplastics. *Bioresource Technology, 362*, 127785. https://doi.org/10.1016/j.biortech.2022.127785

Najafpour, G. D. (2007). Single-cell protein. In G. D. Najafpour (Ed.), *Biochemical engineering and biotechnology* (pp. 332–341). Elsevier. https://doi.org/10.1016/B978-044452845-2/50014-8

Najari, Z., Khodaiyan, F., Yarmand, M. S., & Hosseini, S. S. (2022). Almond hulls waste valorization towards sustainable agricultural development: Production of pectin, phenolics, pullulan, and single cell protein. *Waste Management, 141*, 208–219. https://doi.org/10.1016/j.wasman.2022.01.007

Odu, N., Uzah, G., & Akani, N. (2020). Optimization of citric acid production by *Aspergillus niger* and Candida tropicalis for solid state fermentation using banana peel substrate. *Journal of Life and Bio Sciences Research, 1*(02), 51–60. https://doi.org/10.38094/jlbsr1214

Oliveira, F., Moreira, C., Salgado, J. M., Abrunhosa, L., Venâncio, A., & Belo, I. (2016). Olive pomace valorization by *Aspergillus* species: Lipase production using solid-state fermentation. *Journal of the Science of Food and Agriculture, 96*(10), 3583–3589. https://doi.org/10.1002/jsfa.7544

Onyeaka, H., Anumudu, C. K., Okpe, C., Okafor, A., Ihenetu, F., Miri, T., Odeyemi, O. A., & Anyogu, A. (2022). Single cell protein for foods and feeds: A review of trends. *The Open Microbiology Journal, 16*(1). https://doi.org/10.2174/18742858-v16-e2206160

Ordoñez-Cano, A. J., Ramírez-Esparza, U., Méndez-González, F., Alvarado-González, M., Baeza-Jiménez, R., Sepúlveda-Torre, L., Prado-Barragán, L. A., & Buenrostro-Figueroa, J. J. (2025). Recovery of phenolic compounds with antioxidant capacity through solid-state fermentation of pistachio green hull. *Microorganisms, 13*(1), 35. https://doi.org/10.3390/microorganisms13010035

Ortiz-Sanchez, M., Inocencio-García, P.-J., Alzate-Ramírez, A. F., & Alzate, C. A. C. (2023). Potential and restrictions of food-waste valorization through fermentation processes. *Fermentation, 9*(3), 274. https://doi.org/10.3390/fermentation9030274

Ozzeybek, M., & Cekmecelioglu, D. (2024). Formulation of apple pomace, orange peel, and hazelnut shell mix for co-production of bacterial pectinase and cellulase enzymes by mixture design method. *Biomass Conversion and Biorefinery, 14*(2), 1853–1861. https://doi.org/10.1007/s13399-022-02409-0

Pan, X. R., Shang-Guan, P. K., Li, S. H., Zhang, C. H., Lou, J. M., Guo, L., Liu, L., & Lu, Y. (2025). The influence of carbon dioxide on fermentation products, microbial community, and functional gene in food waste fermentation with uncontrol pH. *Environmental Research, 267*, 120645. https://doi.org/10.1016/j.envres.2024.120645

Patel, A., Mahboubi, A., Horváth, I. S., Taherzadeh, M. J., Rova, U., Christakopoulos, P., & Matsakas, L. (2021). Volatile fatty acids (VFAs) generated by anaerobic digestion serve as feedstock for freshwater and marine oleaginous microorganisms to produce biodiesel and added-value compounds. *Frontiers in Microbiology, 12*, 614612. https://doi.org/10.3389/fmicb.2021.614612

Pau, S., Tan, L. C., Arriaga, S., & Lens, P. N. L. (2024). Lactic acid fermentation of food waste at acidic conditions in a semicontinuous system: Effect of HRT and OLR changes. *Biomass Conversion and Biorefinery, 14*(10), 10979–10994. https://doi.org/10.1007/s13399-022-03201-w

Paz, A., Chalima, A., & Topakas, E. (2021). Biorefinery of exhausted olive pomace through the production of polygalacturonases and omega-3 fatty acids by Crypthecodinium cohnii. *Algal Research, 59*, 102470. https://doi.org/10.1016/j.algal.2021.102470

Pinela, J., Añibarro-Ortega, M., & Barros, L. (2024). Food waste biotransformation into food ingredients: A brief overview of challenges and opportunities. *Food, 13*(21), 3389. https://doi.org/10.3390/foods13213389

Pour, F. H., & Makkawi, Y. T. (2021). A review of post-consumption food waste management and its potentials for biofuel production. *Energy Reports, 7*, 7759–7784. https://doi.org/10.1016/j.egyr.2021.10.119

Rahman, N. A., Ngaini, Z., Farooq, S., Ai Ling, S. C., & Shahren, P. N. J. (2025). Sustainable lactic acid production from agricultural waste: A review of current techniques, challenges and future directions. *Bioresources and Bioprocessing, 12*(1), 81. https://doi.org/10.1186/s40643-025-00923-3

Rajendran, N., & Han, J. (2022). Integrated polylactic acid and biodiesel production from food waste: Process synthesis and economics. *Bioresource Technology, 343*, 126119. https://doi.org/10.1016/j.biortech.2021.126119

Rajput, S. D., Pandey, N., & Sahu, K. (2024). A comprehensive report on valorization of waste to single cell protein: Strategies, challenges, and future prospects. *Environmental Science and Pollution Research, 31*(18), 26378–26414. https://doi.org/10.1007/s11356-024-33004-7

Ray, R. C., Behera, S. S., Awogbemi, O., Sooch, B. S., Thatoi, H., Rath, S., & Aguilar-Rivera, N. (2025). Beyond enzymes and organic acids, solid-state fermentation as an alternative for valorizing fruits and vegetable wastes into novel bio-products in a circular economy: A critical review. *AIMS Microbiology, 11*(2), 462–500. https://doi.org/10.3934/microbiol.2025021

Reihani, S. F. S., & Khosravi-Darani, K. (2019). Influencing factors on single-cell protein production by submerged fermentation: A review. *Electronic Journal of Biotechnology, 37*, 34–40. https://doi.org/10.1016/j.ejbt.2018.11.005

Rodriguez-Perez, S., Cabeza Sanchez, A., & Lopez-Abelairas, M. (2022). Modelling and mathematical optimisation of wastewater treatment in food industries. *Open Research Europe, 2*, 62. https://doi.org/10.12688/openreseurope.14737.1

Roukas, T., & Kotzekidou, P. (2020). Pomegranate peel waste: A new substrate for citric acid production by *Aspergillus niger* in solid-state fermentation under non-aseptic conditions. *Environmental Science and Pollution Research, 27*(12), 13105–13113. https://doi.org/10.1007/s11356-020-07928-9

Rousta, N., Aslan, M., Yesilcimen Akbas, M., Ozcan, F., Sar, T., & Taherzadeh, M. J. (2024). Effects of fungal based bioactive compounds on human health: Review paper. *Critical Reviews in Food Science and Nutrition, 64*(20), 7004–7027. https://doi.org/10.1080/10408398.2023.2178379

Sabater, C., Ruiz, L., Delgado, S., Ruas-Madiedo, P., & Margolles, A. (2020). Valorization of vegetable food waste and by-products through fermentation processes. *Frontiers in Microbiology, 11*, 1–11. https://doi.org/10.3389/fmicb.2020.581997

Sadh, P. K., Kumar, S., Chawla, P., & Duhan, J. S. (2018). Fermentation: A boon for production of bioactive compounds by processing of food industries wastes (by-products). *Molecules, 23*(10), 2560. https://doi.org/10.3390/molecules23102560

Saha, D., & Kumar Das, P. (2024). Perspective chapter: Bioconversion of agricultural and food wastes to vinegar. In A. N. Barros, J. Campos, & A. Vilela (Eds.), *Functional food–Upgrading natural and synthetic sources*. IntechOpen. https://doi.org/10.5772/intechopen.109546

Salafia, F., Ferracane, A., & Tropea, A. (2022). Pineapple waste cell wall sugar fermentation by *Saccharomyces cerevisiae* for second generation bioethanol production. *Fermentation, 8*(3), 100. https://doi.org/10.3390/fermentation8030100

Salazar-López, N. J., Barco-Mendoza, G. A., Zuñiga-Martínez, B. S., Domínguez-Avila, J. A., Robles-Sánchez, R. M., Ochoa, M. A. V., & González-Aguilar, G. A. (2022). Single-cell protein production as a strategy to reincorporate food waste and agro by-products back into the processing chain. *Bioengineering, 9*(11), 623. https://doi.org/10.3390/bioengineering9110623

Salimi, A., Khodaiyan, F., Askari, G., & Hosseini, S. S. (2024). A zero-waste approach towards a sustainable waste management of apple: Extraction of value-added products and their application as edible coating. *Food Hydrocolloids, 147*, 109304. https://doi.org/10.1016/j.foodhyd.2023.109304

Sapmaz, T., Manafi, R., Mahboubi, A., Lorick, D., Koseoglu-Imer, D. Y., & Taherzadeh, M. J. (2022). Potential of food waste-derived volatile fatty acids as alternative carbon source for denitrifying moving bed biofilm reactors. *Bioresource Technology, 364*, 128046. https://doi.org/10.1016/j.biortech.2022.128046

Satheeshkumar, S., Sivagurunathan, P., Muthulakshmi, K., & Uma, C. (2019). Utilization of fruit waste for the production of citric acid by using *Aspergillus niger*. *Journal of Drug Delivery and Therapeutics, 9*(4-A), 9–14. https://doi.org/10.22270/jddt.v9i4-A.3487

Sekoai, P. T., Roets-Dlamini, Y., O'Brien, F., Ramchuran, S., & Chunilall, V. (2024). Valorization of food waste into single-cell protein: An innovative technological strategy for sustainable protein production. *Microorganisms, 12*(1), 166. https://doi.org/10.3390/microorganisms12010166

Serrano, A., Newton, G., Alonso-Fariñas, B., Fermoso, F. G., & Villa-Gomez, D. K. (2020). pH-controlled fermentation of strawberry waste as phenol solubilisation method. *Journal of Cleaner Production, 266*. https://doi.org/10.1016/j.jclepro.2020.121924

Sharma, V., Tsai, M.-L., Nargotra, P., Chen, C.-W., Kuo, C.-H., Sun, P.-P., & Dong, C.-D. (2022). Agro-industrial food waste as a low-cost substrate for sustainable production of industrial enzymes: A critical review. *Catalysts, 12*(11), 1373. https://doi.org/10.3390/catal12111373

Sharma, M., Tellili, N., Kacem, I., & Rouissi, T. (2024). Microbial biopolymers: From production to environmental applications—A review. *Applied Sciences, 14*(12), 5081. https://doi.org/10.3390/app14125081

Signorello, L., Arena, M. P., Brugnoli, M., Romeo, F. V., & Gullo, M. (2025). Valorization of olive mill wastewater by selective sequential fermentation. *Food, 14*(13), 2170. https://doi.org/10.3390/foods14132170

Skiba, E. A., Budaeva, V. V., Ovchinnikova, E. V., Gladysheva, E. K., Kashcheyeva, E. I., Pavlov, I. N., & Sakovich, G. V. (2020). A technology for pilot production of bacterial cellulose from oat hulls. *Chemical Engineering Journal, 383*, 123128. https://doi.org/10.1016/j.cej.2019.123128

Södergren, J., Larsson, C. U., Wadsö, L., Bergstrand, K.-J., Asp, H., Hultberg, M., & Schelin, J. (2022). Food waste to new food: Risk assessment and microbial community analysis of anaerobic digestate as a nutrient source in hydroponic production of vegetables. *Journal of Cleaner Production, 333*, 130239. https://doi.org/10.1016/j.jclepro.2021.130239

Song, L., Cai, C., Chen, Z., Lin, C., Lv, Y., Ye, X., Liu, Y., Dai, X., & Liu, M. (2024). Lactic acid production from food waste: Advances in microbial fermentation and separation technologies. *Bioresource Technology, 414*, 131635. https://doi.org/10.1016/j.biortech.2024.131635

Sorour, M., Ramadan, B., Mehanni, A., & Abdelsabour, N. (2021). Single cell protein production using low quality fruits of some dates. *Archives of Agriculture Sciences Journal, 4*(1), 115–126. https://doi.org/10.21608/aasj.2021.65566.1056

Sun, W., Zhang, Z., Li, X., Lu, X., Liu, G., Qin, Y., Zhao, J., & Qu, Y. (2024). Production of single cell protein from brewer's spent grain through enzymatic saccharification and fermentation enhanced by ammoniation pretreatment. *Bioresource Technology, 394*, 130242. https://doi.org/10.1016/j.biortech.2023.130242

Swetha, T. A., Ananthi, V., Bora, A., Sengottuvelan, N., Ponnuchamy, K., Muthusamy, G., & Arun, A. (2023). A review on biodegradable polylactic acid (PLA) production from fermentative food waste–Its applications and degradation. *International Journal of Biological Macromolecules, 234*, 123703. https://doi.org/10.1016/j.ijbiomac.2023.123703

Taheri, S., & Hosseini, S. S. (2025). Waste not, want not: Comprehensive valorization of fruit and vegetable waste from single-product recovery to zero-waste strategies. *Cleaner Waste Systems, 11*, 100300. https://doi.org/10.1016/j.clwas.2025.100300

Talan, A., Pokhrel, S., Tyagi, R. D., & Drogui, P. (2022). Biorefinery strategies for microbial bioplastics production: Sustainable pathway towards circular bioeconomy. *Bioresource Technology Reports, 17*, 100875. https://doi.org/10.1016/j.biteb.2021.100875

Tang, J., Hu, Z., Pu, Y., Wang, X. C., & Abomohra, A. (2024). Bioprocesses for lactic acid production from organic wastes toward industrialization-a critical review. *Journal of Environmental Management, 369*, 122372. https://doi.org/10.1016/j.jenvman.2024.122372

Thakur, S., Chaudhary, J., Singh, P., Alsanie, W. F., Grammatikos, S. A., & Thakur, V. K. (2022). Synthesis of bio-based monomers and polymers using microbes for a sustainable bioeconomy. *Bioresource Technology, 344*, 126156. https://doi.org/10.1016/j.biortech.2021.126156

Thiviya, P., Gamage, A., Kapilan, R., Merah, O., & Madhujith, T. (2022). Single cell protein production using different fruit waste: A review. *Separations, 9*(7), 178. https://doi.org/10.3390/separations9070178

Tornuk, F., & Akman, P. K. (2025). Recent developments in the valorization of agri-food waste and byproducts by fermentation. *Journal of the Science of Food and Agriculture, 106*(5), 2567–2577. https://doi.org/10.1002/jsfa.70160

Tsouko, E., Kourmentza, C., Ladakis, D., Kopsahelis, N., Mandala, I., Papanikolaou, S., Paloukis, F., Alves, V., & Koutinas, A. (2015). Bacterial cellulose production from industrial waste and by-product streams. *International Journal of Molecular Sciences, 16*(7), 14832–14849. https://doi.org/10.3390/ijms160714832

Uçkun Kiran, E., & Liu, Y. (2015). Bioethanol production from mixed food waste by an effective enzymatic pretreatment. *Fuel, 159*, 463–469. https://doi.org/10.1016/j.fuel.2015.06.101

Uçkun Kiran, E., Trzcinski, A. P., Ng, W. J., & Liu, Y. (2014). Enzyme production from food wastes using a bBiorefinery concept. *Waste and Biomass Valorization, 5*(6), 903–917. https://doi.org/10.1007/s12649-014-9311-x

UN. (2025). *Stop food loss and waste. For the people. For the Planet*. UN. https://www.un.org/en/observances/end-food-waste-day

Upadhyay, A., Kovalev, A. A., Zhuravleva, E. A., Pareek, N., & Vivekanand, V. (2023). Enhanced production of acetic acid through bioprocess optimization employing response surface methodology and artificial neural network. *Bioresource Technology, 376*, 128930. https://doi.org/10.1016/j.biortech.2023.128930

Vicenssuto, G. M., & de Castro, R. J. S. (2020). Development of a novel probiotic milk product with enhanced antioxidant properties using mango peel as a fermentation substrate. *Biocatalysis and Agricultural Biotechnology, 24*, 101564. https://doi.org/10.1016/j.bcab.2020.101564

Vidya, P., Am, A., & Jalalugeen, H. (2018). Optimization and utilisation of various fruit peel as substrate for citric acid production by *Aspergillus niger* isolated from orange and carrot. *The Pharma Innovation Journal, 7*(6), 141–146.

Vijayendra, S. V. N., & Shamala, T. R. (2014). Film forming microbial biopolymers for commercial applications—A review. *Critical Reviews in Biotechnology, 34*(4), 338–357. https://doi.org/10.3109/07388551.2013.798254

Villanueva-Galindo, E., Pérez-Rangel, M., & Moreno-Andrade, I. (2025). Lactic acid production from different sources of organic solid waste: Evaluation of the inoculum type and operational optimization. *Biomass Conversion and Biorefinery, 15*(5), 7707–7718. https://doi.org/10.1007/s13399-024-05563-9

Vong, W. C., Au Yang, K. L. C., & Liu, S. Q. (2016). Okara (soybean residue) biotransformation by yeast *Yarrowia lipolytica*. *International Journal of Food Microbiology, 235*, 1–9. https://doi.org/10.1016/j.ijfoodmicro.2016.06.039

Wang, Q., Feng, K., & Li, H. (2020a). Nano iron materials enhance food waste fermentation. *Bioresource Technology, 315*, 123804. https://doi.org/10.1016/j.biortech.2020.123804

Wang, Q., Li, H., Feng, K., & Liu, J. (2020b). Oriented fermentation of food waste towards high-value products: A review. *Energies, 13*(21), 5638. https://doi.org/10.3390/en13215638

Wang, J., Zeng, A., & Yuan, W. (2022). Succinic acid fermentation from agricultural wastes: The producing microorganisms and their engineering strategies. *Current Opinion in Environmental Science & Health, 25*, 100313. https://doi.org/10.1016/j.coesh.2021.100313

Wang, T., Sun, B., Tang, K., Shen, W., Chen, C., & Sun, D. (2023). Sustainable bacterial cellulose derived composites for high-efficiency hydrogen evolution reaction. *International Journal of Biological Macromolecules, 242*, 125173. https://doi.org/10.1016/j.ijbiomac.2023.125173

Wang, X., Sun, W., Sun, S., Cheng, S., Campos, L. C., & Li, Z. (2025). Process optimization and bioaugmentation for enhanced acetic acid fermentation from kitchen waste. *Journal of Water Process Engineering, 78*, 108690. https://doi.org/10.1016/j.jwpe.2025.108690

Williams, D. L., Schückel, J., Vivier, M. A., Buffetto, F., & Zietsman, A. J. J. (2019). Grape pomace fermentation and cell wall degradation by *Kluyveromyces marxianus* Y885. *Biochemical Engineering Journal, 150*, 107282. https://doi.org/10.1016/j.bej.2019.107282

Williamson, K. M. (2021). *Multivalorizaiton of beverage industry waste streams for the development of bioplastics and food ingredients*. The Ohio State University. https://www.proquest.com/docview/2689207865?fromopenview=true&pq-origsite=gscholar&sourcetype=Dissertations&Theses#

Wu, G., Chong, J. W. R., Khoo, K. S., Tang, D. Y. Y., & Show, P. L. (2025). Upcycling food waste for microalgae cultivation toward lipid production in a closed-loop and system-integrated circular bioeconomy. *Biotechnology for Biofuels and Bioproducts, 18*(1), 74. https://doi.org/10.1186/s13068-025-02679-6

Yadav, J. S. S., Yan, S., Ajila, C. M., Bezawada, J., Tyagi, R. D., & Surampalli, R. Y. (2016). Food-grade single-cell protein production, characterization and ultrafiltration recovery of residual fermented whey proteins from whey. *Food and Bioproducts Processing, 99*, 156–165. https://doi.org/10.1016/j.fbp.2016.04.012

Yang, L., Chen, L., Li, H., Deng, Z., & Liu, J. (2022a). Lactic acid production from mesophilic and thermophilic fermentation of food waste at different pH. *Journal of Environmental Management, 304*, 114312. https://doi.org/10.1016/j.jenvman.2021.114312

Yang, R., Chen, Z., Hu, P., Zhang, S., & Luo, G. (2022b). Two-stage fermentation enhanced single-cell protein production by *Yarrowia lipolytica* from food waste. *Bioresource Technology, 361*, 127677. https://doi.org/10.1016/j.biortech.2022.127677

Ye, L., Bogicevic, B., Bolten, C. J., & Wittmann, C. (2024). Single-cell protein: Overcoming technological and biological challenges towards improved industrialization. *Current Opinion in Biotechnology, 88*, 103171. https://doi.org/10.1016/j.copbio.2024.103171

Yu, P., Tu, W., Wu, M., Zhang, Z., & Wang, H. (2021). Pilot-scale fermentation of urban food waste for volatile fatty acids production: The importance of pH. *Bioresource Technology, 332*, 125116. https://doi.org/10.1016/j.biortech.2021.125116

Zeng, J., Zeng, H., & Wang, Z. (2022). Review on technology of making biofuel from food waste. *International Journal of Energy Research, 46*(8), 10301–10319. https://doi.org/10.1002/er.7868

Zhang, A. Y., Sun, Z., Leung, C. C. J., Han, W., Lau, K. Y., Li, M., & Lin, C. S. K. (2013). Valorisation of bakery waste for succinic acid production. *Green Chemistry, 15*(3), 690. https://doi.org/10.1039/c2gc36518a

Zhang, Z., Wang, Y., Wang, X., Zhang, Y., Zhu, T., Peng, L., Xu, Y., Chen, X., Wang, D., Ni, B.-J., & Liu, Y. (2024). Towards scaling-up implementation of polyhydroxyalkanoate (PHA) production from activated sludge: Progress and challenges. *Journal of Cleaner Production, 447*, 141542. https://doi.org/10.1016/j.jclepro.2024.141542

Zhao, W., Zhang, J., Hou, P., Zhang, G., & Long, Z. (2025). Valorisation of food waste through self-fermentation and photosynthetic bacterial protein production: Efficiency, microbial dynamics and safety assessment. *Bioresource Technology, 436*, 132982. https://doi.org/10.1016/j.biortech.2025.132982

Zhong, C. (2020). Industrial-scale production and applications of bacterial cellulose. *Frontiers in Bioengineering and Biotechnology, 8*. https://doi.org/10.3389/fbioe.2020.605374

Zhou, W., Bergsma, S., Colpa, D. I., Euverink, G.-J. W., & Krooneman, J. (2023). Polyhydroxyalkanoates (PHAs) synthesis and degradation by microbes and applications towards a circular economy. *Journal of Environmental Management, 341*, 118033. https://doi.org/10.1016/j.jenvman.2023.118033

[illegible]

Chapter 14
Next-Generation Protein Foods: Integrating Fungal Fermentation, Traditional Tempeh, and Cultivated Hybrids

Vipasha Sharma, Tarun Kumar, Pallavi Sharma, and Sandip Patil

Abstract As population increases globally and environmental impacts on conventional livestock production become more pronounced, the search for more sustainable protein alternatives has intensified. Proteins produced via fermentation, such as mycoprotein and tempeh, as well as hybrid lab-grown foods, offer a growing solution combining old food technology and new biotechnology. Mycoprotein, produced by fermentation of a fungus, is a complete protein source, including sufficient amounts of all the essential amino acids, dietary fiber, and offers health benefits by improving feelings of fullness, and enhancing metabolic health and cholesterol management. Its production via submerged and solid-state fermentation is environmentally sound, requiring less land and water, and producing lower amounts of greenhouse gas emissions. Similarly, tempeh, an ancient Indonesian food made from fermented soybeans, has nutritionally enhanced properties from the fermentation process which increases digestibility of the protein, improves the bioavailability of beneficial compounds, and diminishes the presence of anti-nutritional factors. Recent advancements in technology, especially with regard to starter cultures, the control of fermentation using metabolomics, and the use of alternative substrates, have greatly enhanced the potential of tempeh for improved function and greater market potential. Additionally, hybrid lab-grown foods, which combine cultivated animal cells with plant and fermentation components, offer a more sustainable options to mimic traditional meat and provide improved nutritional profiles. Although advances in technology have occurred, there still exists challenges in consumer acceptance, regulatory issues, and economically viable large-scale production. In conclusion, hybrid systems and fermentation-derived proteins have pioneered the creation of sustainable, nutritious, and scalable options for proteins derived from animals.

V. Sharma · T. Kumar (✉) · P. Sharma
Department of Biotechnology, Chandigarh University, Mohali, Punjab, India

S. Patil
Department of Haematology and Oncology, Shenzhen Children Hospital, Shenzhen, China

S. Dhiman, S. Kaur (eds.), *Next-Gen Plant Based Fermented Foods: Innovations for a Sustainable Future*, https://doi.org/10.1007/978-3-032-29927-7_14

Keywords Fermenting the future of alternative proteins · Mycoproteins · Tempeh innovations and lab-grown hybrid products

14.1 Introduction

The ongoing increase in world population and concurrent changes in the global food sector necessitates the adoption of multi-angle sustainability measures. Also, conventional animal farming practices have a deficit on the sustainable side, contributing significantly to greenhouse gas emissions, land and water use, and overall consumption (Poore & Nemecek, 2018). There are also measurable challenges including climate change, resource depletion and customer demand. For sustainable, nutritious and ethical food sources, plant-based proteins are being explored as one of the most viable options. Within these, we posit that proteins derived from fermentation combines elements of the prehistoric and the modern in a manner that epitomizes the potential of food of the future (Kaur et al., 2022). Alternative proteins are a highly promising class of materials that are nutritious and environmentally sustainable as a substitute for traditional meat ingredients. Currently, there are plant-based protein products on the market, however, these have low market attractiveness because they also have poor protein solubility and structural complexity. Therefore, to design a food product with bioactivity, it is imperative to relate the protein's structure and processing to the gastro-small intestinal phase of digestion. Ultimately, this is expected to yield the optimal combination of protein and bioavailability and favorable digestion characteristics (Liu et al., 2024). The market for plant proteins is expected to reach 62% by 2025.

Innovation and sustainability are the driving forces behind the rapid growth of proteins from microbial and fermentation sources. Protein from insects is emerging but is stymied by regulatory and consumer acceptance hurdles (Alternative Protein Market Demand & Trends 2025 to 2035, n.d.). Hence, microbial protein is positioned to be a significant player in a fully sustainable system. Even more interesting is the anticipation that scaling up production will be easier than other methods. In addition to addressing challenges related to agricultural residues, lignocellulosic biomass, and industrial by-products, this approach is aligned with the goals of the circular economy (Cedeno et al., 2025), and with advances in fermentation (including precision) technology, mycoprotein, algal protein, and even precision fermented proteins are within reach of sustainable alternative protein sources (Liu et al., 2024). Mycoprotein, and hybrid lab-grown proteins, and novel products related to tempeh (including modernized forms of traditional tempeh products) encapsulate the intersection of traditional food cultures and the latest fermentation technologies, and are the focus of this chapter (Figs. 14.1 and 14.2).

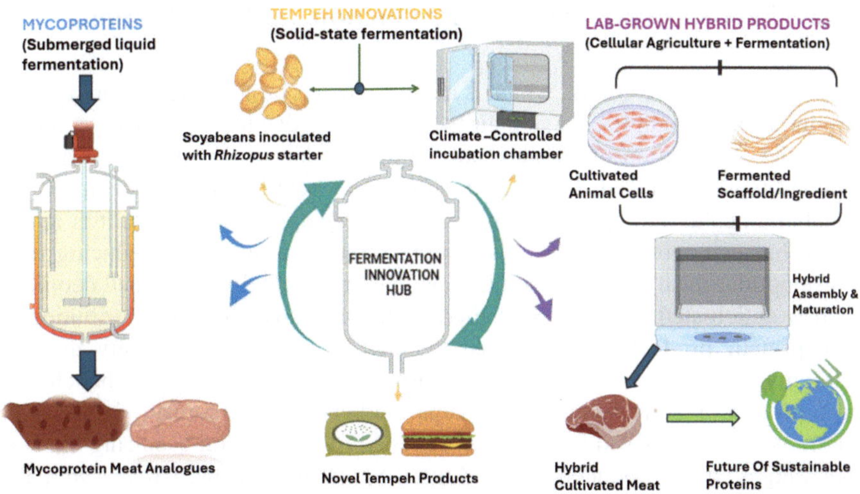

Fig. 14.1 The figure describes three types of alternative meat products. The first panel shows how mycoprotein is produced via submerged liquid fermentation. The second panel shows how tempeh is produced using conventional methods. The third panel shows how lab-grown hybrid meat is produced by using combined cultivated cells of complementary proteins

Feature/Attribute	Mycoproteins	Tempeh Innovations	Lab – Grown Hybrid Products
Core Process	Fermentation (Submerged Liquid Fermentation)	Fermentation (Solid- State)	Cellular Agriculture & Scaffold Fermentation
Key Organism/Cell Type	Filamentous Fungi	*Rhizopus* spp. & Novel Substrates	Animal Cells + Plant/ Microbial Scaffold
Primary Structure/Texture Source	Interwoven Fungal Hyphae Network	Mycelial Binding of Solid Ingredients	Cultured Animal Tissue on Edible Framework
Typical Final Product Form	Mince, Chunks Fillets (Meat Analogues)	Dense Blocks, Burgers (Whole Food Texture)	Structured Whole – Cuts (Steaks, Filets)
Key Benefit/Innovation Focus	Scalability, High Fiber, Muscle- like Texture	Traditional Basis, Gut Health, Diverse Ingredients	Biomimicry, Authentic Taste, Complete Sensory Experience

Fig. 14.2 The figure compares the overview of next generation alternative protein. This figure shows three types of different processes for the production of mycoproteins, tempeh and lab grown hybrid products attributing relevance to the alternative protein innovation

14.2 Mycoproteins: Science, Production, and Applications

Meat substitutes like Quorn, which is made from myoprotein, are becoming more popular as more people change their lifestyle habits due to the growing concerns about the exploitation of animals and the destruction of the planet. Because of mycoprotein's balanced amino acid profile and health benefits like improved metabolic health and appetite control, it is a very appealing option. It is anticipated that mycoprotein will be used more frequently by people reducing meat consumption, meat eating flexitarians, and environmentally concerned people. It is also anticipated that by 2030 there will be a positive net contribution to mycoprotein production from a commercial standpoint for society and the economy. ("Mycoprotein," 2022). Quorn is the commercialized benchmark mycoprotein, readily available in supermarkets. The organism from which Quorn is made is derived from *Fusarium venenatum*, a natural occurring filamentous fungus. It was founded by Dr. Arnold Spicer in Marlow, UK in 1967. The Quorn is grown in maize/wheat-derived high grade carbohydrates utilising an airlift fermenter.

A good example is mycoprotein which contains 6% dietary fibre (Mycoprotein and Health—Denny—2008—Nutrition Bulletin—Wiley Online Library, n.d.). This is high in dietary fibre as many other food sources contain less than 1% and it is also low in fat and saturated fat and has a high-quality protein. In addition to this, mycoprotein contains an unusual type of fibre which is polymeric n-acetyl glucosamine with beta 1,3 and 1,6 glucan. It has been suggested that this type of fibre is responsible for the positive effects seen in clinical studies of mycoprotein and its ability to maintain healthy blood lipids (Turnbull et al., 1990) and increase feelings of fullness (Williamson et al., 2006). Over the past 20 years, Denny et al. (2008) have summarised the clinical studies that have been done on the positive effects of mycoprotein on health. Research continues to be done on the mechanisms that cause feelings of fullness.

Mycoprotein has demonstrated tradeable abilities in relation to lifestyle diseases and has even provided remedies to lifestyle choices of developed nations. Nonetheless, the debate is moving to the global problem of a continually growing population and food (or protein) insecurity. There is also a rising recognition that the food and farming systems are significant contributors to the greenhouse gas problems (Garnett, 2011).

Recent technological developments in the structure of proteins, taste-based design and nutritional enhancement pave the way to natural-mimicking analogues. Fermentation techniques, in particular, stand out as they enable improvements of digestibility, bioavailability and functional properties and sustainability through resource-efficient production (Lima et al., 2022).

14.2.1 Fermentation Processes in Mycoprotein Production

Biomass Fermentation:
The optimal biomass production of mycoprotein is done by using fermentation processes which are explained below.

Submerged Fermentation (SmF): This process is performed in liquid bioreactor (e.g., airlift fermenters) under controlled environment which requires high-energy input particularly for aeration and agitation. This process offers fine control over microbial contamination, ensuring food safety. In SmF, fibrous mass is produced having meat-like texture. Microbes are grown aerobically or partial anaerobic in this process and water is extensively employed for the same (Nevalainen, 2020). In the course of this medium standardization is performed as far as Carbon, Nitrogen, micronutrients, salts, vitamins and minerals are concerned. Submerged fermentation has been stated as more advantageous because it is easier and quick to control the culture conditions (Bakratsas et al., 2023). In the field for growing of fungus preferential use of liquid bioreactors with high aeration (e.g. airlift fermentors) are standard. Optimal parameters (pH ~6.0, temperature 28–30 °C, aeration and agitation) are sustained to obtain maximum biomass and protein production (Majumder et al., 2024; Risner et al., 2023).

Solid-State Fermentation (SSF): This process involves solid substrates such as cereal grains. Uses less water and energy than SmF. Applicable in low volume or decentralized production. Frequently used in applications similar to conventional tempeh.

In this approach the fermentation takes place on solid substrates, and therefore less water is needed. The SSF is proven to have good growth on barley, wheat and oats (in the Fusarium genus). The strategy relies on the activation of the Fusarium mycelium under liquid culture. The liquid media are critical for proper mycelium growth and depend directly on the composition of the media which comprises suitable concentrations of carbon, nitrogen, sugar(s), salts, vitamins and minerals when complex media is being used. Following the optimal growth of mycelium in liquid culture, the media is prepared for solid substrate fermentation, which include cereal grains as the solid substrate and subsequent addition of carbon and nitrogen sources then a combination of vitamins, salts and minerals. It also includes controlled environment such as temperature, pH and humidity (pH ~ 6.0; temperature: 28–30 °C.; aerated—stirred). Importantly, the mass: mass nutrient to mycelial biomass ratio has a crucial effect on achievement of optimum mycoprotein ("Filamentous Fungi Processing by Solid-State Fermentation," 2023; Schweiggert-Weisz et al., 2020).

14.2.2 Downstream Processing

After fermentation, the biomass is then heated to reduce the amount of RNA. Centrifugation to separate solids. Storage and formulated using freezing or drying techniques.

Benefits of mycoprotein sourced from fermentation has the sustainability approach as it requires less land and water, combatting CO_2, suitable for industrial-scale production, high in protein and fiber, but fairly neutral tasting. The product is versatility as it can be applied in meat analogs, dairy alternatives and dry products (Cunha et al., 2025).

14.2.3 Protein Quality, Amino-Acid Composition, Dietary Fibre and Health Effects

Mycoprotein possesses full essential amino acid profile and in some instances better than others plant proteins such as that of soy, pea, rice which confirms it as a high-quality protein source (Ahmad et al., 2022; Lee et al., 2024). Its amino acid score and digestibility generally exceed those of plant proteins (Cao et al., 2025; Landeta-Salgado et al., 2025). The inclusion of all essential amino acids also matches with dietary factors related to the health and metabolism of muscles (Coelho et al., 2020; Derbyshire & Ayoob, 2019).

Mycoprotein intake has several health effects, including increased muscle protein synthesis, reduced cholesterol and glycemic response and enhanced satiety (Ahmad et al., 2022; Coelho et al., 2020; Derbyshire & Ayoob, 2019; Lee et al., 2024). Some of its healthy compound contents such as high fiber and bioactive compounds influence the metabolism systemic health and gut performance (Lee et al., 2024; Boukid et al., 2024). Accumulating evidence backs its involvement in muscular maintenance at any age as well as possibilities for reducing disease risk (Coelho et al., 2020; Derbyshire & Ayoob, 2019).

Its fibrous texture and functionality due to which it is used as meat analogues and an array of the food products increasing their textures and dietary profiles (Kaur et al., 2022; Rohmadi et al., 2025). Researches have reported that protein bars and meat analogues have been developed by incorporating succinylated chickpea protein with acceptable sensory profiles at low incorporation levels (You et al., 2024; Zhang et al., 2024). It is compatible with extrusion and fermentation processes that enable (Liu et al., 2023; Rohmadi et al., 2025) industrial applicability (Cao et al., 2025; Rohmadi et al., 2025).

14.2.4 Environmental Sustainability and Production Efficiency

The environmental footprint of mycoprotein production is low—it uses less land and water than animal and some plant proteins ("Mycoprotein," 2024). Utilization of agro-industrial wastes as substrate improves circularity and reduces expenditures (Hashempour-Baltork et al., 2020). Life cycle assessments have shown very much lower greenhouse emissions than animal (meat and dairy) proteins (Hashempour-Baltork et al., 2020). Mycoprotein production using fungal fermentation has lower environmental impact, when compared with cultivation of plant protein due to less land use, water and lesser carbon emission (Hashempour-Baltork et al., 2020). Agro-industrial wastes utilization and circular bioeconomy methods would also improve its sustainability aspects (Wikandari et al., 2023).

14.2.5 Applications and Food Product Development

Studies have documented the successful utilization of mycoprotein and plant proteins in several food applications, such as meat analogues, protein bars and frozen desserts, achieving an effective compromise between nutritional quality and sensory acceptance (Ng et al., 2024a, b; Rohmadi et al., 2025; You et al., 2024). Further, the texture, flavor and consumer acceptability issues are tackled with blending, fermentation and advanced processing technologies (Kaur et al., 2022; Zhang et al., 2024).

By application of solid-state and submerged fungal fermentation, protein content, amino acid profile, and digestibility plant-based proteins are improved closing nutritional gaps as well as functional properties (de Reu et al., 1993; Wolkers–Rooijackers et al., 2024). New fermented proteins, such as from Shiitake mycelia or fungal consortia, present interesting nutritional and safety profiles (Landeta-Salgado et al., 2025).

Safety assessments reveal both mycoproteins and fermented plant proteins to be safe but, with some potential allergenicity, particularly in individuals sensitive towards fungi or particular plants (Hashempour-Baltork et al., 2020). Consumer acceptance is driven by sensory characteristics, nutritional information, and ethical aspects, highlighting the need to enhance palatability and marketability (Wikandari et al., 2021).

14.3 Tempeh and Its Modern Innovations

Tempeh is one of the oldest fermented food products originating from the island of Java in Indonesia (Ajie et al., 2016; Kurniadi et al., 2017; Teoh et al., 2024; Wahyuono et al., 2021; Wiloso et al., 2019). It is made from soybeans and is highly nutritious, containing a high level of protein (Aaslyng & Højer, 2021; Babu et al., 2009; Mital & Garg, 1990). The food is an inexpensive substitute for meat, and for this reason has become a food staple in many countries including the United States and Europe. During the fermentation process, soybeans are transformed into a white and firm cake-like product (Mital & Garg, 1990; Teoh et al., 2024).

The traditional production of tempeh uses the fungus *Rhizopus oligosporus* (Kurniadi et al., 2017; Mital & Garg, 1990; Teoh et al., 2024; Wahyuono et al., 2021). Other species of Rhizopus, including Rhizopus oryzae and Rhizopus stolonifer, are also used in the production of tempeh (Kameda et al., 2018; Nursiwi et al., 2021). This process alters the nutritional content of the soybeans. *R. oligosporus*, for example, produces enzymes that are able to convert glycosides of isoflavones into isoflavones, which are much more bioavailable, and also lessens the presence of certain anti-nutrients (e.g., phytic acid). This ultimately increases the digestibility of the soybeans (Borges et al., 2016; Kameda et al., 2018; Mital & Garg, 1990; Sutardi & Buckle, 1985a, b; Teoh et al., 2024; Wahyuono et al., 2021).

The traditional methods of producing tempeh are often utilized in home industries in Indonesia and involve non-standardized methods and equipment. The process usually includes the following steps: soaking and dehulling, cooking, inoculation followed by fermentation. During fermentation, the inoculated soybeans are packed (often using banana leaves) which offers a suitable environment for aerobic fermentation and moisture control (Wahyuono et al., 2021). The mold *Rhizopus oligosporus* is able to grow rapidly and form a dense white mycelial mat which binds the soybeans into a compact cake. This fermentation occurs at 30–35 °C for 24–48 h (Mital & Garg, 1990; Teoh et al., 2024). During this fermentation phase, the fungus secretes various enzymes, which causes the breakdown of proteins into distinct and the various and ample amino acids and lipids into free fatty acids, and contributing to the distinct flavor and aroma of the free fatty acids of tempeh (Jeleń et al., 2013; Nowak & Szebiotko, 1992; Teoh et al., 2024).

Traditional tempeh production may also involve a process called "back-slopping," in which a small amount of the soak water from a previously successful fermentation is added to a new batch. This is believed to be important for the attainment of high-quality tempeh because it influences the microbial communities during the soaking step (Yan et al., 2013).

Although *Rhizopus* is the main fermenting agent, the microbial diversity in tempeh is considerable. In addition to this mold, various bacteria and yeasts, such as *Lactobacillus plantarum, Staphylococcus xylosus,* and *Saccharomyces cerevisiae*, participate in the fermentation and affect the various aspects of the process, such as its flavor and safety (Ashenafi & Busse, 1991; Chen et al., 2020; Feng et al., 2005; Rizal et al., 2021; Suwanto, 2021). For instance, LAB may enhance the safety of tempeh by suppressing pathogenic bacteria. The variability of the microbial dynamics, especially yeasts and LAB, tends to differ based on production techniques during tempeh production (Efriwati et al., 2013).

14.3.1 Enhancement of Nutritional Profile

Tempeh is a functional food (Górska et al., 2025) that has an improved nutritional profile. This is especially true when new raw materials are considered, e.g., tarwi and mung bean that display greater nutritional value compared to the standard soybean tempeh (Górska et al., 2025). The variety of the bioactive profiles observed in diverse substrates also reflects that functional foods with selective health-promoting properties could be designed (Górska et al., 2025).

To conclude, the tempeh fermentation process predominantly mediated by *R. oligosporus* is a strong force capable of not only removing anti-nutritional compounds but also maintaining and improving bioactive beneficial properties, toward creation of healthier plant-based foods with more functionalities.

14.3.2 Fermentation Processes in Tempeh Production

The fermentation of tempeh is achieved using a complex mixed culture of *Rhizopus* sp., *lactobacilli*, yeasts, and other bacteria; this combination may exert either synergistic or competitive influences on growth kinetics, proteolysis, and nutritional properties (Pramudito et al., 2021; Rizal et al., 2021; Suwanto, 2021; Wolkers–Rooijackers et al., 2024). Co-fermentation of tempeh with *Propionibacterium freudenreichii* and probiotic yeasts enhances its vitamin B12 and β-glucan content. Fermentation parameters affect the levels of isoflavone aglycone and the antioxidant capacity of the tempeh. Fermentation with a proprietary blend of microorganisms synergistically and optimally bioactive compounds allows the tempeh to achieve its potential as a functional food (Rizal et al., 2021; Signorini et al., 2018). The best multi-microbial starter coupled with the conserved biodiversity of *Rhizopus* isolates enhances fermentation consistency and sensory acceptability (Hartanti et al., 2024; Pramudito et al., 2021; Qibty et al., 2023; Wiesel et al., 1997; Wikandari et al., 2021).

Recent advancements include improved starter formulations, automated fermentation chambers with temperature and humidity controls, and fuzzy logic models to determine optimal inoculum concentration and fermentation duration. Technologies have been created to mitigate issues with the quality and consistency of manufacturing tempeh (Pramesti et al., 2025).

The literature considers the fermentation of tempeh using various other legumes and seeds such as (i) lupin, (ii) sorghum, (iii) chickpea, (iv) broad bean, (v) durian seeds, (vi) velvet bean, and (vii) faba bean. The other raw materials require fermentation to be accompanied by optimizations of both the fermentation conditions and the starter culture. Thus, this provides an additional opportunity to enhance the nutrient profile of the tempeh and diversify its potential market (Devagopalan & Krishnaswamy, 2023; Fernandez Castaneda et al., 2024; Hachmeister & Fung, 1993; Faisal et al., 2025; Reyes-Moreno et al., 2004).

The metabolic profiling or 'metabolomics' of the various phases of fermentation systems shows the dynamics of the constituent sugars, amino acids, organic acids, and isoflavones. This also relates to the interplay of metabolite profiles with metabolic conditions and microbial metabolism. This helps elucidate the intricate biochemistry of transformations affecting the quality and process optimization of tempeh (Kadar et al., 2020; Riyanto et al., 2024). Several sensory attributes (texture, aroma, color, and taste) are linked to the microbial fermentations, the processes and raw materials used.

14.3.3 Innovations in Tempeh

The modern science and cutting-edge technology have led into development of tempeh in various ways,

- **Novel substrates:** The use of chickpeas, lentils, oat, and even sun flower seeds for soy-free tempeh production are investigated owing to the demand from individuals with allergy-friendly diets (Nout & Kiers, 2005).
- **Biofortification:** Addition of Vitamin B12 or iron to tempeh product in order to fulfil the need of modern diet (Ahnan-Winarno et al., 2021).
- **Flavour engineering:** Bitterness is minimized and favourable flavours are maximized through the use of controlled fermentation techniques (The Book of Tempeh—William Shurtleff, Akiko Aoyagi—Google Books, n.d.).
- **Global market adaptations:** Meanwhile, ready-to-eat tempeh snacks, marinated tempeh strips and fusion recipes are gaining ground in Western nations.

Tempeh illustrates one-way ancient wisdom can be synthesized with science and technology to bring a sustainable, healthy food into the modern world.

14.3.4 Hybrid Lab-Grown Foods: Fermentation Meets Cell Culture

Hybrid lab-grown foods are produced by integrating traditional meat elements with novel ingredients. These lab-grown hybrids are found be ethically sound and sustainable towards modern meat alternatives (Efriwati et al., 2013; Kaur et al., 2025; Boukid et al., n.d.).

This novel food category attempts to capture the middle ground limited by animal protein alternatives by incorporating protein sources from plants, microbes, algae, or cultured cells, in addition to animal protein (Villacís-Chiriboga et al., 2025). The inspiration towards their development is due to global demand, environmental concerns, animal welfare and public health issue (Yao et al., 2024).

14.3.5 Definition and Components of Hybrid Lab-Grown Foods

Hybrid lab-grown foods refer to a category of food products that combine fully animal-derived components, e.g. meat cells, with ingredients of plant or fermentation origin (Kaur et al., 2025; Yao et al., 2024; Stanišić et al., 2025).

Essentially, the objective is to achieve good meat by combining the health and environmental advantages of plant-based protein sources and foods with the familiar taste and texture of meat (Kaur et al., 2025; Villacís-Chiriboga et al., 2025).

The main constituents of hybrid lab-grown foods include:

Cultivated (Cell-Based) Ingredients: These include muscle, fat, or connective tissues obtained from animal cells grown in bioreactors, eliminating the need for conventional animal farming and slaughtering (Guan et al., 2025; Kaplan & McClements, 2025; Rasmussen et al., 2024). Recently, the integration of stem cells technology has been employed to produce 'seed cells' which can be utilized for the cultivation of meat. These cells allow for the exponential proliferation and subsequent differentiation of cells into skeletal muscle constructs (Lee et al., 2023; Rasmussen et al., 2024; Strauss et al., 2025).

Plant-Based Ingredients: These typically include proteins from plants such as soy, pea, and wheat proteins. At moderate substitution levels, these plant-based components assist in enhancing texture, juiciness, and nutritional equilibrium (Song et al., n.d.; Kaur et al., 2025; Villacís-Chiriboga et al., 2025).

Fermentation-Derived Ingredients: This category comprises proteins and other functional constituents synthesized via precision fermentation. Specific microorganisms may be modified to manufacture various components, including growth factors, heme substitutes (i.e., soy leghemoglobin), or functional lipids that enhance flavor, color, and nutritional value (Yao et al., 2024).

14.3.6 Production Processes and Technological Pillars

The production of hybrid lab-grown foods is based on numerous interdisciplinary techniques and advanced technologies. In order to address the complexity of natural meat, the co-cultivation of myocytes and adipocytes through tissue engineering are pursued (Ogilvie et al., 2024). They also use tissue engineering to construct meat-like, three-dimensional (3D) structures created with edible scaffolds and microcarriers that promote the growth and differentiation of cells. Further, alginate can be used to improve adhesion for making the suitable scaffold to culture the meat cells (Lee et al., 2023; Moslemy et al., 2023; Ogilvie et al., 2024). In biofabrication, the cells are cultivated, and then incorporated into cell/plant protein or mycelium matrices, using 3D bioprinting, and scaffold-based assembly (Chandrababu & Puthumana, 2024). This is important to accurately reproduce the sensory attributes of meat (Bryant & Barnett, 2020; Ogilvie et al., 2024). In addition, the co-extrusion of the plant-based proteins with meat and fish can be employed to produce fibrous and meat-like structures (Villacís-Chiriboga et al., 2025). This also closes the textural gap of the plant-based alternatives and the meat products (Villacís-Chiriboga et al., 2025). From a sensory standpoint, the binding, flavor release, and the cooking behavior of the hybrid products have to been standardized through scientific and culinary approach. The interactions among the different constituents and their impact on the overall acceptability of a product in terms of taste, aroma, and mouthfeel are the focus of researchers (Kaur et al., 2025; Ogilvie et al., 2024).

14.3.7 Impact on Environment and Health

In order to consider the environmental concern, the traditional meat productions impact on a larger scale due to high emission of greenhouse gases and surplus use of land and water (Ching et al., 2022; Kim et al., 2022; Lanzoni et al., 2024; Lucas et al., 2023; Sinke et al., 2023). However, the lab-grown hybrids are more sustainable and eco-friendlier. The product must be formulated considering the health benefits by reducing saturated fat and cholesterol, increasing fiber content in lab grown hybrid foods (Fig. 14.3) (Kim et al., 2022; Lucas et al., 2023; Sinke et al., 2023).

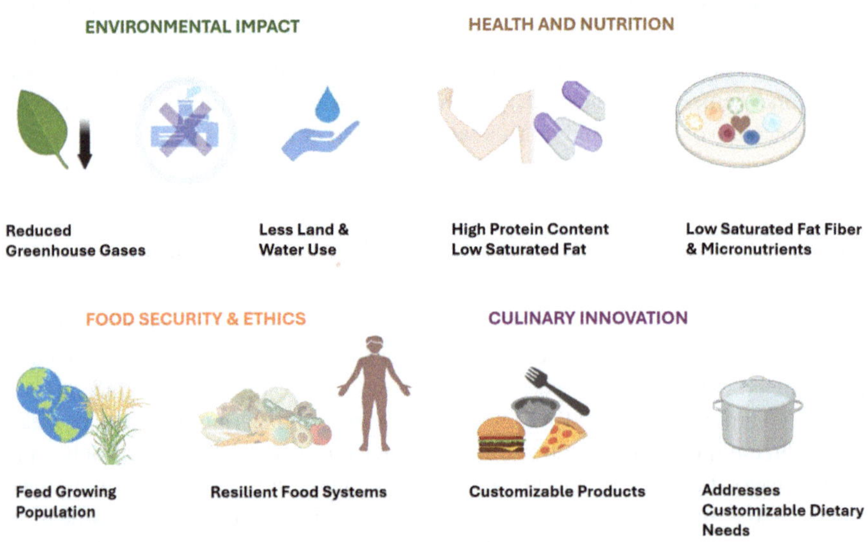

Fig. 14.3 The benefits of using alternative and sustainable protein productions impacting human health as well as environment. The process is also benefiting food security and culinary innovation

14.4 Consumer Acceptance and Challenges

Despite the evident positives regarding hybrid lab-grown foods, the challenges in gaining consumer trust and developing regulatory frameworks are of great concern (Bryant & Barnett, 2020; Gorbunova, 2025; Zhu & Zhang, 2025). Responses to new food technologies and products, including cultured meat and plant-based alternatives, are varied (Bryant & Barnett, 2020; de Oliveira Padilha et al., 2022; Sun et al., 2024). Concerns surrounding the 'naturalness' of lab-grown meat are triggered by food neophobia (Hamlin et al., 2022). The familiar principle suggests that the more novel a food product is, the less likely it is to be accepted (Dai et al., 2024). Social perception of cultured meat is prevalent, but sensory replication is the most significant challenge. It is quite difficult to replicate the sensory characteristics of traditional meat. A blend of meat and meat substitutes (in the range of 20–50%) can improve both the texture and the nutritional quality of the product. A product with more than that level of substitution can develop undesirable volatiles, color changes and off-odors that are unappetizing. The sensory improvements (taste, smell and mouth feel) account for a considerable part of the increasing demand for meat substitutes (Lambert et al., 2024; Stanišić et al., 2025). Economic challenges also exist. The cost and the possibility of large-scale cell cultivation are the most significant obstacles to the industrial cultivation of certain types of cells (Adi et al., 2024; Ching et al., 2022; Moslemy et al., 2023). For large-scale viable food production, economically viable cell cultures and serum-free mediums are essential. Regulatory challenges pertaining to new categories of cultivated and hybrid foods are ongoing. Regulations on the safety, hybrid foods, and marketing and food labeling are important to develop and restrain to regulate and build consumer confidence and to increase available markets (Ching et al., 2022; Regulatory Challenges for Lab-Grown Ingredients, 2024). As in any developing field, the frontiers of IPR and patenting in the field of new technologies are yet to be explored (Zhu & Zhang, 2025).

14.5 Conclusion

Proteins derived through fermentation are helping to revolution sustainable food systems to improve food security, reduce harm to the environment, and increase overall human health. The production of mycoproteins through fungal fermentation is an example of an environmentally sustainable, and versatile, low impact way to grow nutritionally complete proteins for functional foods and substitutes for meat. Tempeh is a traditional fermented food that illustrates how the nutritional and digestibility of plant proteins foods can be enhanced. The use of novel substrates, new microbial consortia, and innovative probiotic food processing can aid further enhancement of the food. Moreover, hybrid lab-grown foods, which consist of a combination of cultivated animal cells and proteins from plants and microorganisms, have the ability to mimic the texture and taste of meat while also reducing the dependence on traditional livestock farming. While fermentation technology and sustainability assessments have progressed, there remain barriers to their implementation, including the economy of scale, regulation, and consumer acceptance. Integration of food science, biotechnology, and sustainability science will be essential to further advance alternative protein systems for the widespread use of these vital systems. The future of fermentation-based proteins and hybrid proteins will help to create food systems that are sustainable.

References

Aaslyng, M. D., & Højer, R. (2021). Introducing tempeh as a new plant-based protein food item on the Danish market. *Food, 10*(11). https://doi.org/10.3390/foods10112865

Adi, P., Mulyani, R., Yudhistira, B., Chang, C.-K., Gavahian, M., & Hsieh, C.-W. (2024). Designing cultivated meat: Overcoming challenges in the production process and developing sustainable packaging solutions. *Trends in Food Science & Technology, 152*, 104675. https://doi.org/10.1016/j.tifs.2024.104675

Ahmad, M. I., Farooq, S., Alhamoud, Y., Li, C., & Zhang, H. (2022). A review on mycoprotein: History, nutritional composition, production methods, and health benefits. *Trends in Food Science & Technology, 121*, 14–29. https://doi.org/10.1016/j.tifs.2022.01.027

Ahnan-Winarno, A. D., Cordeiro, L., Winarno, F. G., Gibbons, J., & Xiao, H. (2021). Tempeh: A semicentennial review on its health benefits, fermentation, safety, processing, sustainability, and affordability. *Comprehensive Reviews in Food Science and Food Safety, 20*(2), 1717–1767. https://doi.org/10.1111/1541-4337.12710

Ajie, F. T., Wicaksono, A., Hendrix, T., & Angwar, M. (2016). Business analysis on sweet condensed tempeh as a functional food. In *2016 Global conference on business, management and entrepreneurship* (pp. 789–794). https://doi.org/10.2991/gcbme-16.2016.147

Alternative Protein Market Demand & Trends 2025 to 2035. (n.d.). Retrieved November 21, 2025, from https://www.futuremarketinsights.com/reports/alternative-protein-market

Ashenafi, M., & Busse, M. (1991). Microbial development during tempeh fermentation from various beans and effect of Lactobacillus plantarum on the natural microflora. *International Journal of Food Science & Technology, 26*(5), 501–506. https://doi.org/10.1111/j.1365-2621.1991.tb01995.x

Babu, P. D., Bhakyaraj, R., & Vidhyalakshmi, R. (2009). *A low cost nutritious food "tempeh"-a review.*. https://www.cabidigitallibrary.org/doi/full/10.5555/20103317586

Bakratsas, G., Polydera, A., Nilson, O., Chatzikonstantinou, A. V., Xiros, C., Katapodis, P., & Stamatis, H. (2023). Mycoprotein production by submerged fermentation of the edible mushroom Pleurotus ostreatus in a batch stirred tank bioreactor using agro-industrial hydrolysate. *Foods, 12*(12), 2295. https://doi.org/10.3390/foods12122295

Borges, C. W. C., Carrão-Panizzi, M. C., Mandarino, J. M. G., da Silva, J. B., Benedetti, S., & Ida, E. I. (2016). Contents and bioconversion of β-glycoside isoflavones to aglycones in the processing conditions of soybean tempeh. *Pesquisa Agropecuária Brasileira, 51*, 271–279. https://doi.org/10.1590/S0100-204X2016000300009

Boukid, F., Fanari, F., & Mefleh, M. (2024). Plant-based fermented foods and microbial ingredients in meat analogs. In *Handbook of plant-based meat analogs* (pp. 169–186). Academic Press. https://doi.org/10.1016/B978-0-443-21846-0.00015-0

Boukid, F., Baune, M. C., Terjung, N., Francis, A., & Smetana, S. (n.d.). The 'meathybrid' concept: Bridging the gap between texture, taste, sustainability and nutrition. *International Journal of*

Food Science and Technology. Oxford Academic. Retrieved March 5, 2026, from https://academic.oup.com/ijfst/article/59/11/8645/7933056?guestAccessKey=

Bryant, C., & Barnett, J. (2020). Consumer acceptance of cultured meat: An updated review (2018–2020). *Applied Sciences, 10*(15), 5201. https://doi.org/10.3390/app10155201

Cao, X., Liu, H., Yang, M., Mao, K., Wang, X., Chen, Z., Ran, M., & Hao, L. (2025). Evaluation of the nutritional quality of yeast protein in comparison to animal and plant proteins using growing rats and INFOGEST model. *Food Chemistry, 463*, 141178. https://doi.org/10.1016/j.foodchem.2024.141178

Cedeno, F. R. P., Olubiyo, O. J., & Ferreira, S. (2025). From microbial proteins to cultivated meat for alternative meat-like products: A review on sustainable fermentation approaches. *Journal of Biological Engineering, 19*(1), 44. https://doi.org/10.1186/s13036-025-00509-9

Chandrababu, A., & Puthumana, J. (2024). CRISPR-edited, cell-based future-proof meat and seafood to enhance global food security and nutrition. *Cytotechnology, 76*(6), 619–652. https://doi.org/10.1007/s10616-024-00645-y

Chen, Y.-C., Hsieh, S.-L., & Hu, C.-Y. (2020). Effects of red-bean tempeh with various strains of Rhizopus on GABA content and cortisol level in zebrafish. *Microorganisms, 8*(9). https://doi.org/10.3390/microorganisms8091330

Ching, X. L., Zainal, N. A. A. B., Luang-In, V., & Ma, N. L. (2022). Lab-based meat the future food. *Environmental Advances, 10*, 100315. https://doi.org/10.1016/j.envadv.2022.100315

Coelho, M. O. C., Monteyne, A. J., Dunlop, M. V., Harris, H. C., Morrison, D. J., Stephens, F. B., & Wall, B. T. (2020). Mycoprotein as a possible alternative source of dietary protein to support muscle and metabolic health. *Nutrition Reviews, 78*(6), 486–497. https://doi.org/10.1093/nutrit/nuz077

Cunha, P. M., Antoniel, E. P., Silva, L. O. S., Maia, G. N., Sydney, A. C. N., Sydney, E. B., Mascarin, G. M., Bicas, J. L., Pierucci, A. P. T. R., Duque-Estrada, P., & Damasio, A. (2025). From fungal cell factories to mycoproteins: Alternative protein production for human diets. *Current Food Science and Technology Reports, 3*(1), 21. https://doi.org/10.1007/s43555-025-00066-8

Dai, X., Leung, A. K. -y., & Chong, M. (2024). The link between people's social perceptions of cultivated meat eaters and their acceptance of cultivated meat. *Asian Journal of Social Psychology, 27*(3), 473–486. https://doi.org/10.1111/ajsp.12609

de Oliveira Padilha, L. G., Malek, L., & Umberger, W. J. (2022). Consumers' attitudes towards lab-grown meat, conventionally raised meat and plant-based protein alternatives. *Food Quality and Preference, 99*, 104573. https://doi.org/10.1016/j.foodqual.2022.104573

Denny, A., Aisbitt, B., & Lunn, J. (2008). Mycoprotein and health. *Nutrition bulletin, 33*(4), 298–310.

de Reu, J. C., Zwietering, M. H., Rombouts, F. M., & Nout, M. J. R. (1993). Temperature control in solid substrate fermentation through discontinuous rotation. *Applied Microbiology and Biotechnology, 40*(2), 261–265. https://doi.org/10.1007/BF00170377

Derbyshire, E., & Ayoob, K.-T. (2019). Mycoprotein: Nutritional and health properties. *Nutrition Today, 54*(1), 7. https://doi.org/10.1097/NT.0000000000000316

Devagopalan, H., & Krishnaswamy, I. (2023). Optimization of growth conditions and protease activity for tempeh production from sorghum. *Journal of Experimental Agriculture International, 45*(11), 58–63.

Efriwati, Suwanto, A., Rahayu, G., & Nuraida, L. (2013). Population dynamics of yeasts and lactic acid bacteria (LAB) during tempeh production. *HAYATI Journal of Biosciences, 20*(2), 57–64. https://doi.org/10.4308/hjb.20.2.57

Faisal, M., Syaubari, Febriana, A., & Mauliza, D. (2025). Utilization of Durian seeds (Durio Zibethinus Murr) and soybean mixtures as an alternative substrate for tempe production with Rhizopus oligosporus. *Rasayan Journal of Chemistry, 18*(01), 22–28. https://doi.org/10.31788/RJC.2025.1819022

Feng, X. M., Eriksson, A. R. B., & Schnürer, J. (2005). Growth of lactic acid bacteria and *Rhizopus oligosporus* during barley tempeh fermentation. *International Journal of Food Microbiology, 104*(3), 249–256. https://doi.org/10.1016/j.ijfoodmicro.2005.03.005

Fernandez Castaneda, L. A., Auer, J., Leong, S. L., Newson, W. R., Passoth, V., Langton, M., & Zamaratskaia, G. (2024). Optimizing soaking and boiling time in the development of tempeh-like products from faba bean (Vicia faba L.). *Fermentation, 10*(8), 407. https://doi.org/10.3390/fermentation10080407

Garnett, T. (2011). Where are the best opportunities for reducing greenhouse gas emissions in the food system (including the food chain)? *Food Policy, 36*, S23–S32. https://doi.org/10.1016/j.foodpol.2010.10.010

Gorbunova, N. A. (2025). Assessment of the consumers' attitude to the alternative meat. Review. *Theory and Practice of Meat Processing, 10*(1), 54–66. https://doi.org/10.21323/2414-438X-2025-10-1-54-66

Górska, K., Pejcz, E., & Harasym, J. (2025). Tempeh and fermentation—Innovative substrates in a classical microbial process. *Applied Sciences, 15*(16), 8888. https://doi.org/10.3390/app15168888

Guan, X., Sun, W., Ma, Z., Du, G., Chen, J., & Zhou, J. (2025). From lab to industry: Technologies and challenges for scaling up bioprocesses in cell-based food production. *Trends in Food Science & Technology, 161*, 105040. https://doi.org/10.1016/j.tifs.2025.105040

Hachmeister, K. A., & Fung, D. Y. C. (1993). Tempeh: A mold-modified indigenous fermented food made from soybeans and/or cereal grains. *Critical Reviews in Microbiology, 19*(3), 137–188. https://doi.org/10.3109/10408419309113527

Hamlin, R. P., McNeill, L. S., & Sim, J. (2022). Food neophobia, food choice and the details of cultured meat acceptance. *Meat Science, 194*, 108964. https://doi.org/10.1016/j.meatsci.2022.108964

Hartanti, A., Kurniawan, B., & Gunawan, A. (2024). Potential of Rhizopus delemar and Rhizopus microsporus as tempeh starters. *Biodiversitas Journal of Biological Diversity, 25*. https://doi.org/10.13057/biodiv/d251120

Hashempour-Baltork, F., Hosseini, S. M., Assarehzadegan, M.-A., Khosravi-Darani, K., & Hosseini, H. (2020). Safety assays and nutritional values of mycoprotein produced by Fusarium venenatum IR372C from date waste as substrate. *Journal of the Science of Food and Agriculture, 100*(12), 4433–4441. https://doi.org/10.1002/jsfa.10483

Jeleń, H., Majcher, M., Ginja, A., & Kuligowski, M. (2013). Determination of compounds responsible for tempeh aroma. *Food Chemistry, 141*(1), 459–465. https://doi.org/10.1016/j.foodchem.2013.03.047

Kadar, A. D., Astawan, M., Putri, S. P., & Fukusaki, E. (2020). Metabolomics-based study of the effect of raw materials to the end product of Tempe—An Indonesian fermented soybean. *Metabolites, 10*(9), 367. https://doi.org/10.3390/metabo10090367

Kameda, T., Aoki, H., Yanaka, N., Kumrungsee, T., & Kato, N. (2018). Production of Isoflavone Aglycone-enriched tempeh with Rhizopus stolonifer. *Food Science and Technology Research, 24*(3), 493–499. https://doi.org/10.3136/fstr.24.493

Kaplan, D. L., & McClements, D. J. (2025). Hybrid alternative protein-based foods: Designing a healthier and more sustainable food supply. *Frontiers in Science, 3*. https://doi.org/10.3389/fsci.2025.1599300

Kaur, L., Mao, B., Beniwal, A. S., Abhilasha, Kaur, R., Chian, F. M., & Singh, J. (2022). Alternative proteins vs animal proteins: The influence of structure and processing on their gastro-small intestinal digestion. *Trends in Food Science & Technology, 122*, 275–286. https://doi.org/10.1016/j.tifs.2022.02.021

Kaur, P., Kaur, R., Sharma, S., & Kaur, S. (2025). Hybrid meat for flexitarian consumers: Technological advancements, diversity in plant proteins and consumer preferences. *Critical Reviews in Food Science and Nutrition*, 1–17. https://doi.org/10.1080/10408398.2025.2564890

Kim, S., Beier, A., Schreyer, H. B., & Bakshi, B. R. (2022). Environmental life cycle assessment of a novel cultivated meat burger patty in the United States. *Sustainability, 14*(23). https://doi.org/10.3390/su142316133

Kurniadi, M., Andriani, M., Sari, I., Angwar, M., Nurhayati, R., Khasanah, Y., & Wiyono, T. (2017). *Nutritional and sensory characteristics of sari tempe formulated from import soybean (glycine max)*. 030098. https://doi.org/10.1063/1.4968351

Lambert, E. G., O'Keeffe, C. J., Ward, A. O., Anderson, T. A., Yip, Q., & Newman, P. L. H. (2024). Enhancing the palatability of cultivated meat. *Trends in Biotechnology, 42*(9), 1112–1127. https://doi.org/10.1016/j.tibtech.2024.02.014

Landeta-Salgado, C., Munizaga, J., González-Troncoso, M. P., Daza-Sanchez, A., Martínez, I., & Lienqueo, M. E. (2025). In vitro bioaccessibility of edible seaweed proteins from the Chilean coast and proteins from the novel seaweed-derived mycoprotein. *Molecules, 30*(1), 165. https://doi.org/10.3390/molecules30010165

Lanzoni, D., Rebucci, R., Formici, G., Cheli, F., Ragone, G., Baldi, A., Violini, L., Sundaram, T. S., & Giromini, C. (2024). Cultured meat in the European Union: Legislative context and food safety issues. *Current Research in Food Science, 8*, 100722. https://doi.org/10.1016/j.crfs.2024.100722

Lee, D.-K., Kim, M., Jeong, J., Lee, Y.-S., Yoon, J. W., An, M.-J., Jung, H. Y., Kim, C. H., Ahn, Y., Choi, K.-H., Jo, C., & Lee, C.-K. (2023). Unlocking the potential of stem cells: Their crucial role in the production of cultivated meat. *Current Research in Food Science, 7*, 100551. https://doi.org/10.1016/j.crfs.2023.100551

Lee, D., Pan, J. H., Kim, D., Heo, W., Shin, E. C., Kim, Y. J., Shim, Y. Y., Reaney, M. J. T., Ko, S.-G., Hong, S.-B., Cho, H. T., Kim, T. G., Lee, K., & Kim, J. K. (2024). Mycoproteins and their health-promoting properties: Fusarium species and beyond. *Comprehensive Reviews in Food Science and Food Safety, 23*(3), e13365. https://doi.org/10.1111/1541-4337.13365

Lima, M., Costa, R., Rodrigues, I., Lameiras, J., & Botelho, G. (2022). A narrative review of alternative protein sources: Highlights on meat, fish, egg and dairy analogues. *Foods, 11*(14), 2053. https://doi.org/10.3390/foods11142053

Liu, F., Li, M., Wang, Q., Yan, J., Han, S., Ma, C., Ma, P., Liu, X., & McClements, D. J. (2023). Future foods: Alternative proteins, food architecture, sustainable packaging, and precision nutrition. *Critical Reviews in Food Science and Nutrition, 63*(23), 6423–6444. https://doi.org/10.1080/10408398.2022.2033683

Liu, Y., Aimutis, W. R., & Drake, M. (2024). Dairy, plant, and novel proteins: Scientific and technological aspects. *Foods, 13*(7), 1010. https://doi.org/10.3390/foods13071010

Lucas, E., Guo, M., & Guillén-Gosálbez, G. (2023). Low-carbon diets can reduce global ecological and health costs. *Nature Food, 4*(5), 394–406. https://doi.org/10.1038/s43016-023-00749-2

Majumder, R., Miatur, S., Saha, A., & Hossain, S. (2024). Mycoprotein: Production and nutritional aspects: A review. *Sustainable Food Technology*. https://doi.org/10.1039/D3FB00169E

Mital, B. K., & Garg, S. K. (1990). Tempeh—Technology and food value. *Food Reviews International, 6*(2), 213–224. https://doi.org/10.1080/87559129009540867

Moslemy, N., Sharifi, E., Asadi-Eydivand, M., & Abolfathi, N. (2023). Review in edible materials for sustainable cultured meat: Scaffolds and microcarriers production. *International Journal of Food Science & Technology, 58*(12), 6182–6191. https://doi.org/10.1111/ijfs.16703

Mycoprotein and health—Denny—2008—Nutrition Bulletin—Wiley Online Library. (n.d.). Retrieved March 1, 2026, from https://onlinelibrary.wiley.com/doi/abs/10.1111/j.1467-3010.2008.00730.x

Nevalainen, H. (Ed.). (2020). *Grand challenges in fungal biotechnology*. Springer International Publishing. https://doi.org/10.1007/978-3-030-29541-7

Ng, C. K. Z., Leng, W. Q., Lim, C. H., & Du, J. (2024a). Physicochemical property characterization, amino acid profiling, and sensory evaluation of plant-based ice cream incorporated with soy, pea, and milk proteins. *Journal of Dairy Science, 107*(12), 10268–10279. https://doi.org/10.3168/jds.2024-25008

Ng, Z. Y., Kee, P. E., Abdullah, R., Lan, J. C.-W., Ling, T. C., Jiang, J.-J., Lim, J. W., & Khoo, K. S. (2024b). Conversion of lignocellulosic biomass waste into mycoprotein: Current status and future directions for sustainable protein production. *Biomass Conversion and Biorefinery.* https://doi.org/10.1007/s13399-024-06130-y

Nout, M. J. R., & Kiers, J. L. (2005). Tempe fermentation, innovation and functionality: Update into the third millenium. *Journal of Applied Microbiology, 98*(4), 789–805. https://doi.org/10.1111/j.1365-2672.2004.02471.x

Nowak, J., & Szebiotko, K. (1992). Some biochemical changes during soybean and pea *tempeh* fermentation. *Food Microbiology, 9*(1), 37–43. https://doi.org/10.1016/0740-0020(92)80060-H

Nursiwi, A., Pertiwi, F., Ishartani, D., Sari, A. M., & Zaman, M. Z. (2021). Substrates and storage time evaluation for preparing tempeh starter from Rhizopus oryzae CBS130145. *IOP Conference Series: Earth and Environmental Science, 828*(1), 012003. https://doi.org/10.1088/1755-1315/828/1/012003

Ogilvie, O. J., Bennie, R. Z., Trlin, H. J. F., Loo, L. S. W., Zhou, H., Jin, A., Oh, J. K., Dobson, R. C. J., Yu, H., & Domigan, L. J. (2024). Interdisciplinary methods for analysing food matrix structures of hybrid cell-based meats: A review. *Food Structure, 39*, 100361. https://doi.org/10.1016/j.foostr.2023.100361

Poore, J., & Nemecek, T. (2018). Reducing food's environmental impacts through producers and consumers. *Science, 360*(6392), 987–992. https://doi.org/10.1126/science.aaq0216

Pramesti, A., Fadhilah, A., Athallah, B. F., Panjaitan, G. O. S., Utami, N. A., Nugraha, P. A., Rohidi, Trianawati, M. L., Zulqisthi, D., Assariy, M. F., & Asabillah, C. H. (2025). Implementation of fuzzy logic in yeast concentration and fermentation time for tempeh quality. *Journal of Applied Science, Technology & Humanities, 2*(3), 349–364. https://doi.org/10.62535/1am4xb61

Pramudito, T. E., Putri, E., Paluphi, E., & Yogiara, Y. (2021). The effect of starter culture on bacterial profile in soybean tempeh. *Food Research, 5*, 380–389. https://doi.org/10.26656/fr.2017.5(1).436

Qibty, M., Yulianto, W. A., & Pujimulyani, D. (2023). Effect of adding probiotic yeast and fermentation time on probiotic tempeh characteristics. *Journal of Food and Agricultural Technology, 1*(1), 9–19. https://doi.org/10.26486/jfat.v1i1.3390

Rasmussen, M. K., Gold, J., Kaiser, M. W., Moritz, J., Räty, N., Rønning, S. B., Ryynänen, T., Skrivergaard, S., Ström, A., Therkildsen, M., Tuomisto, H. L., & Young, J. F. (2024). Critical review of cultivated meat from a Nordic perspective. *Trends in Food Science & Technology, 144*, 104336. https://doi.org/10.1016/j.tifs.2024.104336

Regulatory challenges for lab-grown ingredients. (2024). *Food Science and Technology, 38*(3), 46–47. https://doi.org/10.1002/fsat.3803_12.x

Reyes-Moreno, C., Cuevas-Rodríguez, E., Milán-Carrillo, J., Cárdenas-Valenzuela, O., & Barrón-Hoyos, J. (2004). Solid state fermentation process for producing chickpea (Cicer arietinum L) tempeh flour. Physicochemical and nutritional characteristics of the product. *Journal of the Science of Food and Agriculture, 84*(3), 271–278. https://doi.org/10.1002/jsfa.1637

Risner, D., McDonald, K. A., Jones, C., & Spang, E. S. (2023). A techno-economic model of mycoprotein production: Achieving price parity with beef protein. *Frontiers in Sustainable Food Systems, 7*. https://doi.org/10.3389/fsufs.2023.1204307

Riyanto, R. A., Fukusaki, E., & Putri, S. P. (2024). Metabolite profiling highlights the effect of microbial intervention in the soaking step of tempe. *International Journal of Food Science and Technology, 59*(10), 7414–7425. https://doi.org/10.1111/ijfs.17481

Rizal, S., Kustyawati, M. E., Murhadi, & Hasanudin, U. (2021). The growth of yeast and fungi, the formation of β-Glucan, and the antibacterial activities during soybean fermentation in producing tempeh. *International Journal of Food Science, 2021*(1), 6676042. https://doi.org/10.1155/2021/6676042

Rohmadi, M. I. N., Muhandri, T., Hunaefi, D., & Setiarto, R. H. B. (2025). Meat analog processing technology from non-animal protein to create fibrous structures similar to animal meat: Review. *Current Applied Science and Technology*, e0264052. https://doi.org/10.55003/cast.2025.264052

Schweiggert-Weisz, U., Eisner, P., Bader-Mittermaier, S., & Osen, R. (2020). Food proteins from plants and fungi. *Current Opinion in Food Science, Food Microbiology, Functional Foods and Nutrition, 32*, 156–162. https://doi.org/10.1016/j.cofs.2020.08.003

Signorini, C., Carpen, A., Coletto, L., Borgonovo, G., Galanti, E., Capraro, J., Magni, C., Abate, A., Johnson, S. K., Duranti, M., & Scarafoni, A. (2018). Enhanced vitamin B12 production in an innovative lupin tempeh is due to synergic effects of Rhizopus and Propionibacterium in cofermentation. *International Journal of Food Sciences and Nutrition, 69*(4), 451–457. https://doi.org/10.1080/09637486.2017.1386627

Sinke, P., Swartz, E., Sanctorum, H., van der Giesen, C., & Odegard, I. (2023). Ex-ante life cycle assessment of commercial-scale cultivated meat production in 2030. *The International Journal of Life Cycle Assessment, 28*(3), 234–254. https://doi.org/10.1007/s11367-022-02128-8

Song, Z., Hamill, R. M., & Kerry, J. P. (n.d.). Hybrid plant-meat products – addressing the sustainability debate around processed meat consumption: A review. *Sustainable Food Technology*. RSC Publishing. https://doi.org/10.1039/D5FB00723B. Retrieved March 5, 2026, from https://pubs.rsc.org/en/content/articlehtml/2026/fb/d5fb00723b

Stanišić, N., Delić, N., Cekić, B., Stojiljković, N., Gogić, M., Samolovac, L., & Stajić, S. (2025). Current processing technologies and challenges in hybrid meat production. *Processes, 13*(9). https://doi.org/10.3390/pr13092853

Strauss, B., Kleinsorge, B., & Lotfian, P. (2025). Best practice for cell banking techniques used in cell cultivated products. *FSA Research and Evidence*. https://doi.org/10.46756/001c.144239

Sun, J., Caputo, V., & Taylor, H. (2024). Using machine-learning methods in meta-analyses: An empirical application on consumer acceptance of meat alternatives. *Applied Economic Perspectives and Policy, 46*(4), 1506–1532. https://doi.org/10.1002/aepp.13446

Sutardi, & Buckle, K. A. (1985a). Phytic acid changes in soybeans fermented by traditional inoculum and six strains of Rhizopus oligosporus. *Journal of Applied Bacteriology, 58*(6), 539–543. https://doi.org/10.1111/j.1365-2672.1985.tb01709.x

Sutardi, & Buckle, K. A. (1985b). Reduction in phytic acid levels in soybeans during tempeh production, storage and frying. *Journal of Food Science, 50*(1), 260–263. https://doi.org/10.1111/j.1365-2621.1985.tb13325.x

Suwanto, A. (2021). Dynamics of microbial community during tempeh fermentation. *Biotropia, 28*(1). https://doi.org/10.11598/btb.0.0.0.820

Teoh, S. Q., Chin, N. L., Chong, C. W., Ripen, A. M., How, S., & Lim, J. J. L. (2024). A review on health benefits and processing of *tempeh* with outlines on its functional microbes. *Future Foods, 9*, 100330. https://doi.org/10.1016/j.fufo.2024.100330

The Book of Tempeh—William Shurtleff, Akiko Aoyagi—Google Books. (n.d.). Retrieved November 20, 2025, from https://books.google.co.in/books?hl=en&lr=&id=hdKzFlaqWT8C&oi=fnd&pg=PA6&dq=Shurtleff,+W.+and%E2%80%82Aoyagi,+A..+(1979).+The+book+of+tempeh+(Vol.+1).+Soyinfo+Center.&ots=-Ia3gCAkoe&sig=RO5OMKj-nkMDJbVSe9yZEH2Rvn4&redir_esc=y#v=onepage&q=Shurtleff%2C%20W.%20and%E2%80%82Aoyagi%2C%20A..%20(1979).%20The%20book%20of%20tempeh%20(Vol.%201).%20Soyinfo%20Center.&f=false

Turnbull, W., Leeds, A., & Edwards, G. (1990). Effect of mycoprotein on blood lipids. *The American Journal of Clinical Nutrition, 52*(4), 646–650. https://doi.org/10.1093/ajcn/52.4.646

Villacís-Chiriboga, J., Sharifi, E., Elíasdóttir, H. G., Huang, Z., Jafarzadeh, S., & Abdollahi, M. (2025). Hybrid plant-based meat alternatives structured via co-extrusion: A review. *Trends in Food Science & Technology, 160*, 105013. https://doi.org/10.1016/j.tifs.2025.105013

Wahyuono, S., Lukitaningsih, E., Yuniarti, N., & Wahyuono, S. (2021). *Fermentation of soybean seeds using Rhizopus oligosporus for tempeh production and standardization based on Isoflavones content*. (no. 2021040610). Preprints. https://doi.org/10.20944/preprints202104.0610.v1

Wiesel, I., Rehm, H. J., & Bisping, B. (1997). Improvement of tempe fermentations by application of mixed cultures consisting of Rhizopus sp. and bacterial strains. *Applied Microbiology and Biotechnology, 47*(3), 218–225. https://doi.org/10.1007/s002530050916

Wikandari, R., Kinanti, D. A., Permatasari, R. D., Rahmaningtyas, N. L., Chairunisa, N. R., Sardjono, Hellwig, C., & Taherzadeh, M. J. (2021). Correlations between the chemical, microbiological characteristics and sensory profile of fungal fermented food. *Fermentation, 7*(4), 261. https://doi.org/10.3390/fermentation7040261

Wikandari, R., Tanugraha, D. R., Yastanto, A. J., Manikharda, Gmoser, R., & Teixeira, J. A. (2023). Development of meat substitutes from filamentous fungi cultivated on residual water of tempeh factories. *Molecules, 28*(3), 997. https://doi.org/10.3390/molecules28030997

Williamson, D. A., Geiselman, P. J., Lovejoy, J., Greenway, F., Volaufova, J., Martin, C. K., Arnett, C., & Ortego, L. (2006). Effects of consuming mycoprotein, tofu or chicken upon subsequent eating behaviour, hunger and safety. *Appetite, 46*(1), 41–48. https://doi.org/10.1016/j.appet.2005.10.007

Wiloso, E. I., Sinke, P., Muryanto, Setiawan, A. A. R., Sari, A. A., Waluyo, J., Putri, A. M. H., & Guinée, J. (2019). Hotspot identification in the Indonesian tempeh supply chain using life cycle assessment. *The International Journal of Life Cycle Assessment, 24*(11), 1948–1961. https://doi.org/10.1007/s11367-019-01617-7

Wolkers–Rooijackers, J. C. M., Turner, O., Almekinders, E., & Smid, E. J. (2024). Spatial-temporal distribution of oxygen and its effect on microbial dynamics and vitamin B12 content in lupin tempeh. *LWT - Food Science and Technology (Online), 201*, 116275. https://doi.org/10.1016/j.lwt.2024.116275

Yan, Y., Wolkers-Rooijackers, J., Nout, M. J. R., & Han, B. (2013). Microbial diversity and dynamics of microbial communities during back-slop soaking of soybeans as determined by PCR–DGGE and molecular cloning. *World Journal of Microbiology and Biotechnology, 29*(10), 1969–1974. https://doi.org/10.1007/s11274-013-1349-6

Yao, Y., Li, C., Yuen, J. S. K., Jr., Stout, A. J., & Kaplan, D. L. (2024). Manufacture of hybrid alternative protein food products using a combination of plant-based ingredients, fermentation-derived ingredients, and animal cells. In *Cellular agriculture* (pp. 251–266). Academic Press. https://doi.org/10.1016/B978-0-443-18767-4.00032-9

You, X.-Y., Ding, Y., Bu, Q.-Y., Wang, Q.-H., & Zhao, G.-P. (2024). Nutritional, textural, and sensory attributes of protein bars formulated with mycoproteins. *Foods, 13*(5), 671. https://doi.org/10.3390/foods13050671

Zhang, C., Wu, X., Chen, J., & Zhou, J. (2024). Novel fungal alternative proteins from *Penicillium limosum* for enhancing structural and functional properties of plant-based meat analogues. *Food Chemistry, 444*, 138627. https://doi.org/10.1016/j.foodchem.2024.138627

Zhu, H., & Zhang, Z. (2025). Cultivated meat: Pioneering sustainable solutions for global food challenges. *Food Chemistry International, 1*(1), 11–13. https://doi.org/10.1002/fci2.70007

Chapter 15
Fermentation as a Climate-Resilient Agricultural Practice: An Analysis of National and International Legal Frameworks Advancing SDG 2 and SDG 13

Showkat Ahmad Wani, Sheikh Inam Ul Mansoor, Eshan Singhal, and Sunny Dhiman

Abstract Fermentation, an ancient biotechnology process, has regained prominence as an essential method for promoting sustainable agriculture and enhancing climate resilience in the context of global food security issues. The chapter discusses fermentation not just as an age-old food preservation method but as a novel practice of agriculture with remarkable environmental, economic, and nutritional advantages. The study examines how SDG 2 (Zero Hunger) and SDG 13 (Climate Action); fermentation presents the potential to decrease post-harvest loss, diversify diets, improve soil fertility, and minimize greenhouse gas emissions. By processing agricultural residues into high-value items like biofertilizers, animal feed, and bioenergy, fermentation technologies serve circular economy frameworks and climate-resilient agriculture systems.

This research presents an in-depth examination of the global legal frameworks that regulate and support fermentation in sustainable agriculture. It critically discusses multilateral policy agreements and actions such as the Paris Agreement, the Convention on Biological Diversity (CBD), the International Treaty on Plant Genetic Resources for Food and Agriculture (ITPGRFA), and the Codex Alimentarius as well as their applicability to fermented food systems. The study also examines how the United Nations Framework Convention on Climate Change

S. A. Wani (✉)
School of Law, Alliance University, Bengaluru, Karnataka, India

S. I. U. Mansoor
Symbiosis Law School Hyderabad, Symbiosis International University, Pune, Maharashtra, India
e-mail: sheikh.mansoor@slsh.edu.in

E. Singhal
Mewar University, Gangrar, Chittorgarh, Rajasthan, India

S. Dhiman
Department of Biotechnology, University Institute of Biotechnology, Chandigarh University, Mohali, Punjab, India

S. Dhiman, S. Kaur (eds.), *Next-Gen Plant Based Fermented Foods: Innovations for a Sustainable Future*, https://doi.org/10.1007/978-3-032-29927-7_15

(UNFCCC) and the Food and Agriculture Organization (FAO) enable capacity development, technology transfer, and policy coordination in this industry. Using comparative case studies from Latin America, Africa, and Asia, the chapter demonstrates how various environmental policies influence the adaptation of fermentation-based technologies. It also points to regulatory issues, such as intellectual property, biosafety, and developing countries' technology diffusion barriers.

The chapter supports a unified global approach that connects fermentation with national climate action plans (NAPs) and food security policies, and it highlights the importance of inclusive governance arrangements that acknowledge indigenous knowledge and ensure equitable access to resources. This study places fermentation as a scientific development and a policy imperative, calling on global legal frameworks to acknowledge its contribution in addressing climate change while guaranteeing durable, healthy, and accessible food systems for everyone.

Keywords Food security · Biosafety · Global strategy · Fermentation · Climate-resilient

15.1 Introduction

Food insecurity and climate change are amongst the most alarming and interdependent issues facing the planet in the 21st century. The high temperature, unpredictable climatic patterns, soil erosion, and the increased occurrences of extreme weather have been detrimental to agricultural output, especially in the developing countries that face a high risk of climate change (Hasegawa et al., 2018). Meanwhile, the food system in the world remains the source of about one-third of anthropogenic greenhouse gas emissions, and yet it is not sufficient in ensuring fair access to healthy nutrition (FAO, 2025; Borrego-Ruiz et al., 2024). Given this context, implementing climate-resilient agricultural methods to ehance food security and ecological sustainability has become a priority for global growth and climate governance aligning with Sustainable Development Goal 2 (Zero Hunger) and SDG 13 (Climate Action).

One of the oldest biotechnological processes is fermentation, as it appeared as a beneficial but not properly examined instrument in the contemporary scenario. Historically identified with preserving and enriching the nutritional properties of food, fermentation is also becoming known in the context of sustainability practices in agriculture, such as improving the quality of the soil, utilizing organic waste, making biofertilisers, optimising animal food, and producing renewable bioenergy (El-Mansi, 2018). Fermentation helps turn extra agricultural products into something valuable, which supports the idea of a circular economy and reduces the need for resource-heavy chemical inputs. Such a stance makes fermentation a climate-resilient agricultural practice that has a significant mitigation and adaptation capacity (Zhang et al., 2025).

Fermentation is yet to gain recognition in the legal climate and agricultural systems despite the increased scientific opinion. The various international agreements on climate change, biodiversity, food security, and biosafety usually promote large-scale technological interventions, thus disregarding all low-cost, decentralised and community-based techniques like fermentation. This represents a significant gap regarding fermentation's potential to reduce post-harvest losses, increase nutritional diversity, and lower greenhouse gas emissions throughout the food supply chain, which is central to both SDG 2 and SDG 3 (SDG stands for Sustainable Development Goals). Lack of legal recognition has made governance processes indeterminate and unequal, especially in the developing states (Intergovernmental Panel on Climate Change, 2023).

It is possible that the potential uses of international legal provisions, such as the Paris Agreement, the United Nations Framework Convention on Climate Change (UNFCCC), the International Treaty on Plant Genetic Resources for Food and Agriculture (ITPGRFA), and the Convention on Biological Diversity (CBD), offer quite strong normative facilitation to fermentation-based agriculture systems (Chiarolla, 2017). Similarly, global regulations and markets for fermented food apply the Codex Alimentarius food safety/standards (Adam et al., 2025). Nevertheless, these tools work independently, which constrains consistency and alignment of the policy with fermentation as a climate-resilient approach. Additionally, the implementation usually faces the challenges of intellectual property rights of microbial resources, biosafety regulation, and traditional knowledge access, hence making it difficult to translate fermentation into national climate and food security policies (McManis, 2012).

The invisibility of fermentation that is provided by the law also poses serious concerns of fairness and justice (Onyeaka et al., 2024). The systems of indigenous and traditional knowledge are highly embedded in fermentation practices, especially in Asia, Africa, and Latin America. However, these knowledge systems continue to go unacknowledged as well as unsecured by the international regulatory system, leading to asymmetries in benefit sharing and access to technologies. Without extensive governance systems, fermentation is likely to end up delegitimized under commercial biotechnology as a matter without proper consideration of the rights of the community, biosafety concerns, and ethical protection of human rights (Tamang et al., 2020).

This chapter argues that it is time to redefine fermentation as not only a technological procedure, but also an agriculturally relevant policy as well as capable of enhancing climate resiliency, food security, and sustainable development. The study will locate fermentation at the crossroads of international climate law, governance, food security and biodiversity regulation, allowing it to close existing normative gaps and benefit new scholarship on climate-adaptive food systems. Its main research questions are (1) How do the currently established international and national legal frameworks control and shape fermentation-based agricultural practices? (2) How can fermentation progressively futurize the compound goals of SDG13 and SDG2? (3) What legal and regulatory challenges are there in the way of the fair dissemination of fermentation technologies in the developing world? By answering these questions, the paper demonstrates that fermentation is both a scientific breakthrough and a policy requirement of law, which promotes a harmonised approach to governance, introducing fermentation to the country as a component of climate action plans, security policies, and sustainable global commitments without compromising the biosafety and indigenous knowledge systems.

These approaches to the question include the qualitative, interdisciplinary methodology that consists of the doctrinal legal analysis, comparative policy analysis, and case studies. The doctrine part entails a methodological analysis of the international legal tools that are applicable in the area of climate change, biodiversity, food security, and biosafety, such as the UNFCCC, the Paris Agreement, the CBD, the ITPGRFA and the Codex Alimentarius standards (Tsioumani, 2004). The normative applicability of these tools is mainly determined with regard to executive applicability to agricultural practices based on fermentation (Danelyan & Gulyaeva, 2022). Lastly, the analysis will utilize an analytical model based on the SDG approach to assess the beneficial synergistic impact of fermentation on SDG 2 and SDG 13 (Baky et al., 2025).

15.2 Conceptual Framework in Sustainable Agriculture, Fermentation and Climate Resilience

The multidisciplinary approach to the vision of fermentation as a climate-resilient agricultural practice considers non-discriminatory biotechnology, sustainability science, and legal governance (Chaturvedi et al., 2024). Through the scientific methods of transforming organic substances through the use of microbes, the biochemical process has been deeply engrained in human agricultural- and food-related systems, which have endured throughout civilized history. Recent sustainability demands need a more comprehensive theoretical framework according to which fermentation is seen not only as a food-processing methodology but as a systemic intervention with the potential to achieve a better level of flexibility, structure, and fairness in agri-food systems (Steinkraus, 2018). A reconsideration of the key conditions of climate change shows that adaptability and low-emission agriculture-related methods play a vital role in the sustenance of food production and the reduction of environmental degradation (Ali et al., 2025). The agricultural system based on the ability of the system to forecast, absorb, and recover from climatic stresses without incurring long-term loss in productivity or ecological integrity tends to assess climate flexibility (Mukherji, 2023).

In this context, fermentation contributes to flexibility in various ways. To start with, fermentation technologies can prolong the shelf life of agricultural produce by enhancing post-harvest stability, which helps save the global economy by reducing food loss and waste, both of which are major contributors to greenhouse gases emissions (FAO, 2025). Second, fermentation is a dietary approach that promotes food security in the community because it increases nutritional availability and food security, in addition to improving food safety during the stress caused by climatic conditions (Materia et al., 2021). Third, fermentation activities of microbes form the basis of soil health and nutrient recycling, and this keeps agriculture productive in the face of fluctuating climatic conditions (Sharma et al., 2022).

Fermentation strategic vision is also close to the postulates of sustainable agriculture, including cost-effectiveness, ecological balance, and socio-economic diversity (Lawal et al., 2023). Sustainable agriculture aims to reduce the use of synthetic inputs to achieve a balance between environmental effects and the enhancement of biodiversity while maintaining the sustainable livelihoods of farmers (Altieri & Nicholls, 2020). Biofertilizers and biopesticides made through fermentation provide a feasible alternative to chemical fertilizers and pesticides, thus leading to the decrease of greenhouse gases and soil erosion. Regenerative agricultural practices, bolstered by microbial processes occurring naturally to ferment soil organic compounds, add to soil organic matter and improve water retention and resistance to nutrient deficiency and temperature variation.

Fermentation and climate-resilient agriculture are theoretically connected by the view of a circular economy. This economic model focuses on waste minimization, resource re-use, and even on the regeneration of natural systems, which reverses the linear produce-consume-discard model that describes industrial agriculture (Geissdoerfer et al., 2017). The agricultural residues and biomass are converted by fermentation technologies to produce products that are useful—bioenergy, biofertilizer, compost, and animal feed supplements. Such reuse does not only lessen the emission of methane in the process of decomposing organic waste, but it also creates closed-loop systems that increase the overall organization of resources. Under climate governance, such processes are evaluated based on their role in the mitigation and adaptation process together (Jiao et al., 2025).

Of particular interest is the association between fermentation and mitigation of climate change concerning its ability to achieve emission reduction of greenhouse gases along the agri-food value chain. By replacing nitrous oxide, the most common and one of the most powerful greenhouse gases that are related to agricultural activity, with fermented organic inputs, the emission of nitrous-oxide can be significantly reduced (Mukherji, 2023). Moreover, bioenergy systems based on fermentation, like anaerobic digesters, produce renewable energy and minimize the amount of organic waste. Such cuts complement contributions to enhance the diversity of the farming systems and reduce vulnerability to climate along food delivery lines (Gonzalez, 2011).

At the crossroads of food security and climate elasticity, fermentation is a key factor in the further development of Sustainable Development Goal 2 (García-Franco et al., 2021). Food security does not only involve access to food but also the approach, utilization and time stability of food. Fermented foods fulfill every four-dimensional requirement encompassing enhancement of nutrient bioavailability, food safety and preservation, and no use of energy-consuming refrigeration. In addition, nutritional diversity and culturally suitable food systems are preserved due to fermentation; these are key factors in nutritional stability and overall health of the population in sensitive areas (Chai & Chen, 2025). Fermentation, therefore, is one of the major preservation methods and value additions in food (Marshall, 2022).

At the same time, fermentation serves as an enabler to SDG 13 because it will lead to climate action by encouraging low-carbon agricultural systems and resilience measures. The SDG model requires a set of approaches to be integrated and to have an effect on each other (United Nations, 2015). This integration is seen through the practice of fermentation, which involves the integration of climate mitigation, adaptation, and sustainable livelihoods. The success of this potential, however, will rely on empowering legal and policy frameworks that will make fermentation technologies integrate securely and equally (Mukherjee et al., 2022).

Fermentation technologies have become increasingly placed at the intersection of agricultural innovation and sustainability, but their efficient application will be tied to the quality and design of legal and governance systems. The international environmental and food production policies, such as those developed by the United Nations, make sustainability the focus of climate resilience and ecological sustainability (Clapp & Scott, 2018). In this respect, the fermentation technologies can comply with the overall sustainability agenda, decreasing the emission of greenhouse gases, improving the soil condition, and allowing resource-effective manufacturing. Different schemes like UNFCCC and the Paris Agreement encourage implementation of climate-reducing technology, which indirectly promotes fermentation through low-emission solution focus (Rakshit et al., 2025). However, such international instruments fail to appreciate fermentation technologies explicitly, which leaves a loophole in the policy assistance and strategic implementation advice (Mukherjee et al., 2022). Likewise, bio-governance, which is especially conveyed by the Convention on Biological Diversity (CBD), is based on the importance of microbial resources as a traditional base of biotechnological progress. The CBD's focus on access and benefit-sharing processes, especially the Nagoya Protocol, has complex legal issues regarding the use of microbial strains that are essential for fermentation (Smith et al., 2017). Although these provisions are meant to secure a fair utilization of the genetic resources, they tend to cause an administrative burden on the creation of a legal loophole, especially in cross-border research and development projects. These limitations are critically acute within multinational partnerships, whereby national standards tend to be unsuited and unevenly applied, which deters the strengths of scale and transfer of fermentation-based solutions. Therefore, although it is seen that the future of fermentation and sustainable agricultural structures has been proven to be highly promising, a fragmented system of control still serves as an obstacle to the advancement of these practices. Lack of policies at the international level well-coordinated with technology hinders investment coordination, innovations and institutional adoption (Motarjemi, 2002). To scientists and policymakers, it takes much more than scientific skill to navigate fragmented governance of the law and policies. An integrated and future oriented regulation is vital in unlocking these transformative effects of fermentation on the global food systems and this should be achieved through a more formal recognition of fermentation within the environmental, agricultural and biodiversity policy frames (Buck & Hamilton, 2011). There is a new model of regulation of microbial resources, which provides an all-inclusive approach based on biosafety, human ethics, and the protection of intellectual property rights. At this framework is the acknowledgment that scenes of fermentation rely on complex community of microbes that a huge percentage of which are a part of the indigenous farming ecosystems. The emerging biogenetic resource for sustainable development have led to serious ethical problems like the threat of biopiracy, unfair allocation of financial gains, and the loss of indigenous knowledge of ecology (Nambisan, 2017). Though the intellectual property regimes are meant to create incentives to innovate, they may also act as barriers to accessibility of microbial technologies especially by the smallholder farmers in the less fortunate economic belts. The existence of this duality points to the tremendous necessity of a consistent theoretical framework that can allow harmonizing scientific and technological advancement with ideas of distributive justice, social equality, and eco-friendly environmental governance (Gunduz, 2023).

The use of an inappropriate regulatory paradigm can be treated as a substantial hindrance to the inception of the fermentation technologies in climate-adaptive food production systems. Despite its general acknowledgment as a low-hazard bioprocess, microbial fermentation is often regulated following biosafety concerns that were developed around the notion of genetically modified organisms (GMOs), which constitutes a fundamental misinterpretation of the precautionary principle as well as an overapplication of the principle (Kozarev et al., 2020). Such misclassification of boards creates undue compliance expenses, not only in terms of time and cost but also in terms of finances and resources, for organizations that are innovative in fermentation-based solutions to agriculture. Therefore, such an uncertain and cumbersome terrain not only stifles important investment in research and development but also limits the opportunity to commercially enlarge the sustainable innovations on a large scale. To access this potential and make the policy comply with scientific reality, the establishment of a new risk assessment architecture is essential. Such systems should be strictly evidence-based, on empirical data as opposed to theoretical risk, and designed in such a way that they actively encourage sustainability and the equal distribution of technological advancement (Motarjemi, 2002).

The system of indigenous and traditional knowledge (ITK) is an important but overlooked resource in fermentation, providing local insights and experiences related to ecological and cultural socal interactions (Sharma et al., 2025). In areas like Asia, Africa, and Latin America, to name but a few, fermentation has become a measure of a suitable adaptation to a given environmental situation and the requirements of society. These ancient methodologies have been tested over time and unite the ideas of sustainability to use resources, active ecological management, and properties of resilience to the system that become even more important when overcoming the growing climate-related problems. Nonetheless, modern world governance systems tend to not have formal systems to respect, protect and thoughtfully assimilate this knowledge making it vulnerable to being marginalized, exploited or recontextualized into new commercial commodities out of their cultural and ecological contexts. To the researchers, closing this divide requires a reconsideration of the legal paradigms in a bid to provide just values and preservation of ITK contributes to food systems, biodiversity and climate adaptation (Rai et al., 2024).

To ensure indigenous ways of knowing are genuinely entrenched into the dominating systems of governance, there is an imminent need to reorganize it. This must mark a conscious shift to institutional structures that have been created in partnerships, which legitimize a pluralism of knowledge traditions (Elliott, 2018). This kind of evolution requires going beyond performative acts of inclusion to incorporate real substantive shared-power practices that will take into account the intellectual authority of epistemologies and facilitate equitable custodianship towards both material and cultural history. To make this vision workable, we need a two-fold approach: not just systemic policy changes, but also a transformation of our conceptualization of how various place-based ecological knowledge, particularly of the historically marginalized peoples, can inform world directions towards sustainability (Rai et al., 2024).

Fermentation plays a key role in stopping the intersection of climate adaptation, food sovereignty, sustainable agriculture, and regulatory policy, which will be discussed through this conceptual framework (Ajayeoba & Ijabadeniyi, 2025). One of the core principles of this model is the need to shun the reductionist as well as compartmentalization of policymaking towards a comprehensive governance structure. The new architecture needs to realize the plentiful role played by fermentation on ecological, nutritional, and cultural scales. This approach is echoed in modern sustainability science, which promotes systems thinking and cross-sectoral partnership and policy coherence as the keys to advancing the Sustainable Development Goals (Geissdoerfer et al., 2017). Placing fermentation in the center of this nexus, the framework sheds light on its ability to act both as a source of cultural presence and as an instrument of planetary and his own health.

15.3 Fermentation and SDG 2, SDG 13

The fact that Sustainable Development Goal 2 (Zero Hunger) and Sustainable Development Goal 13 (Climate Action) are interdependent suggests an ever-growing acknowledgment that global food system's fortification and climate resilience development are two concepts that support each other (Chen & Di Lernia, 2022). Climate change issues bring complex challenges to agricultural production, nutrition, and equitable access to food, and agricultural systems are significant sources of anthropogenic atmospheric emissions and ecosystem depletion through unsustainable production (FAO, 2025). The biotechnological approach (fermentation) used in the past, but innovative, is a kind of two-fold approach that can enhance food security and promote the climate action agendas. Fermentation can provide a plausible process to meet the requirement of eliminating food insecurity and decarbonizing food chains as a solution to mitigation and adaptation goals inside a unifying sustainability strategy by optimizing resource-use efficiency and reducing energy-intensive production chains (Galanakis, 2024).

Sustainable Development Goal 2 requires full elimination of hunger and the achievement of universal food security in tandem with the aim of effecting improved levels of nutrition and sustainable agriculture. The fermentation bioprocess is a strategic tool of the advancing agenda, addressing the inherent challenges of the agri-food paradigm specifically the ripple effects of post-harvest loss and malnutrition (Arora & Mishra, 2022). This predicament is scary since in the year 2000, close to one-third of all food that is produced on the planet to be consumed by mankind is discarded or lost (Nellemann, 2009). The burden of this global food loss crisis is differentially borne by the developing world countries in which the degradation of harvested crops is ridiculously amplified by the lack of infrastructure in the preservation and storage of the crop.

In addition to being the traditional preservative used, fermentation expands its importance by enhancing the usefulness of foodstuffs available by increasing the nutritional value and bioavailability of nutritionally essential micronutrients. Biochemical activities triggered by microbes inherent in fermentation raise protein digestibility, enhance levels of bioactive micronutrients, and minimize the antinutritive factors within food matrices, in addition to positively influencing nutritional outcomes. Such physiological benefits are of special concern when addressing the problem of micronutrient shortages and hidden hunger as a form of malnutrition that occurs despite adequate caloric intakes in different geographies. Fermentation adheres to the requirements of Sustainable Development Goal 2, namely, its focus on the development of the nutrient-rich and ecologically friendly food system by promoting culturally flexible and diversified modes of dieting. Such a twin emphasis on nutritional enhancement and sustenance of the system gives significance to the role of fermentation as a key initiative to solve modern food security issues and health statuses in the world.

Transforming agricultural residue to biofertilizers and other organic soil amendments via the microbial fermentation of agricultural residues enables the implementation of the agronomic systems that are ecological. Such biologically produced inputs lead to an alternative that is financially stronger than chemically generated fertilizers, whose production and use is directly linked to high energy consumption and the release of large amounts of nitrous oxide (N_2O), a strong greenhouse gas. In addition to this substitution effect, fermentation-microbial activity is also introduced, which is helpful in improving the health of the soil. It improves physicochemical characteristics of the soil and causes the soil to have a high-water retention capacity, better biogeochemical recycling of vital nutrients, and high resistance to abiotic stress factors in crops. This type of edaphic optimization is crucial to the sustenance of agricultural output over a long-term, with increasing climatic variability threatening. The resulting fermentation technology is, therefore, a strategic tool that can fulfill simultaneously the sustainability requirement of SDG 2 (Zero Hunger) and the climate action requirement of SDG 13 and makes it a key, multi-purpose tool to realize synergistic benefits of both food security and climate responsibility (Arora & Mishra, 2022).

15.4 International Legal and Policy Framework Governing Fermentation-Based Agriculture

The management of agricultural practices based on fermentation is the common feature shared by the complex of international legal policies on climate change, biodiversity, food safety, and food security. Fermentation is not commonly treated as a category that should be regulated but is implicitly controlled by larger regimes that make decisions affecting agricultural innovations, the utilization of microbial resources, climate mitigation, and the health of people in general. It is also important to understand how these frameworks are interconnected to assess the degree to which fermentation can be organised as a climate-resilient agricultural approach that suggests well with the Sustainable Development Goals 2 and 13. This part takes a critical look at three critical areas, namely the international climate change regime, the biodiversity and genetic resource, and the global safety and standards frameworks (Arora & Mishra, 2023).

15.4.1 UNFCCC, Climate Policy Instruments and Climate Change Regime Parts Agreement

The international climate change regime, centred on the United Nations Framework Convention on Climate Change (UNFCCC), offers a general normative framework to promote pathways for climate-resilient and low-emission development. UNFCCC acknowledges the role of agriculture as a cause and effect of climate change in a bid to provide food security through the establishment of adaptation and mitigation efforts soliciting food. Fermentation has not been directly mentioned in the convention; nonetheless, its goals are well aligned with the goal of sustainable development and flexibility of vulnerable sectors as adopted in the convention (Amjath-Babu et al., 2019).

The aligning goals in the Paris Agreement are reinforced by adding stimulus so that Parties must develop mitigation and adaptation plans in line with national conditions and sustainable development priorities (Glanemann et al., 2020). Article 2 highlights the need to protect food security but to manage climate change thus providing normative space in agricultural production such as biofertiliser production and organic waste valorisation which can mitigate emissions against improving agricultural resilience.

The factual instruments in the implementation framework of the Paris framework consist of Nationally Determined Contributions (NDCs) and National Adaptation Plans (NAPs). The development of sustainable agriculture, healthier soils, and reduction of wastes have been listed among the priority issues in the climate strategy of many countries, specifically those in the Global South (Abubakar et al., 2022). Nonetheless, fermentation technologies tend to be covered by other terms, like "climate-smart agriculture", or "organic farming", and there are few policy-specific and specific support. This de facto status limits the availability of climate finance, capacity-building programs and technological market structures that have the potential to accelerate the upcoming task related to the use of fermentation-based solutions.

The mechanisms of technology transfer and capacity-building in the UNFCCC as well as the technology mechanism and the Climate Technology Centre and Network offer possibilities to mainstream fermentation as part of the implementation of climate policy. The objectives of these mechanisms are to support the spread of environmentally sound technologies to developing nations, particularly those that could enhance adaptive capacity and emission intensity. The objectives are best synergised with fermentation technologies which are low-cost and decentralised in nature. However, it depends on national policy prioritisation to be included, thus necessitating legal policy articulation.

15.4.2 Biodiversity and Generic Resources Governance CBD, Nagoya Protocol, ITPGRFA

Bioethics fermentation can be closely connected with the international biodiversity and gene resource law, as it is based on microbial diversity and ancient knowledge of agriculture. The Convention on the Biological Diversity (CBD) provides a holistic system of safeguarding the biological diversity, sustainable utilisation of the elements thereof, and the equitable and just distribution of the resultant benefits of the use of the genetic resources. The microorganisms involved in fermentation are classified under the genetic resources of the CBD, meaning that fermentation practices are subject to access and benefit-sharing requirements.

The Nagoya Protocol brings into effect the access-and-benefit-sharing model of the CBD such that prior informed consent and agreed-upon terms of the utilisation of genetic resources and associated traditional knowledge are required. The protocol has brought about regulatory complexities to fermentation-based innovation though it was meant to prevent biopiracy and enable equity (Thomas Greiber, 2012). Small-scale producers and researchers also have some source of uncertainty on the compliance obligations especially where microbial resources are omnipresent or sourced through traditional food systems. Such regulatory ambiguity makes it less easy to promote innovation and spread fermentation technologies (particularly in a developing economy with a small institutional legal context) (Pereira & Viola, 2021).

The International Treaty on Plant Genetic Resources for Food and Agriculture (ITPGRFA) supplements the CBD by facilitating a multilateral system of accessibility to the plant genetic materials that are imperative in supplying food security. Though the treaty is mainly concerned with plant genetic material, its goals are directly connected with fermented-based agriculture, especially situations, when the production of such fermented products is based on standard types of crops. The emphasis of the treaty on the rights of farmers, the sharing of benefits, sustainable usage presents a normative foundation of fermentation as part of the food security and climate resilient policy. The treaty, however, fails to spell out microbial resources, and this has created a gap in regulation, which goes against the holistic governance of fermentation systems (Tsioumani, 2004).

Another area of overlap between biodiversity governance and safeguarding of indigenous and traditional knowledge is the safeguarding of indigenous and traditional knowledge. An ecological knowledge system, including cultural tradition and fermentation practices that are often passed over generations, can be a part of it. Such knowledge has been realised by the CBD as a key to biodiversity conversion and sustainable use, although it still is not evenly implemented (Pereira & Viola, 2021). Commercialisation of fermented products where recognition and benefit sharing is not properly done threatens to alienate the indigenous people and undermine customs. To solve this issue, a model of governance should be developed: innovation and consideration of social ethics in equal measure; fermentation should be part of sustainable development in an inclusive way.

15.4.3 Food Safety, Standards, and Biosafety-Codex Alimentarius and World Standards

Food safety standards are one of the critical aspects of the law in fermentation to take place, especially in international trade and health of people. Jointly managed by the World Health Organization and the Food and Agriculture Organization, the Codex Alimentarius Commission sets international standards on food and the manufacturing of food items, food guidelines, and food codes of practice to safeguard the healthiness of the consumer and trade activities free of unfair competition. The codex standards apply to fermented foods and ingredients, and they affect the country's policies regarding regulations and market access (Niyigaba et al., 2025).

Whilst codex standards have a twofold role in respect to fermentation. On the one hand, they offer harmonised system, which makes international trade easier and brings consumer confidence to fermented products. Conversely, strict safety and labelling standards could be a problem for the small-scale manufacturers and the traditional fermentation machinery, especially in the developing nations. Entry barriers include compliance costs and technical requirements, which threatens the fermentation-based solutions that are generally aligned with sustainability and climate goals (Rawat et al., 2025).

The regulatory environment of the fermentation technologies is also defined by biosafety governance. Although the method of traditional fermentation is generally considered safe, there is growing interest in the advancement of using specific or engineered microbial strains, which brings up questions with regard to the risk assessment and regulatory concerns. Different nations worldwide adopt biosafety norms, including those informed by the Cartagena Protocol on Biosafety, that emphasis precaution and risk management when using the living modified organisms (Cordonier Segger et al., 2013). When implementing these norms for fermentation, a delicate way should be applied that differentiates between practices that appear low risk when applied in the traditional context and interventions with higher risks when performed through biotechnology. The regulations that are overcautious can lead to discouragement of innovations and implementation of the positive fermentation technologies, but too little control can ruin the trust and safety of the population. Food safety, biosafety, intellectual property, and law further add to the complications of the action or governance of fermentation (Dutta & Das, 2023).

The existence of a patent on the microbial strains and fermentation process can stimulate the research and development, but it can also limit access and affordability especially for the smallholder farmers and the community-based producers (Marcus, 1979). The protection and spreading of fermentation technology are under the influence of the international intellectual property right regime together with the Agreement on Trade-Related Aspects of Intellectual Property Rights. The trade-off between innovation incentives and the interests of the population remains one of the problems of the world food and agriculture regulation.

In general, the issue of fermentation-based agriculture is still insufficiently supported by international laws and policy frameworks. All three regimes claim biodiversity, and food safety provide applicable normative principles, but due to their disintegration, fermentation as a climate-resilient method can hardly be recognized. Settling such gaps would need strong policy coherence, legal clarity, and dynamic systems of governance to respond to the multifunctional character of fermentation. These improvements would increase the ability of international law to underpin fair, adaptable, and sustainable food systems against climate change.

15.5 Role of International Institutions and Policy Coordination

International organisations play a paramount role in influencing the governance of the fermentation-based agricultural practices by establishing coordination of policies, capacity building, and the diffusion of technologies across national boundaries. Since fermentation is a transdisciplinary phenomenon that cuts across agriculture, climate change, biodiversity, and food safety, consistent institutional involvement is critical in terms of incorporating the practice in the global strategies for sustainable development and climate resilience. International organizations like the Food and Agriculture Organization of the United Nations (FAO), and the United Nations Framework Convention on Climate Change (UNFCCC) are bodies and allies of multilateral organizations that carry out the same roles, like bridging international law with the domestic policy of states (Mekouar, 2024).

The FAO is at the forefront of promoting fermentation as a sustainable and climate-resilient food system. The FAO assists member states to embrace sustainable agricultural practices that will improve food security and reduce the risks of climate change through the processes of providing normative guidance, technical assistance, and policy advisory services. FAO's work in agroecology, climate-smart agriculture, and reducing food waste includes collecting successful methods and starting projects that helps share knowledge and build skills, especially in developing countries where fermentation is a traditional practice (Leippert et al., 2020).

The UNFCCC institutions involved in climate governance have made fermentation a common practice by designing mechanisms that are aimed at supporting adaptation, mitigation, and transportation of technologies. Environmentally friendly technological mechanisms like the Climate Technology Centre and Network spread the environmentally friendly technology that champions the low emissions and climate-resilient development strategies. To achieve these goals, fermentation technologies, due to their minimal emissions, low cost, and decentralized character, can be well applicable, but their adoption requires coordination between climate institutions and agricultural agencies at international and national levels. On the same note, monetary initiatives like the Green Climate Fund promise any opportunities for expanding fermentation-based interventions when incorporated in the wider climate and food-security plans (Serli et al., 2023).

The coordination of policies of international policy-making bodies is essential in defeating the discontinuity that is defining the management of fermentation. Institutions focused on biodiversity under the Convention on Biological Diversity, food-safety under the Codex Alimentarius Commission, and climate under UNFCCC generally have different mandates, which results in overlaps and gaps in regulations. More inter-institutional exchange of information will enhance policy coherency by lowering the standards and preserving traditional understanding and equity in the human right to access fermentation technologies (Ruppel, 2022). This kind of coordination is necessary specifically in handling cross-cutting regulations like biosafety, intellectual rights, and sharing of benefits, which have direct impacts on the spread of fermentation-based solutions. In addition, global organizations can play a significant role in positioning fermentation as a part of global development policies such as the 2023 Agenda for Sustainable Development. Through the direct identification of fermentation as a goal-implicating process for the SDG2 and SDG13, international institutional policies and food-security frameworks make fermentation a tool to attain these targets. In that regard, international institutions will also serve as agenda-setting actors, as well as facilitators, and influence policy discourse on sustainable and climate-resilient agriculture (Martin & Vos, 2024).

Altogether, getting institutions involved, followed by policy alignment, is the key to realizing the full potential of fermentation as a practice of climate-resistant agriculture. Enhancing cooperation between the international organisations will help in sealing the loopholes in the existing governance, facilitate inclusive innovation, and make fermentation play significant roles in global climate and food-security goals.

15.6 Comparative Case Studies: Latin America, Africa, Asia

Intra-regional comparative analysis offers a beneficial perspective in which agricultural practice based on fermentation is framed by the varying ecological conditions, socio-economic moves, and legal-policy frameworks. The countries of Latin America, Africa, and Asia sever as important examples due to the strong fermentation capabilities, significant climate exposure, and diverse responses to sustainable agriculture and climate change regulations. Comparisons of these areas can give us insight into the potential transformation ability of fermentation as well as the structural factors that restrict its integration into formal policies of climate and food security (Mukherjee et al., 2022).

15.6.1 Latin America Fermentation, Agroecology and Climate Adaptation

The practices that are rationalized through fermentation in Latin America are strongly associated with the agroecological movements as well as indigenous agricultural systems, which focus on biodiversity protection and soil restoration in addition to community resilience (Lasso García et al., 2024). Locations like Brazil, Mexico, and Peru have traditionally used fermentation in food preparation, soil fixation, and waste treatment, especially in smallholder and indigenous agriculture. Fermented products are used broadly to supplement the soil fertility and crop resistance to climate variability, such as silage, compost tea, and biofertilizer (Altieri & Nicholls, 2020).

On a policy angle, the known national-level responses taken by some Latin American nations against climate change are national agro-ecological and sustainable agriculture policies. The National Policy on Agroecology and Organic Production of Brazil, e.g., implicitly defends the inputs via fermentation because it encourages the application of organic fertilizers and biological means of controlling pests and insects. Though in neither climate legislation nor Nationally Determined Contributions does fermentation receive any specific mention, the practices of fermentation form part of a larger idea of climate adaptation and low-emission agriculture. This indirect recognition allows fermentation to help achieve SDGs 2 and 13, specifically by capturing carbon in the soil, reducing chemical use, and improving food security for rural population.

Nevertheless, there are still regulatory issues. Food safety laws and certification policies, which support large-scale fermented product producers, normally restrict the commercialization of fermented products in Latin America. Also, benefit-sharing requirements of the Convention on Biological Diversity and Nagoya Protocol had generated legal ambiguity in understanding the utilization of microbes, especially where the traditional knowledge overlays with commercial innovation (Reichman, 2019). Although they are designed to enhance equity, these schemes have elaborate compliance conditions that unwillingly sideline small-scale producers and community fermentation facilities.

15.6.2 Community-Based Fermentation and Food Security Resilience: Africa

Fermentation has been recognized as a source of food security, nutrition, and climatic resilience in Africa, especially in resource-limited and rural contexts. The use of fermented foods such as Agri, Injera, Ogi and Kenkey are widely used staples in the continent, as they serve as a relatively cheap source of nutrition that is also more nutritionally beneficial and has a longer shelf life in the conditions of lack of refrigeration (Steinkraus, 2018). Further to food usage, fermentation is also currently used to control soils and to recycle wastes, fermented organic inputs are offered to enhance soil fertility and water retention in areas prone to drought. African states are amongst the most vulnerable to climate change, so adopting agriculture is a top priority.

Sustainable land management plans, climate resilience, and productivity of smallholder agriculture are some of the focuses of regional frameworks, including the African Union Comprehensive African Agricultural Development Programme (Critchley et al., 2023). The fermentation-based practice aligns with these aims by providing inexpensive and locally customizable solutions that reduce reliance on imported agrochemicals. The community-based composting and fermentation programs in countries like Ethiopia and Kenya have recorded a positive response in crop yield and stress to climatic fluctuations.

Despite these advantages, African law lacks proper control over the fermentation process. Food safety standards tend to concentrate on export-based standards at the expense of traditional fermented food eaten in overseas markets. The regulatory loophole may leave the consumers vulnerable to health threats at the same time as restricting the prospects of value-added and market entry. Besides, weak institutionalization capacity and poor policy coordination are barriers towards capturing fermentation as part of the national climate adaptation strategies (Abourabi, 2025). Despite increasing keen awareness of sustainable agriculture and soil management among African NDCs, fermentation as a specific intervention has never been mentioned in a list of interventions in climate finance and technical support.

More challenging to the spread of fermentation technologies in Africa are intellectual property and technology transfer challenges. Proprietary agendas of agricultural inputs and government-sourced models of innovation can peripherally access the local control of food systems and limit local access to indigenous use of fermentation processes and may privilege external models of capitalist productivity and production. To meet these problems, a governance framework that pays attention to fermentation as a social good and helps in community-led innovation in climate and food security frameworks is needed (Nambisan, 2017).

15.6.3 Asia: Traditional Knowledge, Innovation and Complexity in Regulations

Asia is a very diverse country where the ancient tradition of fermentation and the recent application of biotechnology co-exist. The production of fermented foods is very old in countries like India, China, Japan, and Indonesia, where people prepare idli, miso, kimchi, and soybean sauce (Meshram et al., 2024). These are highly cultural practices, which are entrenched in culture and diet regimes, and their contribution to nutrition and food security is high (De, 2025).

Fermentation is another recent trend that has been taken up as a technique to achieve sustainable agriculture, especially in organic and natural farming movements. A particularly good example is India, which through its natural farming campaigns promotes fermentation-based inputs like Jeevamrit and panchagavya to promote the use of fewer chemicals in fertility and improve the soil quality. These methods can be described as consistent with national priorities of sustainable agriculture and climate resilience as manifested in state policy to support organic and natural farming (Lee et al., 2024).

The legal acceptance of these fermentation-derived inputs is rather unclear, though, and regulatory practices tend to be devoid of the standards of safety, quality, and effectiveness. This uncertainty makes fermentation-based solutions nearly unexplorable in scale, which is unclear to the farmers (Resnik, 2024).

Japan and South Korea are countries in East Asia that have managed to incorporate traditional fermentation in the modern regulatory and market mechanisms, especially into the food sector (Chin et al., 2024). Other strong food safety standards coupled with the cultural recognition of the same have facilitated the attainment of the global market presence by the fermented products without compromising the quality or consumer confidence. Nevertheless, soil and climate resilience through agricultural fermentation attract slightly less policy attention, as they are part of a more generalized strategy of promoting industrial technological answers to traditional farming techniques in climate management (Mukherjee et al., 2022). The integration of fermentation operations and intellectual property law is a complicated form of governance throughout Asia. Even though patents and proprietary technologies have spurred innovation in industrial fermentation operations, they also contain microbial resources and cultural knowledge under legal monopoly. This conflict is especially acute in the developing Asian setting, where small-scale farmers often lack access to patented fermentation methods, and their traditional practices are unprotected. The allocation of incentives towards innovation and the equal distribution and sharing of benefits accordingly, therefore, form a key challenge in integrating fermentation into climate-resilient agriculture.

15.6.4 Comparative Inside and Governance Implications

When Latin America, Africa, and Asia are compared, it is evident that there are convergence and divergence in the manner in which fermentation-based agriculture is governed. Fermentation has in every part contributed significantly to food security, climate change adaptation, and sustainability, but it is mostly overlooked in official legislation and policymaking. In countries, where fermentation is amalgamated in the more agroecological or sustainable agriculture movements, including in some Latin America countries, it receives indirect policy support. When fermentations occur in informal settings in Africa, it is highlighted as accessible and resilient; however, these practices face limitations in expanding their activities due to lack of government control and institutional attention. Regulatory complexity and intellectual-property regimes all make up a progressively dual landscape of happenings and commitments in Asia (Candel, 2014).

It is based on these regional experiences that context-sensitive forms of governance that acknowledge that fermentation is entirely a legitimate and climate-resistant practice are required. Incorporating fermentation into national climate action policy, agricultural policy, and food safety policies can be used to promote policy coherence and enable co-benefits both in SDG 2 and SDG 13. Notably, comparative analysis has shown that good governance should strike a balance between standardisation and flexibility, innovation and equity, and also global standards and local knowledge systems.

15.6.5 Governance and Regulatory Issues

Even though fermentation-based agriculture can help improve food security and adapt to climatic changes, it faces major regulatory and governmental challenges that make it hard to get official validation and massive application. Such problems are caused by the faulty legal framework and ambiguity of the regulations as well as structural inequality in the international and national system of governance. This lack of a consistent regulatory method indicatively reflects a concerted trend in environmental and agricultural legislation in giving priority to the industrial-scale, technological solutions as opposed to the decriminalised and traditionally grounded approaches in fermentation (Ledesma-Amaro & Bamezai, 2025).

One of the key regulatory issues is the unclear legal status of fermentation technologies within agricultural and environmental legislation. With inputs like biofertilisation, compost teas, and microbial inoculants carried out through fermentation frequently lying between regulatory lines, or at least across regulatory jurisdictions, there often also exist inconsistencies in the way such inputs are treated. Most of the countries regulate these products by the fertilizer laws that control chemical inputs or through the biosafety framework established to deal with genetically modified organisms. This inconsistency of regulations creates excessive compliance costs on small-scale producers as well as deterring innovation, as it is especially true in developing nations where there is limited institutional capability. Scalability of fermentation as an agricultural practice is climate-resilient and weak because of the absence of specific standards to regulate it (Knorr & Augustin, 2024).

Another problem in governance is intellectual property rights. Although the patents and plant-variety protection are supposed to encourage technological processes, they pose questions on the issues of access, equity, and sharing benefits. Intellectual rights over a strain of microbe or fermentation process will limit the freedom of farmers to either produce or localize inputs and support dependence on commercial producers. This scenario is especially worrisome where the development of fermentation is based on the indigenous and traditional systems of knowledge, which are under-shielded within the current intellectual-property environments. This imbalance between the officially recognised intellectual property regulation and the unofficial knowledge system jeopardizes the shift of the community-based innovation and damages the social sovereignty (De, 2025).

The regulation of fermentation is also harder through biosafety regulation. Even though the traditional fermentation is popularly considered low-risk, regulatory frameworks tend to impose precautionary rules that were formulated in relation to the current biotechnology context, including genetically engineered organisms. Although the Cartagena Protocol on Biosafety concerned mainly the risk assessment and precautions in the transboundary movement of the living modified organisms, its principles are occasionally applied to the fermentation practices that lack scientific distinction. Excessive biosafety regulations have the potential to hinder the spread of fermentation technologies, and inadequate regulation can offer legitimate concerns about food safety and environmental effects. This is because matching a reasonable regulation balance continues to be a challenge (Hukić & Hukić, 2025).

Another governance limitation is technological barriers to diffusion, especially in developing countries. Affordability and flexibility of the fermentation practice are limited by accessibility to finance, technical knowledge, and broad-scale services provision. Technology-transfer programs and other international climate-finance mechanisms do not consider fermentation as a viable option; thus, the issue of fermentation continues to be a bias towards managing context-specific and community-based solutions rather than quantifiable and standardised interventions (Haselip et al., 2015).

Systematic obstacles are a barrier to the involvement of indigenous and small-holder communities, who often hold the most knowledge about fermentation, in formal policy. Traditional knowledge has been acknowledged by the international biodiversity law, but there are weak mechanisms of implementation and benefit-sharing in commercial biotechnology regimes; the most able communities, promoted as the most promising in developing climate-resilient food systems, are left out of the benefit-sharing arrangements (Sooresh et al., 2023). Overall, the regulatory and governance problems facing fermentation-based agriculture illustrate underlying problems in the global environmental and agricultural governance. To handle these issues, it is important that we move on to more adaptive, inclusive, and context-sensitive regulatory framework, which will value fermentation as a legitimate and valuable climate-resilient practice. This reform is necessary to get the maximum out of fermentation with regard to supporting a sustainable food system and combined climate action.

15.7 Policy Recommendations and Conclusion: Fermentation as a Legal and Policy Imperative for Climate Resilient Food Systems

The above discussion has illustrated that fermentation plays a unique but poorly acknowledged role at the nexus of climate governance, food security, and sustainable agricultural development. Fermentation is rarely mentioned in official regulatory discourse because international legal frameworks continue to point to integrated solutions in solving the issues of climate change and hunger. The marginalization is more a product of the form of governance that prioritizes technologies of capacity building and favors industrial technologies based on practices, which are decentralized and culturally necessary to adapt to the ecology. The discussion below summarizes the legal, policy, and comparative lessons that have been established during this work and makes a normative argument of fermentation as an essential element of climate-resilient food systems.

One of the key findings of this study is that fermentation represents the combined rationality behind Sustainable Development Goal 2 (SDG 2) and SDG 13. Fermentation can improve nutritional performance, minimize post-harvest loss, include all other siloed interventions, improve soil health, and reduce greenhouse gas emissions, unlike food security or climate mitigation interventions alone. All these co-benefits make fermentation emerge as a feasible implementation of the nexus approach to SDGs promoted in the United States. However, there is no clear legal audit in climate and agricultural policies, which limits the enjoyment of these benefits, restricting access to climate finance and technology-transfer schemes, as institutional support.

Fermentation, glance-wise, fits the agenda of the UNFCCC and the Paris Agreement, especially their focus on supporting food security and mitigation and adaptation efforts. However, this study shows, the idea of fermentation is rarely mentioned in Nationally Determined Contributions or National Adaptation Plans, which means that promoting low-cost, and scalable solutions that could really help developing economies has been ignored (Augustin, 2024). This disconnect reflects broader challenges in international climate law, where agricultural adaptation solutions often lack specificity and functionality. Including fermentation in the climate-planning tools would improve the consistency of policies and reinforce climate commitment implementation.

The fermentation is complicated even further by the biodiversity and genetic resource regimes. Though the Convention on Biological Diversity and the Nagoya Protocol are aimed at fostering fair access and sharing of benefits, their application has caused regulatory ambiguity in the fermentation procedures based on omnipresent sources of microbes and traditional knowledge (Convention on Biological Diversity, 2011). In comparative analysis, cases has standardized on how over-complex access and benefit-sharing requests could deter innovation and marginalize small-scale production, especially in places where fermentations are undertaken in an informal manner. The same observation highlights why regulatory strategies ought to be modified to separate high-risk biotechnological applications and the low-risk traditional fermentation systems (De, 2025).

Another area of conflict is that of food safety and biosafety systems. The Codex Alimentarius standards and national food-safety regulations are important for protecting public health and trade; however, they unintentionally exclude traditional fermented foods and inputs by imposing compliance requirements that apply only to industrial producers. On the same note, policy bias, such as the imposition of precautionary biosafety norms that are generated to protocols that express fermentation, may result in disproportionately burdensome regulations. These issues indicate a need to have adaptive governance models that would be safe, innovative, and inclusive.

The fact that, the effectiveness of fermentation is highly reliant on the context of governance is strengthened by the comparative analysis across Latin America, Africa, and Asia (Tamang et al., 2020). In agroecological or sustainable-agriculture movements, where the idea of fermentation is integrated, such as in parts of Latin America, it takes advantage of the indirect support of policies (Giraldo, 2025). In Africa, informality helps to increase the level of accessibility but offers existing regulatory gaps that restrict the level of scaling and institutional acceptability. Regulatory fermentation and intellectual regimes develop in Asia and generate a mottled landscape of innovation exclusions. Such regional differences emphasize the need of a context-sensitive legal framework which reflects both respects of local knowledge systems but allows policy integration (Ogunlade et al., 2025).

The issues of equity present themselves as a cross-cutting issue during the study. The indigenous communities and small-scale groups of people maintain fermentation traditions whose inputs continue to be undervalued by formal legal systems. There is a risk of repeating structural inequalities in climate and agriculture governance through the inability to preserve the traditional knowledge and unfairly share benefits. Appreciating of fermentation as a public good and integrating it into inclusive governance systems is hence vital towards promotion of just and sustainable food systems (Resnik, 2024).

With these results in mind, the present research would draw a number of policy implications. To begin with, the fermentation process, as a type of practice in the agricultural sector that is environmentally friendly in response to climate change and promotes biodiversity, must be clearly accepted in international and state legal norms. This would help increase its place within climate-action agendas, agricultural policies and development strategies in accordance with SDG 2 and SDG 13. Second, the regulatory systems of biosafety, food safety and intellectual property should be defibrillated into the characteristic of the low-risk and community-based character of several fermented activities. It will necessitate the varying regulatory provisions that promote innovation as well as protect the human and environmental wellbeing.

Third, the international organizations like the FAO and UNFCCC bodies need to be more active to mainstream fermentation by guiding policy, building capacities and climate-finance. Such institutions are able to shift the priorities of countries and resources distribution by presenting fermentation as a strategy to mitigate climate and adapt to the changes and provide their populations with food. Fourth, the government framework should facilitate the safeguarding of indigenous and traditional knowledge systems acknowledging the core position they hold in enabling the fermentation culture and provision of climate resilience.

Finally, fermentation can be considered a technique to overcome the modern-day climate and food-security issues rather than a primitive or a supportive technique in agriculture. Scientific soundness, environmental friendliness, and socio-cultural entrenchedness make it ideally applicable to the coordinated goal of sustainable growth. This study proposes paradigm shift by putting fermentation in the context of international legal and policy to achieve climate-vulnerable agriculture that embraces low-cost, inclusive, and environmentally based solutions and technological innovation. It is crucial to acknowledge that fermentation is a scientific improvement and a policy necessity in the future of constructing sustainable, balanced and climate-safe food systems.

References

Abourabi, Y. (2025). Environmental governance in Africa: The challenge of food security. In Y. Abourabi, J. Durand de Sanctis, & J.-N. Ferrié (Eds.), *African continental governance: Normative trends and agency challenges*. Palgrave Macmillan. https://doi.org/10.1007/978-3-032-04431-0_5

Abubakar, I. R., Maniruzzaman, K. M., Dano, U. L., AlShihri, F. S., AlShammari, M. S., Ahmed, S. M. S., Al-Gehlani, W. A. G., & Alrawaf, T. I. (2022). Environmental sustainability impacts of solid waste management practices in the global south. *International Journal of Environmental Research and Public Health, 19*(19), 12717. https://doi.org/10.3390/ijerph191912717

Adam, A. K., Sadhu, T., Mondal, B. P., & Pamucar, D. (2025). Low-carbon agricultural practices enhance climate resilience and food security in India. *Discover Sustainability, 7*, 142. https://doi.org/10.1007/s43621-025-01675-y

Ajayeoba, T. A., & Ijabadeniyi, O. A. (2025). *Transforming food for the future: Precision fermentation as a key to sustainability, nutrition, and health* (SSRN Scholarly Paper No. 5119686). SSRN. https://doi.org/10.2139/ssrn.5119686

Ali, S. A., et al. (2025). Holistic approaches to enhancing climate resilience and biodiversity conservation for food security. In B. Pramanick, S. V. Singh, S. Maitra, S. Celletti, & A. Hossain (Eds.), *Climate-smart agricultural technologies* (Sustainability sciences in Asia and Africa). Springer. https://doi.org/10.1007/978-981-96-7699-6_16

Altieri, M. A., & Nicholls, C. I. (2020). Agroecology and the reconstruction of a post-COVID-19 agriculture. *The Journal of Peasant Studies, 47*(5), 881–898. https://doi.org/10.1080/03066150.2020.1782891

Amjath-Babu, T. S., Aggarwal, P. K., & Vermeulen, S. (2019). Climate action for food security in South Asia? Analyzing the role of agriculture in nationally determined contributions to the Paris agreement. *Climate Policy, 19*(3), 283–298. https://doi.org/10.1080/14693062.2018.1501329

Arora, N. K., & Mishra, I. (2022). Current scenario and future directions for sustainable development goal 2: A roadmap to zero hunger. *Environmental Sustainability, 5*(2), 129–133. https://doi.org/10.1007/s42398-022-00235-8

Arora, N. K., & Mishra, I. (2023). Sustainable development goal 13: Recent progress and challenges to climate action. *Environmental Sustainability, 6*(3), 297–301. https://doi.org/10.1007/s42398-023-00287-4

Augustin, M. A. (2024). Innovation in precision fermentation for food ingredients. *Critical Reviews in Food Science and Nutrition, 64*(18), 6218–6238. https://doi.org/10.1080/10408398.2023.2166014

Baky, M. H., Zahran, H. A., Khalifa, I., Nabih, E. K., Zayed, A., Capanoglu, E., & Farag, M. A. (2025). Sustainable valorization of olive by-products: Environmental impact, biotechnological and probiotic approaches for bioactive recovery in alignment with circular bioeconomy. *Critical Reviews in Food Science and Nutrition*, 1–25. https://doi.org/10.1080/10408398.2025.2555406

Borrego-Ruiz, A., González-Domenech, C. M., & Borrego, J. J. (2024). The role of fermented vegetables as a sustainable and health-promoting nutritional resource. *Applied Sciences, 14*(23), 10853. https://doi.org/10.3390/app142310853

Buck, M., & Hamilton, C. (2011). The Nagoya protocol on access to genetic resources and the fair and equitable sharing of benefits arising from their utilization to the convention on biological diversity. *Review of European Community & International Environmental Law, 20*(1), 47–61. https://doi.org/10.1111/j.1467-9388.2011.00703.x

Candel, J. J. L. (2014). Food security governance: A systematic literature review. *Food Security, 6*(4), 585–601. https://doi.org/10.1007/s12571-014-0364-2

Chai, K. F., & Chen, W. N. (2025). Microbes in fermentation: Guardians or threats to food safety? *Journal of Agricultural and Food Chemistry, 73*(33), 20643–20652. https://doi.org/10.1021/acs.jafc.5c07574

Chaturvedi, V., Ghosh, A., Garg, A., Avashia, V., Vishwanathan, S. S., Gupta, D., Sinha, N. K., Bhushan, C., Banerjee, S., Datt, D., Bansal, J., Pathak, M., Dhar, S., Singh, A. K., Khan, N., Rashmi, R. R., Agrawal, S., Agarwal, D., Singh, A., et al. (2024). India's pathway to net zero by 2070: Status, challenges, and way forward. *Environmental Research Letters, 19*(11), 112501. https://doi.org/10.1088/1748-9326/ad7749

Chen, V., & Di Lernia, C. (2022). Climate and carnism: Regulatory pathways towards a sustainable food system. *University of New South Wales Law Journal, 45*, 997–1041. https://heinonline.org/HOL/Page?collection=journals&handle=hein.journals/swales45&id=1016&men_tab=srchresults

Chiarolla, C. (2017). Convention on biological diversity and regulatory law as plant protection. In *Research handbook on intellectual property and the life sciences* (pp. 197–213). Edward Elgar Publishing.

Chin, X. H., Lee, S. Y., Patel, M., & Smith, J. (2024). Enhancing food safety in soybean fermentation through strategic microbial risk assessment. *Heliyon, 10*, e25007. https://doi.org/10.1016/j.heliyon.2024.e25007

Clapp, J., & Scott, C. (2018). The global environmental politics of food. *Global Environmental Politics, 18*(2), 1–11. https://doi.org/10.1162/glep_a_00464

Convention on Biological Diversity. (2011). *Nagoya protocol on access to genetic resources and the fair and equitable sharing of benefits arising from their utilization*. CBD Secretariat. https://www.cbd.int/abs

Cordonier Segger, M.-C., Perron-Welch, F., & Frison, C. (Eds.). (2013). *Legal aspects of implementing the Cartagena protocol on biosafety* (pp. 19–59). Cambridge University Press.

Critchley, W., Harari, N., Mollee, E., Mekdaschi-Studer, R., & Eichenberger, J. (2023). Sustainable land management and climate change adaptation for small-scale land users in Sub-Saharan Africa. *Land, 12*(6), 1206. https://doi.org/10.3390/land12061206

Danelyan, A. A., & Gulyaeva, E. E. (2022). Problems of biosafety in current international law. *Moscow Journal of International Law, 2*, 66–84. https://doi.org/10.24833/0869-0049-2022-2-66-84

De, S. (2025). Living-in-Place in Kavery Nambisan's *The Scent of Pepper*. South Asian Review. Advance online publication. https://doi.org/10.1080/02759527.2025.2504241

Dutta, S., & Das, D. (2023). Bioprocess: Control, management, and biosafety issues. In *Basic biotechniques for bioprocess and bioentrepreneurship* (pp. 355–364). Academic Press.

Elliott, M. (2018). Indigenous resurgence: The drive for renewed engagement and reciprocity in the turn away from the state. *Canadian Journal of Political Science, 51*(1), 61–81. https://doi.org/10.1017/S0008423917001032

El-Mansi, M. (2018). Fermentation microbiology and biotechnology: An historical perspective. In M. El-Mansi, C. F. Bryce, A. L. Demain, & A. H. Salgado (Eds.), *Fermentation microbiology and biotechnology* (4th ed., pp. 3–8). CRC Press.

Food and Agriculture Organization of the United Nations (FAO), International Fund for Agricultural Development (IFAD), United Nations Children's Fund (UNICEF), World Food Programme (WFP), & World Health Organization (WHO). (2025). *The state of food security and nutrition in the world 2025: Addressing high food price inflation for food security and nutrition* (FAO Flagship Report). FAO. https://www.fao.org/publications/fao-flagship-publications/the-state-of-food-security-and-nutrition-in-the-world/en

Galanakis, C. M. (2024). The future of food. *Foods, 13*(4), 506.

García-Franco, A., Godoy, P., de la Torre, J., Duque, E., & Ramos, J. L. (2021). United nations sustainability development goals approached from the side of the biological production of fuels. *Microbial Biotechnology, 14*(5), 1871. https://doi.org/10.1111/1751-7915.13912

Geissdoerfer, M., Savaget, P., Bocken, N. M., & Hultink, E. J. (2017). The circular economy – A new sustainability paradigm? *Journal of Cleaner Production, 143*, 757–768.

Giraldo, O. F. (2025). The agroecological movement. A panoramic view. *The Journal of Peasant Studies, 52*(1), 47–73. https://doi.org/10.1080/03066150.2024.2318463

Glanemann, N., Willner, S. N., & Levermann, A. (2020). Paris climate agreement passes the cost-benefit test. *Nature Communications, 11*(1), 110. https://doi.org/10.1038/s41467-019-13961-1

Gonzalez, C. G. (2011). Climate change, food security, and agrobiodiversity: Toward a just, resilient, and sustainable food system. *Fordham Environmental Law Review, 22*, 493–522. https://www.jstor.org/stable/44175833

Greiber, T. (2012). *An explanatory guide to the Nagoya protocol on access and benefit-sharing* (pp. 83–137). IUCN.

Gunduz, C. P. B. (2023). Eco-friendly and cost-effective methods applied to sustainable food industries. In *Smart food industry: The Blockchain for sustainable engineering* (pp. 50–71). CRC Press.

Hasegawa, T., Fujimori, S., Havlík, P., Valin, H., Bodirsky, B. L., Doelman, J. C., et al. (2018). Risk of increased food insecurity under stringent global climate change mitigation policy. *Nature Climate Change, 8*(8), 699–703. https://doi.org/10.1038/s41558-018-0230-x

Haselip, J., Hansen, U. E., Puig, D., et al. (2015). Governance, enabling frameworks and policies for the transfer and diffusion of low carbon and climate adaptation technologies in developing countries. *Climatic Change, 131*, 363–370. https://doi.org/10.1007/s10584-015-1440-0

Hukić, M., & Hukić, E. (2025). Ethical and biosafety challenges in modern microbiology: Navigating the risks of emerging technologies. In *Spectrum of dual-use technologies: Unforeseen risks versus returns* (pp. 163–182). Springer Nature. https://doi.org/10.1007/978-3-032-00202-0_6

Intergovernmental Panel on Climate Change. (2023). Climate change 2023: Synthesis report – Summary for policymakers. In H. Lee & J. Romero (Eds.), *AR6 synthesis report of the IPCC Sixth Assessment Report*. IPCC. https://www.ipcc.ch/report/ar6/syr/downloads/report/IPCC_AR6_SYR_SPM.pdf

Jiao, F., Zhu, X., Li, C., Xie, H., & Li, H. (2025). Advancing biomass fermentation and composting for global sustainability: A bibliometric perspective (2019–2024). *Discover Sustainability, 6*, 667. https://doi.org/10.1007/s43621-025-01574-2

Knorr, D., & Augustin, M. A. (2024). The future of foods. *Sustainable Food Technology, 2*, 253–265. https://doi.org/10.1039/D3FB00199G

Kozarev, V., Damianova, Z., & Asenova, D. (2020). Genetically modified organisms and the precautionary principle. *D2. 4.1 Intra Case Study Analysis*, 45. https://recipes-project.eu/sites/default/files/2021-03/RECIPES%20D2.4.1%20Intra%20case%20study%20analysis_final.pdf#page=46

Lasso García, C., German Bermúdez, F. A., Vanden Berghe, W., Zurita-Benavides, M. G., & Orellana-Manzano, A. (2024). Fermented beverages among indigenous Latin American societies. *Frontiers in Sustainable Food Systems, 8*, 1390162. https://doi.org/10.3389/fsufs.2024.1390162

Lawal, T. O., Mustapha Abdulsalam, A. M., & Sundararajan, S. (2023). Economic and environmental implications of sustainable agricultural practices in arid regions: A cross-disciplinary analysis of plant science, management, and economics. *International Journal, 10*(3), 3100–3114. https://www.researchgate.net/profile/Abdulsalam-Mustapha/publication/375546915_Economic_and_Environmental_Implications_of_Sustainable_Agricultural_Practices_in_Arid_Regions_A_Cross-disciplinary_Analysis_of_Plant_Science_Management_and_Economics/links/65538470ce88b87031e7581d/Economic-and-Environmental-Implications-of-Sustainable-Agricultural-Practices-in-Arid-Regions-A-Cross-disciplinary-Analysis-of-Plant-Science-Management-and-Economics.pdf

Ledesma-Amaro, R., & Bamezai, S. (2025). *Sustainable use and conservation of fermentation-associated microorganisms within the agrifood system* (pp. 41–45). Food & Agriculture Organization.

Lee, C.-H., Ahn, J., & Son, H.-S. (2024). Ethnic fermented foods of the world: An overview. *Journal of Ethnic Foods, 11*(1), 39. https://doi.org/10.1186/s42779-024-00254-2

Leippert, F., Darmaun, M., Bernoux, M., & Mpheshea, M. (2020). *The potential of agroecology to build climate-resilient livelihoods and food systems*. Food and Agriculture Organization of the United Nations & Biovision. https://doi.org/10.4060/cb0438en

Marcus, I. (1979). Fermentation processes and products: Problems in patenting. In *Microbial technology* (pp. 497–530). Academic Press. https://doi.org/10.1016/B978-0-12-551502-3.50024-4

Marshall, M. (2022). The refrigerator as a problem and solution: Food storage practices as part of sustainable food culture. *Food and Foodways, 30*(4), 261–286. https://doi.org/10.1080/07409710.2022.2124726

Martin, W., & Vos, R. (2024). *The SDGs and food system challenges: Global trends and scenarios toward 2030* (IFPRI Discussion Paper 2237). International Food Policy Research Institute. https://hdl.handle.net/10568/138940

Materia, V. C., Linnemann, A. R., Smid, E. J., & Schoustra, S. E. (2021). Contribution of traditional fermented foods to food systems transformation: Value addition and inclusive entrepreneurship. *Food Security, 13*(5), 1163–1177. https://doi.org/10.1007/s12571-021-01185-5

McManis, C. R. (Ed.). (2012). *Biodiversity and the law: Intellectual property, biotechnology and traditional knowledge*. Earthscan.

Mekouar, M. A. (2024). Food and agriculture organization of the united nations (FAO). *Yearbook of International Environmental Law, 34*(1), yvae031. https://doi.org/10.1093/yiel/yvae031

Meshram, B. D., Shaikh, A., Datir, R. P., & Ranveer, S. (2024). Fermented products of India and abroad: A review. In *Trends & prospects in food technology, processing and preservation* (pp. 361–400). Today and Tomorrow's Printers and Publishers. Retrieved from https://www.researchgate.net/profile/Sg-Ranvir/publication/338764985_Fermented_products_of_India_and_Abroad_A_review/links/5e296e10299bf15216760402/Fermented-products-of-India-and-Abroad-A-review.pdf

Motarjemi, Y. (2002). Impact of small scale fermentation technology on food safety in developing countries. *International Journal of Food Microbiology, 75*(3), 213–229. https://doi.org/10.1016/S0168-1605(01)00709-7

Mukherjee, A., Gómez-Sala, B., O'Connor, E. M., Kenny, J. G., & Cotter, P. D. (2022). Global regulatory frameworks for fermented foods: A review. *Frontiers in Nutrition, 9*, 902642. https://doi.org/10.3389/fnut.2022.902642

Mukherji, A. (2023). *Climate change 2023 synthesis report.*

Nambisan, P. (2017). *An introduction to ethical, safety and intellectual property rights issues in biotechnology*. Academic Press.

Nellemann, C. (Ed.). (2009). *The environmental food crisis: The environment's role in averting future food crises: A UNEP rapid response assessment*. UNEP/Earthprint.

Niyigaba, T., Küçükgöz, K., Kołożyn-Krajewska, D., Królikowski, T., & Trząskowska, M. (2025). Advances in fermentation technology: A focus on health and safety. *Applied Sciences, 15*(6), 3001. https://doi.org/10.3390/app15063001

Ogunlade, C. A., Olaniyan, A. R., Babalola, R. T., Oyefeso, B. O., & Jaiyeoba, K. F. (2025). Processing techniques for bio-based products in the global south. In *Sustainable bioeconomy development in the global south: Volume II bioeconomy techniques* (pp. 39–65). Springer Nature Singapore. https://doi.org/10.1007/978-3-032-00202-0_6

Onyeaka, H., Duan, K., Miri, T., Pang, G., Shiu, E., Pokhilenko, I., et al. (2024). Achieving fairness in the food system. *Food and Energy Security, 13*(4), e572. https://doi.org/10.1002/fes3.572

Pereira, J. C., & Viola, E. (2021). *Climate change and biodiversity governance in the Amazon: At the edge of ecological collapse?* Routledge. https://doi.org/10.4324/9780429296581

Rai, S., Lama, S., & Dhyani, S. (2024). Mainstreaming indigenous and traditional ecological knowledge (TEK) systems in global climate policy for resilient ecosystems and societies. In S. C. Rai & P. K. Mishra (Eds.), *Sacred landscapes, indigenous knowledge, and ethno-culture*

in natural resource management. Advances in geographical and environmental sciences. Springer. https://doi.org/10.1007/978-981-97-4206-6_25
Rakshit, S., Aiswarya, S., & Sahoo, A. K. (2025). Global carbon market: Policy pathways for low carbon emissions in the agriculture sector. In *Agriculture toward net zero emissions* (pp. 61–81). Elsevier. https://doi.org/10.1016/B978-0-443-13985-7.00006-3
Rawat, S., Magar, Z. A. A., Gaikwad, P. S., Negi, T., Rai, S., & Choudhary, P. (2025). Biotechnological horizons in food processing: Future perspectives and challenges. In *Biotechnological innovations in food processing* (pp. 661–707). Springer. https://doi.org/10.1007/978-3-031-96526-5_15
Reichman, J. H. (2019). Why the Nagoya protocol to the convention on biological diversity matters to science and industry everywhere. In *Intellectual property and development: Understanding the interfaces* (pp. 295–314). Springer. https://doi.org/10.1007/978-981-13-2856-5_14
Resnik, D. B. (2024). Biosafety, biosecurity, and bioethics. *Monash Bioethics Review, 42*(1), 1–15. https://doi.org/10.1007/s40592-023-00187-6
Ruppel, O. C. (2022). Soil protection, food security and the nexus between climate governance and trade in agriculture. In *International yearbook of soil law and policy 2020/2021* (pp. 481–520). Springer. https://doi.org/10.1007/978-3-030-96347-7_19
Serli, D., Parrish, R., Lauriola, P., & Zeka, A. (2023). Green climate fund adaptation's interventions on food security: Metrics and health outcomes. *European Journal of Public Health, 33*(Suppl. 2). https://doi.org/10.1093/eurpub/ckad160.1095
Sharma, P., Sangwan, S., Kumari, A., Singh, S., & Kaur, H. (2022). Impact of climate change on soil microorganisms regulating nutrient transformation. In A. Vaishnav, S. Arya, & D. K. Choudhary (Eds.), *Plant stress mitigators*. Springer. https://doi.org/10.1007/978-981-16-7759-5_8
Sharma, K., et al. (2025). Integrating indigenous technical knowledge for sustainable agriculture and horticulture. In *Advances in knowledge acquisition, transfer, and management* (pp. 207–225). IGI Global. https://doi.org/10.4018/979-8-3693-3542-0.ch011
Smith, D., da Silva, M., Jackson, J., & Lyal, C. (2017). Explanation of the Nagoya protocol on access and benefit sharing and its implication for microbiology. *Microbiology, 163*(3), 289–296. https://doi.org/10.1099/mic.0.000425
Sooresh, M. M., Willing, B. P., & Bourrie, B. C. T. (2023). Opportunities and challenges of understanding community assembly in spontaneous food fermentation. *Foods, 12*(3), 673. https://doi.org/10.3390/foods12030673
Steinkraus, K. H. (2018). *Handbook of indigenous fermented foods* (2nd ed.). CRC Press.
Tamang, J. P., Cotter, P. D., Endo, A., Han, N. S., Kort, R., Liu, S. Q., Mayo, B., Westerik, N., & Hutkins, R. (2020). Fermented foods in a global age: East meets west. *Comprehensive Reviews in Food Science and Food Safety, 19*(1), 184–217. https://doi.org/10.1111/1541-4337.12520
Tsioumani, E. (2004). International treaty on plant genetic resources for food and agriculture: Legal and policy questions from adoption to implementation. *Yearbook of International Environmental Law, 15*(1), 119–144. https://doi.org/10.1093/yiel/15.1.119
United Nations. (2015). *Transforming our world: The 2030 agenda for sustainable development*. UN General Assembly.
Zhang, S., Zhang, H., Xie, F., & Wu, D. (2025). Climate change and sustainable agriculture: Assessment of climate change impact on agricultural resilience. *Sustainability, 17*(16), 7376. https://doi.org/10.3390/su17167376

Chapter 16
Sensory Acceptance and Consumer Perception: Navigating Flavor, Texture, and Cultural Preferences in Fermented Foods

Sergio Erick García-Barrón, María Aurora Pintor-Jardines, Claudia Ariadna Acero-Ortega, Anne Gschaedler-Mathis, and Enrique J. Herrera-López

Abstract Fermented plant-based foods (FPBFs) are a dietary alternative due to the different compounds they contain, which are generated during fermentation. Consuming these foods could be beneficial to health, making them a sustainable alternative for reducing food waste and generating added value. It is important to understand consumer perceptions and preferences to identify the physical-chemical, process, labeling, and regulatory parameters to be addressed. Therefore, intrinsic and extrinsic product attributes, along with psychological, situational, sociocultural and biological and physiological factors are studied to determine their influence on consumer perceptions and preferences. The research adopts a transdisciplinary approach based on a holistic conceptual model trying to explain the perception and preference of plant-based fermented products. It considers dynamic and

S. E. García-Barrón · C. A. Acero-Ortega
ESDAI, Universidad Panamericana, Ciudad de México, Mexico
e-mail: sergiogarcia@up.edu.mx; cacero@up.edu.mx

M. A. Pintor-Jardines
Departamento de Biotecnología, Universidad Autónoma Metropolitana Iztapalapa, Ciudad de México, Mexico
e-mail: ma.pintorj@izt.uam.mx

A. Gschaedler-Mathis
Unidad de Biotecnología Industrial, Centro de Investigación y Asistencia en Tecnología y Diseño del Estado de Jalisco, Zapopan, Jalisco, Mexico
e-mail: agschaedler@ciatej.mx

E. J. Herrera-López (✉)
Laboratorio para la Innovación en Bioelectrónica e Inteligencia Artificial, LINBIA, Unidad de Biotecnología Industrial, Centro de Investigación y Asistencia en Tecnología y Diseño del Estado de Jalisco, Zapopan, Jalisco, Mexico
e-mail: eherrera@ciatej.mx

S. Dhiman, S. Kaur (eds.), *Next-Gen Plant Based Fermented Foods: Innovations for a Sustainable Future*, https://doi.org/10.1007/978-3-032-29927-7_16

complex interactions among food, consumers and environment, contributing to their ecological validity. Finally, a section on future perspectives is included since it may be useful for this food segment.

Keywords Plant-based · Fermented foods · Consumer perception · Sensory acceptance

16.1 Introduction

Fermentation is a preservation technique developed throughout human history, along with customs representing symbolic and cultural value (Augustin et al., 2023). This technique has not only contributed to the creation of new products but has enabled the preservation of perishable foods such as milk, meat, and cereals (Anagnostopoulos & Tsaltas, 2019). Furthermore, it accounts for almost a third of global food consumption up to 40% in some populations (Andreu et al., 2022).

Fermentation has been used to generate complex macromolecules such as proteins, carbohydrates and metabolites such as flavonoids, terpenes, among others (Xiang et al., 2019). These compounds not only contribute to ensuring the shelf life of products but also improve their sensory properties, nutritional value, and biological activities against chronic diseases, e.g., through signal regulatory effects, lipid modulators, immunity stimulants, antimicrobials, antiparasitic, or anticarcinogens (Cuamatzin-García et al., 2022; García-Barrón et al., 2025).

In recent years, consumers have become increasingly conscious of their diet and, as a result, their interest in healthy and natural foods, such as fermented foods, has grown (García-Barón et al., 2025). In this sense, the demand for and consumption of fermented plant-based foods (FPBFs) has increased substantially. Likewise, the demand for FPBFs that are sustainable and promote human health is growing rapidly (Jaeger et al., 2023b). Therefore, it is necessary to analyze and understand the perceptions and preferences of FPBF consumers. This would allow us to recognize the factors that impact their behavior and identify the product parameters that need further development (Mandha et al., 2021).

The selection and consumption of food is a complex process because different factors are involved. Therefore, a multidisciplinary approach would be necessary to explain this process holistically (Chen & Antonelli, 2020). In this regard, Mojet (Köster, 2009) proposed a holistic model that considers six different factors that influence eating behavior and food choice: biological and physiological factors, extrinsic product characteristics, intrinsic product characteristics, psychological factors, situational factors, and sociocultural factors. This chapter, therefore, addresses the factors that influence consumer perception and preference for plant-based fermented products. It moreover presents different perspectives derived from the study of these factors.

16.2 Factors Influencing the Perception and Consumption Preferences of Plant-Based Fermented Products

To ensure reliability and facilitate analysis of the involved factors, the following were classified: biological and physiological factors, extrinsic product characteristics, intrinsic product characteristics, psychological factors, situational factors, and sociocultural factors.

16.2.1 Biological and Physiological Factors

Factors such as age and gender performance have a fundamental role in the perception and preference of different food products (Fernqvist et al., 2024). Research has shown that perceptions of food tend to change as people age (Mojet, 2001). Furthermore, Prutkin et al. (2000) point out that, on a physiological level, women tend to be more sensitive than men.

According to Roh et al. (2018) and Greis et al. (2023), there is a relationship between age, perception, and preference for FPBFs. According to Kwak et al. (2017) and Rojas-Rivas et al. (2019), age influences the perception of traditional fermented products (Rojas-Rivas & Cuffia, 2020). Therefore, when seeking to differentiate consumer segments or identify market trends, age should be considered a key factor (Kwak et al., 2017; Lee et al., 2019; Fibri & Frøst, 2020).

Gender is another subfactor studied within the category of FPBFs, as it influences the perception of the appearance and sensory characteristics of products. In line with this, Chijioke et al. (2021) mentioned that women are more involved in food preparation since they tend to be more perceptive. It is therefore necessary to consider gender, as it can influence how a product is prepared. Table 16.1 summarizes the FPBFs subfactors associated with biological and physiological factors.

16.2.2 Extrinsic Product Characteristics

When consumers encounter foods with multiple attributes, selecting one over another becomes a complex task. This is where extrinsic characteristics play a

Table 16.1 List of FPBFs subfactors encountered in biological and physiological factors

Factor	Subfactors	Reference
Biological and physiological	Age	Fibri and Frøst (2020), Greis et al. (2023), Kwak et al. (2017), Lee et al. (2019), Roh et al. (2018), and Rojas-Rivas and Cuffia (2020)
	Gender	Chijioke et al. (2021), Marlapati et al. (2024), and Rojas-Rivas and Cuffia (2020)

decisive role (Symmank, 2018) as they are the main motivators for consumers who are unfamiliar with the sensory properties of the food they want to purchase. These characteristics even cause consumers to generate expectations about what they hope to find in the product they are going to choose (Banovic & Grunert, 2023); an aspect that stands out in new foods or those made with new technologies (Jaeger et al., 2023a; Saint-Eve et al., 2020; Banovic & Grunert, 2023; Fibri & Frøst, 2020). In the specific case of fermented foods of plant origin, secondary characteristics such as consumer health, health claims or healthy food labeling, sustainability claims, nutritional information, product origin and labeling, packaging design, price, and type of label, which are ultimately what can determine the decision to try that food and can influence consumers' purchase intention.

About health claims, Jaeger et al. (2023a) concluded that they influence consumer purchase intent, although their effect tends to carry less weight than the sensory characteristics of the product, suggesting that the health benefits that can be obtained from a food only occur when it meets an adequate sensory standard. Therefore, health claims must be presented clearly and objectively, as factors such as wording, the amount and type of information, framing, and positioning on the packaging can influence consumers. It is also necessary to determine whether consumers pay attention to these claims when deciding whether to purchase the product (García-Barón et al., 2025).

On the other hand, health labels refer to product information that complementarily emphasizes ingredients or properties related to health, such as probiotics, antioxidants, sweeteners, protein content, fiber, or salt reduction (García-Barrón et al., 2025). Additionally, the way in which information about the product or the technology used is presented influences consumer perceptions in terms of health (Banovic & Grunert, 2023; Greis et al., 2023; Jaeger et al., 2023a; Lee et al., 2019; Carneiro et al., 2024). Health benefit information can increase acceptance, but its effectiveness depends on the product and the audience, especially among consumers with different sociodemographic characteristics (Lee et al., 2019). Furthermore, environmental information can sometimes influence willingness to pay for plant-based products (Saint-Eve et al., 2020). However, the effectiveness of these claims has a limit, since sensory attributes tend to have a greater influence on food choice than health claims, so a balance between health and pleasure is necessary (Jaeger et al., 2023a; Saint-Eve et al., 2020).

Among the extrinsic factors, nutritional labeling refers to government-regulated information on labels, providing information about the product, influencing consumer expectations and willingness to pay (Lee et al., 2019; Banovic & Grunert, 2023; Greis et al., 2023; Carneiro et al., 2024). The declared protein content of plant-based fermented products is a key issue (Saint-Eve et al., 2020). In the case of fermented products containing both plant and animal proteins (e.g., milk and peas), it is essential to indicate the percentage of plant protein on the label.

During informed tasting, each dish clearly stated the percentages (25% pea, 50% pea, 75% pea), providing participants with compositional information that influenced their subsequent evaluations, including willingness to pay (WTP) (Saint-Eve et al., 2020). This is consistent with the findings of Brückner-Gühmann et al. (2019), who reported a greater preference for food as its protein content increased. On the other hand, the sodium content of this type of product is not only a concern for consumers from a nutritional point of view, but also in terms of microbial safety, as its quantity can influence microbial growth (Lee et al., 2019; Majhenič et al., 2025).

Another subfactor to consider in FPBFs is origin labeling, since according to Fibri and Frøst (2020), consumers prefer, appreciate, and trust local products because they are familiar with their sensory characteristics. This information can positively influence consumer emotions, as reported by Kwak et al. (2017), who concluded that the origin of traditional products is a key determinant of willingness to pay for this type of food, especially among older consumers.

Packaging is considered the first form of communication with consumers and therefore influences how they perceive themselves (Carneiro et al., 2024). Similarly, the brand is a determining attribute (Carneiro et al., 2024; Kwak et al., 2017), probably because it represents a promise or guarantee of quality based on symbolic value (Keller et al., 2015). In the case of new or unknown products, visual appearance and descriptive information determines factors in consumer behavior (Greis et al., 2023; Jaeger et al., 2023a).

Within the intrinsic factors, price plays a determining role in the perception and preference of fermented products (Saint-Eve et al., 2020; Kwak et al., 2017). As shown by Majhenič et al. (2025), the term "cheap" is often associated with a product category, indicating that price perception is intrinsic to its conceptual image. Thus, the price effect is modulated by contextual information on origin, traditionality, health, or environmental benefits, although a high price may be limited by the sensory quality of the product (Jaeger et al., 2023a).

Conversely, sustainability claims refer to information on environmental or ethical aspects acting as essential factors in food marketing, especially in the case of alternative or innovative plant-based products, as they can influence consumer perception (Jaeger et al., 2023a; Banovic & Grunert, 2023). Furthermore, environmental impact can motivate product selection (Greis et al., 2023; Jaeger et al., 2023a). However, the impact may be limited if consumers do not possess adequate objective knowledge in combination with other variables such as psychographics (Banovic & Grunert, 2023; Carneiro et al., 2024; Fibri & Frøst, 2020). Finally, extrinsic factors are crucial as they determine consumer perception and acceptance. They can generate cognitive expectations about sensory and hedonic characteristics (Greis et al., 2023; Jaeger et al., 2023a). However, the importance of these factors may vary according to the different consumer segments that may be encountered. Table 16.2 summarizes the FPBFs subfactors associated with extrinsic product characteristics.

16.2.3 Intrinsic Product Characteristics

Like other foods, FPBFs are composed of carbohydrates, lipids, proteins, vitamins, and water (Gupta et al., 2022). It is also important to consider that the manufacturing

Table 16.2 List of FPBFs subfactors encountered in the extrinsic product characteristics factor

Factor	Subfactors	Reference
Extrinsic product characteristics	Healthy label	Brückner-Gühmann et al. (2019)
		Lee et al. (2019)
	Nutritional facts	Brückner-Gühmann et al. (2019)
		Saint-Eve et al. (2020)
	Origin label	Fibri and Frøst (2020)
	Type of product label	Banovic and Grunert (2023)
		Kwak et al. (2017)
	Health claims, sustainability claim, type of product label	Jaeger et al. (2023a)
	Sustainability claim, type of product label	Greis et al. (2023)
	Type of product label	Carneiro et al. (2024)

process, raw materials, and other variables influence intrinsic characteristics (Deba-Rementeria et al., 2023; Sangija et al., 2022; Saint-Eve et al., 2020; Bernal-Gil et al., 2020), which are decisive for the identity and nature of foods.

Due to their composition, these types of foods can have a health-promoting effect (Melini et al., 2019), an example of this is pulque (Rojas-Rivas et al., 2019), which has been demonstrated in various studies (Fernqvist et al., 2024); therefore, this information can be used to promote consumption. During the fermentation process, different microorganisms such as bacteria, fungi, and yeasts are responsible for the chemical transformation of different compounds (Marco & Golomb, 2016; Rastogi et al., 2022). To this regard, it is important to consider that microbial type is decisive in sensory characteristics and consumer preferences (Deba-Rementeria et al., 2023; Nemati et al., 2023). Among the different strains used is the genus *Lactobacillus*. According to Mandha et al. (2021), it is possible to use *L. plantarum*, *L. rhamnosus*, *L. casei*, *L. brevis*, and Pediococcus *pentosaccus* in watermelon juice production. Since juices made with *L. brevis* and *Pediococcus pentosaccus* have more acceptance. Similarly, Boeck et al. (2022) evaluated alternative yogurts made with isolated lentil protein and fermented with *L. paracasei* FST 6.1, which were highly accepted, as was the product obtained from the starter culture *S. thermophiles*. The sensory characteristics were a less bitter aftertaste, with a more yogurt-like taste than other strain.

Santos et al. (2025), reported that coconut beverages fermented with *B. bifidum* can influence the level of enjoyment as they present descriptors such as "ideal sweetness", "coconut flavor," and "ideal texture." These results support the application of *B. bifidum* in the formulation of probiotic coconut water beverages as a promising functional matrix. On the other hand, in the development of symbiotic fermented vegan beverages, Valero-Cases et al. (2023) investigated the impact of fermentation with *Lactiplantibacillus plantarum* on the sensory profile and preference of orange-carrot juice blends and consumer acceptance. According to the results, orange-flavored and sweet juices were preferred.

Intrinsic factors include nutritional attributes, which strongly influence the preference and perception of FPBFs. These products are successfully evaluated thanks to the availability of essential proteins and amino acids, efficient digestion of complex carbohydrates, and incorporation of bioactive compounds that promote well-being (Majhenič et al., 2025). These products include non-alcoholic beverages fermented with *Sclerocarya birrea* (Marula), supplemented with biofortified corn and provitamin A, which stand out for their nutritional characteristics and excellent consumer acceptance (Dlamini et al., 2025). Fermented beverages based on cereals and legumes have also been developed, seeking to increase synergistic interactions between microorganisms and plant matrices in order to enhance techno-functional properties, providing consumers with new alternatives (Viretto et al., 2025). Similarly, beverages have been formulated using carrot, orange, and inulin juices fermented with *Lactiplantibacillus plantarum*, an interaction that enhances their antioxidant, bioactive, and probiotic properties (Valero-Cases et al., 2023). Another product highly valued for its protein content and nutritional value is plant-based yogurt (Saint-Eve et al., 2020; Gupta et al., 2022). The partial replacement of animal proteins with plant proteins in yogurts has led to the development of new consumer alternatives, which appear to be well accepted by consumers, without neglecting the importance of nutritional benefits (Marlapati et al., 2024). Hence, nutritional features should be considered as key elements for encouraging the consumption of fermented plant-based foods.

Within intrinsic factors, the processing method can influence product perception, particularly when it comes to those considered "traditional" as the process is critical in product differentiation (Kim & Lee, 2014; Adinsi et al., 2014; Jo & Kim, 2020) as well as influencing the preferences (Roh et al., 2018; Kwak et al., 2017; Fibri & Frøst, 2020; Akissoé et al., 2014; Wu et al., 2024; Santos et al., 2025; Dlamini et al., 2025). In addition to influencing the perception of image (Bernal-Gil et al., 2020; Carneiro et al., 2024). Thus, the information on the processing method is essential to improve preferences, especially among millennial consumers (Fibri & Frøst, 2020). On the other hand, the optimization of the beverage processing method can help to improve the level of taste, particularly off-flavors (Mishra et al., 2024; Cohen et al., 2023). In this regard, Sangija et al. (2022) applied different innovative fermentation methods to an indigenous vegetable to minimize postharvest losses, which improved sensory quality and overall acceptability.

Among the intrinsic factors, raw materials are determinant in perception and preferences, especially in traditional products, since they are often linked to an identity and to a geographic region (Fibri & Frøst, 2020; Adinsi et al., 2014; Akissoé et al., 2014; Kim et al., 2020; Rojas-Rivas & Cuffia, 2020). According to Byeon et al. (2023), within the same region, products considered traditional may present different characteristics due to the raw material. In the same sense, Wu et al. (2024) and Panjapiyakul et al. (2017) obtained similar results for both douchi and attiéké, since in both cases, the sensory characteristics and aromatic profile were associated with the combination of raw material and production process.

Defining the appropriate proportions of raw materials is one of the main challenges, as this can compromise the level of acceptability. In this regard, it is essential to understand the impact of raw materials (Cohen et al., 2023; Diby et al., 2023; Marlapati et al., 2024; Valero-Cases et al., 2023; Viretto et al., 2025). The use of different raw materials is crucial for the development of new products that aim to diversify the offering and ensure sustainability (Mishra et al., 2024; Sangija et al., 2022). However, incorporating and combining different raw materials poses major technological challenges for the sector (Gupta et al., 2022; Greis et al., 2023). According to Saint-Eve et al. (2020), the combined use of raw materials of both plant and animal origin allows for the creation of more sustainable products that are accepted by consumers. Similarly, it is important to consider that the nature and quality of raw materials influence consumer behavior, as in the case of kombucha, where the use of different types of teas has been explored (Dartora et al., 2023), and when seeking to fortify products (Akanni et al., 2024; Dlamini et al., 2025).

Among intrinsic attributes, sensory properties are the most significant factors in the perception and preference of plant-based fermented products. The influence of sensory properties depends on the type of product (Gupta et al., 2022; Diby et al., 2023; Majhenič et al., 2025; Jaeger et al., 2023b; Dlamini et al., 2025). This influence is associated with the age of the population (Greis et al., 2023). Sensory properties are of vital importance in both the reformulation and optimization of new products (Brückner-Gühmann et al., 2019; Deba-Rementeria et al., 2023; Dlamini et al., 2025), especially in terms of differentiation (Dartora et al., 2023). In this regard, Mandha et al. (2021) indicate that terms associated with "natural characteristics" tend to be more appreciated by consumers. One observation is that plant-based yogurts are less acceptable than yogurts made with dairy ingredients, which translate into sensory characteristics that are not typical of this product (Jaeger et al., 2023b). Sensory studies have determined that yogurts with characteristics such as white color, creamy appearance, sweet taste, and texture strongly influence the overall acceptance of plant-based yogurts (Gupta et al., 2022; Greis et al., 2023; Boeck et al., 2022; Jaeger et al., 2023b; Mustapa et al., 2024; Cardello et al., 2024; Höhl & Dreyer, 2025; Marlapati et al., 2024). When it comes to beverages, certain sensory aspects such as color, taste, and smell strongly influence consumers' purchasing decisions (Mishra et al., 2024; Mustapa et al., 2024; Carneiro et al., 2024; Valero-Cases et al., 2023).

When it comes to traditional products, sensory properties are decisive for product identity and preference (Chijioke et al., 2021; Adinsi et al., 2014; Jo & Kim, 2020; Kim & Lee, 2018; Rojas-Rivas & Cuffia, 2020). Therefore, the presence of unexpected or different flavors can affect the level of enjoyment of this type of product (Ambarita et al., 2019). Different studies have pointed out that sensory properties are related to the way a traditional product is consumed (Roh et al., 2018; Kwak et al., 2017; Rojas-Rivas et al., 2019; Kim et al., 2020), which is relevant for identifying the determinants of liking (Roh et al., 2018; Wu et al., 2024; Byeon et al., 2023; Panjapiyakul et al., 2017). Furthermore, identifying the sensory properties responsible for preferences in traditional products is essential when adapting them to different types of consumers (Lee et al., 2019; Fibri & Frøst, 2020; Adinsi et al.,

Table 16.3 List of FPBFs subfactors encountered in the Intrinsic product characteristics factors

Factor	Subfactor	Reference
Intrinsic product characteristics	Microbial type	Nemati et al. (2023)
	Processing method	Bernal-Gil et al. (2020)
	Raw material	Akanni et al. (2024)
	Sensory properties	Brückner-Gühmann et al. (2019), Cardello et al. (2024), Chijioke et al. (2021), Höhl and Dreyer (2025), Jaeger et al. (2023a, b), Kim and Lee (2018), Mustapa et al. (2024), and Park et al. (2020)
	Health, sensory properties	Rojas-Rivas et al. (2019)
	Microbial type, sensory properties	Boeck et al. (2022) and Mandha et al. (2021)
	Nutritional properties, sensory properties	Saint-Eve et al. (2020)
	Nutritional properties, sensory properties	Majhenič et al. (2025) and Lee et al. (2019)
	Processing method, raw material	Sangija et al. (2022)
	Processing method, sensory properties	Jo and Kim (2020), Kwak et al. (2017), Carneiro et al. (2024), and Roh et al. (2018)
	Raw material, sensory properties	Byeon et al. (2023), Dartora et al. (2023), Diby et al. (2023), Greis et al. (2023), Kim et al. (2020), Panjapiyakul et al. (2017), and Rojas-Rivas and Cuffia (2020)

2014; Akissoé et al., 2014). However, such adaptations should not influence the consumer's perception of the product's image (Park et al., 2020). Table 16.3 summarizes the FPBFs subfactors associated with intrinsic product characteristics factors.

16.2.4 Psychological Factors

Within the psychological factor, different sub-factors influence the choice of FPBF (Chen & Antonelli, 2020; Bauer & Reisch, 2018). These sub-factors include behavioral control, which refers to the self-perceived ease or difficulty of carrying out a given behavior and is related to self-efficacy (Ajzen, 1991). In this sense, perceived behavioral control is a significant predictor of the purchase intention of different fermented plant-based products, as shown by (Mustapa et al., 2024; García-Barrón et al., 2025). On the other hand, within psychological factors, emotions and feelings play a significant role (Jaeger et al., 2023b; Dartora et al., 2023; Gupta et al., 2022).

By analyzing emotions and feelings, it is possible to better explain hedonic aspects (Gupta et al., 2022). According to Cardello et al. (2024), this may be due to certain sensory characteristics that can generate positive emotions because they contribute to high activation. In line with this, Jaeger et al. (2023b) point out that there are links between sensory drivers and emotional aspects generated by products. Furthermore, when it comes to traditional fermented foods, the emotion of pride is significantly evoked (Fibri & Frøst, 2020). Likewise, the measurement of emotions and feelings can contribute to the development of new products, packaging, and labels, in addition to obtaining information to generate marketing and commercialization strategies (Gupta et al., 2022; Jaeger et al., 2023b; Carneiro et al., 2024).

Among fermented products, familiarity plays a decisive role in the perception of sensory characteristics, as it allows consumers to differentiate between these products (Mandha et al., 2021). The level of familiarity becomes significant in the case of traditional foods (Roh et al., 2018; Park et al., 2020). Similarly, familiarity must be considered in the development of new products, as it can help dispel potential concerns about new alternatives (Kim et al., 2020).

In contrast, when consumers are unfamiliar with a food, neophobia is a factor to consider. This psychological factor refers to a reluctance to try new or unfamiliar foods (Carneiro et al., 2024; Greis et al., 2023; Pliner & Hobden, 1992). In line with this, Bernal-Gil et al. (2020) point out that the level of neophobia is related to the perception and intention to try a traditional fermented beverage, since the perception of negative characteristics is related to high levels of neophobia. Similarly, products with high levels of neophobia are considered unsafe and unnatural, since such levels of neophobia would be related to memory and food experiences (Fibri & Frøst, 2020). In contrast, Marlapati et al. (2024) reported that information about the composition and nutritional contributions can reduce levels of neophobia in plant-based yogurts. Therefore, informing consumers about product features can reduce levels of neophobia (Banovic & Grunert, 2023). However, the implications of this strategy need to be further explored (Marlapati et al., 2024). Table 16.4 summarizes the FPBFs subfactors associated with psychological factors.

Table 16.4 List of FPBFs subfactors encountered in psychological factors

Factor	Subfactor	Reference
Psychological factors	Behavioral control	Mustapa et al. (2024)
	Emotions and feelings	Cardello et al. (2024), Dartora et al. (2023), Gupta et al. (2022), and Jaeger et al. (2023a, b)
	Familiarity	Kim et al. (2020), Mandha et al. (2021), Park et al. (2020), and Roh et al. (2018)
	Neophobia	Banovic and Grunert (2023), Bernal-Gil et al. (2020), Greis et al. (2023), and Marlapati et al. (2024)
	Emotions and feelings, neophobia	Fibri and Frøst (2020) and Cardello et al. (2024)

16.2.5 Situational Factors

Situational factors include context and point of sale. Food choices are often made in specific situations that determine decisions, i.e., according to context. According to Meiselman (2006) and Onwezen (2018), context is defined as the situational and temporal conditions in which foods are selected or consumed. Thus, the perception and type of consumption of an FPBF depend on the context in which it is consumed, as it can occur at different times of the day (Gupta et al., 2022). Likewise, the type of product, consumption habits, and culture must be considered.

With regard to the point of sale, this can influence the perception of a product's image and quality. This is because consumers may associate a particular context with poor conditions or a negative reputation, which influences their perception of the product (Rojas-Rivas et al., 2019). Therefore, point-of-sale conditions must be improved to prevent them from negatively affecting consumer perception. Table 16.5 summarizes the subfactors of plant based fermented foods associated with situational factors.

16.2.6 Sociocultural Factors

Within the choice and consumption of FPBF, sociocultural factors play a determining role. Among these factors are attitudes, which are the result of positive or negative evaluations that an individual makes of a given process. The outcome is based on the consequences of behavior and the value the individual places on those consequences, influencing the perception and choice of FPBF (Mustapa et al., 2024; Cardello et al., 2024; Kim et al., 2020; Jaeger et al., 2023b; Marlapati et al., 2024). When a negative attitude is expressed, there is less willingness to pay (Jaeger et al., 2023b). Moreover, in traditional products, attitude analysis is vital to know the perception of this type of product (Bernal-Gil et al., 2020), identification of consumer segments (Greis et al., 2023; Fibri & Frøst, 2020). Moreover, attitude would be related to moral satisfaction as people with a positive attitude presented higher plant-based yogurt moral satisfaction (Marlapati et al., 2024). In this sense, it is necessary to develop information strategies to make consumers aware of the benefits and avoid negative attitudes (Banovic & Grunert, 2023).

Another sociocultural aspect that can influence the perception of FPBFs is beliefs (Banovic & Grunert, 2023), especially traditional products, as they are related to health benefits (Rojas-Rivas et al., 2019) and the image of these products (Bernal-Gil

Table 16.5 List of FPBFs subfactors encountered in the situational factor

Factor	Subfactor	Reference
Situational factor	Contcxt	Gupta ct al. (2022)
	Point of purchase	Rojas-Rivas and Cuffia (2020) and Rojas-Rivas et al. (2019)

et al., 2020). Likewise, consumers believe that purchasing power is key in shaping purchase intention (Mustapa et al., 2024). Therefore, in the case of FPBF, beliefs can be considered as a multifactorial sub-factor.

Within FPBFs, consumption habits are multidimensional aspects that relate to sensory and non-sensory attributes such as psychological, preparation methods, and communication (Bernal-Gil et al., 2020).

Cultural influence on the perception and preferences of processed foods, as with other foods, is a function of cultural expectations, intrinsic attributes, and place of residence, demonstrating this cultural influence (Greis et al., 2023; Akissoé et al., 2014; Kim et al., 2020). However, objective knowledge is essential for this (Banovic & Grunert, 2023), so it is desirable to study the mechanisms underlying the cultural factor (Park et al., 2020).

Different works show that educational level can influence the perception of FPBF. In line with this, consumers with a higher educational level tend to be more sensitive in reevaluating traditional FPBF (Rojas-Rivas et al., 2019; Adinsi et al., 2014; Rojas-Rivas & Cuffia, 2020). Similarly, educational level is related to the perception of the level of healthiness (Marlapati et al., 2024).

Finally, within the socio-cultural factor, a group of sub-factors was identified that have been little addressed, such as knowledge. In this regard, Deba-Rementeria et al. (2023) mention that consumers are familiar with the fermentation process and are aware of its effects on sensory characteristics. Another variable identified is values, which can influence the perception and evaluation of a fermented product, as mentioned by Marlapati et al. (2024), who point out that moral and health values can explain the consumption of plant-based yogurts. In line with this, García-Barrón et al. (2025) point out that although health awareness as a consumer value does not explain consumption or consumption intention, it does determine consumer attitudes. Similarly, Rojas-Rivas et al. (2019) point out that cultural and attitudinal values mainly influence young consumers of traditional beverages. Since these sub-factors have been little addressed in the context of FPBFs, it is advisable to consider them in consumer studies to better understand their role within this food category. Table 16.6 summarizes the FPBFs subfactors associated with sociocultural factors.

Table 16.6 List of FPBFs subfactors encountered in sociocultural factors

Factor	Subfactor	Reference
Sociocultural factor	Attitude	Cardello et al. (2024), Fibri and Frøst (2020), and Jaeger et al. (2023a, b)
	Consumption habits	Panjapiyakul et al. (2017)
	Cultural influence	Akissoé et al. (2014) and Park et al. (2020)
	Educational level	Adinsi et al. (2014)
	Knowledge	Deba-Rementeria et al. (2023)

16.3 Perspectives for Plant-Based Fermented Products

Plant-based fermented products have gained attention as sustainable alternative food sources, as they can benefit consumers' health. This has had a positive influence on consumer perceptions and preferences. It is important to note that intrinsic attributes, particularly sensory qualities, are the most influential sub-factors affecting consumer behavior. In contrast, factors related to biological and physiological characteristics, as well as situational aspects, have received comparatively less attention.

Due to this situation, it is necessary to study the role of different contexts on the consumption and preferences of FPBFs, especially in real-life situations, as this has not been sufficiently explored. This reflects the complexity of food choice: extrinsic factors can improve acceptability, but this depends on the sensory quality of the product being acceptable. Within the FPBFs, innovation and development are mainly focused on sustainable and nutritious alternatives. These advances include the valorization of by-products, improvement of fermentation processes, and enhancement of the sensory characteristics of little-known products or those with low levels of acceptance. Therefore, it is necessary to consider aspects such as neophobia, emotions, and the ability of communication strategies to familiarize consumers with FPBFs and the benefits of consuming them.

Likewise, it is necessary to adopt a holistic approach to assess the effect of different types of information on consumer behavior, especially in terms of sustainability. In addition, it is advisable to provide detailed information on processing methods, raw materials, processing methods, and ecological practices (Fibri & Frøst, 2020). Similarly, it is necessary to study aspects such as personality, emotions, gender, and age, as this would help to efficiently tailor products to consumers (Greis et al., 2023).

Although hedonic tests are useful for estimating consumer preferences, they need to be complemented with physiological measures, such as heart rate, and psychological aspects, such as emotions or personality traits, as suggested by Gupta et al. (2022) and Jaeger et al. (2023b). Finally, implementing a multidisciplinary strategy would allow for a comprehensive investigation of the several factors that influence consumer behavior. This would provide a broader understanding of both products and consumers, facilitating improvements in product design, acceptance, and overall consumption (Chijioke et al., 2021).

16.4 Conclusions

This chapter analyzed the various factors that influence the perception and preferences toward fermented plant-based foods (FPBF). Based on the model proposed by Mojet (Köster, 2009), six groups of factors were identified and classified: biological and physiological factors, extrinsic product characteristics, intrinsic product characteristics, psychological factors, situational factors, and sociocultural factors. Among

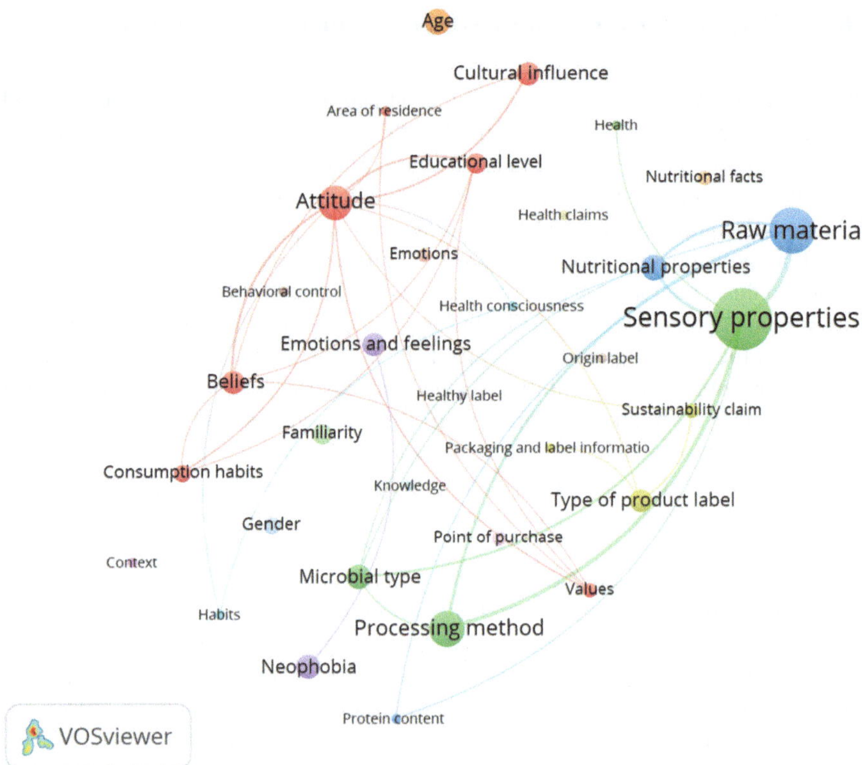

Fig. 16.1 Map network analysis elaborated using the Software VOSviewer, showing the associations among the sub-factors and the factors: situational factors (red), biological and physiological (green), psychological factors (blue), extrinsic product characteristics (yellow), sociocultural factors (purple), and intrinsic product characteristics (cyan)

these, and considering the number of published studies, intrinsic characteristics particularly sensory attributes stand out as a central element in the acceptance of FPBF. Following the network analysis described in García-Barón et al. (2025), it was possible to show that some subfactors are addressed in isolation, so it would be advisable to adopt a holistic approach to understand their possible relationship, Fig. 16.1.

The growing interest in these products reflects a global shift toward options that harmonize health, sustainability, and cultural identity. According to the analysis, consumer perception and preference for FPBFs are due to a complex interaction between biological, psychological, sensory, sociocultural, and contextual factors, which define how these products are assessed and accepted. Although extrinsic cues—such as origin, labeling, packaging, sustainability claims, and price—established initial expectations, it is intrinsic characteristics that eventually determine acceptance, with sensory characteristics being the most important factors influencing preference and consumption.

At the same time, psychological factors such as familiarity, emotions, perceived behavioral control, and food neophobia influence the willingness to try and repeatedly consume FPBF. These elements are modulated by situational and sociocultural contexts that shape the meanings, experiences, and values attributed to these kinds of foods. Traditional products reflect how cultural identities, processing methods, and raw materials determine both the symbolic and sensory dimensions of preferences.

In general, understanding the factors that determine the choice of FPBFs provides valuable insight for the development of products that are nutritious, technologically viable, and emotionally appealing, as well as culturally and sensorially acceptable. Therefore, innovations must balance health benefits with pleasurable sensory experiences. Likewise, they must communicate the product's qualities clearly and credibly, taking consumer characteristics into account. This integrated approach will enable FPBFs to expand their role in sustainable diets and strengthen their presence on the global food scene.

References

Adinsi, L., Akissoé, N. H., Dalodé-Vieira, G., Anihouvi, V. B., Fliedel, G., Mestres, C., & Hounhouigan, J. D. (2014). Sensory evaluation and consumer acceptability of a beverage made from malted and fermented cereal: Case of gowe from Benin. *Food Science & Nutrition, 3*(1), 1–9. https://doi.org/10.1002/fsn3.166

Ajzen, I. (1991). The theory of planned behavior. *Organizational Behavior and Human Decision Processes, 50*(2), 179–211. https://doi.org/10.1016/0749-5978(91)90020-T

Akanni, G. B., Qaku, X. W., Adetunji, A., & Dlamini, B. C. (2024). Consumer acceptability, metabolite profile analysis and storage stability of sorghum-Bambara groundnut *mahewu*. *International Journal of Food Science & Technology, 59*(3), 1363–1374. https://doi.org/10.1111/ijfs.16879

Akissoé, N. H., Sacca, C., Declemy, A.-L., Bechoff, A., Anihouvi, V. B., Dalodé, G., Pallet, D., Fliedel, G., Mestres, C., Hounhouigan, J. D., & Tomlins, K. I. (2014). Cross-cultural acceptance of a traditional yoghurt-like product made from fermented cereal. *Journal of the Science of Food and Agriculture, 95*(9), 1876–1884. https://doi.org/10.1002/jsfa.6892

Ambarita, M., Raes, K., & De Meulenaer, B. (2019). The sensory acceptance and profile of Indonesian sambal-terasi: Impact of terasi type and concentration. *International Journal of Gastronomy and Food Science, 16*, 100149. https://doi.org/10.1016/j.ijgfs.2019.100149

Anagnostopoulos, D. A., & Tsaltas, D. (2019). Fermented foods and beverages. In *Innovations in traditional foods* (pp. 257–291). https://doi.org/10.1016/b978-0-12-814887-7.00010-1

Andreu, M., Saavedra-Coutado, C., Andreu, M., & Saavedra-Coutado, C. (2022). El rol de los fermentos en la sostenibilidad alimentaria. *Nutrición Hospitalaria, 39*(SPE3), 56–59. https://doi.org/10.20960/nh.04313

Augustin, M. A., Hartley, C. J., Maloney, G., & Tyndall, S. (2023). Innovation in precision fermentation for food ingredients. *Critical Reviews in Food Science and Nutrition, 64*, 6218–6238. https://doi.org/10.1080/10408398.2023.2166014

Banovic, M., & Grunert, K. G. (2023). Consumer acceptance of precision fermentation technology: A cross-cultural study. *Innovative Food Science & Emerging Technologies, 88*, 103435. https://doi.org/10.1016/j.ifset.2023.103435

Bauer, J. M., & Reisch, L. A. (2018). Behavioural insights and (un)healthy dietary choices: A review of current evidence. *Journal of Consumer Policy, 42*(1), 3–45. https://doi.org/10.1007/s10603-018-9387-y

Bernal-Gil, N. Y., Favila-Cisneros, H. J., Zaragoza-Alonso, J., Cuffia, F., & Rojas-Rivas, E. (2020). Using projective techniques and food Neophobia scale to explore the perception of traditional ethnic foods in Central Mexico: A preliminary study on the beverage Sende. *Journal of Sensory Studies, 35*(6), e12606. https://doi.org/10.1111/joss.12606

Boeck, T., Ispiryan, L., Hoehnel, A., Sahin, A. W., Coffey, A., Zannini, E., & Arendt, E. K. (2022). Lentil-based yogurt alternatives fermented with multifunctional strains of lactic acid bacteria—Techno-functional, microbiological, and sensory characteristics. *Foods, 11*(14), 2013. https://doi.org/10.3390/foods11142013

Brückner-Gühmann, M., Banovic, M., & Drusch, S. (2019). Towards an increased plant protein intake: Rheological properties, sensory perception and consumer acceptability of lactic acid fermented, oat-based gels. *Food Hydrocolloids, 96*, 201–208. https://doi.org/10.1016/j.foodhyd.2019.05.016

Byeon, Y. S., Heo, J., Park, K. H., Chin, Y., Hong, S.-S., Lim, S.-D., & Kim, S. S. (2023). Consumer preference of traditional Korean soy sauce (Ganjang) and its relationship with sensory attributes and physicochemical properties. *Food, 12*(12), 2361. https://doi.org/10.3390/foods12122361

Cardello, A. V., Llobell, F., Jin, D., Ryan, G. S., & Jaeger, S. R. (2024). Sensory drivers of liking, emotions, conceptual and sustainability concepts in plant-based and dairy yoghurts. *Food Quality and Preference, 113*, 105077. https://doi.org/10.1016/j.foodqual.2023.105077

Carneiro, G. R., Rocha, S., Vitória, M., Barão, C. E., & Pimentel, T. C. (2024). Probiotic almond-fermented beverages processed by ultrasound: Vegan and non-vegan consumer perceptions through packaging. *Food, 13*(13), 1975. https://doi.org/10.3390/foods13131975

Chen, P.-J., & Antonelli, M. (2020). Conceptual models of food choice: Influential factors related to foods, individual differences, and society. *Food, 9*(12), 1898. https://doi.org/10.3390/foods9121898

Chijioke, U., Madu, T., Okoye, B., Ogunka, A. P., Ejechi, M., Ofoeze, M., Ogbete, C., Njoku, D., Ewuziem, J., Kalu, C., Onyemauwa, N., Ukeje, B., Achonwa, O., Forsythe, L., Fliedel, G., & Egesi, C. (2021). Quality attributes of fufu in south-East Nigeria: Guide for cassava breeders. *International Journal of Food Science & Technology, 56*(3), 1247–1257. https://doi.org/10.1111/ijfs.14875

Cohen, G., Sela, D. A., & Nolden, A. A. (2023). Sucrose concentration and fermentation temperature impact the sensory characteristics and liking of Kombucha. *Foods, 12*(16), 3116. https://doi.org/10.3390/foods12163116

Cuamatzin-García, L., Rodríguez-Rugarcía, P., El-Kassis, E. G., Galicia, G., Meza-Jiménez, M. d. L., Baños-Lara, M. d. R., Zaragoza-Maldonado, D. S., & Pérez-Armendáriz, B. (2022). Traditional fermented foods and beverages from around the world and their health benefits. *Microorganisms, 10*(6), 1151. https://doi.org/10.3390/microorganisms10061151

Dartora, B., Crepalde, L. T., Hickert, L. R., Fabricio, M. F., Ayub, M. A. Z., Veras, F. F., Brandelli, A., Perez, K. J., & Sant'Anna, V. (2023). Kombuchas from black tea, green tea, and yerba-mate decocts: Perceived sensory map, emotions, and physicochemical parameters. *International Journal of Gastronomy and Food Science, 33*, 100789. https://doi.org/10.1016/j.ijgfs.2023.100789

Deba-Rementeria, N. S., Paz, A., Estrada, O., & Vázquez-Araújo, L. (2023). Consumer perception and physicochemical characterization of a new product made from lactic acid fermented orange peels. *International Journal of Gastronomy and Food Science, 31*, 100647. https://doi.org/10.1016/j.ijgfs.2022.100647

Diby, S., Deffan, P., Adinsi, L., Bechoff, A., Kanon, L., Bouniol, A., Eric, Y., Deuscher, Z., Bugaud, C., N'Zue, B., & Bomoh, C. (2023). Use of sensory and physico-chemical parameters to understand consumer perception of *attiéké*, a fermented cassava product. *Journal of the Science of Food and Agriculture/Journal of the Science of Food and Agriculture, 104*, 4596–4605. https://doi.org/10.1002/jsfa.13141

Dlamini, N. D., Kolanisi, U., Siwela, M., & Dlamini, N. G. (2025). Development, nutritional evaluation, and consumer acceptability of non-alcoholic fermented beverages made from Sclerocarya birrea (Marula) and provitamins A biofortified maize. *Food and Humanity, 5*, 100640. https://doi.org/10.1016/j.foohum.2025.100640

Fernqvist, F., Spendrup, S., & Tellström, R. (2024). Understanding food choice: A systematic review of reviews. *Heliyon, 10*(12), e32492. https://doi.org/10.1016/j.heliyon.2024.e32492

Fibri, D. L. N., & Frøst, M. B. (2020). Indonesian millennial consumers' perception of tempe – And how it is affected by product information and consumer psychographic traits. *Food Quality and Preference, 80*, 103798. https://doi.org/10.1016/j.foodqual.2019.103798

García-Barón, S. E., Carmona-Escutia, R. P., Herrera-López, E. J., Leyva-Trinidad, D. A., & Gschaedler-Mathis, A. (2025). Consumers' drivers of perception and preference of fermented food products and beverages: A systematic review. *Foods, 14*(5), 713. https://doi.org/10.3390/foods14050713

García-Barrón, S. E., Gonzalez-Hemon, G., Herrera, E. J., Carmona-Escutia, R. P., Leyva-Trinidad, D. A., & Gschaedler, A. C. (2025). Prediction of the consumption of a traditional fermented beverage via an expanded model of the theory of planned behaviour. *British Food Journal, 127*(7), 2381–2398. https://doi.org/10.1108/bfj-07-2024-0724

Greis, M., Nolden, A. A., Kinchla, A. J., Puputti, S., Seppä, L., & Sandell, M. (2023). What if plant-based yogurts were like dairy yogurts? Texture perception and liking of plant-based yogurts among US and Finnish consumers. *Food Quality and Preference, 107*, 104848. https://doi.org/10.1016/j.foodqual.2023.104848

Gupta, M. K., Viejo, C. G., Fuentes, S., Torrico, D. D., Saturno, P. C., Gras, S. L., Dunshea, F. R., & Cottrell, J. J. (2022). Digital technologies to assess yoghurt quality traits and consumers acceptability. *Journal of the Science of Food and Agriculture, 102*, 5642–5652. https://doi.org/10.1002/jsfa.11911

Höhl, K., & Dreyer, J. (2025). Sensory analysis and hedonic evaluation of fermented plant-based products similar to yogurt. Not imitating, but emancipating. *Ernährungs Umschau, 72*(1), 10–17.

Jaeger, S. R., Giacalone, D., Jin, D., Ryan, G. S., & Cardello, A. V. (2023a). Information about health and environmental benefits has minimal impact on consumer responses to commercial plant-based yoghurts. *Food Quality and Preference*, 104820. https://doi.org/10.1016/j.foodqual.2023.104820

Jaeger, S. R., Cardello, A. V., Jin, D., Ryan, G. S., & Giacalone, D. (2023b). Consumer perception of plant-based yoghurt: Sensory drivers of liking and emotional, holistic and conceptual associations. *Food Research International, 167*, 112666. https://doi.org/10.1016/j.foodres.2023.112666

Jo, Y., & Kim, M. K. (2020). Influences of appearance characteristics on consumer acceptance and perception of "gu-soo" in fermented soybean paste (*doenjang*). *Journal of Sensory Studies, 35*(5). https://doi.org/10.1111/joss.12597

Keller, K. L., Parameswaran, A. M. G., & Jacob, I. (2015). *Strategic brand management: Building, measuring, and managing brand equity*. Pearson Education India.

Kim, M. K., & Lee, K.-G. (2014). Correlating consumer perception and consumer acceptability of Traditional Doenjangin Korea. *Journal of Food Science, 79*(11), S2330–S2336. https://doi.org/10.1111/1750-3841.12676

Kim, M. K., & Lee, K.-G. (2018). Defining gu-soo perception in Doenjang (fermented soybean paste) using consumer tests with limited sensory modality and instrumental analysis. *Food Chemistry, 267*, 210–216. https://doi.org/10.1016/j.foodchem.2017.11.035

Kim, M.-R., Chung, S.-J., Adhikari, K., Shin, H., Cho, H., & Nam, Y. (2020). Cross-cultural consumer acceptability for ethnic fermented sauce products: Comparisons among Korean, UAE, and US consumers. *Foods, 9*(10), 1463. https://doi.org/10.3390/foods9101463

Köster, E. P. (2009). Diversity in the determinants of food choice: A psychological perspective. *Food Quality and Preference, 20*(2), 70–82. https://doi.org/10.1016/j.foodqual.2007.11.002

Kwak, H. S., Jung, H. Y., Kim, M. J., & Kim, S. S. (2017). Differences in consumer perception of Korean traditional soybean paste (Doenjang) between younger and older consumers by blind and informed tests. *Journal of Sensory Studies, 32*(6), e12302. https://doi.org/10.1111/joss.12302

Lee, Y., Lee, S. M., & Kim, K. (2019). Investigation on the intrinsic potential of using potassium chloride for partially replacing NaCl in kimchi and the effect of information on consumer acceptance of sodium-reduced kimchi according to age. *Journal of Food Science, 84*(10), 2961–2972. https://doi.org/10.1111/1750-3841.14784

Majhenič, A. Č., Levart, A., Salobir, J., Prevc, T., & Pajk Žontar, T. (2025). Can plant-based cheese substitutes nutritionally and sensorially replace cheese in our diet? *Foods, 14*(5), 771. https://doi.org/10.3390/foods14050771

Mandha, J., Shumoy, H., Devaere, J., Schouteten, J. J., Gellynck, X., de Winne, A., Matemu, A. O., & Raes, K. (2021). Effect of lactic acid fermentation of watermelon juice on its sensory acceptability and volatile compounds. *Food Chemistry, 358*, 129809. https://doi.org/10.1016/j.foodchem.2021.129809

Marco, M. L., & Golomb, B. L. (2016). Fermented foods, lactobacillus, and health. *Microbe, 11*, 349–370.

Marlapati, L., Kinchla, A. J., & Nolden, A. A. (2024). Conjoint analysis study to examine consumer's preferences for hybrid yogurt. *Sustainability, 16*(17), 7460. https://doi.org/10.3390/su16177460

Meiselman, H. L. (2006). The role of context in food choice, food acceptance and food consumption. In R. Shepherd & M. Raats (Eds.), *The psychology of food choice* (pp. 179–199). CABI. https://doi.org/10.1079/9780851990323.0179

Melini, F., Melini, V., Luziatelli, F., Ficca, A. G., & Ruzzi, M. (2019). Health-promoting components in fermented foods: An up-to-date systematic review. *Nutrients, 11*(5), 1189. https://doi.org/10.3390/nu11051189

Mishra, S., Mithul Aravind, S., Said Ajlouni, C., Ranadheera, S., Singh, B. P., & Chakkaravarthi, S. (2024). Optimization of fermentation process and characterization of non-alcoholic functional beverage from pigmented rice varieties. *Biocatalysis and Agricultural Biotechnology, 58*, 103213. https://doi.org/10.1016/j.bcab.2024.103213

Mojet, J. (2001). Taste perception with age: Generic or specific losses in threshold sensitivity to the five basic tastes? *Chemical Senses, 26*(7), 845–860. https://doi.org/10.1093/chemse/26.7.845

Mustapa, M. A. C., Kallas, Z., Silande, C., Gagnaire, V., Jan, G., López-Mas, L., & Aguiló-Aguayo, I. (2024). From taste to purchase: Understanding the influence of sensory perceptions and informed tasting on plant-based product purchases - an extension of the theory of planned behavior. *Journal of Agriculture and Food Research, 16*, 101188. https://doi.org/10.1016/j.jafr.2024.101188

Nemati, V., Hashempour-Baltork, F., Alizadeh, A. M., & Varzakas, T. (2023). Production of traditional Torba yogurt using lactic acid bacteria isolated from fermented vegetables: Microbiological, physicochemical and sensory properties. *Journal of Agriculture and Food Research, 14*, 100850. https://doi.org/10.1016/j.jafr.2023.100850

Onwezen, M. (2018). Including context in consumer segmentation: A literature overview shows the what, why, and how. In *Methods in consumer research* (Vol. 1, pp. 383–400). https://doi.org/10.1016/b978-0-08-102089-0.00015-7

Panjapiyakul, P., Srinin, T., & Kitsawad, K. (2017). The product development of Thai fermented sausage with mixed Tomyum ingredients. *British Food Journal, 119*(10), 2203–2216. https://doi.org/10.1108/bfj-10-2016-0512

Park, H., Ko, J., Lim, J., & Hong, J. (2020). American consumers' perception and acceptance of an ethnic food with strong flavor: A case study of Kimchi with varying levels of red pepper and fish sauce. *Journal of the Science of Food and Agriculture, 100*(6), 2348–2357. https://doi.org/10.1002/jsfa.10106

Pliner, P., & Hobden, K. (1992). Development of a scale to measure the trait of food neophobia in humans. *Appetite, 19*(2), 105–120. https://doi.org/10.1016/0195-6663(92)90014-w

Prutkin, J., Duffy, V. B., Etter, L., Fast, K., Gardner, E., Lucchina, L. A., Snyder, D. J., Tie, K., Weiffenbach, J., & Bartoshuk, L. M. (2000). Genetic variation and inferences about perceived taste intensity in mice and men. *Physiology & Behavior, 69*(1–2), 161–173. https://doi.org/10.1016/s0031-9384(00)00199-2

Rastogi, Y. R., Thakur, R., Thakur, P., Mittal, A., Chakrabarti, S., Siwal, S. S., Thakur, V. K., Saini, R. V., & Saini, A. K. (2022). Food fermentation – Significance to public health and sustainability challenges of modern diet and food systems. *International Journal of Food Microbiology, 371*, 109666. https://doi.org/10.1016/j.ijfoodmicro.2022.109666

Roh, S. H., Lee, S. M., Kim, S. S., & Kim, K.-O. (2018). Importance of applying condiments in a commonly consumed food system for understanding the association between familiarity and sensory drivers of liking: A study focused on Doenjang. *Journal of Food Science, 83*(2), 464–474. https://doi.org/10.1111/1750-3841.14040

Rojas-Rivas, E., & Cuffia, F. (2020). Identifying consumers' profile and factors associated with the valorization of pulque: A traditional fermented beverage in Central Mexico. *Food Science and Technology International, 26*(7), 593–602. https://doi.org/10.1177/1082013220917554

Rojas-Rivas, E., Viesca-González, F. C., Favila-Cisneros, H. J., & Cuffia, F. (2019). Consumers' perception of a traditional fermented beverage in Central Mexico. *British Food Journal, 122*(2), 708–721. https://doi.org/10.1108/bfj-05-2019-0317

Saint-Eve, A., Irlinger, F., Pénicaud, C., Souchon, I., & Marette, S. (2020). Consumer preferences for new fermented food products that mix animal and plant protein sources. *Food Quality and Preference, 90*, 104117. https://doi.org/10.1016/j.foodqual.2020.104117

Sangija, F., Martin, H., & Matemu, A. (2022). Effect of lactic acid fermentation on the nutritional quality and consumer acceptability of African nightshade. *Food Science & Nutrition, 10*, 3128–3142. https://doi.org/10.1002/fsn3.2912

Santos, B. N., Pereira, A. L. F., Fonteles, T. V., & Rodrigues, S. (2025). Development and evaluation of potential probiotic coconut water beverages: Fermentation, storage, and consumer perception. *Processes, 13*(8), 2554. https://doi.org/10.3390/pr13082554

Symmank, C. (2018). Extrinsic and intrinsic food product attributes in consumer and sensory research: Literature review and quantification of the findings. *Management Review Quarterly, 69*(1), 39–74. https://doi.org/10.1007/s11301-018-0146-6

Valero-Cases, E., Frutos, M. J., & Pérez-Llamas, F. (2023). Development of synbiotic vegan beverages: Probiotic viability, sensory profile, consumers' acceptance and functional stability. *International Journal of Food Science & Technology, 58*(5), 2325–2335. https://doi.org/10.1111/ijfs.16361

Viretto, C., Tlais, A. Z. A., Tuccillo, F., Polo, A., Arora, K., Verté, F., Katina, K., Di Cagno, R., & Gobbetti, M. (2025). Maximize the synergistic interactions among microbial consortia and plant-based matrices to design fermented cereal-pulse-based beverages. *Food Research International, 220*, 117045. https://doi.org/10.1016/j.foodres.2025.117045

Wu, Z., Chao, J., Tang, H., Liu, T., Jiang, L., & Liu, Y. (2024). Characterization of key aroma-active compounds in different types of Douchi based on molecular sensory science approaches. *Food Chemistry: X, 21*, 101170. https://doi.org/10.1016/j.fochx.2024.101170

Xiang, H., Sun-Waterhouse, D., Waterhouse, G. I. N., Cui, C., & Ruan, Z. (2019). Fermentation-enabled wellness foods: A fresh perspective. *Food Science and Human Wellness, 8*(3), 203–243. https://doi.org/10.1016/j.fshw.2019.08.003

Proekt, J., Du, B., V. B., Greer, L., [illegible] … [illegible] … Snyder, D. L., [illegible] … McGlone, [illegible] … Blumenthal, T. M. (2009). [illegible] … taste intensity [illegible] … *Physiology & Behavior*, 97(1–2), 165–173. https://doi.org/10.1016/j.physbeh.2009.02.[illegible]

[illegible]

[illegible]

[illegible]

[illegible]

Chapter 17
The Future of Plant-Based Fermented Foods

Sunny Dhiman, Sukhminderjit Kaur, and Showkat Ahmad Wani

Abstract Plant-based fermented foods are undergoing a transformative evolution driven by the convergence of traditional knowledge, modern biotechnology, and sustainability imperatives. This chapter explores the historical development of plant-based fermentation across diverse cultures and its transition toward next-generation products enabled by precision fermentation, microbial engineering, and artificial intelligence. Fermentation enhances nutritional quality, sensory attributes, and functional properties by improving bioavailability, reducing antinutrients, and generating bioactive metabolites, including postbiotics. The chapter further examines global market trends, highlighting increasing consumer demand for clean-label, functional, and plant-based products, alongside regional variations in adoption and growth. Sustainability considerations are central, with plant-based fermented foods offering lower environmental footprints and supporting circular bioeconomy approaches through the valorization of agro-food by-products such as fruit pomace. Emerging innovations, including hybrid plant-microbial systems, designer fermented foods, and net-zero fermentation technologies, are discussed as key drivers shaping future food systems. However, challenges related to scalability, standardization of functional claims, and long-term health validation remain critical barriers. The chapter concludes that integrating fermentation technologies with circular resource utilization and interdisciplinary innovation is essential for establishing plant-based fermented foods as a resilient, sustainable, and scalable component of future global food systems.

Keywords Plant-based fermentation · Circular bioeconomy · Precision fermentation · Postbiotics · Food waste valorization · Sustainable food systems

S. Dhiman (✉) · S. Kaur
Department of Biotechnology, University Institute of Biotechnology, Chandigarh University, Mohali, Punjab, India

S. A. Wani
School of Law, Alliance University, Bangalore, India

S. Dhiman, S. Kaur (eds.), *Next-Gen Plant Based Fermented Foods: Innovations for a Sustainable Future*, https://doi.org/10.1007/978-3-032-29927-7_17

17.1 Introduction

Plant-based fermented foods represent a convergence of traditional food practices and modern biotechnological innovation, offering sustainable solutions to meet growing global food demands. Fermentation enhances the nutritional quality, sensory attributes, and functional properties of plant-derived ingredients, making them more digestible and bioavailable. As consumer interest in health, sustainability, and clean-label products increases, plant-based fermentation is emerging as a key driver in the transformation of future food systems.

17.1.1 Evolution of Plant-Based Fermented Foods Across Cultures

Fermented foods represent one of humanity's oldest biotechnological practices, with evidence suggesting fermentation techniques have been employed for millennia across diverse civilizations worldwide (Muhammed et al., 2025). The fermentation of plant-based materials has evolved organically through generations of cultural knowledge systems, adapting to local agricultural availability, environmental conditions, and generational knowledge transfer. Fermentation began as a fundamental food preservation technique in the pre-refrigeration era, but has transcended this practical purpose to become deeply embedded in cultural identity and dietary practices (Mahajani et al., 2023). Across Asia, Africa, Europe, and the Americas, traditional fermented plant-based foods have maintained their significance, from Korea's kimchi and Japan's miso to India's idli and dosa, African traditional porridges, and Eastern European sauerkraut and kvass (Santa et al., 2025).

The global diversity of these products reflects unique microbial terroirs—distinctive microbial communities shaped by regional climate, ingredients, and traditional production methods. In sub-Saharan Africa alone, there exists one of the world's richest selections of indigenous fermented foods that constitute valuable cultural heritage with significant socio-economic impact (Khayeka-Wandabwa et al., 2024). These traditional practices encode deep ecological knowledge accumulated through centuries, representing sustainable, community-based food systems that provided nutrition and preservation without modern technological infrastructure. Traditional Tunisian fermented foods, Ethiopian injera fermentation, Malaysian tempoyak production, and North-East Indian fermented fish products exemplify how cultures have independently developed sophisticated fermentation systems tailored to their local ecosystems (Alebidi et al., 2026). These ethnomicrobiological traditions demonstrate the universal applicability of fermentation as a food processing and preservation strategy across distinct geographical and climatic contexts.

17.1.2 Traditional vs Next-Generation Fermented Products

A critical distinction has emerged between traditional artisanal fermented foods and what may be termed "next-generation" plant-based fermented products designed through modern biotechnology and precision microbiology. Traditional fermented foods are characteristically produced through spontaneous or mixed-culture fermentations using naturally-occurring microorganisms, often with minimal standardization and significant batch-to-batch variability (Muhammed et al., 2025). These products frequently employ multi-species microbial consortia, where diverse lactic acid bacteria, yeasts, and filamentous fungi interact in complex ecological relationships, producing distinctive sensory profiles and bioactive compounds through spontaneous processes refined over generations.

In contrast, next-generation plant-based fermented foods increasingly employ precision fermentation technologies, rationally designed starter cultures, and biotechnological innovations to achieve consistency, safety, and enhanced functionality (Dhiman et al., 2025). These modern approaches utilize novel starter cultures, strain engineering, CRISPR-based genome editing, and advanced fermentation systems that reshape the nutritional landscape of plant-based foods. The distinction reflects a fundamental shift from heterogeneous, artisanal production toward more controlled, scalable, scientifically-validated processes designed for industrial application while sometimes maintaining traditional sensory appeal. This represents an important technological convergence where traditional fermentation wisdom is being integrated with modern microbiological science, enabling producers to manipulate fermentation outcomes with unprecedented precision while addressing specific nutritional and functional targets (Horlacher et al., 2023).

Next-generation products specifically address historical limitations of plant-based foods through fermentation: improved protein digestibility, enhanced bioactive compound synthesis, reduction of antinutrients (such as phytic acid, tannins, and raffinose-family oligosaccharides), and improved micronutrient bioavailability (Molina et al., 2025). Modern plant-based fermented beverages, dairy alternatives, and meat analogs exemplify this approach, utilizing fermentation to eliminate off-flavors, reduce beany notes, and achieve flavor and texture profiles comparable to conventional animal-based products. These developments have coincided with expanding consumer interest in plant-based diets, creating commercial momentum for fermented plant-based alternatives that were previously niche products.

17.1.3 Growing Relevance in Sustainable Food Systems

Plant-based fermented foods have acquired enhanced significance within global sustainability discourse as research increasingly documents their environmental advantages and role in climate-resilient food systems. The agricultural production of animal-based proteins generates substantial greenhouse gas emissions, with livestock farming accounting for approximately 14.5% of anthropogenic greenhouse gas emissions globally (Solomon et al., 2023). In contrast, comparative life-cycle assessments demonstrate that plant-based foods, particularly when fermented through low-input processes, possess markedly lower environmental footprints. A comparative analysis of burgers illustrates this disparity: the water required to produce beef burger was 1.8 times greater than a hybrid burger and 21 times greater than a plant-based burger, while carbon footprint emissions were approximately 2 times higher for beef compared to hybrid options and 13 times higher than plant-based variants (Dominguez-Lacueva et al., 2025).

Beyond carbon and water metrics, fermented plant-based foods contribute to sustainable food systems through multiple pathways. Fermentation enhances the bioavailability of nutrients from plant-based raw materials that might otherwise pose bioavailability challenges due to antinutrient content, thereby reducing the need for resource-intensive nutrient supplementation or fortification (Abbaspour, 2024). The fermentation process can be implemented at domestic and small-scale levels with minimal technological input, supporting food security in resource-limited contexts. Additionally, fermented plant-based products often utilize underutilized crops, agricultural by-products, and food waste streams as fermentation substrates, promoting circular bioeconomy principles and reducing overall resource consumption (Pal et al., 2024).

17.1.4 Climate Change and Resource Constraints

Climate change poses escalating threats to traditional agricultural systems, and fermented plant-based foods offer adaptation and mitigation strategies. Plant-based proteins derived from legumes, cereals, and pseudocereals demonstrate superior climate resilience compared to animal protein systems, though climate change itself threatens legume productivity, with projected yield reductions between 10–49% for major legume crops by 2050 without adaptation measures (Dave et al., 2024). However, fermented plant-based foods can be produced from diverse crop species and even underutilized indigenous crops, enhancing agrobiodiversity and reducing systemic vulnerability to climate-induced disruptions. Furthermore, the lower environmental footprint of plant-based fermented foods aligns with global climate mitigation targets, contributing to reductions in greenhouse gas emissions from food systems.

Global food systems face unprecedented pressures from climate change and escalating resource constraints that are fundamentally reshaping the future of protein production and dietary patterns. The United Nations projects a global population reaching 10 billion by 2050, coupled with a doubling of protein demand, while simultaneously available arable land continues to decline and freshwater resources face increasing scarcity (FAO, 2025). These converging challenges have catalyzed strategic investment in alternative protein sources and sustainable food technologies, positioning plant-based fermented foods as critical components of climate-adaptive food systems.

The urgency of addressing climate change has generated policy momentum toward dietary transition toward plant-based options. Carbon pricing initiatives, sustainability-oriented dietary recommendations from major health and environmental organizations, and the evolving development of climate policies addressing greenhouse gas emissions in agriculture are contributing to a policy and socio-economic landscape that is increasingly conducive to the expansion of plant-based food markets (Ahmad, 2024). Research indicates that dietary patterns incorporating reduced meat consumption paired with increased plant-based food intake could achieve simultaneous health benefits and substantial environmental impact reduction, with monetized health improvements potentially exceeding environmental damage avoidance values (Springmann et al., 2016).

17.1.5 Rising Vegan, Flexitarian, and Health-Conscious Consumers

Consumer dietary patterns are undergoing rapid and measurable transformation, driven by environmental consciousness, health awareness, and ethical considerations. Research from multiple geographic contexts demonstrates substantial growth in plant-based diet adoption. In Germany, Italy, Spain, and France, over 50% of consumers have reduced red meat intake, with notable portions adhering to vegan diets, particularly among German consumers (Mustapa et al., 2025). In an Australian consumer study analyzing plant-based product consumption, 60% of respondents reported following special diets, including 25% identifying as vegan and 19% as flexitarian (Estell et al., 2021).

The flexitarian dietary pattern characterized by predominantly plant-based eating with occasional meat consumption, has emerged as the most rapidly growing dietary category, particularly among younger generations and in urban populations. Market research suggests that by 2030, more than half of consumers globally may shift to flexitarian dietary patterns (Ahmad, 2024). These dietary transitions are not driven exclusively by environmental concern; consumer motivation research reveals complex, multi-dimensional drivers. Plant-based fermented foods are increasingly regarded as functional foods due to their content of bioactive compounds and microbial metabolites that may influence gut microbiota composition and host health. Consumption of these foods has been associated with improved gut microbial balance, enhanced immune-related responses, and potential benefits for metabolic health, although evidence for direct therapeutic effects on chronic disease outcomes remains an active area of research (Peng et al., 2025).

Despite rapid technological innovation and market growth, development of plant-based fermented foods remains scientifically and structurally fragmented. Traditional knowledge systems, precision biotechnology, sustainability assessment, regulatory frameworks, and clinical validation are advancing unevenly. This fragmentation constrains evidence-based optimization, large-scale implementation, and long-term credibility of fermented plant-based foods as sustainable protein solutions. This chapter advances the hypothesis that the future success of plant-based fermented foods depends on integrating traditional microbial knowledge with precision biotechnology, sustainability metrics, and regulatory harmonization. This chapter identifies strategic research directions necessary to transform plant-based fermentation from a promising innovation into a scientifically validated, environmentally robust, and commercially scalable pillar of future food systems.

17.2 Global Market Trends in Plant-Based Foods

Global food markets are experiencing a shift toward clean-label and minimally processed products, and increasing interest in functionally enhanced foods, reflecting consumer preferences for simpler, more "natural" foods and growing concerns about highly processed industrialized food systems (Tachie et al., 2023; Cao & Miao, 2023). Clean-label preferences—short ingredient lists, the absence of artificial additives, and transparent sourcing—have become central drivers of purchase. Fermented plant-based foods occupy a strategic position within this trend because fermentation inherently enhances stability, flavor, and nutritional functionality without synthetic additives. Consumer perception research demonstrates that ingredient list length and additive type significantly influence perceived healthfulness and purchase intention, with products positioned as additive-free or explicitly plant-based often associated with greater consumer trust and willingness to purchase (Asioli et al., 2017; De Castro Carvalho et al., 2025). Fermentation may therefore offer a competitive advantage by aligning biological processing with consumer expectations for naturalness and authenticity.

Health and well-being motivations further strengthen market demand. Consumers increasingly associate plant-based fermented products with digestive benefits, immune support, and enhanced nutrient bioavailability (Dhiman et al., 2025). Consumer research examining preferences for plant-based milk alternatives indicates that taste, perceived healthfulness, and product labeling information are among the most influential attributes shaping consumer choice (Su et al., 2023). Fermentation can enhance sensory fidelity in plant-based dairy and beverage analogues by reducing undesirable off-flavors and improving texture and mouthfeel through microbial metabolism and protein modification. This differentiation is particularly relevant in the context of increasing scrutiny of highly processed plant-based alternatives, as concerns have been raised regarding the nutritional and processing profiles of products containing multiple additives and structuring agents (Macdiarmid, 2022).

Geographic analysis of plant-based food market expansion reveals distinct regional patterns reflecting cultural dietary traditions, economic development levels, and regulatory environments. The most mature markets exist in Western Europe and

North America, where plant-based product availability is extensive and consumer familiarity substantial. Market research from Germany, Italy, Spain, and France documented that over 50% of consumers in these countries have reduced red meat intake, with varying adoption of plant-based alternatives (Mustapa et al., 2025). However, emerging markets in Asia-Pacific, Latin America, and Africa represent the fastest-growing regions for plant-based fermented products, driven by multiple converging factors.

The Asian market represents the largest opportunity and most dynamic growth region for plant-based fermented foods, reflecting both historical fermentation traditions and contemporary consumer health consciousness. Countries with long-standing fermentation traditions, including South Korea, Japan, and China, exhibit high cultural familiarity with fermented foods, which may facilitate consumer acceptance of fermented plant-based products. Bibliometric analyses of fermented food research indicate that East and Southeast Asia dominate scientific output in fermented fish studies, particularly in countries such as China, South Korea, and Thailand, reflecting both strong culinary traditions and active research capacity in these regions (Mamangkey et al., 2025).

The plant-based milk market in the Asia–Pacific region is expanding rapidly and is projected to be among the fastest-growing segments of the global dairy-alternative industry. This growth is driven by urbanization, rising disposable incomes, and the high prevalence of lactose intolerance in many Asian populations, where intolerance rates can reach 70–90% in some regions (Mordor Intelligence, 2025).

Sub-Saharan Africa represents an emerging opportunity for plant-based fermented foods, particularly through the utilization of indigenous crops and traditional fermentation knowledge. The region has a long history of fermented food production involving cereals, legumes, vegetables, and beverages, which provides a strong cultural foundation for the development of novel fermented products. Recent analyses of plant-based food systems in sub-Saharan Africa highlight the potential for commercial production of fermented foods derived from indigenous crops, including cereals, legumes, and traditional beverages, which could contribute to both nutritional improvement and economic development (Ogutu et al., 2024). However, commercialization remains constrained by structural challenges such as limited processing infrastructure, fragmented supply chains, and evolving regulatory frameworks.

Latin America is witnessing increasing consumer interest in plant-based food alternatives, although fermented plant-based products remain less established than in Asian markets where fermentation traditions strongly shape food consumption patterns. Nevertheless, Latin America possesses diverse indigenous fermentation practices, including traditional beverages such as pulque, tepache, and pozol, which demonstrate the cultural relevance of fermented plant-derived foods and provide opportunities for the valorization and commercialization of indigenous fermentation technologies (Steinkraus, 1997; Tamang et al., 2016). These traditions offer a potential foundation for developing commercially scalable fermented plant-based products tailored to regional consumer preferences. In parallel, the Ukrainian market illustrates how external disruptions can accelerate plant-based food adoption. Under recent crisis conditions, the plant-based milk alternative sector expanded rapidly, with market volumes increasing approximately 13.6-fold between 2019 and 2020, driven by disruptions to animal agriculture supply chains and growing consumer demand for locally produced alternatives derived from domestically available crops (Bochkarev, 2023).

Demographic analyses indicate that younger consumers generally exhibit higher adoption rates of plant-based products compared with older generations. However, recent consumer research has identified a non-linear relationship between age and plant-based purchasing behavior, with the likelihood of purchasing plant-based alternatives declining through mid-life but increasing again after the mid-50s, suggesting emerging opportunities among older consumers motivated primarily by health considerations rather than environmental ideology (Van Oppenraaij et al., 2025). Across regions, urban populations also tend to display higher engagement with plant-based products compared with rural populations, reflecting greater product availability, higher levels of consumer awareness, and more developed food retail infrastructure in urban markets.

17.3 Future Outlook of Plant-Based Fermented Foods: Toward 2035 and Beyond

The trajectory of plant-based fermented foods toward 2035 and beyond increasingly depends on convergence with cellular agriculture and precision biology, integration into climate-neutral food systems, and systematic resolution of key research gaps. Advancing this future requires coordinated innovation spanning bioprocess engineering, environmental sustainability, and rigorous health outcome validation.

17.3.1 Convergence with Cellular Agriculture and Precision Biology

The convergence of fermentation biotechnology with cellular agriculture represents one of the most promising frontiers for creating scalable, sustainable protein systems.

17.3.1.1 Hybrid Plant-Microbial Protein Systems

Hybrid approaches combining microbial fermentation with engineered plant matrices are generating sophisticated designer foods with optimized nutritional and functional properties.

Precision fermentation for food protein production demonstrates extraordinary yield improvements enabling industrial viability (Cleaver & Brock, 2025). Commercial viability of lab-grown food proteins—including animal-free dairy, egg, and collagen—depends on microbial hosts' capacity to export correctly folded proteins into culture medium at gram-per-liter titers. Deliberate, data-driven engineering of secretion efficiency determines whether precision-fermented proteins advance from pilot production to routine industrial manufacture. Yeast and filamentous fungi serve as ideal host organisms due to their potential for high-yield secretion (enabling gram-per-liter titers) and ability to reproduce many eukaryotic post-translational modifications. This convergence of microbial engineering with conventional proteins represents transformative pathway toward scaled production.

Kefir fermentation demonstrates how microbial consortia modulate plant-based systems with targeted health outcomes (Choi et al., 2025). In randomized controlled trial with 28 healthy young adults (18–30 years), kefir consumption increased relative abundance of lactate-producing bacteria including *Bifidobacterium breve*, *Ruthenibacterium lactatiformans*, *Weissella koreensis*, and *Leuconostoc mesenteroides*. Critical finding revealed genus Blautia increased significantly, with species luti and wexlerae showing particular enrichment. These shifts were associated with increases in short-chain fatty acid production pathways, demonstrating how fermentation engineering targets specific microbiota modulation with measurable functional outcomes.

Weissella confusa fermentation of soy yogurt exemplifies designer fermented food development (Liu et al., 2025). Safety evaluation demonstrated *W. confusa* SY628 possessed no hemolytic activity, was sensitive to variety of antibiotics, produced no biogenic amines—indicating extremely low pathogenic risk with enhanced acid and bile tolerance suggesting probiotic properties. Fermented soy yogurt with SY628 strain achieved pH 4.49 after 21 days storage with total acidity 76.77°T and viable count 5.81×10^7 CFU/mL, indicating excellent stability. Critically, cohesiveness, viscosity, elasticity, and modulus in SY628 group significantly exceeded other strains, with stable internal network structure. Total amino acid content reached 308.57 μg/g, with umami-tasting amino acids (aspartate, glutamate) comprising 22.95%. Polyunsaturated fatty acids reached notably high 1818.95 μg/g, with significant nutritional value. Most remarkably, SY628 demonstrated highest number of key flavor components, with 2,3-butanedione and other flavor compounds contributing rich sensory profile. This demonstrates systematic design of fermented plant-based products optimizing simultaneous nutritional, functional, and sensory attributes.

Heyndrickxia coagulans isolated from stingless bee honey demonstrates multifunctional probiotic capacity (Prakit et al., 2025). Two isolates (TBRC-18260, TBRC-18261) underwent comprehensive in vitro analysis assessing traits including acid, bile salt tolerance, self-aggregation, pathogen inhibition. TBRC-18260 and TBRC-18261 exhibited core probiotic characteristics with excellent survivability under acidic conditions and bile salt presence, displaying strong antimicrobial activity against various pathogens and significant GABA production capabilities. Hemolytic activity assessments confirmed non-hemolytic nature with wide antibiotic resistance patterns. These isolates emerged as promising candidates for development of functional foods and dietary supplements—exemplifying how indigenous microbial discovery combined with characterization generates next-generation probiotic strains for precision fermentation.

17.3.1.2 Designer Fermented Foods

Engineered fermentation also enables creation of designer fermented foods with precisely defined functional properties addressing specific health objectives, metabolic phenotypes, and consumer preferences. Integration of omics technologies, artificial intelligence, and synthetic biology accelerates designer food development toward commercialization.

Microbial metabolite engineering generates designer bioactive compounds with targeted health functions. Recent biotechnological advancements in fermentation specifically employing omics, multi-omics, and precision fermentation technologies provide deeper insight into microbial diversity and metabolic pathways. Omics technologies enable deciphering of functional interactions and metabolic capabilities, while precision fermentation enables manipulation of microbial processes to optimize fermentation production of desired attributes in foods. Various sustainable substrates are utilized apart from conventional ones. Notably, artificial intelligence, machine learning, and Internet of Things technologies focus on potential to optimize processes, predict microbial interactions, and improve quality control, supported by relevant successful examples.

AI-driven precision fermentation demonstrates extraordinary yield improvements toward scaled production (Priyadharshini et al., 2025). Evidence indicates 300% yield increases for alt-proteins via AI-CRISPR fusion and 60% reduction in bioreactor failures through reinforcement learning optimization of bioprocess parameters. These quantifiable improvements directly translate to economic viability of fermented protein production at industrial scale. However, critical ethical concerns persist including data privacy risks (72% GDPR non-compliance in health monitoring platforms) and algorithmic biases potentially exacerbating nutritional disparities.

Postbiotic generation through fermentation creates stable, functional metabolites with predictable biological activities (Meena et al., 2025; Asefa et al., 2025). Postbiotics—metabolic products and cellular components of probiotics—possess advantages of being non-living yet retaining biological activity. Different inactivation methods (thermal and non-thermal) directly affect anti-inflammatory and antioxidative functions of postbiotics. Many studies demonstrate that choice of inactivation method substantially affects postbiotic composition and activity. Current applications span food industry, food-medicine integration, pet food, and animal breeding. Future research must optimize inactivation methodologies to maximize efficacy of postbiotics, thereby promoting precise application across various fields.

Fermented whey-based products demonstrate comprehensive designer food potential across multiple health dimensions (Sar et al., 2025). Systematic review of 12 human intervention studies found that fermented whey product consumption was associated with improvements in muscle mass, glycemic control, lipid profiles (notably triglycerides and LDL cholesterol reduction), immune function (increased natural killer cell activity), and reductions in oxidative stress and inflammation. Some studies reported benefits for gastrointestinal and urinary tract health. Health effects were attributed to increased bioavailability of branched-chain amino acids, bioactive peptides, and microbial metabolites including exopolysaccharides and short-chain fatty acids. Most interventions were well tolerated with no serious adverse effects reported.

17.3.2 Role in Climate-Neutral Food Systems

17.3.2.1 Net-Zero Fermentation Facilities

Achieving net-zero emissions in fermentation facilities represents essential requirement for plant-based fermented foods to fulfill their environmental sustainability promise. Transition toward net-zero fermentation requires systematic decarbonization of energy use, waste management, and bioprocess optimization.

Transition toward net-zero emissions in chemical and process industries establishes foundational frameworks applicable to fermentation (Glavič et al., 2023). Review examining transition to net-zero emissions in chemical and process industries highlights core areas of carbon reduction, efficient energy use, and sustainable practices. Cutting-edge technologies including biomass utilization, biotechnology applications, and waste management strategies serve as key drivers of transition. The role of hydrogen as clean fuel is central to revolutionizing these sectors, pointing toward cleaner, greener operations. European Green Deal and Sustainable Development Goals provide clear roadmap and framework advancing sustainability, driving innovation, reducing industry's carbon footprint, with notable contribution to existing knowledge. Alignment with SDGs brings numerous benefits to industries, increasing competitiveness, promoting societal well-being, supporting cross-sector collaboration to achieve shared sustainability goals.

Biowaste-to-hydrogen conversion demonstrates promising pathway for simultaneous waste valorization and energy generation (Abawalo et al., 2025). Systematic review of conversion technologies reveals hydrothermal gasification and dark fermentation as particularly promising for wet biowaste streams, including food waste. Comparative analyses demonstrate bio-based pathways offer significantly lower emissions, lower energy use, and lower pollutant outputs than conventional methods. Future research directions emphasize process integration, catalyst development, and lifecycle assessment for optimizing sustainable hydrogen production from fermentation byproducts and food waste streams.

Recovered ammonia as sustainable energy carrier exemplifies circular nutrient recovery from fermentation streams (La Corte et al., 2025). Ammonia recovered from waste streams such as industrial leachates and organic waste represents unique opportunity to harness sustainable, carbon-free energy resources. Integrating recovered ammonia into energy systems enables industries to reduce dependence on conventional ammonia production (which is carbon-intensive), lower greenhouse gas emissions, and advance circular economy practices. Technologies for recovering ammonia and applications in combustion processes and fuel cells demonstrate practical pathways for capturing value from fermentation byproducts while decarbonizing operations.

Waste-derived fertilizer production from agro-industrial residues closes nutrient cycles while reducing synthetic fertilizer dependence (Hidalgo et al., 2025). Review examining conversion of nutrient-rich streams from livestock excreta, food industry residues, and sewage sludge into effective fertilizers through biological, thermal, and chemical processes enables reducing dependence on fossil-fuel-intensive mineral fertilizers while supporting more circular and resilient cropping systems. Analysis of agronomic performance reveals waste-derived fertilizers achieve nutrient release dynamics, improve soil functions, and support crop response under different management conditions. While promising, implementation requires further efforts in standardization, safety assurance, and farmer engagement.

17.3.2.2 Integration into Circular Urban Food Production

Plant-based fermented foods increasingly integrate into urban food production systems designed for resource efficiency, waste minimization, and localized nutrition security.

Circular bioeconomy business models demonstrate pathways overcoming commercialization barriers (Gatto & Re, 2021). Analysis of emerging bio-based technologies and trends highlights circular economy and bioeconomy concepts as central to European Union policy agendas. Study examining small and medium-sized enterprises (SMEs) innovation reveals that eco-nutraceuticals and microalgae represent most innovative industry applications with leading commercialization models and future perspectives. Research demonstrates that bioeconomy-focused business models, when supported through appropriate financing and incubation environments, successfully transition from laboratory innovation toward market deployment.

Regional-scale energy modeling for climate neutrality establishes frameworks for integrated urban food system planning (Zyśk et al., 2025). Complex energy and greenhouse gas emissions analysis at regional scale employs 3E (Energy-Economy-Environment) optimization models incorporating multiple sectors including energy, industry, agriculture, households, tertiary, transport, and forestry. Modeling covers renewable sources availability, technology and fuel costs, legal environment, and sectoral goals from strategic documents. Study demonstrating optimal pathways toward climate neutrality for coal regions offers methodology applicable to urban food system planning. Integrated modeling across multiple sectors demonstrates essential requirement for food systems to simultaneously address energy efficiency, waste management, and renewable resource integration.

17.3.3 *Key Research Gaps and Innovation Priorities*

17.3.3.1 Standardization of Functional Claims

Establishing rigorous, standardized frameworks for substantiating functional claims of fermented foods represents critical gap limiting regulatory approval and consumer confidence. Current evidence base remains insufficient to support most health claims despite growing observational data.

Prospective evidence linking fermented food consumption with mortality outcomes remains limited (Paveljšek et al., 2025). Comprehensive systematic review assessing association between fermented food consumption and risk of all-cause, cardiovascular, and cancer-related mortality identified 52 cohort studies meeting inclusion criteria. Fermented milk products (including yogurt), chocolate, and fermented soy products (particularly natto) suggested modest inverse association with all-cause and cardiovascular mortality. Cheese was associated with reduced all-cause mortality in some studies but showed inconsistent effect on cardiovascular mortality. Evidence for cancer-related mortality was weaker, although yogurt and fermented milk displayed some protective trends. Single cohort suggested potential reduction in all-cause mortality with fermented vegetable consumption, whereas fermented meat suggested no clear association. Biological plausibility was supported by fermentation-derived compounds including bioactive peptides, polyphenols, isoflavones, natto-kinase, and vitamin K2. However, habitual consumption of certain fermented foods may be associated with modest reductions in mortality risk, but current evidence remains insufficient to support European Food Safety Authority-approved health claims. Randomized controlled trials are essential to demonstrate causality.

Soy product consumption and cognitive health demonstrate complexity of long-term health claims (Yu & Zeng, 2025). Dose-response meta-analysis examining association between soy product consumption and risk of cognitive impairment and major neurocognitive disorders included six prospective cohort studies incorporating 68,691 participants. Results revealed that elevated total soy product consumption was markedly correlated with reduced cognitive impairment. Associated with increased risk observed, though not statistically significant. Meta-analysis indicated 1-gram per day increase in natto consumption demonstrated 8–14% decrease in both cognitive impairment and neurocognitive disorder. Subgroup analysis suggested more pronounced protective effect in individuals experiencing stroke (OR 0.66, 95%CI: 0.53–0.82). However, unfermented soy showed paradoxical increased risk (OR 1.20, 95%CI: 0.83–1.72). Risk of bias showed no significant differences. Dose-response further explained inverse relationship with cognitive decline, with higher doses associated with lower risk. Despite diverse results, heterogeneity remained within both overall and multiple subgroup analyses, highlighting need for standardized methodologies.

Functional food standards and certification frameworks require international harmonization (Ankiel et al., 2025). Review examining role of certification and labeling in ensuring authenticity and sustainability of fermented milk products revealed substantial inconsistencies in defining key terms (organic, probiotic, carbon-neutral), which hinder certification harmonization. Complex labels and allergen declarations reduce clarity and trust, while overlapping or vague eco-labels risk contributing to consumer confusion and skepticism. Despite this, credible certifications still enhance purchase intent. Modern technologies including blockchain traceability, interactive QR codes, and digital product passports offer new transparency mechanisms for verification of sustainability claims and functional properties, though implementation costs and regulatory gaps remain barriers.

17.3.3.2 Scalable Sustainable Substrates

Developing consistent, affordable, sustainable substrate sources for fermentation represents foundational requirement for scaling plant-based fermented food production while maintaining economic viability and environmental benefits.

Circular bioeconomy approaches increasingly focus on valorizing agro-food by-products as substrates for fermentation and other biotechnological processes. Agro-industrial residues such as fruit peels, seeds, and pomace are rich in bioactive compounds, including polyphenols, enzymes, and essential nutrients, which can be converted into value-added products through microbial fermentation. Recent reviews highlight the potential of these materials for sustainable product development, particularly in cosmetic and biotechnological applications, demonstrating their functional and economic value (Krzyżostan et al., 2024). Although many studies focus on non-food applications, the underlying principles of biomass valorization and bioconversion are directly relevant to food fermentation systems. Fermentation-based valorization therefore represents a promising strategy for enhancing the bioavailability and functionality of plant-derived materials within circular bioeconomy frameworks.

Fruit pomace valorization demonstrates practical substrate availability and functional benefits (Tama & Karaś, 2025). Annually, approximately 1.3 billion tons of food is wasted globally, with fruits and vegetables constituting significant portion. Pomace—solid remains after juice extraction—comprises valuable nutrients including fiber, polyphenols, flavonoids, carotenoids, organic acids, vitamins, and minerals. Common sources include apples, grapes, citrus fruits, and berries. Researchers have highlighted potential for use as fermentation substrate and food ingredient. For example, replacing flour in cookies with pomace improves antioxidant content and reduces hardness. Adding grape pomace to gummy candies increases levels of anthocyanins, flavanols, and proanthocyanidins while enhancing texture. Fortifying waffles with raspberry pomace boosts nutritional value and may inhibit enzymes linked to free radical production. As functional ingredient, pomace could help lower risk of cardiovascular disease, diabetes, obesity, and colon cancer. Using pomace in production supports sustainability by reducing waste and improving nutrition, positioning it as key ingredient for sustainable innovation.

Biorefinery approaches maximize value extraction from diverse substrates (Wibisono et al., 2025). Advancing industrial food byproduct management through strategic exploration of interconnections between management strategies, technologies, and byproduct types contributes to overarching goal of waste reduction. Review discusses advantages and limitations of current approaches, offering insights regarding implications for safety, economic feasibility, and environmental sustainability. Key categories examined assist development of targeted strategies for more effective byproduct management. Journey toward circular food systems requiring transformation of waste into resources through technological advancements has progressively enabled industries to repurpose materials once considered waste.

17.3.3.3 Long-Term Health Impact Studies

Establishing rigorous, long-term health outcomes research represents critical gap limiting evidence-based health claims and regulatory advancement for fermented plant-based foods.

Gut microbiota modulation through fermented foods requires longitudinal mechanistic studies (Randeni et al., 2024). Comprehensive review examining triangular relationship among diet, gut microbiota, and inflammation highlights that human gastrointestinal tract hosts complex, dynamic microbial community playing pivotal role in numerous physiological processes including digestion, metabolism, and immune function. Recent research highlights significant impact of diet on microbiota composition and functionality, with consequential effects on host health. Growing evidence links microbiota composition to inflammation, a key factor in many chronic diseases including inflammatory bowel disease, obesity, diabetes, and cardiovascular disease. Beneficial dietary patterns including Mediterranean and plant-based diets promote diverse, balanced microbiota supporting anti-inflammatory pathways. Conversely, Western dietary patterns high in saturated fats and refined sugars are associated with dysbiosis and increased inflammation. However, most fermentation studies remain short-term, with inadequate long-term mechanistic understanding.

Functional gastrointestinal disorders and obesity interconnections reveal gaps in mechanistic understanding (Calcaterra et al., 2024). Narrative review summarizing evidence regarding childhood obesity and functional gastrointestinal disorders reveals shared risk factors related to diet, lifestyle, and gut microbiota impairments. Findings emphasize importance of holistic, multidisciplinary approaches to symptom management. Beyond nutrition education, physical activity, and medical care, complementary intervention strategies including personalized dietary modifications (e.g., low-FODMAP, fiber-enriched diets) prove critical. Modulation of microbiota through probiotics, prebiotics, and integrative support shows significant promise. However, current evidence heterogeneity underscores need for robust longitudinal studies developing standardized protocols maximizing treatment efficacy.

Constipation-predominant irritable bowel syndrome demonstrates dysbiosis patterns requiring long-term intervention studies (Wang et al., 2025). Metagenomic and metabolomic analysis of IBS-C individuals compared with healthy controls revealed no significant overall diversity differences ($p > 0.05$). However, six bacterial species exhibited differential abundance, with relative abundance of beneficial taxa including *Megasphaera elsdenii*, *Bifidobacterium bifidum*, and *Alistipes inops* significantly reduced ($p < 0.05$). Metabolomic profiling demonstrated enriched pathways related to short-chain fatty acid metabolism, with SCFA levels downregulated in IBS-C individuals. Trend toward increased pro-inflammatory metabolite (leukotriene D5) was observed. Critically, these findings offer insights into pathophysiological mechanisms yet require longitudinal intervention studies assessing fermented food efficacy in restoring dysbiotic patterns and ameliorating symptoms.

Neurodevelopmental disorder and gut-brain axis interactions suggest promising intervention targets (Chen et al., 2025). Comprehensive review examining pathological mechanisms underlying autism spectrum disorder, attention-deficit hyperactivity disorder, and epilepsy reveals abnormal activation of key signaling pathways (NF-κB, MAPK, PI3K/AKT/mTOR) strongly associated with gut dysbiosis. Natural compounds with multiple bioactivities can directly and indirectly affect immune and digestive tract function by regulating structure of intestinal microbial community. Emerging evidence suggests fermented foods enriching Akkermansia muciniphila and butyric acid-producing bacteria may provide therapeutic strategy for neurodevelopmental disorder management. However, evidence remains nascent with critical need for long-term mechanistic and clinical studies.

17.4 Conclusions

The rapid evolution of plant-based food systems reflects a convergence of sustainability imperatives, technological innovation, and shifting consumer preferences. Fermentation has emerged as a central enabling technology, enhancing the nutritional quality, sensory attributes, and functional value of plant-based products while supporting the development of postbiotics and other bioactive compounds. At the same time, circular bioeconomy approaches are transforming agro-food by-products, such as fruit pomace and processing residues, into valuable fermentation substrates, reducing waste and creating new opportunities for sustainable product development.

Consumer trends further reinforce this transition, with increasing demand for clean-label, functional, and environmentally responsible foods. Fermentation aligns well with these expectations by offering natural processing pathways that improve flavor, texture, and bioavailability without reliance on extensive additives. Advances in biotechnology, including AI-assisted optimization and precision fermentation, are accelerating innovation and improving process efficiency, positioning plant-based fermented products as scalable and commercially viable solutions.

Despite these advances, challenges remain in ensuring consistent product quality, optimizing raw material utilization, and achieving economic scalability across diverse regional contexts. Addressing these challenges will require interdisciplinary collaboration, integrating food science, biotechnology, and systems-level sustainability approaches.

Overall, the integration of fermentation technologies with circular resource utilization presents a powerful pathway toward more resilient and sustainable food systems. By valorizing underutilized biomass and aligning with evolving consumer demands, plant-based fermentation has the potential to play a transformative role in shaping the future of global food production.

References

Abawalo, M., Pikoń, K., Landrat, M., & Ścierski, W. (2025). Hydrogen production from biowaste: A systematic review of conversion technologies, environmental impacts, and future perspectives. *Energies, 18*(17), 4520. https://doi.org/10.3390/en18174520

Abbaspour, N. (2024). Fermentation's pivotal role in shaping the future of plant-based foods: An integrative review of fermentation processes and their impact on sensory and health benefits. *Applied Food Research, 4*(2), 100468. https://doi.org/10.1016/j.afres.2024.100468

Ahmad, I. (2024). The nexus of plant-based markets, technological advancements, health and safety protocols, and future trajectories. *Journal of Culinary Science & Technology, 22*, 373–385. https://doi.org/10.1080/15428052.2024.2345309

Alebidi, S., Mallek, H., Fragasso, M., Capozzi, V., Abidi, F., Essid, I., Spano, G., & Selmi, H. (2026). Exploring the diversity and applications of lactic acid bacteria from Tunisian traditional fermented foods. *Microorganisms*. https://doi.org/10.3390/microorganisms14020383

Ankiel, M., Halagarda, M., Piekara, A., Sady, S., Żmijowska, P., Popek, S., Pachołek, B., Jefmański, B., Kucia, M., & Krzywonos, M. (2025). Role of certifications and labelling in ensuring authenticity and sustainability of fermented milk products. *Sustainability, 17*(18), 8398. https://doi.org/10.3390/su17188398

Asefa, Z., Belay, A., Welelaw, E., & Haile, M. (2025). Postbiotics and their biotherapeutic potential for chronic disease and their feature perspective: A review. *Frontiers in Microbiomes, 4*, 1489339. https://doi.org/10.3389/frmbi.2025.1489339

Mordor Intelligence. (2025, November 18). Asia-Pacific dairy alternatives market forecasts 2030. Mordor Intelligence. https://www.mordorintelligence.com/industry-reports/asia-pacific-dairy-alternatives-market

Asioli, D., Aschemann-Witzel, J., Caputo, V., Vecchio, R., Annunziata, A., Næs, T., & Varela, P. (2017). Making sense of the "clean label" trends: A review of consumer food choice behavior and discussion of industry implications. *Food Research International, 99*, 58–71. https://doi.org/10.1016/j.foodres.2017.07.022

Bochkarev, D. O. (2023). Trends and prospects for the development of the ukrainian market of plant-based milk substitutes in the war and postwar period. *Economic Innovations, 25*(1(86)), 34–43. https://doi.org/10.31520/ei.2023.25.1(86).34-43

Calcaterra, V., Cena, H., Loperfido, F., Porri, D., Basilico, S., Gazzola, C., Ricciardi Rizzo, C., Conti, M. V., Luppino, G., Wasniewska, M. G., & Zuccotti, G. (2024). Functional gastrointestinal disorders and childhood obesity: The role of diet and its impact on microbiota. *Nutrients, 17*(1), 123. https://doi.org/10.3390/nu17010123

Cao, Y., & Miao, L. (2023). Consumer perception of clean food labels. *British Food Journal, 125*(2), 433–448. https://doi.org/10.1108/BFJ-03-2021-0246

Chen, L., Li, Z., & Fan, Y. (2025). Neurodevelopmental disorders and gut-brain interactions: Exploring the therapeutic potential of pycnogenol through microbial-metabolic-neural networks. *Frontiers in Cellular and Infection Microbiology, 15*, 1601888. https://doi.org/10.3389/fcimb.2025.1601888

Choi, Y., Keum, G. B., Kang, J., Doo, H., Kwak, J., Kim, H., Chae, Y., Lee, S., Yang, H., Kim, S., Sun, X., Kim, H. B., & Yoo, S. J. (2025). Evaluation of kefir consumption on gut microbial diversity in a healthy young population using full-length 16S rRNA sequencing. *Frontiers in Microbiology, 16*, 1587831. https://doi.org/10.3389/fmicb.2025.1587831

Cleaver, A., & Brock, J. (2025). Harnessing fungal secretion systems for precision fermentation of food proteins. *ACS Synthetic Biology, 14*(11), 4153–4162. https://doi.org/10.1021/acssynbio.5c00521

Dave, K., Kumar, A., Dave, N., Jain, M., Dhanda, P. S., Yadav, A., & Kaushik, P. (2024). Climate change impacts on legume physiology and ecosystem dynamics: A multifaceted perspective. *Sustainability, 16*(14), 6026. https://doi.org/10.3390/su16146026

De Castro Carvalho, G., Carneiro, J. D. D. S., Rocha, R. A. R., & Pereira, E. (2025). Plant-based trends: Consumer's perception and factors that influence the purchase of plant-based products. *British Food Journal, 127*(9), 3396–3426. https://doi.org/10.1108/BFJ-08-2024-0830

Dhiman, S., Kaur, S., Thakur, B., Singh, P., & Tripathi, M. (2025). Nutritional enhancement of plant-based fermented foods: Microbial innovations for a sustainable future. *Fermentation, 11*(6), 346.

Dominguez-Lacueva, P., Beriain, M. J., & Aldaya, M. (2025). A comparative analysis of the water and carbon footprints of hybrid, plant-based, and animal-based burgers. *Renewable Agriculture and Food Systems, 40*, e18. https://doi.org/10.1017/S1742170525100112

Estell, M., Hughes, J., & Grafenauer, S. (2021). Plant protein and plant-based meat alternatives: Consumer and nutrition professional attitudes and perceptions. *Sustainability, 13*(3), 1478. https://doi.org/10.3390/su13031478

Food and Agriculture Organization of the United Nations. (2025, December 1). Better land, soil and water management key to feeding 10 billion people, FAO warns. FAO Newsroom. Retrieved February 25, 2026, from https://www.fao.org/newsroom/detail/better-land%2D%2Dsoil-and-water-management-key-to-feeding-10-billion-people%2D%2Dfao-warns/en

Gatto, F., & Re, I. (2021). Circular bioeconomy business models to overcome the valley of death. A systematic statistical analysis of studies and projects in emerging bio-based technologies and trends linked to the sme instrument support. *Sustainability (Basel, Switzerland), 13*(4), 1899. https://doi.org/10.3390/su13041899

Glavič, P., Pintarič, Z. N., Levičnik, H., Dragojlović, V., & Bogataj, M. (2023). Transitioning towards net-zero emissions in chemical and process industries: A holistic perspective. *Processes, 11*(9), 2647. https://doi.org/10.3390/pr11092647

Hidalgo, D., Martín-Marroquín, J. M., Corona, F., & Verdugo, F. (2025). Waste-derived fertilizers: Conversion technologies, circular bioeconomy perspectives and agronomic value. *Agronomy, 15*(9), 2167. https://doi.org/10.3390/agronomy15092167

Horlacher, N., Oey, I., & Agyei, D. (2023). Learning from tradition: Health-promoting potential of traditional lactic acid fermentation to drive innovation in fermented plant-based dairy alternatives. *Fermentation, 9*(5), 452. https://doi.org/10.3390/fermentation9050452

Khayeka-Wandabwa, C., Choge, J. K., Linnemann, A., & Schoustra, S. E. (2024). Linking fermented foods to microbial composition and valorisation: Blueprint for Kenya. *Food Reviews International*. (Print). https://doi.org/10.1080/87559129.2024.2355992

Krzyżostan, M., Wawrzyńczak, A., & Nowak, I. (2024). Use of waste from the food industry and applications of the fermentation process to create sustainable cosmetic products: A review. *Sustainability, 16*(7), 2757. https://doi.org/10.3390/su16072757

La Corte, D., Maddaloni, M., Vahidzadeh, R., Domini, M., Bertanza, G., Ansari, S., Marchionni, M., Tola, V., & Artioli, N. (2025). Recovered ammonia as a sustainable energy carrier: Innovations in recovery, combustion, and fuel cells. *Energies, 18*(3), 508. https://doi.org/10.3390/en18030508

Liu, M., Li, X., Tao, Y., Zhao, L., & Zhang, X. (2025). Safety evaluation of Weissella confusa SY628 and the effect of its fermentation on the taste and quality of soy yogurt. *Frontiers Media*. https://doi.org/10.3389/fmicb.2025.1567399

Macdiarmid, J. I. (2022). The food system and climate change: Are plant-based diets becoming unhealthy and less environmentally sustainable? *Proceedings of the Nutrition Society, 81*(2), 162–167. https://doi.org/10.1017/S0029665121003712

Mahajani, K., C.s., V., Sahu, R., & Nayak, P. (2023). Cultural and regional influences on the development and acceptance of fermented functional foods. *Journal of Food and Biotechnology*. https://doi.org/10.51470/fab.2023.4.2.06

Mamangkey, J., Naibaho, F. G., Purba, I. R., Hibatullah, A. Y., & Soa, C. A. P. (2025). Trends and knowledge gaps in fish-based fermented foods: A bibliometric analysis from the last decade (2015–2024). *International Journal of Science, Technology & Management, 6*(6), 1061–1070.

Meena, K. K., Joshi, M., Gupta, L., & Meena, S. (2025). Comprehensive insights into postbiotics: Bridging the gap to real-world application. *Food Nutrition, 1*(2), 100024. https://doi.org/10.1016/j.fnutr.2025.100024

Molina, G. E. S., Ras, G., Da Silva, D. F., Duedahl-Olesen, L., Hansen, E. B., & Bang-Berthelsen, C. H. (2025). Metabolic insights of lactic acid bacteria in reducing off-flavors and antinutrients in plant-based fermented dairy alternatives. *Comprehensive Reviews in Food Science and Food Safety, 24*(2), e70134. https://doi.org/10.1111/1541-4337.70134

Muhammed, Y., Minervini, F., & Cavoski, I. (2025). From ancient fermentations to modern biotechnology: Historical evolution, microbial mechanisms, and the role of natural and commercial starter cultures in shaping organic and sustainable food systems. *Foods*. https://doi.org/10.3390/foods14244240

Mustapa, M. A. C., Baba, Y., Baishakhy, S. D., & Kallas, Z. (2025). Evolving appetites: Current evidence and future perspectives in terms of meat substitutes in europe. *Food Science & Nutrition, 13*(1), e4753. https://doi.org/10.1002/fsn3.4753

Ogutu, F. O., Okiko, G., Wanjala, G., Luvitaa, S., Obong'o, B. O., Vriesekoop, F., & Munialo, C. D. (2024). Unlocking the potential of plant-based foods in sub-Saharan Africa: A review of the opportunities and challenges. *International Journal of Food Science and Technology, 59*(8), 5326–5342. https://doi.org/10.1111/ijfs.17327

Pal, P., Singh, A. K., Srivastava, R. K., Rathore, S. S., Sahoo, U. K., Subudhi, S., Sarangi, P. K., & Prus, P. (2024). Circular bioeconomy in action: Transforming food wastes into renewable food resources. *Food, 13*(18), 3007. https://doi.org/10.3390/foods13183007

Paveljšek, D., Pertziger, E., Fardet, A., Panagiotakos, D. B., Savary-Auzeloux, I., Adamberg, S., Peñas, E., Frias, J., Ntantou, A., Diamantoglou, I., Domínguez-Soberanes, J., Louis, S., Chassard, C., Praćer, S., Vergères, G., & Matalas, A. (2025). A systematic review of prospective evidence linking non-alcoholic fermented food consumption with lower mortality risk. *Frontiers in Nutrition, 12*, 1657100. https://doi.org/10.3389/fnut.2025.1657100

Peng, B., Nie, P., & Xu, H. (2025). Plant-based fermented foods: Classification, biochemical transformations, and health benefits. *Fermentation, 11*(7), 364. https://doi.org/10.3390/fermentation11070364

Prakit, B., Chaiyod, R., Khongkool, K., Chanasit, W., & Lertworapreecha, M. (2025). Multifunctional probiotic and safety attributes Heyndrickxia coagulans isolated from stingless bee honey. *Annals of Microbiology, 75*(1), 3. https://doi.org/10.1186/s13213-025-01791-0

Priyadharshini, D., Muthuvel, I., Saraswathy, S., Kavitha, P. S., & Jegadeeswari, V. (2025). Precision to plate: AI-driven innovations in fermentation and hyper-personalized diets. *Frontiers in Nutrition, 12*, 1659511. https://doi.org/10.3389/fnut.2025.1659511

Randeni, N., Bordiga, M., & Xu, B. (2024). A comprehensive review of the triangular relationship among diet–gut microbiota–inflammation. *International Journal of Molecular Sciences, 25*(17), 9366. https://doi.org/10.3390/ijms25179366

Santa, D., Huch, M., Stoll, D. A., Cunediolu, H., Priidik, R., KarakaBudak, B., Matalas, A., Pennone, V., Girija, A., Arranz, E., Syrpas, M., Mukherjee, A., Cotter, P. d., Mojsova, S., Chassard, C., Pracer, S., Vergres, G., & Ktt, M. (2025). Health benefits of ethnic fermented foods. *Frontiers in Nutrition*. https://doi.org/10.3389/fnut.2025.1677478

Sar, T., Bogovic Matijasic, B., Danilovic, B., Gamero, A., Gandía, M., Krausova, G., Martínez-Villaluenga, C., Peñas, E., Bagherzadehsurbagh, E., Cemali, Ö., Santa, D., Künili, I. E., Kesenkas, H., Chassard, C., Pracer, S., Vergères, G., & Ergün, B. G. (2025). A systematic review of health promoting effects of consumption of whey-based fermented products on adults. *Frontiers in Nutrition, 12*, 1651365. https://doi.org/10.3389/fnut.2025.1651365

Solomon, T., Gupta, V., & Ncho, C. M. (2023). Balancing livestock environmental footprints with forestry-based solutions: A review. *Ecologies, 4*(4), 714–730. https://doi.org/10.3390/ecologies4040047

Springmann, M., Godfray, H. C. J., Rayner, M., & Scarborough, P. (2016). Analysis and valuation of the health and climate change cobenefits of dietary change. *Proceedings of the National Academy of Sciences of the United States of America, 113*(15), 4146–4151. https://doi.org/10.1073/pnas.1523119113

Steinkraus, K. H. (1997). Classification of fermented foods: Worldwide review of household fermentation techniques. *Food Control, 8*(5–6), 311–317. https://doi.org/10.1016/S0956-7135(97)00050-9

Su, W., Zhang, Y. Y., Li, S., & Sheng, J. (2023). Consumers' preferences and attitudes towards plant-based milk. *Food, 13*(1), 2. https://doi.org/10.3390/foods13010002

Tachie, C. Y. E., Nwachukwu, I. D., & Aryee, A. N. A. (2023). Trends and innovations in theformulation of plant-based foods. *Food Production, Processing and Nutrition.* https://doi.org/10.1186/s43014-023-00129-0

Tama, A., & Karaś, M. (2025). The health-promoting potential of fruit pomace and its application in the confectionery industry. *Applied Sciences, 15*(10), 5790. https://doi.org/10.3390/app15105790

Tamang, J. P., Watanabe, K., & Holzapfel, W. H. (2016). Review: Diversity of microorganisms in global fermented foods and beverages. *Frontiers in Microbiology, 7*. https://doi.org/10.3389/fmicb.2016.00377

Van Oppenraaij, S. L., Putker, M., Van Schaik, A., Weijs, P. J. M., & Verlaan, S. (2025). Perspectives of adults aged 55+ on plant-based diets rich in protein. *Scientific Reports, 15*(1), 31537. https://doi.org/10.1038/s41598-025-17087-x

Wang, Y.-L., Xu, X.-Q., Long, Y.-Y., & Cheng, Y.-L. (2025). Characterization of gut microbiota and metabolites in individuals with constipation-predominant irritable bowel syndrome. *Frontiers in Microbiology, 16*, 1617288. https://doi.org/10.3389/fmicb.2025.1617288

Wibisono, D. A. S., Saw, C.-Y., Wu, T.-Y., & Chau, C.-F. (2025). Advancing industrial food byproduct management: Strategies, technologies, and progress in waste reduction. *Processes, 13*(1), 84. https://doi.org/10.3390/pr13010084

Yu, J., & Zeng, H. (2025). Association of high consumption of soy products with the risk of cognitive impairment and major neurocognitive disorders: A systematic review and dose-response meta-analysis. *Frontiers in Nutrition, 12*, 1635844. https://doi.org/10.3389/fnut.2025.1635844

Zyśk, J., Wyrwa, A., Raczyński, M., Pluta, M., & Suwała, W. (2025). Regional-scale energy modelling for developing strategies to achieve climate neutrality. *Energies, 18*(7), 1787. https://doi.org/10.3390/en18071787

Index

S. Dhiman, S. Kaur (eds.), *Next-Gen Plant Based Fermented Foods: Innovations for a Sustainable Future*, https://doi.org/10.1007/978-3-032-29927-7

MIX
Papier aus verantwortungsvollen Quellen
Paper from responsible sources
FSC® C105338
FSC
www.fsc.org

If you have any concerns about our products,
you can contact us on
ProductSafety@springernature.com

In case Publisher is established outside the EU,
the EU authorized representative is:
Springer Nature Customer Service Center GmbH
Europaplatz 3, 69115 Heidelberg, Germany

Printed by Libri Plureos GmbH
in Hamburg, Germany